BASELINE GEOSCIENCE DATA GENERATION FOR SUSTAINABLE EXPLORATION OF NATURAL RESOURCES

BASELINE GEOSCIENCE DATA GENERATION FOR SUSTAINABLE EXPLORATION OF NATURAL RESOURCES

O.P. Mishra
Formerly National Centre for Seismology,
Ministry of Earth Sciences (MoES),
Government of India, New Delhi, India

D.C. Naskar
Formerly National Centre for Seismology,
Ministry of Earth Sciences (MoES),
Government of India, New Delhi, India

ELSEVIER

Elsevier
Radarweg 29, PO Box 211, 1000 AE Amsterdam, Netherlands
125 London Wall, London EC2Y 5AS, United Kingdom
50 Hampshire Street, 5th Floor, Cambridge, MA 02139, United States

ISBN: 978-0-443-45221-5

For information on all Elsevier publications visit our website at https://www.elsevier.com/books-and-journals

Publisher: Peter Llewellyn
Acquisitions Editor: Jennette McClain
Editorial Project Manager: Rahul Basak
Production Project Manager: M. Balakrishnan
Cover Designer: S. Bakyalakshmi

Typeset by Integra Software Services

Contents

About the authors

Dr. O.P. Mishra is a distinguished fellow of the National Academy of Sciences, India, with over 35 years of experience in the field of Applied Geophysics, Seismology, and Disaster Risk Management Science. He formerly served at the National Centre for Seismology (NCS), Ministry of Earth Sciences (MoES), as its Scientist and Head/Director during September 2014 - February 2026. He earned his MSc Tech from the Indian Institute of Technology (Indian School of Mines) [IIT (ISM)] Dhanbad, one of the premier institutes of Earth Sciences in the world. Dr. Mishra obtained a PhD/DSc from the Geodynamics Research Centre (GRC), Ehime University, Japan, and a postgraduate diploma in management (PGDMA) from AIMA, New Delhi. He worked with different organizations in India and abroad in different capacities, namely the Geological Survey of India for about 20 years where he applied almost all geophysical methods for natural resource exploration and contributed toward high-resolution electrical and seismic imaging of subsurface strata to understand the deep-seated resources and tectonic intricacies of the study region. He also served at Geodynamics Research Centre, Ehime University, where he conducted cooperative research on 3D seismic tomography for different geotectonic settings of the world. Dr. Mishra has held key positions, including the Head/Director of the SAARC Disaster Management Centre and Foreign Guest Professor at GRC, Japan. A prolific researcher, he has published a series of research papers and supervised several researchers at the graduate and doctoral levels. He has authored several textbooks, covering geophysical methods and geothermal energy exploration as strong classroom materials for students and researchers. His accolades include the National Mineral Award (2008) by the Government of India, and the Anni Talwani Gold Medal (2015) and Decennial Award (2025) by the Indian Geophysical Union (IGU). He is a life member of several esteemed professional bodies of different countries, including AGU, and of other professional bodies of the world.

Dr. D.C. Naskar was a formerly consultant at the National Centre for Seismology (NCS), Ministry of Earth Sciences (MoES), Government of India, New Delhi, where he mentored early-career scientists and plans seismological studies during July 2022 – March 2026. He holds a PhD in Science from Jadavpur University and an MSc in Physics from Calcutta University. His career began in July 1989 with the Geological Survey of India (GSI), where he served as the Director (Geophysics) until his retirement in January 2021. Over his 32 years at GSI, he gained extensive experience in geophysical mapping, leading surveys across India in mineral exploration, groundwater studies, and environmental research. He also served as the Chief Consultant (Geophysics) at DMT Pvt. Solutions Ltd, and worked in earlier roles with South Eastern Railway and the West Bengal State Electricity Board. Notable projects include a kimberlite exploration with De Beers and a GPS survey for the 2004 Indonesian earthquake's impact in the Andaman Nicobar

Islands. Dr. Naskar has authored a series of books, besides several research publications to his credit. He received awards of recognition, including the H.S. Pareek Award for his research on coal science. He is a life member of the American Geophysical Union (AGU) and the European Association of Geoscientists and Engineers (EAGE).

Foreword

The growing global demand for natural resources, coupled with increasing environmental concerns and the need for sustainable development, has placed geosciences at the forefront of responsible resource exploration. In this context, the book *Baseline Geoscience Data Generation for Sustainable Exploration of Natural Resources* by O.P. Mishra and D.C. Naskar emerges as a timely and comprehensive contribution to the scientific and the natural resource management community. This book offers an integrated perspective on geophysical, geospatial, and remote sensing techniques, emphasizing their critical role in generating reliable baseline data for informed decision-making in mineral and energy-based resource exploration.

The strength of this volume lies in its systematic and multidisciplinary approaches. Beginning with fundamental concepts of geophysics and the evolution of mineral exploration, the authors skilfully guided readers through a wide spectrum of modern exploration techniques. These include potential field methods, geoelectrical and electromagnetic surveys, seismic investigations, and advanced remote sensing technologies. Each chapter is thoughtfully structured, balancing theoretical foundations with practical applications and real-world case studies drawn from diverse geological settings across the globe.

One of the notable aspects of this book is its strong emphasis on the integration of geological, geophysical, geochemical, and geospatial data within a unified framework, reflecting the current direction of exploration geoscience, where interdisciplinary methods are essential for improving accuracy, reducing uncertainty, and minimizing environmental impact. The inclusion of geographic information systems, geographic positioning systems, drone-based surveying, and artificial intelligence further highlights the authors' forward-looking vision, aligning the text with emerging technological trends that are reshaping the industry.

The book also stands out for its global perspective in which several case studies on the national geophysical mapping programs from various countries are incorporated, providing valuable insights into international practices, challenges, and innovations. This makes the text particularly relevant not only for students and researchers but also for professionals, policymakers, and stakeholders involved in resource management worldwide.

Equally commendable is the attention given to sustainability and ethical considerations. The authors recognize that exploration activities must be conducted responsibly, with due regard for environmental protection, social impact, and long-term resource management. The discussion on ethical practices, data standardization, and international collaboration reinforces the importance of adopting a holistic and transparent approach in modern geoscience endeavors.

Another distinguishing feature of this work is its pedagogical value. The inclusion of practical exercises, programming applications using Python, R, and MATLAB, and hands-on case studies enhances its utility as a textbook for academic institutions. It not only equips students with theoretical knowledge but also develops their analytical and computational skills, which are essential in today's data-driven geoscience landscape.

In summary, this book represents a significant and much-needed addition to the literature on mineral exploration and geoscientific data analysis. Its comprehensive coverage, clarity of presentation, and emphasis on sustainability make it a valuable resource for a global audience. The esteemed publisher, Elsevier, has done justice by bringing the commendable and insightful contributions of authors with its great dedication to the scholarly efforts of authors.

I am confident that this volume will serve as a key reference for students, educators, researchers, and industry professionals and will contribute meaningfully to advancing sustainable exploration practices in the different parts of the globe.

Mrinal K. Sen
Professor and Shell Foundations Centennial Chair in Geophysics, The University of Texas at Austin, United States

Preface

The sustainable exploration of natural resources in the 21st century demands a holistic, strategic, and science-driven geoscientific approach that integrates diverse techniques and cutting-edge technological innovations. With global demand for natural resources rapidly increasing—driven by industrial growth, population expansion, and the worldwide shift toward green energy—there is a critical need to generate reliable baseline geoscience data through advanced, multidisciplinary tools. Such data serve as the foundation for informed decision-making in resource planning, environmental stewardship, and long-term sustainability. This book, *Baseline Geoscience Data Generation for Sustainable Exploration of Natural Resources*, is a timely and comprehensive contribution aimed at addressing this global imperative. It is designed to serve as an essential reference for geoscientists, researchers, students, and policymakers involved in the systematic investigation and sustainable management of Earth's subsurface resources.

Authored by two highly experienced geoscientists—**Dr. O.P. Mishra**, former Director of the National Centre for Seismology, Ministry of Earth Sciences, Government of India, and **Dr. D.C. Naskar**, former Director (Geophysics) of the Geological Survey of India and current Consultant at NCS—this book presents a consolidated framework of modern geoscientific methods, both theoretical and applied. Dr. Mishra brings with him over **35 years of expertise** in applied geophysics, seismology, and disaster risk science, his significant contributions to high-resolution seismic imaging and tectonic research, besides his outstanding contributions in the field of geophysical exploration with different organizations, including the Geological Survey of India. Dr. Naskar, with more than **35 years of professional experience**, offers deep insights into geophysical mapping, mineral exploration, and remote sensing, gained through his extensive work across diverse geological terrains in India and international collaborations. Together, their combined leadership and academic rigor provide a strong foundation for the themes explored in this volume.

In **Chapter 1**, we introduce the concept of **global geophysical mapping (GGM)** as a foundational and transformative tool in the domain of mineral exploration. The chapter explores how GGM integrates geophysical, geological, and geospatial datasets to generate region-wide resource intelligence. It outlines the importance of such large-scale initiatives in identifying prospective mineral zones, improving resource estimation, and reducing uncertainty. The global relevance of GGM is demonstrated through international case examples and emerging initiatives, emphasizing its strategic role in national resource planning and policy development. **Chapter 2** offers a comprehensive overview of both **traditional and modern geophysical techniques** used in baseline geoscience data generation. It provides conceptual clarity on the principles, instrumentation, and data types associated with electrical, magnetic, gravity, seismic, and electromagnetic methods. Emphasis is placed on the transition from single-method exploration to integrated, multisensor approaches, highlighting how the synergy between methods enhances subsurface imaging, improves geological modeling, and leads to more accurate interpretations—paving the way for informed and sustainable exploration.

In **Chapter 3**, the focus shifts to **potential field surveying**, encompassing both **magnetic and gravity methods**. The chapter details the physical principles, data acquisition strategies, and data processing workflows for these techniques, particularly in regional reconnaissance and tectonic studies. It emphasizes their utility in delineating large-scale geological structures, lithological boundaries, and concealed mineralized zones. The chapter further explores recent advances in airborne and satellite-based magnetic-gravity surveys and how modern processing tools (e.g., Euler deconvolution, upward continuation, and 3D inversion) have enhanced the resolution and reliability of subsurface interpretations. **Chapter 4** explores the principles and applications of **geoelectrical surveying**, including techniques like **resistivity imaging**, **vertical electrical sounding (VES)**, and **induced polarization (IP)**. These methods are widely applied in mineral exploration, groundwater prospecting, environmental assessments, and geo-engineering investigations. The chapter delves into practical survey design, field execution, and interpretation strategies, showcasing case examples from varied geological terrains. Special focus is placed on 2D and 3D electrical resistivity tomography (ERT), which has revolutionized near-surface imaging in complex and heterogeneous zones.

In **Chapter 5**, we examine **electromagnetic (EM) surveying techniques**, which are highly effective in detecting conductive minerals such as sulfides, graphite, and saline groundwater. The chapter discusses time-domain and frequency-domain EM methods, airborne and ground-based systems, and their applications in both reconnaissance and detailed exploration. Advanced topics such as EM inversion, depth resolution, and noise handling are addressed, along with case studies that illustrate the successful application of EM methods in structurally complex and mineral-rich terrains. **Chapter 6** transitions to **remote sensing techniques**, presenting an in-depth look at how satellite-based and aerial multispectral, hyperspectral, NISAR applications, and thermal imaging are used in lithological mapping, alteration mineral detection, and lineament analysis. The chapter highlights how remote sensing provides cost-effective, large-area synoptic views that complement ground geophysics. Practical applications in mineral targeting, structural interpretation, and environmental monitoring are discussed using real-world datasets and image processing techniques such as supervised classification, band rationing, and principal component analysis (PCA).

Chapter 7 focuses on **geospatial surveying techniques**, specifically the integration of **geographic information systems (GIS)** and **global positioning system (GPS)** technologies. It explains how spatial data is acquired, stored, analyzed, and visualized in GIS environments to support decision-making in mineral exploration. The chapter covers digital cartography, terrain modeling, spatial overlays, and proximity analysis. Practical examples include GIS-based mineral prospectivity modeling and GPS-integrated field mapping, highlighting how geospatial tools facilitate precise survey planning and real-time navigation in exploration projects. In **Chapter 8**, we explore the innovative use of **drone-based geoscientific surveying techniques**, which have become increasingly valuable for rapid, high-resolution data collection in inaccessible or rugged terrains. The chapter elaborates on drone-mounted sensors such as optical cameras, LiDAR, multispectral and thermal imagers, and their integration with GIS for 3D terrain modeling and anomaly detection. Operational aspects, data accuracy considerations, and case studies demonstrate how drones are reshaping the landscape of preliminary mineral assessments and environmental monitoring. **Chapter 9** covers **seismic surveying techniques**, presenting both high-resolution shallow seismic methods and deep seismic reflection/refraction surveys. The chapter introduces wave propagation theory, data acquisition configurations, and interpretation strategies for mapping subsurface layering, faults, and mineralized structures. It also addresses seismic tomography and passive seismology for tectonic and crustal studies. The role of seismic surveys in hazard assessment, particularly in seismically active and resource-bearing zones, is highlighted with Indian and global case applications.

Chapter 10 extends the discussion to the oceanic domain through **marine mineral exploration and the deep ocean mission**. This chapter reviews technologies such as marine magnetometry, seafloor mapping, marine seismic surveys, and remotely operated vehicles (ROVs) for exploring polymetallic nodules, gas hydrates, and hydrothermal sulfide systems. India's Deep Ocean Mission is presented as a case of national strategic exploration, aligned with global best practices. Environmental considerations and deep-sea mining regulations are also discussed to promote responsible ocean resource utilization. **Chapter 11** offers a **comparative analysis of National Geophysical Mapping Programs** from different countries, including the United States, Australia, Canada, China, and India. It reviews their strategic approaches, coverage areas, technologies adopted, and policy frameworks. This global perspective underscores the importance of coordinated, government-led initiatives in building open-access geoscientific databases and fostering collaboration between research institutions, industry, and policymakers. **Chapter 12** presents **case studies and applied insights** from real-world exploration campaigns. These include multimethod surveys in mineralized belts, exploration workflows in challenging terrains, and lessons learned from integrating geological, geophysical, and geochemical data. Each case highlights methodological challenges, solutions, and exploration outcomes, illustrating the value of holistic, interdisciplinary approaches in successful resource discovery.

Chapter 13 discusses the planning and strategic execution of **baseline geoscientific data acquisition**, focusing on survey design, instrumentation selection, data quality assurance, and prioritization strategies based on geological models and exploration goals. It also examines the integration of multiple survey methods, the role of stakeholder collaboration, and cost-effective resource deployment to ensure comprehensive yet efficient baseline data generation. In **Chapter 14**, the emphasis is on **hands-on practical applications**, including code-based examples and computational workflows for geophysical and remote sensing data analysis. Readers are introduced to data processing platforms such as Python, QGIS, and Geosoft Oasis Montaj, with guided exercises on anomaly enhancement, filtering, and inversion modeling. This chapter bridges theory and practice, equipping users with skills to process and interpret real exploration data. Finally, **Chapter 15** explores **emerging technologies in geoscience**, with a focus on artificial intelligence (AI), machine learning (ML), and geostatistics. It examines how these tools are being used for predictive modeling, mineral prospectivity mapping, and big data analytics in exploration. The chapter also emphasizes ethical considerations, including environmental sustainability, indigenous rights, and responsible data use, underscoring the need for a balanced approach between innovation and social responsibility in resource development.

Chapter 16 describes a structured compilation of key equations, concepts, and tools essential for geoscientific exploration. It brings together foundational and advanced methodologies—ranging from gravity, magnetic, electrical, seismic, and remote sensing techniques to emerging applications of AI, ML, and geostatistics. Designed as a practical reference, this chapter also includes unit conversions, software tools, visual aids, and integrated workflows to support accurate data interpretation and modeling. It aims to serve as a quick-access resource for geoscientists, students, and exploration professionals working across disciplines.

This book is a culmination of our collective experience in geophysical exploration and geoscientific research. It reflects not only the evolution of methods and tools but also the growing need for a responsible and informed approach to natural resource management. We trust that this work will serve as a critical reference and learning resource across academia, research institutions, government agencies, and the mineral exploration industry.

Dr. O.P. Mishra
Former Director National Centre for Seismology,
Ministry of Earth Sciences,
Government of India, New Delhi

Dr. D.C. Naskar
Former Director (Geophysics), GSI Consultant
National Centre for Seismology, New Delhi

Acknowledgment

Authors profusely express sincere gratitude to Dr. Jitendra Singh, Hon'ble Minister of Earth Sciences & Department of Science and Technology, Government of India for his kind inspiration and motivation to share our experience, expertise and lessons learnt among the users of diverse categories involved in the exploration of natural resources to make this piece of our contribution as a reference material for sustainable exploration of minerals and critical minerals, ground water, hydrocarbons, and renewable energy sources. Invaluable suggestions of the Hon'ble Minister to highlight the importance of digital data integration, spatial mapping, and advanced analytical tools associated with artificial intelligence and machine learning in the book is gratefully acknowledged, which form the backbone of informed decision-making for exploration, land-use planning, hazard mitigation, and sustainable utilization of natural resources.

The authors wish to express their sincere and profound gratitude to Dr. M. Ravichandran, Secretary, Ministry of Earth Sciences, Government of India, for his visionary leadership and unwavering support, which have provided the institutional guidance and encouragement necessary for scientific research and its dissemination of fundamental knowledge of the subject through the publication of this book.

The authors are deeply grateful to the Geological Survey of India (GSI) for providing the foundational support, infrastructure, and resources during our association with GSI, which made this publication possible as GSI has a proven track record of applying integrated geophysical methods for baseline geoscience data generation. Beyond the technical support, the collaborative spirit and mentoring environment within the GSI have continually inspired us to strive for excellence and maintain the highest standards of scientific inquiry throughout the preparation of this work.

The authors also extend their sincere gratitude to the National Centre for Seismology (NCS), Ministry of Earth Sciences (MoES), Government of India, for fostering an environment conducive to publishing the text and reference book for students and researchers. The support, facilities, and collaborative opportunities available at NCS have been invaluable in refining the methodologies, interpretation, and presentation of the material compiled in this book. The constructive interactions with colleagues and peers at NCS have contributed greatly by enhancing the analytical depth and quality of this book.

This work reflects the collective support, inspiration, and guidance received from different institutions, colleagues, and mentors who have contributed to the completion of this work. Finally, we most humbly acknowledge our family members for their unwavering support, patience, and encouragement during the preparation of this book till its finalisation. We are deeply grateful for their enduring patience.

We are especially grateful to Mr. Hari Dass Sharma, Technical Assistant at the NCS-MoES, for his meticulous assistance in preparing and enhancing the figures and other visual materials, which have significantly improved the clarity, accuracy, and presentation of this book. His dedication and minute attention ensured effective communication of the complex scientific data, making the book more accessible and informative to readers. Time-to-time support in refining figures by Mr. Ravikant Mahto, Research Associate, NCS-MoES, is also acknowledged. The authors would like to acknowledge peer reviewers of the original proposal of the book. Colleagues, mentors, and members of the broader scientific community are also acknowledged whose thoughtful feedback and constructive suggestions have enriched the quality of the book.

Finally, the authors understand that despite all efforts, if errors or shortcomings may remain with this edition, it is the sole responsibility of the authors to rectify them in the subsequent edition of the book.

We also sincerely thank the publisher Elsevier and its entire editorial and production team for their professionalism and sincere effort to translate this contribution into an excellent book to fruition.

O.P. Mishra
D.C. Naskar

Synopsis

Baseline Geoscience Data Generation for Sustainable Exploration of Natural Resources offers a comprehensive, interdisciplinary, and future-focused exploration of the methods, technologies, and strategies necessary for the responsible discovery and development of mineral and natural resources. Authored by Dr. O.P. Mishra, former Director of the National Centre for Seismology, Ministry of Earth Sciences, Government of India, and Dr. D.C. Naskar, former Director (Geophysics) at the Geological Survey of India and consultant to MoES, the book draws upon decades of institutional expertise and field practice to present a cohesive framework for integrating advanced geoscientific data into sustainable exploration workflows.

Designed for a global readership of geoscientists, researchers, policy architects, and advanced students, the book navigates the reader through the full spectrum of baseline data generation—from fundamental principles to applied case studies and emerging technologies. It emphasizes the convergence of geophysical, geospatial, geochemical, and remote sensing techniques in building a scientifically sound and environmentally responsible resource exploration paradigm.

Beginning with foundational chapters, the book details the evolution of geophysical exploration methods—including resistivity, induced polarization, gravity, magnetics, gamma-ray spectrometry, and seismic techniques—and their application across diverse geological settings. It systematically explains airborne, borehole, waterborne, and marine geophysical techniques, establishing the importance of multidisciplinary data acquisition and interpretation.

Chapters on gravity and magnetic surveying provide insight into the role of potential field methods in detecting ore bodies such as iron, gold, copper, and geothermal systems. The text expands into electrical and electromagnetic methods like magnetotellurics, time-domain EM, and induced polarization, with specific attention to their role in subsurface characterization. Ground penetrating radar (GPR) is introduced as a complementary tool for shallow investigations.

Remote sensing is treated both practically and theoretically, with strong emphasis on spectral analysis, satellite imagery, sensor technology, and image processing techniques for mineral prospectivity analysis. These are paired with real-world case studies to demonstrate the application of Earth observation data in exploration geology. Further, the book highlights the synergy of GIS and GPS technologies for spatial analysis, mineral system modeling, and exploration targeting. A notable feature is the extensive treatment of drone-based geoscientific surveys, including unmanned aerial vehicle integration with magnetic, GPR, seismic, and electromagentic systems—underscoring their role in enhancing resolution, reducing costs, and improving field efficiency.

The seismic chapters offer a detailed account of wave propagation, seismic reflection techniques, and modern crustal imaging in mineral exploration. Case studies span hard rock terrains, shale gas fields, and offshore zones, including the Tropicana Gold Mine and marine gas fields, demonstrating seismic methods as powerful tools for subsurface mapping and resource prediction. The section on marine mineral exploration and deep ocean missions investigates polymetallic nodules, seafloor massive sulfides (SMS), and Rare Earth Elements (REEs)-rich muds. The book delves into the engineering, environmental, and legal dimensions of deep-sea mining, including a discussion of the High Seas Agreement (2023) and Blue Economy frameworks. Advanced technologies such as controlled-source electromagnetics, self-potential mapping, and GIS-based marine exploration are contextualized through global examples from Japan, India, and Alaska.

A globally comparative chapter evaluates national geophysical mapping programs, offering insight into approaches adopted by India, the United States, Canada, Australia, China, and several European and African countries. The role of national geological surveys, data standardization, public-private partnerships, and international collaboration is explored in detail. Case studies from India's GSI-led programs showcase the integration of multidisciplinary data for resource targeting across different tectonic regimes.

Several chapters are devoted to case studies, offering rich, contextual analysis of exploration programs. Indian examples include gold and base metal discoveries in the Cuddapah Basin, Kerala coastal deposits, and prospects in Tamil Nadu and Rajasthan. Globally, case studies span Southeast Asia, Sweden, Australia, and the United States—illustrating how integrated geophysical and geospatial surveys are unlocking critical mineral potential in varied geological terrains.

A dedicated chapter on data acquisition planning discusses the strategy behind collecting and integrating geological, geochemical, and geophysical datasets. Case studies from South Africa, Australia, the Andes, and Namibia illustrate how modern exploration campaigns optimize data acquisition, leverage remote technologies, and prioritize environmental and social responsibility. In bridging theory and application, the book includes a robust module on hands-on data analysis, featuring programming exercises using Python, R, and MATLAB for geophysical, remote sensing, and geospatial data interpretation. These exercises, coupled with tutorials and real-world datasets, empower learners to build exploration models and interpret complex geoscientific information.

Emerging technologies form a critical component of the book's forward-looking perspective. Artificial intelligence and machine learning are explored for their transformative roles in geophysical interpretation, prospectivity mapping, uncertainty quantification, and seismic imaging. Case studies from Brazil, Tibet, Xinjiang, Canada, and Africa show how AI/ML is driving more accurate, scalable, and ethical decision-making in exploration. Ethical dimensions—including environmental impact, fair sourcing, community benefit, transparency, and long-term sustainability—are integrated throughout this discussion, asserting that innovation must be accompanied by responsibility.

The final chapter provides a dictionary of equations and geophysical formulas—a technical reference that supports the practical application of concepts presented throughout the book. In summary, this book presents a holistic, globally relevant, and technically rigorous treatment of baseline geoscience data and its pivotal role in advancing sustainable exploration of natural resources. It fuses classical geophysical principles with the latest technological innovations, enriched by diverse case studies, practical exercises, and ethical foresight. With its dual focus on scientific depth and applied relevance, the book is an essential resource for professionals, students, and policymakers aiming to balance mineral development with environmental stewardship and socioeconomic equity in a rapidly evolving global landscape.

CHAPTER

1

Global geophysical mapping (GGM): A catalyst for mineral exploration

1.1 Introduction

Mineral exploration utilizes a broad array of methodologies and techniques, often in combination, to identify and assess potential mineral deposits across the globe. These methods include geological mapping, which involves detailed analysis of surface geology to understand the distribution of different rock types and structures (Mishra and Naskar, 2025; Naskar et al., 2024). Geochemical surveys play a critical role by examining the concentration of specific elements or minerals in soil and rock samples, with the presence of certain "indicator minerals" often signaling the possibility of a larger deposit beneath the surface.

Geophysical surveys are another essential tool, utilizing advanced technology to measure the physical properties of rocks and soils. These surveys are designed to identify subsurface features that could indicate the presence of mineral deposits. Remote sensing and satellite imagery have become increasingly significant in the modern era, allowing geologists to capture images across various wavelengths of light, which help identify minerals at the surface and, in some cases, beneath vegetation or other surface cover. Drilling is a direct exploration method that involves extracting cylindrical core samples from the earth to analyze rock types and mineral grades. Trenching and pitting, which are surface excavation techniques, also provide valuable insights by exposing bedrock for detailed sampling and examination.

On a global scale, geophysical surveys are fundamental in mineral exploration, guiding all stages of the process with clearly defined objectives. These surveys rely on both passive methods (which utilize natural earth fields) and active methods (which involve artificially energizing the earth), interpreting the resulting responses to reveal geological variations indicative of mineral deposits. Predicting mineral prospectivity is a complex task that requires the integration of geological, geophysical, geochemical, and remote sensing data (Sun et al., 2019). The form and distribution of mineral deposits are closely linked to these data, which reflect the signatures of the intricate processes involved in ore formation. Access to high-quality baseline data is crucial for improving the effectiveness of global mineral exploration.

Baseline Geoscience Data Generation for Sustainable Exploration of Natural Resources
https://doi.org/10.1016/B978-0-443-45221-5.00001-1

By leveraging advanced technologies and comprehensive data integration, the exploration community can enhance its ability to identify and evaluate mineral resources worldwide, ensuring more efficient and sustainable exploration efforts.

With the accelerating pace of global development, the demand for minerals has surged, driven by the need for key commodities such as gold, copper, iron, and other essential resources. Mineral exploration, however, presents significant challenges, particularly when vast, often rugged areas must be surveyed. In many cases, minerals are hidden beneath complex geological formations, making their identification and extraction even more difficult. Deposits may be deep, small, scattered, or of low grade, with only a minimal physical contrast between the host rock and the targeted mineral, making detection more challenging. Given these complexities, it is critical to select the right investigation techniques to effectively address these challenges. Geophysical methods, in particular, are often the most suitable choice. These techniques rely on measuring the response of geological materials to physical forces, with the goal of identifying anomalies that may indicate the presence of mineral deposits worth further exploration. In geophysical exploration, parameters such as density, magnetic susceptibility, and electrical conductivity are essential, as they provide the contrast needed to differentiate between the host rock and the target mineral. While elastic properties also play a role, their use in exploration is more limited.

As countries worldwide pursue green goals and transition to renewable energy, there is an increased focus on the search for minerals critical to emerging technologies, particularly those used in electronics and battery production. This has spurred a global push to explore the Earth's crust for specific minerals that are essential to the development of new technologies. As a result, the exploration community faces the dual challenge of improving the efficiency of exploration techniques while also meeting the growing demand for minerals that will enable the next generation of sustainable technologies.

For decades, geophysical instruments such as magnetometers, gravimeters, electromagnetic (EM) systems, resistivity, self-potential (SP), and induced polarization (IP) equipment have played a crucial role in mineral exploration. These tools offer extensive coverage of both horizontal and vertical dimensions, making them invaluable in exploring large areas. On a regional scale, geophysical surveys are highly effective in validating geological concepts and generating evidence maps that outline key mineralization controls, such as craton margins, mobile belt contacts, large igneous provinces, crustal-scale lineaments, and mantle plumes, as well as flood basalts and younger intrusives (Naskar et al., 2020). At more detailed exploration stages, geophysical prospecting can directly detect ore bodies when significant contrasts in physical properties exist, such as in deposits of iron oxide-copper-gold, skarn, massive sulfides, barytes, and chromite (Kaur et al., 2018).

Globally, mineral exploration is typically focused on metallic deposits, though it also covers bulk minerals, fossil fuels, and groundwater. The majority of exploration efforts are directed toward metallic deposits, using gravity, magnetic, and radiometric methods at the regional scale, followed by various electrical and electromagnetic techniques in more detailed investigations. The confidence level in interpreting geophysical data varies by method, with most techniques providing indirect evidence of mineralization by mapping favorable host rocks and geological environments, such as carbonatites, kimberlites, porphyritic intrusions, and hydrothermal alteration zones. Additionally, geophysical surveys are effective in

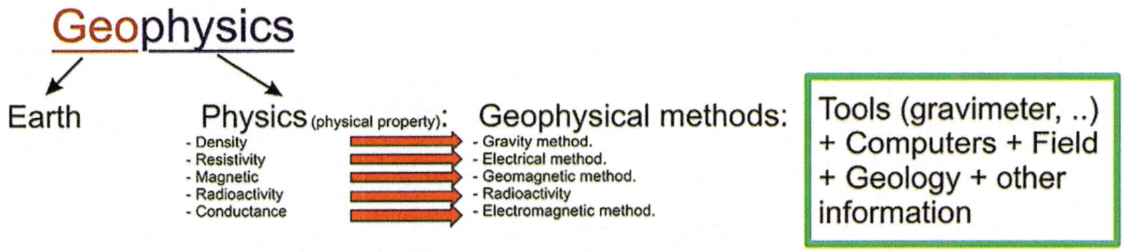

FIG. 1.1 Introduction to geophysics. *United Arab Emirates University (UAEU) & after: EIT Raw Materials, MinEX Consulting (2014), Dentith and Mudge (2014), https://em.geosci.xyz content physical properties electrical conductivity electrical conductivity values Open Access (https://em.geosci.xyz/).*

identifying structural controls of mineralization, including fault zones, fractures, formation contacts, and shear zones.

Exploration is not limited to discovering new deposits but also includes the re-evaluation of known or previously surveyed areas, often driven by advancements in technology or rising commodity prices, which can render previously uneconomic deposits viable. This re-evaluation may also involve expanding mineral resources at active mine sites or reprocessing waste material. After initial geophysical surveys, more detailed investigations typically follow, often targeting specific areas with borehole geophysics, coring, and physical sampling. The core of geophysical methods lies in the analysis of energy propagation through geological media. Integrating various geophysical techniques ensures greater accuracy and reliability, playing an essential role throughout all stages of global mineral exploration (Sundararajan, 2012).

1.1.1 What is geophysics?

Geophysics is a major branch of Earth science that applies the principles and methods of physics to study the Earth (Fig. 1.1). Traditionally, the term geophysics refers to the solid Earth applications, which include studying the Earth's shape, gravitational and magnetic fields, electromagnetic fields, internal structure, composition, dynamics, and their surface expressions in processes such as plate tectonics, magma generation, volcanism, and rock formation (Sheriff, 1991). Key aspects of geophysics include: Georadar—Dielectric constant; Seismics—Elastic constants, density; Spontaneous Polarization (SP)—Oxidation potential, hydrogen concentration; Induced Polarization (IP)—Electrochemical properties of electronically conducting particles in rock pores; Rock-specific physical properties. To study these properties, specific geophysical methods are employed, supported by computational tools and technology.

1.1.2 Objective of geophysical surveys

The primary aim of a geophysical survey is to gather information about the geological structure or buried objects, such as their existence, size, and shape, by examining the distribution of physical properties (Fig. 1.2).

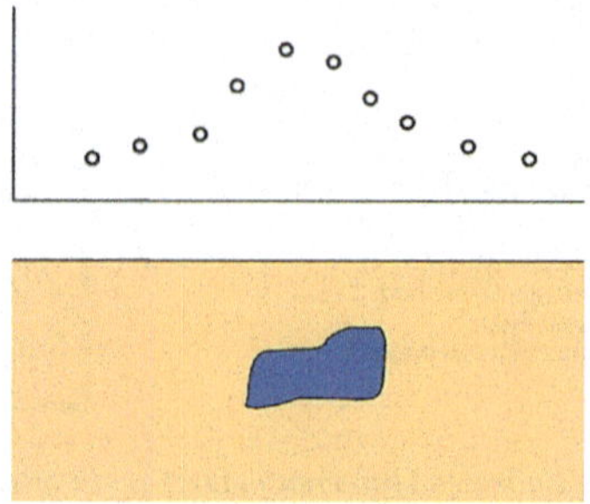

FIG. 1.2 Detection and analysis of geophysical signals. *United Arab Emirates University (UAEU) & after: EIT Raw Materials, MinEX Consulting (2014), Dentith and Mudge (2014), https://em.geosci.xyz content physical properties electrical conductivity electrical conductivity values Open Access (https://em.geosci.xyz/).*

1.1.2.1 *Forward and inverse problems*

Geophysical methods rely on studying the propagation of various physical fields within the Earth. Traditional geophysical data analysis involves developing geological models and comparing them with observational data. Numerical modeling of geophysical data for a given set of parameters is often referred to as a forward problem. Solving the forward problem allows us to predict geophysical data based on specific geological structures.

In the past, a simple "layer cake" ground model was often used to generate synthetic data, which would then be compared with field measurements. The model would be adjusted manually until the synthetic and actual data matched. This labor-intensive approach, known as forward modeling, was feasible only for simple geologies (e.g., 1D).

However, forward modeling has largely been replaced by automatic numerical inversion processing, which directly converts measurements into spatial property distributions (1D, 2D, and 3D) without manual intervention. Inversion is a key aspect of modern geophysical data processing and interpretation, offering significant improvements over forward modeling. It enables the display of the spatial distribution of geophysical properties through tomographic images. The inverse problem in geophysical exploration involves determining subsurface model parameters by approximating the medium type, typically treated as a least-squares optimization problem. This optimization minimizes the difference between observed and predicted data (Fig. 1.3), expressed as (Virieux and Operto, 2009; Tarantola, 1984):

$$\delta \mathrm{d} = \mathrm{d}_{\mathrm{obs}} - \mathrm{d}_{\mathrm{pred}} \tag{1.1}$$

1.1.3 Geophysics in mineral exploration

Exploration geophysics is a powerful tool used to directly detect mineralization by measuring the physical properties of subsurface materials. For instance, density contrasts can be measured between dense iron ore and lighter silicate host rocks, or electrical conductivity contrasts can be detected between conductive sulfide minerals and resistive silicate rocks. A general overview of geophysical methods in mineral exploration includes the following: The

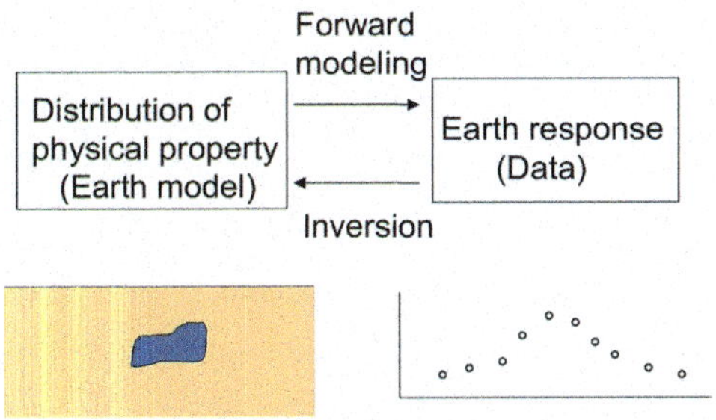

FIG. 1.3 Forward and inverse challenges in geophysical exploration. *United Arab Emirates University (UAEU) & after: EIT Raw Materials, MinEX Consulting (2014), Dentith and Mudge (2014), https://em.geosci.xyz content physical properties electrical conductivity electrical conductivity values Open Access (https://em.geosci.xyz/).*

science of generating indirect information about subsurface geological structures and anomalous physical and chemical properties of rocks at depth using geophysical instruments; **Geophysical exploration methods** include: Surface-based methods, Airborne methods (e.g., drones, airplanes), In-mine methods (e.g., tunnels, open pits), Borehole-based methods; Geophysical models can be constructed in 1D, 2D, 3D, and 4D (time); Additional applications include water resource studies, environmental impact assessments, and monitoring.

A key aspect of mineral exploration geophysics is the identification of physical property contrasts. In recent years, there has been increased recognition of the need for a deeper understanding of these physical properties. As a result, there has been a growing push to collect more data and use it more effectively. Several reasons contribute to this trend: geophysicists seek more physical property data to improve survey planning and interpretation, while geologists use these properties to gain additional insights into geology that might be difficult to detect with traditional or more costly methods, such as hand specimen examination, thin section analysis, lithogeochemistry, or assays (Table 1.1). If a drill hole intersects a geophysical target, physical property measurements of the core or surrounding rocks can verify whether the target was reached and provide data to model the target's response (Smith, Shore, and Rainsford, 2012).

1.1.4 The role of geophysics in mineral exploration

Geophysics has long been a crucial tool in mineral exploration, assisting in the identification and delineation of mineral deposits and playing a vital role in reducing geological uncertainty in mining operations. Haile and Atsbaha (2014) highlight that geophysical techniques are routinely integrated into geological investigations to map subsurface geological structures. As noted by Frasheri et al. (1995), the most commonly employed geophysical methods in mineral exploration include electrical, electromagnetic, gravimetric, magnetic, and seismic techniques.

Shallow geophysics, which focuses on investigating geological structures in the upper crust, is frequently utilized in modern mining practices (Zhou et al., 2014; Sun et al., 2017). Resistivity and induced polarization (IP) methods are particularly useful for exploring deposits of sulfides, metals, and graphite (Yusefi et al., 2017). Moreira et al. (2016) demonstrated

TABLE 1.1 Qualitative evaluation of geophysical method applicability.

Geophysical method	Air or ground	Application	Mineral
			Diamonds
Magnetic	Air	Geological framework	Highly effective
		Direct targeting	Highly effective
	Ground	Geological framework	Highly effective
		Direct targeting	Highly effective
			Lode gold
	Air	Geological framework	Highly effective
		Direct targeting	Generally ineffective
	Ground	Geological framework	Highly effective
		Direct targeting	Moderately effective
			VMS deposits
	Air	Geological framework	Highly effective
		Direct targeting	Highly effective
	Ground	Geological framework	Highly effective
		Direct targeting	Highly effective
			MVT lead-zinc deposits
	Air	Geological framework	Highly effective
		Direct targeting	Generally ineffective
	Ground	Geological framework	Moderately effective
		Direct targeting	Generally ineffective
			SEDEX deposits
	Air	Geological framework	Highly effective
		Direct targeting	Moderately effective
	Ground	Geological framework	Highly effective
		Direct targeting	Moderately effective
			Porphyry copper deposits
	Air	Geological framework	Highly effective
		Direct targeting	Highly effective
	Ground	Geological framework	Highly effective
		Direct targeting	Highly effective

Geophysical method	Air or ground	Application	Mineral
			Uranium deposits
	Air	Geological framework	Highly effective
		Direct targeting	Generally ineffective
	Ground	Geological framework	Highly effective
		Direct targeting	Generally ineffective
			Olympic dam type deposits
	Air	Geological framework	Highly effective
		Direct targeting	Highly effective
	Ground	Geological framework	Highly effective
		Direct targeting	Highly effective
			Magmatic Ni–Cu-PGEs deposits
	Air	Geological framework	Highly effective
		Direct targeting	Moderately effective
	Ground	Geological framework	Moderately effective
		Direct targeting	Moderately effective
			Diamonds
Electromagnetic	Air	Geological framework	Moderately effective
		Direct targeting	Highly effective
	Ground	Geological framework	Highly effective
		Direct targeting	Highly effective
			Lode gold
	Air	Geological framework	Moderately effective
		Direct targeting	Moderately effective
	Ground	Geological framework	Moderately effective
		Direct targeting	Generally ineffective
			VMS deposits
	Air	Geological framework	Moderately effective
		Direct targeting	Highly effective
	Ground	Geological framework	Moderately effective
		Direct targeting	Highly effective
			MVT lead-zinc deposits
	Air	Geological framework	Generally ineffective
		Direct targeting	Generally ineffective

Continued

TABLE 1.1 Qualitative evaluation of geophysical method applicability—cont'd

Geophysical method	Air or ground	Application	Mineral
	Ground	Geological framework	Moderately effective
		Direct targeting	Moderately effective
			SEDEX deposits
	Air	Geological framework	Moderately effective
		Direct targeting	Moderately effective
	Ground	Geological framework	Moderately effective
		Direct targeting	Moderately effective
			Porphyry copper deposits
	Air	Geological framework	Generally ineffective
		Direct targeting	Generally ineffective
	Ground	Geological framework	Generally ineffective
		Direct targeting	Generally ineffective
			Uranium deposits
	Air	Geological framework	Highly effective
		Direct targeting	Highly effective
	Ground	Geological framework	Highly effective
		Direct targeting	Highly effective
			Olympic dam type deposits
	Air	Geological framework	Generally ineffective
		Direct targeting	Moderately effective
	Ground	Geological framework	Generally ineffective
		Direct targeting	Moderately effective
			Magmatic Ni-Cu-PGEs deposits
	Air	Geological framework	Highly effective
		Direct targeting	Highly effective
	Ground	Geological framework	Highly effective
		Direct targeting	Highly effective
			Diamonds
Electric	Ground	Geological framework	Generally ineffective
		Direct targeting	Moderately effective

Geophysical method	Air or ground	Application	Mineral
			Lode gold
	Ground	Geological framework	Generally ineffective
		Direct targeting	Moderately effective
			VMS deposits
	Ground	Geological framework	Moderately effective
		Direct targeting	Highly effective
			MVT lead-zinc deposits
	Ground	Geological framework	Moderately effective
		Direct targeting	Highly effective
			SEDEX deposits
	Ground	Geological framework	Generally ineffective
		Direct targeting	Generally ineffective
			Porphyry copper deposits
	Ground	Geological framework	Generally ineffective
		Direct targeting	Highly effective
			Uranium deposits
	Ground	Geological framework	Generally ineffective
		Direct targeting	Moderately effective
			Olympic dam type deposits
	Ground	Geological framework	Moderately effective
		Direct targeting	Moderately effective
			Magmatic Ni-Cu-PGEs deposits
	Ground	Geological framework	Highly effective
		Direct targeting	Highly effective
			Diamonds
Gravity	Air	Geological framework	Moderately effective
		Direct targeting	Highly effective
	Ground	Geological framework	Highly effective
		Direct targeting	Highly effective
			Lode gold
	Air	Geological framework	Moderately effective
		Direct targeting	Generally ineffective

Continued

TABLE 1.1 Qualitative evaluation of geophysical method applicability—cont'd

Geophysical method	Air or ground	Application	Mineral
	Ground	Geological framework	Moderately effective
		Direct targeting	Generally ineffective
			VMS deposits
	Air	Geological framework	Moderately effective
		Direct targeting	Highly effective
	Ground	Geological framework	Moderately effective
		Direct targeting	Highly effective
			MVT lead-zinc deposits
	Air	Geological framework	Moderately effective
		Direct targeting	Moderately effective
	Ground	Geological framework	Moderately effective
		Direct targeting	Highly effective
			SEDEX deposits
	Air	Geological framework	Moderately effective
		Direct targeting	Moderately effective
	Ground	Geological framework	Moderately effective
		Direct targeting	Highly effective
			Porphyry copper deposits
	Air	Geological framework	Moderately effective
		Direct targeting	Generally ineffective
	Ground	Geological framework	Moderately effective
		Direct targeting	Moderately effective
			Uranium deposits
	Air	Geological framework	Moderately effective
		Direct targeting	Generally ineffective
	Ground	Geological framework	Highly effective
		Direct targeting	Generally ineffective
			Olympic dam type deposits
	Air	Geological framework	Moderately effective
		Direct targeting	Moderately effective
	Ground	Geological framework	Moderately effective
		Direct targeting	Moderately effective

Geophysical method	Air or ground	Application	Mineral
			Magmatic Ni-Cu-PGEs deposits
	Air	Geological framework	Highly effective
		Direct targeting	Highly effective
	Ground	Geological framework	Highly effective
		Direct targeting	Highly effective
			Diamonds
Radiometric	Air	Geological framework	Moderately effective
		Direct targeting	Generally ineffective
	Ground	Geological framework	Generally ineffective
		Direct targeting	Generally ineffective
			Lode gold
	Air	Geological framework	Highly effective
		Direct targeting	Moderately effective
	Ground	Geological framework	Highly effective
		Direct targeting	Moderately effective
			VMS deposits
	Air	Geological framework	Highly effective
		Direct targeting	Moderately effective
	Ground	Geological framework	Moderately effective
		Direct targeting	Moderately effective
			MVT lead-zinc deposits
	Air	Geological framework	Moderately effective
		Direct targeting	Generally ineffective
	Ground	Geological framework	Generally ineffective
		Direct targeting	Generally ineffective
			SEDEX deposits
	Air	Geological framework	Highly effective
		Direct targeting	Generally ineffective
	Ground	Geological framework	Moderately effective
		Direct targeting	Generally ineffective
			Porphyry copper deposits

Continued

TABLE 1.1 Qualitative evaluation of geophysical method applicability—cont'd

Geophysical method	Air or ground	Application	Mineral
	Air	Geological framework	Highly effective
		Direct targeting	Highly effective
	Ground	Geological framework	Highly effective
		Direct targeting	Highly effective
			Uranium deposits
	Air	Geological framework	Highly effective
		Direct targeting	Highly effective
	Ground	Geological framework	Highly effective
		Direct targeting	Highly effective
			Olympic dam type deposits
	Air	Geological framework	Highly effective
		Direct targeting	Highly effective
	Ground	Geological framework	Highly effective
		Direct targeting	Highly effective
			Magmatic Ni-Cu-PGEs deposits
	Air	Geological framework	Moderately effective
		Direct targeting	Generally ineffective
	Ground	Geological framework	Moderately effective
		Direct targeting	Generally ineffective
			Diamonds
Seismic	Ground	Geological framework	Moderately effective
		Direct targeting	Moderately effective
			Lode gold
	Ground	Geological framework	Moderately effective
		Direct targeting	Generally ineffective
			VMS deposits
	Ground	Geological framework	Moderately effective
		Direct targeting	Moderately effective
			MVT lead-zinc deposits
	Ground	Geological framework	Moderately effective
		Direct targeting	Moderately effective

Geophysical method	Air or ground	Application	Mineral
			SEDEX deposits
	Ground	Geological framework	Moderately effective
		Direct targeting	Moderately effective
			Porphyry copper deposits
	Ground	Geological framework	Moderately effective
		Direct targeting	Moderately effective
			Uranium deposits
	Ground	Geological framework	Moderately effective
		Direct targeting	Moderately effective
			Olympic dam type deposits
	Ground	Geological framework	Moderately effective
		Direct targeting	Moderately effective
			Magmatic Ni-Cu-PGEs deposits
	Ground	Geological framework	Moderately effective
		Direct targeting	Moderately effective

Note: VMS: Volcanogenic massive sulfide, MVT: Mississippi valley type, SEDEX: Sedimentary exhalative, Porphyry copper deposit: Porphyry copper deposits are copper ore bodies that are formed from hydrothermal fluids that originate from a voluminous magma chamber several kilometers below the deposit itself, Olympic dam type deposit: Iron oxide associated copper-uranium-gold-silver-rare earth element deposit, located 520 km north—northwest of Adelaide in central South Australia.
Source: United Arab Emirates University (UAEU)—The 16th Arab International Conference of Mineral Resources, UAE.

how resistivity and IP were essential in modeling a gold deposit, while Coelho et al. (2020) assessed the potential of resistivity tomography as a prospecting tool for supergene ore. Martins (2016) showcased how resistivity and IP methods support morphological modeling in limestone mines.

Additionally, Martins (2016) illustrated that resistivity and IP methods are invaluable in small-scale mining by providing essential geological modeling data quickly and at a low cost. This approach has been shown to reduce the number of unproductive drill holes by 30–50%, making exploration more cost-effective. Moreira et al. (2019) further advocated for resistivity tomography in mineral exploration due to its speed and cost-reducing benefits.

Incorporating geophysical methods throughout all stages of a small-scale mining project can significantly decrease geological uncertainty. These methods help identify and delineate mineral deposits while providing crucial data for mine planning and decision-making. Dentith and Mudge offer comprehensive descriptions of the most commonly used geophysical methods in mineral exploration, particularly in their application to diverse mineral deposits. The growing global demand for mineral commodities has spurred increased use of geophysical technologies in exploring various types of mineral deposits. A schematic illustration of geophysical surveying used for detecting mineralization and mapping contacts between different rock types is shown in Fig. 1.4.

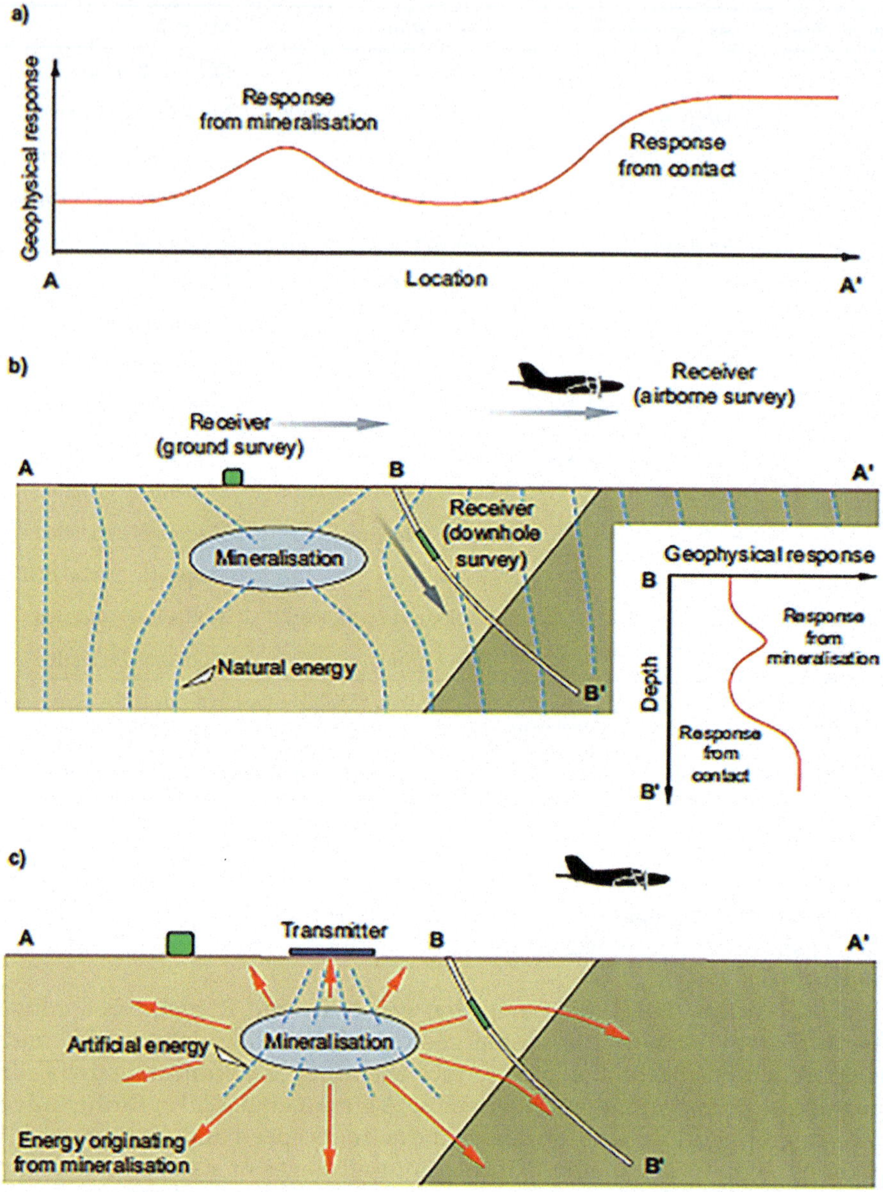

FIG. 1.4 Schematic representation of geophysical signatures on the surface through modern technological devices. *Source: Dentith and Mudge (2014).*

1.2 The evolution of mineral exploration: From ancient searches to modern technologies

Mineral exploration has evolved significantly from ancient times when early humans relied on surface clues and manual digging to find metals like gold, copper, and iron. Over the centuries, knowledge of geology, mineralogy, and geography expanded, leading to more systematic approaches. The Industrial Revolution introduced mechanized tools and basic geological mapping. In the modern era, exploration has become highly advanced, using technologies such as satellite imaging, geophysical surveys, remote sensing, and AI-driven data analysis. This evolution has made exploration more accurate, efficient, and sustainable, enabling access to deeper and previously hidden mineral resources.

1.2.1 Introduction to mineral exploration

Mineral exploration is the systematic process of identifying and assessing the extent of mineral deposits within the Earth's crust, aiming to locate ore bodies that can be mined profitably. This vital activity drives the mining industry by providing the raw materials essential for modern society, including those used in electric vehicles, renewable energy technologies, and everyday products (Moore, 2023). The exploration process involves multiple stages, beginning with initial research and data analysis, and advancing to detailed drilling and sampling. Each stage generates valuable information that helps exploration companies decide whether to proceed with mining operations.

1.2.2 Early beginnings of mineral exploration

The origins of mineral exploration trace back to the earliest human civilizations, which began searching for stones and metals to create tools. Over time, scientific methods were applied, especially during the Middle Ages, leading to more organized exploration. The Industrial Revolution further intensified exploration efforts, driven by the rising demand for minerals like coal and iron.

1.2.3 Technological advancements in exploration

In the 20th century, technological advancements, such as satellite imagery and geophysical surveys, revolutionized the field, enabling more precise and efficient exploration practices. Today, mineral exploration has become a sophisticated, multidisciplinary endeavor, utilizing cutting-edge technology to discover new mineral resources that are critical for sustaining modern society.

1.2.4 The role of mineral exploration in the mining industry

As the foundation of the mining industry, which supports many global economies, mineral exploration is crucial for several reasons. It stimulates economic growth by generating substantial revenue from mineral extraction, supports technological progress by providing minerals needed for advanced technologies, and informs strategic planning for both governments and industries by mapping the availability of key resources.

1.2.5 Stages of mineral exploration

The process of mineral exploration is complex, unfolding through distinct stages, each of which is crucial for ultimately locating viable mineral deposits. It begins with reconnaissance, where broad surveys are conducted to identify areas with potential mineralization. Target generation follows, where specific areas with promising mineral deposits are pinpointed for further investigation. Permitting is a critical stage in which the necessary legal and environmental clearances are obtained before exploration can proceed. Surface exploration techniques are then employed to gather more detailed geological data. Early-stage exploration involves detailed mapping, sampling, and geophysical surveys to assess the mineral potential of the area. Core drilling is conducted to extract rock samples from beneath the surface for analysis. Resource modeling then helps estimate the size, grade, and distribution of the mineral deposit, which leads to de-risking, where potential economic and environmental risks are evaluated. Finally, production decision-making occurs, where companies decide whether to proceed with mining operations based on the collected data, analysis, and feasibility studies.

1.3 Geophysical techniques in mineral exploration: Tools and approaches for effective targeting

Innovative approaches to mineral targeting rely on a range of geophysical investigation techniques that can address common challenges encountered during the exploration phase of a mining project. These methods offer a nondestructive and cost-effective means of gaining valuable insights into subsurface conditions. For prospecting and exploration, techniques such as induced polarization (IP), electromagnetic (EM) surveys, resistivity, and self-potential (SP) are essential, with ground-penetrating radar (GPR) and seismic surveys also providing critical data, particularly in borehole investigations.

It's important to note that, in some cases, the primary goal is not to directly detect the mineral itself but rather to understand the geological setting in which it is likely to be found. For example, gold deposits are challenging to detect directly, but the surrounding geological environment, characterized by specific electrical properties like conductivity, resistivity, and chargeability, can help pinpoint areas of potential interest. Geophysical modeling is crucial in this process, offering generalized but nonunique solutions that help determine the geometry (shape, size, etc.) of subsurface sources. Using advanced mathematical tools, the spatial location and depth of these sources can be estimated with a high degree of accuracy, helping to guide further exploration efforts.

1.3.1 Resistivity and induced polarization (IP) in geophysical mapping for mineral exploration

Resistivity and Induced Polarization (IP) surveys are frequently conducted together, as the same equipment can be used to measure both parameters without significantly increasing field acquisition time. The combination of resistivity and IP provides complementary data that enhances the interpretation of subsurface conditions, which is particularly useful in mineral exploration targeting specific deposits. Measurements can be taken from the surface or using a combination of surface and borehole electrodes, improving resolution and providing a clearer definition of three-dimensional targets.

1.3.2 Transient electromagnetic (TEM) surveys in global geophysical mapping

Transient Electromagnetic (TEM) surveys are particularly effective for mapping conductive mineral deposits, such as high-grade polymetallic sulfides (Cu, Zn, Pb ± Au, Ag) or Ni–Cu (±Co) in mafic rocks, as well as gold deposits associated with high-grade pyrrhotite or pyrite mineralization. TEM is also useful for determining the depth to bedrock, mapping variations in bedrock type, and identifying changes within the bedrock. Recent advancements in processing software have enhanced the ability to extract induced polarization (IP) effects from TEM data, opening up additional opportunities for using TEM systems in mineral exploration.

1.3.3 Self-potential (SP) surveys in geophysical mapping for mineral exploration

Self-potential (SP) surveys can be conducted using similar equipment to that used for IP and resistivity measurements, with some specific requirements for the measuring electrodes. SP measurements are particularly useful for exploring sulfide mineralization, as these formations often contain base metals like Cu, Zn, Ni, and Pb.

1.3.4 Ground penetrating radar (GPR) in geophysical mapping for subsurface exploration

Ground Penetrating Radar (GPR) can be an invaluable tool in specific exploration applications, offering essential data to inform further investigations. GPR measurements are useful for determining the depth to bedrock, conducting borehole investigations to identify geological contacts in the bedrock, and gathering information on fractures or rock quality, such as fractures in marble or clay layers in limestone. Additionally, GPR is effective for exploring fluvial deposits of gold and diamonds, as well as beach deposits containing titanium and iron-rich heavy minerals. It can also be used to detect and track mineral-rich veins, major fault zones, and in the exploration of lateritic nickel deposits.

1.3.5 Gravity surveys in global geophysical mapping for mineral exploration

Positive gravity anomalies are linked to shallow bodies with high density, while gravity lows are associated with shallow bodies that have low density. Consequently, deposits of high-density minerals like chromite, hematite, and barite result in gravity highs, while low-density deposits such as halite, weathered kimberlite, and diatomaceous earth cause gravity lows. The gravity method can also help estimate the total anomalous mass (ore tonnage) responsible for the observed anomaly. In the context of mineral exploration and environmental studies, gravity surveys can assist in identifying various lithologies, structural features, and, in some cases, even ore bodies themselves (Wright, 1981). However, small anomalous bodies, such as underground workings, are difficult to detect unless they are located at shallow depths.

1.3.6 Magnetic surveys in geophysical mapping for mineral exploration

The magnetic method relies on detecting small variations in magnetic mineral content, such as magnetic iron and iron-titanium oxide minerals (including magnetite, titanomagnetite, titanomaghemite, and titanohematite) as well as some iron sulfide minerals like pyrrhotite and greigite, present in rocks. Magnetic anomalies can result from primary igneous or sedimentary processes that define the magnetic mineralogy, or they may arise from secondary alterations that either introduce or remove magnetic minerals. In the context of mineral exploration and geoenvironmental studies, secondary effects in rocks hosting ore deposits, particularly those associated with hydrothermal systems, are significant (Criss and Champion, 1984; Hanna, 1969). Magnetic surveys can highlight zones of past hydrothermal activity.

1.3.7 Gamma ray spectrometry in geophysical mapping for mineral exploration

Gamma ray methods (Durrance, 1986; Hoover et al., 1992) utilize scintillometry to detect the presence of natural radioelements such as potassium, uranium, and thorium, with multi-channel spectrometers offering precise measurements of individual radioelement concentrations. These methods are widely used in uranium exploration due to their ability to provide direct detection. Among these radioelements, thorium is generally the least mobile and exhibits geochemical behavior similar to zirconium. Like uranium, thorium content tends to increase in felsic rocks and generally rises with higher alkalinity. Gamma ray spectrometry, by directly measuring natural radioelements, also provides geoenvironmental data on radiation dose and radon potential. As uranium and/or potassium are often concentrated in or near certain deposits, their presence can help indirectly assess the likelihood of hazardous material release from ore or waste piles. In the presence of sulfide minerals, oxidation processes can accelerate the mobilization of uranium.

1.3.8 Thermal methods in geophysical mapping for mineral exploration

Thermal methods encompass two distinct techniques: (a) borehole or shallow probe methods for measuring thermal gradients, which are valuable on their own, and when combined with knowledge of thermal conductivity, can provide an estimate of heat flow, and (b) airborne or satellite-based measurements, which are used to determine the Earth's surface temperature and the thermal inertia of surface materials, as well as the thermal infrared radiation emitted from the Earth's surface. Thermal noise includes factors such as topography, variations in thermal conductivity, and intrinsic endothermic and exothermic sources.

1.3.9 Seismic surveys in mineral exploration

Seismic techniques have been relatively underused in mineral exploration at the deposit scale, primarily due to their high cost and the challenges involved in acquiring and interpreting data in complex, faulted, and altered igneous terrains. However, shallow seismic surveys, which utilize more affordable sources and smaller survey areas compared to regional surveys, can be cost-effective for certain geoenvironmental studies in the near subsurface. Reflection seismic methods provide detailed structural information, whereas refraction methods deliver accurate depth estimates to lithologies with varying acoustic impedance. The refraction method, in particular, has been used to map low-velocity alluvial deposits that may contain gold, tin, or sand and gravel. Seismic methods are also used in geoenvironmental studies, including assessing the structure, thickness, and hydrology of tailings, as well as evaluating the extent of acid mine drainage surrounding mineral deposits (Hack, 1982; Steeples and Miller, 1990; Darracott and McCann, 1986).

Seismic investigations are particularly useful for mapping geological structures that host mineral deposits. In some cases, velocity contrasts detected by seismic methods can help identify large ore bodies. Additionally, seismic techniques can be used to estimate the depth to bedrock, which is important for assessing the "rippability" of the bedrock and its potential impact on extraction strategies and costs.

1.4 An overview of geophysical methods in mineral exploration

Geophysical methods play a crucial role in mineral exploration and have significantly advanced over time in terms of equipment design, method development, and the precision of data acquisition, processing, and interpretation. Each geophysical method is based on contrasts in the physical properties of the subsurface rocks and the target being explored. Since every geophysical method is better suited for specific types of mineral deposits and has limitations on the depth of exploration, relying on a single method often does not provide reliable or satisfactory results (Table 1.2). This can be attributed to several factors, including data quality, noise influences, geophysical field parameters, the choice of interpretation tools, and the inherent limitations of the method itself. As a result, the success of geophysical surveys can be enhanced by using multiple methods in combination, followed by the selection of appropriate processing and interpretation techniques. This approach, known as an integrated geophysical strategy, ensures higher accuracy and more reliable results. It is important to note that geophysical modeling typically offers generalized, nonunique solutions regarding the geometry (such as shape and size) of subsurface sources, while mathematical tools can help accurately estimate the spatial location and depth of these sources. This chapter highlights the key features of various geophysical methods used in mineral resource exploration.

1.5 Driving forces behind mineral exploration: An overview

Mineral exploration is driven by a combination of economic, technological, and societal factors. Growing global demand for metals and minerals—especially for use in renewable energy, electronics, and infrastructure—fuels the need to discover new resources. Technological advancements enable exploration in previously inaccessible areas, while geopolitical factors and supply chain security also play key roles. Environmental regulations and sustainability goals are increasingly shaping exploration strategies, pushing the industry toward more responsible and innovative practices.

TABLE 1.2 Geophysical surveying techniques and their relevant parameters for different types of deposits (Al-Sayid, 2019).

Method	Measured parameters	Operative physical property	Suitable deposit type
Seismic	Travel time of reflected and refracted seismic waves	Density and elastic mode velocity	Oil and gas, layered sedimentary basin
Gravity	Spatial variation in the strength of Earth's gravitational field	Density contrast between the surrounding host rocks	Massive sulfides, chromite, salt dome, barite, kimberlite, concealed basin
Magnetic	Spatial variation in the strength of the geomagnetic field	Magnetic susceptibility	Magnetite, ilmenite, pyrrhotite-rich sulfides
Electrical			
Resistivity	Earth's resistance	Electrical conductivity	Groundwater, sulfide ore
Induced polarization (IP)	Polarization voltage/ frequency development of ground resistance	Electrical capacitance	Large sulfide dissemination, graphite
Self-potential (SP)	Electrical potential	Electrical conductivity	Sulfide veins, graphite, groundwater
Electromagnetic (EM)	Response to EM radiation	Electrical conductivity and inductance	Sulfide ore, graphite deposits
Radiometric	Gamma radiation	Gamma ray	Thorium, uranium, radium
Borehole geophysics and Mise-a-la-Masse	Down-hole probe	All types	Continuity of mineralization and dip

1.5.1 Global distribution of ore deposits

Ore refers to natural rock or sediment that contains one or more valuable minerals concentrated above background levels, typically metals that can be mined, processed, and sold profitably (Jenkin et al., 2014; Neuendorf et al., 2011). An ore deposit is an economically significant accumulation of minerals within a host rock, distinct from a mineral resource in that it contains minerals in high enough concentrations to be economically viable. Ore deposits are typically named based on their location or after their discoverers. Minerals of interest in ore deposits are usually oxides, sulfides, silicates, or native metals such as copper or gold (Wills and Finch, 2015).

Mineral discoveries worldwide have included a broad array of valuable resources such as coal, iron ore, uranium, base metals, and precious metals like gold. Over the years, the global landscape of mineral exploration has evolved, with significant changes in the time required for discoveries to be transformed into operational mines. In the 1950s, approximately 50% of deposits became mines within 15 years of their discovery. However, by the 2000s, this rate had dropped to <10%, reflecting the increasing complexity of modern mineral exploration and development.

Minerals are essential for modern life, with many used in the production of renewable energy technologies. For instance, copper is a key component in wind turbines, and silver plays an important role in solar panels. Additionally, critical minerals like lithium, nickel, and cobalt are indispensable for the production of lithium-ion batteries, which are crucial for the growing electric vehicle (EV) industry. A battery-electric vehicle, for example, uses approximately 83 kg of copper, compared to just 9 kg in a conventional internal combustion engine vehicle.

Moreover, minerals such as calcium, phosphorus, iron, and zinc are commonly found in rocks and soil, playing an integral part in various geological processes and serving as fundamental components for both biological systems and industrial applications. Scientists have identified over 3000 different types of minerals, with quartz being the most common mineral on Earth and basalt being the most common rock.

In conclusion, ore deposits are vital to the global economy, particularly with the increasing demand for minerals essential for energy production, electronics, and emerging technologies. Understanding the global distribution of these ore deposits is crucial for future exploration and resource management strategies.

1.5.2 The need for minerals and exploration in a circular economy

A circular economy is a model of resource use and consumption that focuses on sharing, leasing, reusing, repairing, refurbishing, and recycling materials and products for as long as possible (Fig. 1.5). This approach aims to address global issues such as climate change,

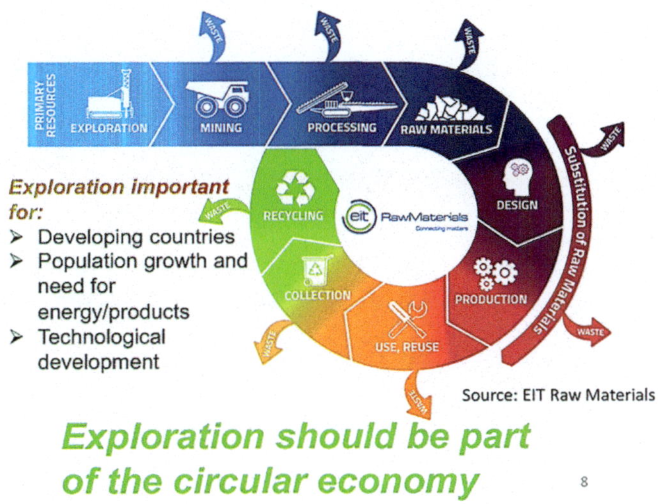

FIG. 1.5 Circular economies for minerals and exploration. *[Source: United Arab Emirates University (UAEU), European Institute of Innovation & Technology (EIT) Raw Materials].*

biodiversity loss, waste, and pollution by promoting the design and implementation of the three core principles of the model. The key principles necessary for the shift to a circular economy are: eliminating waste and pollution, keeping products and materials in continuous use, and restoring natural systems (Pomykała et al., 2017).

1.5.3 Exploring deeper: Advancements in the exploration of basemetal deposits beneath deeper cover—global discoveries from 1900 to 2013

Deep underground mining is increasingly being developed due to the depletion of shallow mineral deposits. One of the primary benefits of deep underground mining is its reduced environmental impact compared to shallow mining methods. As a result, mineral exploration is increasingly dependent on geophysical techniques that can provide crucial subsurface information for locating future ore deposits. Consequently, it is becoming more essential for exploration geologists to stay informed about advanced geophysical methods and their potential roles in aiding the discovery of mineral resources. To achieve this, exploration geologists must acquire a solid understanding of geophysical methods. The mineral exploration industry is gradually moving into deeper cover, although the rate of depth increase is relatively slow. On average, base metal deposits are found at twice the depth as gold deposits. Additionally, brownfields exploration typically occurs at depths two to three times greater than greenfields exploration (Source: MINEX Consulting) (Fig. 1.6).

1.5.4 Geophysical data acquisition, processing, and interpretation

Data acquisition is an essential stage in the establishment of geophysical test sites (GTS), which are critical for analyzing the geophysical responses, signals, and behaviors of the buried materials (Omeiza et al., 2023). Different geophysical techniques, such as gravity,

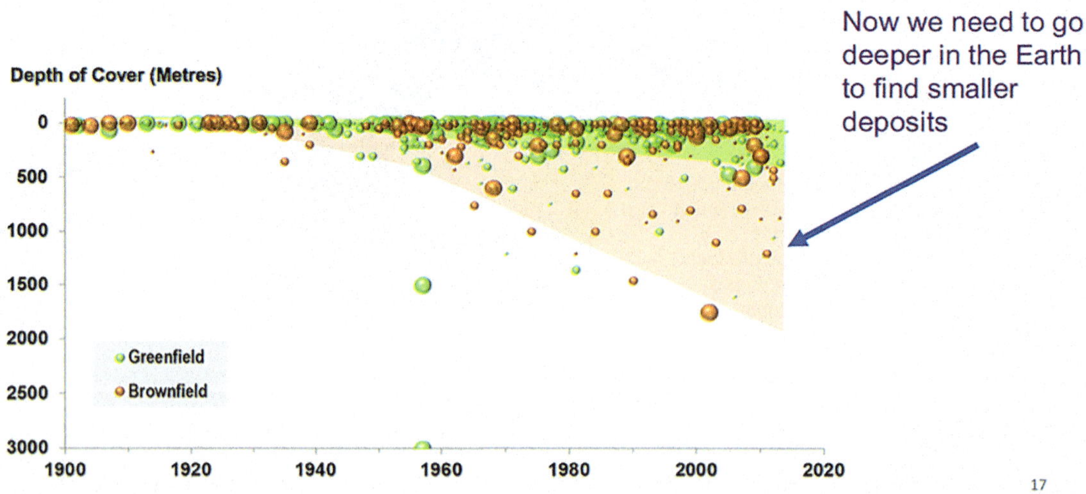

FIG. 1.6 Schematic representation of mineral depths in the shallow subsurface. *Source: United Arab Emirates University (UAEU); MinEX Consulting.*

magnetics, seismic refraction and reflection, ground-penetrating radar, electrical resistivity depth sounding and profiling, very low frequency, electromagnetic induction, spontaneous potential, ground motion monitoring, and telluric measurements, are employed for data acquisition at GTS sites. This data helps in studying anomalies generated by various targets and their response to the applied geophysical methods.

A significant step in the data acquisition process involves converting the analog signals from measuring instruments into digital formats that can be manipulated and analyzed using computer software (Daryono et al., 2019; Zhao, 2021; Wilson, 2020). The processed data can be stored in repositories for future reference, validation, testing, and calibration of geophysical instruments.

Geophysics involves measuring signals, which are processed and analyzed before interpretation and presentation through forward modeling and inverse modeling. Typically, a geophysical method measures data that is influenced by a "geophysical" mass property, such as seismic velocity, when the travel time of seismic waves is measured. These properties are generally controlled by lithology and rock mass conditions. The selection of geophysical methods should prioritize those that respond most sensitively to the variability of geophysical mass properties relevant to the specific civil engineering or exploration problem.

However, geophysical data processing and interpretation can be challenging due to the heterogeneous and nonunique (uncertain) nature of subsurface materials (Omeiza et al., 2023). The physical effects measured may not always be interpretable as originating from a single, unique source at a specific depth due to the potential for multiple sources at varying depths to produce similar physical effects (Hill, 2000; Giao et al., 2003; Huizar-Alvarez et al., 2003).

1.5.5 Principal stages of a geophysical program in mineral exploration

From defining the objectives of the geophysical survey(s) to providing an interpretation of the subsurface geology, the process involves various stages to ensure accurate and reliable results (Fig. 1.7).

1.5.6 The nature of geophysical responses

The general characteristics of a geophysical response (Fig. 1.8) and how these change with variations in: (a) Sign of the physical property contrast (ΔP), (b) Magnitude of the contrast, (c) Depth of the source, and (d) Shape of the source. Understanding the anomalies generated by various subsurface targets and their responses to different geophysical techniques in various subsoil types is crucial to near-surface geophysical investigations. To enhance subsurface information, academic platforms such as geophysical test sites (GTS) have been developed to improve near-surface geophysical investigations. Several significant studies have described GTS as an ideal platform to enhance knowledge of data acquisition and interpretation methods, offering insights into subsurface parameters (Ahmed et al., 2021; Foti, 2017; Giao and Vichalai, 2006; Rosado-Fuentes, 2021).

However, geophysical applications may not always be capable of detecting all subsurface materials due to their depths and sizes. Some targets are either too deep or too small to be detected and resolved. Consequently, mapping depth anomalies or signals generated by such

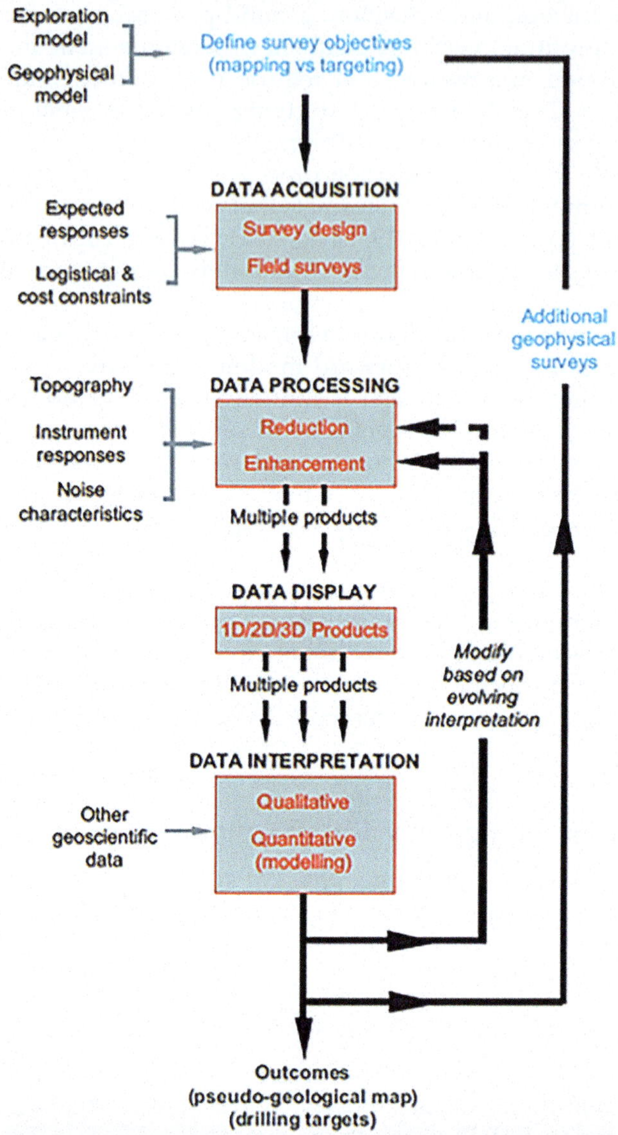

FIG. 1.7 Geophysical data acquisition, processing, and interpretation. *Source: Dentith and Mudge (2014).*

targets can be difficult or impossible if their physical properties are similar to those of the surrounding materials (Telford et al., 1990). Therefore, it is essential for geologists and geophysicists to collaborate effectively as mineral exploration increasingly moves toward deeper targets in the search for the next generation of ore deposits.

Gravity and magnetic geophysical methods are passive techniques. They do not rely on controlled sources; instead, they focus on detecting naturally occurring variations in the Earth's gravity and magnetic fields. Due to this, these surveys have been used for various

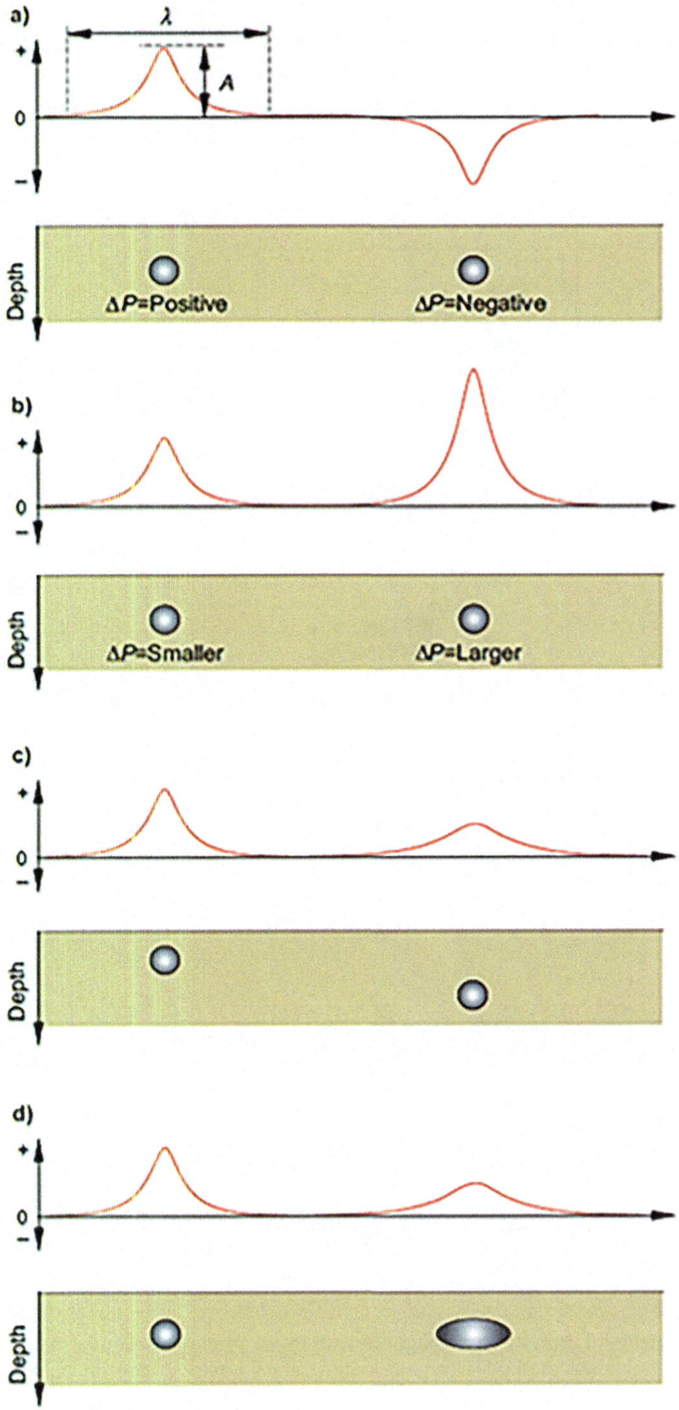

FIG. 1.8 Schematic of geophysical responses. *Source: Dentith and Mudge (2014).*

military purposes, such as the detection of submarines and volatile unexploded munitions (Saibi, 2018). The gravity method, in particular, is widely used to detect subsurface density anomalies and geological structures. The interpretation of gravity data can be applied to mineral and oil exploration, as well as regional geological studies. The successful and effective use of gravity methods relies on fast forward modeling and stable inversion tools to accurately image subsurface density structures (Fig. 1.9A–C).

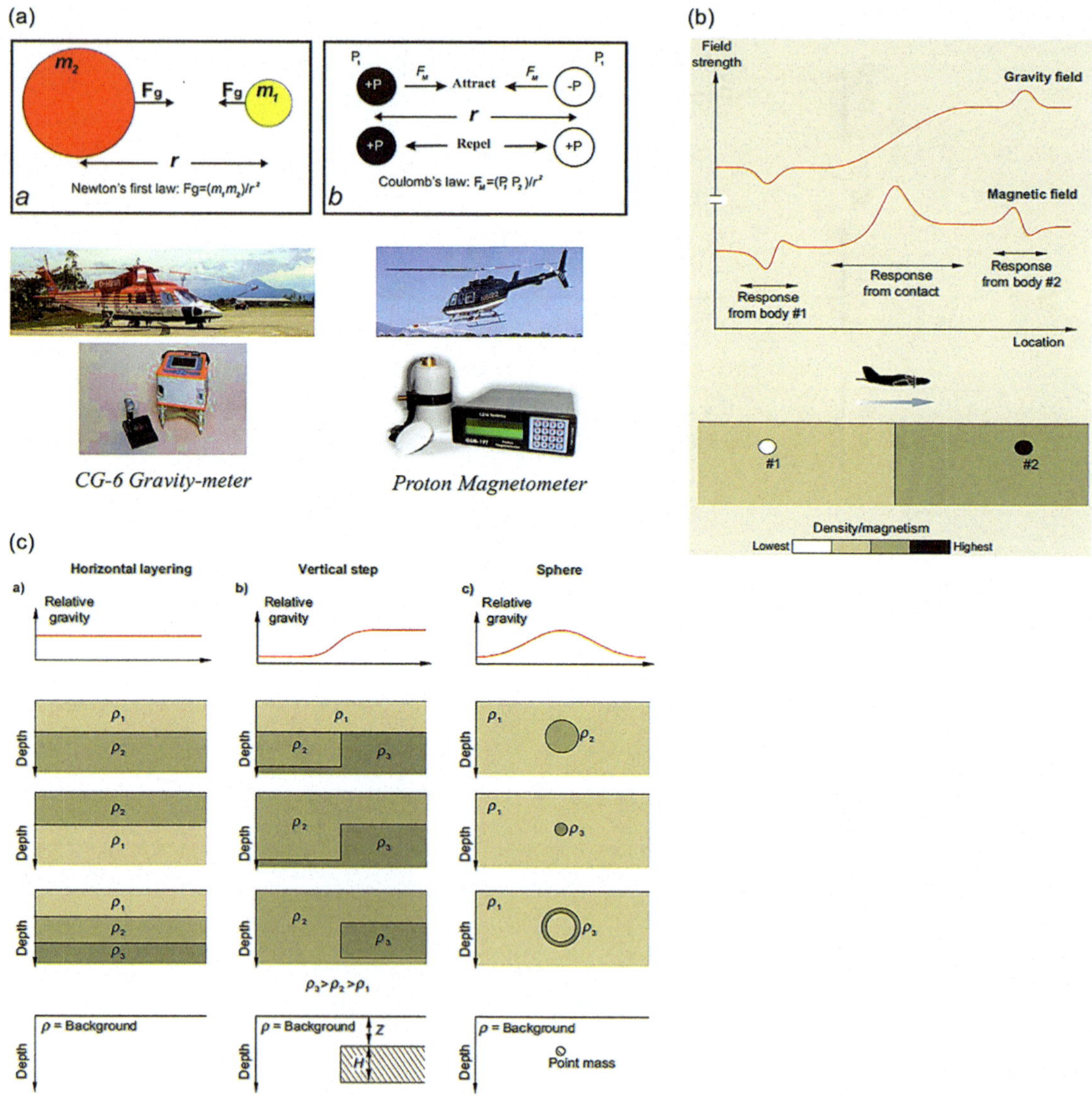

FIG. 1.9 (A) Equipment for gravity and magnetic survey in mineral resource exploration. (B) Geophysical responses in gravity and magnetic methods in regard to mineral resource exploration. (C) Equivalent gravity models for three common geological sections: (a) gravity anomaly, (b) some possible density cross sections that produce the gravity anomaly, and (c) equivalent model. *Source: Dentith and Mudge (2014).*

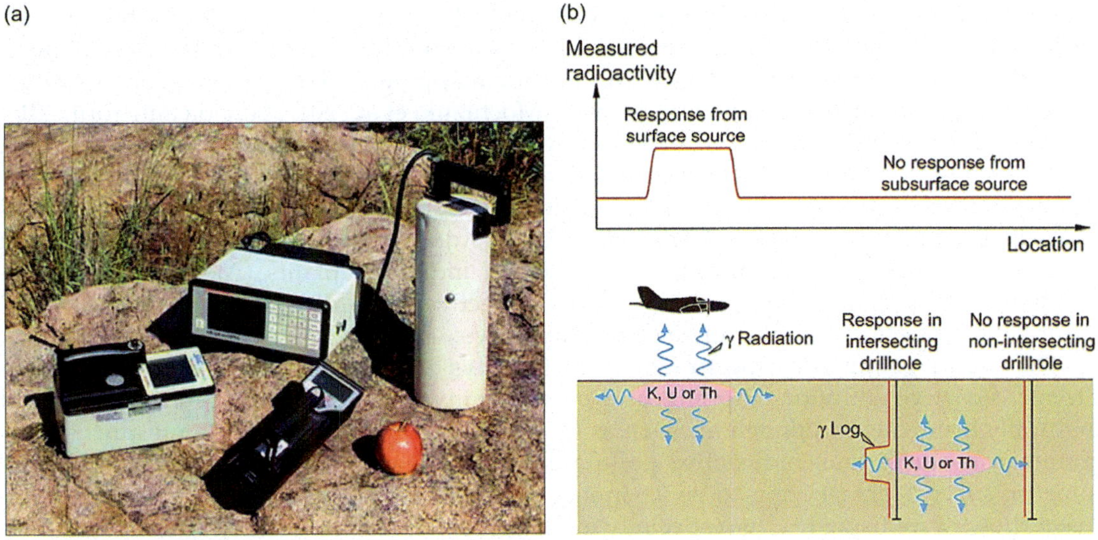

FIG. 1.10 (A) Modern g-ray spectrometers. (B) Schematic illustration of the radiometric method. *Source: Dentith and Mudge (2014).*

Survey and data-gridding parameters are selected to capture all expected relevant anomaly wavelengths from the source being investigated, while minimizing costs. At a relatively low cost compared to seismic surveys, gravity and magnetic techniques can provide geophysical coverage of large and small areas. In many regions of Canada, relatively sparse yet valuable magnetic and gravity grids are available free of charge from the federal government. Calibrated with geological knowledge, gravity and magnetic anomalies can yield an indirect but extremely useful picture of lateral changes in rock composition and structural patterns (Lyatsky, 2010).

Gamma-ray surveys are increasingly being employed in a variety of applications, such as estimating erosion (Tyler et al., 2001), mapping mine tailings (Koomans, van der Wal and Kamsma, 2014), precision agriculture (Egmond et al., 2018; Piikki et al., 2015; Söderström et al., 2016), characterizing sediments (Anjos et al., 2007; De Meijer, 1998), and identifying contamination of heavy metals or organic micro-compounds in sediments and soils (Dentith and Mudge, 2014; van der Graaf et al., 2007; Söderström and Eriksson, 2013) (Fig. 1.10A).

Depending on the required spatial resolution and the area to be mapped, gamma-ray measurements can be carried out using ground-based methods, typically by off-road vehicles, or airborne methods, using helicopters or airplanes. Ground-based surveys are often preferred when high spatial resolution is necessary, as the detector is positioned close to the surface, providing a small footprint (Dentith and Mudge, 2014). However, airborne measurements are useful for large-area mapping, though they come with the drawback of reduced resolution due to the height and speed of the aircraft. Nonetheless, airborne methods offer advantages, particularly in accessing challenging terrains such as wet, rocky, dangerous, or densely vegetated areas (Fig. 1.10B).

Unmanned Aerial Vehicle (UAV)-borne gamma-ray measurements combine the advantages of both ground-based and airborne approaches. UAVs can fly over areas that are inaccessible by vehicles while maintaining proximity to the ground, enabling high spatial resolution radiometric data collection. Additionally, UAVs are useful in radioactive or hazardous areas that would pose a health risk to human operators. Recent studies have demonstrated the application of UAV-based surveys for mapping 137Cs concentrations resulting from the Fukushima Dai-Ichi nuclear disaster (Mochizuki et al., 2017; Sanada and Torii, 2015; Martins, 2016; Tang et al., 2016) and locating lost radioactive sources (Gong et al., 2019). UAV-based radiometric surveys have moved beyond the proof-of-concept phase and are now routinely used. Advanced algorithms are being developed to address inversion challenges and identify radiation hotspots (Zhang et al., 2018). Additionally, UAVs are being considered as a tool for monitoring radioactive materials in industrial plants (Aleotti et al., 2019). Remediation operations in areas with high radioactive contamination typically use small scintillation detectors, which can be carried by UAVs.

The use of gamma-ray spectrometers in geophysical applications differs significantly from radioactive pollution mapping, as it primarily involves determining concentrations of naturally occurring radionuclides such as 40 K, 238U, and 232Th, which emit much lower gamma-ray fluxes than radioactive pollutants. Despite these challenges, UAV-borne radiometric surveys are proving to be an innovative and effective technique. Various studies have shown promising results, especially in the characterization of legacy uranium and iron mines (MacFarlane et al., 2014; Martin et al., 2015; Pirttijarvi and Oy, 2016; Šálek et al., 2018).

Electrical and electromagnetic methods play a pivotal role in studying the Earth's subsurface. These methods are powerful tools in environmental and geotechnical investigations (Pellerin, 2002). Techniques originally developed for deeper applications, such as mining, geothermal, and crustal studies, are adapted for shallow targets by increasing frequencies, reducing decay times, and using smaller array configurations. A crucial factor in their effectiveness is the use of dense station spacing to minimize spatial aliasing and the acquisition of high-quality data to resolve small subsurface features. New instruments focus on providing continuous or dense measurements with high precision, accompanied by interpretational methods that can quickly process large datasets. Continuous electrical and time-domain electromagnetic systems have been developed for geological mapping in hydrological investigations, employing one-dimensional inversion routines that are both fast and reliable. On a smaller scale, electrical systems have been successfully used in archaeology studies. Additionally, electromagnetic systems working near or above the upper limits of the quasi-static approximation are proving effective in mapping subsurface infrastructure beneath conductive, clay-rich cover, where ground-penetrating radar (GPR) would be ineffective.

Induced polarization (IP) and resistivity systems, utilizing multiplexing techniques, do not provide continuous measurements but enable relatively rapid data collection and dense sampling. These systems are applied in various fields, including landfill and contaminant characterization, verifying the integrity of engineered subsurface structures, and monitoring infiltration in the vadose zone. Conductivity and resistivity are key properties of materials that describe how easily they allow the flow of electric current. The resistivity of a material is the inverse of its conductivity, with higher resistivity indicating lower conductivity, and vice versa. Here's a short overview of typical ranges for conductivity and resistivity in some common minerals, rock types, and near-surface materials:

Granite has a resistivity range of 10^3 to 10^5 Ohm-m and low conductivity, as it is typically an insulating rock with relatively low water content. *Sandstone* has a resistivity of 10^1 to 10^3 Ohm-m and moderate conductivity, depending on porosity and mineral content, with water saturation significantly affecting conductivity. *Basalt* shows a resistivity of 10^2 to 10^4 Ohm-m and has moderate conductivity depending on mineral composition. *Limestone* has a resistivity range of 10^2 to 10^4 Ohm-m and generally low conductivity, varying with the amount of pore water and impurities. *Clay* exhibits a resistivity from 10^{-1} to 10^2 Ohm-m and high conductivity, especially when saturated with water, due to the presence of charged particles and ions. *Saltwater (brine)* has a resistivity of 0.1 to 10 Ohm-m and very high conductivity, as saltwater is an excellent conductor due to its high ion content. *Shale* typically has a resistivity of 10^1 to 10^3 Ohm-m and moderate conductivity affected by water content and mineral composition. *Soil* has a resistivity of 10^1 to 10^3 Ohm-m, and its conductivity varies widely depending on moisture, organic content, and mineralogy, with wet soils being more conductive. *Coal* generally exhibits a resistivity range of 10^1 to 10^3 Ohm-m and moderate conductivity, with variation due to impurities and moisture content.

A general trend observed is that metals (like copper, silver, or gold) have extremely low resistivity and high conductivity, while nonmetallic minerals (like quartz) generally have higher resistivity and lower conductivity. Sedimentary rocks show considerable variation depending on porosity and mineral content, with clays being highly conductive when wet. This range reflects how the presence of water, ions, and porosity significantly affects the resistivity and conductivity of geological materials.

The electromagnetic spectrum and the frequency bands (Fig. 1.11) used by various geophysical methods include AFMAG (Audiofrequency Magnetics), AMT (Audiofrequency Magnetotellurics), ASTER (Advanced Spaceborne Thermal Emission and Reflection Radiometer), CSAMT (Controlled Source Audiofrequency Magnetotellurics), FDEM (Frequency-domain Electromagnetics), GPR (Ground Penetrating Radar), HyMap (Hyperspectral Mapper), IP (Induced Polarization), Landsat MSS (Landsat Multispectral

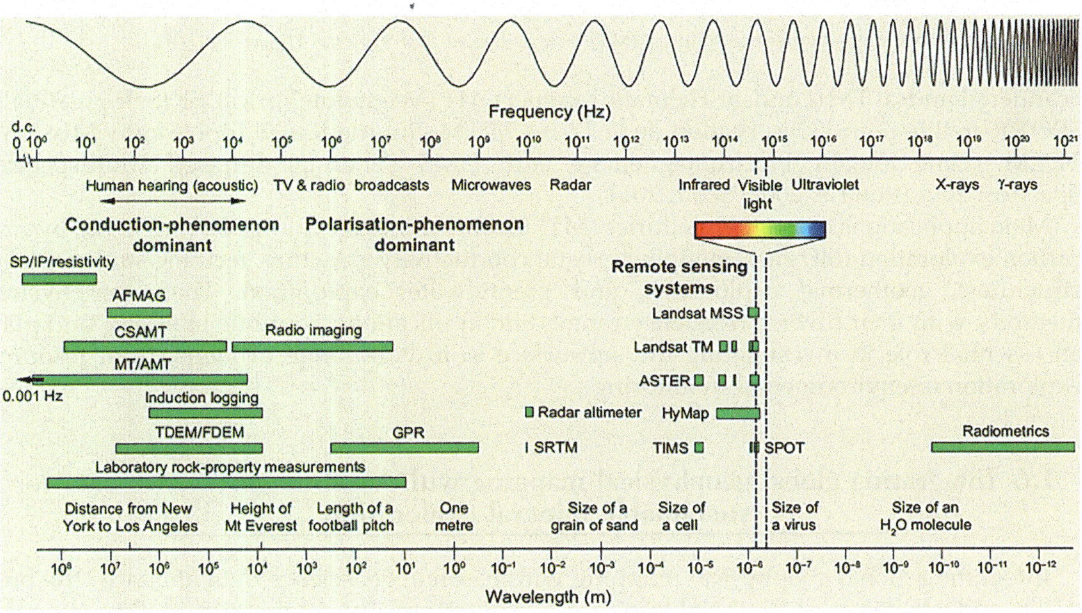

FIG. 1.11 The electromagnetic spectrum and the frequency bands used by the various geophysical methods. *Source: Dentith and Mudge (2014).*

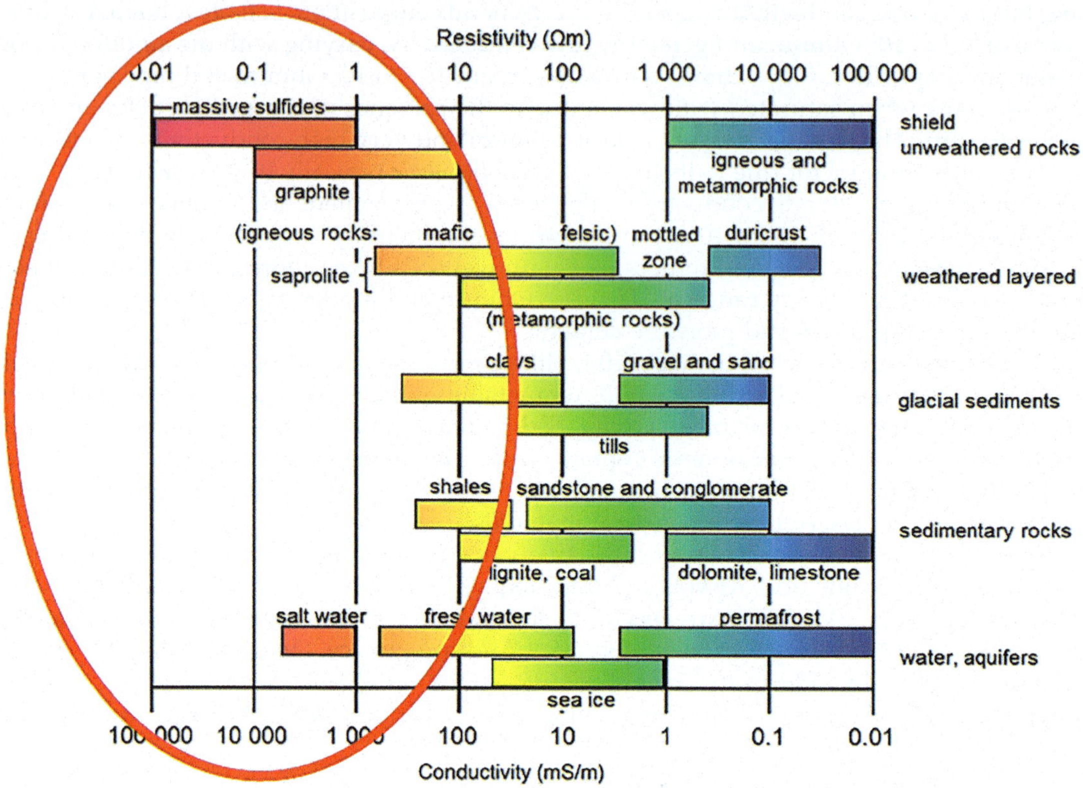

FIG. 1.12 Applications of magnetotellurics (MT). *Source: United Arab Emirates University (UAEU).*

Scanner), Landsat TM (Landsat Thematic Mapper), MT (Magnetotellurics), SP (Self-potential), SPOT (Satellite pour l'Observation de la Terre), SRTM (Shuttle Radar Topography Mission), TDEM (Time-domain Electromagnetics), and TIMS (Thermal Infrared Multispectral Spectroscopy) (Lowrie, 2007; Scott, 2014).

Main applications of magnetotellurics (MT) include mineral exploration (Fig. 1.12), hydrocarbon exploration (oil/gas), mapping crustal conductivity structure, tectonic studies (deep structures), geothermal exploration, and groundwater exploration. These geophysical methods, with their diverse frequency ranges and applications, continue to evolve and play an essential role in investigating the subsurface in a wide range of fields, from resource exploration to environmental monitoring.

1.6 Integrating global geophysical mapping with baseline geoscience data for sustainable mineral exploration

Integrating global geophysical mapping with baseline geoscience data enhances the precision and efficiency of sustainable mineral exploration. By combining satellite imagery, gravity, magnetic, and radiometric data with geological, geochemical, and structural maps, explorers can better identify subsurface anomalies and potential ore deposits. This integrated approach supports data-driven targeting, reduces exploration risk, and promotes environmentally responsible decision-making. It also aids in understanding deep crustal structures that control mineralization processes. Ultimately, this synergy accelerates discovery while minimizing ecological impact and exploration costs.

1.6.1 Detailed explanation

Global Geophysical Mapping (GGM) refers to the systematic collection and analysis of geophysical data at a global or regional scale, utilizing techniques such as magnetometry, gravity surveys, seismic imaging, and electromagnetic methods. These geophysical methods help to understand the Earth's subsurface structure, composition, and properties, which are critical in identifying potential mineral deposits and other natural resources.

However, the true value of GGM is realized when it contributes to baseline geoscience data, which serves as the foundation for all exploration and resource management activities. Baseline geoscience data consists of a comprehensive set of geological, geophysical, geochemical, and geospatial information that provides insights into the Earth's natural resources and environmental conditions. This data is crucial for making informed decisions throughout the exploration phase and ensuring that sustainable practices are followed.

1.6.2 Link to sustainable exploration

Sustainability in mineral exploration requires a careful balance between resource extraction and environmental stewardship. By integrating GGM with baseline geoscience data, explorers can better understand the geological context of mineral deposits while also identifying areas of ecological sensitivity, geohazards, or other environmental concerns. This integrated data helps in:

1. *Identifying potential exploration zones*

GGM provides large-scale geophysical maps that can pinpoint areas of interest for mineral exploration. This allows companies to focus their efforts on regions with high potential, minimizing unnecessary disturbance to other areas. Such a focused approach promotes the sustainable exploration of resources.

2. *Reducing environmental impact*

The information from GGM, when integrated with geospatial and environmental baseline data, helps in identifying areas where extraction would have minimal environmental disruption. Additionally, it aids in the planning of exploration methods and the implementation of mitigation strategies to protect local ecosystems.

3. *Efficient resource management*

Accurate geophysical maps, supported by baseline geoscience data, ensure that exploration activities are well-targeted, preventing over-exploitation or wasteful extraction methods. This contributes to the long-term sustainability of resources by reducing the risk of resource depletion or contamination.

4. Monitoring and compliance

Baseline data, including geophysical data, is essential for postexploration monitoring and ensuring compliance with environmental regulations. This data can be used to track any potential changes to the environment that may result from exploration activities, ensuring that corrective actions are taken when necessary.

1.7 Conclusions

In conclusion, Chapter 1 has provided an in-depth exploration of the pivotal role of geophysics in modern mineral exploration. Geophysics, through its advanced tools and methodologies, has revolutionized the way minerals are discovered and extracted, significantly enhancing the efficiency, precision, and sustainability of exploration efforts. From ancient techniques to cutting-edge technologies, the evolution of mineral exploration has been shaped by the continuous refinement of geophysical methods, which now serve as indispensable tools for identifying mineral deposits beneath the Earth's surface.

The integration of global geophysical mapping with baseline geoscience data offers a sustainable approach to mineral exploration, allowing for better-targeted and environmentally responsible exploration activities. As we move forward, the driving forces behind mineral exploration, such as the growing demand for raw materials in the context of a circular economy and the quest for deeper ore bodies, continue to underscore the importance of geophysical surveys in the search for untapped resources.

With advancements in geophysical data acquisition, processing, and interpretation, coupled with a deeper understanding of ore deposit distribution and the complexities of Earth's geology, the future of mineral exploration looks promising. By leveraging these cutting-edge techniques, exploration efforts can be optimized, ensuring that mineral resources are discovered and extracted in a manner that is not only efficient but also aligned with global sustainability goals.

CHAPTER

2

Overview of geophysical techniques for baseline geoscience data

2.1 Overview of geophysical exploration methods, emphasizing the importance of a multidisciplinary approach in baseline geoscience data collection

Geophysics is the study of the Earth's physical properties through measurements recorded at or near the surface. These measurements are crucial for understanding subsurface conditions and have become indispensable in the exploration of natural resources such as minerals, oil, and gas, as well as in addressing geological and environmental challenges (Naskar et al., 2022; Haldar, 2018; Abdelrahman et al., 2003; Khalil et al., 2014). Geophysical techniques leverage the physical properties of rocks and materials to create methods that provide valuable insights into the Earth's interior (Fig. 2.1). The fundamental concepts and applications of different geophysical methods have been discussed comprehensibly for easy understanding of their diverse applicability by users for the exploration of natural resources (Mishra and Naskar, 2024, 2025 and 2026).

These methods can be broadly classified based on their application environments: on land, within boreholes or wells, and in surface water bodies. Furthermore, geophysical methods are categorized into active and passive techniques. Passive methods measure natural fields, such as gravity, magnetism, and self-potential (SP), while active methods involve the introduction of external sources, such as seismic waves, electrical resistivity, and induced polarization (IP), to probe subsurface physical properties.

Radiometric surveying, a key technique under the nuclear methods category, measures the natural gamma radiation emitted by radioactive isotopes in geological formations. This method is crucial for identifying mineral deposits, particularly those containing uranium, thorium, and potassium. Radiometric techniques are commonly applied in both surface and airborne surveys, and are often used in combination with other methods, such as seismic and electromagnetic techniques, to provide a comprehensive view of the subsurface. By measuring the distribution of radioactive elements, radiometric methods help identify areas with potential mineral resources.

Geophysical methods are also categorized based on the specific environment in which they are applied. These include surface geophysics, borehole geophysics, and waterborne geophysics. Additionally, methods can be specialized for unique contexts, such as marine

Baseline Geoscience Data Generation for Sustainable Exploration of Natural Resources
https://doi.org/10.1016/B978-0-443-45221-5.00017-5

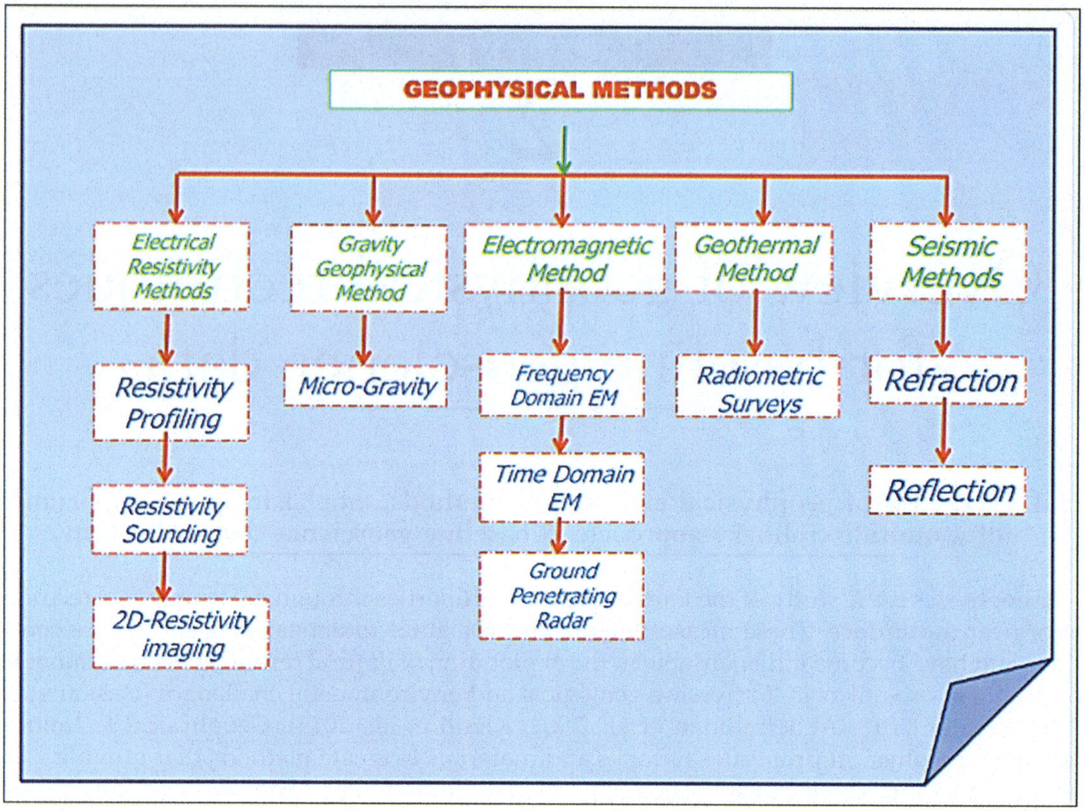

FIG. 2.1 Overview of geophysical techniques and a list of widely used methods.

geophysics for sea-based surveys or aero-geophysics for air-based surveys. Reconnaissance surveys, often conducted from the air, offer significant advantages in terms of speed and efficiency, making them a valuable tool in early-stage resource exploration.

The following sections provide detailed explanations of the major categories of geophysical methods, along with descriptions of the underlying physical principles and their specific applications. These include:

- Surface Geophysical Methods
- Borehole Geophysical Methods
- Waterborne Geophysical Methods

In addition to discussing the various techniques, the following topics will be explored in depth:

- The physical properties measured by geophysical methods.
- Inversion approaches for data processing and analysis.

Table 2.1 lists various geophysical methods, categorized by type (active and passive) and the environment in which they are applied (surface, borehole, and waterborne). This table provides a comprehensive overview of commonly employed geophysical methods,

TABLE 2.1 Overview of common geophysical methods and their applications.

Geophysical method	Type	Application environment	Description
Gravity Method	Passive	Surface	Measures variations in the Earth's gravitational field to detect subsurface density changes.
Magnetic Method	Passive	Surface	Measures variations in the Earth's magnetic field caused by variations in the magnetic properties of rocks.
Self-potential (SP) method	Passive	Surface	Detects natural voltage differences in the Earth's subsurface due to electrochemical processes.
Seismic refraction	Active	Surface, borehole	Measures the travel time of seismic waves through the subsurface to determine layers and their properties.
Seismic reflection	Active	Surface, borehole	Uses reflected seismic waves to image subsurface structures and detect resource deposits.
Electrical resistivity	Active	Surface, borehole, waterborne	Measures the resistivity of subsurface materials to map geological formations and identify fluids.
Induced polarization (IP)	Active	Surface, borehole	Measures the delayed response of the subsurface to an applied electrical current to detect minerals.
Electromagnetic (EM) method	Active	Surface, borehole, waterborne	Measures the conductivity of subsurface materials to locate variations in mineralization and groundwater.
Ground penetrating radar (GPR)	Active	Surface, borehole	Uses radar pulses to image the subsurface, detecting objects, structures, or variations in materials.
Borehole geophysics (wireline)	Active	Borehole	Measures physical properties within a borehole to assess formations encountered during drilling.
Microgravity	Passive	Surface	Detects small changes in gravity to locate variations in subsurface density, useful in mineral exploration.
Marine seismic	Active	Marine/ waterborne	Uses seismic waves in water to study the seabed and subsurface geology, often for oil and gas exploration.
Aero-geophysics	Active	Airborne	Uses airborne equipment to measure geophysical properties of large areas quickly and efficiently.
Borehole seismic	Active	Borehole	Measures seismic waves in boreholes for detailed subsurface imaging and resource exploration.

organized by both method type and application environment. It serves as a useful reference for understanding the diverse range of techniques used in different geophysical surveys. Fig. 2.2 presents a flow chart outlining the data description process and the sequential methodology steps involved in geophysical mapping. This visual guide illustrates the key stages of data collection, processing, and analysis, offering a structured approach to applying geophysical techniques for sustainable exploration and management. The flow chart provides an overview of how baseline geoscience data can be integrated to enhance our understanding of subsurface properties, ultimately supporting more effective resource exploration and management strategies.

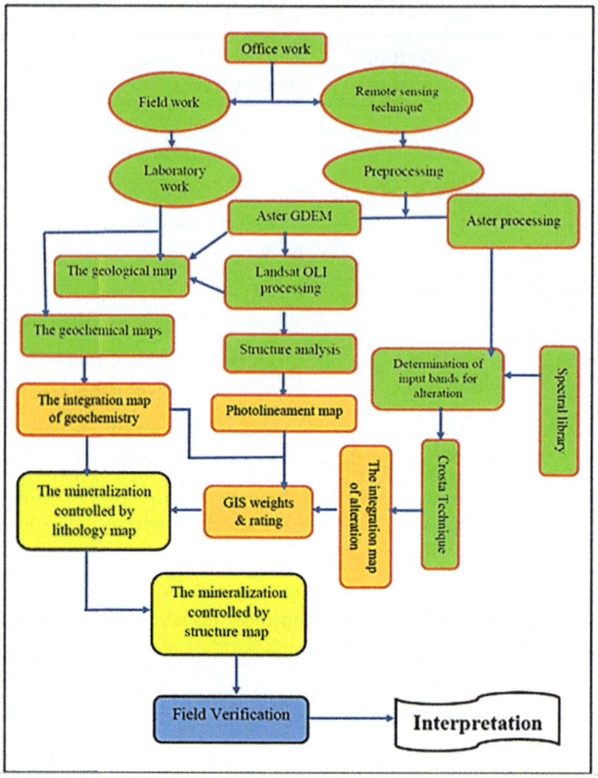

FIG. 2.2 Flow chart simplifying data description and methodology steps. *Source: Arnous et al. (2021). https://s100.copyright.com/CustomerAdmin/PLF.jsp?ref=093fd189-e9da-476e-952e-254624b12f53*

2.2 Surface geophysical exploration methods

Surface geophysical surveys are widely employed to investigate specific land areas (i.e., sites) with the most appropriate geophysical techniques suited to the survey's objectives. The choice of method depends on the hypothesized physical properties of the target material. Surface geophysical methods aim to collect data that can image the subsurface to a depth of investigation (DOI), which is determined by factors such as the method used, instrumentation, and site characteristics.

One of the key advantages of surface geophysical techniques is that they generally require little to no penetration into the ground. Sensors such as electrodes or geophones are typically placed or driven no deeper than 12 to 24 inches (~30 to 60 cm) below the surface. Due to this noninvasive or minimally invasive nature, these methods are cost-effective, require less time, and can be applied with minimal disturbance to the site.

Surface geophysical methods can be broadly categorized into five main groups based on the physical principles they rely on. Each of these categories includes commonly used techniques for shallow subsurface investigations, particularly in environmental studies. The following sections provide an overview of each category, highlighting its applications and typical uses.

2.2.1 Potential field methods

- **Gravity Method**
- **Magnetic Method**

Gravitational and magnetic fields obey Laplace's equation, making both gravity and magnetic fields types of potential fields. Consequently, the gravity and magnetic methods are classified as passive geophysical methods, as they measure naturally occurring potential fields without applying any external energy or force. The gravity method measures variations in the Earth's density, while the magnetic method detects changes in the Earth's magnetic field.

2.2.2 Electrical methods

- **Electrical Resistivity**
- **Induced Polarization (IP)**
- **Complex Resistivity**
- **Self-Potential (SP)**

Electrical methods detect subsurface electrical properties by measuring naturally occurring (passive) or artificially introduced (active) voltage potentials, electrical currents, and/or electromagnetic fields. These properties are measured either galvanically, through direct electrode connections to the ground, or inductively, by inducing electromagnetic fields through the subsurface. Subsurface electrical properties such as resistivity can indicate variations in the composition, structure, and extent of subsurface materials. Electrical resistance and resistivity, fundamental to Ohm's Law ($V = IR$), can be influenced by various factors, such as porosity, water content, composition, salinity of pore water, and grain size distribution.

For instance, in most shallow subsurface environments, electric current conduction primarily occurs through the pore water, which can become a conductive electrolyte when mixed with chemical salts. Resistivity values, however, can overlap across different geological formations, which complicates direct interpretation (Table 2.2). Factors such as porosity, water content, and mineral composition need to be carefully considered when interpreting resistivity data. In some electrical methods, such as induced polarization, solid components (such as clay minerals and metallic content) may also be measured.

This table, based on data from Keller and Frischknecht (1966) and Telford et al. (1976), provides typical resistivity ranges for various natural materials. However, as resistivity values often overlap between different materials, careful analysis of the factors influencing resistivity is necessary for accurate geological interpretation.

2.2.3 Electromagnetic (EM) methods

- **Frequency Domain Electromagnetic (FDEM)**
- **Time Domain or Transient Electromagnetic (TDEM)**
- **Ground Penetrating Radar (GPR)**
- **Very Low Frequency Electromagnetic (VLF)**

TABLE 2.2 Resistivity values for common Earth materials.

Material	Typical resistivity range (Ohm·m)
Air	10^{13} to 10^{16}
Sand	10^{1} to 10^{3}
Clay	10^{-1} to 10^{1}
Granite	10^{3} to 10^{6}
Limestone	10^{2} to 10^{4}
Shale	10^{-2} to 10^{2}
Water (fresh)	10^{-2} to 10^{2}
Water (salt)	10^{-1} to 10^{1}
Coal	10^{0} to 10^{3}
Iron ore	10^{-1} to 10^{2}

Note: Resistivity values can vary significantly due to geological and environmental factors such as porosity, composition, water content, and salinity.

Electromagnetic methods detect the electrical properties of the subsurface by inducing EM energy and measuring the response of the Earth's materials. The EM instruments transmit a time-varying electric current through a transmitter loop, generating a magnetic field. If a conductive body or material is present nearby, an electromotive force (voltage) is generated within it due to the magnetic field (Faraday's Law). This leads to the flow of electrical current in the subsurface conductors and a secondary magnetic field that is detected by a receiver coil. The data from this process is used to map the electrical properties of subsurface materials. Electromagnetic properties, such as electrical conductivity, dielectric permittivity, and magnetic susceptibility, control the magnitude of the received EM signal, providing important insights into subsurface characteristics.

2.2.4 Seismic methods

- **Seismic Refraction**
- **Seismic Reflection**
- **Passive Seismic (Horizontal-to-Vertical Spectral Ratio—HVSR)**

Seismic methods are effective for imaging large subsurface areas and have been widely applied in both deep Earth studies and natural resource exploration. Seismic technology has been extensively used in hydrocarbon exploration and environmental investigations, providing high-resolution images of the shallow subsurface, typically from tens to hundreds of meters deep. Seismic methods rely on the density contrasts between materials, which affect the propagation velocity of seismic waves. As seismic energy passes through the subsurface, it interacts with geological layers, and variations in material density cause seismic waves to refract or reflect. Geophones are used to detect this energy and analyze the subsurface structure. Research on seismic wave velocities for various minerals aids in interpreting the seismic data to determine subsurface materials (Christensen, 1990).

2.2.5 Nuclear methods

Nuclear methods in geophysical exploration refer to techniques that measure naturally occurring radiation emitted from the Earth's surface. One of the most widely applied nuclear methods is **radiometric surveying** (gamma ray spectrometry), which detects and maps radioactive emanations (gamma rays) originating from rocks, soils, and other geological formations.

2.2.5.1 Radiometric surveying

Radiometric surveying is the technique used to measure natural radiation emitted from the Earth's surface. This method detects and maps the radioactive emanations (gamma rays) originating from rocks and soils. Gamma radiation is produced through the natural decay of isotopes such as uranium ($^{238}_{92}U$), thorium ($^{232}_{90}Th$), and potassium ($^{40}_{19}K$). The radiometric method can detect these elements at or near the Earth's surface. Common radioactive minerals include uraninite (^{238}U), monazite, thorianite (^{232}Th), rubidium ($^{87}_{37}Rb$) found in granite-pegmatite, feldspar ($^{4\circ}K$), muscovite, and sylvite in acidic igneous rocks, as well as radiocarbon ($^{14}_{6}C$).

Radioactive decay generates heat, which influences thermal conditions within the Earth. Additionally, the rate of decay for certain elements serves as a critical tool for geochronology, helping scientists date the formation of rocks. The demand for nuclear fuels, as well as the search for associated nonradioactive deposits like titanium and zircon, has made radiometric surveys a crucial method for mineral exploration.

Isotopes are atoms of the same element with the same number of protons but different numbers of neutrons. Some isotopes are unstable and undergo spontaneous decay, releasing energy and particles. Radioactivity refers to the disintegration of atomic nuclei, emitting particles such as alpha (α) particles (helium nuclei), beta (β) particles (electrons emitted from neutrons), and gamma (γ) rays, which are a form of pure electromagnetic radiation (Duval, 1980).

While alpha, beta, and gamma particles may all be emitted during radioactive decay, only gamma rays can travel a significant distance and are therefore useful for exploration. Gamma rays, having the highest energy, can penetrate several hundred feet through the atmosphere but only a few inches (10–60 cm) into the Earth's surface. Higher altitudes (above 300 m) may interfere with radiometric data, so for accurate detection, radioactive elements or their decay products must be present at the Earth's surface. The energy of gamma rays is characteristic of the radioactive element from which they originated.

Radiometric measurements can be conducted using various methods, such as airborne instruments (airplanes, helicopters, drones), ground-based systems (vehicle-mounted or handheld devices), or borehole tools. The measurement principle remains consistent across these setups. However, radiometric surveys cannot be conducted offshore or in permafrost regions due to the attenuation of the signal by water.

To detect common radioactive elements like uranium, thorium, and potassium, scintillometry (gamma ray spectrometry) is employed. Multichannel spectrometers are used to measure the abundances of specific radioelements. This method is particularly valuable for uranium exploration, as it can directly detect the element. Among these three radioelements, thorium is the least mobile and behaves similarly to zirconium.

Radiometric surveys are ideal for mapping large, flat areas. Although the technique has numerous applications, it is primarily used in geological mapping. Changes in lithology or soil types often correlate with variations in the concentrations of radioelements. The method is effective in detecting mineral deposits directly, including potassium alterations typically associated with hydrothermal ore deposits. Radiometric surveys are also used in uranium and thorium exploration, heat flow studies, and environmental mapping, such as delineating surface drainage features.

2.2.5.2 *Principles of radioactivity*

Radioactivity refers to the disintegration of atomic nuclei through the emission of energy and particles of mass. In geosciences, the application of radioactivity is based on understanding the physical properties of radiation sources and our ability to detect them through remote sensing techniques. Atoms are the smallest units of matter with unique chemical properties. An atom consists of a nucleus, which is surrounded by electrons. The nucleus contains positively charged protons and uncharged neutrons. The size of an atom is on the order of 10^{-10}m, while the diameter of its nucleus is around 10^{-15} meters. Protons and neutrons each have a mass of 1.67×10^{-27} kg, and electrons have a mass of 9.11×10^{-31} kg. The elementary charge of an electron is 1.602×10^{-19} Coulombs.

The number of protons in a nucleus defines an element's atomic number Z. The total number of protons and neutrons in the nucleus is called the mass number A. Atoms of the same element with the same atomic number but different numbers of neutrons are known as isotopes. These isotopes are represented by their chemical symbol and mass number, for example, ${}_{Z}^{A}X$, where Z is the atomic number and A is the mass number (A = Z + N, where N is the number of neutrons). While isotopes have identical chemical properties, they exhibit different physical properties.

When atoms have the same number of protons and neutrons, they are referred to as nuclides. Some isotopes, however, have excess energy in their nuclei and are unstable. These unstable isotopes undergo a process called nuclear decay, where they disintegrate into more stable nuclei, often of a different isotope. This disintegration is accompanied by the emission of particles or energy, a phenomenon known as nuclear radiation or radioactivity. The isotopes that undergo radioactive decay are called radionuclides.

The rate of radioactive decay can be described by the following equation:

$$N = N_0 e^{-\lambda t} \tag{2.1}$$

where N is the number of atoms remaining after time t, N_0 is the number of atoms initially present, λ is the decay constant (s^{-1}), and e is Euler's number (approximately 2.71828).

The half-life ($T_{\frac{1}{2}}$) of a radionuclide, the time it takes for half of the initial amount of the substance to decay, is related to the decay constant by:

$$T_{\frac{1}{2}} = \frac{0.693}{\lambda} \tag{2.2}$$

From the decay equation, the activity A of a radionuclide is given by:

$$A = \frac{dN}{dt} = -\lambda N = -\lambda N_0 e^{-\lambda t} = A_0 e^{-\lambda t} \tag{2.3}$$

where A_0 is the initial activity, A is the activity after time t, λ is the decay constant, and N is the number of remaining atoms.

Thus, the activity also decreases exponentially with the same half-life. The activity A represents the number of disintegrations (or decays) occurring per unit time, often measured in Becquerels (Bq), where 1 Bq equals one disintegration per second. The unit Curie (Ci) is used to measure the activity of a radionuclide, with 1 Ci corresponding to 3.7×10^{10} disintegrations per second.

Radioactive decay is independent of external factors such as temperature, pressure, and chemical state of the material, meaning the rate of decay is a constant characteristic of the radionuclide.

2.2.5.3 *Types of radioactive decay*

Ernest Rutherford's experiments on the interaction of radiation with magnetic and electric fields led to the discovery of different types of radioactive decay (Fig. 2.3). These types of decay involve the emission of various particles or energy, transforming the parent isotope into a different element or isotope.

The major types of radioactive decay are as follows:

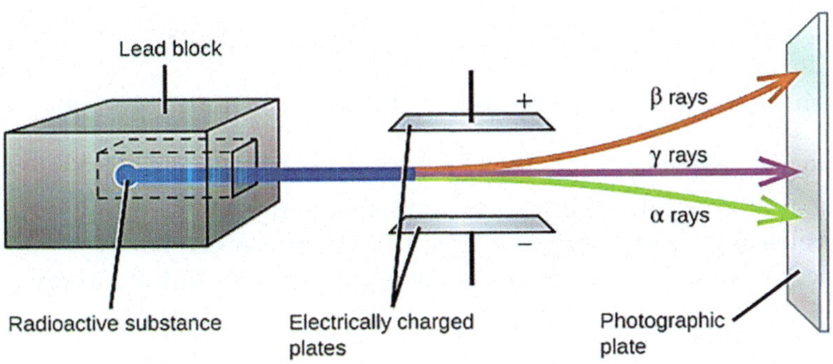

FIG. 2.3 Behavior of alpha, beta, and gamma particles in an electric field. *Source: University of Central Florida (UCF)—Chapter 19, 19.3 Radioactive Decay. https://pressbooks.online.ucf.edu/chemistryfundamentals/chapter/radioactive-decay-2*

2.2.5.3.1 Alpha decay

In alpha decay, a nucleus loses an alpha particle, which consists of 2 protons and 2 neutrons. This results in a decrease in the atomic number by 2 and the mass number by 4. A common example is the decay of uranium-238:

$$^{238}_{92}\mathrm{U} \rightarrow {}^{234}_{90}\mathrm{Th} + {}^{4}_{2}\mathrm{He} \tag{2.4}$$

The emitted alpha particle is a helium nucleus.

2.2.5.3.2 Beta minus (β^-) decay

Beta-minus decay involves the emission of a beta particle, which is identical to a negatively charged electron. In this type of decay, a neutron in the nucleus is converted into a proton, increasing the atomic number by 1 while the mass number remains unchanged. A typical example is:

$$^{234}_{90}\mathrm{Th} = {}^{234}_{91}\mathrm{Pa} + \beta^{-1} \tag{2.5}$$

This process increases the atomic number by one (from thorium to protactinium).

2.2.5.3.3 Beta plus (β^+) decay (positron emission)

Beta-plus decay, or positron emission, is less frequent and involves the emission of a positron (the antimatter counterpart of the electron) and a neutrino. In this process, a proton is converted into a neutron, causing the atomic number to decrease by 1 while the mass number remains the same. An example is:

$$^{11}_{6}\mathrm{C} = {}^{11}_{5}\mathrm{B} + e^{+} + \nu \tag{2.6}$$

In this reaction, carbon-11 decays into boron-11, emitting a positron and a neutrino.

2.2.5.3.4 Electron capture

Electron capture occurs when an orbital electron (usually from the innermost shell) is absorbed into the nucleus, where it combines with a proton to form a neutron. This process decreases the atomic number by 1. The missing electron position is filled by another electron, and characteristic X-rays are emitted. A typical example is:

$$^{40}_{19}\mathrm{K} + {}^{0}_{-1}e = {}^{40}_{18}\mathrm{Ar} \tag{2.7}$$

Here, potassium-40 captures an electron and transforms into argon-40.

2.2.5.3.5 Spontaneous fission

Spontaneous fission occurs when heavy atoms, like uranium or plutonium, split into two lighter nuclei, releasing energy and neutrons. This process results in the formation of smaller nuclei and is accompanied by the release of a significant amount of energy.

2.2.5.3.6 Gamma radiation

After a radionuclide undergoes any of the above decays, the resulting daughter nucleus may still be in an excited state. It can release this excess energy in the form of gamma rays (high-energy electromagnetic radiation). An example is:

$$^{234}_{91}\text{Pa} \rightarrow {}^{234}_{91}\text{Pa} + \gamma \tag{2.8}$$

This emission does not change the atomic or mass number but represents a release of energy.

2.2.5.3.7 Decay equations and atomic number changes

The decay equations can be summarized as follows:

- Alpha Decay:

$$^{A}X \rightarrow {}^{A-4}Y + \alpha\,(\text{alpha emission}),\ Z_Y = Z_X - 2 \tag{2.9}$$

- Beta Decay:

$$^{A}X \rightarrow {}^{A}Y + \beta^{-1}\,(\text{beta emission}),\ Z_Y = Z_X + 1 \tag{2.10}$$

- Electron Capture:

$$^{A}X \rightarrow {}^{A}Y + \text{photon}\,(\text{electron capture}),\ Z_Y = Z_X - 1 \tag{2.11}$$

Fig. 2.4: Summary of types of radioactive decay [Source: University of Central Florida (UCF)—Chapter 19, 19.3 Radioactive Decay].

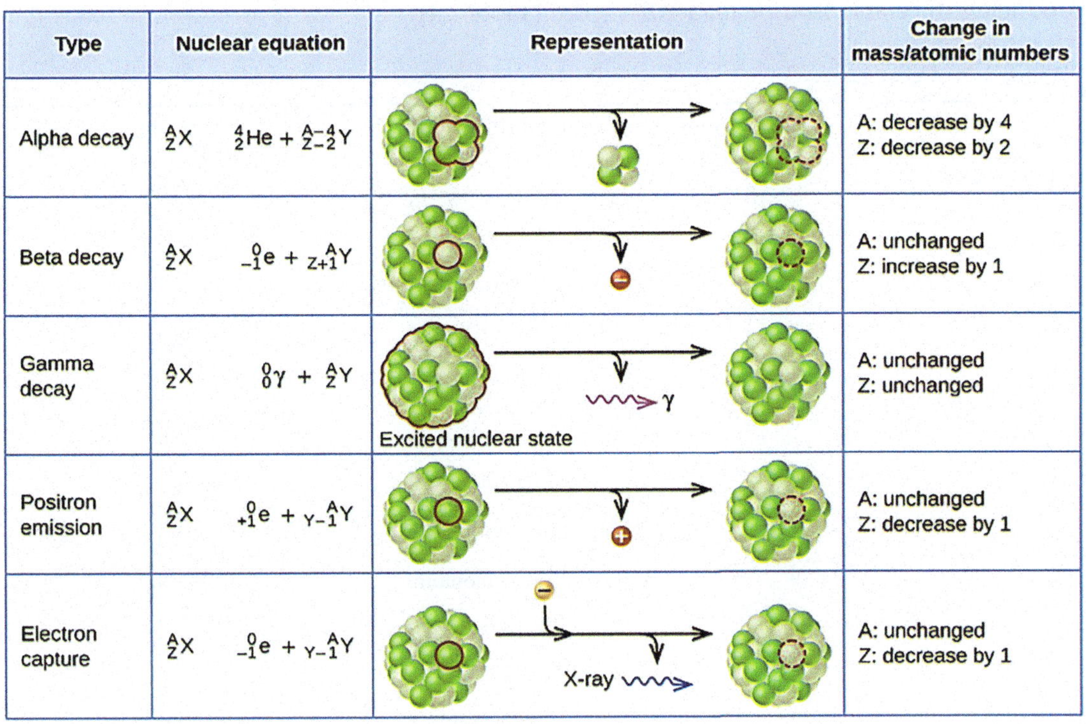

Type	Nuclear equation	Representation	Change in mass/atomic numbers
Alpha decay	$^{A}_{Z}X$ $\quad ^{4}_{2}He + {}^{A-4}_{Z-2}Y$		A: decrease by 4 Z: decrease by 2
Beta decay	$^{A}_{Z}X$ $\quad ^{0}_{-1}e + {}^{A}_{Z+1}Y$		A: unchanged Z: increase by 1
Gamma decay	$^{A}_{Z}X$ $\quad ^{0}_{0}\gamma + {}^{A}_{Z}Y$	γ Excited nuclear state	A: unchanged Z: unchanged
Positron emission	$^{A}_{Z}X$ $\quad ^{0}_{+1}e + {}^{A}_{Y-1}Y$		A: unchanged Z: decrease by 1
Electron capture	$^{A}_{Z}X$ $\quad ^{0}_{-1}e + {}^{A}_{Y-1}Y$	X-ray	A: unchanged Z: decrease by 1

FIG. 2.4 Types of decay, along with their equations and changes in atomic and mass numbers. *https://pressbooks.online.ucf.edu/chemistryfundamentals/chapter/radioactive-decay-2*

2.2.5.3.8 Multiple decay modes

Some radionuclides can decay through more than one mode. For example, bismuth-212 can undergo both beta decay and alpha decay, with 66% of disintegrations occurring via beta emission and 34% via alpha emission. However, the half-life of the radionuclide remains the same, regardless of the decay mode.

2.2.5.3.9 Radioactive decay series

Radioactive decay often occurs in a chain (or series), where a parent isotope decays into several daughter isotopes, which are also radioactive, until a stable isotope is eventually formed. The series reaches equilibrium when the activities of all radionuclides in the series become equal. This equilibrium is expressed as:

$$\lambda_1 N_1 = \lambda_2 N_2 = \lambda_3 N_3 = \ldots = \lambda_i N_i \tag{2.12}$$

Here, λ_i represents the decay constants, and N_i represents the number of atoms for each radionuclide in the series.

2.2.5.3.10 Examples of natural decay series

Uranium-238 decay series: Uranium-238 ($^{238}_{92}U$) decays through a series of steps, eventually producing the stable isotope lead-206 ($^{206}_{82}Pb$) (Table 2.3).

Uranium-235 decay series: Uranium-235 ($^{235}_{92}U$) decays through a series of steps, eventually producing the stable isotope lead-207 ($^{207}_{82}Pb$) (Table 2.4).

TABLE 2.3 Decay series.

Parent nuclide	Decay mode	Product of decay	Half-life
$^{238}_{92}U$ (uranium)	–	–	$4.468 \times 10^9 y$
$^{238}_{92}U$	α	$^{234}_{90}Th$ (thorium)	24.1 d
$^{234}_{90}Th$	β^-	$^{234}_{91}Pa$ (protactinium)	1.18 min
$^{234}_{91}Pa$	–	$^{234}_{91}Pa$	6.7 h
$^{234}_{91}Pa$	β^-		$2.48 \times 10^5 y$
$^{234}_{92}U$	α	$^{230}_{90}Th$	$7.52 \times 10^4 y$
$^{230}_{90}Th$	α	$^{226}_{88}Ra$ (radium)	1602 y
$^{226}_{88}Ra$	α	$^{222}_{86}Rn$ (radon)	3.825 d
$^{222}_{86}Rn$	α	$^{218}_{84}Po$ (polonium)	3.05 min
$^{218}_{84}Po$	α β^-	$^{214}_{82}Pb$ (lead) $^{218}_{85}At$ (astatine)	26.8 min 2 s
$^{218}_{85}At$	α	$^{214}_{83}Bi$ (bismuth)	19.7 min

Parent nuclide	Decay mode	Product of decay	Half-life
$^{214}_{83}Bi$	β^-	$^{214}_{84}Po$	164 μs
	α	$^{210}_{81}Tl$ (thallium)	1.32 min
$^{210}_{81}Tl$	β^-		~22 y
$^{210}_{82}Pb$	β^-	$^{210}_{83}Bi$	5.02 d
$^{210}_{83}Bi$	β^-	$^{210}_{84}Po$	138.3 d
	α	$^{206}_{81}Tl$	4.19 min
$^{206}_{81}Tl$	β^-	$^{206}_{82}Pb$	Stable

Note: y stands for year, d stands for day, min stands for minute, s stands for second, and μs stands for microsecond (= s).
Radiological Health Handbook (1970); Ivanovich and Harmon (1982).

TABLE 2.4 Decay series.

Parent nuclide	Decay mode	Product of decay	Half-life
	–	–	7.13×10^8 y
$^{235}_{92}U$	α	$^{231}_{90}Th$	25.64 h
$^{231}_{90}Th$	β^-	$^{231}_{91}Pa$	3.43×10^4 y
$^{231}_{91}Pa$	α	$^{227}_{89}Ac$(actinium)	22 y
$^{227}_{89}Ac$	β^-	$^{227}_{90}Th$	18.17 d
	α	$^{223}_{87}Fr$(francium)	21 min
	β^-	$^{223}_{88}Ra$	11.68 d
$^{223}_{88}Ra$	α	$^{219}_{86}Rn$	3.92 s
$^{219}_{86}Rn$	α	$^{215}_{84}Po$	1.83 ms
$^{215}_{84}Po$	α	$^{211}_{82}Pb$	36.1 min
$^{211}_{82}Pb$	β^-	$^{211}_{83}Bi$	2.16 min
$^{211}_{83}Bi$	β^-	$^{211}_{84}Po$	0.52 s
	α	$^{207}_{81}Tl$	4.79 min
$^{207}_{81}Tl$	β^-	$^{207}_{82}Pb$	Stable

Note: ms stands for millisecond (= 10^{-3} s).
Radiological Health Handbook (1970); Ivanovich and Harmon (1982).

Thorium-232 decay series: Thorium-232 ($^{232}_{90}Th$) undergoes a decay series that eventually produces the stable isotope lead-208 ($^{208}_{82}Pb$) (Table 2.5).

The decay of uranium-238 and thorium-232 does not involve gamma rays directly; however, gamma ray emissions from their daughter products are used to estimate their concentrations. The process of radioactive decay, including alpha decay, beta decay, electron capture, and spontaneous fission, results in the transformation of unstable parent isotopes into more stable daughter isotopes. The decay process is often accompanied by the emission of radiation and can occur in a series, eventually leading to stable isotopes. These decay chains are crucial in fields such as geochronology and radiometric dating.

2.2.5.4 *Disequilibrium in radioactive process*

Disequilibrium occurs when one or more decay products in a radioactive decay series are either completely or partially removed or added to the system. Thorium typically remains in equilibrium in nature, and there are no disequilibrium issues with potassium. However, disequilibrium is common in the uranium decay series, where it can occur at several stages in the ^{238}U decay chain. For example, ^{238}U can be selectively leached relative to ^{234}U, and ^{234}U can be selectively leached compared to ^{238}U. Additionally, ^{230}Th and $^{226}_{88}Ra$ (radium) can be

TABLE 2.5 Decay series.

Parent nuclide	Decay mode	Product of decay	Half-life
$^{232}_{90}Th$	–	–	1.39×10^{10} y
$^{232}_{90}Th$	α	$^{228}_{88}Ra$	5.75 y
$^{228}_{88}Ra$	β^-	$^{228}_{89}Ac$	6.13 h
$^{228}_{89}Ac$	β^-	$^{228}_{90}Th$	1.913 y
$^{228}_{90}Th$	α	$^{224}_{88}Ra$	3.64 d
$^{224}_{88}Ra$	α	$^{220}_{86}Rn$	55.6 s
$^{220}_{86}Rn$	α	$^{216}_{84}Po$	0.145 s
$^{216}_{84}Po$	α	$^{212}_{82}Pb$	10.64 h
$^{212}_{82}Pb$	β^-	$^{212}_{83}Bi$	60.5 min
$^{212}_{83}Bi$	β^-	$^{212}_{84}Po$	304 ns
	α	$^{208}_{81}Tl$	3.1 min
$^{208}_{81}Tl$	β^-	$^{208}_{82}Pb$	Stable

Note: ns stands for nanosecond ($= 10^{-9}$ s).
Radiological Health Handbook (1970); Ivanovich and Harmon (1982).

removed from the decay chain, and $^{222}_{86}Rn$ (radon gas) is mobile and can escape from soils and rocks into the atmosphere. The restoration of equilibrium can take days, weeks, or even millions of years, depending on the half-lives of the involved radioisotopes.

Disequilibrium in the uranium decay series is a significant source of error in gamma ray spectrometry. Uranium concentration estimates often rely on the measurement of $^{214}_{83}Bi$ and $^{214}_{82}Pb$ isotope abundances, which occur later in the radioactive decay chain and may not be in equilibrium with uranium. As a result, uranium concentration estimates are usually reported as "equivalent uranium" (eU), based on the assumption of equilibrium. Although the thorium decay series is almost always in equilibrium, thorium concentrations are also often reported as "equivalent thorium" (eTh).

2.2.5.5 Interaction of gamma rays with matter

Radiation consists of a flux of elementary particles and energy quanta, which can be classified based on their physical characteristics and energy. Alpha radiation is a flux of positively charged alpha particles, with an initial energy of several MeV and an initial velocity on the order of 10^7 m/s. Alpha particles are highly ionizing but have a low penetration range in matter, being absorbed by about 10^{-2} m of air and 10^{-5} m of rock. The energy of alpha particles is discrete and specific to the radionuclide.

Beta radiation consists of a flux of electrons with a continuous energy spectrum, reaching a maximum energy depending on the radionuclide. Beta particles can approach the speed of light. The penetration range of beta particles depends on their initial energy, and for a 2 MeV beta particle, the range is about 8 m in air and 1 cm in water. As beta radiation passes through matter, it loses energy via ionization and generates electromagnetic radiation known as bremsstrahlung. Positrons combine with electrons to produce two gamma quanta of 511 keV each.

Gamma radiation is a form of electromagnetic radiation with a discrete energy (E), frequency (ν), and wavelength (λ), traveling at the speed of light (c). These quantities are related by the equation:

$$E = h\nu = \frac{hc}{\lambda} \tag{2.13}$$

where h is Planck's constant (6.6261×10^{-34} Js) and c is the speed of light. Electromagnetic radiation with energy <40 keV is categorized as X-rays, while gamma rays have energies greater than 40 keV. Gamma rays interact with matter through three main processes: the photoelectric effect, Compton scattering, and pair production.

The probability of a photon interacting with matter is expressed by the cross-section σ (m^2), which depends on the photon energy (E) and the composition of the matter. Fig. 2.5 shows the relationship between scattering and absorption processes, photon energy, and the atomic number of the absorbing material. For gamma rays of natural terrestrial origin (up to 2.615 MeV) and for materials such as rock, water, and air, Compton scattering is the dominant interaction process. Gamma rays typically lose energy through successive Compton scattering events, eventually leading to absorption via the photoelectric effect (ICRU, 1994).

As gamma rays interact with matter, their intensity decreases with distance from the source. The absorption of gamma rays is described by linear attenuation coefficients (μ, m^{-1}) or mass attenuation coefficients (μ/ρ, m^2/kg). The attenuation of gamma rays in a narrow

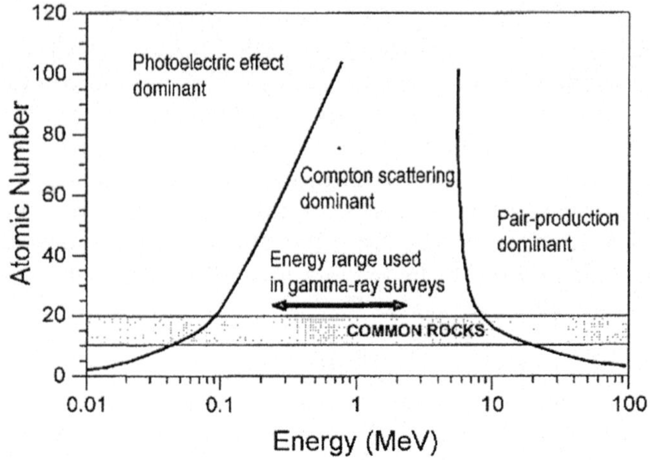

FIG. 2.5 The interaction of gamma rays with matter. *Source: IAEA-TECDOC-1363, 2003). https://www.iaea.org/about/terms-of-use*

beam can be modeled by an exponential function. The range of gamma rays from natural radionuclides is approximately 700 m in air, up to 0.5 m in rock, and a few centimeters in lead. Since gamma rays are the most penetrating form of natural and man-made radiation, they are widely used in radiation environment studies.

2.2.5.6 *Area survey meters*

The instrumentation used for radioactivity measurements has continuously improved with the rapid advancement of new technologies. Some of the earliest radioactivity measurements were made using ionization chambers and electrometers for borehole measurements in 1935 (Howell and Frosch, 1939). Shortly thereafter, Geiger-Muller tubes sensitive to gamma radiation were developed (Evans and Mugele, 1936) and were quickly applied to geological problems worldwide (Brewer, 1937; Kurbatov, 1937; Landsberg and Klepper, 1939). By the late 1940s, portable Geiger counters weighing about 10 kg became available. At the same time, research into scintillation detectors showed that thallium-activated sodium-iodide crystals produced the greatest sensitivity to gamma radiation (Hofstadter, 1949). In 1949, R.W. Pringle and K.I. Roulston of the University of Manitoba developed a portable scintillation counter that could function as a spectrometer to distinguish between uranium and thorium in ores by identifying gamma rays with unique energies from each decay series. A commercial version of the instrument became available in 1950 (Brownell, 1950; Pringle et al., 1950).

Radiation instruments used for survey monitoring include both gas-filled detectors and solid-state detectors, such as scintillator or semiconductor detectors. With the invention of the transistor in the early 1950s, instruments for radioactivity measurements became lighter, more compact, and more advanced. In the late 1950s and early 1960s, portable gamma-ray spectrometers were developed, allowing for in-situ measurements of potassium (K), equivalent uranium (eU), and equivalent thorium (eTh) concentrations (Schneider and Schwerdtel, 1960; Adams et al., 1959; Adams and Lowder, 1964). By 1968, gamma-ray spectrometry was also used in aerial surveys (Adams and Gasparini, 1970; Rogers and Adams, 1969a, 1969b; Darnley and Fleet, 1968). During the 1970s, portable instruments became even more compact and versatile (Darnley, 1972; Darnley, 1993).

Field and laboratory methods primarily rely on the ionizing properties of radiation and instruments that convert this radiation into electrical signals. Various detectors, such as ionization chambers, proportional counters, Geiger-Muller tubes, scintillation counters, semiconductor detectors, thermoluminescence detectors, and mechanical and chemical track detectors, are used to monitor and quantify α, β, γ, and neutron radiation in the environment (Fig. 2.6).

A Geiger-Muller (GM) counter consists of a gas-filled tube with a metal cylinder (the cathode) and a thin conductive wire (the anode) along the tube's axis. The tube is typically filled with argon gas mixed with halogen vapor. GM counters, which range from 2 to 30 cm in length and 1 to 4 cm in diameter, operate with an applied voltage of several hundred volts. These counters exploit the progressive growth of ionization in a strong electric field between the anode and cathode. When a photon interacts with the cathode, it releases an electron that is directed into the GM tube. The resulting ionization amplifies the signal, generating an electric current between the electrodes, which creates a voltage pulse at the anode output. The gas ionization multiplication coefficient is about 10^6, but the output pulse is not proportional to the gamma ray energy absorbed. GM counters have a low detection efficiency (<2%) and a dead time of approximately 10^{-4} s.

Scintillation counters consist of a scintillator and a photomultiplier. When a gamma photon interacts with the scintillation crystal, it produces scintillations, which are visible light

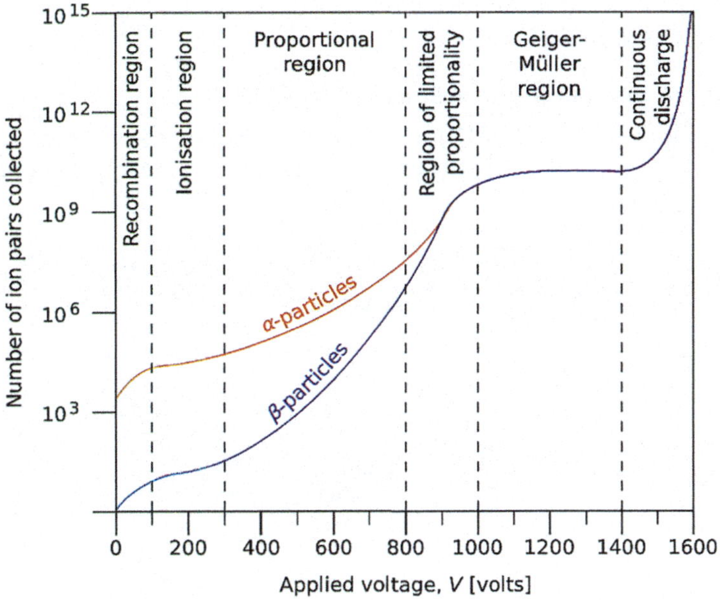

FIG. 2.6 Ion pairs collected vs. applied voltage in a gaseous ionization chamber. *Source: Oxford Physics 2009, C4: Particle Physics Major Option Particle Detectors. https://www.iaea.org/about/terms-of-use*

photons that eject electrons from the photocathode of the attached photomultiplier. These electrons multiply progressively at the dynodes of the photomultiplier, and the resulting electron cloud strikes the anode, inducing a negative voltage pulse with amplitude proportional to the incident photon's energy. Scintillation counters are widely used in gamma-ray spectrometry, with thallium-activated sodium iodide (NaI (Tl)) crystals being the most common detectors in field surveys. These crystals have a detection efficiency of up to 100% for low-energy gamma rays, but slightly less for high-energy gamma rays. The dead time is around 10^{-7} s, and the energy resolution for ^{137}Cs at 662 keV ranges from 7 to 10%. Total count instruments have an energy discrimination threshold, above which all gamma rays are recorded. Changes to this threshold can affect the instrument's response, so reference radiation sources are used to adjust the gain periodically. Total count measurements estimate overall radiation levels.

Gamma-ray spectrometers extend the scintillometer by utilizing the direct proportionality between the energy of an incoming gamma ray and the pulse amplitude at the detector output. Fig. 2.7 shows a block diagram of a gamma-ray spectrometer. After amplification and digitization, the pulse amplitudes are analyzed, and the spectrometer produces an energy spectrum of detected radiation. Since individual radionuclides emit specific gamma-ray energies, the spectra can be used to identify the radiation source. Gamma-ray spectrometers are classified as "integral" or "differential." Integral spectrometers record pulses above a discrimination threshold, while differential spectrometers record pulses within a given amplitude interval or channel. Wider energy intervals, comprising several channels, are referred to as energy windows. Modern analysers may use up to 256 or 512 channels, with each channel representing several keV. To avoid energy spectrum drift, gamma-ray spectrometers should have amplitude gain stabilization. Further descriptions of detectors and radiometric devices can be found in Adams and Lowder (1964), IAEA (1989), ICRU (1994), IAEA (1979), IAEA (1991), and Rossi and Staub (1949).

2.2.5.7 *Surface nuclear magnetic resonance (SNMR)*

Nuclear magnetic resonance (NMR) is another technique used under **Nuclear Methods** to quantify the free water in the subsurface. This method can be applied both on the surface

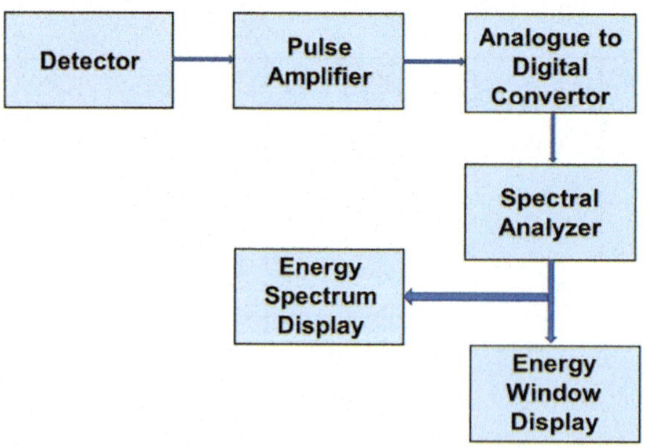

FIG. 2.7 Block diagram of a gamma ray spectrometer.

and in boreholes and is especially useful for mapping the presence of water and hydrogen ions in the subsurface. NMR is widely used in **environmental geophysics** to detect water sources and assess water-related conditions.

2.3 Borehole geophysical methods

Borehole geophysics involves the collection and analysis of geophysical data obtained from boreholes, wells, or test holes. These methods provide direct access to subsurface properties, offering in situ, continuous records of critical formation characteristics such as lithology, fluid content, porosity, permeability, and fracture properties. Borehole geophysical data often complement surface geophysical surveys, providing a more detailed and comprehensive understanding of subsurface conditions.

In the context of resource exploration, particularly oil and gas, borehole geophysics is essential for assessing reservoir properties. It is used to estimate key parameters such as porosity, permeability, fluid saturation, and formation composition. Similarly, these methods are highly valuable in groundwater studies, including the assessment of supply, recharge, contamination monitoring, and remediation efforts. Borehole techniques also support subsurface modeling and are frequently employed in the analysis of water wells, such as for determining water quality and construction details (Khan, 2000).

Beyond oil, gas, and groundwater applications, borehole geophysics is critical for other geotechnical and environmental studies. It helps determine the lithology, bed thickness, fracture content, and well construction integrity, such as evaluating well screens and casings. This data is essential for well planning, stratigraphic interpretation, and the lateral extrapolation of information from core samples to better understand subsurface conditions over larger areas.

A primary output of borehole geophysics is the well log, which records the variation of physical properties along the length of the borehole. These logs are collected using a cable-suspended tool (or "sonde") that moves up and down the borehole, with the depth of the tool precisely tracked by a winch with a depth counter. The system typically consists of a power supply, a recording unit, and processing modules that digitize and process the data, allowing for detailed analysis and corrections using specialized software.

Borehole geophysical techniques are similar to surface methods, but they offer several advantages, such as higher measurement resolution and precise depth referencing. They can also study a volume of formation beyond the disturbed zone of the borehole, often covering a larger area than core sampling alone. Due to the complexity of subsurface formations, multiple logs are usually recorded simultaneously to capture a more complete picture. The integration of various logs and imaging techniques allows for more accurate interpretations of subsurface conditions, as the different methods complement each other by highlighting different physical properties.

Overall, borehole geophysics provides valuable insights into subsurface properties and processes, making it indispensable in resource exploration, groundwater management, environmental studies, and geotechnical assessments. These methods can be categorized into several groups, each with various subtypes, depending on the specific geophysical properties being measured and the depth of investigation required.

2.3.1 Electrical borehole geophysics

Electrical borehole geophysical methods, including borehole resistivity (both normal and focused), self or spontaneous-potential (SP) measurements, single-point resistance (SPR) logging, and induced polarization (IP), are crucial for understanding subsurface electrical properties. These techniques are widely applied in hydrogeological, contamination, and environmental studies to investigate the electrical characteristics of subsurface materials. By measuring the electrical currents or voltage potentials, either naturally occurring or artificially introduced, these methods help reveal the electrical structure of formations encountered by the borehole.

The resistivity of earth materials is one of the key electrical properties investigated in borehole geophysics. Resistivity refers to how strongly a material resists the flow of electrical current, and this property is directly influenced by factors such as mineral composition, structure, porosity, permeability, saturation, and the chemical composition of pore fluids. It is important to note that electrical resistivity methods are typically based on the relationship between electrical current (injected into the subsurface) and voltage potential (measured during the survey), allowing for data collection at different depths and scales.

Subsurface resistivity varies due to changes in the composition and physical properties of the earth materials, such as variations in mineral content, clay composition, and fluid saturation. The presence of electrolytic pore fluids (e.g., salt-containing groundwater) within rock pores is the primary conductor of electrical current. However, certain minerals, such as clays, can significantly affect the resistivity measurements due to their unique conductive properties. In interpreting resistivity data, it is essential to account for these factors, which can influence the apparent resistivity of different formations.

Electrical resistivity methods are highly sensitive to spatial variations within geological formations, allowing the differentiation of materials that have distinct hydrological or geological characteristics. By applying these methods, it is possible to identify zones of varying permeability, porosity, and fluid content, which are crucial for understanding subsurface conditions, especially in groundwater and contamination studies.

2.3.2 Electromagnetic borehole geophysics

Electromagnetic borehole geophysical methods, such as borehole electromagnetic induction, magnetic susceptibility borehole logging, and borehole-radar reflection logging, are based on the principles of electromagnetic induction. These techniques utilize the phenomenon whereby a voltage potential is generated across a conductive material when subjected to a time-varying magnetic field. Borehole electromagnetic geophysical tools induce electromagnetic energy into the surrounding subsurface formations and measure the response of these materials to the applied electromagnetic waves.

At the core of electromagnetic borehole tools is a time-varying alternating current (AC) introduced into an internal transmitter coil. This creates a magnetic field with the same frequency and phase as the alternating current. The primary magnetic field extends into the earth materials surrounding the borehole, with the depth of penetration depending on the specifications of the tool and the environmental conditions present within the subsurface. As the magnetic field interacts with conductive materials in the subsurface, it induces an electromotive force (EMF) that generates a secondary electromagnetic wave.

The flow of electrical current within subsurface conductors influenced by the primary magnetic field creates a secondary magnetic field, which is detected by a receiver coil within the borehole instrument. The system records the returning signal, which is then compared to the original transmitted signal. By processing and interpreting these signals, it is possible to evaluate the variations in electrical conductivity and magnetic susceptibility of the materials surrounding the borehole. These measurements offer valuable insights into the subsurface properties, including variations in material composition and structure along the borehole's length.

In electromagnetic borehole geophysics, the contrast in dielectric permittivity, which refers to the ability of materials to store electrical energy, plays a significant role in electromagnetic wave reflection. This principle is exploited to generate high-resolution images of the formations intersected by the borehole. By analyzing these reflections, geophysicists can create detailed representations of subsurface conditions, allowing for the identification and characterization of geological features and the assessment of material properties within the borehole's surrounding formations.

2.3.3 Nuclear borehole geophysics: An overview of techniques for subsurface exploration

Nuclear borehole geophysics is a crucial technique for examining the subsurface of the Earth by utilizing principles derived from nuclear physics. This field studies the behavior and interactions of atomic nuclei under both natural and externally applied conditions. The primary objective of nuclear borehole geophysics is to assess the physical and chemical properties of formations intersected by boreholes, providing essential data for geophysical exploration and management.

Various nuclear borehole techniques offer insights into the material properties surrounding the borehole, including density, porosity, and chemical composition. Methods such as **gamma-gamma density logging, nuclear magnetic resonance (NMR)**, and **neutron borehole logging** focus on estimating the physical properties of the rock formations, such as porosity and density. Additionally, techniques like **natural gamma** and **spectral gamma logging** are used to identify specific chemical elements, crucial for differentiating lithologies (rock types) (Telford et al., 1990).

2.3.3.1 *Radiation types in nuclear borehole geophysics*

Radiation utilized in nuclear borehole geophysics can be categorized into **ionizing** and **nonionizing** radiation. Ionizing radiation, which includes gamma rays and neutron radiation, is capable of breaking chemical bonds and ionizing atoms. These types of radiation are associated with most nuclear borehole methods and carry enough energy to interact with atomic structures, facilitating the measurement of various subsurface properties. In contrast, **nonionizing radiation** is primarily relevant in the context of nuclear magnetic resonance (NMR) methods, which involve the behavior of protons in hydrogen nuclei without producing ionization (Nelson, 2001).

2.3.3.2 *Passive vs. Active nuclear borehole methods*

Nuclear borehole techniques can be further divided into **passive** and **active** methods:

2.3.3.2.1 Passive methods

These include **natural gamma** and **spectral gamma logging**, which detect radiation emitted by naturally occurring radioactive elements—primarily **potassium (K)**, **thorium (Th)**, and **uranium (U)**, along with their decay products. These elements are inherently present in the Earth's crust and provide valuable data on lithological differences (Knox et al., 2016).

2.3.3.2.2 Active methods

These techniques, such as **gamma-gamma density logging** and **neutron logging**, require an external radiation source. They measure the interaction between the emitted radiation and the surrounding rock or fluid, providing valuable information about the material's properties, such as density, porosity, and fluid content (Benoit et al., 2013).

2.3.3.3 *Innovative techniques in nuclear borehole geophysics*

One of the most innovative techniques in nuclear borehole geophysics is **Borehole Nuclear Magnetic Resonance (BNMR)**, inspired by magnetic resonance imaging (MRI) technology. BNMR is used to measure the behavior of hydrogen protons in water or hydrocarbons within the rock formations. Unlike traditional methods, NMR does not use ionizing radiation and offers distinct advantages, such as high resolution for detecting fluid properties and saturation levels. This makes it a powerful tool in the exploration of fluid reservoirs (Esmaili et al., 2016).

2.3.3.3.1 Advantages and applications of nuclear borehole methods

One of the significant advantages of nuclear borehole methods—especially those involving gamma and neutron radiation—is their ability to penetrate deep into rock formations. This capability is particularly important for comprehensive geophysical surveys, enabling the detection of variations in subsurface material properties that might otherwise remain undetected by other methods. Nuclear methods are useful for investigations in any borehole casing, including those in hard-to-reach or highly heterogeneous geological settings (Keller et al., 2017).

2.3.3.3.2 Historical development of radiometric methods

The development of radiometric methods in geophysical exploration can be traced back to the discovery of **uranium** in 1789 by German chemist Martin Klaproth. This discovery laid the foundation for understanding radioactive elements. In 1896, French physicist Henri Becquerel discovered **radioactivity**, which led to the study of nuclear physics and the development of techniques for measuring radioactivity (Telford et al., 1990).

Following Becquerel's discovery, early radiation detectors were developed during the early 20th century. These detectors allowed for the measurement of radiation levels, which spurred further advancements, including the invention of **portable field instruments**. In the 1940s, the advent of **scintillation detectors** significantly enhanced the sensitivity of radiation measurements (Nelson, 2001).

In 1947, the United States, Canada, and the former USSR initiated **airborne radiometric surveys**, marking the beginning of the widespread application of radiometric techniques in **mineral exploration**. These surveys, along with ground-based gamma-ray spectrometry, helped in the in-situ estimation of key radioelements, such as potassium, uranium, and thorium. These surveys provided crucial data for geological mapping and resource exploration (Esmaili et al., 2016).

Throughout the 1960s and 1970s, advancements in **multichannel analysers**, **digital data recording**, and improvements in **semiconductor detectors** refined the precision of radiometric measurements. The integration of advanced computing systems and data processing technologies further revolutionized geophysical exploration methods, leading to the high-efficiency techniques used today (Benoit et al., 2013).

2.3.3.3.3 Radiation sources in nuclear borehole geophysics

Radiation sources used in nuclear borehole geophysics can be classified into **natural** and **artificial** categories:

Natural radiation Natural radiation is emitted by various rocks, especially those containing radioactive elements like **uranium (U)**, **thorium (Th)**, and **potassium (K)**. These elements are abundant in rocks such as **granites**, **shales**, and **clays**. Additionally, **radon gas**, a decay product of uranium, also contributes to natural radiation levels and poses a significant environmental concern. **Cosmic radiation** contributes to the natural background radiation levels on Earth (Telford et al., 1990).

Artificial radiation Artificial sources of radiation include radiation leakage from **nuclear reactors** and the use of **radioactive isotopes** for various applications. For instance, radioactive tracers are used in **industrial** and **environmental studies** to measure liquid flow, while **nuclear medicine** uses isotopes for diagnostic imaging and therapy (Knox et al., 2016). Artificial sources typically emit higher energy levels compared to natural radiation sources. Both natural and artificial radiation sources are essential in geophysical exploration. Understanding these sources is crucial for accurately interpreting radiometric data in subsurface studies and ensuring the effective application of nuclear borehole geophysics (Esmaili et al., 2016).

In conclusion, nuclear borehole geophysics plays a vital role in modern geophysical exploration. The continued advancement of these techniques provides more detailed and reliable data that support sustainable resource management, environmental monitoring, and exploration. As technology evolves, the ability to assess subsurface materials with higher precision will continue to benefit industries like mining, oil, gas, and environmental science (Knox et al., 2016; Keller et al., 2017).

2.3.3.4 Ground radiometric methods

Ground radiometric methods, including portable and car-borne gamma ray spectrometry, are employed for both regional and detailed mapping surveys to estimate surface concentrations of radioelements. Borehole gamma ray spectrometry is used to obtain information about lithology and mineralization in subsurface geological formations. Similarly, marine gamma ray spectrometry is utilized to map the seafloor and assess radioelement concentrations.

2.3.3.5 Portable gamma ray spectrometry

Portable gamma ray spectrometry has been utilized since the 1960s for applications such as uranium exploration, geological mapping, and environmental studies. Portable gamma ray spectrometers used for natural radionuclide mapping typically monitor energy windows centered on the 1461 keV ($^{40}_{19}K$), 1765 keV ($^{214}_{83}Bi$), and 2615 keV ($^{208}_{81}Tl$) photo peaks to estimate concentrations of K, U, and Th, respectively (IAEA, 1989). Several methods have been developed for estimating the concentration of nuclear fallout radionuclides (Grasty and Cox, 1997), most of which use both low and high energy windows for each radioisotope. Multichannel methods have also been developed.

The response of a portable gamma ray spectrometer is influenced by the size, location, and geometry of radioactive sources. Meaningful measurements along a traverse can only be obtained if the source-detector geometry remains consistent across all observations. The same source-detector geometry used for spectrometer calibration should be applied in the field. The sampling time required for a measurement depends on the radioactivity of the source and the precision needed. However, because the K, U, and Th spectra overlap, determining the necessary counting time for a specified precision in these radionuclide estimates is more complex. For example, a 350 cm^3 NaI (Tl) detector may require a sampling time of 2 min for highly radioactive rocks and 6 min for low radioactivity rocks as a reasonable compromise.

To estimate K, U, and Th concentrations in rocks or soils, the recorded count rate is related to the concentrations of the radioelements by the following equation:

$$n_i = S_{iK}C_K + S_{iU}C_U + S_{iTh}C_{Th} + n_{iBG} \tag{2.14}$$

where n_i is the count rate in the ith energy window (c/s), S_{ij} is the spectrometer's sensitivity for detecting the jth element in the ith energy window (c/s per unit concentration of the jth element), c_j is the concentration of the jth element (% K, ppm U, ppm Th), and n_{iBG} is the background count rate in the ith energy window (c/s). The concentrations are expressed in % K, ppm eU, and ppm eTh. Ratios of radioelement concentrations (Th/U, Th/K, U/K) are often used to minimize the impact of terrain geometry on concentration estimates. Anomalous ratios can indicate mineralization and rock alteration. Estimates of radioelement concentrations may also be converted to dose rates in air or exposure rates. Careful record-keeping of field measurement conditions helps to eliminate false anomalies.

2.3.3.6 Car-borne gamma ray spectrometry

Car-borne gamma ray spectrometry systems have a field of view similar to portable systems rather than airborne ones. The operational speed of a car-borne system is approximately 1/5 to 1/10 that of an airborne system. As a result, the resolution of data collected along a car-borne traverse is much higher than that of an airborne flight line. However, the data is more susceptible to local variations, often considered "geological noise," such as soil composition, moisture content, surface water, transported vs. in-situ materials, weathering, and outcrop exposure. Topographic effects, even in moderate terrain, can significantly impact car-borne data, particularly when road cuts expose fresh bedrock. Trees and buildings may also introduce noise, depending on the survey's objectives. Additionally, pockets of radon can cause localized effects.

The detectability of anomalies is influenced by both the detector volume and the vehicle speed, which together determine the signal-to-noise ratio and spatial wavelength resolution. To assist the operator, audio and/or visual alarms are often incorporated into the system to indicate the detection of a significant anomaly, allowing for immediate follow-up. Variations in background levels in car-borne systems may be higher than in airborne systems due to the collection of dust, mud, and other debris inside and outside the vehicle.

2.3.3.7 Borehole gamma ray spectrometry

Gamma ray spectrometry is one of several techniques used to acquire measurements in boreholes after drilling, helping to characterize the physical properties of the intersected geology. Borehole surveys often involve multi-parameter measurements using various sensors to provide complementary datasets. Gamma ray spectrometry is part of a broader suite of nuclear techniques used in borehole logging, especially for petroleum exploration. For example, density (gamma-gamma) logging systems use a gamma ray source, such as cesium-137 ($^{137}_{55}Cs$), to generate back-scattered gamma rays. These rays reflect the electron density of the surrounding rock, allowing for the determination of bulk density. Neutron logging systems, which emit neutrons, are used to assess porosity, liquids (e.g., oil, gas, water, and brine), and certain lithologies (e.g., calcium, silicon). Killeen (1997b) discusses the use of nuclear sources (gamma rays, neutrons) and the measurement of resulting emissions (gamma rays, neutrons, X-rays) for mineral exploration, including density logging, heavy element detection, and the identification and grading of ore zones.

A typical borehole system consists of a probe (tool) with detectors (passive system) or both sources and detectors (active system) that is lowered into the borehole (Killeen, 1997a). The probe is connected to a surface recording system by a cable that transmits power to the probe (or via batteries) and sends data to the surface (or stores it in the probe's memory). A winch and pulley assembly is used to lower and raise the probe at a predetermined speed for continuous logging, or to hold it stationary for measurements, with a depth counter incorporated. Multiparameter surveys may require probes with multiple instrumentation sets or multiple logging sessions with different probes. Gamma ray spectrometer borehole probes typically use NaI (Tl) or CsI (Tl) detectors measuring 25 mm × 76 mm or smaller (Killeen, 1997b), about one-third the volume of hand-held detectors. In boreholes, the detectors are surrounded by the source and provide optimal 4π sampling geometry, thus avoiding topographic effects.

The gamma ray spectral probe measures the full gamma ray spectrum, from which the total count and windows for K, U, and Th are extracted, similar to ground and airborne systems. These probes have largely replaced total count scintillometry systems. Measurements are made continuously as the probe moves up or down the borehole, typically at a speed of 3 m/minute (Killeen, 1997b). The lateral penetration adjacent to a borehole ranges from 10 to 30 cm, depending on the rock's density. The geological applications of borehole gamma ray spectrometry are similar to those of ground and airborne methods, but are generally more focused on advanced mineral exploration projects. Direct detection of ore zones is limited to radioactive minerals, though indirect detection (e.g., potassium-enriched alteration associated with gold mineralization) and identifying host rock phases (e.g., kimberlite) are also important applications. As with other borehole techniques, lithologies and horizons can be traced from hole to hole.

2.3.3.8 Laboratory gamma ray spectrometry

Laboratory analysis of potassium (K), uranium (U), thorium (Th), and radium (Ra) in rock samples complements field surveys in uranium exploration, radiometric mapping, and mineral prospecting. In a laboratory setting, a NaI (Tl) scintillation detector is shielded from background radiation using lead. A multichannel analyser is used to detect, analyze, and display gamma rays emitted by rock samples. For more advanced applications, high-resolution germanium (Ge) laboratory spectrometers are employed. These instruments are useful for identifying environmental isotopic pollution, detecting unknown gamma radiation sources, and studying radioactive equilibrium and the distribution of radionuclides in soil and rock samples.

A typical laboratory scintillation gamma ray spectrometer includes a NaI (Tl) scintillation detector, with a minimum size of 76 × 76 mm NaI (Tl), a preamplifier, a multichannel analyser (1024–4096 channels), a stabilized high-voltage power supply, a display unit, and a data recording system. The sample and detector are enclosed in 6–10 cm of lead shielding to reduce background radiation. A 5 mm thick brass plate is placed inside the lead shielding to minimize the effects of induced X-ray radiation. The energy resolution of the detector can be assessed using the ^{137}Cs peak at 662 keV. Noise levels in the 10–30 keV energy range should be experimentally determined and eliminated by adjusting the instrument's lower energy discriminator.

The four-component analysis of U, Th, Ra, and K is based on measurements within four specific energy windows (i = 1, 2, 3, 4). The count rates per unit mass are related to the concentrations by the following equation:

$$n_i = S_{iU}C_U + S_{iTh}C_{Th} + S_{iRa}C_{Ra} + S_{iK}C_K \tag{2.15}$$

where n_i is the background-corrected count rate per unit mass in the iii-th energy window (c/s per g), S_{ij} is the sensitivity of the spectrometer for detecting the jth element in the i-th energy window (counts/s per unit mass and concentration), and C_j represents the concentration of the jth element. The relative concentrations of uranium and radium in a sample can be used to estimate the radioactive equilibrium between these isotopes.

2.3.4 Seismic borehole geophysics

Seismic borehole geophysics involves the use of seismic waves for the detailed exploration of subsurface structures. In this approach, seismic sources (transmitters) and/or receivers (such as geophones or hydrophones) are positioned within a borehole to acquire measurements. Seismic waves, which are acoustic energy waves, propagate through the Earth and interact with subsurface materials as they travel from their origin (e.g., an earthquake focus or a shot point). When these waves pass through geological formations, the material particles oscillate around their equilibrium positions. Depending on the nature of the seismic wave, these oscillations cause the material to compress or stretch, with the material returning to its original state after the wave has passed. This phenomenon—elastic deformation—forms the basis of seismic borehole methods, which are governed by the principles of wave propagation in elastic media (Paillet and Cheng, 1991).

The velocity of seismic waves is influenced by a range of factors, including the density, elasticity, lithology, porosity, permeability, fluid saturation, and composition of the pore fluids within the formation. These factors collectively determine the speed at which seismic waves travel through the subsurface. By analyzing the travel times of seismic waves, it becomes possible to interpret the characteristics, identity, and structure of subsurface materials. Seismic data can therefore provide insights into the geologic properties of a given area, such as the type of rock, its porosity, and the fluid content in the pores.

As seismic waves propagate through the Earth, their amplitude gradually decreases, a phenomenon known as **attenuation**. This loss of energy occurs through various mechanisms such as **absorption, scattering, reflection**, and **refraction**—processes that typically depend on the contrasts between different subsurface materials. Attenuation rates can be influenced by the seismic wave frequency, which is inversely related to the depth of penetration. In general, seismic energy is absorbed more rapidly in loosely consolidated formations, whereas it tends to dissipate more slowly in denser, more consolidated rocks (Paillet and Cheng, 1991). Thus, attenuation patterns can offer valuable clues about the material properties of subsurface layers.

In seismic borehole geophysics, the four primary seismic wave properties that are measured include **velocity**, **amplitude**, **attenuation**, and **frequency**. These properties are essential for interpreting subsurface conditions. Compared to surface seismic methods, borehole seismic techniques generally use higher-frequency seismic waves. This results in improved **spatial resolution**, which is further enhanced by the shorter distance between the seismic source and the receiver within the borehole. As a result, seismic borehole methods provide more detailed data than traditional surface-based seismic surveys, enabling geoscientists to discern finer variations in subsurface characteristics.

2.3.4.1 Key techniques in seismic borehole geophysics

Two widely used techniques in seismic borehole geophysics are **sonic borehole logging** and **vertical seismic profiling (VSP)**:

Sonic borehole logging

This method measures the travel time of seismic waves as they pass through the borehole and surrounding materials. The sonic tool emits sound waves into the formation and records the time it takes for the waves to travel from the source to the receiver. This data is used to determine the **seismic velocity** of the formation, which can help characterize the geological properties of the subsurface, including lithology and porosity.

Vertical seismic profiling (VSP)

VSP involves the use of both seismic sources at the surface and receivers placed within the borehole. By recording the arrival times of seismic waves at different depths within the borehole, VSP provides a high-resolution profile of subsurface structures. This method is particularly effective in characterizing the vertical variations in seismic velocity and identifying geological features such as faults, fractures, and layers of differing material properties.

Both methods are crucial for enhancing the resolution of seismic imaging and providing more precise subsurface data, which are vital for resource exploration and geotechnical applications. In summary, seismic borehole geophysics offers powerful tools for investigating the subsurface, allowing for the detailed characterization of geological formations. Through the use of techniques such as sonic logging and vertical seismic profiling, seismic borehole methods provide valuable data on seismic wave velocities, attenuation, and other subsurface properties. These methods offer higher resolution and more accurate data than surface seismic techniques, making them essential for advancing geophysical exploration and management.

2.3.4.2 *Physical and imaging borehole methods*

Physical and imaging borehole methods provide high-resolution data essential for geological and hydrogeological studies, offering valuable insights into borehole conditions and surrounding formations. These methods are integral to baseline geoscience data generation, helping to map the subsurface, assess formation properties, and inform site models. A borehole caliper measures the borehole diameter, detecting irregularities that can indicate potential zones of collapse or instability. Borehole deviation measures vertical and horizontal shifts from the intended path, aiding in the accuracy of well placement and interpretation, particularly in directional drilling. The acoustic televiewer (ATV) uses sound waves to produce detailed images of the borehole wall, revealing fracture features, lithology, and geometry, especially in water-filled boreholes. The optical televiewer (OTV) uses light to capture high-resolution images in clear borehole environments, helping identify geological structures like fractures and bedding planes. Formation microresistivity imaging (FMI) measures resistivity changes to generate images that characterize lithological features, fractures, and fluid flow paths, which are critical for understanding porosity and saturation. Lastly, a borehole video camera visually inspects the borehole wall and casing, offering a straightforward approach for observing geological features and fluid conditions (Prensky, 1999). Together, these methods contribute to comprehensive baseline geoscience data that can be used for geological mapping, hydrogeological modeling, and environmental assessments.

2.3.4.3 *Fluid column and flow borehole methods*

Fluid column and flow borehole methods are essential tools for understanding subsurface fluid behavior and properties. These methods provide critical data for hydrogeological studies and are also useful in the context of mineral exploration, where groundwater flow and fluid characteristics can influence site assessments and exploration models. Fluid-temperature and fluid-conductivity logging measure the temperature and electrical conductivity of the borehole fluid. These parameters can reveal the presence of different water types, identify fluid flow patterns, and assist in determining thermal anomalies. Conductivity can also help infer the salinity of water, providing insights into the geochemical properties of the surrounding formation. Active line source (ALS) logging involves introducing a known signal, such as heat or electrical current, into the borehole fluid and measuring how the fluid responds. This technique helps estimate hydraulic properties of the formation, such as flow velocity and thermal conductivity, which are valuable in understanding groundwater movement and identifying flow paths within the formation. Vertical flowmeter logging measures the vertical velocity of fluid within the borehole. This method helps map groundwater flow within the surrounding formation, providing key insights into the flow structure, transmissivity, and hydraulic conductivity of the aquifer. It is particularly useful for identifying high-permeability zones and understanding how groundwater moves through different layers. Single borehole tracer (SBT) logging introduces a tracer (chemical or radioactive) into the borehole fluid and tracks its movement. This method is critical for understanding groundwater flow patterns, including the identification of zones with varying permeability, and is useful in estimating parameters like hydraulic conductivity and effective porosity.

Together, these fluid column methods generate baseline geoscience data crucial for mineral exploration. For instance, understanding the movement of fluids can reveal the potential for mineralization in aquifers and help identify areas where mineral deposits may be present. Additionally, the estimation of hydraulic parameters such as transmissivity and conductivity is essential for constructing site models and understanding how mineralized fluids may migrate through the subsurface. These data play a key role in the exploration, environmental assessment, and modeling of mineral deposits.

The integration of physical, imaging, and fluid column methods provides detailed and complementary datasets that are essential for understanding subsurface conditions. These methods help not only map geological structures and fracture systems but also assess the behavior of groundwater within the formation, enabling better decision-making in hydrogeological and environmental studies.

2.4 Waterborne geophysical methods

Waterborne geophysical methods have gained significant importance in environmental, hydrogeological, and geological studies. Traditionally applied in marine seismic surveys and resource extraction, these methods are now increasingly adapted for high-resolution imaging in near-surface investigations and environmental monitoring. The fundamental geophysical principles behind waterborne methods are similar to those used in land or borehole applications, with necessary adjustments to account for factors such as the submersion of instruments in water, the electrical conductivity of surrounding water (whether fresh, saline, or brackish), and signal coupling challenges.

Waterborne geophysical methods offer a versatile range of applications, including debris detection, sub-bottom geological mapping, and the study of groundwater-surface water interactions. These techniques are particularly valuable in scenarios where traditional land-based methods would be challenging or impractical, such as in riverbeds, lakes, and coastal areas.

Key waterborne geophysical methods include:

Continuous Seismic Profiling (CSP): This technique provides detailed subsurface imaging by measuring seismic wave reflections. It is especially useful for mapping the stratigraphy and geological layers beneath water bodies, such as rivers, lakes, and offshore areas.

Continuous Resistivity Profiling (CRP): By measuring variations in electrical resistivity, CRP helps delineate subsurface structures, including sediment composition, groundwater flow patterns, and contamination zones.

Waterborne Ground Penetrating Radar (WGPR): WGPR uses radar waves to detect and map subsurface features such as buried objects, geological structures, and groundwater-related anomalies.

Fibre Optic Distributed Temperature Sensing: This method detects temperature variations along the waterborne survey path, providing insight into subsurface flow patterns or thermal anomalies that may be indicative of groundwater movement.

Waterborne Magnetic Surveying: Magnetic surveys in water bodies detect variations in the Earth's magnetic field, which are useful for locating magnetic materials or anomalies, aiding in mineral exploration or archaeological investigations.

Waterborne Transient Electromagnetics (WTEM): This method maps variations in conductivity in the subsurface, often applied in hydrogeological studies to assess groundwater resources and contamination.

The growing use of waterborne geophysical methods underscores their critical role in advancing baseline geoscience data for sustainable exploration and management. These techniques contribute significantly to geophysical mapping, providing reliable data that supports environmental protection, resource management, and sustainable exploration strategies.

2.4.1 Marine gamma-ray spectrometry

Marine gamma ray spectrometry, while limited by the attenuation of gamma rays in water, has made significant advancements due to innovative detector designs that enable gamma ray measurements in both shallow and deep waters. Jones (1999) provides an overview of seabed exploration using radiometric methods. Most studies focus on geological mapping of seafloor outcrops at depths up to 6 km, with additional surveys conducted around active hydrothermal vents.

There are two types of detectors used in marine gamma ray spectrometry. In both cases, detectors are attached to a ship or submersible by a cable, transmitting data to an onboard spectrometer that analyzes the pulses received from the seafloor. One system involves a detector housed in a pressure-resistant casing, lowered onto the seafloor for spot measurements. The second system is a towed detector that moves just centimeters above the seafloor, requiring innovative designs to prevent snagging on outcrops.

Gamma ray attenuation in water is about 1000 times that in air, but only half that in solid materials. For a photon energy of 1.76 MeV, a seafloor clearance of 0.175 m provides an equivalent of 150 m of aircraft terrain clearance. Therefore, to be effective, measurements are best taken at or near the seafloor (within 20 cm). The relatively slow ship speed (1.0 to 2.5 m/s) and counting times of 100 to 500 s help improve the signal despite the challenging operating conditions. Background measurements are taken several meters above the seafloor or within the water column at the beginning and end of a traverse when the sensor is lowered or raised.

The detector assembly itself contributes most of the background radiation, including cosmic and gamma rays. In deep water, radon contribution is typically minimal as bottom waters sweep away emanations from the seafloor, except near active hydrothermal vents or areas of gas and water seepage. Tides and currents on the continental shelf generally suffice to remove nuclear fallout, such as from tests, accidents, or waste disposal. However, seabed emissions may increase near waste disposal sites if materials have been absorbed by nearshore sediments.

Instruments may be calibrated using a uniform seafloor with known radioelement concentrations, though this is often impractical. Instead, spectrometers are typically calibrated on concrete pads using the same procedures as for airborne spectrometers. Elemental count rates are determined using sensitivity coefficients derived during calibration or by measuring a ground-based test line with the marine system. Presentation of the data in profile form is generally more suitable than gridded images unless the survey follows a dense, regular pattern of lines.

2.5 Airborne radiometric methods

2.5.1 Introduction

Airborne radiometric methods are widely used to measure the natural gamma radiation emitted by naturally occurring radioactive elements in the Earth's surface. These elements—primarily potassium, uranium, and thorium—decay over time and emit gamma radiation at characteristic energy levels, which can be detected and analyzed. The energy spectra of gamma radiation emitted by their isotopes or daughter elements, such as K-40 (potassium), Bi-214 (uranium), and Tl-208 (thorium), provide valuable information about the concentrations of these elements in the Earth's surface materials.

In geophysical radiometrics, the most significant method involves measuring gamma radiation to map the distribution of radioactive elements in the subsurface. Airborne radiometric surveys use gamma-ray spectrometers mounted in aircraft to detect and record the natural radioactivity of near-surface rocks and soils. These surveys are particularly useful for geological mapping, as they help identify areas rich in minerals such as granite, gneiss, shale, and basalt, which are associated with higher concentrations of potassium, uranium, and thorium.

An example system used by the **BGR** (Federal Institute for Geosciences and Natural Resources) involves a spectrometer with five crystals totaling 21 liters in volume and weighing about 100 kg. The detector and associated electronics are mounted in a helicopter, and the system records a complete gamma spectrum every second. The data collected are processed to determine ground concentrations of potassium (in %), uranium, and thorium (in ppm) (Fig. 2.8). In 2013, BGR upgraded from the GR-820 spectrometer to the modern RS-500 spectrometer, featuring the RSX-5 crystal detector for more precise measurements.

2.5.2 Survey methodology

Airborne geophysical surveys using gamma-ray spectrometers are generally carried out along regular flight lines. The spacing between these flight lines varies depending on the survey's objectives. For detailed geological and environmental mapping, flight lines are typically spaced between 50 m and 400 m apart. For regional geochemical baseline surveys, the spacing may be up to 1 km or more. The aircraft flies at a constant nominal altitude above the ground, which typically ranges from 40 m to 100 m. Helicopters can fly lower than fixed-wing aircraft, enabling greater flexibility in topographically complex areas.

In addition to flight lines, complementary perpendicular lines, known as tie lines, are often flown. These are usually spaced approximately five times further apart than the flight lines. The spacing between flight lines represents a balance between the desired data resolution and the cost of the survey. While flying height is related to line spacing, it is also constrained by safety and operational considerations. Fixed-wing aircraft typically fly at speeds between 50 and 60 m/s, whereas helicopters fly slower, around 25 to 30 m/s.

Gamma ray spectrometric data are usually collected in conjunction with other remote sensing data, such as magnetic field measurements. This integrated approach ensures the optimal collection of data across multiple geophysical methods, enhancing the survey's overall utility. Gamma ray spectrometric data are recorded at 1-second intervals, with a

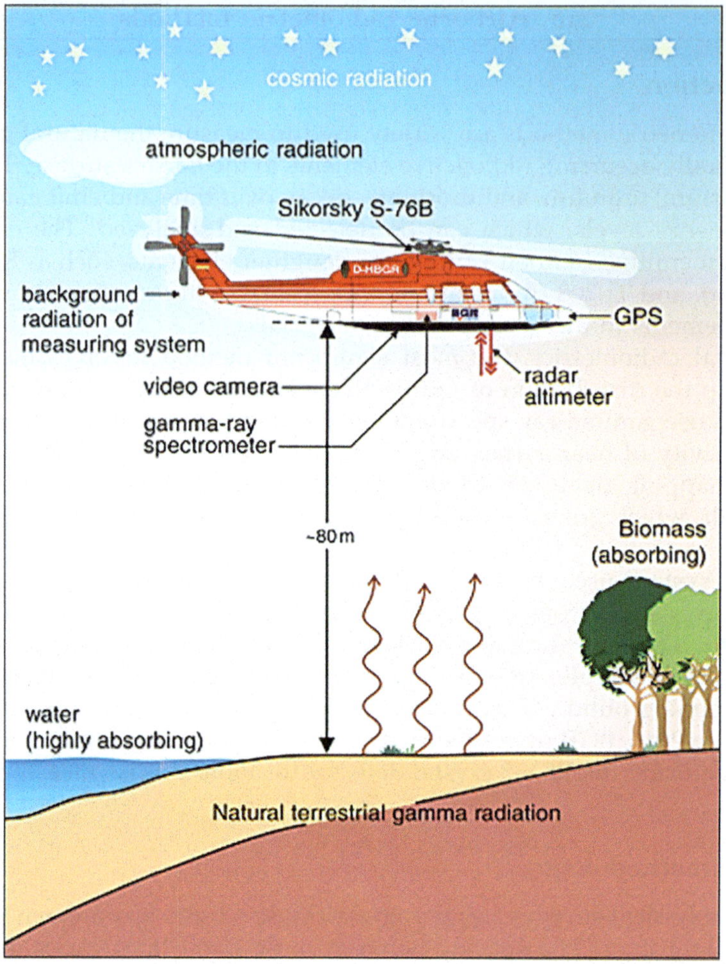

FIG. 2.8 Schematic representation of the gamma ray spectrometry used in airborne radiometric surveys, measuring the gamma spectrum of the upper soil layers from the air. *Source: BGR helicopter, BGR—Airborne radiometrics. https://www.whymap.org/EN/Themen/GG_Geophysik/Aerogeophysik/Aeroradiometrie/aeroradiometrie_inhalt_en.html*

fixed-wing aircraft covering approximately 55 m during each interval. However, the radiation signal decays exponentially with distance from the source, and this can reduce the signal-to-noise ratio as the aircraft moves farther from the ground.

Airborne gamma ray surveys can also be affected by rugged or rough terrain. To counter this, surveys can be conducted in "drape mode," where the aircraft's altitude adjusts to follow the terrain, maintaining a consistent height above the surface. Key corrections must be applied to airborne gamma ray data to ensure its accuracy. These include background correction, stripping correction, height correction, and sensitivity correction for converting measurements into elemental concentrations. Calibration for these corrections is typically empirical. Data processing is a crucial step in airborne radiometric surveys as it removes nongeological influences, converts raw data into estimates of radioelement concentrations, and ensures the integration of new and old datasets for regional radiation assessments.

2.6 Optimizing gamma-ray spectrometers for UAV-borne surveys

The integration of gamma-ray spectrometers into unmanned aerial vehicles (UAVs) has revolutionized geophysical surveys, providing advantages such as increased flexibility, reduced operational costs, and the ability to access remote or challenging terrains. UAVs equipped with gamma-ray spectrometers are increasingly being used in a variety of geophysical applications, including environmental monitoring, geological mapping, and resource exploration. To optimize the use of these spectrometers on UAV platforms, several key factors must be considered:

2.6.1 Sensor calibration

Proper calibration is critical for ensuring accurate gamma radiation readings, especially given the varying altitudes and terrains UAVs can access. Calibration ensures that the data collected from different UAV flights is consistent and reliable, even in remote or harsh environments.

2.6.2 Weight and power constraints

UAVs have limited payload capacities, so the size and weight of the gamma-ray spectrometer must be minimized without sacrificing the sensitivity needed for reliable data collection. As UAV payload capacities improve, more powerful and sensitive spectrometers can be integrated, but weight remains a key consideration in optimizing UAV performance.

2.6.3 Flight path design

UAVs offer precise control over flight altitude, which is advantageous for ensuring consistent gamma-ray data acquisition. The design of the flight path—balancing factors such as flight speed, altitude, and line spacing—plays a crucial role in ensuring optimal spatial resolution and data quality. Precise flight planning is essential for maximizing the effectiveness of UAV-borne gamma-ray spectrometry.

2.6.4 Data processing

Given the large volumes of data generated during UAV surveys, advanced algorithms and onboard processing systems are required to manage the data efficiently. Real-time analysis during the flight can provide immediate feedback, while postflight processing ensures that data is corrected and calibrated for accurate interpretation.

2.6.5 Terrain adaptation

UAVs have the ability to fly in drape mode, which allows them to maintain a constant altitude over varying topography. This is particularly useful for reducing signal distortions caused by uneven terrain and ensuring that gamma-ray measurements remain consistent, even in complex landscapes.

Recent advancements in heavy-duty UAVs have made it possible to integrate large gamma-ray spectrometers weighing several kilograms. These systems, now available at affordable prices, are capable of mapping naturally occurring radionuclides such as ^{40}K, ^{238}U, and ^{232}Th with high spatial resolution. The primary advantage of UAV-borne gamma-ray spectrometers is their ability to be deployed in areas that are difficult or impossible to reach with traditional methods, such as remote mountain ranges, dense forests, or rugged terrains.

However, due to the relatively low activity of naturally occurring radionuclides, optimizing UAV flight parameters is crucial. Longer integration times, slower flight speeds, or the use of larger detectors are often necessary to effectively map the spatial distribution of elements such as potassium (^{40}K), uranium (^{238}U), and thorium (^{232}Th). This raises the important question: What is the minimum practical detector size required to measure these radionuclides effectively from UAV platforms? By addressing these optimization strategies, UAV-borne gamma-ray spectrometry can significantly enhance geophysical exploration, environmental monitoring, and geological mapping, providing valuable data that supports sustainable exploration and resource management.

2.7 Geophysical properties

To gain a comprehensive understanding of the theory and science behind geophysical methods, it is essential to grasp the key geophysical properties that underpin these techniques. The following terms related to geophysical properties are crucial for this understanding: Density, Porosity, Magnetic Susceptibility, Electrical Conductivity and Resistivity, Dielectric Permittivity, Electromagnetic Properties of Matter, Electromagnetic Signal Attenuation (Skin Depth), Seismic Velocities (Vp, Vs), and Seismic Reflectivity.

Density: Density (σ) is the intrinsic unit mass of a material, and is defined as the mass of one cubic meter with units of kilogram per cubic meter (kg/m^3) in SI units and with units of gram per cubic centimeter (gm/cc) in cgs units. One gm/cc is equivalent to 1000 kg/m^3. Porosity, saturation, and density are related as a function of porosity φ as:

$$\sigma_{\mathrm{f}} = \sigma_{\mathrm{m}}(1-\varphi) + 1000S\varphi \tag{2.16}$$

where f is the formation, m is the matrix, and S is the fractional water saturation. A water density of 1000 kg/m^3 is assumed.

Porosity: Porosity, saturation, and density are related as a function of porosity φ as:

$$\sigma_{\mathrm{f}} = \sigma_{\mathrm{m}}(1-\varphi) + 1000S\varphi \tag{2.17}$$

where f is the formation density, m is the matrix density, and S is the fractional water saturation. A water density of 1000 kg/m^3 is assumed.

Magnetic susceptibility: Magnetic susceptibility, k, is a measure of the ability of a material to be magnetized and is related to magnetic permeability (μ) by:

$$\mu = \mu_o(1 + k) \quad (2.18)$$

where μ_0 is the magnetic permeability of free space, which is $4\pi \times 10^{-7}$. Ferromagnetic materials have the largest k and contain iron, nickel, cobalt, and many of their alloys. For example, magnetite is a ferromagnetic material with one of the highest magnetic susceptibility values. In environmental waste sites, the largest sources of magnetic anomalies are iron and steel. The unit of magnetic susceptibility, k, in the SI system is dimensionless. Magnetic permeability (μ) has units of Henry per meter. The geomagnetic field (B) has units of force per magnetic pole or Tesla. The practical unit of geomagnetism is the nanotesla (nT), or gamma. Most environmental magnetic surveys are measured in nT.

Electrical conductivity and resistivity: Electrical conductivity (σ) is a measure of the ability of the material to conduct an electrical current. The units of conductivity are Siemens per meter (S/m), or more commonly milliSiemens per meter (mS/m). The Siemen, which is the unit of conductance, is the reciprocal of the Ohm, the unit of resistance. The units of conductivity are sometimes given as mho/meter or millimho/meter. Resistivity (ρ) is the inverse of conductivity ($\rho = \frac{1}{\sigma}$). The units of resistivity are Ohm-meter (Ω·m). The electrical conductivity of earth materials is influenced by metal content (e.g., sulfides), porosity, clay content, permeability, and saturation. All metal objects of interest at environmental sites typically have a very large conductivity contrast with surrounding materials. This contrast can usually be detected with electrical and electromagnetic methods. If metals are present, their electrical conductivity is governed by Ohmic conduction via electron propagation, like in electrical wiring. In the absence of metals, formation conductivity is related to the volume and conductivity of the water in earth materials and is effectively electrolytic conduction. Groundwater conducts electrical current via ions; therefore, the conductivity of groundwater depends strongly on the total dissolved solids and the chemistry of the groundwater. Archie's Law (Archie, 1942) describes this relationship within a porous, clay-free medium with a nonconducting matrix.

$$\rho_e = a\varphi^{-m}S^{-n}\rho_w \quad (2.19)$$

where ρ_e = resistivity of the formation as geophysically measured electrical resistivity; ρ_w = resistivity of the water; a, m, and n = empirical constants value dependent on the matrix type; φ = fractional pore volume (i.e., porosity); and S = the fraction of pores containing water (i.e., the electrolyte). The inverse of this can be used for conductivity. Archie's Law allows the calculation of various geophysical and hydrogeologic parameters depending on known measurements. For example, porosity is estimated if other parameters are known via geophysical and hydrogeological data.

If clays and/or shales are present, these hydrated minerals with high porosities and low permeabilities may not be very electrically conductive, but their surface properties cause an excess of cations in the pore fluid immediately adjacent to the clay surfaces. This results in high conductivity near the clay surfaces, which can dominate the overall conductance if the pore water conductivity is low. Waxman and Smits (1968) can be investigated for more on this topic. To conduct electricity a rock or volume of sediment, with a nonconducting matrix must be permeable as well as porous. Mazac et al., 1985 present a summary of the relationship between Darcy's law for fluid flow and Ohm's law for electrical current flow. Archie's law predicts an inverse relationship between resistivity (rho) and (effective) porosity (phi), which determines hydraulic conductivity (k). However, for some materials hydraulic conductivity increases and porosity decreases with grain size, leading to a direct relationship between ρ and k. Mazac et al., 1985 can be referenced for an in-depth analysis of this relationship.

Dielectric permittivity: Dielectric permittivity relates charge separation to an applied electric field. That is, materials with no ions or electrons (i.e., free charges) may transmit electrical current when a voltage is applied. Dielectric permittivity describes the charge displacement in these materials when a voltage potential is applied. This mobile charge is bound to the molecules of the material (e.g., the positive charge on the nucleus and the negative electrons of the electron shells). An applied electrical field polarizes this charge distribution and separates the positive and negative charges in opposite directions. Permittivity is sometimes expressed relative to the permittivity of free space, ε_o, and in terms of the dielectric constant κ (κ = 1 for air and κ = 80 for water).

$$\varepsilon = \kappa\varepsilon_o \tag{2.20}$$

Water (κ = 80) clearly dominates the permittivity of any rock-water or sediment—water mixtures and is a function of the porosity of the rock or sediment. Permittivity is the primary factor influencing the speed of electromagnetic radiation in earth materials at GPR frequencies. A contrast in GPR velocity, which would be due to κ contrast, would then produce reflections of electromagnetic energy within the earth. As a result, κ is the major influence on ground penetrating radar measurements. Additionally, κ is strongly frequency dependent in parts of the electromagnetic frequency spectrum and is considered only at ground penetrating radar (GPR) frequencies, in the range 100 to 1000 MHz. These relationships are shown below. The velocity, V, of electromagnetic waves through a medium of permittivity ε and magnetic permeability μ is:

$$V = \frac{1}{\sqrt{\mu\kappa}} \tag{2.21}$$

The reflection coefficient, R, the ratio of the reflected to incident EM signal amplitude for an EM signal traveling from medium 1 toward medium 2 (i.e., a contrasting permittivity) is:

$$R = \frac{\sqrt{\kappa_1} - \sqrt{\kappa_2}}{\sqrt{\kappa_1} + \sqrt{\kappa_2}} \tag{2.22}$$

Electromagnetic properties of matter: Any geophysical method that relies on electromagnetic (EM) energy and EM wave propagation is measuring or responding to the EM properties of matter. The EM properties of matter include: Electrical conductivity (σ), or its inverse, electrical resistivity (ρ); Relative magnetic permeability (μ); Relative dielectric permittivity (ε)—also called dielectric constant.

Electromagnetic signal attenuation (skin depth): In electromagnetic (EM) methods, the electrical conductivity of the earth plays a pivotal role in the EM signal penetration and depth resolution that can be obtained. High electrical conductivity removes, or attenuates, energy from the EM wave through the work done by moving electrical charge. High-conductivity materials effectively absorb the EM energy and limit or attenuate its flow, or penetration, deeper into the earth. This attenuation of the EM signal is exponential with depth and is described by the skin depth relationship. Skin depth is the depth at which the EM signal attenuates to $\frac{1}{e}$ or approximately a third. Skin depth is typically measured in meters and is related to the frequency (*f*) of the EM wave and the electrical conductivity (σ) of the media (i.e., geologic materials) through which the EM wave is propagating.

$$\text{Skin depth} = 500\sqrt{\frac{\rho}{f}} = \sqrt{\frac{1}{\sigma f}} \tag{2.23}$$

Skin depth is inversely proportional to conductivity (σ) and frequency (*f*). Higher electrically conductive earth materials and higher frequency EM waves result in shallow EM wave penetration. The frequency of the wave presents a trade-off for resolution and depth. The higher the frequency, the higher the resolution and lower penetration (smaller skin depth), while the lower the frequency, the deeper the signal penetration (larger the skin depth) and lower resolution.

Seismic velocities: In environmental geophysics, seismic energy sources are typically a small explosion or a hammer blow, which generate three fundamental types of elastic waves: P (primary, compressional) waves, S (secondary, shear) waves, and surface waves. The P and S waves propagate through the body of the Earth, while the surface waves can exist only close to the free surface. Since P and S waves are predominantly used in environmental geophysics, surface waves are not included in this discussion.

P waves have a particle motion in the direction of propagation, while particles under the influence of S waves move transverse to the direction of propagation. S waves cannot propagate through liquids because liquids have no shear rigidity. P wave velocities are typically 50% higher than S waves. V_p and V_s are governed by the elastic moduli [Young's modulus (E), Poisson's ratio (ν), and Bulk modulus (k)] and the density (ρ) of geologic material.

$$\text{P wave velocity} = V_p = \sqrt{\frac{k + \frac{4}{3}\mu}{\rho}} \tag{2.24}$$

$$\text{S wave velocity} = V_s = \sqrt{\frac{\mu}{\rho}} \tag{2.25}$$

where, ρ = density; k = bulk modulus; and μ = shear modulus. Seismic velocities are usually expressed in SI units of meters per second. Occasionally, seismic velocities are expressed as kilometers per second or meters per millisecond.

Seismic reflectivity: In seismic reflection, the layers or bands observed in a seismic section are termed reflections. These reflections in the seismic data result from contrasts in the seismic impedance (density × seismic velocity) across a geologic boundary. For a seismic wave intersecting a layer or boundary at normal incidence, the ratio of the reflected to incident amplitude is the acoustic impedance (R), which defines the strength or amplitude of the seismic reflection.

$$R = \frac{\rho_2 V_2 - \rho_1 V_1}{\rho_2 V_2 + \rho_1 V_1} \tag{2.26}$$

Here's a table format that outlines key geophysical properties for baseline geoscience data generation, which is critical for sustainable management (Table 2.6):

This table presents a concise overview of key geophysical properties and their relevance to various sustainable management applications, such as resource exploration, environmental monitoring, and infrastructure planning.

TABLE 2.6 Geophysical properties for baseline geoscience data generation in sustainable exploration and management.

Geophysical property	Description	Units	Relevance to sustainable exploration management
Density	Mass per unit volume of material. It is used to identify different rock types and mineral content.	kg/m^3 or g/cm^3	Helps in assessing resource potential (e.g., mineral deposits), groundwater storage, and land stability.
Porosity	Fraction of a rock's volume that is void space. It indicates how much fluid a material can store.	Percentage (%)	Important for water and hydrocarbon storage, soil permeability, and assessing aquifer capacity.
Magnetic susceptibility	The degree to which a material can be magnetized in the presence of a magnetic field.	SI units (dimension less)	Useful for identifying mineral deposits and mapping geological structures in exploration.
Electrical conductivity	A measure of how well a material can conduct electricity.	Siemens per meter (S/m)	Vital for mapping groundwater contamination, soil salinity, and characterizing subsurface materials.
Electrical resistivity	The inverse of conductivity, indicating how strongly a material resists electric current.	Ohm-meters (Ω·m)	Key for groundwater studies, environmental monitoring, and identifying ore bodies.
Dielectric permittivity	Measure of how much a material can store electrical energy in an electric field.	Farads per meter (F/m)	Important for subsurface imaging techniques, including ground-penetrating radar (GPR).
Electromagnetic properties	How materials interact with electromagnetic fields, influencing wave propagation.	Varies (depends on frequency)	Crucial for remote sensing, soil moisture estimation, and environmental monitoring.
Electromagnetic signal attenuation (skin depth)	Distance over which an electromagnetic wave decreases to 1/e of its initial intensity in a conductive medium.	Meters (m)	Impacts subsurface imaging, such as in metal detection, and provides insights into material composition.
Seismic velocities (Vp, Vs)	The speed at which seismic waves travel through the Earth, dependent on the medium's properties.	Meters per second (m/s)	Used in geotechnical studies, subsurface structure mapping, and assessing earthquake risk.
Seismic reflectivity	The ability of subsurface interfaces to reflect seismic waves, based on contrasts in impedance.	Dimension less (ratio)	Essential for hydrocarbon exploration, mineral exploration, and understanding subsurface geology.

2.8 Inversion in geophysics

Geophysics involves the study of the Earth's subsurface by collecting and analyzing physical property measurements recorded at or near the surface. Geophysicists use a variety of remote sensing techniques to explore the subsurface, with each method targeting specific physical properties. For instance, electrical and electromagnetic methods focus on resistivity, while seismic methods measure density and elastic properties. Geophysics complements traditional in situ geochemical or hydrogeological sampling by providing spatial and temporal data that cannot be obtained through direct sampling alone.

In geophysics, inversion plays a key role in processing and interpreting measurements to infer geological properties such as the presence of oil, minerals, or water content. The inversion process seeks to estimate geophysical parameters, often represented as **X** (the model), using a limited set of measurements **d** (the data). These model parameters are typically arranged in a grid (1D, 2D, or 3D) and represented as a vector:

$$\bar{X} = [X_I, \ldots, X_m]^T \tag{2.27}$$

where $\mathbf{X}_i$ represents the resistivity or seismic arrival time at grid node **i**, and **m** is the total number of parameters to estimate. The measurements **d** are represented as a vector of voltages or arrival times:

$$\hat{d} = \left[d_I, \ldots\ldots, d_n\right]^T \tag{2.28}$$

where **n** is the number of measurements collected. The relationship between the model **X** and the data **d** is typically expressed as:

$$GX = d \tag{2.29}$$

Here, **G** is the forward operator that describes the physical relationship in the problem. For example, if the Earth's resistivity structure (**X**) is known, applying **G** will provide the potential measurements (**d**) in the gridded space. Similarly, if the velocities of P-wave propagation (**m**) are known, **G** will give the arrival times at the geophones (**d**). Calculating the forward operator often involves solving partial differential equations (PDEs).

The inversion process estimates the model **X** from the data **d** using the equation:

$$X = G^{-1}d \tag{2.30}$$

where G^{-1} is the inverse operator, completing the process of converting geophysical measurements into meaningful geological information (Fig. 2.9). However, inverse problems are challenging due to key issues like Existence, Uniqueness, and Instability. Often, no model perfectly fits the observed data. Even when solutions exist, they may not be unique, even with an infinite number of data points. Additionally, computing the inverse solution can be unstable—small variations in the measurements may cause significant changes in the estimated model.

Another critical challenge is noise in the data, which can result from random measurement errors and approximations made during the numerical solution process (such as discretizing continuous functions). This noise, along with simplifications of the true physical model, can complicate the inversion process.

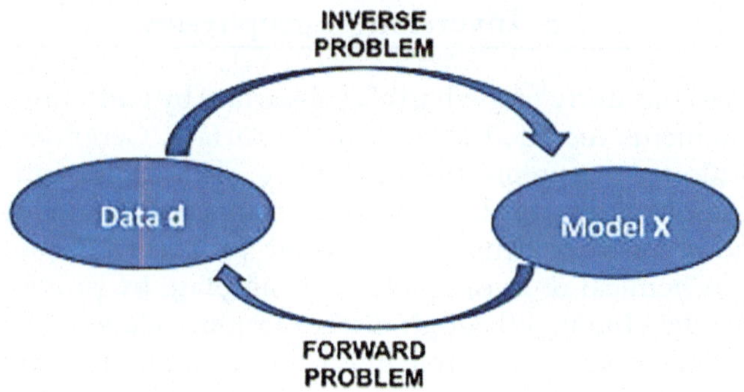

FIG. 2.9 The forward and inverse problem. *Source: United States Environmental Protection Agency (US EPA). https://www.epa.gov/*

Given these complexities, the inverse problem can be reformulated as:

$$GX = d + e \tag{2.31}$$

where **e** represents random noise. The inverse problem is then transformed into an optimization problem, where the goal is to find a model that minimizes the error **e**. There are several optimization methods, generally divided into two categories: gradient-based and stochastic-based. For many geophysical problems, which are often nonlinear, the transition from data to model typically cannot be achieved in a single step. However, it is possible to linearize the problem using a Taylor expansion:

$$G(X + dX) = G(X) + \frac{\partial G(X)}{\partial X} + O(dX)^2 \tag{2.32}$$

where $\mathbf{O(\Delta X)^2}$ represents higher-order terms that can often be neglected. This suggests that while the relationship between the model and the data may be nonlinear, the changes in the model and data can be approximated as linear. As a result, nonlinear problems can be solved iteratively, starting from an initial model and updating it incrementally until the misfit is sufficiently small. This iterative process is commonly depicted in a flowchart (Fig. 2.10).

Various algorithms have been proposed to solve the inverse problem (whether linear or nonlinear), typically focusing on minimizing an error norm relative to the model norm. A popular gradient-based approach is **least squares smoothness-constrained inversion**, which seeks a model that fits the data well while also being smooth. Geological observations often suggest that geological formations are layered, with minimal sharp changes between neighboring spatial points. The optimization problem is therefore formulated as:

$$S = \| GX - d \|^2 + \lambda^2 \| CX \|^2 \tag{2.33}$$

where the first term ensures that the recovered model fits the observed data, and the second term enforces smoothness in the model. The parameter λ controls the trade-off between fitting the data and ensuring smoothness, while **C** represents the second derivative of the model (Constable et al., 1987). The solution to this optimization problem can be found using an iterative update equation (Occam's update equation):

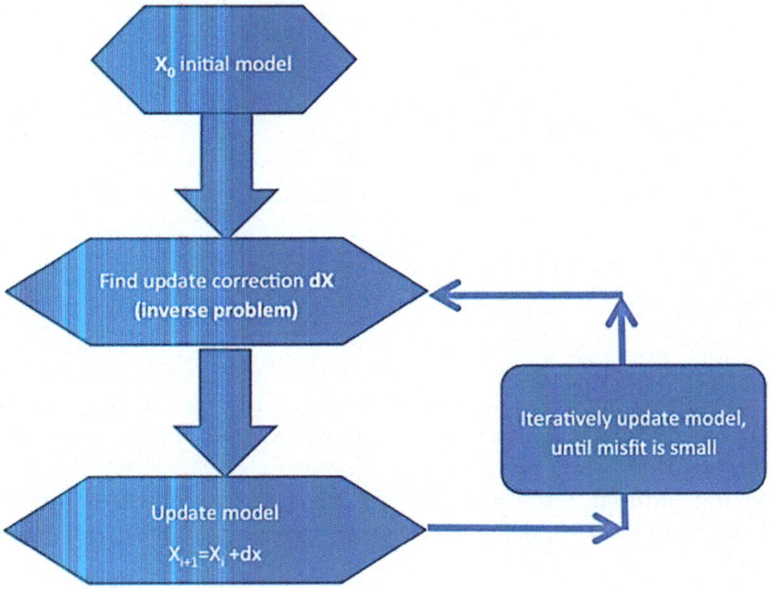

FIG. 2.10 The inverse model flowchart. *Source: United States Environmental Protection Agency (US EPA). https://www.epa.gov/*

$$X_{i+I} = (J^T J + \lambda C^T C)^{-I} J^T [G(X) - d + J^T X] \tag{2.34}$$

This iterative procedure gradually refines the model, balancing the data fit and the smoothness of the model, ultimately providing a solution that is both physically plausible and mathematically stable. In the context of inverse problems, the iteration number is denoted as **i**, and the update to the model is represented as **ΔX**. The Jacobian matrix **J**, or sensitivity matrix, describes the relationship between the model and the data, while **C** is the smoothness matrix that ensures the model remains smooth during the inversion process. The Lagrange parameter λ plays a critical role in controlling the smoothness effect on the model. If λ is too small, the inversion can become unstable, whereas if it is too large, the model may lack resolution.

Finding the optimal value of λ is crucial but not unique. The process typically begins with an analysis of the inverse equations, often involving the **Singular Value Decomposition (SVD)** of the matrix **G**. For an **m** × **n** matrix **G**, the SVD decomposition is given by:

$$G = USV^T \tag{2.35}$$

where **U** is an **m** × **n** unitary matrix, **S** is an **m** × **n** diagonal matrix containing nonnegative singular values along its diagonal, and **V** is an n × **n** unitary matrix. The diagonal elements of **S**, denoted as S_{ii}, are called the singular values of **G**. Due to the rank deficiency of **G**, some of the singular values in **S** may be very small or even zero, leading to an unstable inverse solution. This instability is one of the primary challenges in solving inverse problems.

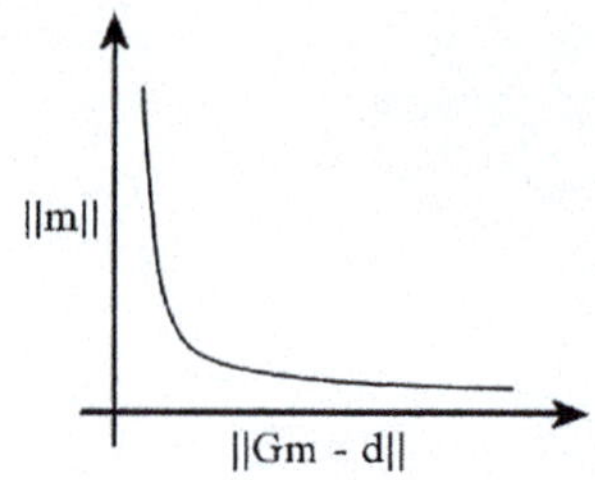

FIG. 2.11 The L-curve as a tool for finding the optimum Lagrange value. *Source: United States Environmental Protection Agency (US EPA). https://www.epa.gov/*

To address this, more sophisticated methods trial different values of the Lagrange parameter and plot the corresponding values of ||**X**|| (the model norm) and ||**G X—d**|| (the data misfit). This plot often forms an L-shaped curve (Fig. 2.11), where the corner of the **L** indicates the optimal value of λ.

Determining the optimal λ requires solving the forward problem for each trial value, which can be computationally expensive. Two common techniques to determine the trade-off parameter are:

1. Starting with an empirical value for λ (e.g., λ = 0.15) and gradually adjusting it with each iteration.
2. Using the Active Constraint Balancing (ACB) technique for space-domain regularization, which assigns varying values of the trade-off parameter based on the resolution matrix R:

$$R = (J^T J)^{-I} J^T J \tag{2.36}$$

Parameters with high resolution are given smaller trade-off values, while poorly resolved parameters are assigned larger values. The **Backus-Gilbert spread function** (Menke, 1984) is used to assess the resolving power of different parameters, with large values indicating poor resolution and small values indicating good resolution. The spread function for the **i-th** parameter is calculated as:

$$SP_i = \sum_{j=1}^{N} \{W_{ij}(1 - S_{ij})R_{ij}\}^2 \tag{2.37}$$

where **N** is the number of parameters, $\mathbf{W_{ij}}$ is a weighting factor based on the spatial distance between parameters **i** and **j**, and $\mathbf{S_{ij}}$ is a matrix accounting for the regularization applied during the inversion. The spread function is a valuable tool for resolution and sensitivity analysis, helping to assess the quality of the inversion process.

2.9 Conclusion and summary of geophysical techniques

Geophysical techniques are vital noninvasive methods used to measure and interpret the physical properties of the Earth's subsurface. These techniques play a crucial role in a variety of fields, such as resource exploration, environmental monitoring, civil engineering, and hazard assessment. By analyzing how subsurface materials interact with different energy waves (e.g., seismic, electromagnetic, and gravitational), geophysicists can map geological structures, locate natural resources, and monitor environmental changes.

The principles behind these methods typically rely on wave propagation, electromagnetic fields, and gravitational forces. The selection of a geophysical technique depends on the target material's physical properties, depth of investigation, and the resolution required. These methods are particularly valuable in natural resource exploration, enabling noninvasive, cost-effective, and high-resolution data collection over large areas.

2.9.1 Widely used geophysical methods

2.9.1.1 Seismic methods

Principle: Seismic methods involve generating energy waves, usually through controlled explosions or mechanical sources, and analyzing the reflected waves to infer subsurface structures.

Applications: Oil and gas exploration, mineral exploration, geotechnical studies, earthquake research, and civil engineering.

Common Techniques:

Reflection seismology: Maps subsurface geological layers by analyzing the reflection of seismic waves at boundaries between different rock types.

Refraction seismology: Measures refracted seismic waves to determine seismic velocity and layer depth.

Microseismic monitoring: Detects small seismic events to monitor reservoir activity or fault movements.

Surface-wave methods: Analyze the speed of surface seismic waves (Love and Rayleigh waves) to assess soil properties and site conditions.

2.9.1.2 Electrical and electromagnetic methods

Principle: These methods measure the electrical conductivity or resistivity of subsurface materials or the response of materials to electromagnetic fields.

Applications: Hydrogeology, environmental site assessments, archaeological investigations, and mineral exploration.

Common Techniques:

Electrical resistivity tomography (ERT): Measures the resistivity of the ground by injecting current and measuring potential differences at various points. It's used to map groundwater, soil contamination, and mineral deposits.

Induced polarization (IP): Measures the delayed response of the ground to an electrical current, often used in mineral exploration to detect ore bodies.

Ground-penetrating radar (GPR): Uses high-frequency electromagnetic waves to map shallow subsurface structures. It is often used in environmental and archaeological investigations.

Magnetotellurics (MT): Measures natural variations in the Earth's electromagnetic field to map deep geological structures.

2.9.1.3 *Magnetic methods*

Principle: Magnetic surveys measure variations in the Earth's magnetic field caused by the magnetic properties of subsurface materials.

Applications: Mineral exploration (e.g., detecting ore bodies), geologic mapping, and monitoring of geohazards.

Common Techniques:

Magnetic gradiometry: Measures the difference in magnetic field strength at multiple points to map geological structures, fault zones, and mineral deposits.

Total magnetic intensity (TMI): Measures the total magnetic field to detect anomalies in the Earth's magnetic field caused by underlying magnetic materials.

2.9.1.4 *Gravitational methods*

Principle: Gravity surveys measure variations in the Earth's gravitational field caused by differences in subsurface density.

Applications: Mineral exploration (e.g., locating dense ore bodies), structural geology, and subsurface void detection.

Common Techniques:

Gravity surveying: Measures small variations in gravitational attraction caused by density contrasts in the Earth's subsurface. It's commonly used to detect large-scale geological structures like faults, salt domes, and mineral deposits.

2.9.1.5 *Ground-based geophysical methods*

Principle: These methods involve direct or indirect measurement of physical properties of the ground in situ, typically at shallow depths.

Applications: Environmental monitoring, geotechnical studies, archaeological investigations.

Common Techniques:

Seismic refraction: Used to determine subsurface layer velocities and densities, important for site assessments and groundwater studies.

Electrical resistivity: Maps variations in resistivity to identify subsurface features such as water tables, contamination plumes, and mineral deposits.

Microgravity: Measures very small gravitational anomalies to detect voids, mineral deposits, or changes in ground density.

2.9.1.6 *Radiometric methods*

Principle: Measures the natural radioactivity of subsurface materials to identify specific mineral types and environmental features.

Applications: Environmental studies, mineral exploration (e.g., uranium and thorium), and geological mapping.

Common Techniques:

Gamma-Ray spectrometry: Detects gamma radiation emitted by naturally occurring radioactive elements like uranium, thorium, and potassium in the subsurface.

Alpha and Beta counting: Measures the intensity of alpha and beta radiation from radioactive minerals for environmental and resource assessment.

TABLE 2.7 Overview of geophysical techniques and their applications.

Geophysical method	Principle	Primary applications
Seismic methods	Reflection and refraction of seismic waves to image subsurface structures	Oil and gas exploration, earthquake studies, geotechnical engineering
Electrical resistivity	Measures resistance to electric current to infer subsurface properties	Hydrogeology, environmental monitoring, mineral exploration
Ground penetrating radar (GPR)	Electromagnetic wave propagation to image shallow subsurface structures	Archaeology, environmental studies, infrastructure inspection
Magnetometry	Measures variations in the Earth's magnetic field	Mineral exploration, geological mapping, fault detection
Gravitational methods	Measures variations in gravitational field due to density differences	Geological mapping, mineral exploration, detection of voids
Induced polarization (IP)	Measures delayed response of the ground to electric current	Mineral exploration, groundwater studies
Magnetotellurics (MT)	Measures natural electromagnetic variations to image deep structures	Deep geological mapping, resource exploration
Gamma-ray spectrometry	Detects natural radioactivity from subsurface materials	Mineral exploration, environmental monitoring

2.9.2 Summary of key geophysical methods

Geophysical methods are integral tools in noninvasive exploration and assessment of subsurface conditions. These techniques enable scientists and engineers to gain valuable insights into the Earth's properties without the need for direct sampling. They are essential in sustainable resource management, offering critical data for the exploration of natural resources, assessing environmental risks, and ensuring safe infrastructure development.

By integrating seismic, electrical, electromagnetic, magnetic, gravitational, and radiometric methods, geophysicists can build detailed models of subsurface structures. The combination of surface, borehole, airborne, and waterborne techniques offers flexibility, allowing for targeted and efficient exploration, even in challenging environments.

These methods contribute to the sustainable management of natural resources by reducing the need for invasive exploration methods and minimizing the environmental footprint of resource extraction. Geophysical techniques are also invaluable in monitoring and mitigating natural hazards, making them essential tools for environmental stewardship and geotechnical safety. Table 2.7 provides a concise overview of the methods discussed, highlighting their applications and contributions to geoscientific exploration.

CHAPTER

3

Advancing potential field surveying for sustainable exploration

3.1 Gravity surveying methods for sustainable exploration and management

The gravity method investigates anomalies in the Earth's gravitational field caused by changes in subsurface densities. These density variations (density contrasts) result from the presence of a causative body (target source) within the surrounding rocks. Rock density is a key geophysical parameter influenced by both mineral composition and porosity. Table 3.1 provides examples of the density values for various minerals and rocks. The gravity method is versatile and can be applied across land, air, and marine environments. Gravity anomalies are indicative of anomalous densities within the Earth's interior. This method has diverse applications in fields such as mineral and hydrocarbon exploration, cave detection, geothermal studies, and archaeological investigations (Naskar et al., 2012; Essa, 2012; Hinze, 2013; Deng et al., 2016; Mishra and Naskar, 2024).

The primary goal of gravity interpretation is to locate and characterize the buried mineral source, focusing on parameters like density contrast, depth, and shape (Seguin et al., 1994; Essa, 2014; Biswas, 2015; Mehanee and Essa, 2015). Several approaches are employed to interpret gravity data (Kilty, 1983; Stavrev, 1997; Abdelrahman et al., 2006; Abdelrahman and Essa, 2013). More recently, three-dimensional modeling and inversion of gravity data have led to more precise results. Additionally, newer non-conventional methods, such as particle swarm optimization, very fast simulated annealing, genetic algorithms, forced neural networks, and differential evolution algorithms, are increasingly being utilized to enhance gravity survey interpretation.

Newton's Law of Gravitation states that the force of gravitational attraction between two point masses is given by:

$$F = \frac{Gm_1m_2}{r^2} \tag{3.1}$$

where F denotes the gravitational force, G is the universal gravitational constant ($6.672\times10^{-11}\,Nm^2kg^{-2} = 6.672\times10^{-11}\,m^3kg^{-1}s^{-2}$), m_1 and m_2 are masses of two attracting bodies in kg, and r is the distance between them in m. This Eq. (3.1) enables us to calculate the gravitational force the Earth is pulling, e.g., a rock on the Earth's surface:

Baseline Geoscience Data Generation for Sustainable Exploration of Natural Resources
https://doi.org/10.1016/B978-0-443-45221-5.00007-2

TABLE 3.1 Density and magnetic susceptibility of some minerals and rocks.

Mineral/rock	Density (gm/cc)	Magnetic susceptibility (10^{-6} SI)
Gold	19.28	–0.14
Chromite	4.80	3000–120,000
Iron	7.87	3,900,000
Copper	8.90	–9.63
Graphite	2.16	–80 to 200
Hematite	5.26	500–40,000
Magnetite	5.20	1,000,000–5,700,000
Pyrite	5.01	35–5000
Sphalerite	4.08	–31 to 750
Rock salt	2.5–2.6	-0.01×10^{-3}
Dolomite	2.87	–10 to 940
Granite	2.64	0–50,000

Source: Essa and Munschy (2019).

$$F = \frac{GM_E m_r}{R_E^2} \tag{3.2}$$

where M_E is the mass of the Earth, m_r is the mass of the rock, and R_E is the radius of the Earth. It is inconvenient to use and measure the gravitational force, since it depends on masses of both bodies M_E and m_r. Dividing both sides of Eq. (3.2) by m_r we get:

$$\frac{F}{m_r} = \frac{GM_E}{R_E^2} \tag{3.3}$$

According to Newton's second law of motion, the force is computed as mass × acceleration, and we can transform Eq. (3.3) into:

$$g = \frac{GM_E}{R_E^2} \tag{3.4}$$

defining the acceleration caused by the Earth. The acceleration g is called the "acceleration due to gravity" or "acceleration of gravity." In geophysical applications, accelerations are of concern rather than forces. The value of g on the Earth's surface is 9.81 m/s^2. For a spherical, non-rotating, homogeneous Earth, g is everywhere the same. Mass of a homogeneous Earth (ME) = 5.977×10^{24} kg and mean radius of the Earth (spherical Earth) (R_E) = 6371 km. The unit of acceleration of gravity—1 cm/s^2—is also referred to as Gal, in honor of Galileo, who was the first to measure its value. The gravitational field at a point is measured in milligal (mGal).

$$1\text{Gal} = 1\text{cm}/\text{s}^2(10^{-2}\text{m}/\text{s}^2) = 10^3\text{mGal}$$

$$1\text{mGal} = 10^{-5}\text{m}/\text{s}^2$$

$$1\mu\text{Gal} = 10^{-6}\text{mGal} = 10^{-8}\text{m}/\text{s}^2\ (\text{precision of gravimeter for geotechnical surveys}).$$

Gravity unit: 10 gu = 1 mGal. Mean gravity around the Earth = 9.81 m/s^2 = 981,000 mGal. The gravity field is reassuringly simple, unipolar, and almost perfectly vertical. The common unit of gravity measurement in exploration geophysics is mGal. These units refer to acceleration due to gravity, and the average value at the Earth's surface is around 9.8 m/s^2. By comparison, anomalies in mineral and oil exploration seldom exceed a few hundred mGal. Density is the physical property of interest for a gravity survey. The Earth's interior is not homogeneous, which implies that density must increase with depth in the Earth. Density is an intrinsic property of a material and is measured in mass per unit volume (kg/m^3 in SI and gm/cm^3 or g/cm^3 in cgs units). The modern gravimeters are capable of readings with a precision of 0.001 mGal.

Gravitational potential field

The gravitational potential field is conservative (i.e., the work to move a mass in this field is independent of the way). The first derivative of U in a direction gives the component of gravity in that direction (Fig. 3.1).

$$\nabla U = \vec{g} = \frac{\vec{F}}{m_2} \text{ with } \nabla U = \frac{\partial U}{\partial x}\vec{i} + \frac{\partial U}{\partial y}\vec{j} + \frac{\partial U}{\partial z}\vec{k}$$

where $\hat{i}$, $\hat{j}$, and $\hat{k}$ are the unit vectors.

$$U = \int_{\infty}^{R} \vec{g}\cdot\vec{r}dr = Gm_1\int_{\infty}^{R}\frac{dr}{r^2} = -\frac{Gm_1}{R} \tag{3.5}$$

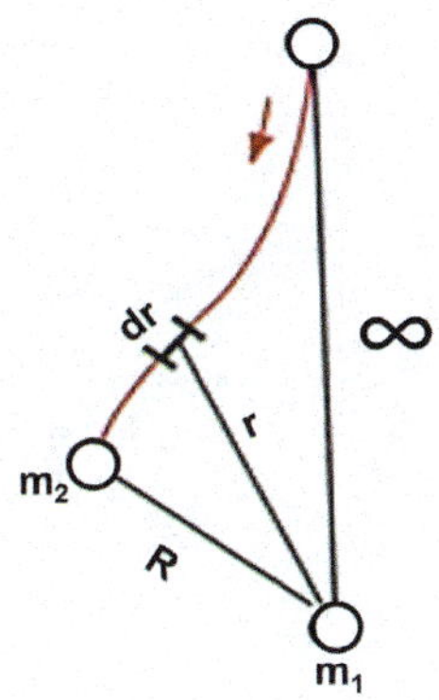

FIG. 3.1 Calculation of gravitational potential at a given point. *Source: Personal.*

gives the gravitational potential of a point mass m_1. Acceleration may be computed by calculating the derivative of potential. Using Cartesian coordinates (x, y, z), the acceleration will have components (g_x, g_y, g_z). These may be computed by calculating, separately, the derivatives of the potential with respect to x, y, and z:

$$g_x = -\frac{\partial U}{\partial x},\ g_y = -\frac{\partial U}{\partial y},\ g_z = -\frac{\partial U}{\partial z} \tag{3.6}$$

Until now, only the gravitational acceleration and potential of point masses have been considered. A solid body may be considered to be composed of numerous small particles, each of which exerts a gravitational attraction at an external point P (Fig. 3.2A). By integrating over the volume of the body, its gravitational potential at P can be calculated. At a point in the body with coordinates (x, y, z), let the density be ρ(x, y, z), and let its distance from P be r(x, y, z) as in Fig. 3.2B. The gravitational potential of the body at P is

$$U = -G\iiint \frac{\rho(x,y,z)}{r(x,y,z)}dxdydz \tag{3.7}$$

The integration readily gives the gravitational potential and acceleration at points inside and outside a hollow or homogeneous solid sphere. The values outside a sphere at a distance r from its center are the same as if the entire mass (M) of the sphere were concentrated at its center.

$$U = -\frac{GM}{r},\ g = -\frac{GM}{r^2}\hat{r} \tag{3.8}$$

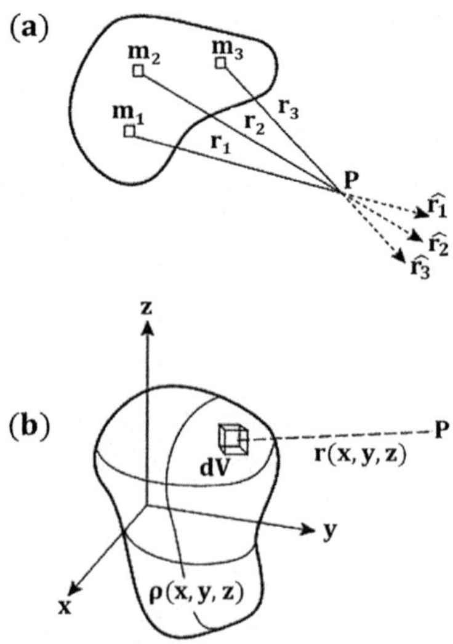

FIG. 3.2 (A) Gravitational attraction of individual particles in a solid body at an external point. (B) Calculation of gravitational potential for a continuous mass distribution. *Source: Personal.*

In this equation $\hat{r}$ is a unit vector in the direction of increase in coordinate r.

Measurement of the gravity component

Gravimeters effectively respond only to the vertical component of gravity of an anomalous mass. Consider the gravitational effect of an anomalous mass δg, with horizontal and vertical components δg_x and δg_z respectively, on the local gravity field g (Fig. 3.3). It can be shown that θ is usually insignificant since $\delta g_z \ll g$. Solving the rectangle of forces:

$g + \delta g = [(g + \delta g_z)^2 + \delta g_x{}^2]^{½} = [(g^2 + 2g\delta g_z + \delta g_z{}^2 + \delta g_x{}^2]^{½}$. Terms in δ^2 are insignificantly small and can thus be ignored. Binomial expansion then gives

$$g + \delta g = g + \delta g_z$$

$$\delta g = \delta g_z \tag{3.9}$$

Therefore, measured perturbations in gravity effectively correspond to the vertical component of the attraction of the causative body (Fig. 3.3). A discrepancy between the corrected, measured gravity and the theoretical gravity is called a gravity anomaly. It arises because the density of the Earth's interior is not homogeneous as assumed. If the Earth were an ideal oblate spheroid of uniform density, then the gravity measured at every point on its surface would be given precisely by a simple algebraic expression. However, the Earth has a rugged surface and non-uniform composition, which distorts its gravitational field (Radogna et al., 2004).

Measurement of gravity anomaly: Point mass

This gravity anomaly can reveal the presence of subsurface structures of unusual density. For example, a mass of dense ore below the surface will give a positive anomaly due to the increased gravitational attraction of the ore. Different theoretical models will predict different values of gravity, and so a gravity anomaly is always specified with reference to a particular model. The Bouguer, free-air, and isostatic gravity anomalies are each based on different theoretical corrections to the value of gravity. From Newton's law, the gravitational potential of a point mass (Fig. 3.4).

$$\Delta g_r = \frac{Gm}{r^2}$$

$$\Delta g = \Delta g_z = \frac{Gm}{r^2}\cos\theta = \frac{Gm(z' - z)}{r^3} \tag{3.10}$$

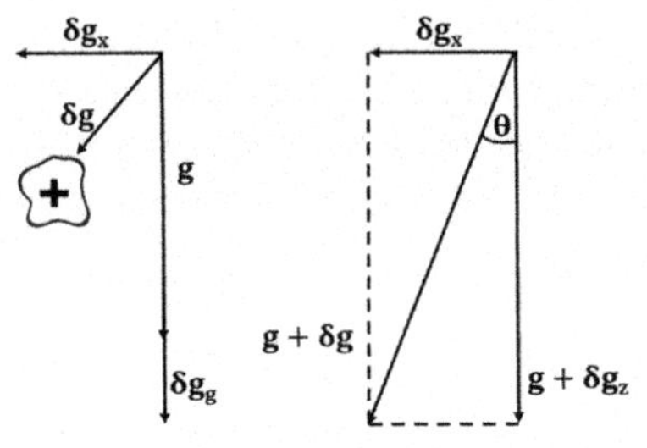

FIG. 3.3 Measurement of gravity components—gravity measures only the vertical component. *Source: Personal.*

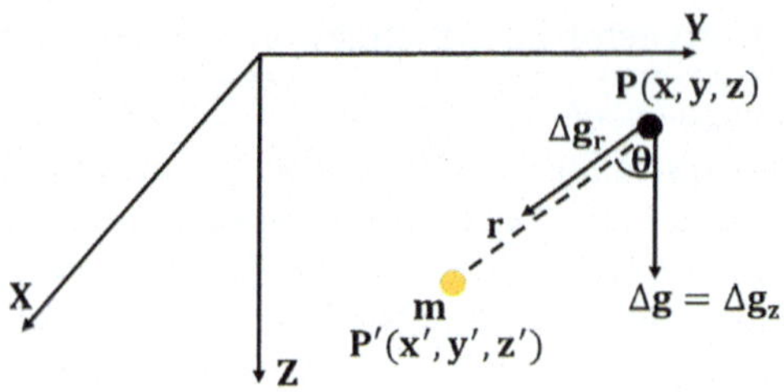

FIG. 3.4 Calculation of gravitational anomaly for a point mass in a Cartesian coordinate system. *Source: Personal.*

Measurement of gravity anomaly: Irregular shape

Until now, we have considered only the gravitational acceleration and potential of point masses. A solid body may be considered to be composed of numerous small particles, each of which exerts a gravitational attraction. From Eq. (3.10), $\Delta g = \frac{Gm(z'-z)}{r^3}$, the gravity anomaly for an irregularly shaped body (Fig. 3.5) for $\delta m = \rho \delta x' \delta y' \delta z'$. Therefore, $\delta g = \frac{G\rho(z'-z)}{r^3}\rho \delta x' \delta y' \delta z'$ with ρ the density (g/cm^3) and

$$r = \sqrt{(x'-x)^2 + (y'-y)^2 + (z'-z)^2}$$

For the whole body,

$$\Delta g = \sum\sum\sum \frac{G\rho(z'-z)}{r^3}\delta x' \delta y' \delta z'$$

If $\delta x'$, $\delta y'$ and $\delta z'$ approaches zero then

$$\Delta g = \int\int\int \frac{G\rho(z'-z)}{r^3} dx'dy'dz' \tag{3.11}$$

Therefore, the gravitational anomaly can be efficiently computed. The direct problem in gravity is straightforward: Δg is found by summing the effects of all elements that make up the body.

Derivatives—horizontal and vertical derivatives of gravity

Generally, the derivative grids pick up the subtle variations in density/magnetization contrasts that are otherwise masked or faintly reflected in the original grids. The derivatives can be computed along the horizontal or vertical direction. Normally, the vertical component of gravity g_z is measured. The horizontal components of gravity g_x, and g_y are small and not measured (Fig. 3.6). The spatial rate of change of gravity (in the x-direction) is written mathematically as $\frac{\partial g}{\partial x}$. This horizontal derivative can help map edges of buried changes in density.

However, when data were gathered over an area, it is more common to employ the so-called *total horizontal derivative*, which is the square root of the sum of the squared x and y horizontal derivatives. The rate of change of gradients, written as $\frac{\partial^2 g}{\partial x^2}$, can also be

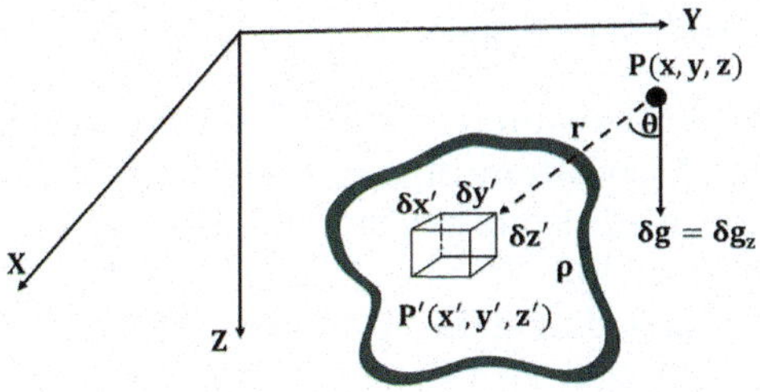

FIG. 3.5 Gravity anomaly for an irregularly shaped body. *Source: Personal.*

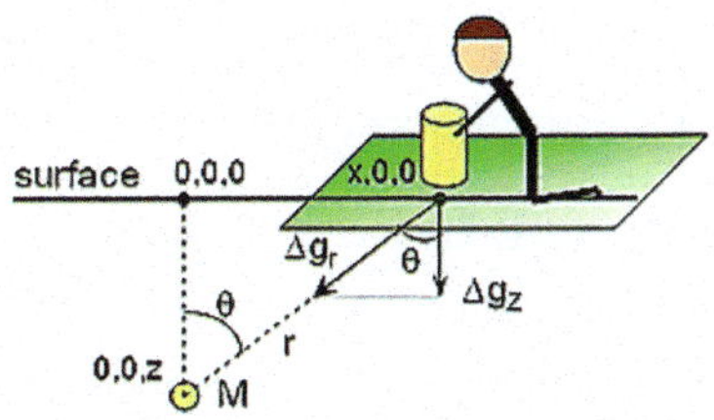

FIG. 3.6 Measured gravitational acceleration. *Source: https://gpg.geosci.xyz/.*

estimated. The second vertical derivative is often considered useful. But what is the first vertical derivative? It is the vertical rate of change of gravity, and it is not easy to find unless the gravity is measured everywhere at two elevations. However, gravity (a potential field) follows Laplace's equation, which states that the squared gradient of gravity equals 0:

$$\frac{\partial^2 g}{\partial x^2} + \frac{\partial^2 g}{\partial y^2} + \frac{\partial^2 g}{\partial z^2} = 0 \tag{3.12}$$

The $\frac{\partial^2 g}{\partial x^2}$ and $\frac{\partial^2 g}{\partial y^2}$ terms can be obtained directly from a Bouguer anomaly map, so the $\frac{\partial^2 g}{\partial z^2}$ term can be derived. This last term describes how fast the "vertical rate of change in gravity" is changing. However, the success of this process depends upon adequate data spacing. These types of calculations are very efficient in the Fourier domain.

The rate of change of gravity with distance is designated as the gradient. In fact, there are units for gravitational gradient ($\frac{\partial g}{\partial x}, \frac{\partial g}{\partial y}, \frac{\partial g}{\partial z}$)—the Eotvos Unit (EU), which is 10^{-6} mGal/cm.

Gravity of the Earth: Shape of the Earth—spheroid

The Earth is inhomogeneous and it rotates. Rotation causes the Earth to be an oblate spheroid. The best-fitting spheroid is called the reference spheroid, and gravity on this surface is given by the International Gravity Formula (IGF), 1967:

$$g_\varphi = 9.780318\left(1 + 5.3024\times10^{-3}\sin^2\varphi + 5.9\times10^{-6}\sin^2 2\varphi\right) \quad (3.13)$$

where φ is geographic latitude. This model is used in gravimetry.

In 1980, the International Association of Geodesy adopted a Geodetic Reference System (GRS80) in which the reference ellipsoid has an equatorial radius (R_e) equal to 6378 km and a polar radius (R_p) equal to 6356 km. The polar radius of the Earth is ~22 km less than the equatorial radius, which means that g is ~0.4% less at the equator than at the pole. At the equator, g is 5300 mGal (Foulger and Peirce, 2007). Compared to the best-fitting sphere, the spheroid is flattened ($R—R_p$) by about 14.3 km at each pole and the equator bulges ($R_e—R$) by about 7.2 km. The polar flattening f is defined as the ratio (Fig. 3.7).

$$f = \frac{R_e - R_p}{R_e} \quad (3.14)$$

A recent best estimate of the flattening (eccentricity) is $f = \frac{1}{298.247}$ (Mohr and Taylor, 2005; McCarthy and Petit, 2004; Groten, 2004; Lowrie, 2007). R = Radius of the spherical Earth.

Normal gravity

The International Reference Ellipsoid is the standardized reference figure of the Earth. The theoretical value of gravity on the rotating ellipsoid can be computed by differentiating the gravity potential U_g is called the geopotential (where $U_g = U_G - \frac{1}{2}\omega^2 r^2\theta$, where U_g = potential of gravity, U_G = gravitational potential of the oblate ellipsoid notation used in MacCullagh's formula, ω = angular velocity, linear velocity v = ωr, where r = radius vector, a point rotating with the Earth at a distance r from its center, θ = ωt = radius vector makes an angle with x axis, t = time; Lowrie, 2007). This yields the radial and transverse components of gravity, which are then combined to give the following formula for gravity normal to the ellipsoid:

$$g_n = g_e\left(1 + \beta_1 \sin^2\lambda + \beta_2 \sin^2 2\lambda\right) \quad (3.15)$$

is known as the normal gravity formula. The theoretical gravity at a given latitude is given by the normal gravity formula Eq. (3.15). The constants, g_e, β_1, and β_2 defined in 1980 for the Geodetic Reference System (GRS80), are still in common use. The constant β_2 is about 1000 times smaller than β_1. β_1 and β_2 are constants dependent on the shape and speed of rotation of the Earth. If we drop second-order terms and use λ = geographic latitude = 90° (see Figure 2.23, Lowrie, 2007), the value of normal gravity at the pole is $g_p = g_e (1 + \beta_1)$, so by rearranging and retaining only first-order terms,

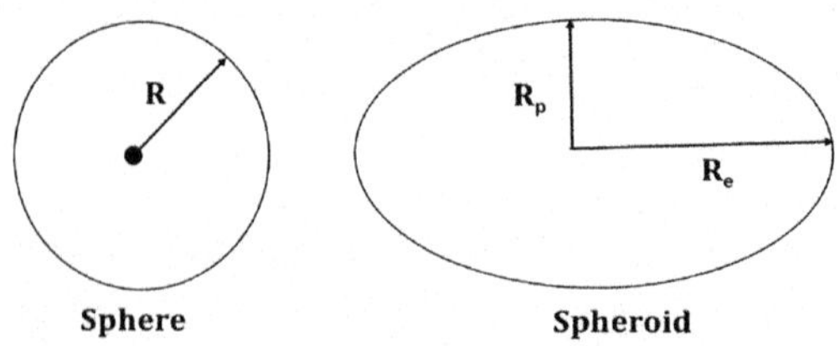

FIG. 3.7 Shape of the Earth—from spherical model to ellipsoid due to rotation. *Source: Personal.*

$$\frac{g_p - g_e}{g_p} = \frac{5}{2}m - f \tag{3.16}$$

where m = equatorial acceleration ratio = $\frac{1}{288}$ (Mohr and Taylor, 2005; McCarthy and Petit, 2004; Groten, 2004; Lowrie, 2007). This expression is called Clairaut's theorem. An increase in gravity from the equator to the pole of approximately $5.186 \times 10^{-2}\,m/s^2$, or 5186 mGal, is observed. Here, g_p = gravity at the pole, and g_e = gravity at the equator. The Earth's ellipsoidal shape, rotation, irregular surface relief, and internal mass distribution cause gravity to vary from place to place over its surface.

The geoid

The geoid is an equipotential surface, meaning it is a surface where the gravitational potential is constant. The direction normal to this surface defines the vertical or plumb-line direction, while the plane tangent to the equipotential surface at any point defines the horizontal at that location. It represents the mean sea level and is perpendicular to the plumb line at all points. On land, the geoid corresponds to the level that water would reach in canals connecting the seas. The geoid is a conceptual surface that is distorted due to variations in the distribution of mass. It bulges upward over land (positive geoid undulation, indicating mass excess) and dips downward in the oceans (negative geoid undulation, indicating mass deficit) as shown in Fig. 3.8 (Foulger and Peirce, 2007).

3.1.1 Gravity measurement

Measurements of gravity are of two types. The first corresponds to the determination of the absolute magnitude of gravity at any place; the second consists of measuring the change in gravity from one place to another. In geophysical studies, especially in gravity prospecting, it is necessary to measure accurately the small changes in gravity caused by underground structures. In national gravity surveys, the relative variations determined with a gravimeter may be converted to absolute values by calibration with absolute measurements made at selected stations.

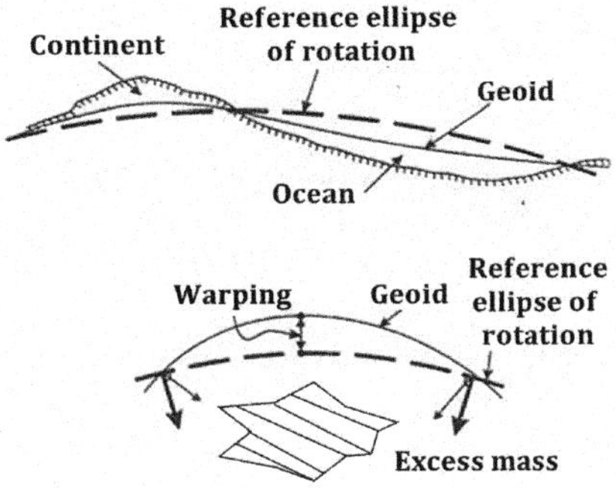

FIG. 3.8 Relationship between the geoid, spheroid, topography, and anomalous mass distribution. *Source: Personal.*

3.1.1.1 *Absolute measurement of gravity*

The classical method of measuring gravity is with a pendulum (compound or reversible), first described by Henry Kater in 1818, and allows more exact measurements. The period of the instrument is given by

$$T = 2\pi\sqrt{\frac{I}{mgh}} = 2\pi\sqrt{\frac{L}{g}} \tag{3.17}$$

where I is the moment of inertia of the pendulum about a pivot, h is the distance of the center of mass from the pivot, and m is the mass of the pendulum. Knowing the length L from Kater's method obviates knowledge of I, m, and h. The sensitivity of the compound pendulum is found by differentiating Eq. (3.17). This gives

$$\frac{\Delta g}{g} = -2\frac{\Delta T}{T} \tag{3.18}$$

To achieve a sensitivity (precision) of approximately 1 mGal, it is necessary to measure the period with an accuracy of around 0.5 μs. The equipment used for this purpose was quite bulky but remained the primary method for absolute gravity measurements until the 1950s. Today, modern techniques for determining the acceleration of gravity rely on observing the motion of falling objects. For an object dropped from an initial position z_0 with an initial velocity u, the equation of motion describes its position z at time t as:

$$z = z_0 + ut + \frac{1}{2}gt^2 \tag{3.19}$$

The absolute value of gravity is determined by fitting a quadratic curve to the recorded data of position vs. time. One limitation of the free-fall method is the influence of residual air molecules remaining in the evacuated chamber. This issue is mitigated in the rise-and-fall method. In the original version of the rise-and-fall method, a glass sphere was shot vertically upwards and allowed to fall back along the same path. Timing devices positioned at two different levels recorded the times it took for the ball to pass during both its upward and downward motion. The distance h between the two timing levels (approximately 1 m) was measured with high accuracy using optical interferometry. Let T_1 represent the time the sphere spends above the first timing level, and T_2 the time spent above the second level. Additionally, let z_1 and z_2 denote the distances from the zenith level to the respective timing levels.

The corresponding times of fall are $t_1 = \frac{T_1}{2}$ and $t_2 = \frac{T_2}{2}$. Then, $z_1 = \frac{1}{2}g\left(\frac{T_1}{2}\right)^2$, with a similar expression for the second timing level. Their separation is

$$h = z_1 - z_2 = \frac{1}{8}g\left(T_1^2 - T_2^2\right) \tag{3.20}$$

The following elegantly simple expression for the value of gravity is obtained:

$$g = \frac{8h}{\left(T_1^2 - T_2^2\right)} \tag{3.21}$$

Sensitivity and accuracy are comparable to those of the free-fall method.

3.1.1.2 *Relative measurement of gravity: The gravimeter*

In principle, a gravimeter is a highly sensitive balance that operates based on the direct application of Hooke's law. When a mass mmm is suspended from a spring of length l, it causes the spring to stretch by an amount (l + δl). The extension δl is proportional to the restoring force of the spring, which in turn is related to the value of gravity, as described by the following equation:

$$F = mg = -k\delta l,\ \text{or},\ g = -\frac{k}{m}\delta l \tag{3.22}$$

or,

$$g = -\frac{4\pi^2}{T^2}\delta l\ \text{with}\ T = 2\pi\sqrt{\frac{m}{k}}\ \text{for one period} \tag{3.23}$$

The constant k represents the elastic constant of the spring. A gravimeter is initially calibrated at a known location. When gravity differs at another location, the spring's extension changes, and this change can be used to calculate the variation in gravity. This type of gravimeter, based directly on Hooke's law, is called a *stable* gravimeter. However, it has been largely replaced by more sensitive, *unstable,* or astatized types, which are designed so that an additional force acts in the same direction as gravity, opposing the spring's restoring force.

If the natural length l of the spring is minimized, ideally approaching zero, Eq. (3.22) shows that the restoring force becomes proportional to the physical length of the spring, rather than its extension. The zero-length spring, first introduced in the LaCoste-Romberg gravimeter, is now a standard feature in modern gravimeters. These springs are typically of the helical type. During the manufacturing process of a zero-length spring, the helical spring is given an extra twist, causing it to naturally uncoil. The operation of a gravimeter is depicted in Fig. 3.9. Gravimeters are lightweight, durable, and portable, with sensitivities (precision) ranging from about 0.01 to 0.001 mGal.

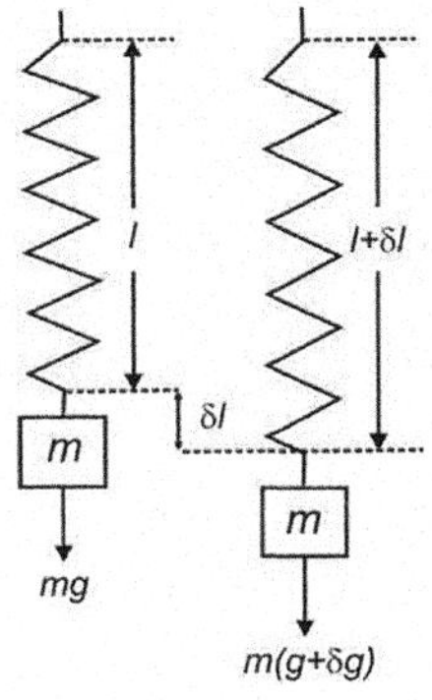

FIG. 3.9 Principle of operation of an unstable (astatic) gravimeter—spring extension (δl) due to change in gravitational pull (δg). *Source: Saibi, 2018—Open Access; INTECH. https://www.intechopen.com/chapters/57238.*

There are two types of gravity instruments: an absolute and a relative gravimeter. An absolute gravimeter is a highly precise tool applied to reservoir monitoring and subsidence studies, usually used in microgravity studies. The commonly used gravimeter in geophysical exploration applications is the relative gravimeter. The LaCoste and Romberg, and Scintrex instruments are the most commonly used gravimeters (Fig. 3.10). The Geological Survey of India (GSI) is at present deploying CG-5 gravimeters of Scintrex make, and the procedures involved in setting up the gravimeter are given in the operating manual, which shall be followed strictly in all respects. These gravimeters are calibrated in the factory itself. The main calibration constants are determined on the Orangeville calibration line of Toronto, which was established and maintained by the Geological Survey of Canada. State-of-the-art DGPS system, which has the capacity to measure both position location and elevation within a short period of observation time.

3.1.2 Surveying of gravity

When a gravimeter is set up at a specific location and monitored for an hour or more, repeated readings typically show smooth variations over time, with changes amounting to several hundredths of a mGal. This variation is primarily due to *instrumental drift*, caused by imperfections in the spring's elasticity (known as creep). In modern gravimeters, this effect is small and can be compensated for by applying a *drift correction*, which is determined by repeatedly taking measurements at the same station throughout the day.

During these measurements, the gravimeter is also influenced by Earth and ocean tides.[a] While the solid Earth responds to the gravitational pull of the Sun and Moon similarly to the

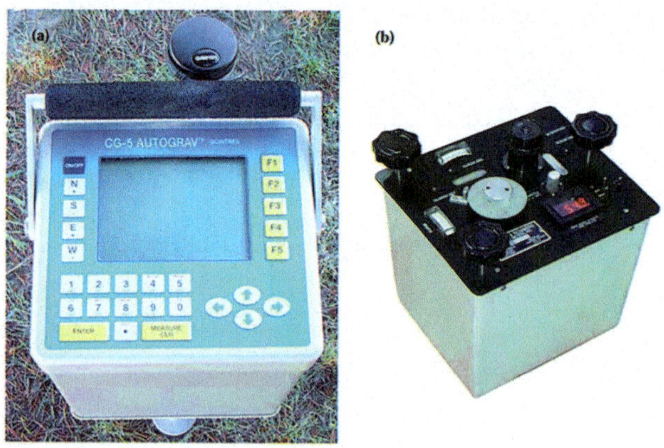

FIG. 3.10 Some commonly used gravity meters (A) Scintrex CG-5 Autograv (B) Lacoste & Romberg gravity meter. *Source: Scintrex. https://scintrexltd.com.*

[a] Tidal forces are inversely proportional to the cube of the distance between a body of mass M, which is producing the field, and the body of mass m: $F_t = \frac{2GMmr}{R^3}$ where R is the distance between the two bodies and $r \ll R$ is the distance from the centre of the reference body along the axis. This force causes the seas to rise and fall according to the lunar rhythm. The tide effect causes temporal variation of gravity due to Earth's rotational deformation. The amplitude of the gravitational effect of the Earth tides is up to 0.3 mGal throughout the day at a fixed point on Earth.

oceans, the resulting movements are much smaller. The Sun and Moon also exert a pull on the gravimeter and its components, causing additional effects that can impact gravity readings. These tidal-induced changes in gravity typically occur with periods of about 12 h. Earth tide effects can be managed by taking repeated readings at the same station, just as with instrument drift. To correct for tidal effects, it is necessary to know the exact time of each measurement.

Gravimeter measurements are typically taken at a network of stations, with the spacing between them depending on the survey's purpose. In environmental studies, a high-resolution gravity survey of a small area requires stations to be placed just a few meters apart. Conversely, in regional gravity surveys aimed at identifying hidden geological structures of potential commercial interest, stations may be spaced several kilometers apart. One common method to improve survey accuracy is to designate one station as a reference point (base) and periodically revisit it throughout the day, a practice known as loops. This base station can also be used to link relative gravity differences to an *absolute gravity* value, referencing the Swiss International Gravity Standardization Network (IGSN) (Morelli et al., 1971). After correcting for errors, the *observed gravity* (g_{obs}) is calculated. The standard survey design under the NGPM in the Geological Survey of India (GSI) is shown in Table 3.2.

In addition to recording gravity and magnetic data in the instrument, it is recommended to also maintain a record in the field notebook. The standard format for data recording in the field notebook is provided in Table 3.3.

The process of determining the relative difference between the established base (Morelli et al., 1974) and the field base is conducted systematically. At least three sets of readings are taken at both the field base and the IGSN base, with the results recorded. The average difference is then calculated to determine the value of the field base. An example is provided in Table 3.4.

TABLE 3.2 Standard gravity survey design.

Survey stage	Scale of survey	Station interval	Traverse interval	Contour interval	Remarks
National geophysical mapping programme (NGPM)	1:50,000	One station in 2.5 sq km (1–2) km	Random distribution along track/ road	1–2 mGal	Auto level for leveling
				1–2 mGal	DGPS for leveling
				1–2 mGal	With DGPS in Himalayan/inaccessible rugged terrain

TABLE 3.3 Gravity data record sheet.

Instrument no. and make: Traverse ID: Area with Toposheet no.: Date: Name of the Observer:
Field season (FS): Reading time: District: State:

Sl. no.	Base/station ID	Time	Reading (mGal)	Longitude (°)	Latitude (°)	Geology	Remarks/ landmarks

TABLE 3.4 Gravity base tie.

Sl. no.	Station	Time	Reading	Difference (mGal)	Drift (mGal)	Corrected difference
Drift = —0.05/180 mGal/min						
1	Base 1	07.30	736.5			
2	Base 2 (IGSN 1971)	08.00	740.5	740.5–736.5 = 4.0	–0.008	4–0.008 = 3.992
3	Base 1	08.30	736.515	736.515–736.5 = 0.015	–0.016	
4	Base 2	09.00	740.525	740.525–736.515 = 4.01	–0.025	4.01–0.025 = 3.985
5	Base 1	09.30	736.525	736.525–736.515 = 0.01	–0.033	
6	Base 2	10.00	740.540	740.540–736.525 = 4.015	–0.041	4.015–0.041 = 3.974
7	Base1	10.30	736.55	736.55–736.5 = 0.05	–0.05	3.983 is the average difference value of the two bases in mGal

Readings from Sl. no. 1 to 3, 3 to 5, and 5 to 7 form 3 sets of readings for the *Base Tie*.

3.1.3 Factors affecting gravity

Gravitational acceleration measured at any point is influenced by five factors, all related to either mass (M), distance (r), or both. These effects must be accounted for in the data through corrections, as explained further in the section on data reduction.

1. **Latitude**: Gravity varies by about 5000 mGal from the equator to the poles, with the highest values at the poles. The rate of change in gravity with respect to latitude is greatest at 45° latitude, where it is approximately 0.8 mGal/km.
2. **Elevation**: Changes in elevation (i.e., changes in distance, r) have a significant effect on gravity measurements. Modern instruments can detect elevation changes of just a few centimeters. For example, the difference in gravity between sea level and the top of Mount Everest is about 2000 mGal.
3. **Slab Effect**: Elevation changes usually involve moving through rocks and soil, not just air (except in airborne surveys). The gravitational attraction of these materials counteracts the effect of elevation changes. To correct for this, a factor of $0.0419 \times h \times \rho$ mGal is subtracted, where h is the elevation in meters and ρ is the density of the intervening material in gm/cc. This correction is known as the Bouguer correction.
4. **Topography**: The gravitational effect of nearby topographic features (such as hills or valleys) can be significant but is generally no >1.0 mGal. Although applying these corrections can be laborious, they are important when steep topography is present near measurement locations.
5. **Earth Tides**: Tidal effects can cause gravity variations of up to 0.3 mGal. These are typically accounted for by taking multiple measurements at a single station (base station) over the course of the survey.

6. Lateral Density Variations:

- Large-scale structural anomalies, such as those found in basin and range geology, can cause gravity variations of 100–500 mGal.
- Targets for oil exploration, like salt domes, may result in anomalies around 10 mGal.
- In mineral exploration, ore bodies typically produce anomalies of about 1 mGal.
- Geotechnical applications may detect anomalies as small as 20 microGal (μGal).
- **Effects of a Moving Platform**: When the instrument is in motion during measurements, the acceleration from the motion of a rotating sphere must be considered. These effects can be particularly significant in aircraft. Even the slight movement of a ship on calm seas, with mild swells, can introduce substantial measurement errors. As mentioned earlier, all data sets must be corrected for these effects, which are explained in more detail in a later section.

3.1.4 Earth's gravity field and data correction methods

Since the Earth is not a perfect, homogeneous sphere, gravitational acceleration varies across its surface. Its magnitude is influenced by five factors: latitude, elevation, surrounding topography, Earth tides, and subsurface density variations. In geophysical prospecting, the focus is typically on the latter, which is generally much smaller than the effects of latitude and elevation changes. The process of eliminating unwanted components is commonly referred to as gravity reduction.

3.1.4.1 Calibration of the gravimeter

This is a number provided by the manufacturer that translates scale readings into mGal.

$$\text{Calibration} = \text{actual reading} + \text{drift} - \text{base reading} = g_{\text{station}} - g_{\text{base}}$$

3.1.4.2 Latitude correction

The rotation of the Earth and its equatorial bulge produce variations in the gravitational value with latitude. The variation of the gravity value with latitude is given by the equation (1980 formula, Moritz, 1980; IAG (International Association of Geodesy), 1980).

$$g_\varphi = 978.0327(1 + 0.0052790414\sin^2\phi - 0.0000232718\sin^4\phi + 0.0000001262\sin^6\phi)\,\text{Gal} \quad (3.24)$$

where g_φ is the gravity value at latitude φ. The theoretical gravity value corresponding to the latitude of a particular point, as obtained from the above formula, is removed from the observed gravity, also referred to as station gravity.

$$^{b}\text{Latitude correction} = g_{\text{station}} - g_\varphi = 0.8140\sin 2\lambda \ \text{mGal/km} \quad (3.25)$$

[b]From Eq. (3.15), $\frac{\partial g_n}{\partial \lambda} = g_e(\beta_1 \sin 2\lambda + \beta_2 \sin 4\lambda)$. After converting $\partial\lambda$ from radians to kilometers and neglecting the β_2 term, the latitude correction (Δg_{lat}) is $0.8140\sin 2\lambda$ mGal/km of N–S displacement.

The latitude gradient represents the change in gravity with N–S distance from a base station, i.e., the centrifugal force changes only in the N–S direction, not in the W–E, and latitude (φ) is negative in the southern hemisphere (low latitude).

3.1.4.3 *Free-air correction (FAC)*

The gravity field decreases with the inverse square of the distance from the source, and it is essential to apply corrections for elevation differences between stations to ensure all field measurements are referenced to a common datum, typically mean sea level (MSL). In the Free-Air Correction (FAC), the material between the datum and the station is not considered (Fig. 3.11). Since gravity (g) decreases with height, this effect is corrected by adding a value. Now, differentiate Eq. (3.4).

$$\frac{\Delta g_{FA}}{\Delta R} = -2G\frac{M_E}{R^3} = -\frac{2g}{R}\text{mGal / m} \tag{3.26}$$

On substituting the Earth's radius (6371 km) for R and the mean value of gravity (981,000 mGal) for g, the value of Δg_{FA} is found to be 0.3086 mGal/m.

3.1.4.4 Bouguer correction (BC)

The Bouguer Correction (BC) accounts for the attraction of the material between the station and the datum plane, i.e., msl (Fig. 3.12). This effect was ignored during the free-air corrections. Hence, a slab is added with an average density of surrounding rocks—the Bouguer correction (Saibi, 2018):

$$\frac{\Delta g_B}{\Delta R} = 2\pi G\rho = 0.04192\rho \text{ mGal / m} \tag{3.27}$$

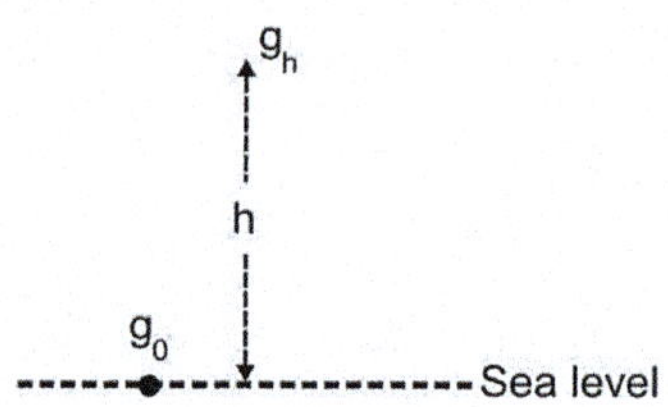

FIG. 3.11 Free-air correction effect. *Source: Saibi, 2018—Open Access; INTECH. https://www.intechopen.com/chapters/57238.*

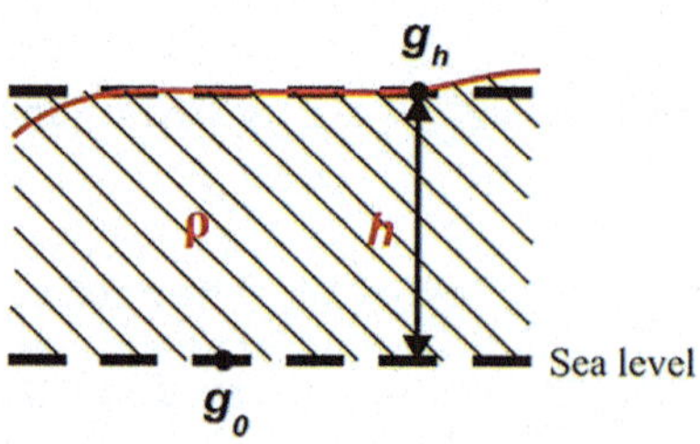

FIG. 3.12 Bouguer effect. *Source: Saibi, 2018—Open Access; INTECH. https://www.intechopen.com/chapters/57238.*

where ρ is the density of the *Bouguer slab* and h is the thickness of the rock slab. It is also known as the formula for an *infinite slab* of rock. The Bouguer correction is subtracted on land (as the attraction due to the mass of the rock between the datum and the observation point must be removed from the observed gravity), but at sea, it must be added to account for the lack of mass between the sea floor and sea level.

The correction is equivalent to the replacement of the water layer by material of a specified rock density ρ_r. In this case:

$$BC_{sea} = 2\pi G(\rho_r - \rho_w)\,\text{mGal/m},\ \rho_w = \text{density of sea water}$$

The free-air and Bouguer corrections are often combined into an *elevation correction*:

$$\frac{\Delta g_E}{\Delta R} = \frac{\Delta g_{FA}}{\Delta R} - \frac{\Delta g_B}{\Delta R} = (0.3086 - 0.04192\rho) = 0.1967\,\text{mGal/m} \tag{3.28}$$

assuming, ρ = density of Bouguer slab for crustal rocks = 2.67 g/cc.

The data format for a simple software application that corrects for free-air, Bouguer slab, and for normal variation of gravity is given in the following Table 3.5.

3.1.4.5 *Terrain correction (TC)*

This correction takes care of surface irregularities in the vicinity of the station. Topographic undulations reduce the gravity effect because of upward attraction (hills) and lack of downward attraction in the case of valleys (Fig. 3.13) (Musset and Khan, 2000). Hence, the

TABLE 3.5 Data format for data reduction using GP-Macros.

Sl. no.	Longitude (°)	Latitude (°)	Elevation (msl) in m	Station gravity (1971) in mGal	Theoretical gravity (1980) in mGal	Bouguer anomaly in mGal	Free-air anomaly in mGal

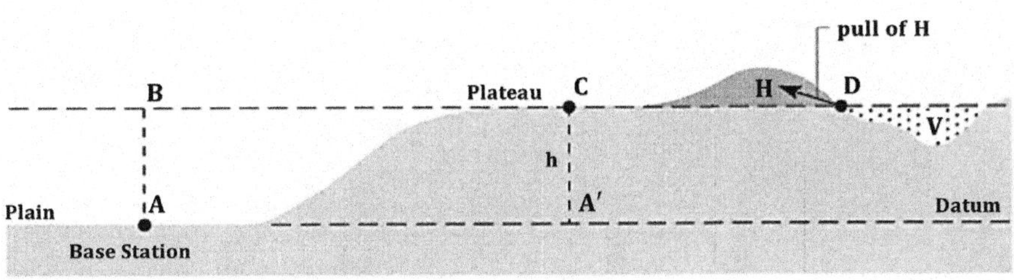

FIG. 3.13 Topographic corrections: Free-air effect (left), Bouguer effect (center), and terrain effect (right). The datum represents an arbitrary reference plane of constant elevation to which all measurements are adjusted. The elevation corrections are made in three steps: Free-air correction, Bouguer correction, and terrain correction. *Source: Personal Gravity, Expl.ravity.*

correction factor is always positive. It is a very cumbersome process to apply the terrain correction. This correction is normally ignored in surveys under project NGPM, unless it is very essential, when the anomaly sought (signal) is of the same order as the topographic noise (Fig. 3.14). Descriptions of the terrain conditions at the point of observation may be noted, as they are required during terrain corrections (Hammer, 1939).

There are various methods for calculating terrain corrections. The most common approach involves dividing the area around the gravity stations into zones, estimating the average altitude of each zone, and then computing the gravity effect for each zone (Hammer, 1939). One simple method is to print the zoning chart (Fig. 3.15) on a transparent sheet and overlay it onto the topographic map.

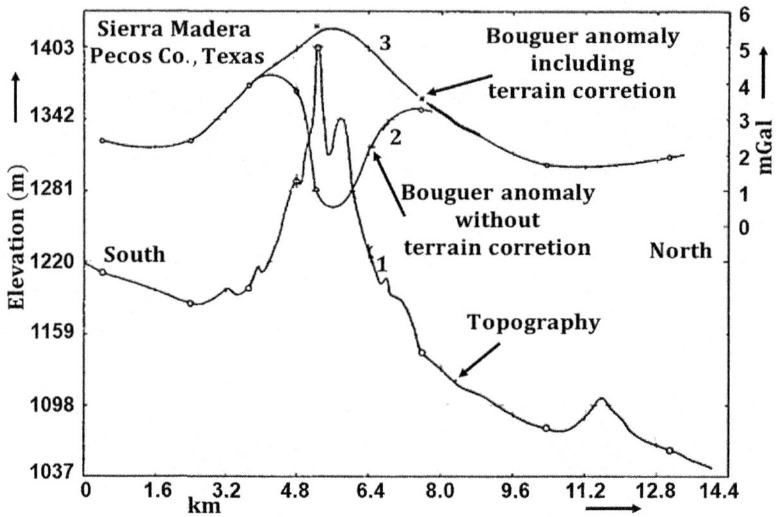

FIG. 3.14 Gravimeter profile across Sierra Madera, Pecos County, Texas, highlighting the importance of terrain corrections. *Source: Modified.*

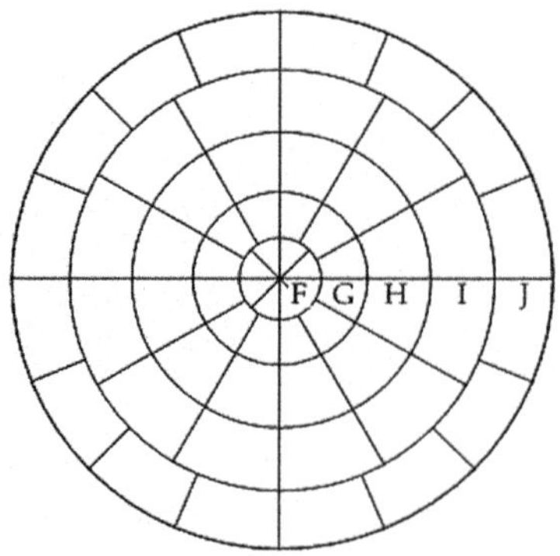

FIG. 3.15 Zone chart for use in evaluating terrain corrections at gravity stations. *Source: Taha and Hussein Muhammed (2023); Hammer (1939). https://doi.org/10.21203/rs.3.rs-3764006/v1.*

Finally, the total terrain correction for the current station is obtained by summing the effects from all compartments and zones. This method of calculating terrain corrections can be quite labour-intensive. As a result, modern approaches typically use computer programs that calculate the corrections based on the Digital Elevation Model (DEM) of the terrain.

3.1.5 Gravity anomalies

The gravity anomaly at a point may be defined as the difference between the observed gravity and the theoretically computed one for the point using the International Gravity Formula:

$$\delta g = \delta g_{obs} - g_{\phi} \quad (3.29)$$

To make the observed and theoretical gravity values comparable, the reduction steps are adopted, each step of reduction resulting in a particular type of anomaly.

3.1.5.1 *Free-air anomaly (FAA)*

The anomaly mainly represents the topographic features superposed over the effects of the subsurface. Therefore,

$$\text{FAA} = \text{observed gravity} + \text{free air correction} - \text{theoretical gravity} = g_{obs} + \text{FAC} - g_{\phi} \quad (3.30)$$

The free-air anomaly is thought of as removing all the mass above sea level into an infinitesimally small thin layer, and then measuring the gravity. FAA is generally used for marine surveys and investigation of deep mass distribution, e.g., the theory of isostasy.

3.1.5.2 *Bouguer anomaly (BA)*

It is defined as $\text{BA} = g_{obs} - g_{model}$, where $g_{model} = g_{\varphi}$–FAC + BC–TC, and g_{model} = model for an on-land gravity survey. Therefore,

$$\text{BA} = g_{obs} + \text{FAC}(-,+)\text{BC} + \text{TC} - g_{\phi} \quad (3.31)$$

The Bouguer anomaly, being free from the effects of material between the observation point and the datum, reflects variations in the mass distribution beneath the datum. Therefore, a zero Bouguer anomaly (BA) indicates that the Earth's mass distribution perfectly matches the ideal spherical Earth model. However, elevated continental regions typically show a negative BA, indicating a lower-than-normal rock density in these areas. Conversely, oceanic regions often exhibit a positive BA, suggesting a higher-than-normal rock density in the oceanic crust. The variations caused by lateral inhomogeneities are referred to as the Bouguer anomaly or simply the gravity anomaly.

3.1.5.3 *Isostatic anomaly (IA)*

These are the anomalies obtained after correcting the BA for the isostatic behavior of the crust (mass balance, i.e., equilibrium between the Earth's crust and upper mantle, e.g., mountain belts on the Earth's surface, are compensated for at depth). This correction is made by taking into account the surface topographic mass distribution. Therefore,

$$\text{IA} = g_{obs} + \text{FAC} - \text{BC} + \text{TC} + \text{Isostatic correction} - g_{\varphi} \quad (3.32)$$

A zero isostatic anomaly indicates complete compensation, while positive anomalies suggest under-compensation, and negative anomalies indicate overcompensation. Isostatic anomalies are generally of limited significance in prospecting, apart from helping to identify potential zones of strain accumulation and tectonic activity.

3.1.6 Rock densities

The use of gravity for prospecting requires density contrasts to be used in interpretations. Rock densities vary very little, the least of all geophysical properties. Most rocks have densities in the range 1.5–3.5 g/cc, with extreme values up to 4 g/cc in massive ore deposits. In sedimentary rocks, density increases with depth and age, i.e., compaction and cementation. In igneous rocks, density increases with basicity, so granites tend to have low densities, and basalts show high densities. Density is related to velocity using Gardner's relationship:

$$\rho = \alpha v^{\beta} \tag{3.33}$$

where ρ = bulk density, v = P wave velocity in ft/s, $\alpha = 0.23$, $\beta = 0.25$. No single density value is completely satisfactory for a large area, but there are a number of techniques available to estimate the density of rocks in situ (Nettleton, 1939). The underlying goal in these techniques is to minimize the correlation between Bouguer gravity anomalies and elevation. The density of a rock is dependent on both its composition and porosity.

In *direct measurement*, the sample is weighed in air and water. Dry and saturated samples are measured. Applying Archimedes principle

$$\rho_r = \frac{W_a}{W_a - W_w} \tag{3.34}$$

where ρ_r = density of rock sample, W_a = weight in air, W_w = weight in water.

Using a *borehole gravimeter* is only possible if a borehole is available in the formation of interest. The density in the interval between the measurements is calculated using the equation:

$$g_1 - g_2 = 0.3086h(\text{FA term}) - 4\pi G\rho h(2 \times \text{Bouguer term})$$

where g_1 and g_2 are two measurements at points in the borehole separated by a vertical distance h. Twice the Bouguer term must be used because the slab of rock between the two points exerts downward pull at the upper station and an upward pull at the lower station. Thus

$$\rho = \frac{0.3086h - \Delta g}{4\pi Gh} \tag{3.35}$$

The *borehole density logger* (gamma-gamma logger), which uses a gamma ray source (Co60) and a Geiger counter, measures only the scattered gamma radiation. This method works well in solid rock formations but is less effective in soft, washed-out formations. Its maximum penetration is around 15 cm, with an effective sampled volume of approximately 0.03 m^3, which can be problematic if this small volume is not representative of the surrounding formation. The accuracy is about 1% of the density, but the borehole log tends to be irregular, requiring averaging over tens of meters to obtain values suitable for gravity reduction.

Nettleton's method involves conducting a gravity survey over a topographic feature and reducing the data using various density values. The density that results in an anomaly with the least correlation to the topography is chosen. This method has the advantage of determining bulk density, rather than just the density of a small sample (Lowrie, 2007; Nettleton, 1939). However, it has some drawbacks: only near-surface rocks are sampled, which may be weathered, and the topographic feature could consist of a different rock type than the surrounding area, possibly existing due to these differences (Fig. 3.16).

Rearranging the Bouguer equation as if the variation in gravity over the area is small, we may write:

$$BA = BA_{ave} + \delta BA$$

where BA = Bouguer anomaly at station, BA_{ave} = average BA over whole area, δBA = small increment of BA. The standard Bouguer anomaly equation is

$$BA = g_{obs} - g_{\phi} + FAC - BC + TC$$

thus:

$$G_{obs} - g_{\varphi} + FAC = BA_{ave} + \delta BA + BC - TC = \rho(0.04191h - \frac{TC}{2000}) + BA_{ave} + \delta BA$$

$$G_{obs} - g_{\phi} + FAC = \rho(0.04191h - \frac{TC}{2000}) + BA_{ave} + \delta BA \tag{3.36}$$

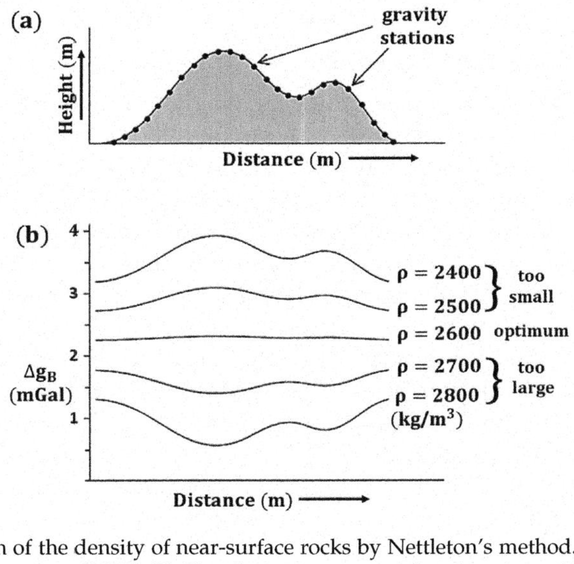

FIG. 3.16 Determination of the density of near-surface rocks by Nettleton's method. (A) Gravity measurements are made on a profile across a small hill. (B) The data are corrected for elevation with various test values of the density. The optimum density gives minimum correlation between the gravity anomaly (Δg_b) and the topography. *Source: Personal.*

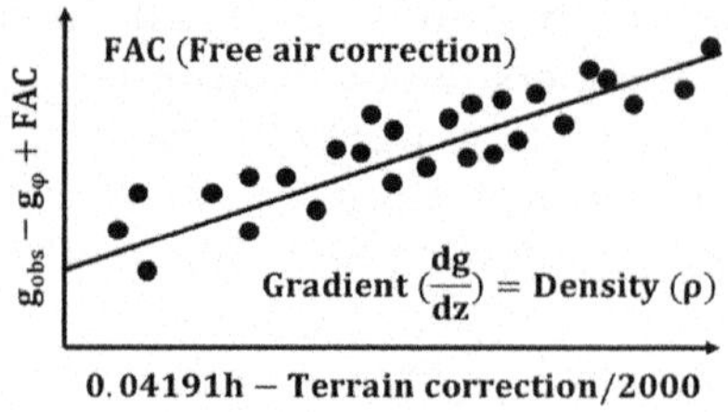

FIG. 3.17 Rearranging the Bouguer equation. *Source: Personal.*

For the Hammer chart using ρ = 2000 kg/m^3, and if δBA is small, then the above equation takes the form y = mx + c. If the line is plotted with (Fig. 3.17)

$$G_{obs} - g_{\phi} + FAC: 0.04191h - \frac{TC}{2000}$$

It should yield a data distribution in the form of a scatter about a straight line. A line can be fitted to the data using the least squares method, with the gradient representing the density (ρ). An example where density played a crucial role is the sulphur exploration at Orla, Texas. In this case, rock density in the region was measured both from samples and borehole data. The predominant lithologies in the area included limestone, dolomite, sand, gypsum, salt, and anhydrite (Foulger and Peirce, 2007).

3.1.7 Interpretation of gravity anomalies

3.1.7.1 Regional and residual anomalies

A gravity anomaly arises from the inhomogeneous distribution of density within the Earth. Consider a subsurface body with a density of ρ, surrounded by rocks with a density of ρ_0. The difference, $\Delta\rho = \rho - \rho_0$, is known as the density contrast between the body and the surrounding material. If the body has a higher density than the host rock, the density contrast is positive; if it has a lower density, the contrast is negative. The horizontal extent of an anomaly is often referred to as its apparent wavelength. This wavelength serves as an indicator of the depth of the anomalous mass: large, deep bodies produce broad (long-wavelength), low-amplitude anomalies, while small, shallow bodies generate narrow (short-wavelength), sharp anomalies.

Typically, a map of Bouguer gravity anomalies displays superimposed anomalies from multiple sources. Long-wavelength anomalies, caused by deep density contrasts, are termed regional anomalies. These are important for understanding the large-scale structure of the Earth's crust beneath major geographical features such as mountain ranges, oceanic ridges, and subduction zones. Short-wavelength residual anomalies, on the other hand, are linked to shallow anomalous masses that could be of interest for commercial exploration. Interpreting these residual anomalies requires geological knowledge. In eroded shield areas, such as Canada or Scandinavia, anomalies with very short wavelengths may indicate near-surface mineralized bodies. In sedimentary basins, short- or intermediate-wavelength anomalies may be associated with structures linked to petroleum or natural gas reservoirs.

Distinguishing between regional and local anomalies is a crucial step in interpreting gravity maps. This analysis can be done by selecting profiles across certain structures or by examining the two-dimensional distribution of anomalies on a gravity map. Various techniques are used to decompose gravity anomalies into their individual components, ranging from simple visual inspection (Fig. 3.18) of anomaly patterns to more advanced mathematical methods such as polynomial representation, Fourier series, anomaly enhancement, and filtering (Mariita, 2009).

3.1.7.2 Applications of the gravity method in geosciences

Gravity surveys can be classified into two types: (1) regional surveys and (2) local surveys. The choice between these two depends on the scale of the geological target being studied. For fault and regional geological surveys, a regional gravity survey with a large grid size is typically used. In contrast, for local geological studies, such as cavity investigations, active fault detection, or the examination of specific geological features, denser gravity surveys with smaller grid sizes and closer spacing between stations are employed.

3.1.7.3 Geology and structural geology

Gravity data provides valuable insights into subsurface geological structures, including faults, rock intrusions, dykes, and sills. It can also help in the study of basin structures, grabens, horsts, and salt intrusions (such as salt domes—structures formed by the upward movement of salt, which has lower density, magnetic susceptibility, and higher seismic velocity and resistivity than surrounding sediments). The low density of salt causes it to rise over time in a diapir, piercing through overlying strata and forming a dome-like shape. These features are crucial for oil and gas exploration. The key physical property involved is rock density, and the density contrast between different rocks generates gravity anomalies, which can be easily detected if the gravity data is of sufficient quality (Fig. 3.19).

Gravity measurements are highly sensitive to noise from external sources, such as vehicles, human activity (including walking), machinery vibrations, nearby seismic operations, and

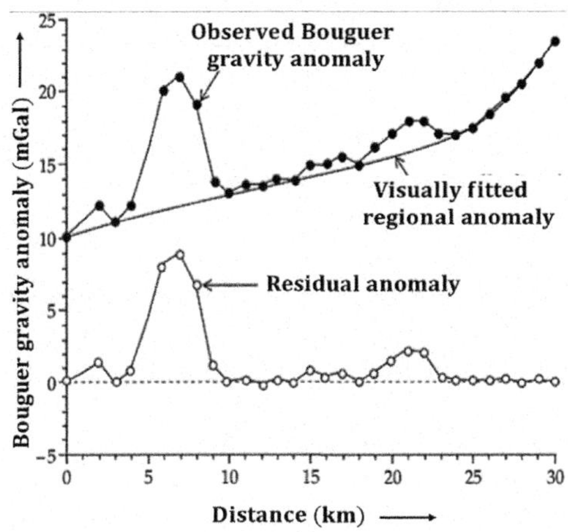

FIG. 3.18 Representation of the regional anomaly on a gravity profile by visually fitting the large-scale trend with a smooth curve. *Source: Personal.*

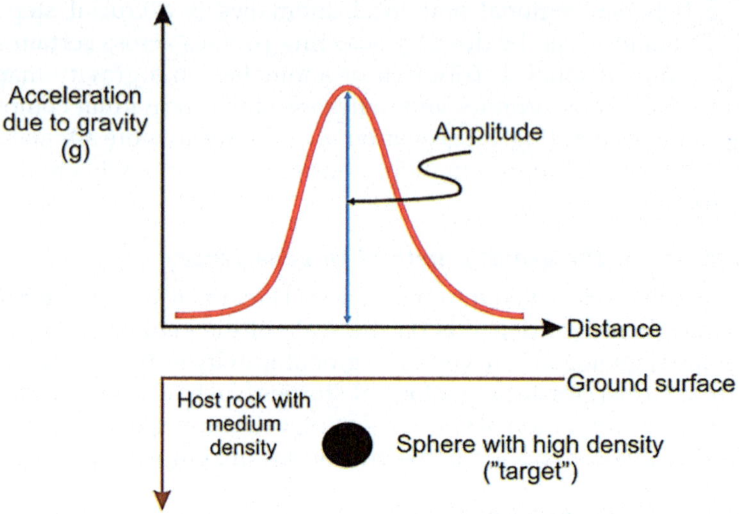

FIG. 3.19 Example of gravity anomaly over a buried sphere with a higher density than the surrounding rocks. *Source: Saibi, 2018. https://www.intechopen.com/chapters/57238.*

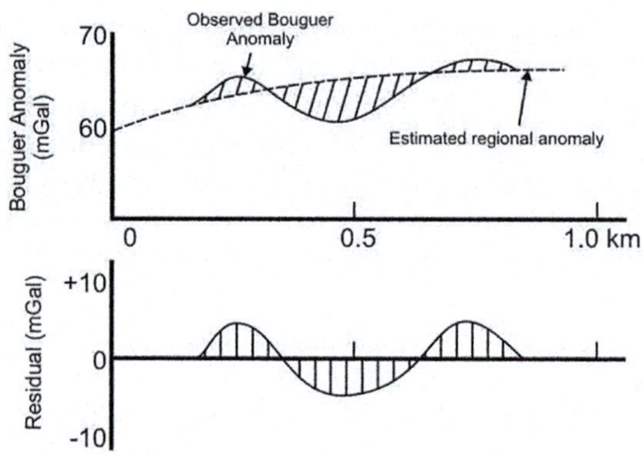

FIG. 3.20 Removal of residual gravity from a regional profile. *Source: Saibi, 2018. https://www.intechopen.com/chapters/57238.*

even earthquakes. Therefore, it is important for the geophysicist or user to be vigilant during field measurements to ensure accurate gravity data.

Bouguer anomaly maps display both regional and residual (local) anomalies (Fig. 3.20). Regional anomalies, which have long wavelengths, are typically caused by deep crustal features, while residual anomalies, characterized by short wavelengths, are associated with shallower structures. Several filtering techniques, such as band-pass filters, can help separate regional and residual gravity anomalies. Power spectrum analysis is also useful for determining the number of geological layers and their depths, as well as for removing noise from the data.

Regional anomalies, characterized by long wavelengths, are typically associated with deep crustal features. In contrast, residual (local) anomalies, which have shorter wavelengths, are linked to shallower structures. Various filtering methods, such as band-pass filters, can be used to differentiate between regional and residual gravity anomalies. Additionally, power spectrum analysis is helpful in determining the number of geological layers and their depths, as well as in removing noise from the data.

3.1.8 High precision structural mapping using advanced gravity processing methods: A case study from the north region of Cameroon

The North Cameroon region has garnered significant attention from researchers due to previous geophysical and geodynamic studies conducted in the area (Kamguia et al., 2005; Tokam et al., 2010; Mouzong et al., 2014; Ghomsi et al., 2022). Kamguia et al. (2005) estimated the sediment thickness and Moho depth beneath the Garoua sedimentary basin using ground gravity data. Tokam et al. (2010) later analyzed the crustal structure beneath the Garoua Rift, revealing a thin crust with a thickness ranging from 26 km to 31 km, based on seismic data inversion. In addition, Mouzong et al. (2014) conducted a structural study in the northern part of Cameroon (Benue Trough) using classical processing techniques, such as the horizontal gradient. Their study identified deep faults primarily trending in the NW-SE direction (Kamto et al., 2023). Furthermore, the region's vulnerability has driven further research aimed at better understanding and exploiting its natural resources. A detailed and accurate mapping of the geological structures (such as lineaments and faults) in this region would greatly benefit mining and water resource exploration. Additionally, major infrastructure projects in the area require a comprehensive understanding of the primary and deep faults present.

The main objective of this study is to create a high-precision map of geological features, including lineaments and faults, using advanced gravity processing methods. The techniques employed in this study include the tilt angle of horizontal gradient (TAHG), improved logistic filter (ILF), fast sigmoid edge detector filter (FSF), and an arcsine function-based edge detection filter (ASF). Euler deconvolution was used to estimate the position and depth of gravity anomalies in the region. Through these advanced edge detection methods, a precise map of structural lineaments has been created. The deepest density anomalies in the region, primarily oriented in the NW-SE direction, correlate well with the extension of the Central and West African rift system. The edge detection techniques and Euler deconvolution both identified a network of superficial lineaments around longitude 13°N and latitude 8°N, located between depths of 0 and 4 km. Crustal features trending NW-SE in the northeastern part of the region suggest that the area has undergone uniform tectonic activity. This study plays a crucial role in enhancing the understanding of the fracture network configuration, which is essential for the more efficient exploitation of mining and groundwater resources in the region.

3.1.9 Microgravity survey for mineral exploration and case studies

The interpretation of gravity changes over space or time requires knowledge of the universal gravitational constant (G) and the free-air gradient (FAG), which describes how gravity varies with height. Gravity measurements are typically expressed in cgs units. Gravity variations are measured in Gals (sometimes written as "gals"), where 1 Gal equals 1 cm/s^2 or 10^{-2} m/s^2. Modern instruments can measure up to 10^{-9} Gal, and since variations are usually hundreds of times larger, microgravity measurements are commonly expressed in microgals (1 μGal = 10^{-8} m/s^2). The term "microgravity" is generally used to distinguish measurements in the range of 1–500 microGals (0.001–0.5 milliGals) from those in geophysical prospecting, like Bouguer anomalies, which typically range from 500 microGals to 100,000 microGals (0.5–100 milliGals).

Microgravity surveying, which detects small gravity signals from subsurface features on the scale of a few microgals, is widely used in commercial applications such as mineral exploration, environmental studies, and the detection of subsurface voids, mine workings, sinkholes, and other underground hazards (Hinze, 1990; Nabighian et al., 2005; Tuckwell et al., 2008). It is particularly useful for detecting deeper voids, especially where ground conditions complicate the use of other active geophysical methods, such as Ground Penetrating Radar (GPR), whose depth of investigation is often limited by signal attenuation in conductive ground (Boddice et al., 2022).

In a conventional (static) gravity survey, measurements are taken at several field stations and at a base or reference station. A Bouguer anomaly map is then produced by comparing the gravity values at the field stations to those at the base station. Before this, several corrections are applied to account for differences in latitude, height, and the gravitational effect of material between stations. Once these corrections are made, data are usually expressed as relative gravity differences with respect to a local base station or an international gravity reference station. This technique is commonly used to identify subsurface density structures and has been applied extensively in volcanic regions to map caldera infill thickness, magma feeder pipe dimensions, and the extent of hydrothermal alteration (Rymer, 1996; Yokoyama and Tajima, 1957; Rymer and Brown, 1989). These anomalies typically range from tens to hundreds of milliGals in magnitude and can have lateral extents of several kilometers.

Microgravity surveys are also expressed relative to a base or reference station. In addition to spatial gravity variations, temporal changes in gravity are measured. A set of gravity differences is usually recorded for field stations relative to the base station and then compared with values obtained in subsequent surveys, which may occur days, months, or even years later. The changes in gravity observed over time are typically much smaller than Bouguer anomalies, typically ranging from tens to hundreds of microGals. A microgravity survey is typically conducted over a small area with the goal of detecting local anomalies (Ahmad et al., 2004). In a microgravity survey, measurements are typically made with an accuracy of a few microgals, and station spacing is typically a few meters (1–10 m). With advanced instruments such as the CG3M (Scintrex) Autograv with high precision (<0.05 mGal) (Gozlan and Cull, 2019) and the CG-5 (Scintrex) Autograv with 0.001 mGal sensitivity (Srinivasarao et al., 2022), gravity changes in the microgal range can now be measured accurately. Microgravity surveys are a promising geophysical method for detecting and mapping subsurface cavities (Butler, 1991). These surveys can detect structures such as near-surface cavities, variations in overburden, and lithologic contrasts in bedrock (Kaufmann, 1996).

3.1.9.1 Exploration of rare metal bearing pegmatites—a case study from Marlagalla, Mandya district, Karnataka

The Atomic Minerals Directorate (AMD) for Exploration and Research is actively exploring rare metals like columbite-tantalite, spodumene, and beryl minerals from weathered pegmatites and their derived gravels in the Marlagalla-Allapatna region of the Nagamangala schist belt, Mandya, Karnataka (Srinivasarao et al., 2022). A microgravity survey in the Marlagalla area has helped identify pegmatite bodies, some of which contain spodumene mineralization. These pegmatites are located within amphibolite, part of the Nagamangala schist belt of the Sargur Group in the southern Western Dharwar Craton. The difference in density between the pegmatite (2.58–2.61 g/cc) and the host rock amphibolite (3.09–3.14 g/cc) creates a detectable contrast, which allows the identification of pegmatites within the amphibolite. The residual gravity map revealed low gravity anomalies ranging from −5 to −200 μGal, indicating the presence of pegmatite bodies. The First Vertical Derivative (FVD) filtered microgravity map further identified the precise locations of the pegmatites. High gravity zones (3–20 μGal/m) on the FVD map correspond to the amphibolite host rock, while gravity lows (–36 to 3 μGal/m) are attributed to the pegmatites. Six major pegmatite bodies, trending in the N10°E direction with widths of ~10–20 m, were identified. The geometry of these bodies was further analyzed using 2D forward modeling of the microgravity data. Boreholes drilled based on the survey results intercepted the pegmatites within the amphibolite. This technique has proven effective for exploring pegmatites in the contrasting amphibolite environment. Additionally, the proposed model will aid in the exploration of rare metals and lithium-bearing pegmatites.

3.1.9.2 Fosterville in an attempt to delineate major fault structures associated with gold mineralisation in central Victoria

Gravity surveys are rarely used as a primary tool for mineral exploration, except in a few notable cases, such as Olympic Dam (Gozlan and Cull, 2001). Gravity data typically takes time to collect, and its sensitivity is often limited to locating major structures. However, with the advent of new digital gravity meters, productivity has improved, and fine details can now be detected at the microgravity scale. As a result, high-definition microgravity surveys were conducted at Fosterville to identify regional trends linked to faulting and to locate anomalies associated with dense sulphide ore bodies for gold mineralization in Central Victoria (Gozlan and Cull, 2001).

The data were collected with an average station spacing of 25 m along 22 traverse lines, which were spaced about 45 m apart, running east-west across the Fosterville fault-line. Microgravity surveys successfully detected deep sulphide deposits at Fosterville. The survey procedures and data reduction techniques were precise enough to resolve anomalies with an accuracy better than 0.05 mGal. However, despite this accuracy, the fault-line and shallow oxide mineralization zones were not directly observed. Only small variations in density could be linked to alteration and weathering from fluid flows in the main structural elements at Fosterville. Therefore, directly detecting the primary sulphides at Fosterville may be more effective than focusing on fault-line location for exploration.

A magnetic survey was also conducted to supplement the gravity data. This survey identified several northeast-trending linear contacts that were interpreted as subsurface fractures. The intersection of these fractures with the Fosterville Fault was seen as a potential pathway for auriferous fluids, which could explain the oxide and sulphide mineralization at Rehe's Pit. Based on independent reports, the spatial frequency components of the magnetic image were analyzed, and the characteristic magnetic signatures of the rocks in the area were accurately determined.

3.1.9.3 *Gold and silver prospection using magnetic, radiometry, and microgravity methods in the kısladag province of western Turkey*

High-resolution magnetic and gravity data are often useful in large-scale studies to define structures directly related to mineralization (Whiting, 1986; Isles et al., 1989; Henley and Adams, 1992). On a smaller scale, geophysical data help outline the regional geologic framework by identifying major compositional boundaries or structural zones that may be favorable for mineralization (Jaques et al., 1997; Leclair et al., 1997; Gunn et al., 1997a, b; Moore et al., 1998). Geological and geophysical exploration can lead to the detection of gold and silver deposits (Locke and De Ronde, 1987; Irvine and Smith, 1990; Grauch et al., 1995; John et al., 1999; Morrell, 2004). Integrated magnetic and radiometric surveys have been particularly useful in exploring gold mineral deposits (Allis, 1990; Irvine and Smith, 1990; Feebrey et al., 1998). Gold minerals often occur in zones of quartz adularia-sericite-pyrite alteration, surrounded by propylitic and argillic alteration zones in low-sulfidation epithermal systems (Christie and Brathwaite, 2003; Yuningsih et al., 2012; Andreeva et al., 2013). Gold and silver mineralized zones are also found in caldera-type geological structures (Groves et al., 2003; Robert et al., 2005; Andreeva et al., 2013).

In western Turkey, the Ulubey area of Usak features two calderas: the Gedik caldera to the south and the Kısladag caldera to the north. Ercan et al. (2014) studied the Kısladag caldera, located southwest of Kısladag Hill. This site is approximately 1.5 to 2 km in length (north-south) and 3 km in width (east-west). A combination of geological, geochemical, and geophysical methods was used to explore the regional geologic framework and identify favorable areas for mineralization. Volcanic hills in the center of the Kısladag caldera are of particular interest. Methods such as magnetics, magnetic total field gradient, radiometry, microgravity, and microgravity gradient were used to investigate the terrain and boundaries of gold and silver deposits in the area. The mineralized zones observed were characterized by low magnetization, low radiation, and high-density contrast. These zones were found beneath a magnetized cap rock. Drill hole locations were determined at varying depths. After completing the geophysical and geochemical exploration, an open-cut mine was established. This mine produces about 5 to 6 tons of gold annually, with depths ranging from 50 m to 225 m.

3.1.10 Marine gravity

On ships, gyrostabilized platforms are necessary because external accelerations, such as the vessel's roll, pitch, and yaw, can reach up to 100,000 mGal. Long-term averaging and damping suspension systems help reduce vertical acceleration errors to 1 mGal. However, cross-coupling errors occur in beam-supported mass instruments, as the circular vertical motion creates torque that doesn't average out during the up-and-down cycle of floating over waves. The Bell marine gravimeter solves this problem by being axially symmetric, eliminating cross-coupling. It can detect anomalies with wavelengths of 1–2 km, but it is expensive.

3.1.11 Airborne gravity

Airborne gravity measurements face more challenges than those on moving platforms. The best resolution currently achievable is around 0.1 mGal. The Eotvos effect can cause errors of 18 to 25 mGal at speeds of 200 knots ($\approx 370\text{km}/\text{hr}$) if there is a 1% error in velocity or heading, so autopilots are necessary. Since 1998, airborne instruments have improved to the point where they can acquire data with enough resolution for resource-scale projects. One example is the use of airborne gravity mapping to study salt domes for petroleum exploration. More information on this can be found by searching online using keywords such as "gravity," "airborne," and "petroleum exploration."

Airborne gravity gradiometry is a new frontier in gravity data acquisition. In the early 2000s, BHP, one of the world's largest mining companies, developed the first airborne gravity gradiometer for mineral exploration. Satellite-derived gravity is also used to precisely characterize the shape of Earth and other planets.

An example of a field procedure that uses aircraft but is not airborne gravity is helicopter-borne long-line systems. This method involves an instrument suspended from a long wire, which is placed on the ground for each measurement. It offers several advantages: no need for cut lines, easier handling of rugged terrain, and smoother handling of instruments compared to a crew and truck. Additionally, helicopter-borne GPS is accurate and works well in forested or rugged areas. The helicopter's position is tracked with differential GPS, and ground-reflecting laser altimeters measure the distance between the helicopter and the instrument. Laser range-finding helps achieve more accurate terrain measurements at each station.

3.2 Airborne gravity gradiometry in mineral deposit exploration

3.2.1 Introduction

As global demand for mineral resources continues to rise due to population growth, urbanization, and industrialization, the mining industry must discover more resources in the Earth's crust to meet the need for metals and raw materials. However, the discovery of new ore deposits, particularly through greenfield exploration, has been declining in recent years due to more challenging conditions, such as thick overburden, isolated districts, declining ore grades, and social and environmental issues. As a result, there is an increased need for technological innovation in mineral exploration to ensure the sustainability of non-renewable resources.

Recent breakthroughs in exploration technologies have highlighted the intrinsic value of these advancements. Specific examples of these technological innovations include induced polarization (IP), airborne electromagnetics (AEM), airborne gravity gradiometry (AGG), spectral remote sensing, global navigation satellite systems (GNSS), unmanned survey platforms, and neural networks/deep learning (Okada, 2022). Initially, Lockheed Martin Corporation, a defense subcontractor for the United States Government, focused on the defense applications of gravity gradiometry. However, there is now growing interest in applying this advanced technology to strategic commercial applications. Fig. 3.21 illustrates the various uses of gravity gradiometry in different fields, reflecting its importance in geophysical mapping for mineral exploration (Veryaskin, 2018).

There are two commercially available airborne gravity gradiometry (AGG) instruments derived from the original full tensor gradient (FTG): the FTG developed by Lockheed Martin

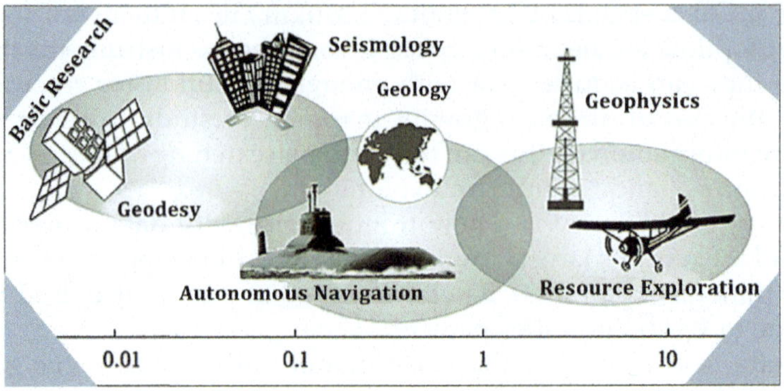

FIG. 3.21 Gravity gradiometry applications compared on the sensitivity scale that is required to be reached for a particular application. Those on the far left are 1000× more sensitive than those on the far right. *Source: Modified.*

(USA) and the FALCON developed by BHP Billiton (Australia). To operate in small survey aircraft for mineral exploration, BHP Billiton, in collaboration with Lockheed Martin, developed the FALCON system. This system uses a larger diameter gravity gradient instrument (GGI) with its axis of rotation nearly vertical (van Leeuwen, 2000; Lee, 2001). This development is considered the first breakthrough in AGG technology.

Before AGG, airborne gravity was explored for resource detection. Airborne gravity requires an additional non-inertial sensor, such as GPS, to separate platform acceleration from gravity measurements (Bruton, 2000). van Kann, 2004 showed that even the giant Olympic Dam deposit was below the instrumental noise level of airborne gravity instruments, meaning it was undetectable. However, all major mineral deposits are well above the noise level of AGG systems. Thus, the introduction of AGG marked a significant breakthrough, as it allowed for the detection of density anomalies caused by mineral deposits from the air for the first time. This technological advancement has played a crucial role in potential field methods used for geophysical mapping in mineral exploration.

Gravity gradiometry is used by oil and mineral prospectors to measure subsurface density by detecting the rate of change in gravitational acceleration caused by underlying rock properties. This data helps build a map of subsurface anomalies, which can then be used to more accurately locate oil, gas, and mineral deposits. It is also used to map water column density, which helps locate submerged objects and determine water depth (bathymetry).

Physical scientists use gravimeters to determine the exact size and shape of the Earth and to support gravity compensation in inertial navigation systems. >2.5 million-line kilometers of airborne and marine surveys have been conducted using this technique (DiFrancesco et al., 2009). These surveys highlight gravity anomalies linked to geological features like salt diapirs, fault systems, reef structures, and kimberlite pipes. Other applications of gravity gradiometry include tunnel and bunker detection (Martin, 2013) and the recent Gravity Field and Steady-State Ocean Circulation Explorer (GOCE) mission, which aims to improve the understanding of ocean circulation.

In addition, spaceborne gravity gradiometers have been under development for many years. The success of the GOCE space mission (van der Meijde et al., 2015) has sparked new ideas and approaches for gravity gradiometry in exploration on the Moon (Carroll, 2011),

Mars (Cuperus et al., 2009), and other planets. Compact and ultra-sensitive gravity gradiometers designed for space missions are expected to be available soon for both national space agencies and private space enterprises, advancing the potential field methods in planetary exploration.

3.2.2 Theoretical aspects of gravity gradiometry survey

Gravity gradiometry is the study of variations (anomalies) in the Earth's gravity field by measuring the spatial gradient of gravitational accelerations. The gradients of the Earth's gravity field can be estimated from a simple map of gravitational acceleration (g_z) by calculating the first and second horizontal derivatives. In some cases, it is beneficial to measure all three gradients (g_z, g_x, g_y). The gravity gradient represents the rate at which gravitational acceleration changes over space. Gravitational acceleration, g, is the gradient of the gravitational potential, φ, as shown below. This technique plays a key role in potential field methods used in geophysical mapping for mineral exploration and other applications.

$$g = \nabla\varphi = \left(\frac{\partial\varphi}{\partial x}, \frac{\partial\varphi}{\partial y}, \frac{\partial\varphi}{\partial z}\right) = (g_x, g_y, g_z) \tag{3.37}$$

The gradient of the acceleration, g, gives the gravity gradient tensor field, T, (Telford et al., 1990) defined as

$$T = \nabla g = \begin{bmatrix} \frac{\partial g_x}{\partial x} & \frac{\partial g_x}{\partial y} & \frac{\partial g_x}{\partial z} \\ \frac{\partial g_y}{\partial x} & \frac{\partial g_y}{\partial y} & \frac{\partial g_y}{\partial z} \\ \frac{\partial g_z}{\partial x} & \frac{\partial g_z}{\partial y} & \frac{\partial g_z}{\partial z} \end{bmatrix} \tag{3.38a}$$

Or

$$T = \nabla g = \nabla\nabla\varphi = \begin{bmatrix} \frac{\partial^2\varphi}{\partial x^2} & \frac{\partial^2\varphi}{\partial x\partial y} & \frac{\partial^2\varphi}{\partial x\partial z} \\ \frac{\partial^2\varphi}{\partial y\partial x} & \frac{\partial^2\varphi}{\partial y^2} & \frac{\partial^2\varphi}{\partial y\partial z} \\ \frac{\partial^2\varphi}{\partial z\partial x} & \frac{\partial^2\varphi}{\partial z\partial y} & \frac{\partial^2\varphi}{\partial z^2} \end{bmatrix} = \begin{bmatrix} T_{xx} & T_{xy} & T_{xz} \\ T_{yx} & T_{yy} & T_{yz} \\ T_{zx} & T_{zy} & T_{zz} \end{bmatrix} \tag{3.38b}$$

The gravitational potential is a smooth function that satisfies Laplace's equation in areas without sources. As a result, the gravity gradient tensor is symmetric and traceless. The smooth nature of the potential leads to symmetry, meaning $T_{xy} = T_{yx}$, $T_{xz} = T_{zx}$, and $T_{yz} = T_{zy}$. The Laplacian property results in a zero trace, meaning $T_{xx} + T_{yy} + T_{zz} = 0$ (Blakely, 1995). Therefore, there are only five independent components (T_{xx}, T_{xy}, T_{xz}, T_{yy}, T_{yz}) at any observation point.

In theory, the maps of these five components on the observation surface are linearly dependent, and one component contains all the necessary information. However, due to the discrete nature of measurements and the presence of noise in the data, the additional component maps still provide useful information. In practical applications, as many components as possible are used to maximize the available data, enhancing the accuracy of geophysical mapping for mineral exploration and other applications.

Gravity gradiometry is a valuable alternative to traditional vertical gravity (g_z) measurements in certain applications. The commonly sparse collection of vertical gravity measurements limits the lateral resolution and the ability to fully characterize a target or structure. Gravity surveys are now a standard part of exploration for hydrocarbons and minerals. Both scalar gravity and gravity gradient surveys are commonly conducted on land, sea, and in the air.

Compared to ground and airborne gravity, gravity gradiometry can potentially provide a more detailed characterization of the target body. The increased data density makes gravity gradiometry preferable to ground gravity surveys. It also has a higher signal-to-noise ratio, making it superior to airborne gravimetry. Gravity gradient surveys offer improved lateral resolution compared to gravity surveys (either ground or airborne) at the same observation locations because of the contribution from the horizontal components, which show how the gravity field changes laterally.

The high-frequency content in gravity gradient measurements results from the decay of the field, which is inversely proportional to the distance cubed. In contrast, gravity measurements have more signal power at low frequencies, making them more sensitive to regional signals and deeper sources. Table 3.6 and Fig. 3.22 compare the g_z and T_{zz} responses from a point source. The unit of measurement for gravity gradient is the Eotvos (E_o), named after Lorand Eotvos. One E_o equals 0.1 mGal/km (10^{-4} mGal/m), or 10^{-9} s^{-2}. Similar to gravity data, gravity gradiometry data requires several reduction steps to remove contributions unrelated to the underlying geological signal. The gravity gradient has units of square reciprocal time, expressed as s^{-2}. This technique plays a crucial role in advancing potential field methods for geophysical mapping in mineral exploration.

TABLE 3.6 Comparison to gravity g_z and G_{zz}.

	Gravity (g_z)	Gravity gradient (T_{zz} or G_{zz})
Signal	$\frac{GMz}{(r^2+z^2)^{\frac{3}{2}}} \times 10^5$ (mGal)	$\frac{GM(r^2-2z^2)}{(r^2+z^2)^{\frac{5}{2}}} \times 10^9$ (Eo)
Peak signal (r = 0)	$\frac{GM}{z^2} \times 10^5$	$\frac{2GM}{z^3} \times 10^9$
Full width at half maximum	1.53 z	≈z
Wavelength (λ)	3.07 z	2z

Source: Markcsc (2011).

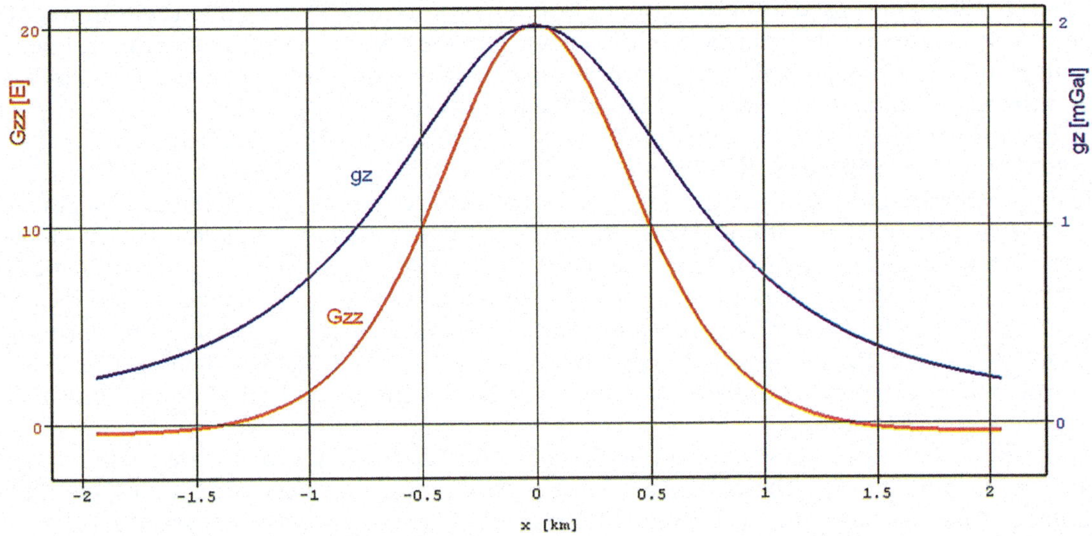

FIG. 3.22 Vertical gravity and gravity gradient signals from a point source buried at 1 km depth. *Source: From Wikimedia Commons, the free media repository, Author Markcsc's own work, 2011. https://commons.wikimedia.org/wiki/File:Gravity-and-gravity-gradient-signals2.jpg.*

The decay rate generally causes a higher signal contribution from shallower targets and a smaller signal contribution from deeper ones. As a result, gravity gradiometry provides better resolution for near-surface features compared to vertical gravity measurements. This makes it a valuable tool in geophysical mapping for mineral exploration, offering enhanced characterization of shallow geological structures.

3.2.3 Current gravity gradiometers

The use of gravity gradiometry in the petroleum industry has steadily increased, but its major breakthrough came in onshore mineral exploration. By October 2005, Falcon systems had flown 1 million-line kilometers of surveys, almost exclusively for mineral exploration. Since then, many more kilometers have been flown, building on the data from Air-FTG and FTGe systems in operation. Airborne gravity gradiometry has become a key method in mineral exploration, allowing explorers to generate high-resolution maps that show both lateral and vertical changes in rock density. These maps are often used as qualitative geological maps to identify geological units, faults, and contacts. Advances in computer technology have also enabled sophisticated numerical inversion programs, which create simulated geological models of the Earth. Below are some examples of mineral exploration targets that gravity gradiometry can help detect.

The primary motivation for using gravity gradiometers in mineral exploration was diamond exploration. Kimberlite pipes, which intrude surrounding rocks, often preserve low-density portions at shallow depths. These can be easily detected as negative gravity anomalies. While conventional ground-based gravimeters can measure these anomalies, BHP

Billiton in the 1990s used airborne gravity gradiometry to detect kimberlite pipes in remote and swampy regions of northern Canada. This technology was soon applied in other regions with difficult terrain, as well as in more accessible areas where the speed and convenience of airborne acquisition provided a competitive advantage (Veryaskin, 2018).

In addition to diamonds, gravity gradiometry is effective for detecting iron oxide copper-gold (IOCG) mineralization. IOCG deposits have diverse geophysical signatures, particularly in their magnetic and gravity responses. For example, the Olympic Dam deposit has a gravity anomaly of about 17 mGal but only a 1000 nT magnetic anomaly, while the Ernest Henry deposit has a 2 mGal gravity anomaly and a 7000 nT magnetic anomaly (Rutter and Esdale, 1985; Esdale et al., 1987, 2003; Reeve, 1990). A combination of airborne magnetics and gravity gradiometry is an effective strategy for targeting IOCG deposits. In addition to verifying known IOCG deposits, a significant new discovery was made in the Candelaria copper belt in Chile, through a strategic alliance between Far West Mining and BHP Billiton (Dransfield and Milkereit, 2007).

Massive sulphide nickel-copper deposits, often found in dense host rocks, are another target for gravity gradiometry. These deposits contain very dense sulphides, which are detectable using gravity measurements. The ability to conduct gravity surveys from an airborne platform makes it possible to detect nickel deposits in remote, difficult-to-access locations. Similarly, volcanogenic massive sulphide (VMS) copper and gold deposits, which consist of dense sulphides, can be detected using gravity gradiometry. The first airborne gravity gradiometry survey was flown over the Bathurst Camp in Canada, a region rich in VMS deposits.

Gravity gradiometry is also an excellent tool for iron ore exploration, where massive haematite and banded iron deposits are often denser than the surrounding host rocks. This technique has been successfully applied to iron ore exploration in Michigan (USA), South Australia, Western Australia, Brazil, and other locations. Similarly, lead and zinc deposits, often found in dense ore bodies within low-density sedimentary rocks, are ideal targets for gravity gradiometry. For example, the Cannington silver-lead-zinc deposit in Australia was identified through Falcon anomalies, and there is potential for discovering other large deposits like the Broken Hill lead-zinc mine (Veryaskin, 2018).

Gravity gradiometry can also detect low-density anomalies, such as coal, which has a very low density compared to surrounding rock formations. This technique has been used to search for coal, as well as petroleum, in southeast Australia and other regions. Additionally, paleochannels, which can host economic gold or hydrocarbon deposits, present low-density features that are well-suited for detection using gravity gradiometry.

Another key advantage of gravity gradiometry is its ability to map geological features, especially those hidden beneath surface cover. The ability to map faults, weathering profiles, and lithologies is particularly valuable in remote or rugged terrains where traditional mapping is challenging or impossible. This technique has proven useful in gold exploration in weathered greenstone belts in Western Australia and West Africa, as well as in porphyry copper exploration in regions like Mongolia and the Andes.

In summary, many mineral and petroleum deposits have characteristics such as contrasting densities that can be detected through gravity measurements. While gravity measurements have been in use for over a century, gravity gradiometry has expanded the capabilities of this technique, allowing surveys to be conducted more quickly and efficiently over large areas, including over water and in rugged or remote terrain. Gravity gradiometry, along with other geophysical, geological, and geochemical methods, is an important tool for mineral and hydrocarbon exploration. Continued advances in technology will further enhance the sensitivity, resolution, and effectiveness of gravity gradiometry, making it even more valuable for future exploration projects.

3.2.3.1 *Platforms*

The development of operational gravity gradiometers began in the 1980s by Bell Aerospace (now Lockheed Martin) for various applications. A key project between BHP (now BHP Billiton) and Lockheed Martin led to the creation of the Airborne Gravity Gradiometer (AGG), based on Lockheed's advanced technology at the time. In October 1999, Sander Geophysics conducted the world's first airborne gravity gradiometer survey for BHP Billiton at the Bathurst Camp in New Brunswick. Since that first survey, the number of active gravity gradiometer systems has increased. These systems have been used in both oil and gas surveys, including marine applications, but the majority have been focused on mineral exploration. Airborne gravity gradiometers have proven valuable for both direct detection and geological mapping of a wide range of mineral deposits.

The choice of aircraft plays an important role in the success of these surveys. Dransfield (2007) highlighted that the dynamics of the aircraft directly affect both noise levels and survey productivity due to the limited ability to reject aircraft-induced motion. Other factors also depend on the aircraft chosen. For instance, spatial resolution is inversely related to the aircraft's speed, meaning slower aircraft provide better resolution. Additionally, the gravity gradient varies inversely with distance, so flying at a lower altitude improves signal strength. Operational and safety requirements also impose a minimum speed for aircraft. Airborne gravity gradiometers can be operated from different types of platforms, including fixed-wing aircraft like Cessna Grand Caravans, airships such as the Zeppelin LZ N07, and helicopters like the Eurocopter AS350-B3. Each platform is suited for specific conditions, depending on the terrain and survey requirements.

3.2.3.1.1 Fixed-wing platform

Most airborne gravity gradiometry surveys using Falcon, Air-FTG, and FTGeX systems are conducted in single-turbine-engine Cessna Grand Caravans. These aircraft typically fly at a survey ground speed of 55–65 m/s (~120 knots) with a ground clearance of 80–100 m in gentle terrain. The Cessna Grand Caravan is the most cost-effective option among the three platforms, providing the fastest coverage of large survey areas. It is also capable of carrying any of the current gravity gradiometers.

3.2.3.1.2 Airship platform

One of the Air-FTG systems has been used for surveys in Botswana for De Beers, using the Zeppelin airship, as described by Hatch et al. (2006a). The airship flies at 16 m/s (32 knots) with a ground clearance of 80 m. Due to the high elevations and temperatures in Botswana, the airship's lift capability is limited, meaning surveys are only conducted at night. The airship can carry any gravity gradiometer, and its main advantage is its low acceleration levels due to high inertia, which leads to minimal dynamic gradient noise.

3.2.3.1.3 Helicopter platform

The Falcon system has also been installed in a Eurocopter AS350-B3 helicopter (Boggs et al., 2007). This helicopter flies at 30 m/s (60 knots) and has typical ground clearances ranging from 25 m to 60 m. Like the airship, the helicopter is fully laden when carrying a gravity gradiometer, which limits its use over high-elevation terrain in hot temperatures. However, successful surveys have been completed in the Canadian Arctic, including surveys with a frequency domain EM system. The AS350-B3 helicopter can only carry the lightweight digital AGG, as other gravity gradiometers are too heavy. The key advantage of the helicopter platform is its ability to fly lower and slower, offering higher resolution and sensitivity, making it ideal for detailed mapping of small, near-surface features.

3.2.3.1.4 Cross-platform comparison

It is useful to compare the performance of gravity gradiometers across different platforms that travel at varying speeds. Murphy et al. (2007) propose using noise power density in the wavenumber domain to compare performance. This method involves squaring the noise density and multiplying it by the survey speed. A comparison using data from Murphy et al. (2007), Hatch et al. (2006b), and Boggs et al. (2007) is provided for the Air-FTG systems in both a Caravan and an airship, as well as the Falcon systems in a Caravan and a helicopter (Fig. 3.23). The comparison does not account for the helicopter's advantage of flying lower, nor does it include operational, safety, and cost considerations (Dransfield, 2007).

The Air-FTG system is generally the most sensitive to aircraft motion, leading to higher error and variability when mounted on a fixed-wing aircraft. However, this sensitivity is fully compensated for when the system is mounted on an airship, which provides a very stable and slow platform. The Falcon system, on the other hand, is much less sensitive to aircraft motion and performs better than the Air-FTG in the same fixed-wing aircraft. The heliborne

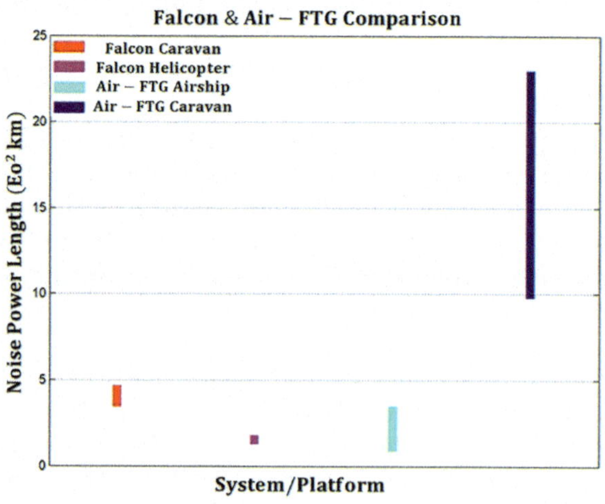

FIG. 3.23 Comparison of error levels across four airborne gravity gradiometer installations in terms of noise power density. This comparison is presented in wavelength form, derived by dividing the bandwidth by the aircraft velocity, to directly relate to the signals of interest. A lower noise power density indicates better sensitivity and resolution. *Source: Modified.*

Falcon system has a noise power density similar to the airship-borne Air-FTG. Additionally, the helicopter-based Falcon has the advantage of flying lower than the other platforms, providing much greater sensitivity to near-surface geology.

3.2.4 Applications of airborne gravity gradiometry in mineral exploration

Airborne gravity gradiometry has become a vital tool in the exploration of various mineral commodities and deposit types. It is utilized both for direct detection of mineral deposits and for enhancing geological mapping. This section provides an overview of some key applications, highlighting their significance in advancing geophysical mapping techniques for sustainable exploration and management.

3.2.4.1 Targeting coal exploration in Latrobe Valley, Australia, using airborne gravity gradiometry

Airborne gravity gradiometry has proven effective in mapping coal seams in the Latrobe Valley, located in southeast Australia. Mahanta (2003) describes how the coal seam is identified by a vertical gravity gradient low, which diminishes where the seam is exposed along its southern edge. As the seam dips gently northwest beneath gravel cover, the gravity signal decreases gradually. Coal seams with thicknesses ranging from 30 m to 50 m and dips slightly below 10° are typically detectable. The Falcon AGG system can successfully identify coal seams greater than 10 m thick at dips exceeding 10°, as demonstrated by Mahanta (2003). This technique offers significant advantages in advancing mineral exploration, contributing to sustainable geophysical mapping and resource management.

3.2.4.2 Targeting copper (IOCG) exploration in Candelaria, Chile, using airborne gravity gradiometry

Airborne gravity gradiometry has proven to be an effective method for detecting Iron Oxide Copper-Gold (IOCG) mineralization, exemplified by the Olympic Dam deposit. Falcon AGG surveys have successfully identified significant deposits such as Ernest Henry and Prominent Hill, and similar surveys have been conducted in regions like the King George Falcon survey and the Wernecke Mountains of Yukon. Notably, the Candelaria copper belt in Chile became a focal point for exploration through this technique, leading to the discovery of the Santo Domingo Sur deposit.

The Candelaria Project, initiated in 2002 through a strategic alliance between Far West Mining and BHP Billiton, focused on exploring the highly prospective IOCG deposits in northern Chile's Candelaria copper belt, spanning nearly 1200 km from Santiago to Antofagasta. In late 2002, a comprehensive 10,700-line-km Falcon airborne gravity gradiometry survey was conducted, covering 5145 sq km across eight blocks of the belt. This survey successfully identified over 70 target areas with distinct gravity anomalies.

From February 2003 to May 2005, reverse circulation drilling was conducted on 18 target areas, leading to the discovery of IOCG mineralization in three of them. The first drilling results, announced in July 2003, revealed a 60 m interval with 2.5% copper and 0.33 g/t (grams per ton) gold, confirming the effectiveness of gravity gradiometry in locating valuable deposits. The discovery of the Santo Domingo Sur deposit, located in the southern part of the survey area, proved to be even more significant. By May 2005, Far West Mining completed its 100% earn-in on the project, with BHP Billiton retaining a 2% net smelter return royalty. This case highlights the critical role of airborne gravity gradiometry in advancing mineral exploration in challenging terrains and fostering the discovery of major mineral deposits.

3.2.4.3 Targeting diamond exploration (kimberlites and palaeochannels) using airborne gravity gradiometry

Airborne gravity gradiometry has emerged as a vital tool in diamond exploration, particularly for identifying primary sources of diamonds such as kimberlites and lamproites, which occur independently of surface geology. While aeromagnetic and airborne electromagnetic methods have traditionally been used in diamond prospecting, the availability of high-resolution airborne gravity data since 1999 has greatly increased the demand for gravity gradiometry in diamond exploration. By October 2005, BHP Billiton's Falcon systems had flown over 1 million-line kilometres, with more than half of these dedicated to diamond exploration. Similarly, Bell Geospace's Air-FTG systems have conducted numerous surveys for diamonds, with one system committed to diamond exploration in Botswana.

In addition to primary diamond sources, airborne gravity gradiometry can also be applied to detect secondary sources, such as palaeochannel deposits. These deposits, formed by ancient river systems, often have distinct density contrasts with the surrounding host rocks, making them detectable using gravity gradiometry techniques. Here are some notable examples of diamond discoveries facilitated by airborne gravity gradiometry:

- **South Africa**: In 2001, a Falcon survey west of Kimberley identified a palaeochannel interpreted to be draining the Finsch Diamond Mine. A joint venture with Tawana Resources commenced drill-testing the gravels in the channel, revealing that significant quantities of diamond-bearing gravels had travelled downstream from the heavily eroded Finsch kimberlite. The survey also identified several new kimberlites (Tawana Resources NL, Annual Report, 2006).
- **Canada**: A Falcon survey in the vicinity of the Ekati mine led to the discovery of the W09 crater facies kimberlite. Drilling of a Falcon-generated target resulted in the recovery of micro-diamonds from this kimberlite, which has a surface expression measuring 100 m by 200 m (Dransfield and Walker, 2005).
- **Australia**: In northwest Australia, two new kimberlite pipes, Persephone and Niobe, were discovered near the Aries pipe, following airborne gravity gradiometry surveys (Isles and Moody, 2004). Additionally, in January 2005, Gravity Diamonds in Melbourne, Australia, announced the discovery of the diamondiferous Abner kimberlite, traced to an airborne gravity gradient anomaly (Dransfield and Walker, 2005).
- **Ekati Mine, Canada**: A survey of the BHP Billiton Ekati tenement in northwest Canada, which included 136 known kimberlite pipes at the time, found that 55% of the kimberlites were identified as anomalies in the gravity gradiometer data. The survey led to the discovery of three new kimberlites, including the diamondiferous Impala pipe (Dransfield and Walker, 2005).
- **Feynman System at Ekati**: Following successful tests in 2006 at Bulgary Ridge, New York (Boggs et al., 2007), the helicopter-borne digital AGG system known as Feynman began production surveying at Ekati. This system successfully detected over 90% of the

known kimberlite pipes in areas previously surveyed by Falcon, showcasing its effectiveness in identifying diamond deposits (Dransfield and Walker, 2005).

These examples demonstrate the substantial contribution of airborne gravity gradiometry to diamond exploration, allowing for more precise identification of both primary kimberlite and secondary palaeochannel diamond sources, and enhancing exploration in regions with established deposits.

3.2.4.4 Targeting copper-zinc (VMS) deposits using airborne gravity gradiometry in exploration

Volcanogenic massive sulphide (VMS) deposits are recognized for their significant gravity response, primarily due to the high density of the mineralized sulphide bodies (Walker and Mannard, 1974; Grant and West, 1965; Fritz and Sheehan, 1984). This distinctive gravity signal has made ground gravity surveys a standard tool for detecting VMS deposits. However, airborne gravity gradiometry has emerged as an advanced technique that offers even higher sensitivity and resolution, making it increasingly valuable in VMS exploration.

The first airborne gravity gradiometry survey targeting VMS deposits was flown over the Bathurst Camp in northern New Brunswick, Canada, which includes the Heath Steele and Stratmat deposits (Dransfield et al., 2001a). This ground-breaking survey demonstrated the potential of airborne gravity gradiometry for detecting VMS deposits, particularly in areas where ground gravity surveys had already been utilized.

The Stratmat deposit, characterized by narrow mineralized lenses associated with gabbroic intrusives, showed that much of the gravity signal could be attributed to the dense intrusives rather than the mineralization itself. In contrast, the Heath Steele deposit, with its larger volume of mineralization, produced a gravity signal more directly related to the VMS deposit. This clear correspondence between gravity highs and mineralized zones was evident in the survey data (Dransfield, 2007), highlighting the capability of airborne gravity gradiometry to distinguish between different geological features and refine exploration targets.

Through the use of airborne gravity gradiometry, exploration for VMS deposits has been significantly enhanced. The technique allows for a high-resolution map of subsurface density variations, helping geologists to more accurately locate and evaluate VMS mineralization, even in challenging terrains. This advancement in geophysical methods has played a crucial role in improving the efficiency and success of VMS exploration, contributing to more sustainable exploration practices and the development of new mineral resources.

3.2.4.5 Targeting other minerals using airborne gravity gradiometry survey

Airborne gravity gradiometry has proven to be an effective tool for exploring a variety of mineral types and deposit styles beyond the traditional targets of oil and gas. Its high sensitivity to density contrasts makes it especially useful for identifying dense mineral deposits, such as iron ore, copper, gold, and other base and precious metals.

Iron Ore Exploration: Airborne gravity gradiometry has demonstrated its utility in iron ore exploration, particularly for massive haematite deposits, which present strong gravity signals due to their high density. For instance, the *Republic Falcon* survey in northwest Michigan, USA, successfully mapped banded iron formations using gravity gradients (Dransfield et al., 2001b). Similarly, a Falcon survey over the Middleback Ranges in South

Australia provided clear delineations of iron mineralization (Lee et al., 2001). The *Air-FTG* gravity gradiometer has also been employed in Brazil's *Quadrilatero Ferrifero*, mapping structures associated with iron ore mineralization (Mataragio et al., 2006).

Base and Precious Metals: Airborne gravity gradiometry has further applications in detecting base metals like zinc, lead, and copper, as well as precious metals such as gold. The Cannington silver-lead-zinc deposit was detectable through airborne gravity gradiometry, as shown by Christensen et al. (2001). A detailed survey over the *Broken Hill* lead-zinc mine also demonstrated the ability of gravity gradiometry to detect mineralization at depth, leading to new zinc discoveries at the Goldfinger target (Lane, 2006; Anderson et al., 2006).

A model study of gold deposits in the greenstone belts of Western Australia's Yilgarn Craton suggested that gravity gradiometry could be particularly useful for detecting the low-density weathered zones commonly associated with these gold deposits (Bayat, 2007). Additionally, Glass Earth applied the *Air-FTG* system in New Zealand as part of an airborne geophysical mapping program for gold exploration.

Nickel Exploration: In the *West Musgraves* region of Western Australia, Dransfield (2007) demonstrated the use of airborne gravity gradiometry to map gabbronorite intrusions that host massive nickel-sulphide (NiS) mineralization at the Neebo-Babel deposits. Since the magnetization of these intrusives is weak and variable, gravity gradiometry provides a distinct advantage in detecting these structures, especially in areas with difficult ground access.

Porphyry Copper and Other Targets: BHP Billiton has also used Falcon systems for regional mapping in Mongolia as part of a porphyry copper exploration program (BHP Billiton Annual Review, 2006). These examples highlight the flexibility of airborne gravity gradiometry in identifying diverse mineralization types across different geological settings. The ability to detect both high-density ore bodies, such as iron and nickel, and subtle variations in mineralized zones makes it a powerful tool for advancing mineral exploration and resource management.

3.2.4.6 *Targeting geothermal exploration using airborne multigeophysics: Kutcharo caldera, Teshikaga area, Japan*

Ishikawa et al. (2020) presented a significant advancement in regional geothermal exploration through the application of multiple geophysical techniques in the Teshikaga area, Japan. Their study utilized data from airborne gravity gradiometry (AGG) with the HeliFalcon system, as well as airborne electromagnetic and magnetic data from the Helitem system. The combined use of these methods allowed for a detailed exploration of geothermal resources beneath the surface.

One of the key findings of the study was that 3D inversion of AGG data enabled a precise interpretation of the deep subsurface structure, revealing geological features such as intrusions and calderas, which are often associated with geothermal activity. Additionally, the resistivity and magnetism data from the relatively shallow subsurface highlighted a conductive anomaly that overlapped with a low magnetic anomaly, which suggested hydrothermal alteration related to geothermal resources. This ability to identify areas of hydrothermal alteration is crucial in geothermal exploration, as these areas often indicate the presence of valuable geothermal reservoirs.

The findings by Ishikawa et al. (2020) also have important implications for hydrothermal deposit exploration, as many of the features of geothermal resources overlap with those found in hydrothermal systems. The ability to detect these anomalies with airborne geophysical techniques offers significant advantages for exploring both geothermal and hydrothermal deposits.

A major breakthrough in the application of AGG was the introduction of the combined use of airborne gravity and gravity gradiometry systems. Traditional AGG systems are highly effective for detecting small-scale gravity anomalies, such as those caused by mineral deposits. However, they are limited when it comes to capturing long-period gravitational field changes. To overcome this limitation, the full-spectrum FALCON system, which combines AGG with airborne gravity, was developed for practical use. This combination enhances the ability to detect both small and large-scale anomalies, significantly improving the sensitivity and resolution of geophysical surveys (Chen et al., 2017).

In conclusion, the integration of multiple airborne geophysical techniques, including AGG, electromagnetic, and magnetic data, has proven highly effective for geothermal exploration in Japan's Teshikaga area. The combination of AGG with airborne gravity systems marks a significant step forward in geophysical survey capabilities, offering a comprehensive approach for detecting both geothermal and hydrothermal resources.

3.2.4.7 Direct detection of buried ore bodies using airborne gravity gradiometry: A case study in South Africa

The routine application of gravity gradiometry for the detection of buried orebodies, which are not exposed at the surface, is now becoming a common practice in the mineral exploration industry. This method is being used across a variety of commodities and orebody types, ranging from banded iron ore formations to volcanogenic massive sulfide (VMS) style polymetallic deposits. The innovative approach of airborne gravity gradiometry was first pioneered by BHP Billiton and Lockheed Martin through the development of the Falcon technology (Lee, 2001). This advancement has led to the deployment of additional systems based on Lockheed Martin instruments, with their capabilities expanding over time.

In South Africa, efforts to enhance exploration technologies have extended beyond gravity gradiometry to include the development and deployment of magnetic gradiometry systems with similar objectives. These systems, such as those from the Institute of Photonic Technology (IPHT), are being used in conjunction with gravity gradiometry to improve the detection and mapping of subsurface mineralization. The integration of these two techniques allows for more comprehensive and accurate identification of potential ore deposits.

Recent developments have also seen a reduction in the detectable wavelengths from airborne gravity gradiometry systems. Over the past year, the detectable wavelengths have decreased to <200 m, and it is anticipated that by 2011, systems will be capable of detecting wavelengths below 100 m. Additionally, anomaly signal strengths of <2 Eotvos for gravity gradiometry will become viable, further enhancing the sensitivity and resolution of these surveys (Fitzgerald et al., 2011).

These advancements in geophysical techniques promise to significantly improve the effectiveness of mineral exploration, particularly in regions with complex or deeply buried ore bodies. As technology continues to evolve, the use of gravity gradiometry and magnetic gradiometry will play an increasingly important role in the discovery of new mineral resources, contributing to more efficient and sustainable exploration practices.

3.2.4.8 3D inversion of gravity gradiometry data for iron ore exploration in Minas Gerais, Brazil: A case study

Martinez et al. (2013) presented a case study applying 3D inversion of gravity gradiometry data to enhance iron ore exploration in Minas Gerais, Brazil. The high-density contrast of the ore bodies in this region generates distinct gravity anomalies that are well captured by airborne gravity gradiometry. The study focused on a 20 km^2 subarea within a larger survey, demonstrating the effectiveness of gravity gradiometry and inversion techniques in accurately characterizing mineral deposits.

The authors applied a stepwise inversion approach, starting with the inversion of the vertical gravity gradient component (T_{zz}), followed by joint inversion of T_{zz} along with the horizontal derivatives (T_{xz}, T_{yz}), and finally, the inversion of all five independent components of the gravity gradient tensor. They found that while the T_{zz} component alone was sufficient to produce geologically plausible and interpretable results, the inclusion of horizontal derivative components improved the resolution of the density model, resulting in more precise delineation of the iron ore formations.

The case study illustrates the significant potential of gravity gradiometry and 3D inversion techniques in enhancing mineral exploration, particularly for iron ore deposits. By utilizing multiple data components and advanced inversion algorithms, the study successfully demonstrated how airborne gravity gradiometry can be employed to accurately delineate ore bodies, contributing to more efficient and sustainable exploration practices.

3.2.5 Applications of potential field surveying in oil and gas exploration

While this book primarily focuses on global geophysical mapping for mineral exploration, it is valuable to briefly examine the applications of gravity gradiometry in oil and gas exploration. Airborne gravity gradiometry offers several key advantages over traditional methods, including significantly lower noise levels, higher resolution data, faster survey coverage, and reduced accessibility challenges compared to surface gravimetry. These attributes make gravity gradiometry particularly useful in areas where seismic data is limited or when mapping large areas to better target expensive seismic surveys.

Airborne gravity gradiometry has already demonstrated its effectiveness in various oil and gas exploration applications. For instance, Rose et al. (2006) reported the successful mapping of an Eocene channel in the Gippsland Basin, Australia's primary domestic oil source, using Falcon gravity gradiometry. Similarly, O'Brien et al. (2005) used FTG data in conjunction with seismic data to perform a 3D inversion, which accurately mapped the base of the K2 salt body in the Gulf of Mexico to depths of 7000 m. In the Papua New Guinea fold belt, a region with challenging terrain and limited surface access, Nelson et al. (2004) highlighted the successful use of airborne gravity gradiometry for structural mapping, proving its utility in remote and difficult environments.

Further examples include the Cliffs oil field in Western Australia, where gravity data clearly delineated a horst block controlling the oil accumulation (Dransfield, 2007). Additionally, in the Bonaparte Gulf off the northwest coast of Australia, Nexus conducted a Falcon survey that successfully mapped salt diapirs, a critical feature in hydrocarbon exploration (Dransfield and Walker, 2005). The survey not only confirmed the presence of a known salt diapir but also identified several other potential targets with similar gravity signatures (Dransfield, 2007).

These applications illustrate the important role that airborne gravity gradiometry plays in oil and gas exploration, helping to enhance exploration efficiency, reduce costs, and provide valuable geological insights for sustainable exploration and management.

3.3 Magnetic methods in advancing potential field surveying for sustainable exploration and management

The magnetic method involves studying the distribution of magnetic minerals in the Earth's upper crust. It is commonly used to estimate crustal thickness or to constrain temperatures within the crust, based on the Curie isotherm (the temperature at which minerals lose their magnetic properties) (Nabighian, 2005). The Curie point method can provide valuable insight into the presence of hot rock masses in the crust. When rocks are heated above a certain temperature, typically a few hundred degrees Celsius, they lose their ferromagnetic properties. Under favorable conditions, the depth to this demagnetization level can be determined with reasonable accuracy.

In geophysical exploration, magnetic measurements detect variations in the magnetic field caused by lateral changes in the magnetization of the crust. These variations can lead to anomalous regions, which indicate structural or lithological contrasts beneath the surface. The Earth's magnetic field induces magnetization in the crust's constituent rocks, with the direction of the induced magnetic vector typically aligning with the Earth's magnetic field. Magnetic data can be collected from the surface, from the air, or even from borehole environments.

Iron ore deposits are classified as strongly magnetic (e.g., magnetite and hematite from magmatic or hydrothermal origins, and contact metamorphic deposits) or weakly magnetic (e.g., hematite and siderite in hydrothermal filtration deposits). Magnetite and hematite are highly magnetic and are easily detectable using magnetic surveys. Similarly, manganese and chromium ores, along with minerals like pyrrhotite and ilmenite, are also highly magnetic and detectable through magnetic prospecting. On the other hand, weakly magnetic ore bodies generate low-magnitude anomalies, making them challenging to detect using magnetic surveys alone. In such cases, high-precision ground surveys and an integrated approach using multiple geophysical methods are essential.

Magnetometers are commonly used for a range of applications, including locating buried tanks and drums, identifying faults and geologic contacts, mineral exploration, geothermal exploration, mapping buried utilities and pipelines, and archaeological studies.

3.3.1 Basic theory

If two magnetic poles of strength m_1 and m_2 are separated by a distance r, a force, F, exists between them. If the poles are of the same polarity, the force will push the poles apart, and if they are of opposite polarity, the force is attractive and will draw the poles together. The equation for F is the following:

$$F = \frac{m_1 m_2}{4\pi\mu r^2} \tag{3.39}$$

where μ is the magnetic permeability of the medium separating the poles; m_1 and m_2 are pole strengths, and r is the distance between them.

3.3.2 Magnetic units

The magnetic flux lines between two poles per unit area is the flux density B (and is measured in weber/m^2 = Tesla). B, which is also called the "magnetic induction," is a vector quantity. The unit of Tesla is too large to be practical in geophysical work, so a sub-unit called a nanotesla (1 nT = 10^{-9} T) is used instead, where 1 nT is numerically equivalent to 1 gamma in c.g.s. units (1 nT = 10^{-5} gauss).

The magnetic field can also be defined in terms of a force field, which is produced by electric currents. This magnetizing field strength H is defined, following Biot-Savart's Law, as being the field strength at the centre of a loop of wire of radius r through which a current I is flowing such that $H = \frac{I}{2r}$. Consequently, the units of the magnetizing field strength H are amperes per metre (A/m). The ratio of the flux density B to the magnetizing field strength H is a constant called the absolute magnetic permeability ($\mu = \frac{B}{H}$) and expressed in weber/Am (or Henry/m). H simply describes how B is modified by the magnetic polarization (or magnetization).

The magnetic moment is the vector joining the two poles —m and +m at a distance l apart (M = mlr) and expressed in Am^2. The intensity of induced magnetization is defined as the magnetic moment per unit volume of material ($I = \frac{M}{V}$) and expressed in A/m. The intensity of magnetization induced in a rock is k times the Earth's magnetic field ($k = \frac{I}{H}$), where k is the susceptibility of the rock. Magnetic susceptibility k in the SI system is a dimensionless constant ($k_{SIunit} = 4\pi k_{cgsunit}$) and is a measure of the extent to which a material can be magnetised and is related to magnetic permeability (μ) by:

$$\mu = \mu_o (1 + k) \tag{3.40}$$

where μ_o is the magnetic permeability in vacuum and ($\mu_0 = 4\pi \times 10^{-7}$ weber/Am). The relative magnetic permeability is defined as ($\mu_r = \frac{\mu}{\mu_0}$). Ferromagnetic materials have the largest k and contain iron, nickel, cobalt, and many of their alloys. For example, magnetite is a ferromagnetic material with one of the highest magnetic susceptibility values. In environmental waste sites, the largest sources of magnetic anomalies are iron and steel.

3.3.3 Physical properties

The primary component of the Earth's magnetic field originates from convection of liquid iron in the outer core of the Earth, and the field strength is on the order of ~50,000 nT. Additionally, the Earth's magnetosphere is influenced by diurnal variations and solar winds. The remaining component of the Earth's magnetic field is due to magnetized materials in the upper Earth's crust.

$$1\,\text{Tesla} = 10^9\,\text{nT} = 10^4\,\text{gauss} = 10^9\,\text{gamma}$$

The magnetic susceptibility is the physical property that defines the magnetic characteristics of a rock, i.e., how easily the material can be magnetized. Magnetic susceptibility relies on the volume percent content of ferromagnetic minerals such as magnetite. Table 3.7 outlines typical magnetic susceptibilities for various rock types.

3.3.4 Remanent magnetization in advancing potential field surveying for sustainable exploration and management

When a magnetic material possesses relatively high susceptibilities, or when the inducing magnetic field is strong, the material may retain a portion of its induced magnetization even after the external magnetic field has dissipated. This retained magnetization is referred to as *remanent magnetization*. In the context of solid Earth geophysics, remanent magnetization is used to trace the movement of continents and ocean basins as a result of plate tectonics.

TABLE 3.7 Typical magnetic susceptibilities of rocks.

Rock type	Susceptibility (k)
Altered ultrabasics	10^{-4} to 10^{-2}
Basalt	10^{-4}
Gabbro	10^{-4} to 10^{-3}
Granite	10^{-5} to 10^{-3}
Andesite	10^{-4}
Rhyolite	10^{-5} to 10^{-4}
Metamorphic rocks	10^{-6} to 10^{-4}
Most sedimentary rocks	10^{-6} to 10^{-5}
Limestone and chert	10^{-6}
Shale	10^{-5} to 10^{-4}

Source: Naskar et al. (2020); Lichoro (2016).

The only reliable method to measure remanent magnetization is by extracting rock samples and conducting laboratory analysis, a process that is both time-consuming and costly. Due to this limitation, in exploration geophysics, it is often assumed that the observed magnetic field lacks a remanent magnetic component. While this assumption simplifies data interpretation, it may not always be accurate, and neglecting remanent magnetization can potentially lead to biased conclusions or misinterpretations.

3.3.5 Magnetic field nomenclature

In gravity studies, gravitational acceleration is measured in units of milligals (mGal) using a gravimeter, and it always acts in a downward direction. However, unlike gravity, the magnetic field does not have a straightforward direction of action. To address this, Earth scientists have developed specific terminology to describe the magnetic field at any given point on the Earth's surface. At any location on the surface, the magnetic field, denoted as F*, has both a strength and a direction. The following terms are used to characterize the direction of the magnetic field (see Fig. 3.24).

Declination (D)
The angle between north and the horizontal projection of F. This value is measured positive through east and varies from 0 to 360 degrees.

Inclination (I)
The angle between the surface of the Earth and F. Positive inclinations indicate F is pointed downward, negative inclinations indicate F is pointed upward. Inclination varies from −90 to 90 degrees.

Magnetic equator
The location around the surface of the Earth where the Earth's magnetic field has an inclination of zero (the magnetic field vector F is horizontal). This location does not correspond to the Earth's rotational equator.

Magnetic poles
The locations on the Earth's surface where the magnetic field has an inclination of either +90° or −90° (meaning the magnetic field vector F is vertical) do not coincide with the Earth's north and south poles. The vertical component of the Earth's magnetic field varies with latitude, ranging from about 30,000 nT at the magnetic equator to 60,000 nT at the magnetic poles, and 47,000 nT in Switzerland.

The magnetic field, F, includes contributions from several sources: The Earth's main magnetic field (the inducing field), the induced magnetization from crustal sources, and external sources such as temporal variations in the Earth's magnetic field. When measuring the magnetic field on the Earth's surface, we detect the magnetization caused by both the main field and induced fields.

The main magnetic field refers to the portion of the Earth's magnetic field generated within its core. It is the dominant part of the magnetic field and is responsible for inducing magnetization in the crustal rocks, which is of interest in exploration applications.

Fig. 3.25 illustrates a basic example of how magnetics relates to geology. Consider a buried dyke with a susceptibility of 0.001, surrounded by sedimentary rocks that have no magnetic

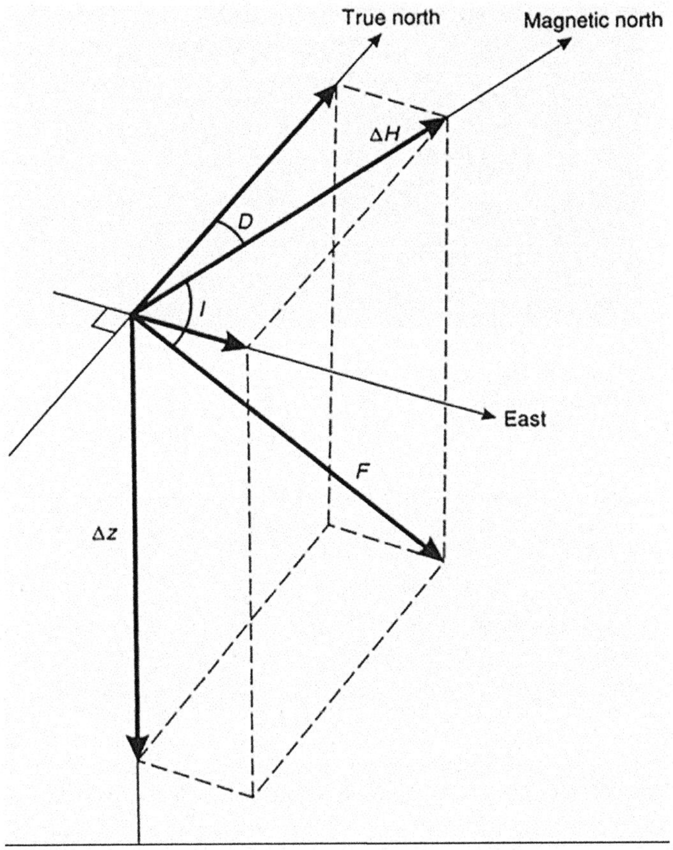

FIG. 3.24 Geomagnetic elements. *Source: Personal.*

susceptibility. The dyke is 3 m wide, 5 m deep, and trends to the northeast. To locate the dyke, geophysicists could measure the magnetic field strength along an east-west line. As the survey approaches the dyke, the induced magnetic field from the dyke would begin to be detectable, in addition to the Earth's main magnetic field. By analyzing the variations in the magnetic field, the location and potentially the dimensions of the dyke could be determined. Much like a gravitational anomaly associated with a dense body, the magnetic anomaly around the dyke is concentrated in the immediate vicinity and rapidly decreases as one moves further from the dyke.

3.3.6 Temporal variations of the Earth's magnetic field

Similar to the gravitational field, the magnetic field also changes over time. When discussing these temporal variations, it is helpful to categorize them into three types based on their frequency and origin. However, while magnetic polarity reversals are significant for understanding Earth's history, they will not be covered in this discussion. Instead, we will focus on the following three types of temporal variations:

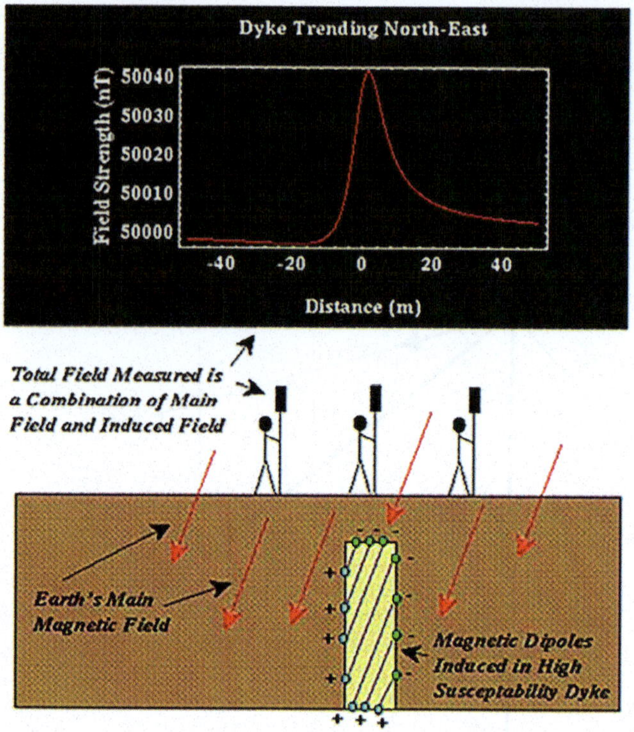

FIG. 3.25 A simple example—magnetics and geology show the total field measured is a combination of the main field and induced field. *Source: EBS 309: Geofizik Carigali (Kamar Shah Ariffin)—Magnetik.*

Secular variations

These long-term variations in the main magnetic field, which occur over several years, are believed to be caused by fluid movement in the Earth's outer core. Since these variations change gradually compared to the duration of a typical exploration magnetic survey, they are unlikely to interfere with data processing.

Diurnal variations

These variations in the magnetic field occur daily and are linked to changes in the Earth's external magnetic field. The fluctuation can be around 20 to 30 nT per day, and it is important to account for this when conducting exploration magnetic surveys.

Magnetic storms

At times, magnetic activity in the ionosphere can sharply increase, often associated with heightened sunspot activity. During such magnetic storms, the magnetic field becomes highly irregular and unpredictable, with amplitude changes reaching up to 1000 nT. It is advisable to avoid conducting exploration magnetic surveys during these storms.

3.3.7 Instrumentation for magnetic surveys in advancing potential field surveying for sustainable exploration and management

Magnetometers are non-mechanical instruments designed to measure the amplitude (strength) of the magnetic field or its specific components. The first significant developments in these instruments occurred during World War II, when Fluxgate magnetometers were created for submarine detection. Since then, other types of magnetometers, such as Proton Precession and Alkali-Vapor magnetometers, have been developed.

The Fluxgate magnetometer, initially designed for use in low-flying aircraft to detect submarines, is now also used for borehole measurements. The transducers from this device are commonly found in electronic compasses and laboratory equipment for measuring remanent magnetization. From 1945 to 1985, much airborne magnetic surveying was conducted using fluxgate detectors, and portable hand-held devices were used for vertical-component ground measurements between 1965 and 1985. A schematic of the fluxgate magnetometer is shown in Fig. 3.26.

The fluxgate magnetometer operates based on a magnetic saturation circuit design. It uses two parallel bars made of ferromagnetic material, which are placed close to each other. The magnetic susceptibility of these bars is high enough that even the weak magnetic field of the Earth can drive the bars to near magnetic saturation. Each bar is wound with a primary coil, but the direction of winding is reversed between the two bars. An alternating current (AC) flows through these primary coils, creating a large and varying magnetic field in each coil. A secondary coil surrounds both the ferromagnetic cores and the primary coils. The varying magnetic fields produced by the primary coil induce a measurable voltage in the secondary coil. This configuration allows the fluxgate magnetometer to measure the strength of any

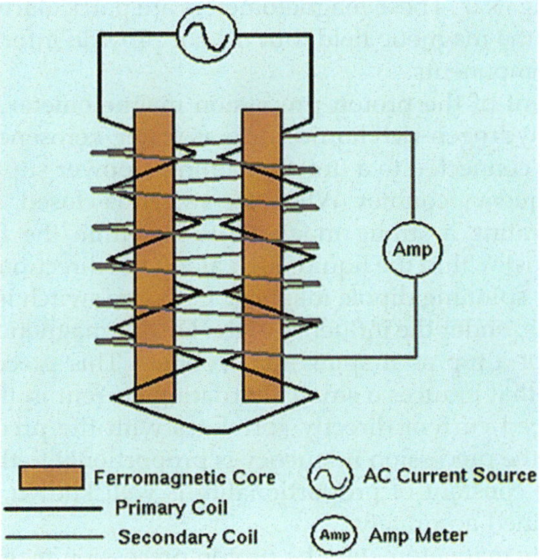

FIG. 3.26 Fluxgate magnetometer. *Source: EBS 309: Geofizik Carigali—Magnetik (Kamar Shah Ariffin).*

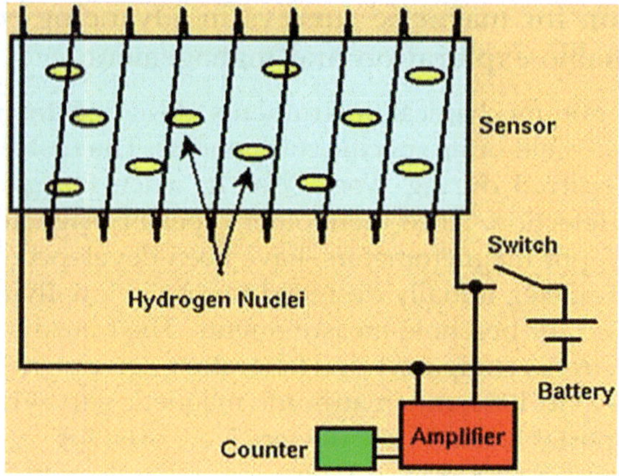

FIG. 3.27 Proton precession magnetometer. *Source: EBS 309: Geofizik Carigali—Magnetik (Kamar Shah Ariffin).*

component of the Earth's magnetic field by adjusting the orientation of the instrument to align the cores with the desired magnetic field component. Fluxgate magnetometers are highly sensitive, capable of measuring the magnetic field strength to within 0.5 to 1.0 nT. These instruments are relatively simple to construct, making them cost-effective (typically between $5000 and $10,000). Unlike gravimeters, fluxgate magnetometers show minimal drift over time.

For land-based magnetic surveys, the proton precession magnetometer is the most commonly used instrument. In contrast to the fluxgate magnetometer, the proton precession magnetometer only measures the total amplitude (strength) of the Earth's magnetic field, often referred to as total field measurements. A schematic of the proton precession magnetometer is illustrated in Fig. 3.27. These magnetometers are particularly effective for measuring the overall intensity of the magnetic field, but do not provide information on the directionality of the magnetic components.

The sensor component of the proton precession magnetometer consists of a cylindrical container filled with a hydrogen-rich liquid, such as water, kerosene, or alcohol, surrounded by a coil. This sensor is connected to a unit containing a power supply, an electronic switch, an amplifier, and a frequency counter. When the switch is closed, a DC current is directed through the coil, generating a strong magnetic field within the fluid-filled cylinder. The hydrogen nuclei (protons) within the liquid align along the direction of the applied magnetic field, behaving like tiny spinning dipole magnets. Once the switch is opened and the current is turned off, the protons, under the influence of the Earth's magnetic field, begin to precess—similar to the motion of a top as it spins and wobbles. This precession generates a time-varying magnetic field that induces a small alternating current in the surrounding coil. The frequency of this induced current directly correlates with the precession frequency of the hydrogen nuclei. Since the precession frequency is proportional to the strength of the Earth's magnetic field, and the constant of proportionality is well known, the total magnetic field strength can be determined accurately.

Similar to the fluxgate magnetometer, the proton precession magnetometer is straightforward to construct, making it an affordable tool for magnetic surveys (typically costing between $5000 and $10,000). It can measure the total magnetic field strength with a sensitivity down to 0.1 nT. Additionally, proton precession magnetometers, like fluxgate models, exhibit minimal instrument drift over time. One of the key advantages of proton precession magnetometers is their ease of use and reliability. Unlike other instruments, they do not require precise levelling or orientation—just a high-angle positioning relative to the Earth's magnetic field is sufficient.

In the absence of external magnetic interference, the magnetic field recorded by the proton precession magnetometer consists of two components: the main magnetic field and the anomalous magnetic field, which is caused by magnetic induction from crustal rocks or remanent magnetization in the Earth's crust.

3.3.8 Planning of surveys and survey design

In the context of advancing potential field surveying for sustainable exploration and management, magnetic surveys are typically conducted alongside gravity surveys. As a result, the survey design principles outlined in Table 3.2 for gravity surveys are also relevant to magnetic surveys. However, due to the distinct characteristics of the magnetic field being measured, certain adjustments are necessary, as highlighted in Table 3.8.

The magnetic surveys in regional geophysical mapping are carried out in conjunction with gravity surveys, and hence the survey design aspects depicted in Table 3.2 are also applicable for magnetic surveys. However, some modifications are required because of the different nature of the measured field and are presented in Table 3.8.

3.3.9 Magnetic data acquisition and corrections—overview

In the realm of advancing potential field surveying for sustainable exploration and management, magnetic data acquisition and the associated corrections play a critical role. Proper data collection, along with the necessary adjustments, ensures the accuracy and reliability of the magnetic measurements, which are essential for informed decision-making in resource exploration and management.

Ground magnetic measurements are typically conducted using portable instruments at regular intervals along straight, parallel lines that cover the survey area. Due to their cost-effectiveness, magnetic surveys generally consist of areal data distributions (two-dimensional surveys) rather than single-line data (one-dimensional surveys). The interval between measurement stations along the lines is often smaller than the spacing between the lines. A single operator usually manages the magnetometer, although additional personnel may be required for grid layout, surveying, or the buddy system.

TABLE 3.8 Magnetic survey design in NGPM project.

Survey stage	Scale of survey	Station interval	Traverse interval	Contour interval	Remarks
National Geophysical Mapping Programme (NGPM)	1:50,000	One station in 2.5 sq km with station interval of 1–2 km	Random distribution along track/road	50 nT	

Magnetic surveys can be affected by intense electromagnetic interference from man-made sources, and ferrous materials near the magnetometer can distort data. For accurate readings, metallic objects such as large belt buckles must be removed, and the compass should be kept >3 m away from the magnetometer. To ensure the instrument is functioning properly, a final test involves immobilizing the magnetometer and taking readings while the operator moves around the sensor. If the readings do not change by >1 or 2 nT, the system is considered stable. For highly precise surveys, operator effects must remain under 1 nT.

Although most magnetometers are designed to operate in environments with significant 60 Hz electromagnetic and radio-frequency fields, extremely low-frequency fields from direct current equipment or large alternating current switches may still cause issues. To minimize the influence of these factors, the sensor should be operated at an elevated height to avoid interference from soil magnetite near the ground. In areas with magnetite-bearing rocks, sensor heights of up to 4 m have been employed to reduce near-surface magnetic effects.

To create accurate magnetic anomaly maps, temporal variations in the Earth's magnetic field during the survey period must be considered. Diurnal drift typically causes variations of a few tens of nT throughout the day, but larger fluctuations, up to hundreds or thousands of nT, can occur during magnetic storms. Surveys should be avoided during severe magnetic storms. Diurnal drift corrections are achieved by making frequent repeat measurements at a base station, which should be situated at least 100 m from large metal objects, roads, and 500 m from power lines when possible. Data from field stations are then corrected for temporal variations by assuming a linear change between base station readings. Continuously recording magnetometers at fixed base stations can also monitor these temporal changes, and field data can be corrected by subtracting variations recorded at the base site, assuming accurate time synchronization between the base and field locations.

Corrections for spatial variations in the strength of the Earth's main magnetic field are referred to as geomagnetic corrections. A widely used method for accounting for these variations is to utilize models of the Earth's main magnetic field, such as the International Geomagnetic Reference Field (IGRF). The IGRF models are regularly updated every five years, through the Definitive Geomagnetic Reference Field (DGRF), to reflect secular variations. By knowing the latitude and longitude of a point on the Earth's surface, the total field strength of the Earth's main magnetic field can be calculated. For the necessary normal correction, the most recent IGRF epoch should be selected.

It is important to note that no elevation correction is applied in ground magnetic surveys, as variations in total field strength with elevation are <0.015 nT per meter. This variation is generally considered negligible. The terrain correction, which would require knowledge of the magnetic properties of topographic features, is also rarely applied due to its complexity.

Magnetic anomalies are calculated as the difference between the diurnal corrected observed total field magnetic values and the normal field (IGRF value), i.e., the observed field minus the theoretical field (IGRF value). The IGRF values at a point can be obtained using the IGRF module of Geosoft software.

Once all corrections are applied, magnetic survey data are typically presented as individual profiles or contour maps. To produce accurate anomaly maps, it is essential to account for temporal changes in the Earth's magnetic field during the survey period. Anomalies caused by cultural features, such as railroads, pipelines, and bridges, are often identified through field observations and maps showing these features. A regional gradient can help describe a residual magnetic anomaly.

Interpreting magnetic data is generally more complex than interpreting gravimetric data, as magnetic anomalies are influenced by multiple factors, such as susceptibility, remanent magnetization, and its orientation. Finding a unique model is challenging because the same anomaly can often be explained by different combinations of magnetic bodies and parameters. Additionally, it is crucial to remember that magnetic fields diminish more rapidly with distance from the source ($B \propto \frac{1}{R^3}$), so caution should be exercised when interpreting local anomalies. A typical data recording sheet for field magnetic surveys in the NGPM program is presented in Table 3.9.

A typical data reduction sheet for the field magnetic surveys in the NGPM programme is presented below in Table 3.10.

3.3.10 Magnetic data processing

The processing techniques for magnetic data closely mirror those used for gravity data, including quality checks, analysis, management, and archival. In addition to these standard methods, magnetic data often require specialized techniques, such as Reduced to Pole (RTP) and Reduced to Equator (RTE), to account for regional variations and improve anomaly interpretation. A thorough understanding of the density and susceptibility values of the rocks and minerals in the survey area is crucial for producing meaningful interpretations and performing quantitative analyses. Key parameters like magnetic inclination, dip, declination, and the ambient (normal) field values are essential prerequisites for interpreting magnetic anomalies and developing models (Naskar et al., 2021). To process gridded magnetic data and separate anomalies of interest, the following steps should be followed using Geosoft software:

TABLE 3.9 Data recording sheet for total field magnetometer in NGPM project.

Station ID	Longitude (°)	Latitude (°)	Time	Field reading (nT)	Diurnal correction (nT)	Diurnal corrected field (nT)	Geology
Base							

TABLE 3.10 Data reduction sheet for total field magnetic data in geophysical mapping project (NGPM).

Station ID	Longitude (°)	Latitude (°)	Elevation (m)	Diurnal corrected field (nT)	IGRF normal field epoch (nT)	IGRF inclination	Declination	Magnetic anomaly (col. 5 minus col. 6) (nT)	Geology
Base									

1. **Grid File Preparation:** The appropriate gridding method must be selected, with careful consideration of grid cell size, search radius, and blanking distance, based on data density and distribution.
2. **Magnetic Anomaly Analysis:** The variations in magnetic anomalies should be examined using different contour intervals to ensure the grid effectively represents the features of interest.
3. **Nonlinear Filtering:** Nonlinear filters should be applied to the gridded magnetic data to eliminate single-point anomalies, which may arise due to undersampling or noise from cultural objects.
4. **Radially Averaged Power Spectra (RAPS):** RAPS should be generated for the magnetic grid to create a plot showing wave number (inverse of wavelength) against log (power).
5. **Power Spectra Interpretation:** Careful analysis of the power spectra in selected areas is needed to draw regression lines that correspond to different geological layers with distinct magnetization characteristics. The slope of these lines indicates the depth of interfaces between layers, and the wavelength ranges of these layers should be recorded.
6. **Filter Selection:** The analysis of wavelength ranges will guide the choice of appropriate cut-off wavelengths for high-pass, low-pass, and band-pass filters in the processing software.

It is theoretically understood that the minimum wavelength present in a gravity or magnetic grid is 2dx, where dx represents the grid spacing. This information is essential for accurately applying filtering techniques during magnetic data processing.

3.3.10.1 Regional residual separation

One of the most challenging aspects of working with gravity and magnetic data is the separation of regional and residual components. This can be accomplished by following the steps outlined below.

Wavelength filtering

a) **High-Pass Filter:** The high-pass filter allows high-frequency anomalies above a certain cut-off value to pass through, highlighting shallow, narrow features with limited depth extent. Essentially, this process emphasizes residual anomalies while eliminating the influence of regional and deeper features.
b) **Low-Pass Filter:** This filter is used when the goal is to remove the effects of shallow features while preserving the responses from regional or deeper geological structures.
c) **Band-Pass Filter:** As the name implies, the band-pass filter allows only a specific range of wavelengths to pass. It is applied when the user aims to analyze and evaluate the response of geological features located at intermediate depths.

Upward continuation

The upward continuation technique is another method used to eliminate responses from shallow sources and focus on the effects of deeper features. By continuing the magnetic anomaly or Bouguer gravity anomaly grids upward, typically by a multiple of the grid spacing, low-amplitude noise can be suppressed. The upward continued grid is then subtracted from the original grid to obtain the residual grid. This process is one approach for separating regional and residual components.

Trend surface analysis

Trend surface analysis is an analytical method used to separate observed data into regional and residual components. In this approach, polynomials of varying orders are generated to represent the anomaly surface under study. The polynomial that most accurately fits the regional surface for the given gravity or magnetic data is selected. This polynomial is then used for further qualitative and quantitative analysis of the regional features. The value of the polynomial at each point is subtracted from the observed magnetic or gravity value to obtain the residual, which more accurately represents the area of interest.

Downward continuation

The downward continuation method is used to enhance the responses from shallow features that may otherwise be obscured by regional responses or gradients. The downward continued grid highlights the "local anomaly" or residual anomaly, which represents relatively shallow features, focusing on the local features of interest.

3.3.10.2 *Derivatives*

Derivative grids are effective at highlighting subtle variations in density or magnetization contrasts that may be masked or faintly observed in the original grids. These derivatives can be calculated in either the horizontal or vertical direction.

Horizontal derivative maps

The horizontal derivative is defined as the rate of change in the gravity or magnetic anomaly per unit distance in the X or Y direction. These maps emphasize magnetization or density contrasts, helping to identify lithological contacts or faults.

Vertical derivative

The vertical derivative represents the variation of the potential field in the vertical direction and is useful for highlighting shallow features.

3.3.10.3 *Analytic signal*

The Analytic Signal (AS) is a function that incorporates the three gradients of the potential field in the X, Y, and Z directions, calculated as the square root of the sum of the squares of the derivatives. This helps to identify the boundaries of an anomalous body.

$$AS = \sqrt{(dx^2 + dy^2 + dz^2)} \tag{3.41}$$

where dx and dy are the horizontal derivatives, and dz is the vertical derivative. The analytic signal, whether applied to profile or spatial data, is highly effective in delineating the boundaries of the causative sources.

3.3.10.4 *Reduction to pole*

Reduction to the pole (RTP) is a technique that recalculates observed Total Field Magnetic Anomaly data as though the inducing magnetic field had a 90° inclination. This transformation converts dipolar magnetic anomalies into monopolar anomalies, centered directly over their causative bodies, which simplifies interpretation. In RTP, the symmetric "highs" are directly over the body, and the maximum gradient of the anti-symmetric dipolar anomalies aligns precisely with the edge of the body. However, pole reduction becomes challenging at low magnetic latitudes (below 21°N) because North-South bodies do not produce detectable induced magnetic anomalies at zero geomagnetic inclination. Additionally, pole reduction is not effective in regions with significant remanent magnetization effects.

3.3.10.5 *Reduced to equator (RTE)*

Magnetic data collected at latitudes below 21°N is not required for generating RTE maps. However, under certain conditions, RTE can be applied to data from higher latitudes for qualitative analysis.

3.3.11 Interpretation of magnetic data

Similar to gravity data, the interpretation of magnetic data is carried out in two stages: qualitative interpretation and quantitative interpretation. However, interpreting magnetic data can be more complex for several reasons, and caution must be exercised throughout the process:

- Unlike gravity anomalies, which produce the same distribution regardless of latitude or orientation of the anomalous mass, magnetic anomalies are highly influenced by the strike of the target and its latitude. This is because the inducing magnetic field varies from the equator to the poles.
- Remanence can also complicate interpretation.
- If magnetization is purely induced, for North-South striking bodies at the equator, the Earth's magnetic field magnetizes the body along its strike without creating any effective magnetization, resulting in no magnetic anomalies.
- The magnetic method may fail to detect anomalies in such cases, even if there is a high susceptibility contrast in the target.
- Ensuring mutual consistency between gravity and magnetic interpretations helps minimize ambiguities in the analysis.

3.3.11.1 *Qualitative interpretation*

The steps used in the qualitative interpretation of gravity data are also applicable to magnetic data. Additionally, the following techniques are commonly employed:

- Delineating zones of high and low contour density.
- Identifying linear gradient zones.
- Describing general trends, including orthogonal trends.
- Identifying and marking bipolar, high-dominant, and low-dominant anomaly zones to understand the nature of the causative source.
- Recording the overall variation of magnetic anomalies in the study area, including the minimum and maximum values.
- Recognizing subtle magnetic variations (often without visible signs) and expected signatures associated with typical ore environments, such as alteration zones.
- Documenting magnetic discontinuities, shifts, and breaks in the anomaly pattern.
- Identifying fault zones, characterized by breaks in the anomaly pattern and shifts in contouring Naskar et al., 2021).

3.3.11.2 *Quantitative interpretation*

The quantitative interpretation process begins with the identification of key features on maps during the qualitative phase, marking the transition to the quantitative stage. Refining the locations involves determining depth values (z), with various methods used to estimate the depth to the tops of anomalous magnetic bodies. These methods include slope measurement techniques, as well as analytic approaches such as the Euler and Werner methods. Gravity and magnetic modeling, incorporating both forward and inversion approaches, play a crucial role in determining the location in the x, y, and z dimensions. Any causative body of gravity and magnetic anomalies must be assessed and visualized in relation to the geological feasibility

TABLE 3.11 Flow chart: gravity surveys as applicable in the NGPM project.

Gravity surveys	
1. Planning at the headquarters stage	Discussions on area, geology, plan of action, etc.
	Calibration and standardization of instruments
2. Planning of survey and quality control	Selection of *Base* Station
	Location and spacing of gravity–magnetic (GM) stations
	Evaluation of data quality
	Precautions while making observations
3. Data processing techniques	Reduction of gravity data: tide correction, drift correction, latitude correction, elevation correction, terrain correction
	Data processing: free air anomaly, Bouguer anomaly
4. Interpretation procedure and software	Regional–residual separation methods: continuation methods, vertical derivative, polynomial fitting, filtering technique, finite element method, spectral analysis of data, horizontal gradient map, apparent density map, analytic signal map
	Qualitative analysis in the field
	Quantitative analysis using Geosoft
5. Geometrical and physical parameters of a causative body	Forward modeling
	Inversion
6. Presentation of data and maps	Bouguer gravity map
	Residual gravity map
	Regional gravity map
	Frequency filtered maps
	Analytic signal map
	Any other map which is relevant for interpretation
	2D geological model along the gravity profile
	Location map
7. Preparation of technical report	Maps and image

TABLE 3.12 Flow chart: magnetic surveys as applicable in the NGPM project.

Magnetic surveys	
1. Planning at the headquarters stage	Discussions on area, geology, plan of action, etc.
	Calibration and standardization of instruments
2. Planning of survey and quality controls	Selection of *Base* Station
	Location and spacing of GM stations
	Evaluation of data quality
	Precautions while making observations
3. Data processing techniques	Reduction of magnetic data: diurnal correction, geomagnetic correction
	Data processing: total intensity map, magnetic anomaly map
4. Interpretation procedure and software	Regional–residual separation methods: continuation method, vertical derivative, polynomial fitting, filtering technique, finite element method, spectral analysis, reduction to pole, reduction to equator, Euler deconvolution, horizontal gradient, apparent susceptibility, analytic signal
	Qualitative analysis in the field
	Quantitative analysis using Geosoft
5. Geometrical and physical parameters of a causative body	Forward modeling
	Inversion
6. Presentation of data and maps	Magnetic contour/image map
	Residual magnetic map
	Regional magnetic map
	Frequency filtered maps
	Analytic signal map
	Any other map which is relevant for interpretation
	2D geological model along magnetic profile
	Location map
7. Preparation of technical report	Maps and images

of the model. The format for Geophysical Mapping (GPM) is outlined in the flow diagrams presented in Tables 3.11 and 3.12, which are essential for ensuring uniformity and maintaining the quality of the output for the National Geophysical Mapping (NGPM) project.

3.3.12 Structural fabric and crustal architecture analysis using gravity and magnetic geophysical surveys: A case study from the Dharwar, North Canara, and Shimoga Districts of Karnataka, India

A regional geophysical survey utilizing gravity and magnetic methods was conducted over a 3600 sq km area of the Peninsular Gneissic Complex (PGC) and the Dharwar Supergroup, aimed at delineating subsurface structural features related to mineralization (Naskar et al., 2023). The Bouguer anomaly map revealed a negative gravity anomaly with amplitudes ranging from −105 mGal to −82 mGal, with a peak negative value of 23 mGal. The gravity map highlights high gravity responses over the Dharwar Supergroup, in contrast to the PGC, suggesting the presence of high-density rock types such as chlorite schist, Banded Iron Formation (BIF), amphibolite, and argillite. These rock types have higher densities than the granitoids and sediments of the cratonic segments. Positive local anomalies on the residual gravity maps are likely due to subsurface anticlines in sedimentary rocks with NW-SE, N–S, and NE-SW orientations, which may have formed during the Dharwar orogeny. Negative anomalies on the residual gravity maps may correspond to synclines within the sedimentary rocks.

Magnetic anomaly values (total field reduced to equator) ranged from about −42 nT to +292 nT. Broad zones of high-intensity anomalies in the central part of the study area were attributed to the combined effect of the shallow PGC basement and Dharwar Supergroup rocks. A prominent NW-SE-trending formational contact in the southern region reflects the boundary between chlorite schist of the Dharwar Supergroup and granite gneiss of the PGC, which is consistent with the gravity data. The Chandragutti magnetic low over the granitoid of the PGC is also associated with a Bouguer gravity low. This granitoid may exhibit a sharp increase in rare metal content of magmatic origin within the surrounding PGC.

Radially averaged power spectra of the gravity and magnetic anomalies suggest potential layering at ensemble average depths of 5.8–7.2 km, 2.7 km, and 1.3–1.5 km, respectively. Euler depth solutions for the gravity anomalies show a NW-SE trending linear structure in the southern part of the study area, occurring at variable depths (1–4 km), which likely represents the deeper section of the PGC-Dharwar Supergroup contact. 2D modeling reveals a significant low in the NW part of the profile over the Chandragutti granitoid, extending to a depth of 10 km. The model also suggests undulation in the basement (Fig. 3.28).

Hard rock geology presents mineral explorers with a diverse array of lithologic, metamorphic, and structural features, which can host a wide range of mineral deposits and generate highly complex geophysical anomaly patterns (Langenberg and Nielsen, 1982; Langenberg, 1983; Sprenke et al., 1986). In regions with a complex polyphase history of deformation, igneous activity, and metamorphism, developing a simple geophysical model can be a challenging task. Each locality and deposit is unique, requiring detailed field mapping and varied geophysical surveys to understand the subsurface. Seismic methods, typically designed for stratified rocks with relatively simple structures, may not always be applicable in such settings (Nettleton, 1971). Mineral explorers, therefore, depend on a broad spectrum of geophysical techniques, including magnetic, gravity, electrical, electromagnetic, and radiometric methods. Above all, these techniques are used in conjunction with a deep understanding of hard rock geology, as well as the results of local field mapping and exploration drilling (Lyatsky, 1996; Lyatsky et al., 2005).

3.3.13 Geophysical investigations in portions of Kolar, Dharmapuri, Chittoor, and Thiruvannamalai districts, Karnataka and Tamil Nadu—a case study

Regional geophysical surveys using gravity and magnetic methods have been conducted over a 2160 sq km area in the Obvious Geological Potential (OGP) category under the National Geophysical Mapping (NGPM) program. The primary objective was to identify

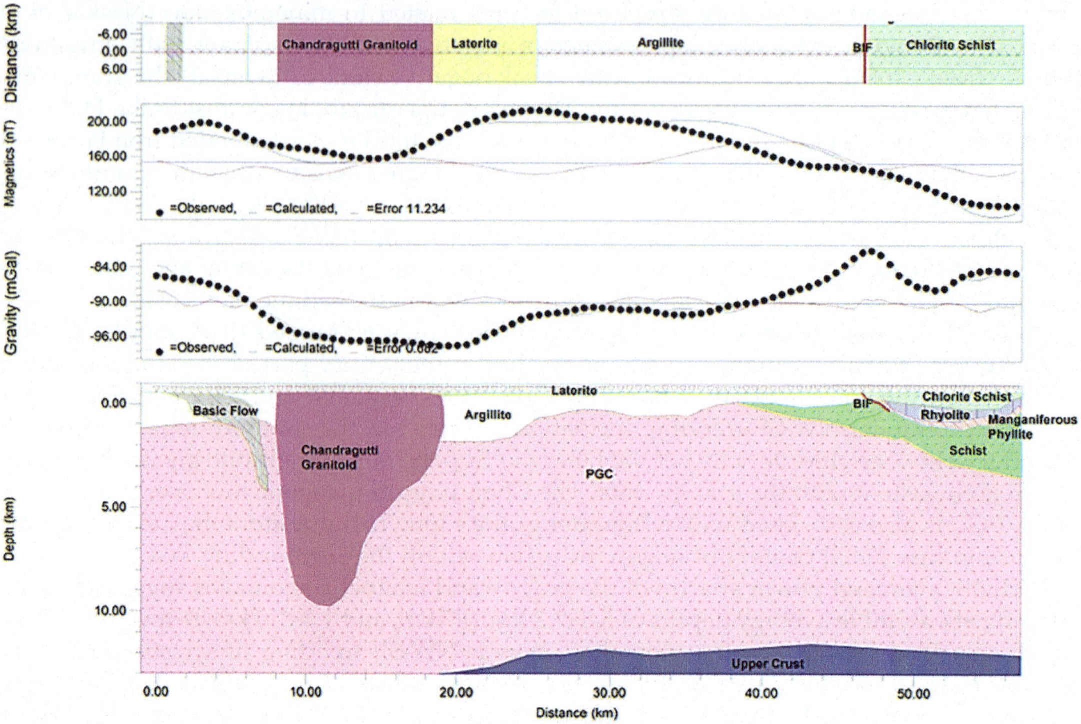

FIG. 3.28 2D gravity and magnetic modeling of Uttar Kanara, Shimoga, and Dharwar districts, Karnataka. Horizontal distance in km. PGC: Peninsular Gneissic Complex. *Source: Naskar et al. (2023).*

subsurface structures and highlight areas for future detailed studies. The Bouguer anomaly (BA) map revealed a negative gravity anomaly with an amplitude of 43 mGal, ranging from −41 mGal in the northern and eastern parts to −84 mGal in the western part. The surveyed region is characterized by high gravity anomalies in the northern and eastern sections, while the southern and central areas show low to medium gravity anomalies. An elongated residual high gravity feature to the east of the Kolar Gold Field, trending nearly north-south and flanked by residual gravity lows, is interpreted as a basement high, which could be significant for mineralization in the region. A north-south trending residual gravity high in the central area is likely associated with faulted contacts, where deformation of the crustal layers in the form of faults and folds is evident over the Peninsular Gneissic Complex (PGC) rocks.

The magnetic anomaly (total field) map displays bipolar magnetic anomalies, ranging from −179 nT to +396 nT. The northern and eastern parts show high-intensity magnetic anomalies, while the central part is characterized by smaller bipolar anomalies. The radially averaged power spectrum of the gravity field indicates three interfaces at depths of 5.6 km, 3.1 km, and 1.6 km, while the magnetic field shows two interfaces at depths of 3.7 km and 2.0 km. These interfaces, revealed by both gravity and magnetic data, are interpreted as the depth range of the contact between the PGC and Kolar Group. Euler depth solutions, both linear and curvilinear, suggest geological or faulted contacts, with major contacts/faults

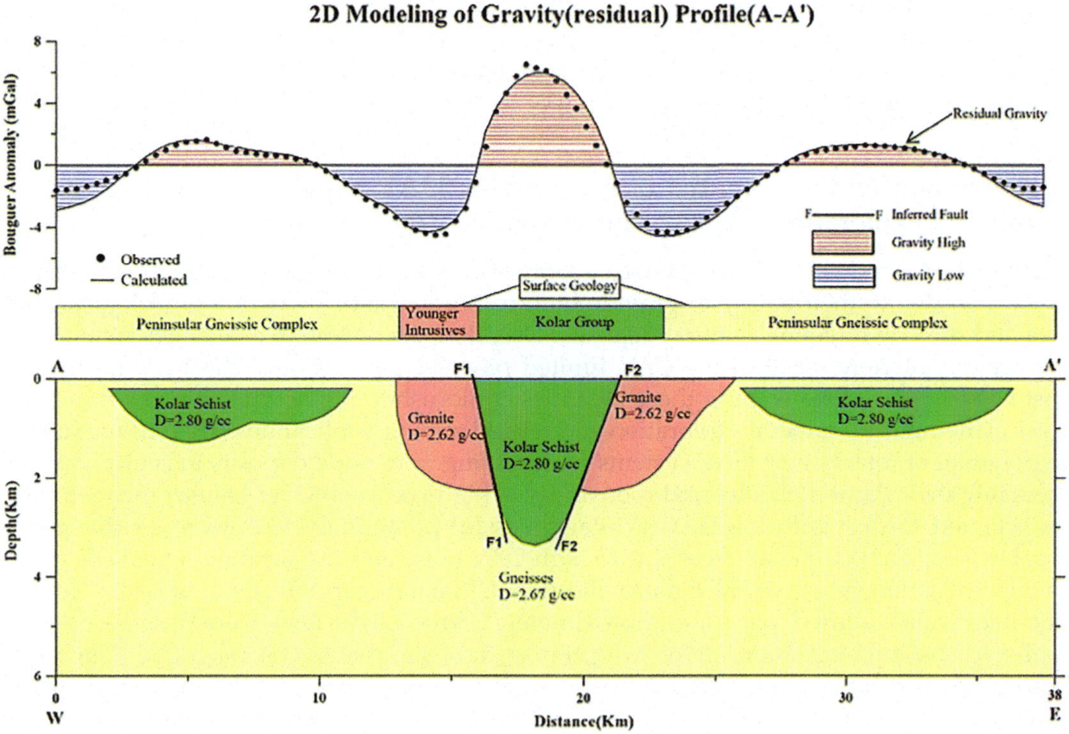

FIG. 3.29 2D gravity model along profile AA' in parts of Kolar, Dharmapuri, Chittoor, and Thiruvannamalai districts, Karnataka and Tamil Nadu. *Source: Naskar et al. (2021).*

occurring at depths of <1000 m due to the sharp contrast in density and susceptibility of the subsurface rocks.

To further investigate the crustal structure, a gravity profile (AA′) was selected to cut across key geological features. The profile was analyzed using 2D modeling software (GEOSOFT), and the observed and modeled curves showed minimal deviation, indicating a good fit (Fig. 3.29). The analysis identified different subsurface layers based on spectral analysis results, the density of measured rock samples, and surface geological constraints. The 2D model along profile AA′ is presented, with the surface geology, density of crustal rocks, faults, tectonic subdivisions, and interpreted crustal layers to aid understanding. Three subsurface layers with varying densities of 2.62 g/cc (granite), 2.67 g/cc (PGC), and 2.80 g/cc (Kolar Group) were inferred. A high gravity signature is associated with the Kolar Schist Belt in the western, central, and eastern parts, while a low gravity signature is found over the granite in the central area, which is covered by gneisses throughout. Two faults, F1–F1 and F2–F2, are interpreted based on the steep gradient of the gravity anomaly at the contact between granite and the Kolar Schist Belt on either side. The Kolar Schist Belt is characterized by a high gravity value of about 6.5 mGal, with an estimated width of 5.3 km and a depth of 3.3 km. The granite on either side of the Kolar Schist Belt exhibits a gravity low of about 4.3 to 4.4 mGal, with widths of 3 km and 4.7 km and depths of 2.2 km and 2.3 km, respectively. The sharp rise in gravity on either side of the high corresponds to the inferred schist belt at near-surface depths. This model is well-resolved, constrained by average depth values derived from radial spectrum analysis and surface geological data, providing a coherent interpretation of the subsurface structure.

3.3.14 Exploring the potential of rotary-wing UAV-based magnetic surveys for mineral exploration: A case study from central Sweden

Unmanned aerial vehicle (UAV)-based geophysical surveys are gaining attention in mineral exploration due to their ability to provide high-resolution data for defining geological structures and directly targeting mineral deposits. Despite their potential, challenges such as electromagnetic interference from the UAV, limited payload capacity, and short flight durations must be addressed. Overcoming these obstacles could usher in a new era of UAV-based geophysical methods for mineral exploration and related mining applications. This study examines the potential of rotary-wing UAV systems for targeting iron-oxide deposits in central Sweden, leveraging their flight flexibility and robustness. A high-precision Over Hauser magnetometer was adapted for use with the UAV. While successful preliminary tests were conducted, the actual experiment faced challenges due to high UAV noise and the strong magnetic field from the mineralization, which delayed data collection. Ultimately, around 20 km of total field magnetic data were acquired, covering approximately 2 km^2. Flight lines were designed perpendicular to the mineralization strike to optimize data sampling. The magnetic data clearly identified two distinct mineralized zones, one rich in magnetite and the other in hematite, located only 50–100 m apart. Historical low-altitude (30 m) fixed-wing aeromagnetic data were compared with the UAV data, showing consistency in delineating the mineralized areas. This demonstrates the potential of UAV-based magnetic surveys for mineral exploration in challenging geological and logistical settings, advancing the capabilities of potential field surveying methods for sustainable exploration and management.

3.4 The magnetic gradient tensor: Properties and applications in source characterization for advancing potential field surveying in sustainable exploration and management

A magnetometer is an instrument equipped with a single sensor that measures magnetic flux density, whereas a magnetic gradiometer consists of pairs of magnetometers with sensors separated by a fixed distance. The Earth's magnetic field at any location is the vector sum of two components: the primary field generated by the Earth's liquid core dynamo and the crustal field from the magnetic minerals in local rocks. This vector has both orientation and amplitude, and even slight changes in any direction can influence its characteristics. When a highly magnetic rock formation is introduced into a homogeneous host, the local magnetic vector is altered. In three-dimensional space, this results in nine spatial gradients (3×3) that form a tensor, which defines the anomalous field.

Scott Hogg, through his company Scott Hogg & Associates, based in Toronto, Canada, has been instrumental in advancing the use of airborne magnetic gradiometers in the mining industry. The development of the first airborne magnetic gradiometers was driven by the oil and gas exploration sector, which sought to use Euler's equation with the measured vertical gradient to estimate the depth to magnetic sources. In the 1960s, Aeroservice introduced both helicopter-based and fixed-wing towed bird systems for magnetic surveys. Interest in Euler's theory for magnetic analysis was renewed in the mining industry two decades later, and it remains a key foundation for many modern interpretation methods.

In Canada, the Geological Survey of Canada (GSC) developed a vertical magnetic gradiometer in the 1970s, leading to a national mapping program that began in the early 1980s. Canadian airborne survey contractors responded by developing a range of fixed-wing and helicopter-based systems. The mining industry's interest in airborne gradient measurement stemmed primarily from the increased spatial resolution and detailed imagery it provided, particularly for detecting small anomalies on the flanks of larger features. Vertical gradient maps, derived from simple filtering of total field data, were found to offer similar benefits to measured gradients, at a lower cost, leading to reduced commercial demand for measured vertical gradients.

In 1983, Geometrics introduced a horizontal gradiometer, marking a significant development in airborne magnetic surveys. This innovation, coupled with techniques developed by Nabighian and Hansen to derive pseudo total field residuals from horizontal gradients, further advanced magnetic data interpretation. Although the Geometrics survey division ceased operations for a time, the concept of horizontal gradient measurement was revived by researchers like Nelson of the NRC and later implemented by companies like De Beers and Fugro (formerly Geodass) in Botswana. The early 1990s saw the introduction of the first 3-axis gradiometer by Fugro, which provided a range of mapping and compilation services based on gradient data.

In Canada, Terraquest became the first company to offer horizontal gradient measurement, followed by Goldak, which pioneered the first full 3-axis fixed-wing gradiometer. Today, almost all airborne contractors provide horizontal gradient systems, with several now offering full 3-axis configurations for simultaneous vertical and horizontal gradient measurements. The growing interest in magnetic gradient measurements reflects their potential to significantly improve the accuracy and resolution of magnetic maps. Vertical gradient maps, analytic signal maps, and Euler-based methods, which rely on gradient information, have proven to be valuable tools in geological interpretation and mineral exploration. Both calculated and measured gradient data have contributed to notable advancements in magnetic survey methodologies, advancing the field of sustainable exploration and management.

3.4.1 Measurement of the magnetic gradient tensor: Advancements in magnetic methods for sustainable exploration and management

Prospecting for Earth's magnetic field anomalies is an essential tool in various applications, such as mineral exploration (Macnae, 1979), detection of unexploded ordnance (Barrow and Nelson, 1998; Nelson and McDonald, 2001; Munschy et al., 2007), and archaeological surveys (Clark, 1996; Linzen et al., 2009). These applications often demand high spatial resolution and precise magnetic field measurements to ensure accurate inversion and 3D modeling. In this regard, the use of high-sensitivity vector magnetometers and tensor gradiometers that measure all components of the gradient tensor has become increasingly advantageous.

Christensen and Rajagopalan (2000) highlight that the next major advancement in magnetic exploration is likely to involve the measurement of the gradient tensor. This development can significantly improve the interpretability of magnetic surveys, especially in regions where anomaly patterns are influenced by remanence or occur at low magnetic latitudes. While the benefits of total field gradiometry are widely recognized, total magnetic intensity (TMI) and its gradients are not considered potential fields, unlike the tensor components, which possess desirable mathematical properties. The key distinction between full tensor gradiometry and total field gradiometry lies in the production of more detailed and quantitatively interpretable maps and 3D models, as opposed to simple anomaly detection. Total field gradient surveys are common (Hood, 1981), but while they share advantages such as reducing the need for base stations and regional corrections, they do not offer the same level of precision or mathematical properties as tensor gradients.

Magnetic surveys remain the most cost-effective and widely used geophysical mapping tool, especially in hard rock environments, with growing importance in hydrocarbon exploration (Schmidt and Clark, 2000). To achieve the high precision required for gradient measurements, superconducting quantum interference devices (SQUIDs) are recognized as the most suitable sensors (Foley et al., 1999; Foley and Leslie, 1998). SQUIDs are highly sensitive to minute changes in magnetic flux, detecting only changes perpendicular to the superconducting loop. While these devices are essentially variometers, they can serve as vector sensors, making them ideal for gradient measurement applications. High-temperature SQUIDs (HTSs), which operate at liquid nitrogen temperatures (–197°C), address some of the logistical challenges associated with using liquid helium, and future advancements may eliminate the need for cryogens altogether through developments in micro miniature Joule-Thomson and non-magnetic Stirling cryocoolers (Zimmerman, 1981).

The sensitivity of SQUIDs is on the order of 100 nT (10^{-13} Tesla), and it is anticipated that gradiometer sensitivity should exceed 0.01 nT/m, with a baseline of 0.1 m. These small, low-power devices, with dimensions of just a few centimeters, may find future applications in down-hole surveys or unmanned aerial vehicles (UAVs). SQUIDs also offer a rapid sampling rate, allowing for the detection of high-frequency noise, such as aircraft interference, which can be effectively filtered out. This is a notable advantage over total field magnetometers, which typically have slower sampling rates that can cause compensation issues. While deploying SQUIDs in aircraft and down-hole presents unique challenges, such as ensuring platform stability and developing slim, robust instruments for down-hole use, SQUIDs offer promising solutions to these issues.

In conclusion, the integration of magnetic gradient tensor measurements and the use of SQUID technology represents a significant advancement in magnetic methods for potential field surveying. These innovations hold considerable promise for improving the accuracy, resolution, and interpretability of geophysical surveys, contributing to more sustainable exploration and management practices.

In the real world, the magnetic gradient tensor is a 3 × 3 s-order tensor:

$$\begin{bmatrix} \frac{\partial B_x}{\partial x} & \frac{\partial B_x}{\partial y} & \frac{\partial B_x}{\partial z} \\ \frac{\partial B_y}{\partial x} & \frac{\partial B_y}{\partial y} & \frac{\partial B_y}{\partial z} \\ \frac{\partial B_z}{\partial x} & \frac{\partial B_z}{\partial y} & \frac{\partial B_z}{\partial z} \end{bmatrix} \tag{3.42}$$

where x, y, and z refer to an orthogonal coordinate system and B = (B_x, B_y, B_z) denotes the magnetic field and its vector components. In practice, we only need to know five of the elements because essentially both the divergence and the curl of the field are zero, so that the tensor is symmetric and traceless (Vrba, 1996).

In practice, we only need to know five of the components. Because the divergence of the magnetic field is zero, i.e.

$$\nabla \cdot B = \frac{\partial B_x}{\partial x} + \frac{\partial B_y}{\partial y} + \frac{\partial B_z}{\partial z} = 0 \tag{3.43}$$

This means that the gradient tensor is traceless (in linear algebra, the trace is the sum of the diagonal elements, so a traceless matrix has the diagonal elements sum to **zero**), and only two of the diagonal terms are required. In addition, in the absence of currents and any significant time variations in electrical fields, the curl of the field is also zero,

$$\nabla \times B = \begin{bmatrix} i & j & k \\ \frac{\partial}{\partial x} & \frac{\partial}{\partial y} & \frac{\partial}{\partial z} \\ B_x & B_y & B_z \end{bmatrix} = 0 \tag{3.44}$$

This implies that the gradient tensor is symmetric since the three orthogonal components of the curl are zero:

$$\frac{\partial B_y}{\partial z} - \frac{\partial B_z}{\partial y} = 0 \tag{3.45}$$

$$\frac{\partial B_x}{\partial z} - \frac{\partial B_z}{\partial x} = 0 \tag{3.46}$$

$$\frac{\partial B_x}{\partial y} - \frac{\partial B_y}{\partial x} = 0 \tag{3.47}$$

Therefore, only three off-diagonal terms are required. Fig. 3.30 depicts the arrangement of SQUID sensors for the detection of axial (diagonal) and transverse (off-diagonal) gradients (Schmidt and Clark, 2006).

The detection of axial (diagonal) gradients requires two separate superconducting quantum interference devices (SQUID) sensors, whereas each transverse (off-diagonal) gradient can be measured using a single planar sensor. This simplification significantly streamlines the system design. Tensor gradiometry has shown considerable potential in various applications, including aeromagnetic surveys with wide line spacings, such as those conducted over sedimentary basins, environmental surveys, and defense applications like submarine and unexploded ordnance detection, as well as down-hole magnetic measurements. Stolz et al., 2006 Stolz et al. (2006) introduced the Full Tensor Magnetic Gradiometry (FTMG) system, which records all components of the magnetic field, including independent tensor gradient components, with high resolution.

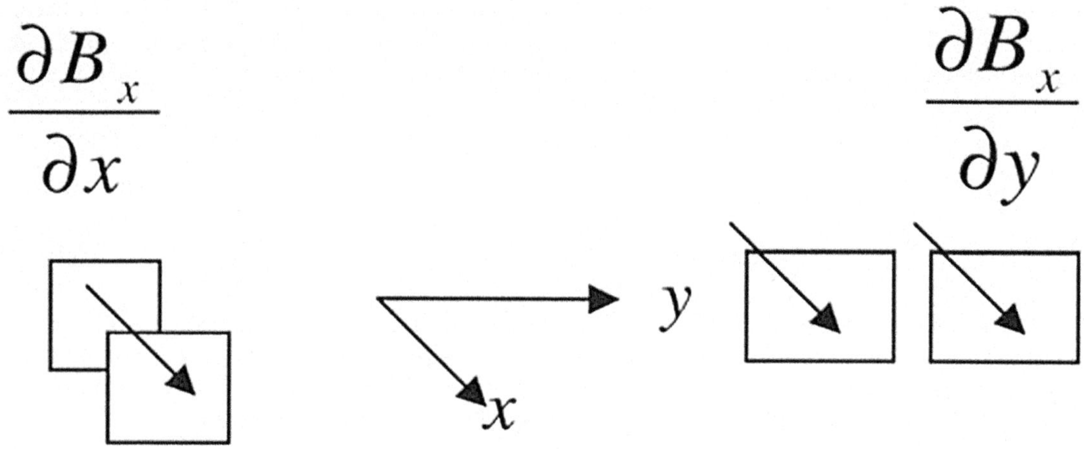

FIG. 3.30 Conceptual arrangements of SQUID sensors for detection of axial and transverse gradients. *Source: Personal.*

Gradient measurement, especially using FTMGs, offers distinct advantages over gradient components derived from total magnetic intensity (TMI) measurements (Rudd et al., 2022). The development of more robust and aerodynamically optimized survey birds has significantly reduced system noise. This reduction in noise enhances sensitivity and accuracy, thereby increasing confidence in the resulting survey data. While the system has primarily been applied to diamondiferous kimberlite exploration, it has also demonstrated success in a range of mineral exploration targets, including gold, nickel, and iron ore. These improved data sets contribute to more accurate geological interpretations across survey areas. Other potential applications for FTMG surveys include infrastructure mapping, unexploded ordnance detection, and the compensation of electromagnetic data in marine environments. The JeSSY STAR instrument, developed at the Institute of Photonic Technology (IPHT) in Jena, is a notable example of this technology.

One key advantage of gradient measurements is their relative insensitivity to orientation. Since gradients mainly arise from anomalous sources, with a relatively low background geomagnetic gradient, they are less influenced by the orientation of the sensor. This contrasts with the measurement of field vectors, which are highly sensitive to orientation due to the dominance of the Earth's core field. As Christensen and Rajagopalan (2000) argue, this makes gradient tensor measurements particularly valuable. Consequently, gradient measurements are best suited for airborne applications. Over the past three decades, airborne magnetic surveys have seen significant improvements in both data acquisition and image processing techniques, making them a routine part of exploration programs, often preceding geological mapping efforts.

Despite treating the magnetic field as a scalar in the past, advances in vector measurement techniques—whether through direct measurements or mathematical manipulation—promise further improvements in magnetic survey accuracy. For example, the total magnetic intensity (TMI) could be adjusted to more accurately reflect a true potential field. By incorporating the gradient tensor into magnetic surveying, further enhancements in sensitivity, resolution, and interpretability are anticipated, driving more sustainable exploration and management practices.

In environmental and defense applications, as well as in the study of compact geological sources such as iron ore copper-gold (IOCG) deposits or alteration zones, isolated dipoles frequently appear. Analytical inversion algorithms for the magnetic gradient tensor are available for such scenarios, offering valuable insight into source characterization (Frahm, 1972; Wynn et al., 1975; Wilson, 1985). Among these, Wilson's (1985) algorithm is one of the most straightforward to implement. However, all these inversion methods are subject to a fourfold ambiguity due to the fact that multiple locations and polarities can generate the same magnetic gradient tensor. Typically, three of the solutions are "ghost solutions," and one corresponds to the true source. The treatment of these ambiguities depends on the specific application, such as post-processing geophysical surveys or real-time tracking of dipole sources. One way to resolve the ambiguity is by knowing whether the source lies above or below the observer, immediately halving the possible solutions. Additionally, if the components of the anomalous magnetic field are known—whether through direct measurement or derived from the total magnetic intensity (TMI)—a unique solution can be determined.

In practical geophysical surveys, anomalies are rarely isolated, which can complicate the use of dipole or other simple geometric model inversion methods. A more general approach for such cases is Euler deconvolution, which has been adapted for magnetic surveys (Zhang et al., 2000). Euler's method, initially developed for gravity surveys, can be extended to magnetics and provides significant advantages over earlier methods discussed by Heath et al. (2003). This approach relies on a "structural index" that encapsulates the source geometry, with values ranging from 0 (for a uniformly magnetized, extensive feature such as a sill) to 3 (for isolated dipoles like IOCG deposits or unexploded ordnance (UXO)). Higher indices, such as 4 for quadrupoles, can also be used, but they are rare in nature. By determining the geometry of the source, the location and magnetic properties can be constrained further by analyzing the eigenvalues and eigenvectors of the magnetic gradient tensor at key locations near the source.

Remarkably, only a few tensor observations are typically needed to derive all determinable information about a source. While there is inherent non-uniqueness in certain scenarios—such as determining the radius of spherical sources or the along-strike component of magnetization for large bodies—critical details about the source can still be extracted. For example, the dip and magnetic inclination of dyke-like bodies may not be fully resolvable, but other attributes, such as the magnetic intensity and depth, can still be constrained (Schmidt et al., 2004). Once the strike of the source is known, the tensor components at each station can be transformed into a coordinate system with x along the profile, y parallel to the strike, and z pointing downward. In this system, elements involving y become relatively small compared to values of B_{xx}, B_{xz}, and B_{zz}, with $B_{xx} \approx -B_{zz}$.

For 2D sources, it can be shown that the eigenvalues of the magnetic gradient tensor are given by (Schmidt et al., 2004):

$$\lambda = \pm\sqrt{\left[(B_{xx})^2 + (B_{xz})^2\right]} = \pm\sqrt{\left[(B_{zz})^2 + (B_{xz})^2\right]} \tag{3.48}$$

These eigenvalues are rotational invariants of the tensor and are independent of the magnetization direction of the source. A unimodal curve of $|\lambda|$ vs. x can indicate a relatively thin sheet. The peak of the eigenvalue, which corresponds to the 2D total gradient or analytic signal amplitude, provides an approximation of the sheet's center (x_0). The half-width at half-maximum (HWHM) of the eigenvalue curve offers an estimate of the depth (h) of the sheet. Additionally, the magnetization-thickness (Jt) product can be estimated from the maximum eigenvalue and the depth. The signs of the components of effective magnetization (J_x, J_z) are determined by the signs of B_{xx} and B_{xz} in the vicinity of the eigenvalue peak, which indicates the quadrant where the effective magnetization lies. This method of tensor gradiometry offers significant advances in magnetic surveying, providing greater detail and accuracy for sustainable exploration and management of mineral resources.

3.4.2 Advantages of determining gradients in general

The advantages of magnetic gradient surveys include:

- Better resolution of shallow features and closely spaced sources
- Better definition of structural features
- Suppression of regional anomalies due to deep sources
- Mapping of subvertical contacts
- Tighter anomalies around compact sources
- Easier detection and delineation of pipe-like sources
- Constraining the local strike direction (not vertical TMI gradients)
- Determination of which side of a line the source lies (not vertical TMI gradients)
- Common-mode rejection of geomagnetic variations
- Relative insensitivity to rotation noise
- Constraining interpolation between flight lines (not vertical TMI gradients) (important as all surveys are somewhat aliased across flight lines)
- Less importance of IGRF corrections (usually unnecessary)
- Direct indication of the Euler structural index when combined with measurements of the field
- Higher resolution than conventional TMI surveys—can be offset against survey height, allowing somewhat higher, therefore considerably safer, flying.

3.4.3 UAV-towed scalar magnetic gradiometry: A case study on iron oxide copper-gold mineralization in Nautanen, Arctic Sweden

Scalar magnetic surveying utilizing unmanned aerial vehicle (UAV) platforms is gaining increasing traction in geophysical applications, particularly for mineral exploration. Although only a few studies have explored UAV-towed scalar field measurements, and even fewer have focused on towed scalar difference measurements or gradients, significant advancements are being made. Dossing et al. (2023) explore the potential and benefits of employing UAV-towed scalar transverse horizontal difference (THD) measurements, showcasing its effectiveness for mineral exploration.

This study presents UAV-towed gradiometry bird data from the Nautanen region in northern Sweden, a known site for iron oxide copper-gold (IOCG) mineralization. The survey covered an area of 2.5 km^2, with an average line spacing of 30 m and a constant flight altitude of 30 m above the ground. The UAV collected high-quality scalar total field and THD data with a dynamic noise level of around ± 0.05 nT, demonstrating the precision achievable with this platform. The collected magnetic data were compared with traditional ground-based magnetic surveys, revealing a strong correlation between magnetic anomaly highs and lows across the survey area. This provided valuable structural insights based on subtle magnetic features identified in both total magnetic intensity anomalies and THD data.

The results of this study underscore the potential of UAV-towed systems to replace conventional airborne surveys, offering a more cost-effective solution for geophysical exploration. However, conducting magnetic surveys using UAV platforms comes with its own set of challenges. UAVs typically fly at speeds of 10–12 m/s, which, when coupled with wind disturbances, can affect the 3D attitude of the towed gradiometry bird during the survey. To address this, advanced processing techniques are essential to obtain the accurate gradient data intended for analysis. In contrast, challenges related to wind disturbances are less prominent in traditional airborne surveys, where flight speeds tend to be much higher than the wind speed.

This study highlights how UAV-towed magnetic gradiometry is advancing the potential of field surveying techniques for sustainable exploration and management, particularly in mineral exploration and resource mapping, and offers a promising alternative to traditional airborne magnetic surveys.

3.4.4 Magnetic and gravity gradiometry framework for iron oxide-apatite and iron oxide copper-gold deposits: Advancing potential field surveying for sustainable exploration and management in Southeast Missouri, United States

High-resolution airborne magnetic and gravity gradiometry data provide a critical geophysical framework for assessing the exploration potential of concealed iron oxide deposits in the Mesoproterozoic basement rocks of southeast Missouri (McCafferty et al., 2016). These advanced geophysical techniques are employed to calculate mineral prospectivity for iron oxide-apatite (IOA) ± rare Earth element (REE) and iron oxide copper-gold (IOCG) deposits, delivering crucial insights into the subsurface.

The geophysical data successfully delineate the geophysical signatures of all known iron oxide deposits in the region and identify several previously unrecognized prospective areas. The airborne data were further processed through inversion techniques, generating three-dimensional models of density and magnetic susceptibility over four concealed deposits: Pea Ridge (IOA ± REE), Boss (IOCG), Kratz Spring (IOA), and Bourbon (IOCG).

The Pea Ridge susceptibility model reveals a vertically extensive magnetic source that can be traced to depths exceeding 2 km. A smaller, shallow density source within the Precambrian basement partially coincides with the magnetic anomaly. In contrast, the Boss model identifies a large (625 m wide), vertically extensive stock with coincident dense and magnetic rock, extending >2600 m across the shallow Precambrian paleosurface. The Kratz Spring deposit, characterized by a smaller volume of iron oxides, exhibits lower density and magnetic susceptibility compared to the other deposits. A prospective area located south of Kratz Spring shows the largest volume of dense, non-magnetic rock, suggesting the presence of hematite-dominant lithology extending from the Precambrian surface to depths greater than 2 km.

The Bourbon deposit is defined by a large, bowl-shaped volume of high-density, high-magnetic susceptibility rock, suggesting that the iron mineralization is vertically restricted to the upper portions of the Precambrian basement. To support the evaluation of these prospects and three-dimensional models, an extensive statistical summary of density and apparent magnetic susceptibility data is provided, based on several hundred samples taken from the deposits, altered wall rocks, and unaltered country rocks.

These advancements in magnetic and gravity gradiometry underline the importance of integrating high-resolution geophysical data in exploring and managing mineral resources, providing valuable tools for sustainable exploration and resource management strategies.

3.5 Conclusion

The chapter titled "Advancing Potential Field Surveying for Sustainable Exploration" in the proposed Elsevier publication "Baseline Geoscience Data for Sustainable Exploration and Resource Management" explores the application of gravity and magnetic methods in the context of sustainable resource exploration. These potential field surveying techniques, which measure variations in the Earth's gravitational and magnetic fields, are essential for detecting subsurface structures, mineral deposits, and geological anomalies, often with minimal environmental disturbance.

The chapter emphasizes that gravity and magnetic surveys are non-invasive, cost-effective, and capable of covering large areas, making them highly suitable for regional-scale exploration. Gravity methods are particularly effective in identifying density contrasts, which can indicate the presence of mineral deposits or geological features such as faults and basin structures. Magnetic surveys, on the other hand, are invaluable for mapping ferrous mineral deposits, volcanic structures, and understanding the overall geology of an area.

The chapter also highlights the advancements in data processing and interpretation techniques that have significantly enhanced the accuracy and resolution of gravity and magnetic surveys. By integrating these methods with other geophysical techniques and geological data, a more comprehensive understanding of the subsurface can be achieved, leading to more informed and efficient exploration decisions.

In conclusion, the chapter advocates for the continued advancement and integration of gravity and magnetic methods in sustainable exploration practices. These techniques provide crucial baseline data for identifying and managing resources while minimizing environmental impacts. When used effectively, potential field surveys contribute to responsible exploration and resource management, ensuring that future generations can benefit from these resources.

CHAPTER

4

Geoelectrical surveying for sustainable exploration and management

4.1 Direct current resistivity methods for baseline geoscience data

Geoelectrical surveying methods are valuable tools for mineral exploration, particularly in assessing the electrical properties of rocks and minerals (Naskar and Rai, 2018). These properties, such as natural electrical potentials, electrical conductivity (or its inverse, resistivity), and dielectric constant, are key in electrical prospecting. Among these, electrical conductivity is the most significant, while the other properties play a lesser role. Natural or spontaneous electrical potentials in the subsurface are typically driven by electrochemical or chemical activity.

Electrical methods involve various measurements that assess the effects of electrical current flow through the Earth. These measurements can include current flow, electrical potential (voltages), and electromagnetic fields. Key electrical techniques commonly employed include direct current (DC) resistivity, induced polarization (IP), and self-potential (SP).

Since most rock-forming minerals are insulators, electrical current primarily flows through rocks via ions in pore water, making the conduction process electrolytic rather than electronic. However, metals and metallic sulfides conduct electricity efficiently through electron flow. As such, geoelectrical methods are especially effective in mineral exploration, particularly in areas where metallic targets or sulfide ores are present. The conductivity of ore deposits depends significantly on the presence of materials like sulfides or graphite, making electrical methods critical for identifying these resources and advancing sustainable exploration and management strategies.

The electrical resistivity of a geological unit or target is measured in Ohm-m and is influenced by factors such as porosity, permeability, water saturation, and the concentration of dissolved solids in pore fluids within the subsurface. Electrical resistivity methods are designed to measure the bulk resistivity of the subsurface, similar to electromagnetic methods. The key distinction between the two is how electrical currents are introduced or induced within the Earth. In electrical resistivity methods, current is injected into the ground through surface electrodes, whereas in electromagnetic methods, currents are generated by the application of time-varying magnetic fields (Yusuf, 2016).

Baseline Geoscience Data Generation for Sustainable Exploration of Natural Resources
https://doi.org/10.1016/B978-0-443-45221-5.00021-7

The fundamental principle behind resistivity methods is Ohm's Law (V = IR) (Parasnis, 1962), where R represents the resistance of the conductor. The unit of resistance is the Ohm (Ω). The inverse of resistance is the conductance of the circuit, measured in reciprocal Ohms (Ω^{-1}), also known as Siemens (S). According to Halliday and Resnick (2015), the resistance (R) of a wire is directly proportional to its length (L) and inversely proportional to its cross-sectional area (A), expressed as:

$$R \propto \frac{L}{A} \text{ or } R = \frac{\rho L}{A} \tag{4.1}$$

Here, ρ is the constant of proportionality, known as electrical resistivity, which is a characteristic of the material and independent of its shape and size (Zohdy, 1989). The inverse of resistivity (ρ) is the conductivity of the material, denoted by σ. The unit of resistivity is Ohm-meter (Ωm), and the unit of conductivity is the reciprocal Ohm-meter ($\Omega^{-1}m^{-1}$).

Highly conductive materials allow electrical current to pass through easily and, therefore, have low resistivity. Electrical current can flow through materials via three modes: electronic, dielectric, or electrolytic conduction. Electronic (Ohmic) conduction occurs in metals and crystals, dielectric conduction in insulators, and electrolytic conduction in liquids. Some crystals, like silicates, exhibit semiconductor behavior, with resistivity that is higher than conductors but lower than insulators.

Since most rock-forming minerals are insulators, electrical current is primarily carried through rocks by the movement of ions in pore waters. This highlights porosity as the key factor controlling the resistivity of rocks—resistivity increases as porosity decreases (Kearey et al., 2002).

Archie's Law provides an equation that defines the relationship between the resistivity of a porous rock and the fluid saturation factor. This law is particularly applicable to rocks and sediments with low clay content. It assumes that electrical conduction occurs through the fluids filling the pores of the rock, with the relationship between pore-water resistivity (ρ_w) and bulk resistivity (ρ_0) expressed through the formation factor (F). Bottom of Form

$$F = \frac{\rho_0}{\rho_w} \tag{4.2}$$

The formation factor is related to porosity (φ) by this equation

$$F = \alpha\varphi^{-m} \tag{4.3}$$

where, φ= Porosity, α = Constant ranging from 0.47 to 2.2, m = Cementation constant ranging from 1.3 to 2.6. Combining **Eqs.** (4.2) and (4.3) we have that

$$\rho_0 = \alpha\rho_w\varphi^{-m}S^{-n} \tag{4.4}$$

which is Archie's Law. If a scenario is considered where the current sink is a finite distance from the source, as presented in (Fig. 4.1).

Absolute potential is usually difficult to monitor in a survey; the potential difference ΔV between the electrode spacing is thus measured as for a semi-infinite homogeneous and isotropic half-space:

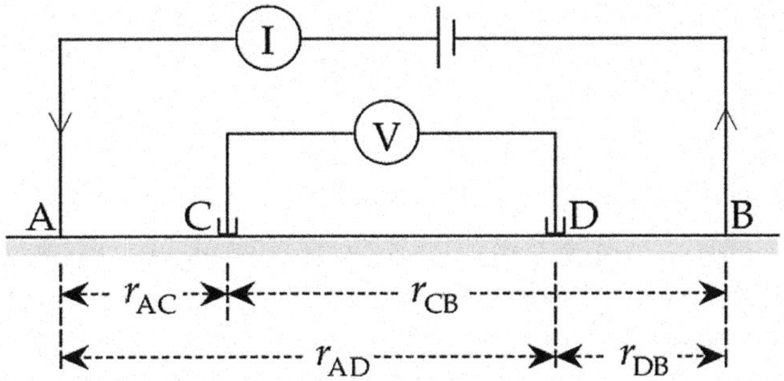

FIG. 4.1 General four-electrode configuration for resistivity measurement, consisting of a pair of current electrodes (A, B) and a pair of potential electrodes (C, D). *Source: Naskar et al. (2002)—Personal.*

$$\Delta V = V_C - V_D = \frac{\rho I}{2\pi}\left[\left(\frac{1}{r_{AC}} - \frac{1}{r_{CB}}\right) - \left(\frac{1}{r_{AD}} - \frac{1}{r_{DB}}\right)\right] \quad (4.5)$$

Therefore;

$$\rho = \frac{2\pi\Delta V}{I}\frac{1}{\left(\frac{1}{r_{AC}} - \frac{1}{r_{CB}}\right) - \left(\frac{1}{r_{AD}} - \frac{1}{r_{DB}}\right)} \quad (4.6)$$

where the ground is homogeneous, the resistivity from the above equation would be constant, independent of either electrode spacing or location. When the subsurface is heterogeneous, the resistivity will vary with relative positions of the electrodes. Computed values are known as the *apparent resistivity* (ρ_a), which is a function of the inhomogeneity.

$$\rho = K\frac{\Delta V}{I} \quad (4.7)$$

where $K = \frac{2\pi}{\left(\frac{1}{r_{AC}} - \frac{1}{r_{CB}}\right) - \left(\frac{1}{r_{AD}} - \frac{1}{r_{DB}}\right)}$ is the geometric factor, and it is independent of electrode configuration used for any electrical resistivity survey. Usually, a complete homogeneous and isotropic Earth medium of uniform resistivity is assumed. For a continuous current flowing in an isotropic and homogeneous medium, the *current density* J (in A/m^2) is related to the electric field E (in V/m) through Ohm's Law:

$$J = \sigma E \quad (4.8)$$

Four metallic electrodes are inserted into the ground and connected to a resistivity meter (Terrameter) via conductive wires. The two outer electrodes, A and B, supply current to the rock formation, while the two inner electrodes, M and N, measure the potential difference. However, the measurement of apparent resistivity in soils and rocks depends on depth and location. The resistivity of soils is influenced by a complex combination of factors, including porosity, permeability, ionic content of the pore fluids, and clay mineralization. Other factors that affect the resistivity of Earth materials may also involve the movement of free electrons or ions in the solid, which can be particularly significant when dealing with certain minerals like graphite, magnetite, or pyrite.

4.1.1 Apparent resistivity and its implications

In an idealized scenario where the ground is a perfectly uniform conducting half-space, the current flow lines follow a dipole pattern, and the resistivity measured using a four-electrode configuration corresponds to the true resistivity of the half-space. However, in real-world conditions, the resistivity is influenced by various lithologies and geological structures, making it highly heterogeneous. This complexity is not accounted for in the four-electrode method, which assumes the ground is uniform. As a result, the measured value represents the apparent resistivity of an equivalent uniform half-space, which typically does not reflect the actual resistivity of any specific area within the subsurface.

4.1.2 Electrode configurations and their applications in resistivity measurements

In resistivity surveys, four electrodes are typically placed at varying positions, and several electrode configurations are used to record field data, each suited to different geological conditions. The most common conventional arrays include Wenner, Schlumberger, Wenner-Schlumberger, dipole-dipole (Naskar and Das, 2009) azimuthal dipole-dipole, equatorial dipole-dipole, pole-pole, pole-dipole (three-electrode), and multiple gradient arrays (Loke et al., 2013). These arrays, along with their respective geometric factors, are illustrated in (Figs. 4.2A–I).

The Wenner and Schlumberger arrays are particularly sensitive to vertical variations in subsurface resistivity beneath the center of the array but are less responsive to horizontal variations (Aizebeokhai, 2010). These arrays have moderate depths of investigation and generally exhibit strong signal strength, which is inversely related to the geometric factor used in calculating apparent resistivity values. However, their main limitation is the relatively poor horizontal coverage as electrode spacing increases. The Wenner array is often preferred in noisy environments due to its high signal strength, but it is less sensitive to three-dimensional (3D) subsurface structures (Dahlin and Loke, 1997).

The dipole-dipole array, on the other hand, is highly sensitive to resistivity variations beneath each dipole pair, making it ideal for detecting horizontal variations. It is especially useful for mapping vertical structures such as dykes and cavities but performs poorly for horizontal structures like sills or sedimentary layers. Additionally, the dipole-dipole array is the most responsive to 3D structures among the common electrode arrays (Dahlin and Loke, 1997). Its depth of investigation is generally shallower than the Wenner array, but it provides better horizontal data coverage. The main drawback is that the signal strength decreases as the distance between dipole pairs increases. Hallof (1964) introduced the concept of plotting apparent resistivity obtained from a collinear dipole-dipole array on a pseudo-section, as shown in Fig. 4.2D.

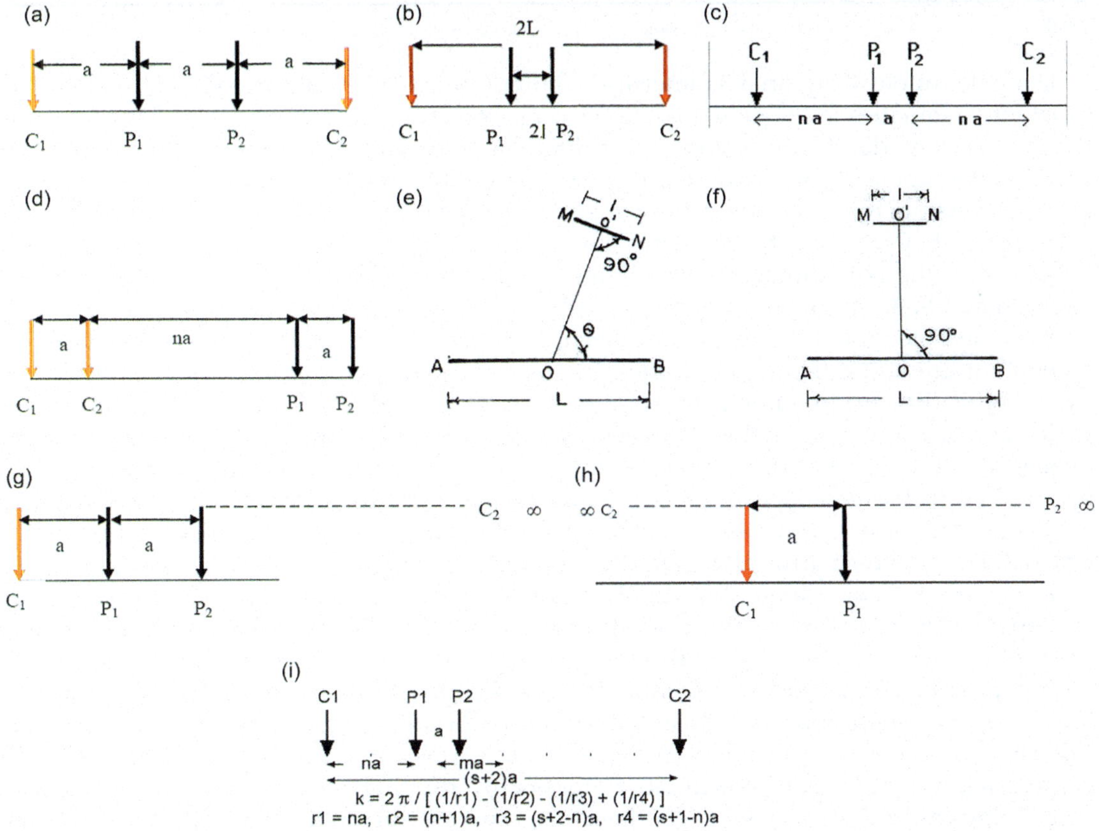

FIG. 4.2 (A) Conventional Wenner array, $C_1P_1 = P_1P_2 = P_2C_2 = a$, geometric factor $K = 2\pi a$ and $\rho_a = 2\pi \mathbf{a}\frac{\Delta \mathbf{V}}{\mathbf{I}}$. (B) Conventional Schlumberger array, $C_1C_2 = 2L$, $P_1P_2 = 2l$, geometric factor $K = \frac{\pi \mathbf{L}^2}{2l}$ **and** $\rho_a = \frac{\pi \mathbf{L}^2}{2l}\frac{\Delta \mathbf{V}}{\mathbf{I}}$. (C) Conventional Wenner-Schlumberger array, C1P1 = P2C2 = na, P1P2 = a, n = unit of dipole length P1P2, geometric factor $K = \pi an(n + 1)$ and $\rho_a = \pi \mathbf{an}(\mathbf{n}+1)\frac{\Delta \mathbf{V}}{\mathbf{I}}$. (D) Conventional dipole-dipole array, $C_1C_2 = P_1P_2 = a$ = dipole length, $C_2P_1 = na$, geometric factor $K = \pi an(n + 1)(n + 2)$ and $\rho_a = \pi \mathbf{an}(\mathbf{n}+1)(\mathbf{n}+2)\frac{\Delta \mathbf{V}}{\mathbf{I}}$. (E) Conventional azimuthal dipole-dipole array, AB = current dipole = L, MN = potential dipole = l, OO′ = r (dipole separation), θ = dipole angle, geometric factor $K = \frac{2\pi \mathbf{r}^3}{\mathbf{Ll}\sin\theta}$ **and** $\rho_a = \frac{2\pi \mathbf{r}^3}{\mathbf{Ll}\sin\theta}\frac{\Delta \mathbf{V}}{\mathbf{I}}$. (F) Conventional equatorial dipole-dipole array, AB = current dipole = L, MN = potential dipole = l, OO′ = r (dipole separation), geometric factor $K = \frac{2\pi \mathbf{r}^3}{\mathbf{Ll}}$ **and** $\rho_a = \frac{2\pi \mathbf{r}^3}{\mathbf{Ll}}\frac{\Delta \mathbf{V}}{\mathbf{I}}$. (G) Conventional pole-dipole array, $C_1P_1 = a$, $P_1P_2 = a$, $P_2C_2 = \infty$, **geometric factor** $\mathbf{K = 4\pi a}$ **and** $\rho_a = 4\pi \mathbf{a}\frac{\Delta \mathbf{V}}{\mathbf{I}}$. (H) Conventional pole-pole array, $C_1C_2 = \infty$, $\mathbf{C_1P_1 = a}$, $\mathbf{P_1P_2 = \infty}$, **geometric factor** $\mathbf{K = 2\pi a}$ **and** $\rho_a = 2\pi \mathbf{a}\frac{\Delta \mathbf{V}}{\mathbf{I}}$. (I) Conventional multiple gradient array, C1P1 = na, P1P2 = a, geometric factor

$$K = \frac{2\pi}{\frac{1}{C1P1} - \frac{1}{P1C2} - \frac{1}{C1P2} + \frac{1}{P2C2}}$$

. Note that for the multiple gradient array, the total array length is a (s + 2) and the distance between the center of the potential dipole pair P1P2 and the center of the current pair C1 C2 is given by ma. *Source: Naskar et al., 2008—Personal.*

The pole-dipole array, an asymmetrical configuration, can result in asymmetric apparent resistivity anomalies in the pseudosection over symmetrical structures, potentially affecting the inversion model. It offers good horizontal coverage and stronger signal strength compared to the dipole-dipole array, and it is less susceptible to telluric noise than the pole-pole array. Reversing the electrode arrangement and repeating measurements can correct the asymmetrical effects, but this increases survey time, as the number of data points doubles. Despite this, the pole-dipole array remains useful when the survey requires higher signal strength and better horizontal coverage.

The pole-pole array consists of one current electrode and one potential electrode, with the second current and potential electrodes theoretically placed at infinite distances. This configuration is difficult to implement in practice, and it is highly susceptible to telluric noise, which can degrade data quality. However, it offers the widest horizontal coverage and the deepest depth of investigation, albeit with the poorest resolution. Subsurface structures can be smeared in the inversion model, making the pole-pole array less effective for detailed analysis. Nonetheless, when small electrode spacing and good horizontal coverage are prioritized, the pole-pole array is a reasonable option.

The pole-pole, pole-dipole, and dipole-dipole arrays are typically used in profiling mode to map both lateral and vertical variations in resistivity. The data is plotted at the intersection point of two 45° lines descending from the current and voltage electrodes or dipoles.

Two primary modes of investigation are associated with these electrode configurations. The Wenner configuration is well-suited for lateral profiling, where the four electrodes are moved stepwise along a profile while maintaining constant inter-electrode distances. This configuration is most effective for locating steeply dipping contacts between rocks with contrasting resistivities, as well as for detecting good conductors such as mineralized dykes, which may represent potential ore bodies. In contrast, Vertical Electrical Sounding (VES) is used to measure resistivity variations with depth (Naskar and Das, 2009). This method is especially useful for determining the depth and resistivity of flat-lying layered rock structures, such as sedimentary beds, or for determining the depth to the water table. The Schlumberger configuration is most commonly employed in VES investigations. The midpoint of the array is kept fixed, and the distance between the current electrodes is progressively increased. A one-dimensional subsurface model is then used to interpret the measurements (Fig. 4.3A). Fig. 4.4 presents an example of data from a sounding survey along with a potential interpretation model.

Despite its limitations, 1D resistivity sounding surveys have been widely used due to two key factors. First, early field equipment lacked the capability for more data-intensive 2D and 3D surveys (Figs. 4.3B and 4.3C). Second, practical computer tools to handle the more complex 2D and 3D models were not available. However, with recent advancements in multielectrode resistivity surveying instruments (Griffiths et al., 1990) and fast computer inversion software (Loke, 1994; Loke et al., 2013), 2D and 3D electrical surveys are now commercially feasible and have become standard practices.

4.1.3 Depth of investigation and apparent resistivity behavior

To better understand the relationship between apparent resistivity and electrode spacing, Fig. 4.5 illustrates a hypothetical Earth model along with corresponding apparent resistivity

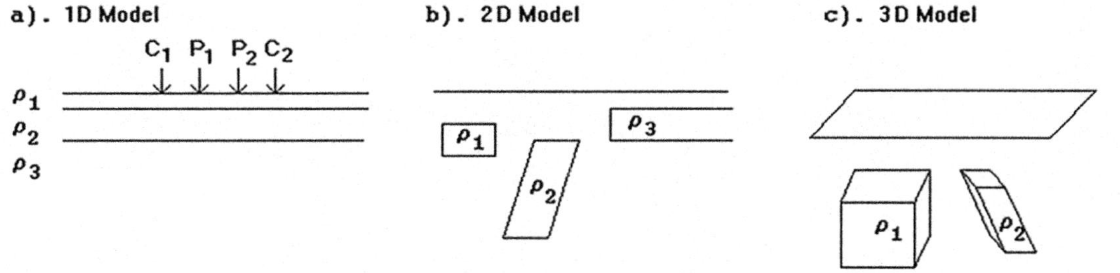

FIG. 4.3 The three different models used in the interpretation of resistivity measurements. *Source: Loke (1997, 1999, 2000).*

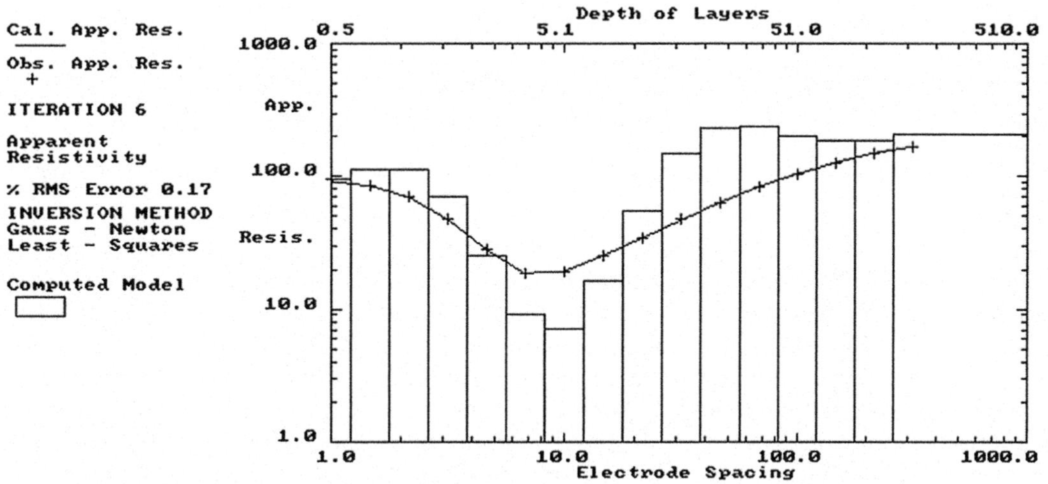

FIG. 4.4 A typical 1D model used in the interpretation of resistivity sounding data for the Wenner array. *Source: Loke (1997, 1999, 2000).*

curves. The model consists of a surface layer with resistivity ρ_1, underlain by a basement layer of resistivity ρ_n that extends downward indefinitely. Intermediate layers with varying thicknesses and resistivities may also be present. Whether using the Wenner or Schlumberger electrode spacing "a," the apparent resistivity vs. spacing curves will share a similar general shape, although they will not perfectly align.

For small electrode spacings, the apparent resistivity closely approximates the resistivity of the surface layer. As the electrode spacing increases, the apparent resistivity approaches that of the basement layer. Thus, every apparent resistivity curve has two asymptotes, represented by the horizontal lines $\rho_a = \rho_1$ and $\rho_a = \rho_n$, which the curve approaches at large electrode spacings. Curve A in the diagram shows a situation where an intermediate layer has a resistivity greater than ρn. Curve B reflects a two-layer scenario or a case where resistivity increases from the surface down to the basement. Curve C represents a situation where the intermediate layer has a resistivity lower than ρ_1.

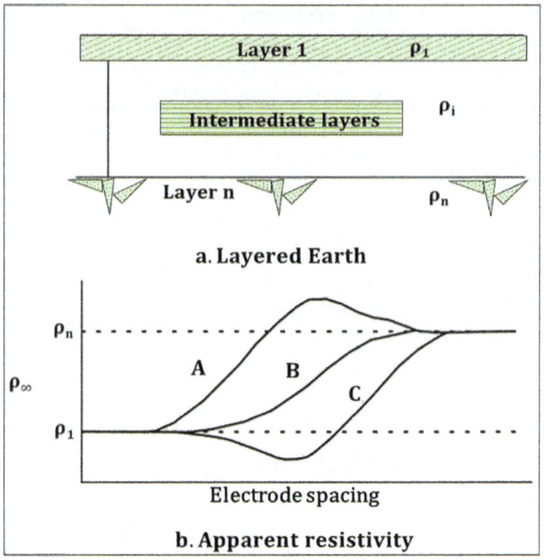

FIG. 4.5 Asymptotic behavior of the apparent resistivity curves at very small and very large electrode spacings. *Personal.*

Unfortunately, neither the maximum of curve A nor the minimum of curve C directly represents the true resistivity values of the intermediate layers. However, these values may be close to the true resistivities if the layers are thick enough. The depth of investigation in a resistivity survey will always be less than the electrode spacing. Typically, to ensure sufficient data collection, the maximum electrode spacing should be at least three times the depth of interest. A useful field guide is to plot the apparent resistivity curve as the survey progresses (Fig. 4.5B) and monitor whether the curve has reached its asymptotic phase, which indicates that the investigation has sufficiently covered the depth of interest.

4.1.4 Sources of noise in geoelectrical measurements

Several factors can introduce noise and affect the accuracy of voltage and current measurements used to calculate apparent resistivities. These sources of interference must be carefully managed to ensure reliable results. The primary sources of noise include:

Electrode Polarization: When a metallic electrode, such as copper or steel, comes into contact with an electrolyte like groundwater (which is not a saturated solution of its own salts), it can generate a measurable contact potential. This potential can contribute significantly to the total measured voltage. To mitigate this, two solutions are available: using nonpolarizing electrodes or periodically reversing the current flow in the current electrodes. Alternatively, employing a slowly varying alternating current (AC) at a few cycles per second can help minimize polarization effects.

Telluric Currents: Naturally occurring electrical currents flow through the Earth and are known as telluric currents. These currents can produce a measurable voltage across the potential electrodes even when no current is being applied through the current electrodes.

To counteract this, the current can be periodically reversed, or a slowly varying AC current can be used to reduce the effects of telluric interference.

Nearby Conductors: Electrical surveys can be disrupted by nearby conductors that make direct contact with the ground, such as buried pipes or chain-link fences. These conductors, due to their low resistivity, will attract current, acting as current sinks. As a result, the current flows preferentially along these conductors instead of through the Earth, effectively short-circuiting the resistivity measurements.

Low Resistivity at the Near Surface: Just as nearby conductors can act as current sinks, a low-resistivity near-surface layer can also impede the flow of current deeper into the Earth. For instance, a highly conductive layer, such as a perched water table, can prevent the current from penetrating more deeply into the subsurface, leading to incomplete or distorted data.

Near-Electrode Geology and Topography: Local variations in geology or water content near the electrodes can cause significant changes in resistivity measurements, particularly in the near-surface. Additionally, rugged topography can influence current flow, concentrating it in valleys and dispersing it on hills, which can result in uneven measurements and data inconsistencies.

Induced Current in Measurement Cables: The current flowing through the cables connecting the current source to the current electrodes can induce a current in the voltage measurement cables, producing a false voltage reading. This noise can be minimized by ensuring that the current-carrying cables are kept physically separate from the voltage cables, preferably by a distance of one to two meters.

By understanding and addressing these potential sources of noise, more accurate and reliable geoelectrical measurements can be obtained, contributing to better interpretation of subsurface conditions.

4.1.5 Instruments and measurements in geoelectrical surveys

The methods and instrumentation used in resistivity surveys are fundamentally based on the use of direct current (DC), primarily because DC enables a deeper investigation of the subsurface compared to alternating current (AC). DC also avoids the complexities introduced by the inductive and capacitive effects of the ground, which result in frequency-dependent resistivity measurements.

In a typical setup, direct current (I) is applied through current electrodes, commonly steel stakes about 50 cm in length, which are driven into the ground to establish a good electrical contact. If there is high contact resistance, which is often the case in dry or rocky soils, the issue can sometimes be alleviated by pouring saltwater around the electrodes to improve conductivity. Alternatively, in areas with particularly hard or rocky ground, aluminum foil sheets buried in shallow depressions or small mounds of wetted Earth can be used to ensure a reliable connection.

Once the current is injected, a potential difference (V) is measured across potential electrodes. Ideally, no current should flow between these potential electrodes, which is typically achieved using a high-impedance operational amplifier or, in older systems, a null-balancing galvanometer. In some instruments, separate "sending" and "receiving" units are used for current and potential measurements.

Power for resistivity meters is generally supplied by dry cell batteries in smaller instruments used for shallow surveys, typically up to a few meters deep. For deeper investigations, especially in mineral exploration, larger systems often rely on petrol or diesel engine-powered electric generators, which can weigh several hundred kilograms. These systems are frequently used for more complex surveys such as induced polarization (IP) surveys (White et al., 2001). The current transmitter is kept separate from the receiving units that measure the potential signal, and voltages of 90 V to several hundred volts may be applied across the current electrodes during engineering surveys (Fig. 4.6).

Special considerations are necessary when using DC to mitigate issues like electrode polarization and telluric currents. Nonpolarizing electrodes, such as porous ceramic pots containing a copper electrode and a saturated copper sulphate solution, can help address these challenges. These electrodes are particularly useful in rocky environments where it is difficult to drive traditional metal stakes into the ground. To ensure good contact with the ground, the pot electrodes should be placed on a cleared surface, possibly aided by adding water if the ground is dry. The electrolyte in the pots must be checked and replenished periodically to maintain saturation levels, ensuring consistent measurement.

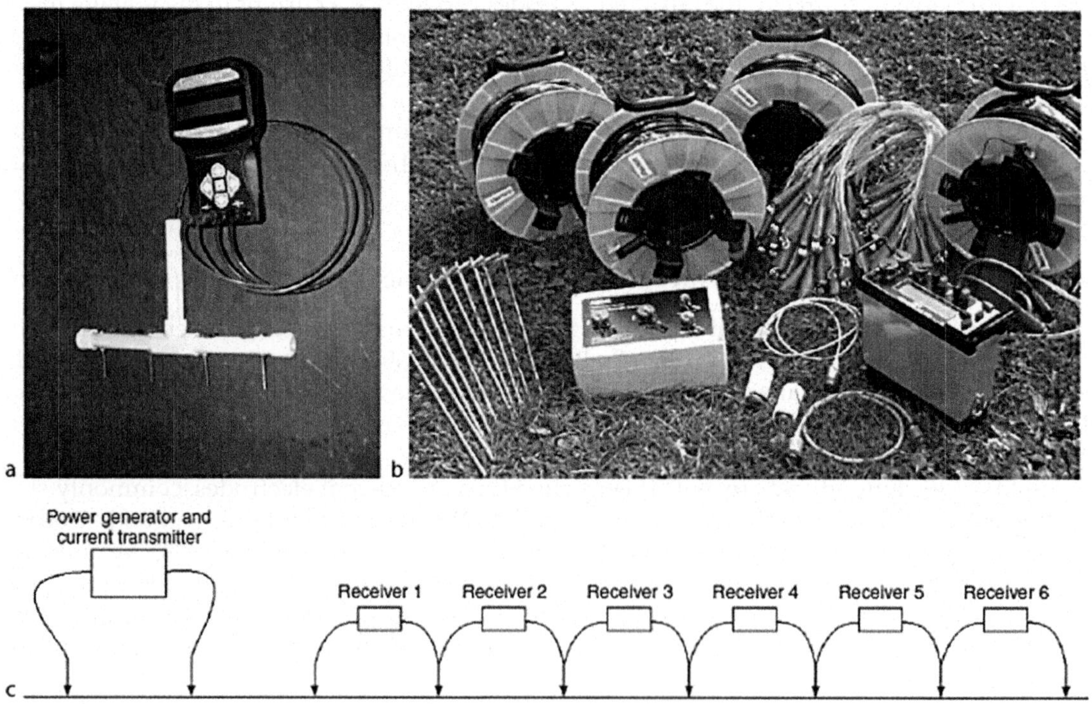

FIG. 4.6 Instrumentation, electrical resistivity–different classes of resistivity instruments. (A) A lightweight portable system (courtesy of Landviser LLC). The distance between the electrodes in the above picture is 5 cm. (B) A typical system used for shallow environmental and engineering surveys (courtesy of Torleif Dahlin and Abem Instruments AB.), and (C) a schematic diagram of a system for deep surveys. *Source: British Geological Survey. https://www.bgs.ac.uk/legal-and-policy/terms-of-use/*

Modern resistivity instruments are equipped with features such as data averaging or stacking, which help improve signal quality in noisy environments and enhance the signal-to-noise ratio for weak signals. The voltage signal is typically in the millivolt range, though modern systems using advanced signal-processing techniques like signal stacking can resolve even smaller voltages. The signal is digitized using a 24-bit analog-to-digital converter (ADC), ensuring precise measurements.

In addition to the direct resistivity measurements, natural self-potential (SP) voltage, which arises from variations in soil chemistry or geomagnetic field changes, can also be present in the ground. This SP voltage can fluctuate over time, but modern resistivity meters are equipped with mechanisms to "buck out" or remove these SP signals. Furthermore, instruments often incorporate notch filters (e.g., 50 Hz, 60 Hz, 16.7 Hz) to eliminate noise from electric power lines or other infrastructure, such as electrified railway systems, which can interfere with the measurements.

These advanced instruments and techniques contribute significantly to geoelectrical surveying, enabling more accurate, reliable, and efficient data collection for baseline geoscience studies and sustainable exploration and management of subsurface resources.

4.1.6 Vertical electrical sounding (VES) in geoelectrical surveys

Vertical Electrical Sounding (VES) is commonly used in geoelectrical surveys to investigate subsurface resistivity variations with depth, especially in two-layer situations. A typical example is when a conductive overburden overlays a resistive basement, which is frequently encountered in environmental studies, where a conductive water table is present beneath drier, more resistive soil or rocks. Prior to the development of portable computing technology, two-layer models were often interpreted using characteristic curves, which offer insights into how the current flow behaves at boundaries between different layers (Naskar et al., 2022).

At the boundary between two layers, current lines behave similarly to optical or seismic rays and are subject to similar laws of reflection and refraction. For instance, if θ represents the angle between a current line and the normal to the interface, the electrical "law of refraction" can be expressed as:

$$\frac{\tan\theta_1}{\tan\theta_2} = \frac{\rho_2}{\rho_1} \tag{4.9}$$

where ρ_1 and ρ_2 represent the resistivities of the two layers. In a set of characteristic curves, the apparent resistivity ρ_a is normalized by the resistivity ρ_1 of the upper layer, and the electrode spacing is expressed as a multiple of the layer thickness. A family of characteristic curves is then calculated for different ratios of $\frac{\rho_2}{\rho_1}$ (Fig. 4.7). The resistivity contrast between the two layers can be conveniently expressed by a k-factor (reflection coefficient), defined as:

$$\mathrm{k} = \frac{\rho_2 - \rho_1}{\rho_2 + \rho_1} \tag{4.10}$$

The k-factor varies between −1 and +1 as the resistivity ratio $\frac{\rho_2}{\rho_1}$ ranges from 0 to infinity. The characteristic curves are plotted as full logarithmic graphs on a transparent overlay, and these are then compared with field data to identify the best-fitting curve. For a three-layer

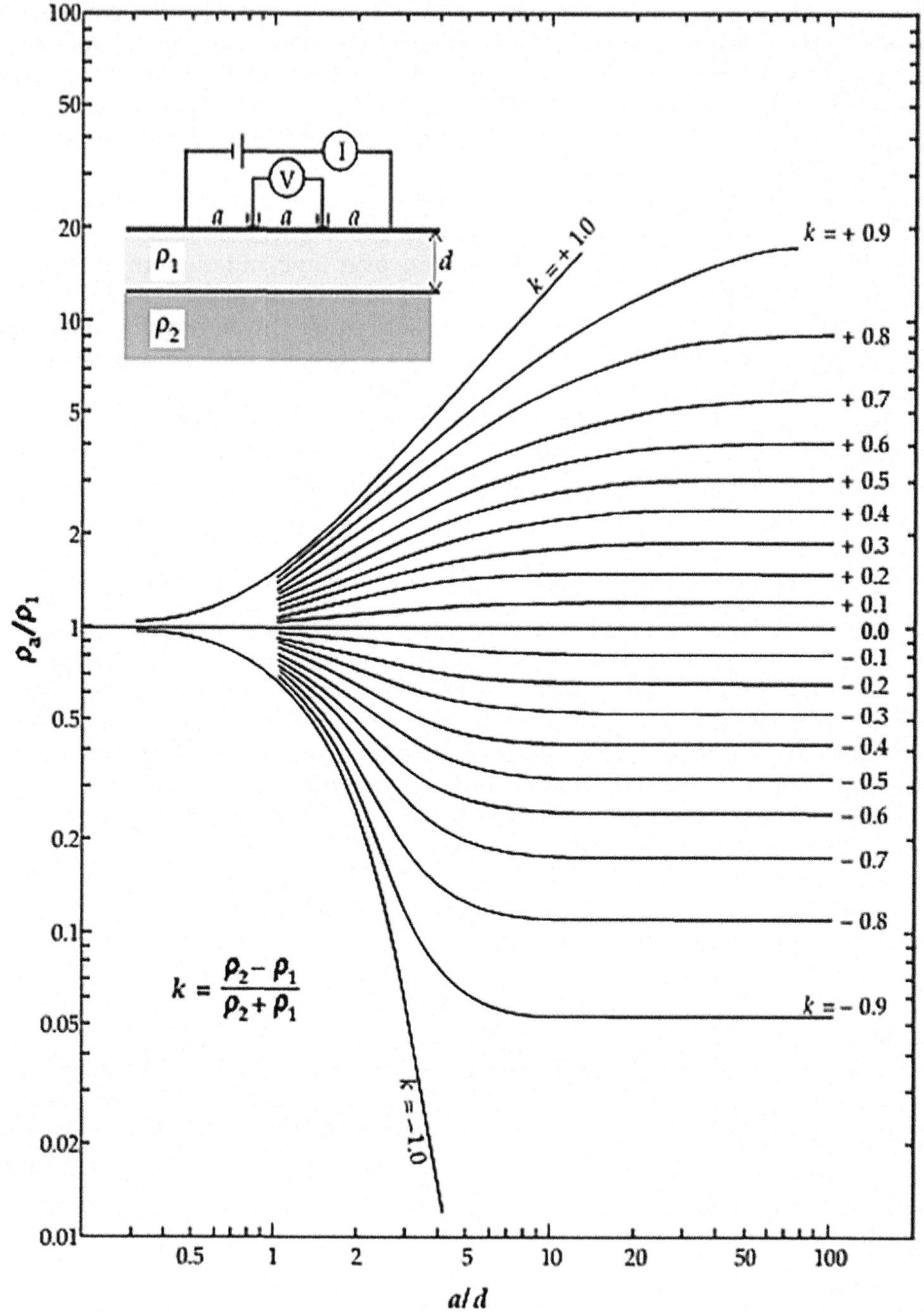

FIG. 4.7 Characteristic curves of apparent resistivity for a two-layer structure using the Wenner array; parameters are defined in the inset. *Source: Lowrie (2007). https://plsclear.com/Assist/Inbox.aspx?sid=1&isarchived=0&tid=1&id=2*

structure, the apparent resistivity curve typically exhibits one of four standard shapes (K, H, A, or Q type), determined by the vertical resistivity sequence in the layers (Fig. 4.8).

Once the observed resistivity profile is identified as being of K, H, A, or Q type, the next step involves performing a one-dimensional inversion of the field data. This inversion procedure requires iterative calculations, which would be very time-consuming without the use of modern computer systems. The inversion method assumes the theoretical response of a multi-layered ground, with each layer characterized by its resistivity and thickness. These parameters must be determined through the inversion process.

By using modern computing techniques, the inversion procedure can be quickly repeated until the discrepancies between the observed and theoretical curves are smaller than a predefined threshold value. The inversion process effectively matches the observed and theoretical curves automatically, offering a detailed model of the subsurface.

The one-dimensional inversion analysis accommodates variations in resistivity and layer thickness with depth. Since the response of a vertically layered structure has an analytical solution, efficient inversion algorithms can be applied. However, recent advancements have introduced methods that also account for lateral heterogeneities in the subsurface. The response of two- or three-dimensional structures requires numerical solutions, which are typically based on finite-difference or finite-element techniques. As the number of unknown

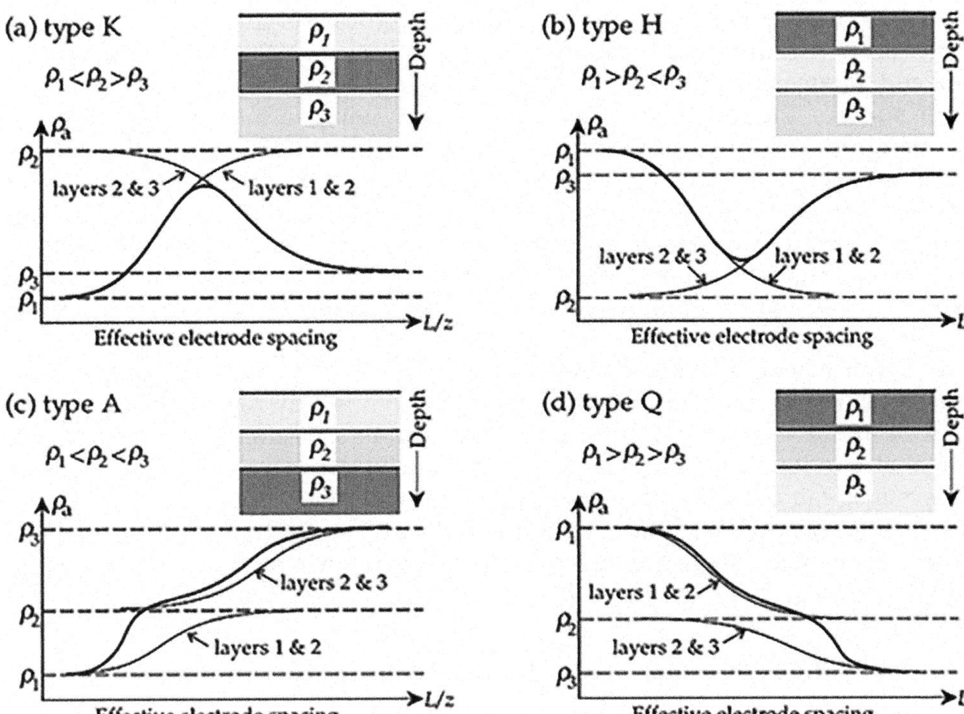

FIG. 4.8 The four common shapes of apparent resistivity curves for a layered structure consisting of three horizontal layers. *Source: Lowrie (2007). https://plsclear.com/Assist/Inbox.aspx?sid=1&isarchived=0&tid=1&id=2*

parameters increases with the complexity of the model, the computational challenges of the inversion process also become more significant.

4.1.7 Advancements in multielectrode and multichannel systems for geoelectrical surveying

The development of multielectrode systems over the past two decades has revolutionized resistivity surveying, significantly expanding the scope of electrical methods in geophysical exploration. The advent of 2D and 3D resistivity tomography has opened new frontiers in subsurface imaging, allowing for more detailed and accurate representations of geological structures. Prior to the 1990s, electrical resistivity methods were primarily applied to resistivity sounding, profiling, and mapping surveys (Telford et al., 1990), with quantitative interpretation mostly confined to one-dimensional (1D) models that assumed horizontal layering of the subsurface (Koefoed, 1979).

The introduction of multielectrode systems enabled the execution of 2D imaging surveys, which provide a much more comprehensive view of subsurface features (Dahlin, 2001). These systems allow for routine, high-resolution imaging and are particularly useful for complex geological scenarios. A variety of electrode arrays have been employed in 2D geoelectrical resistivity surveys, each suitable for different subsurface conditions. The most commonly used conventional arrays include Wenner, dipole-dipole, pole-pole, and pole-dipole configurations. Some of the pioneering works in 2D geoelectrical surveys were conducted using the Wenner array (Griffiths and Turnbull, 1985; Griffiths et al., 1990; Oldenburg and Li, 1999; Olayinka and Yaramanci, 2000).

In these surveys, the resistivity of the 2D model is assumed to vary both vertically and laterally along the survey line but remains constant perpendicular to the line. Unlike the traditional four-electrode system, multielectrode systems involve the use of 25 or more electrodes, which are connected to the resistivity meter via a multicore cable (Fig. 4.9). The resistivity meter, equipped with an internal microcomputer, automatically selects the appropriate set of four electrodes for each measurement through a switching circuit. These systems are highly versatile and programmable, allowing for virtually any electrode array configuration to be used.

Recent advancements include the development of multichannel systems for multielectrode arrays, which significantly reduce survey time. While only two electrodes can be used as current electrodes at any given time, the voltage can be measured between different pairs of potential electrodes. Commercial systems with 4 to 10 channels are widely available, and research systems can accommodate >100 channels (Stummer and Maurer, 2001). This capability has greatly enhanced the efficiency and flexibility of geoelectrical surveys, making them more effective for a wide range of geophysical applications in exploration and environmental monitoring.

4.1.8 Full-waveform systems in geoelectrical surveys

In the pursuit of extracting more detailed information from resistivity data and enhancing the signal-to-noise ratio, significant advancements have been made in the development of systems that record the full waveform of the potential signal (Friedel and Jacobs, 1998;

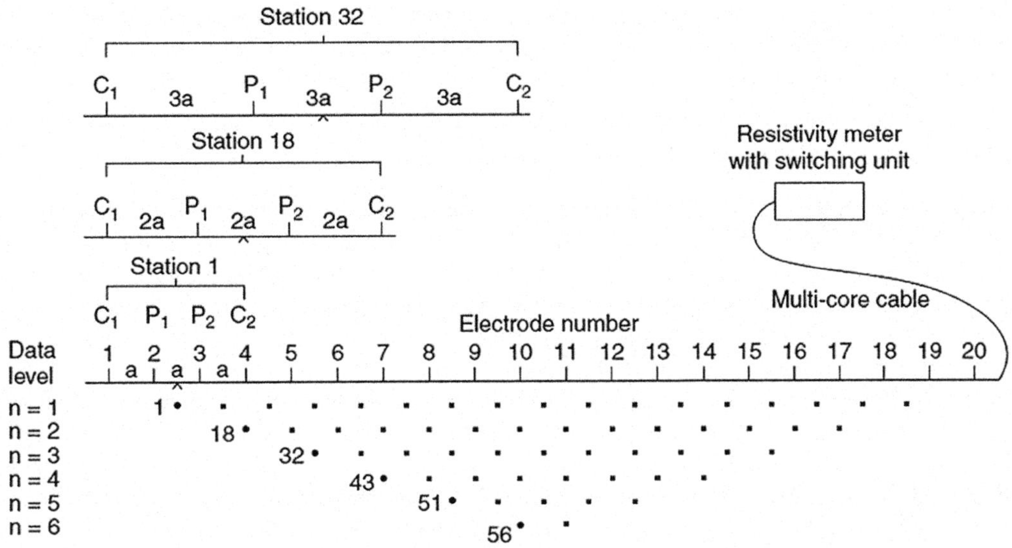

FIG. 4.9 Instrumentation, electrical resistivity. Schematic diagram of a multielectrode system used for a 2D electrical survey and an example sequence of measurements used to build up a pseudosection using the Wenner array. *Source: British Geological Survey. https://www.bgs.ac.uk/legal-and-policy/terms-of-use/*

Storz et al., 2000; Matthews and Zonge, 2003; Rowston et al., 2003; Zhu et al., 2007). This approach is particularly beneficial in surveys that require deep penetration, such as regional investigations or crustal studies, where conventional methods may not provide sufficient data. Additionally, it is also the preferred method for targets that require large dipole offsets, such as studies involving volcanoes or other geologically complex features.

Full-waveform systems offer a significant advantage by capturing the entire waveform of the signal rather than just the apparent resistivity at a single moment. This allows for more sophisticated signal processing and data interpretation techniques, resulting in improved resolution and accuracy in characterizing subsurface features. This technique is indispensable when dealing with challenging environments or when high-resolution data is essential for comprehensive geophysical analysis.

4.1.9 Towed systems for dynamic geoelectrical measurements

Multichannel, multielectrode systems have been successfully adapted for continuous resistivity profiling in waterborne surveys, using floating electrodes that are connected to a cable pulled by a boat (Loke and Lane, 2005; Mansoor and Slater, 2007; Goto et al., 2007). In these systems, data acquisition is coordinated through a personal computer (PC) that integrates the resistivity meter with a GPS unit and water depth sounder, enabling dynamic and real-time monitoring of subsurface conditions.

Similarly, continuous measurement systems have been developed for land surveys. These systems often employ cylindrical steel electrodes arranged in-line (Sorensen, 1996) or spiked wheels designed to maintain constant galvanic contact with the soil (Panissod et al., 1998;

Dabas, 2009). These innovations provide a valuable tool for continuous, high-resolution resistivity data collection, improving efficiency in field surveys and enhancing the quality of geoelectrical data for both environmental and exploration purposes.

4.1.10 Capacitively coupled systems for geoelectrical surveys

In regions with highly resistive surface materials or paved areas, conventional resistivity meter systems, which inject a galvanic current into the ground, often become impractical. To overcome these challenges, capacitively coupled systems offer an alternative. These systems utilize an oscillating, non-grounded electric dipole to generate current flow within the ground, and a second similar dipole is employed to measure the resulting potential distribution at the surface (Kuras et al., 2006).

There are two primary configurations used in both commercial and research instruments of this type. The first configuration, known as the line antenna type, features cylindrical transmitters and receivers towed behind an operator (Fig. 4.10A). This configuration provides measurements that are similar to those obtained from a galvanically coupled in-line dipole-dipole array (Moller, 2001), with investigation depths ranging from 1 m to 20 m, depending on the length of the transmitters and receivers and the distance between them. The second configuration, referred to as the electrostatic quadrupole, uses flat metallic conductors arranged in an equatorial dipole-dipole setup (Fig. 4.10B) (Panissod et al., 1998; Kuras et al., 2007). This configuration is typically employed for surveys with depths of up to a few meters.

Both configurations are dynamic systems in which the array setup remains fixed, but the entire configuration is moved laterally during the survey to ensure comprehensive coverage. These systems facilitate rapid 2D and 3D surveys with dense lateral coverage, making them effective tools for geoelectrical surveys in areas where traditional methods are not feasible (Kuras et al., 2007).

4.1.11 Automated resistivity monitoring systems for geoelectrical applications

Automated PC-controlled resistivity monitoring systems have been developed to detect transient phenomena such as water seepage from dams, landslides, and solute transport (Oldenborger et al., 2007; Sjodahl et al., 2008; Supper et al., 2008; Kuras et al., 2009). These monitoring systems have become increasingly sophisticated, offering continuous, real-time data collection capabilities. They are now being deployed remotely with permanently installed electrode arrays, telemetric control, and seamless data transfer systems. The integration of automated data management systems allows for the efficient handling of large data volumes generated through time-lapse data acquisition (Ogilvy et al., 2009).

These advanced systems provide critical insights into subsurface changes over time, supporting sustainable exploration and management efforts by enabling the monitoring of dynamic geoelectrical processes in real-time.

4.1.12 Computer interpretation in geoelectrical surveying

Following a geoelectrical survey, the raw resistivity measurements are processed into apparent resistivity values, a process that is typically supported by software included with

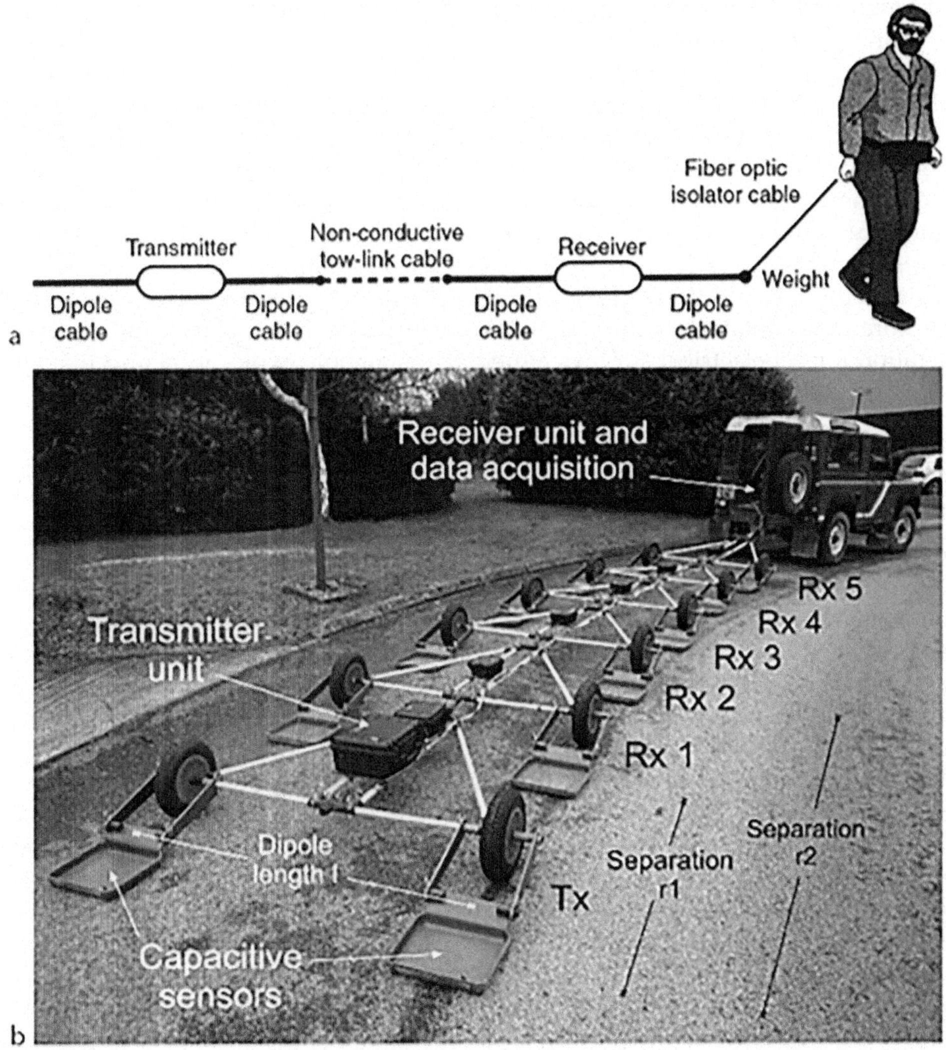

FIG. 4.10 Instrumentation, electrical resistivity. Types of capacitively coupled systems. (A) Line antenna (courtesy of Geometrics Inc.) and (B) plate-wire capacitive sensors (electrostatic quadrupole) arranged in a towed array. *Source: British Geological Survey, NERC. https://www.bgs.ac.uk/legal-and-policy/terms-of-use/*

commercial multielectrode systems. In computer interpretation, the goal is to convert these apparent resistivity values into a resistivity model section that can be used for geological interpretation. This stage of analysis is typically carried out by field engineers, geologists, or geophysicists who may not be familiar with the intricacies of geophysical inversion theory.

Programs such as RES2DINV are designed to perform this conversion in an automatic and user-friendly manner, with minimal input required from the operator. The software comes with a set of default parameters that guide the inversion process, providing reliable results in most cases. However, certain parameters can be adjusted to refine the inversion and achieve more accurate models.

One of the challenges in geophysical inversion is the problem of non-uniqueness, where multiple models can explain the same set of apparent resistivity data. To narrow down the range of potential models, assumptions about the subsurface geology are incorporated into the inversion process. In most surveys, some geological information is available, which can help inform these assumptions. For instance, if it is known that the subsurface features exhibit gradational boundaries, such as pollution plumes or a weathered layer over bedrock, a smoothness-constrained inversion method (deGroot-Hedlin and Constable, 1990) can be used. This method, which is the default in the RES2DINV program, helps generate models that are more representative of the true subsurface structure, improving the accuracy of the interpretation for geoelectrical surveys aimed at sustainable exploration and management.

4.1.13 Overview of resistivity survey in the context of geoelectrical surveying techniques: Contributing to baseline geoscience data for sustainable exploration and management

The Resistivity Survey is a widely used geoelectrical technique that measures the electrical resistance of the subsurface to the flow of an applied electric current. The method involves injecting current into the ground through electrodes and measuring the resulting potential difference at the surface. The resistivity of the subsurface materials is inferred based on how well they conduct electricity. This technique is sensitive to variations in the composition, porosity, and water content of the soil or rock, making it an essential tool for characterizing subsurface conditions.

In the context of sustainable exploration and management, resistivity surveys contribute significantly to baseline geoscience data by providing valuable insights into the distribution of water, minerals, and contaminants beneath the Earth's surface. For example, resistivity surveys are highly effective in locating groundwater resources, mapping aquifers, and assessing soil salinity. The method is also used in environmental monitoring to detect contamination plumes from industrial activities, landfills, or wastewater, helping to prevent further environmental degradation.

In mineral exploration, resistivity surveys are often employed to identify ore bodies, especially in areas where metallic minerals, such as copper, gold, or sulfide deposits, exhibit contrasting resistivity characteristics compared to surrounding materials. The ability to detect variations in resistivity helps reduce the need for extensive drilling and minimizes environmental disruption, ensuring a more sustainable and efficient exploration process.

For sustainable resource management, resistivity surveys play a crucial role in providing data that helps balance resource extraction with environmental preservation. By identifying areas with high water content, for instance, resistivity surveys can guide decisions regarding water management and conservation efforts. Additionally, this method helps identify geotechnical anomalies, such as voids or weak zones, that could affect infrastructure projects like dams, tunnels, or pipelines.

The non-invasive nature of resistivity surveys makes them particularly valuable in environmental studies, as they allow for comprehensive subsurface profiling without the need for intrusive drilling or excavation. By mapping the electrical properties of the subsurface, resistivity surveys provide a detailed understanding of geological features, which can be used to inform decisions related to sustainable land use, resource extraction, and environmental protection.

Overall, resistivity surveys play a vital role in sustainable exploration and management by providing baseline geoscience data that enhances the understanding of subsurface conditions. By supporting more efficient resource exploration, environmental monitoring, and infrastructure development, resistivity surveys contribute to informed decision-making and the sustainable management of natural resources.

4.2 Induced polarization (IP) method: A geoelectrical technique for baseline geoscience data in sustainable exploration and resource management

Induced polarization (IP) is a geophysical phenomenon characterized by a delayed voltage response or low-frequency resistivity dispersion when electrical current is applied to Earth materials. This effect is most prominent in rocks containing metallically conductive minerals such as graphite and specific sulphides. The relationship between IP and valuable sulphide ore deposits has established it as a vital tool in mineral exploration (Naskar and Kulkarni, 2019). The discovery and initial reporting of the induced polarization phenomenon by Conrad Schlumberger paved the way for its widespread adoption in exploration geophysics.

Between 1955 and 1960, numerous mining companies focused on the development of IP technology, with each creating its own equipment to leverage the method's capabilities. In India, the first IP surveys were conducted in 1966 under the "Operation Hard Rock" initiative, which later led to the establishment of the Airborne Mineral Survey Exploration (AMSE) division within the Geological Survey of India (GSI). The full-scale implementation of IP surveying in India commenced in 1970, following the importation of specialized equipment from Canadian mining geophysical companies. This marked a significant milestone in the application of geoelectrical surveying techniques in the region, contributing valuable baseline geoscience data essential for sustainable exploration and management of mineral resources.

In a four-electrode resistivity survey utilizing commutated direct current, the sequence of positive and negative current flow is interspersed with periods when the current is turned off. This results in a box-like waveform for the inducing current (Fig. 4.11A). When the current is interrupted, the voltage across the potential electrodes does not immediately drop to zero. Instead, it undergoes an initial abrupt decrease to a fraction of its steady-state value, followed by a slow decay over several seconds (Fig. 4.11B). Conversely, when the current is re-applied, the potential initially rises sharply and then gradually approaches the steady-state value. These slow changes in the voltage signal—both decay and growth—are attributed to induced polarization (IP), which occurs due to two similar phenomena related to the rock structure: membrane polarization and electrode polarization.

Membrane polarization is a characteristic of electrolytic conduction and arises from differences in the ability of ions within pore fluids to migrate through a porous rock. Minerals within the rock typically carry a negative surface charge, attracting positive ions in the pore fluid. These ions accumulate on the mineral grain surfaces and extend into the adjacent pores, partially obstructing them. When an external voltage is applied, positive ions are able to pass

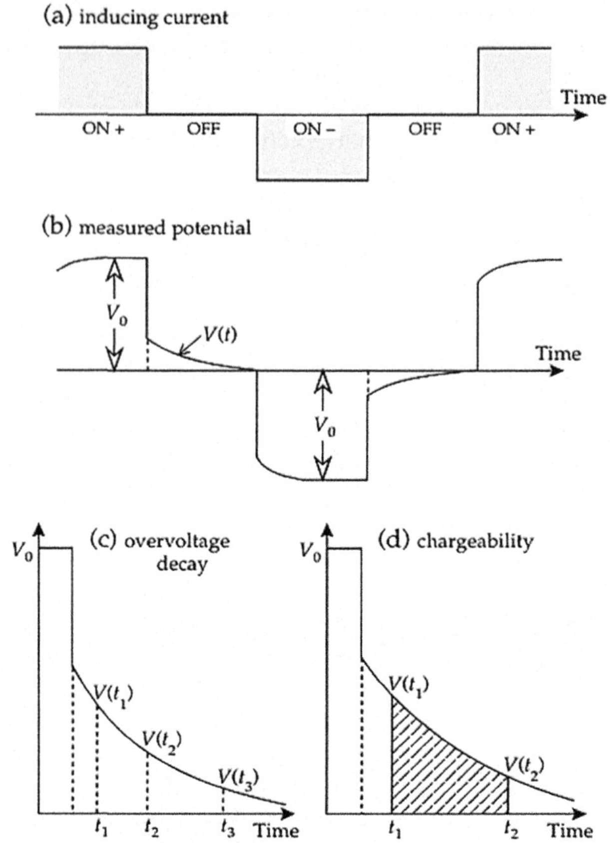

FIG. 4.11 Illustration of IP-related potential decay and its effects on waveform and overvoltage characteristics: (A) Decay of potential following primary current interruption, (B) impact of IP decay time on potential waveform for square-wave input current, (C) overvoltage behavior, (D) metal factor. *Source: Lowrie (2007). https://plsclear.com/Assist/Inbox.aspx?sid=1&isarchived=0&tid=1&id=2*

through the "cloud" of accumulated positive charge, but negative ions accumulate in the pores unless the pore size is large enough to bypass the blockage. This creates a membrane-like effect that selectively allows the passage of certain ions. This phenomenon is especially prominent in rocks containing clay minerals, as these grains are relatively highly charged and readily adsorb ions. The ionic build-up occurs quickly after the voltage is applied, and when the current is switched off, the ions drift back to their initial positions.

Electrode polarization, a similar effect, occurs when ore minerals are present. In this case, metallic grains conduct charge through electronic conduction, while electrolytic conduction takes place in the surrounding pore fluid. However, the flow of electrons through the metal is much faster than the movement of ions in the electrolyte, causing opposite charges to accumulate on opposite surfaces of the metallic grain. This build-up blocks the flow of ions through the pore fluid, creating an overvoltage that persists for some time after the external current is applied. The magnitude of this effect is proportional to the concentration of metallic minerals in the rock. Once the current is turned off, the accumulated ions disperse and the overvoltage decays gradually (Kearey et al., 2002; Glover, 2015).

The two effects responsible for induced polarization are difficult to distinguish at the measurement level. The most common field method used in an induced polarization (IP) survey is based on the dipole-dipole array, where the current electrodes form a transmitter pair and the potential electrodes form a receiver pair. The steady-state voltage (V_0) is recorded and compared with the amplitude of the residual voltage V(t) that decays over time after the current is interrupted (Fig. 4.11C). The ratio of V(t) to V_0 is expressed as a percentage, and this ratio decays over a period of 0.1–10 s after the current is switched on and off. By sampling the decay curve at multiple points, the shape of the curve and the area under it can be analyzed (Fig. 4.11D). The area under the decay curve, expressed as a fraction of the steady-state voltage, is referred to as the chargeability (M), a key parameter in the analysis of induced polarization data (Siegel, 1959).

$$M = \frac{1}{V_0} \int_{t_1}^{t_2} V(t)\,dt \tag{4.11}$$

The chargeability M has the dimensions of time and is typically expressed in seconds or milliseconds. It is the most commonly used parameter in induced polarization (IP) studies.

Induced polarization (IP) influences the duration of the potential decay time. When the decay time is shorter than the duration that the inducing current is turned off, successive half-cycles of the potential will not interfere with each other. However, in the presence of a disseminated conductor, the decay time increases, resulting in overlap and distortion of the half-cycles. This effect becomes more pronounced at higher signal frequencies. As a result, the ratio $\frac{V(t)}{V_0}$ increases, giving the false impression of a better conductor than what is actually present. In other words, the apparent resistivity decreases as the frequency increases.

This frequency dependence of the IP effect is also relevant in resistivity surveys conducted with alternating current (Fig. 4.12). To exploit this phenomenon, apparent resistivity is measured at two low frequencies, typically denoted as f and F (f < F). Commonly, f might range from 0.05 Hz to 0.5 Hz, and F could range from 1 Hz to 10 Hz. The apparent resistivity at these two frequencies, ρ_f and ρ_F, allows the definition of a frequency effect (FE) as (Kemna, 2000; Reynolds, 1997):

$$FE = \frac{\rho_f - \rho_F}{\rho_F} \tag{4.12}$$

The ratio FE is often multiplied by 100 to express it as a percentage (PFE) (Wait, 1959a). In the absence of an IP effect, resistivity remains the same at both frequencies. A higher FE or PFE indicates greater induced polarization in the ground. At frequencies above 10 Hz, mutual inductance between the primary and detection circuits can create unwanted potentials, which can be minimized through field procedures or analytical methods. The presence of metallic conductors is quantified by a similar parameter, the metallic factor (MF), which reflects the conductivity difference at the two frequencies (Marshall and Madden, 1959; Telford et al., 1976).

$$MF = A(\sigma_F - \sigma_f) = A\left(\rho_F^{-1} - \rho_f^{-1}\right) = A\left(\frac{\rho_f - \rho_F}{\rho_f \rho_F}\right) \tag{4.13}$$

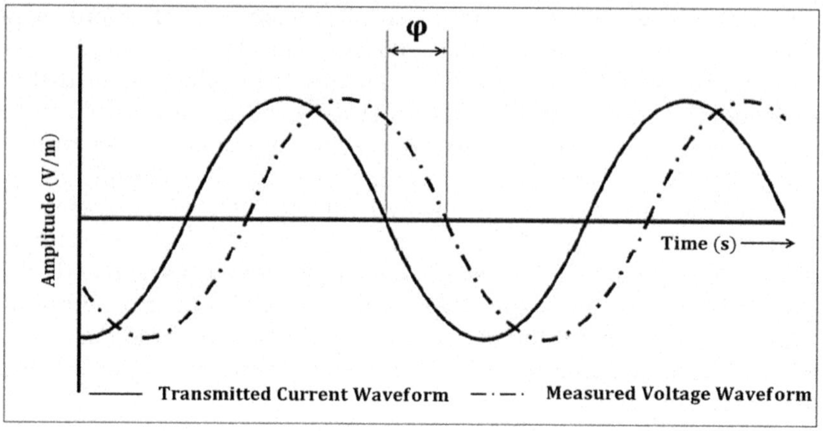

FIG. 4.12 Illustrations of frequency domain Induced Polarization (IP): Visual representations of IP responses in the frequency domain, highlighting key characteristics and interpretations. *(Modified).*

The constant A is equal to $2\pi \times 10^5$; the units of the metallic factor (MF) are those of conductivity (i.e., $\Omega^{-1}m^{-1}$ or Sm^{-1}).

An Induced Polarization (IP) survey typically includes both lateral profiling and vertical sounding using the expanding spread technique. This method utilizes a dipole-dipole array, which is effective for mapping lateral variations in subsurface properties, such as resistivity, chargeability, or metal factor (MF), along the survey profile. In dipole-dipole profiling, measurements are taken at different positions along the profile to capture lateral changes in subsurface properties. In contrast, dipole-dipole sounding focuses on exploring vertical variations by progressively increasing the distance between the transmitter and receiver electrodes. The values of apparent resistivity (ρ_a), frequency effect (FE), or metal factor (MF) obtained from each measurement are plotted beneath the midpoint of the array, at the intersection of two lines inclined at 45°, resulting in a pseudo-section plot (Fig. 4.13A). However, it is important to note that the values shown in the pseudo-section do not represent the true value of the parameter at the indicated depth. For instance, a measurement of apparent resistivity reflects an equivalent half-space beneath the array rather than the actual subsurface conditions at that specific depth.

To better visualize the variation of the IP parameter beneath the profile, a two-dimensional image is created by contouring the results (Fig. 4.13B). This pseudo-section provides a convenient yet artificial representation of anomalous conductors. However, it does not accurately reflect their true lateral or vertical extent. Further investigation, such as exploratory drilling, is often required to understand the nature of these anomalous regions more precisely (Telford et al., 1990).

Resistivity anomalies generally arise from continuous conductors, such as groundwater or massive ore bodies. However, disseminated mineralization within a rock may not produce significant resistivity anomalies. The IP method is especially effective for detecting disseminated concentrations of conductive ore minerals, making it highly valuable in base-metal exploration, where large, low-grade ore bodies can be of commercial interest. In addition to mineral exploration, the IP effect is influenced by the rock's porosity and saturation, making

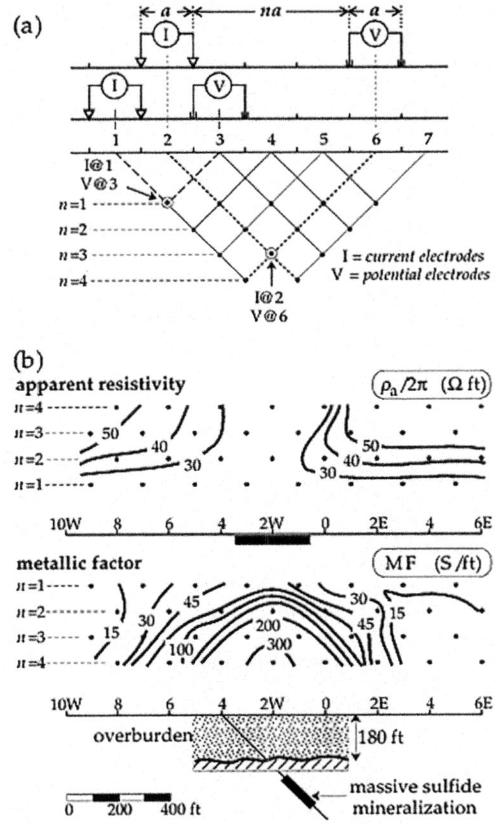

FIG. 4.13 Construction and interpretation of pseudo-sections in IP surveys: (A) Pseudo-section construction for a dipole-dipole IP survey, with measured parameters plotted at the intersection of 45° lines extending from transmitter and receiver mid-points. (B) Pseudo-sections of apparent resistivity and metallic factor for an IP survey over a sulfide orebody. *Source: Lowrie (2007); Telford et al. (1990). https://plsclear.com/Assist/Inbox.aspx?sid=1&isarchived=0&tid=1&id=2*

it useful for applications in groundwater exploration and various environmental assessments. The following tables (Tables 4.1, 4.2, and 4.3 from Telford et al., 1976; Kiberu, 2002) provide a very general guide to possible chargeabilities of materials. One reason why in-situ chargeabilities tend to appear lower than laboratory values is that field measurements involve currents that flow through large volumes of mixed materials.

4.2.1 IP field procedures, data visualization, and interpretation

The IP method involves applying a current through two current electrodes (C) and measuring the voltage across two potential electrodes (P). The field setup is similar to that of resistivity surveys. IP measurements are often taken in drillholes to log near-hole properties and explore deeper layers, as well as in underground surveys. In mining, IP helps estimate metallic mineral grades and identify areas of better mineralization. There is potential to

TABLE 4.1 The values in this table involved more commonly used charging and integration times of 3 s and 0.02–1.0 s, respectively.

Material type	Chargeability (ms)
Groundwater	0
Alluvial	1–4
Gravels	3–9
Precambrian volcanics	8–20
Precambrian gneisses	6–30
Schists	5–20
Sandstones	3–12
Argillites	3–10
Quartzites	5–12

TABLE 4.2 Charging and discharging times (i.e., transmitter on and off times) were about one minute each, which is much longer than field survey systems; therefore, values are larger than field measurements.

Material type	Chargeability (ms)
20% sulfide	2000–3000
8–20% sulfide	1000–2000
2–8% sulfide	500–1000
Volcanic tuffs	300–800
Sandstone, siltstone	100–500
Dense volcanic rocks	100–500
Shale	50–100
Granite, granodiorite	10–50
Limestone, dolomite	10–20

distinguish between electrode polarization in metallic minerals and membrane polarization in clays. Common electrode arrays include gradient, pole-dipole, and dipole-dipole, with the choice depending on resolution needs, overburden conditions, and ore body orientation. IP surveys can be conducted in vertical sounding, horizontal profiling, or 2D/3D tomography modes, using either time-domain or frequency-domain techniques. (Table 4.4) compares commonly used electrode configurations for IP surveys.

The dipole-dipole array is favored by field geophysicists for its ability to resolve anomalies, explore greater depths, and minimize EM coupling effects, despite the challenge of low signal-to-noise ratios. For shallow sources, smaller electrode spacing enhances the apparent IP effect, but larger spacings may miss small, significant anomalies. IP data are typically

TABLE 4.3 Chargeability of minerals at 1% concentration in the samples (charging and integration times as per Table 4.1).

Material type	Chargeability (ms)
Pyrite	13.4
Chalcocite	13.2
Copper	12.3
Graphite	11.2
Chalcopyrite	9.4
Bornite	6.3
Galena	3.7
Magnetite	2.2
Malachite	0.2
Hematite	0.0

TABLE 4.4 Showing the comparison of common IP arrays.

	Performance		
Characteristics	Dipole-dipole	Three electrode	Gradient
Magnitude of response	Good	Fair	Poor
Dip information	Poor	Fair	Good
Depth of exploration	Good	Good	Good
Horizontal resolution	Fair	Poor	Good
Target location	Good	Fair	Good
Freedom from overburden irregularities	Fair	Good	Poor
Freedom from EM coupling	Good	Fair	Poor
Depth estimation	Fair	Fair	Poor
Signal-to-noise ratio	Poor	Fair	Good
Labor needed	Poor	Fair	Good
Speed of coverage	Fair	Fair	Good

Coggon (1973).

presented as profiles, plotting chargeability or metal factor against station location, or on contour plots to highlight anomaly zones. The pseudosection, primarily used for frequency-domain data with the dipole-dipole array, provides an estimate of anomaly depth, though it doesn't represent the actual subsurface geoelectric section (Marshall and Madden, 1959). Modifications to pseudosections have been proposed for more realistic anomaly displays

(Appa Rao and Sarma, 1981). Qualitative interpretation of IP results from profiles or pseudosections can provide information on anomaly location, extent, and depth, based on their sharpness, symmetry, and amplitude. For quantitative analysis, computer modeling is commonly used, with forward and inverse solutions as primary methods. In forward modeling, the interpreter selects a model that best fits the field data, while inverse modeling directly analyzes the data to derive subsurface properties. Various modeling techniques have been discussed in the literature (Coggon, 1971; Aiken and Hastings, 1973). Despite advancements in computer modeling, geological intuition and experience remain crucial for accurate interpretation. Refer to Fig. 4.14 and Table 4.4 for further details.

4.2.2 Limitations of IP

The limitations of conventional IP surveys are apparent. For weak responses, it is often difficult to distinguish anomalies caused by water-filled shear zones from those due to mineral deposits. Similarly, it is challenging to differentiate chargeability or PFE anomalies from uneconomic metallic conductors and those associated with economic ones. Another limitation, common to all electrical methods, is that responses from causative sources beneath highly conductive overburden are typically masked and go undetected in IP surveys. To address these challenges, significant research and development efforts have led to important advancements in IP methodology, which are briefly summarized below.

4.2.2.1 Magnetic induced polarization (MIP) method: A geoelectrical technique for baseline geoscience data in sustainable exploration and resource management

The Magnetic Induced Polarization (MIP) method measures the variation in induced polarization and resistivity of the Earth by assessing the magnetic field associated with galvanic current flow, as opposed to the electric field used in traditional IP or Electric Induced Polarization (EIP) methods. The traditional method of measuring induced polarization (IP) involves using electric fields created by passing a current through the ground. In this method, current flows into the Earth through two electrodes (C1 and C2). The resulting voltage in the Earth is measured across two other electrodes (P1 and P2), placed at a specific distance from the current electrodes. The IP characteristics are determined by comparing the time or shape

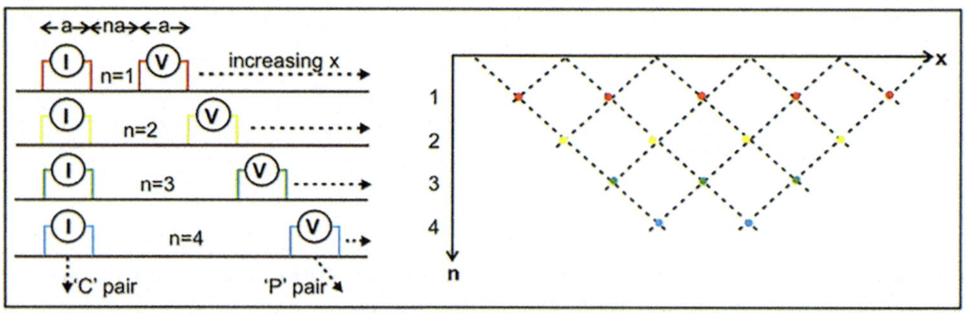

FIG. 4.14 The dipole-dipole array—Left: Electrode positions on the ground; Right: Plotting positions for each x, a, and n combination. *Source: Reynolds (2011). https://marketplace.copyright.com*

differences between the current and the measured voltage. This method, called the "EIP" method, focuses on measuring electric fields (Fig. 4.15). The EIP method has been valuable to the mining industry for the past three decades, contributing to many mineral discoveries (Seigel, 1974). However, like all exploration methods, it has its limitations. One of the biggest challenges is the "masking effect" caused by extreme differences in resistivity (Seigel and Howland-Rose, 1983).

There are significant differences between MIP and EIP in terms of field practice, mathematical theory, and field results. For example, the MIP method is not affected by horizontal layering and reflects only lateral variations in electrical properties. It is also capable of detecting anomalous electrical bodies even beneath highly conductive surface layers. This makes MIP particularly useful in areas with highly conductive overburden, such as saline regions or weathered rock, as found in Australia. MIP responses are often more complex and varied than those seen in EIP measurements, with polarity reversals being common, whereas they are rarely observed in EIP. MIP employs high-sensitivity component magnetometers as sensors, which are smaller in size compared to the electric dipole sensors used in EIP, providing better geometric resolving power. The MIP response is determined by solving Poisson's equation and a magnetostatic problem in terms of potentials using a finite-volume algorithm. It depends on the chargeability difference ($m—m_0$) and the relative conductivity ($\frac{\sigma}{\sigma_0}$), where m_0 and σ_0 are constant values.

The inversion process for m can be approached in three ways: (1) inverting without constraints and adding a constant to the result, (2) inverting with positivity constraints, or (3) working with "lnm." The inversion is framed as an optimization problem, minimizing the objective function while ensuring the model accurately fits the data. A Gauss-Newton method is used to compute the model perturbation at each iteration, and the system is solved using a conjugate gradient least squares method. Depth or distance information requires a depth or sensitivity-based weighting. Positivity constraints are essential, but since MIP responses depend on chargeability differences, the inversion setup can be flexible. Synthetic model studies reveal that a large conductivity ratio between a target and its host strongly affects the MIP response. A ratio difference greater than two orders of magnitude can make the MIP response undetectable, while ratios between 0.1 and 10 have limited impact on recovered chargeability. The inversion algorithm is demonstrated using data from Binduli, Australia (Chen and Oldenburg, 2003).

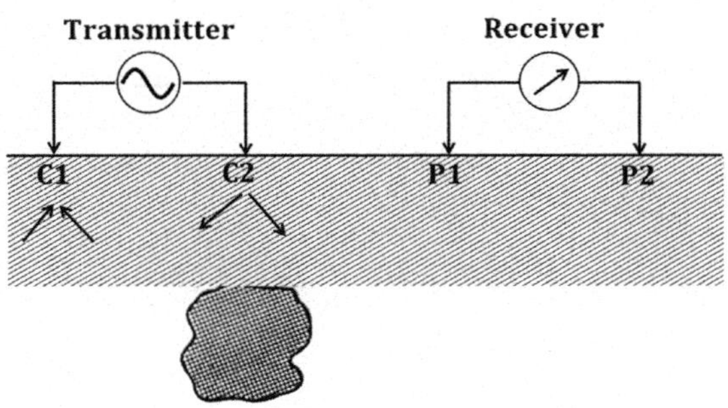

FIG. 4.15 Electric Induced Polarization (EIP) Measurement Setup. *Modified.*

In an MIP survey, an external current is introduced into the ground through a pair of current electrodes (C1 and C2), connected by a U-shaped wire (Fig. 4.16). The magnetic field generated by the current flow in both the wire and the ground can be measured using a magnetometer placed on the surface or in boreholes (Chen and Oldenburg, 2003). Since the frequency of the transmitter current is typically low, usually below 4 Hz, inductive effects are ignored. As a result, the problem is treated as a steady-state situation, where the electric field (E) and magnetic field (H) follow the relevant electromagnetic (EM) equations.

4.2.2.2 Field data example of MIP survey in the Binduli project, 12 km West of Kalgoorlie, Western Australia, by Placer Dome Asia Pacific and Croesus Mining

MIP (Magnetic Induced Polarization) provides a significant advantage over EIP (Electrical Induced Polarization) in conductive environments, where EIP struggles to measure voltages. This is illustrated in the schematic in (Fig. 4.17), studied by Seigel and Howland-Rose (1990). In conductive overburden areas, surface-applied current has difficulty penetrating to energize deeper targets, making both MIP and EIP challenging. However, positioning the source

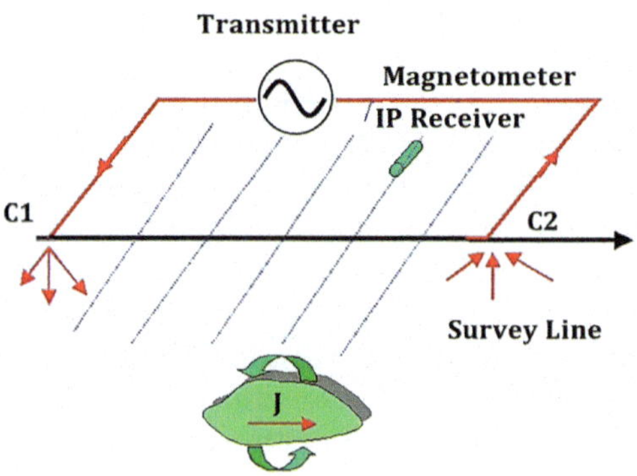

FIG. 4.16 Schematic of the surface MIP survey setup with current electrodes C1 and C2. *Modified.*

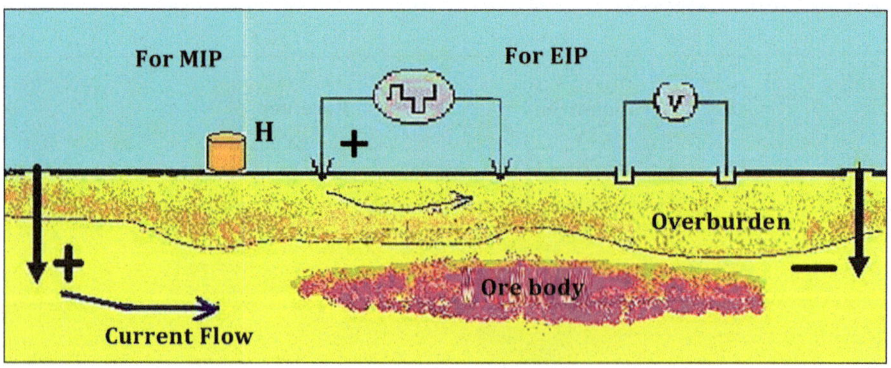

FIG. 4.17 Schematic diagram illustrating the difference between MIP and EIP in a conductive overburden scenario, with MIP current electrodes positioned below the overburden in boreholes. *Modified.*

electrodes below the overburden allows current to directly reach the target. While EIP is still hindered by the overburden's masking effect, MIP, by measuring magnetic signals, faces no such issue.

The principle was demonstrated at the Binduli Project, 12 km west of Kalgoorlie, Western Australia. The Binduli deposit, known for hosting medium to high-grade gold, has produced over 1 million ounces of gold. Mining has focused on low-grade vein deposits and high-grade Eastern Contact Mineralization (ECM) at the Centurion pit. Due to the conductive overburden, including salt lakes, EIP was ineffective, so the MMR/MIP survey aimed to identify larger zones of ECM-like, polarizable mineralization.

A field survey was conducted at the Binduli Project, Western Australia, using the MMR/MIP method to explore mineralization beneath conductive overburden. Four dipoles were used for the source electrodes, and data were collected along 33 survey lines, with 275 data points. The survey area covered approximately 800 m by 800 m, with current electrodes positioned at 8400E (600 N and 2200 N), up to 90 m below the surface. A significant RPS anomaly was identified at the east side of profile 8600E, with a peak magnitude of 2.6 degrees, indicating more polarizable material than the host rock. A 3-D model was created, and the inversion process recovered chargeability values in 31,416 cells, with results from both MMR and a uniform conductivity model showing similar outcomes. The final chargeability model (Fig. 4.18) revealed a significant anomaly at approximately 100 m depth, potentially associated with ECM-style ore. This case underscores the value of baseline geoscience data generation using MMR/MIP for exploration in areas with challenging conductive overburden.

4.2.2.3 Cole-cole model: A framework for understanding induced polarization (IP) responses in geoelectrical baseline data for sustainable exploration and resource management

The Cole-Cole model (Cole and Cole, 1941; Keller and Frischknecht, 1966), as demonstrated by Pelton et al. (1978), provides an effective representation of the complex resistivity (both real and imaginary components) of polarized rock formations. This approach enables

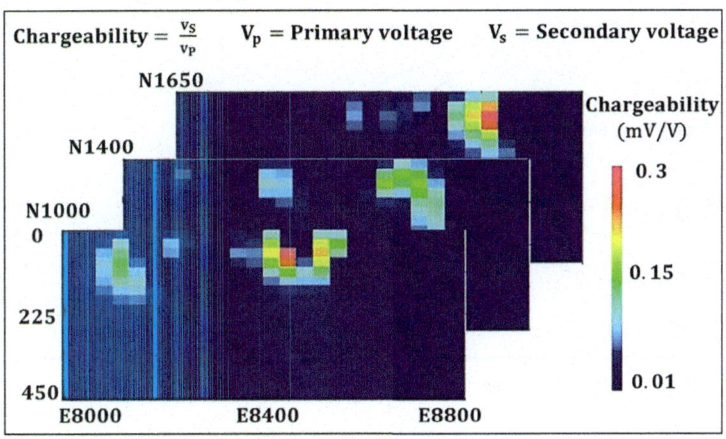

FIG. 4.18 Three cross-sections of the recovered chargeability 3-D model from the Binduli Project Area. *Modified.*

phase-coupled spectral IP responses, which help to improve the signal-to-noise ratio and minimize the impact of electromagnetic coupling effects. Consequently, the IP method is classified as a galvanic electrical technique. This frequency-dependent complex resistivity is depicted in Fig. 4.19, and is described by the following formula:

$$\rho(\omega) = \rho_{DC}\left[1 - m\left\{1 - \frac{1}{1+(-i\omega\tau)^{C}}\right\}\right] \tag{4.14}$$

Here, ρ_{DC} represents the DC resistivity (Ohm-m), ω is the angular frequency (rad/sec), and the time constant τ (tau) determines the frequency at which the peak response in the imaginary resistivity occurs. The dimensionless intrinsic chargeability, m, characterizes the intensity of the induced polarization (IP) effect, while the relaxation parameter C describes the magnitude of this effect. These phenomena are illustrated graphically in Fig. 4.19 (Fu, 2011; Emond, 2007).

Fig. 4.20 illustrates the resistivity behavior of the Cole-Cole model transformed into the time domain by varying different parameters. The resistivity increases over time, with the induced polarization (IP) effect occurring at later times as the time constant (τ) is increased. As chargeability (m) increases, the difference between the amplitude of the IP effect at early and later times becomes more pronounced. Additionally, a higher value of the relaxation parameter (C) results in a steeper resistivity curve over time.

Ongoing research in IP focuses on improving the discrimination between different metallic mineral species, removing electromagnetic coupling effects, enhancing signal-to-noise ratios, and exploring magnetic-induced polarization effects. Mathematical modeling, particularly of various geometric shapes of polarizable bodies, has proven to be a valuable tool for interpreting IP data. Additionally, advances in IP instrumentation are incorporating large-scale integrated circuits and field-based computers for data processing, analysis, and interpretation. The future of IP surveying looks promising, as it remains one of the most effective

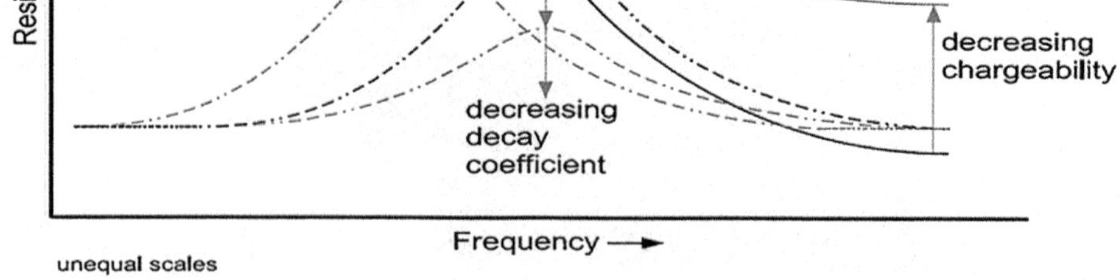

FIG. 4.19 Frequency domain behavior of complex resistivity in the Cole-Cole model: Effects of chargeability (m), time constant (τ), and decay coefficient (C). *Source: Emond et al. (2006). mlei.fu.rice@gmail.com*

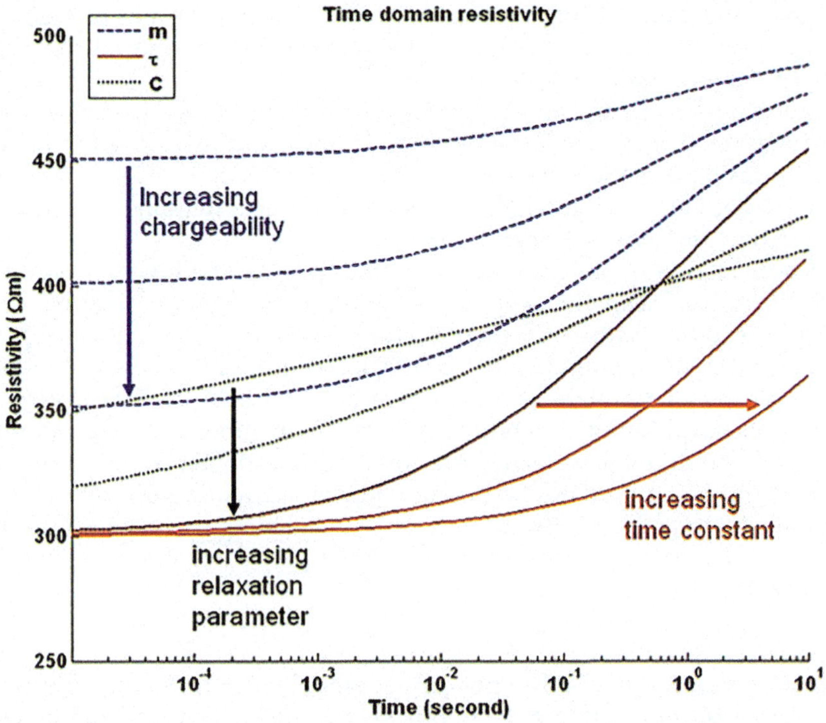

FIG. 4.20 Time domain behavior of the Cole-Cole model: Effects of chargeability (m), time constant (τ), and decay coefficient (C). *Source: Fu (2011). mlei.fu.rice@gmail.com*

geophysical methods for detecting small amounts of metallic minerals in concealed mineral deposits, playing a vital role in sustainable exploration and management.

4.2.3 Overview of induced polarization (IP) method in the context of geoelectrical surveying techniques: Contributing to baseline geoscience data for sustainable exploration and management

The Induced Polarization (IP) method is a powerful geoelectrical surveying technique that measures the delayed voltage response of the subsurface to an applied electrical current. This method is particularly useful for detecting materials that exhibit capacitive effects, such as metallic minerals, clay-rich zones, and groundwater contaminants. The IP method works by inducing an electric field in the ground and then measuring the decay of the potential difference after the current is turned off, providing valuable information about subsurface properties.

In the context of sustainable exploration and management, the IP method contributes significantly to baseline geoscience data by offering insights into the mineral composition, water content, and pollution levels of the subsurface environment. It is especially effective in mapping and characterizing ore bodies, particularly those containing sulfide minerals, which are known to exhibit strong IP responses. By identifying zones of high chargeability and resistivity, IP surveys help pinpoint potential mineral deposits, supporting more efficient and sustainable exploration practices.

The IP method is also valuable in environmental applications, such as monitoring groundwater contamination, detecting landfill leachate plumes, and assessing the integrity of geotechnical structures. In these scenarios, the ability to identify chargeable materials allows for the detection of pollutants or anomalies in the subsurface, helping to guide remediation efforts while minimizing environmental impact.

In sustainable exploration and management, the IP method allows for non-invasive data collection, reducing the need for drilling and minimizing surface disturbance. It complements other geophysical techniques, such as resistivity and self-potential (SP), to provide a more comprehensive understanding of subsurface conditions. The combination of these methods enhances resource management, environmental protection, and exploration efficiency, ensuring that exploration efforts are conducted with minimal disruption and maximum precision. Overall, the IP method contributes to a more responsible and informed approach to resource exploration and environmental monitoring, providing crucial baseline data that supports sustainable practices in the geosciences. By enabling early-stage detection of valuable resources and environmental hazards, IP surveys play a key role in balancing economic development with environmental stewardship.

4.3 Natural potentials and currents in geoelectrical surveying: Contributing to baseline geoscience data for sustainable exploration and management

In the context of geoelectrical surveying, **natural potentials** and **currents** refer to the naturally occurring electrical phenomena in the Earth's subsurface, which can be measured and analyzed to gather crucial baseline geoscience data. These measurements provide valuable insights into subsurface conditions without the need for artificial energy sources or invasive exploration techniques, making them integral to sustainable exploration and resource management.

4.3.1 Self-potential (SP) and its mechanisms in geoelectrical investigations: SP surveying methods as a key tool for sustainable exploration and management

Self-potential (SP) refers to natural voltage differences in the Earth's subsurface caused by electrochemical and mechanical processes. SP is used in geoelectrical studies to address environmental issues, groundwater flow, and mineral exploration. It provides insights into subsurface structures, mineral deposits, groundwater movement, and contamination. SP is commonly applied in base metal exploration to detect massive ore bodies, unlike the IP method, which is used for disseminated ores. It is an affordable and quick method, often used as an initial step in base metal exploration.

In India, the Geological Survey of India (GSI) frequently uses SP for rapid reconnaissance to find sulfide mineralization in areas such as Saladipura (Rajasthan), Kalyadi (Karnataka), Dhani-Basri and Todi ka Bas area, Dausa and Alwar districts, Rajasthan (Naskar and Kulkarni, 2019; Naskar et al., 2018), and Ambaji (Gujarat), as well as for graphite exploration in regions like Trivandrum (Kerala) and Khammam (Andhra Pradesh). SP is also increasingly used in groundwater, geothermal studies (Chandra, 2016; Corwin and Hoover, 1979; Fitterman and Corwin, 1982; Anderson, 1984), environmental hazard mapping (Linde and Revil, 2007), earthquake prediction (Corwin and Morrison, 1977), and engineering applications, including dam seepage mapping and fault detection.

In the oil and gas industry, SP is used in borehole evaluations to detect permeable beds, estimate water salinity, and measure clay content. It is particularly useful in conductive mud environments. The method is inexpensive and simple, making it ideal for routine mineral, groundwater, and civil engineering investigations. SP is considered one of the most cost-effective surface geophysical techniques.

4.3.1.1 Mechanisms of self-potential

Self-potential (SP) refers to naturally occurring voltage differences in the Earth's subsurface, which arise from various mechanisms. One primary cause of SP is the electrokinetic or streaming potential mechanism. This occurs when an electrolyte, such as groundwater, flows through narrow pores in rocks, creating a potential difference. This phenomenon is influenced by factors like the resistivity and viscosity of the electrolyte, the pressure driving the flow, and the interaction between the liquid and solid surfaces (Lyklema, 1995; Marshall and Madden, 1959; Revil, 1999).

In addition to streaming potential, many SP anomalies have an electrochemical or mineral/mineralization potential origin. Variations in ion concentration in groundwater can lead to an electric diffusion potential, driven by concentration gradients. A similar effect occurs when metallic electrodes interact with an electrolyte, generating a contact potential. The combination of diffusion and Nernst potentials results in the self-potential anomalies typically observed in the millivolt range (Stoll et al., 1995).

In mineral exploration, SP measurements are employed to detect ore bodies, as anomalies associated with certain mineral deposits, such as sulfide ores, are often negative, with magnitudes reaching several hundred millivolts. Several models have been proposed to explain these anomalies. The galvanic model initially suggested that ore bodies functioned like a voltaic cell, with oxidation at the surface generating a potential difference (Sato and Mooney, 1960). However, this model has largely been rejected in favor of the more widely accepted redox potential theory. According to this theory, differences in the oxidation states above and below the water table generate the observed self-potential anomalies (Naudet et al., 2003; Revil et al., 2009) (Fig. 4.21).

Several mechanisms of SP are employed in various practical applications. For instance, the streaming potential mechanism is used to detect leaks in earthen dams (Butler and Llopis, 1990) and to locate caverns in karst terrains (Vichabian and Morgan, 2002; Chandra, 2016; Roy, 2022). The redox potential mechanism is applied to identify contaminant plumes (Nyquist and Corry, 2002; Naudet et al., 2003; Revil et al., 2009), while the thermoelectric effect is utilized to map the front of burning coal fires (Revil et al., 2013).

SP measurements play a crucial role in geophysical surveys for various purposes, including locating mineral deposits, monitoring groundwater contamination, and studying subsurface water flow. This non-invasive method is particularly valuable in environmental monitoring, especially for detecting contamination from waste sites or leaking pipelines. While the exact origins of self-potential anomalies are still not entirely understood,

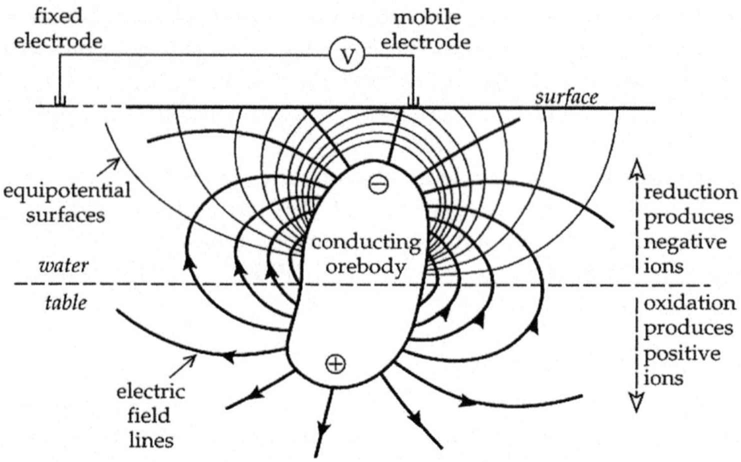

FIG. 4.21 A schematic model illustrating the origin of the self-potential anomaly in an orebody, driven by variations in oxidation potential above and below the water table. *Source: Lowrie (2007). https://plsclear.com/Assist/Inbox.aspx?sid=1&isarchived=0&tid=1&id=2*

SP remains an essential tool in geophysics for exploring subsurface structures and addressing environmental challenges.

4.3.1.2 *Self-potential (SP) surveying*

The equipment required for an SP (self-potential) survey is quite simple. It includes a sensitive high-impedance digital voltmeter to measure the natural voltage between two electrodes placed in the ground. Regular metal stakes are unsuitable as electrodes because electrochemical reactions between the metal and the moisture in the ground can cause incorrect charges, interfering with the small natural potentials. To prevent this, non-polarizable electrodes are used. These electrodes typically consist of a metal rod submerged in a saturated solution of its own salt, such as a copper rod in copper sulphate solution, contained in a ceramic pot. The pot allows the electrolyte to slowly leak through its porous walls, ensuring good electrical contact with the ground.

There are two common field methods for SP surveys (Fig. 4.22). The gradient method (Revil and Jardani 2013), where the components of electric field gradient ($E = -\nabla V$), uses electrodes placed at a fixed distance of about 10 m. The voltage difference between the electrodes is measured, and the pair is moved forward along the survey line. The total potential at each station is found by summing the voltage differences. While non-polarizable electrodes are used to reduce polarization effects, small errors still occur with each measurement. These errors can accumulate over multiple measurements. To reduce this, the "leapfrog" technique is used, where the leading and trailing electrodes are swapped between measurements. The main downside of the gradient method is the cumulative error, but it is practical because only a short length of wire is needed for each measurement.

The total field method uses a fixed base electrode and a mobile measuring electrode. This method measures the total potential directly at each station (Corry et al., 1983). The main challenge with this method is that the long wire connecting the electrodes must be unwound

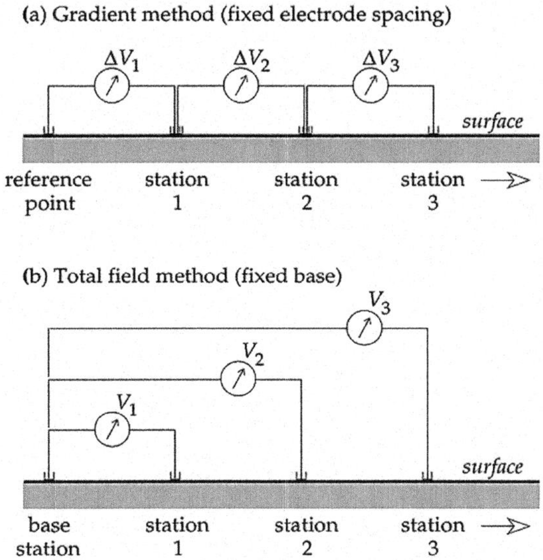

FIG. 4.22 Field techniques for measuring self-potential: (a) Gradient method and (b) total field method. In the gradient method, the total potential (V) at a station is determined by summing the previous potential differences. In the total field method, the total potential (V) is measured directly at each station. *Source: Lowrie (2007). https://plsclear.com/Assist/Inbox.aspx?sid=1&isarchived=0&tid=1&id=2*

and rewound as the electrodes are moved, but this method results in less cumulative error and provides higher quality data. The total field method is generally preferred unless the terrain is difficult. The survey process for both methods involves measuring potential at different stations along a profile. As with gravity and magnetic surveys, the data is mapped (Fig. 4.23), and anomalies are interpreted based on their shape. Often, simple geometric models are used to interpret the anomalies.

For example, a polarized sphere might be used to model circular anomalies, while a horizontal line or cylinder might represent elongated anomalies. Point sources are often used to model SP anomalies, and more complex anomalies can be modeled by combining different sources and sinks.

4.3.1.3 Error correction in self-potential surveys

Self-potential (SP) measurements are prone to errors and noise, arising from various sources. These include the bioelectric effect from vegetation, telluric currents from the Earth's natural electrical activity, and interference from anthropogenic sources such as electric locomotives, high-voltage power lines, and other infrastructure. Additionally, topographic effects can alter the electric polarity of the potential field due to surface undulations, and electrode-related noise, particularly drift errors, can occur over time due to electrochemical reactions.

To minimize drift errors, repeat measurements at the same location are essential. The drift should be kept under 2 mV throughout the day. This is part of the SP survey protocol, which involves conducting repeat measurements at reference points or base stations to ensure

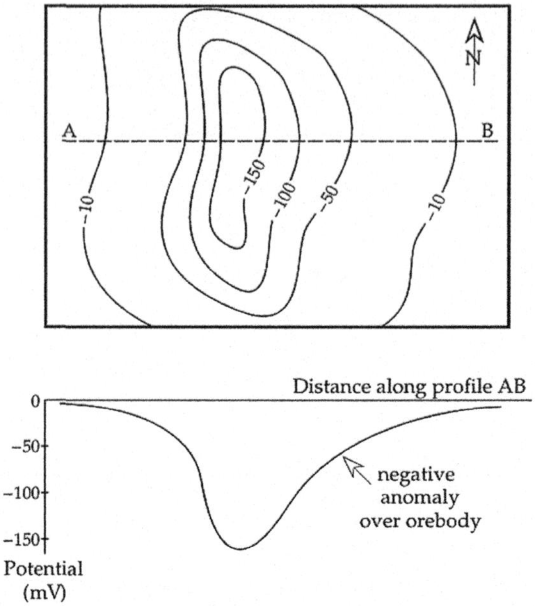

FIG. 4.23 Hypothetical contour lines of a negative self-potential anomaly over an orebody. The asymmetry of the anomaly along the profile AB indicates that the orebody is dipping toward point A. *Source: Lowrie (2007). https://plsclear.com/Assist/Inbox.aspx?sid=1&isarchived=0&tid=1&id=2*

accuracy. Measurements should be made in a closed-loop fashion, particularly around several tie points, to reduce systematic errors. A precision of about 1 mV is usually sufficient.

In on-land measurements, electrodes are typically buried <0.5 m deep and soaked in water to minimize contact resistance and maintain electrolytic contact. Field notes should document not only the voltage, electrode spacing, and station position but also environmental factors such as soil condition, nearby infrastructure, and bodies of water. The reference electrode should be placed in a flat region, far from flowing water, to reduce noise from external sources.

To further minimize the impact of extraneous noise, repeated measurements should be taken in a loop and averaged. Drift errors can be corrected by distributing and adjusting them across the entire dataset of measurements.

4.3.1.4 Interpretation of self-potential data

In self-potential (SP) surveys, measurements are taken at discrete stations along a profile, and the potential at each point results from the combination of multiple electric fields. The key field of interest is the direct current (DC) field, which includes the SP field and possible cultural noise, while time-varying fields (such as telluric currents and other cultural noise) are not relevant and should be avoided. Assessing and minimizing noise in SP data is critical, with methods for estimating noise levels proposed by Roy (2015, 2018).

Several theoretical, experimental, and field studies have contributed to the understanding and interpretation of SP anomalies, including works by GSI researchers (Paul, 1965; Roy and

Chowdhury, 1959; Naskar et al., 2018). Numerous graphical and numerical methods have been developed to interpret SP anomalies, including characteristic points (DeWitte, 1948; Yungul, 1950), curve matching (Meiser, 1962), least-squares (Abdelrahman et al., 1997), gradient analysis (Abdelrahman et al., 1997), nonlinear modeling (Guptasarma, 1983), and Fourier analysis (Atchuta Rao et al., 1982).

To interpret SP anomalies, data are mapped and contoured, with interpretations based on geometric shapes. Visual inspection of anomalies can reveal the elongation and orientation of ore bodies. For example, circular anomalies can be modeled using a polarized sphere, while elongated anomalies can be modeled with a polarized cylinder (El-Araby, 2004). The top of an ore body is typically located beneath the minimum potential. These interpretation methods, often based on simple geometric models, help in understanding subsurface features, including ore bodies and geological contacts.

4.3.2 Overview of self-potential method in the context of geoelectrical surveying techniques: Contributing to baseline geoscience data for sustainable exploration and management

The Self-Potential (SP) method is a valuable geoelectrical surveying technique that measures naturally occurring voltage differences in the Earth's subsurface, which are typically caused by electrochemical, electrokinetic, or thermoelectric effects. Unlike other geophysical methods that rely on external energy sources, SP utilizes the natural potential variations within the ground, making it a non-invasive and cost-effective tool for subsurface exploration.

The method plays a significant role in sustainable exploration and management by contributing to the baseline geoscience data necessary for understanding natural subsurface processes. SP surveys are particularly useful in detecting mineral deposits, identifying ore bodies, and monitoring environmental conditions such as groundwater flow, contamination, and subsurface fluid movement. By interpreting SP anomalies, geophysicists can infer the presence of mineralization, contamination plumes, or even natural resources like geothermal energy.

SP measurements are sensitive to external factors such as topographic features, anthropogenic noise, and electrochemical electrode effects, requiring careful noise reduction and error correction protocols. Techniques like closed-loop measurements, drift monitoring, and careful electrode placement help enhance the accuracy of SP data. Additionally, SP methods are commonly used in conjunction with other geoelectrical techniques, such as resistivity and induced polarization, to provide a comprehensive understanding of the subsurface.

In the context of sustainable exploration and management, SP surveys contribute valuable data to environmental monitoring, resource management, and site remediation efforts. They allow for non-destructive mapping of subsurface features, ensuring minimal disturbance to the environment while providing crucial insights for responsible resource exploration and management strategies. Overall, the SP method, by providing baseline geoscience data, plays a pivotal role in guiding sustainable exploration practices, supporting environmental conservation, and facilitating the management of subsurface resources.

4.3.3 Telluric currents: A key geoelectrical surveying technique for contributing to baseline geoscience data in sustainable exploration and management

Ultraviolet radiation from the Sun ionizes the molecules of air in the Earth's thin upper atmosphere. These ions accumulate in distinct layers, forming the ionosphere at altitudes ranging from approximately 80 km to 1500 km above the Earth's surface. Electric currents in the ionosphere result from the systematic motion of ions, influenced by various factors such as daily and monthly tides, seasonal variations, and periodic fluctuations in ionization associated with the 11-year sunspot cycle. These currents generate magnetic fields with corresponding frequencies, which are detectable at the Earth's surface and can be analyzed from long-term, continuous records of geomagnetic activity, spanning from fractions of a second (geomagnetic pulsations) to several years (Fig. 4.24). The fluctuating magnetic fields, in turn, induce telluric currents—electric currents that flow horizontally through the Earth's crust and mantle.

The distribution of telluric current density is influenced by variations in the resistivity of the horizontal conducting layers. At shallow crustal depths, the flow of these currents is affected by subsurface structures that create resistivity contrasts, potentially caused by geological formations or mineralized zones. The equipment used for measuring telluric current density is relatively simple. It consists of a pair of non-polarizable electrodes separated by a distance L, typically ranging from 10 m to 100 m. The potential difference V between the electrodes is measured using a high-impedance voltmeter, allowing for the calculation of the electric field E at the midpoint between the electrodes as

$$E = \frac{V}{L} \tag{4.15}$$

Using Ohm's Law

$$E = \rho J \tag{4.16}$$

and assuming the telluric current flows through a conducting rock layer with resistivity ρ_1, the telluric current density J at each measurement station along a profile is given by:

$$J = \frac{V}{\rho_1 L} \tag{4.17}$$

Since the direction of the telluric current is not initially known, two pairs of electrodes oriented perpendicular to each other are used. One pair is aligned north-south, while the other is east-west. To account for temporal variations, an orthogonal pair of electrodes is established at a fixed base station outside the surveyed area. Another orthogonal pair is moved across the survey area. The potential differences measured across each electrode pair in both the mobile and base arrays are recorded simultaneously for several minutes at each measurement station. By correlating these records, temporal variations in the direction and intensity of the telluric currents can be accounted for and removed. The deflection of telluric currents by resistive subsurface structures, as depicted in Fig. 4.25, is an idealized representation.

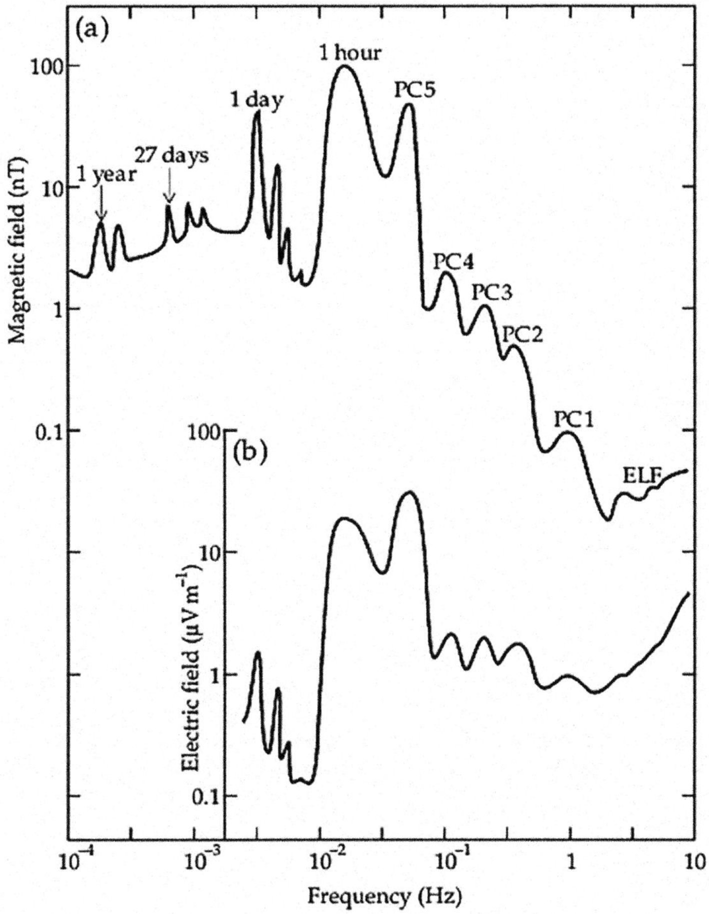

FIG. 4.24 (A) Frequency spectrum of natural variations in the horizontal intensity of the geomagnetic field, and (B) Corresponding spectrum of induced electric field fluctuations, computed for a model Earth with uniform resistivity of 20 Ω-m. *Source: Lowrie (2007); Serson (1973). https://plsclear.com/Assist/Inbox.aspx?sid=1&isarchived=0&tid=1&id=2*

It assumes an infinite resistivity ρ_2 in the core of the anticline. In practice, the current is not entirely diverted through the more conductive layer, and some current flows through the more resistive layer as well. Thus, the resistivity ρ_1 in **Eq.** (4.17) does not correspond directly to the resistivity of the good conductor but rather represents an apparent resistivity that reflects a combination of both ρ_1 and ρ_2. This apparent resistivity is not the true resistivity of either layer but a result of the measurement configuration. This method of telluric current measurement plays a crucial role in providing baseline geoscience data for sustainable exploration and management of natural resources, offering insights into subsurface structures and resistivity variations that are vital for geological studies and mineral exploration.

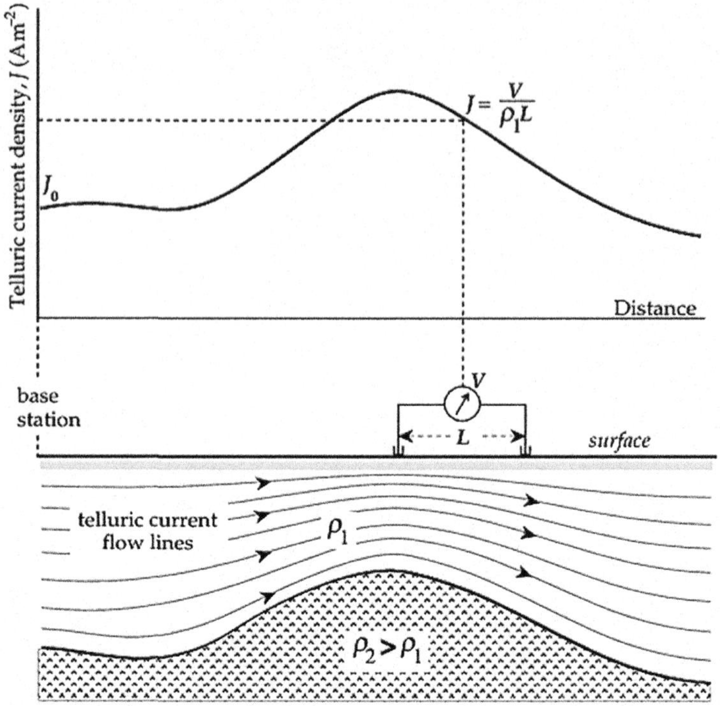

FIG. 4.25 The telluric current lines are altered by variations in the thickness of a conductive layer overlying a more resistive structure (bottom). The telluric current density (top) is determined by measuring the voltage between a pair of electrodes with a fixed separation at the surface. *Source: Lowrie (2007); Robinson and Coruh (1988). https://plsclear.com/Assist/Inbox.aspx?sid=1&isarchived=0&tid=1&id=2*

4.4 Conclusions

The chapter titled "Geoelectrical Surveying for Sustainable Exploration and Management" in the proposed Elsevier publication "Baseline Geoscience Data for Sustainable Exploration and Resource Management" underscores the critical role of geoelectrical techniques in exploring and managing subsurface resources sustainably. These methods, which measure electrical properties such as resistivity and conductivity, are essential for understanding groundwater systems, mineral deposits, and subsurface structures.

The chapter highlights the adaptability of geoelectrical surveying techniques, such as resistivity imaging and induced polarization, in diverse geological settings. It emphasizes their effectiveness in delineating aquifers, assessing soil salinity, and identifying mineralized zones, while stressing the importance of accurate calibration, survey design, and data integration for reliable results. Furthermore, the use of geoelectrical methods ensures minimal environmental disruption, making them suitable for sustainable exploration.

In conclusion, the chapter advocates for the application of geoelectrical surveying as a cornerstone of environmentally conscious resource management. By providing detailed insights into subsurface conditions with minimal ecological impact, these techniques align with the principles of sustainable exploration. The integration of geoelectrical data with other geoscience methods enhances decision-making, ensuring resource exploration and management practices that are both efficient and environmentally responsible.

CHAPTER

5

Electromagnetic surveying techniques

5.1 Theory of electromagnetic induction in geophysical surveys

The development and application of Electromagnetic (EM) methods in mineral exploration date back to the early 1920s. Sheard et al. (2005), Vallee et al. (2011), and Smith (2014) provide detailed reviews of EM methods, focusing on the period prior to 2012. The principle behind EM detection in mineral exploration lies in the varying electrical resistivities of ore bodies and surrounding rocks, which can differ by several orders of magnitude (Guo et al., 2020). This contrast in resistivity, particularly with conductive sulphide ores, makes EM methods especially powerful for identifying ore bodies in various geological environments. As a result, EM methods are commonly applied in airborne, onshore, offshore, and borehole surveys, with continuous improvements driven by both industry and academia.

EM methods—whether airborne or ground-based—are among the most widely used geophysical techniques in mineral exploration. These methods utilize either natural or controlled sources to transmit EM waves into the earth and measure the returned EM fields. The magnetotelluric (MT) method, introduced by Cagniard in 1953, became significant for mineral and hydrocarbon exploration as MT technology advanced and modern data processing techniques emerged. During the 1950s and 1960s, controlled-source frequency domain electromagnetic (FDEM) sounding gained widespread application in mineral exploration, as described by Frischknecht (1967). In addition, to improve the detection of shallow exploration targets, the audio-magnetotelluric (AMT) method was introduced by Berdichevsky et al. (1969). Since the natural signals for both MT and AMT methods are weak, controlled sources are often applied to induce the necessary currents, resulting in the controlled-source audio-magnetotelluric (CSAMT) method (Goldstein, 1971). Both MT and CSAMT methods are widely used for mineral exploration, and their penetration depth is frequency-dependent.

Induced polarization (IP) is another important method for mineral exploration, particularly because of the chargeability of certain mineral deposits (Naskar et al., 2018). The IP method is often combined with MT/CSAMT for more effective ore body detection. Over the past decade, the frequency domain EM method has also seen significant improvements. He (2010) introduced two new frequency-based EM methods: pseudo-random electromagnetic (PSEM) and wide-field electromagnetic (WFEM), both of which define apparent resistivity through the complete analytical expressions of horizontal electrical and vertical magnetic

Baseline Geoscience Data Generation for Sustainable Exploration of Natural Resources
https://doi.org/10.1016/B978-0-443-45221-5.00013-8

field sources in a homogeneous half-spacing. Wait (1956) introduced the transient electromagnetic (TEM) method, which measures electromagnetic signals in the time domain using a loop. Building on TEM, Strack et al. (1990) applied the long-offset transient electromagnetic (LOTEM) method for geophysical investigations of the Earth's crust. More recently, the multichannel transient electromagnetic (MTEM) method (Di et al., 2020; Li et al., 2019), developed from LOTEM, offers a broader dynamic range, allowing for measurements at much smaller offsets than LOTEM (Wright et al., 2001). To address the detection of shallow exploration targets like coal deposits, the short offset transient electromagnetic method (SOTEM) was introduced by Xue et al. (2013). The study by Naskar and Lahiri (2014) demonstrates the effectiveness of the magnetotelluric (MT) method in mapping both shallow and deep crustal structures, which is crucial for coal exploration in the Birbhum coalfield of West Bengal, India. Through the application of MT surveys, the study successfully identified variations in the subsurface resistivity, providing valuable insights into the depth and extent of coal deposits in the region.

Magnetotelluric imaging was employed as a key geophysical tool to understand the coal-bearing formations, emphasizing the potential for detailed coal exploration. The study effectively mapped resistivity anomalies associated with coal layers, offering a noninvasive method for exploring the coal resources in the Birbhum coalfield. These findings underscore the utility of MT surveys in coal exploration by providing a high-resolution image of subsurface structures, which aids in the identification of coal seams in the region.

EM methods, both time and frequency domain, are invaluable not only in the discovery of new mineral deposits and the extension of reserves but also for structural mapping of the Earth, oil and gas exploration, environmental and engineering applications, and groundwater investigation. Advances in sensors, satellite and drone-based platforms, electronics, data processing, inversion techniques, and the application of machine learning have significantly improved the resolution and detectability of EM methods in recent years (Zhou et al., 2022).

5.2 Electromagnetic induction theory

Electromagnetic (EM) surveys conducted at frequencies below 50 kHz are based on the principle of electromagnetic induction. In this method, an alternating magnetic field is generated by a coil or cable, which induces electric currents in conductors. While the conductivity of rocks and soils is generally too low to support significant induction currents, the presence of a good conductor triggers the formation of eddy currents. These eddy currents, in turn, generate secondary magnetic fields that overlay the primary magnetic field, and these secondary fields can be measured at the ground surface (Fig. 5.1A). The various EM induction field methods share a common principle: a transmitter coil generates the primary alternating magnetic field, while a receiver coil measures both the primary signal and the secondary signal resulting from the eddy currents induced in the conductor (Fig. 5.1A).

In the conductor, the magnetic field experiences a phase shift due to the material's conductivity. This phase shift, expressed as [equal to $\frac{z}{\text{skindepth}(\delta)}$], leads to a phase difference

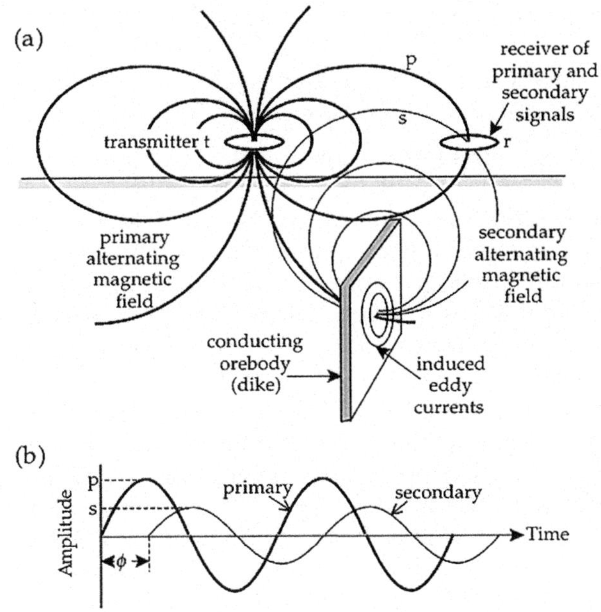

FIG. 5.1 (A) Illustration of primary and secondary fields in the horizontal loop induction method of electromagnetic exploration for shallow orebodies. (B) Amplitudes and phases of the primary (p) and secondary (s) fields. *Source: Lowrie et al. (2007).*

(φ) between the secondary and primary signals detected by the receiver (Fig. 5.1B). While the theory of EM induction can be complex, even in relatively simple situations, a basic understanding can be achieved by applying concepts from electrical circuit theory. This approach involves using complex numbers, where $i = \sqrt{-1}$, to represent the current systems in the transmitter, receiver, and conductor as simple sinusoidal loops.

The transmitter current induces a primary voltage V_p in the receiver, in which the eddy currents in the conductor also induce a secondary voltage V_s. The ratio of V_s to V_p in the receiver is

$$\frac{V_s}{V_p} = -\frac{M_{tc}M_{cr}}{M_{tr}L}\left(\frac{\beta^2 + i\beta}{1+\beta^2}\right) \tag{5.1}$$

where $\beta = \dfrac{\omega L}{R}$ is the *response parameter* of the conductor, M_{tc} = mutual inductance between transmitter and conductor, M_{cr} = mutual inductance between conductor and receiver, M_{tr} = mutual inductance between transmitter and receiver, L = self-inductance of the conductor, and R = resistance of the conductor. The function in parentheses in **Eq.** (5.1) is a complex number, so the voltage ratio (or *response* of the measuring system) can be written P + iQ. The real part P has the same phase as the primary signal and is called the *in-phase component* of the response. The imaginary part Q is 90° out of phase with the primary signal [i.e., if the primary signal is ~cosωt, the imaginary part is ~sinωt = cos $(\frac{\pi}{2} - \omega t)$]; it is called the *quadrature* component.

5.3 Frequency domain airborne electromagnetic method

The basic equations governing electromagnetic (EM) induction bear a resemblance to the diffusion equation, specifically the three-dimensional heat flow equation, expressed as:

[$\frac{\partial T}{\partial t} = \kappa\left(\frac{\partial^2 T}{\partial x^2} + \frac{\partial^2 T}{\partial y^2} + \frac{\partial^2 T}{\partial z^2}\right)$; three-dimensional heat flow equation]

This similarity classifies EM induction as a diffusive method. Diffusive techniques, such as gravity, magnetic field, geothermal, and seismic surface wave methods, respond to a volumetric average of a specific physical parameter and do not provide detailed information on its distribution. Therefore, EM induction provides an average electrical conductivity value for a given volume, with a higher resolution than potential field methods, but still limited in showing fine-scale details.

EM induction is particularly well-suited for airborne electromagnetic (AEM) surveys, which were originally conducted using fixed-wing aircraft but are now more commonly carried out with helicopters. Helicopters are better suited to adapt to rugged terrain and can fly closer to the ground. The transmitter and receiver coils are typically mounted in fixed positions on the aircraft or in a towed "bird," following configurations similar to those used in airborne magnetometer surveys. Modern frequency-domain AEM systems use small transmitter and receiver coils, typically about half a meter in diameter. The transmitter coil can be oriented horizontally or vertically, while the receiver coil is positioned in a way that maximizes coupling. A variety of geophysical methods are often employed simultaneously during airborne surveys, and in some configurations, the transmitter may be in the airplane, while the receiver is located in the bird or even another airplane. This increased separation between the transmitter and receiver enhances the depth penetration of the survey.

An example of a helicopter-borne geophysical system operated by the German Federal Institute for Geosciences and Natural Resources (BGR) is shown in Fig. 5.2. This system is equipped with sensors for collecting five-frequency electromagnetic, magnetic, and gamma-ray spectrometry data, along with altimeters and positioning systems (Kirsch, 2006). Because the secondary field induced by the conductor is much smaller than the primary field, the primary field is typically "bucked out," and the relative secondary field is measured in parts per million (ppm).

AEM flight patterns usually consist of parallel profiles that traverse the terrain. Fixed-wing aircraft typically fly at about 100 m above ground level, while helicopters fly at altitudes of around 30 m. Given the altitude-dependence of AEM data, procedures such as tie-line leveling and AEM leveling (Huang and Fraser, 1999) are commonly applied to parameters like apparent resistivity and apparent depth, which are less sensitive to changes in sensor altitude. Additionally, noise from external sources, such as radio transmitters or power lines, is often filtered or interpolated to remove it from the data. Inductive effects from buildings, electrical installations, or strongly magnetized underground sources should not be removed during initial data processing. These effects, particularly at low frequencies, may manifest as conductive or resistive features outlining the locations of man-made structures or magnetized sources. If needed, these effects can be addressed in later stages of data processing after thorough inspection, and small data gaps can be interpolated for smoother resistivity variations.

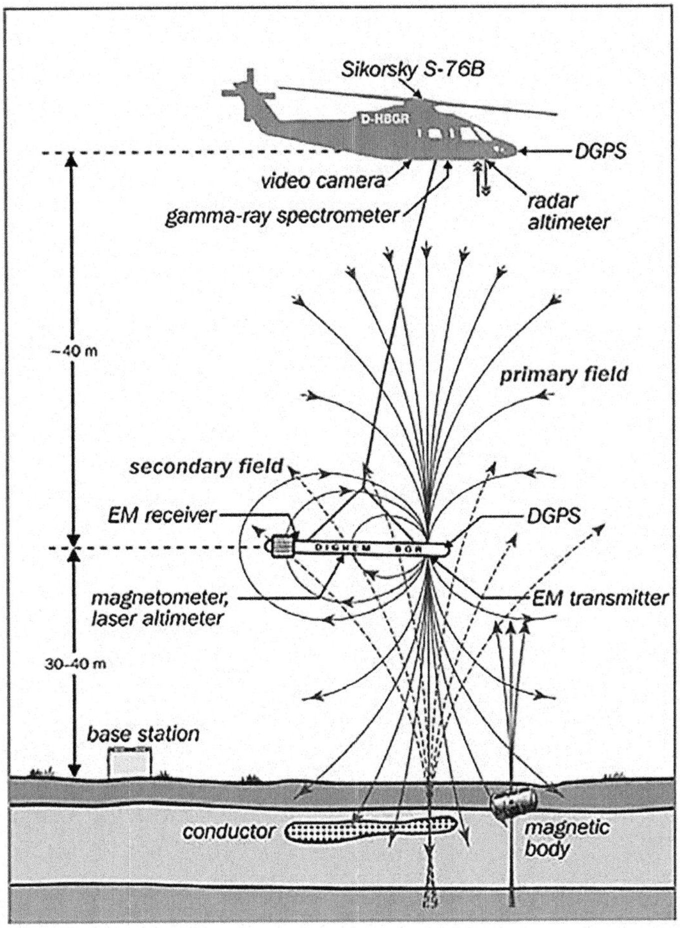

FIG. 5.2 BGR's helicopter-borne geophysical system: Electromagnetic, magnetic, GPS, and laser altimeter sensors are housed by a "bird," a cigar-shaped 9 m long tube, which is kept at about 30–40 m above ground level. The gamma-ray spectrometer, additional altimeters, and the navigation system are installed in the helicopter. The base station records the time varying parameters diurnal magnetic variations and air pressure history (Modified).

AEM survey results are typically presented as maps and vertical resistivity sections (VRS). The maps display parameters like apparent resistivity and centroid depth or resistivities derived from 1D inversion results for specific model layers at various depths or elevation levels. However, for a more accurate interpretation of the AEM data, additional information, such as drilling data, is often necessary to complement the electromagnetic induction results.

Frequency-domain electromagnetic (EM) measurements are well-suited for high-resolution surveys, especially when the targets are located at shallow depths, generally up to about 100 m. However, for deeper targets, ground-based or airborne time-domain measurements are more appropriate. Fixed-wing systems are commonly used for reconnaissance surveys in flat terrain because they have greater range than helicopter-borne systems and are generally more cost-effective. However, fixed-wing systems are less flexible and have lower resolution capabilities compared to helicopters.

Frequency-domain systems that use natural sources, such as audio-frequency magnetic (AFMAG) systems, are typically not practical for detailed EM surveys due to limitations in signal strength and depth penetration. These systems are more suited to broader reconnaissance or large-scale surveys rather than high-resolution exploration of specific targets.

5.4 Frequency domain ground-based electromagnetic technique

When a potential conductivity anomaly is identified from airborne surveys, it is common practice to investigate the anomaly further using a ground-based electromagnetic (EM) induction method. The transmitter for ground-based surveys may be a long cable or a large horizontal loop placed on the ground surface, or it may be a small coil (typically 1 m in diameter), oriented either vertically or horizontally. The receiver is typically a similar small coil, which can be used to detect various aspects of the secondary signal, including its direction, intensity, or phase.

In its simplest application, the tilt of the coil about a horizontal axis is used to measure the dip angle of the combined primary and secondary fields at the receiver. This method allows for the location, outlining, and, to some extent, depth determination of a conductor. However, the dip-angle method captures only part of the information available in the secondary signal and does not provide detailed information about the electrical properties of the conductor. As shown in **Eq.** (5.1), the electrical properties of the conductor influence the phase and relative amplitudes of the in-phase and quadrature components of the secondary signal, relative to the primary. Therefore, methods that measure the phase component of the EM signal provide a more detailed interpretation of the conductor's characteristics.

5.4.1 Slingram and ground conductivity meters

Originally, horizontal orientation transmitter and receiver coils were used for exploration purposes, leading to the term Horizontal Loop Electromagnetic (HLEM) systems, also known by the Swedish names *Slingram* or *Ronka*. In these systems, both coils are operated at a fixed distance from each other during the survey. The depth penetration of the system can be controlled by adjusting either the frequency or the coil separation, and these two factors are often coupled. High-frequency systems, specifically designed for mapping shallow underground structures (e.g., Geonics EM 31 and EM 38), often combine the transmitter and receiver coils into a single instrument, allowing for one-man operation. Operation with vertical coils is also possible. Some instruments, such as Maxmin (Apex) with frequencies ranging from 110 to 3550 Hz (Scott, 2014), are not limited to coplanar coil orientations. Unlike geoelectrical methods, no galvanic contact with the ground is required, meaning measurements can be performed on sealed terrains.

A typical response of the in-phase and quadrature signals to a steeply dipping and highly conducting fracture zone is shown in Fig. 5.3. The depth of penetration is determined by the skin depth, which is influenced by factors such as the frequency and conductivity (Kirsch, 2006). If the coil separation is smaller than the skin depth, the system is said to be

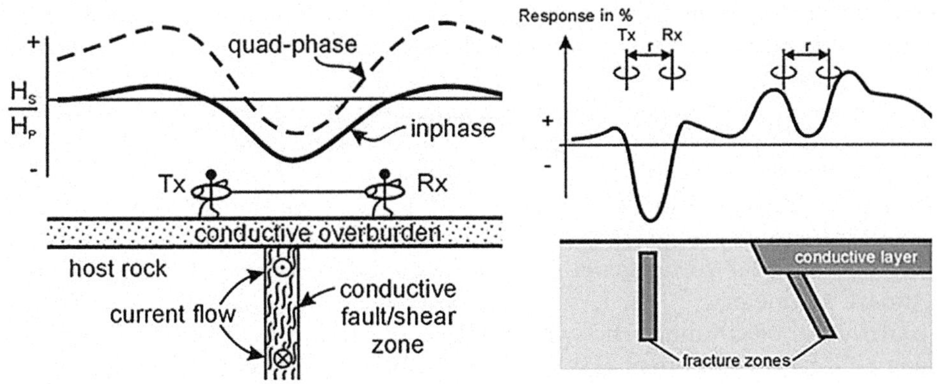

FIG. 5.3 Left: Slingram response over a highly conductive fracture zone. The offset at the quadrature response is due to the conductivity of overburden (McNeill, 1990), right: influence of a good conductive layer on the Slingram response (Grissemann and Ludwig, 1986). Examples showing the influence of dipping layers and conducting sheets (Modified).

operating at low induction numbers (McNeill, 1990). In this case, the in-phase component of the signal becomes very small, and the quadrature component is directly related to the conductivity of the ground, allowing for a direct reading of the mean ground conductivity.

A suitable function for describing this relationship is the dimensionless parameter:

$$\alpha = \mu\sigma\omega s l \tag{5.2}$$

where σ is the conductivity, s is the width of the conductor, l is the coil separation, and ω is the frequency of the EM system. Model response curves for different values of α are directly applicable to the interpretation of real conductors measured in the field (Lowrie, 2007).

The most common application of EM induction methods is lateral profiling, typically conducted along traverses that are perpendicular to the geological strike of suspected conducting bodies, such as dikes. Ground-based EM methods can also be used for vertical sounding, applying the same principles as resistivity methods to determine the conductivities of horizontal layers. The greater the separation between the transmitter and receiver, the deeper the maximum depth at which conductors can be analyzed.

An important form of vertical EM sounding is the magnetotelluric (MT) method, which takes advantage of the penetrative ability of low-frequency signals from natural sources, such as the external geomagnetic field. From the sensitivity function R(Z), the contribution of layers below a depth z, normalized by the coil separation, can be calculated (McNeill, 1980). However, it should be noted that the resolution of these measurements is poor for highly resistive layers (Kirsch, 2006).

5.4.2 Very low frequency and radio magnetotelluric

Very low frequency (VLF) electromagnetic (EM) techniques utilize the signals from very long wavelength radio transmitters, which are globally distributed and primarily used for marine navigation and communication. These transmitters operate in the frequency range of

15–25 kHz and have power outputs ranging from 100 to 1000 kW (McNeill, 1990; McNeill and Labson, 1991). In the context of radio transmission, these frequencies are considered very low.

VLF EM surveys can be conducted using instruments such as the Geonics EM16 or the Crone Radem. Additionally, many magnetometers offer the option of conducting VLF measurements simultaneously. Another option is to use a local transmitter, the TX27, available from Geonics in Toronto. The TX27 system uses a long wire grounded at both ends, positioned about a kilometer away from the survey area, at right angles to the suspected conductor orientation. This setup generates a reasonably uniform VLF field over an area of several square kilometers.

The transmitted electromagnetic wave consists of electric field components (Ex and Ez) and a magnetic field component (Hy) (Fig. 5.4). The electric field component (Ex) induces current flow in the ground, which in turn creates an additional magnetic field component (Hz). This leads to a tilted magnetic field vector in the vicinity of a good conductor, with a crossover point at the center of the conductor (Kirsch, 2006). The tilt angle is measured using a radio receiver, with its antenna oriented toward the direction of the maximum received signal. Tilt angle data can be plotted as profiles, with the crossovers indicating narrow conducting zones. If the data density is high enough to create a contour map, the crossovers are converted into peak values for clearer visualization. This can be achieved by applying an averaging filter, such as the Fraser filter, to the data (Hutchinson and Barta, 2002).

The presence of a shear zone can produce a strong VLF anomaly, even if the zone itself does not have very low resistivity. This phenomenon, known as current channeling, enhances the response of features with significant strike length, meaning that many VLF anomalies are generated by geological features that do not necessarily contain sulfides. As a result, VLF

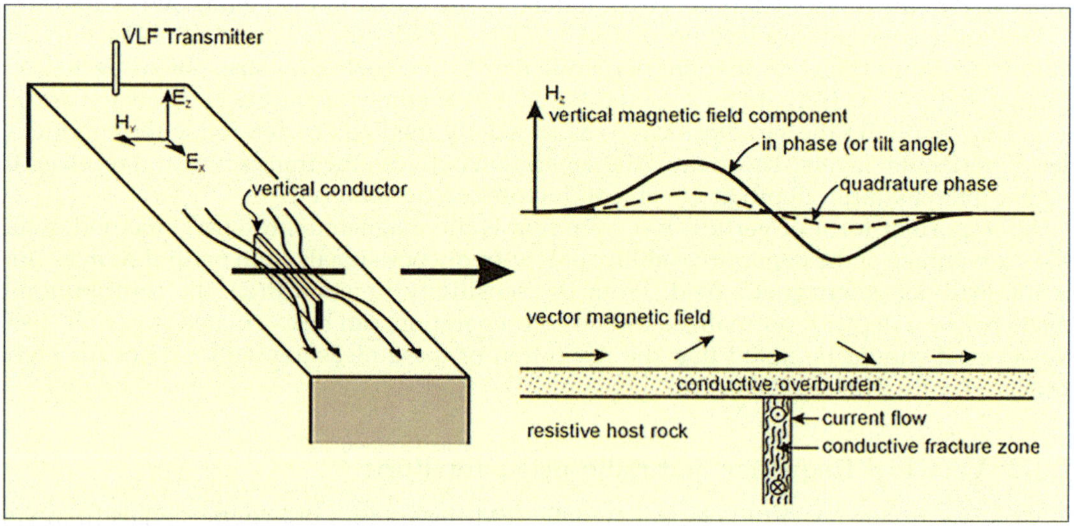

FIG. 5.4 Left: Field components from a remote VLF transmitter and electric current flow in the ground, right: electric field component along the profile indicated in the left picture (Modified).

surveys are better suited for mapping geological structures rather than identifying conductors that could host economic mineralization. The VLF method is often used in conjunction with other follow-up techniques, such as DC resistivity (Benson et al., 1997) and downhole logging (Michaud and Covel, 1998).

The Radio Magnetotelluric (RMT) method is an extension of the VLF technique, applied to higher frequencies (Candansayar and Tezkan, 2006). In RMT, the magnetic field (HHH) is measured by an antenna, while the electric field (EEE) is measured using an electrode pair oriented toward the radio transmitter. The electrode spacing is typically in the range of 1–5 m. Radio transmitters are selected from either the VLF frequency range (VLF-R) or the radio-frequency range of 10–300 kHz (RMT). As the penetration depth of the radio signal is frequency-dependent, a multifrequency data set from VLF-R or RMT can be inverted to model the resistivity-depth structure. For inversion, the apparent resistivity $\rho_a(f)$ and the phase shift $\varphi(f)$ (between E and H) are used.

5.5 Magnetotelluric sounding

Magnetotelluric (MT) sounding is a natural-source electromagnetic (EM) method that utilizes the fluctuating electromagnetic fields originating from the ionosphere. These fields are partly reflected at the Earth's surface and are subsequently reflected off the conducting ionosphere, creating multiple cycles of reflection. As a result, the fields eventually acquire a strong vertical component and can be treated as vertically propagating plane waves with a broad frequency spectrum. These electromagnetic fields penetrate the Earth's surface, inducing telluric electric currents, which then generate secondary magnetic fields.

The telluric currents are detected using two pairs of electrodes, usually oriented in the north-south and east-west directions (denoted as Ex and Ey). Three components of the magnetic field are measured: the vertical component (Hz) and two horizontal components (Hx and Hy), which are parallel to the respective telluric components (Ex and Ey) (Fig. 5.5). This setup allows for the measurement of the electromagnetic field in multiple directions, providing detailed information about the subsurface.

The MT method is capable of yielding conductivity information from much greater depths than artificial-source induction methods. It is widely applied in the exploration of petroleum and deep zones of mineralization in the upper crust. By utilizing long periods in the range of 10–1000 s, the MT method is particularly useful for investigating the structure of the Earth's crust and upper mantle, making it an essential tool for geological and geophysical studies of deep Earth structures (Naskar et al., 2012).

Maxwell's equations are used to define the fundamental law of the electromagnetic field (Kaufman and Keller, 1983), which have been lucidly explained by Mishra and Naskar (2024, 2025) where, E, H, B, J, and D are represented in vector in the following equations:

$$\nabla \times E = -\frac{\partial B}{\partial t}, \text{Faraday's Law} \tag{5.3a}$$

$$\nabla \times H = J + \frac{\partial D}{\partial t}, \text{Ampere's Law} \tag{5.3b}$$

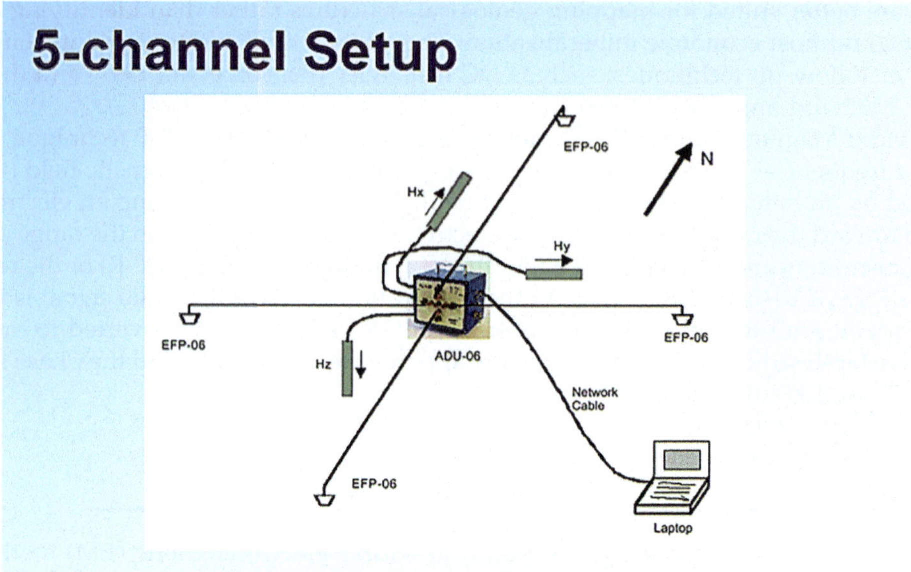

FIG. 5.5 Magnetotelluric (MT) 5-channel setup with electric field (E_x and E_y) and magnetic field (H_x, H_y, and H_z) components. *Source: Naskar and Lahiri (2014).*

$$\nabla.B = 0, \text{Gauss's Law for magnetism} \tag{5.3c}$$

$$\nabla.D = \rho, \text{Gauss's Law} \tag{5.3d}$$

The expressions for Faraday's and Ampere's laws are initially decoupled. However, in the presence of a linear, isotropic medium the material equations can be introduced:

$$D = \varepsilon E \tag{5.4a}$$

$$B = \mu H \tag{5.4b}$$

Furthermore, currents in the presence of an electric field can only flow if the media has a conductivity "σ" other than zero (Ohm's Law):

$$J = \sigma E (\text{Vector form of Ohm Law}) \tag{5.5}$$

where E = electric field (V/m), H = magnetic field (A/m), B = magnetic flux density (Tesla), D = electric displacement (C/m^2), J = current density (A/m^2), μ = magnetic permeability (H/m), ε = dielectric permittivity (F/m), ρ = electric charge density of free charges (C/m^3), and σ = electrical conductivity (S/m).

Using the material **Eqs.** (5.4a) and (5.4b) and Ohm's Law (Eq. 5.5), the Maxwell **Eqs.** (5.3a)–(5.3d) take on the form:

$$\nabla.B = 0 \tag{5.6a}$$

$$\nabla . E = \frac{\rho}{\varepsilon} \tag{5.6b}$$

$$\nabla \times E = -\frac{\partial B}{\partial t} \tag{5.6c}$$

$$\nabla \times B = \mu\sigma E + \mu\varepsilon\frac{\partial E}{\partial t} \tag{5.6d}$$

Eqs. (5.6a)–(5.6d) can be rearranged to yield an expression for extracting information about the subsurface. Taking the curl of **Eq.** (5.6d) yields:

$$\nabla \times (\nabla \times B) = \nabla .(\nabla . B) - \nabla^2 B = \nabla \times \left(\mu\sigma E + \mu\varepsilon\frac{\partial E}{\partial t}\right) = -\mu\sigma\frac{\partial B}{\partial t} - \mu\varepsilon\frac{\partial^2 B}{\partial t^2} \tag{5.7}$$

Since ∇.B = 0, **Eq.** (5.7) simplifies to the equation of telegraphy, which describes the diffusive wave characteristics in a lossy (σ > 0) medium:

$$\nabla^2 B = \mu\left(\sigma\frac{\partial B}{\partial t} + \varepsilon\frac{\partial^2 B}{\partial t^2}\right) \tag{5.8a}$$

and similarly,

$$\nabla \times (\nabla \times E) = \nabla .(\nabla . E) - \nabla^2 E = 0 - \nabla^2 E$$

where in a homogeneous half-space and layered (1D) Earth there are no free charges, so $\nabla . E = 0$.

$$-\nabla^2 E = \nabla \times (\nabla \times E) = \nabla \times \left(-\frac{\partial B}{\partial t}\right) = \nabla \times \left(-\mu\frac{\partial H}{\partial t}\right) = \left\{-\mu\left(\sigma\frac{\partial E}{\partial t} + \varepsilon\frac{\partial^2 E}{\partial t^2}\right)\right\}$$

$$-\nabla^2 E = -\mu\left\{\left(\sigma\frac{\partial E}{\partial t} + \varepsilon\frac{\partial^2 E}{\partial t^2}\right)\right\} \tag{5.8b}$$

$$\nabla^2 E = \mu\left(\sigma\frac{\partial E}{\partial t} + \varepsilon\frac{\partial^2 E}{\partial t^2}\right)$$

The left side of **Eqs.** (5.8a) and (5.8b) describes the variation of the B- and E-field components of the electromagnetic wave in space. The right side describes its variation with time and so its frequency dependence. In Cartesian coordinates

$$\nabla^2 = \frac{\partial^2}{\partial x^2} + \frac{\partial^2}{\partial y^2} + \frac{\partial^2}{\partial z^2}$$

Assuming a time dependence $e^{i\omega t}$ for B (i.e., $B = e^{i\omega t}$) (with ω = 2πf the angular frequency in Hz, $f = \frac{1}{T}$= frequency of oscillation in s^{-1}, T = period of oscillation in s, and $i = \sqrt{-1}$= imaginary unit of complex number), **Eqs.** (5.8a) and (5.8b) yields the Helmholtz equation in the frequency domain:

$$(\nabla^2 - k^2)B = 0 \tag{5.9}$$

where $k^2 = i\omega\mu\sigma - \omega^2\mu\varepsilon$ = wave number of the EM wave. The variables μ, ε and σ are the properties depending entirely on the material through which the electromagnetic waves propagate. For homogeneous materials, ε and μ are constants, while for inhomogeneous materials, they depend on time and position within the material. In the case of isotropic material, μ and ε are scalars, and for anisotropic materials, they are represented by tensors. In a conductive medium, $\sigma \gg \omega\varepsilon$, the wave number k (in m^{-1}) represents the quasi-static approximation, $k \approx \sqrt{i\omega\mu\sigma}$. In the wave equation, the speed of the wave is described as the inverse square root of the coefficient of the time derivative; then, both fields have the same speed of $c = \frac{1}{\sqrt{\mu\varepsilon}} = 3\times10^8\,m/s$, the speed of light. The results above completely describe the electric and magnetic fields in a homogeneous medium.

Following the method used above, the diffusion equation for the E field can be derived in similar fashion:

$$(\nabla^2 - k^2)E = 0 \tag{5.10}$$

The form of these equations (Eqs. (5.9) and (5.10)) is reminiscent of the one-dimensional equation of diffusion or heat conduction equation $\frac{\partial T}{\partial t} = \kappa \frac{\partial^2 T}{\partial z^2}$, where κ (= $\frac{k}{\rho C_p}$, here k = the constant of proportionality is the thermal conductivity in $Wm^{-1}\,K^{-1}$, C_p = specific heat at constant pressure in $Jkg^{-1}K^{-1}$, and ρ = density of the material in kg/m^3) is called the *thermal diffusivity*; it has the dimensions m^2/s, and here, T = absolute temperature in Kelvin (°K).

Eqs. (5.9) and (5.10) are fundamental for understanding the MT method in comparison to other geophysical techniques. The diffusive behavior of electromagnetic waves in the frequency range of MT places this method between geophysical exploration techniques governed by the wave equation (ground penetrating radar, seismics), and the potential methods (gravity, magnetics, DC resistivity). **Eqs.** (5.9) and (5.10) are not only dependent on the conductivity σ and angular frequency ω, but are also depth-dependent B=B(z). Therefore, the diffusion equations have the following solution:

$$E = E_0 e^{-ikz} + E_1 e^{ikz} \tag{5.11a}$$

$$B = B_0 e^{-ikz} + B_1 e^{ikz} \tag{5.11b}$$

where B_0, B_1, E_0, E_1 are the electromagnetic fields at the surface of the earth. B and E have to vanish for $z \to \infty$, and therefore the second term of **Eqs.** (5.11a) and (5.11b) vanishes. Thus, the solution in **Eqs.** (5.11a) and (5.11b) simplifies to:

$$E = E_0 e^{-ikz} = E_0 e^{-i\beta z} e^{-\beta z} \tag{5.12a}$$

$$B = B_0 e^{-ikz} = B_0 e^{-i\beta z} e^{-\beta z} \tag{5.12b}$$

with $k^2 = i\omega\mu\sigma$. The first factor of the **Eqs.** (5.12a) and (5.12b) is the wave amplitude, the second factor (imaginary exponential) is depth variation, and the third is exponential decay.

The wave number $k=\sqrt{i}\sqrt{\omega\mu\sigma}=\sqrt{\omega\mu\sigma}e^{\frac{i\pi}{4}}=(1+i)\sqrt{\frac{\omega\mu\sigma}{2}}$. The inverse of the real part of "k" is also known as the *skin depth or penetration depth* (δ).

$$\delta=\frac{1}{\text{Rek}}=\sqrt{\frac{2}{\omega\mu\sigma}} \tag{5.13}$$

Since the magnetic permeability "μ" does not vary substantially in the earth **Eq.** (5.13) can be approximated as a formula for the skin depth (exploration depth) in (m):

$$\delta=503\sqrt{\frac{\rho}{f}} \tag{5.14}$$

where ρ = apparent resistivity in Ωm, and f = the signal frequency in Hz, which is reciprocal of period T in s. At this depth the magnetic field is attenuated to e^{-1} (~37%) of its value outside the conductor (Vozoff, 1991). The skin depth is dependent on the resistivity of the body and the frequency of the field. Magnetic field decays to 1% at a depth z = 5d (where d is called the skin depth) and to 0.1% at z = 7d, effectively limiting the practical depth of exploration with the induction method (Lowrie, 2007).

Apparent resistivity and phase

Substituting **Eqs.** (5.4a), (5.4b), and (5.5) in **Eqs.** (5.3a)–(5.3d) and assuming the harmonic dependence of the oscillating EM fields, $e^{i\omega t}$ (i.e., $B=e^{i\omega t}$), leads to the following relationships for a homogeneous isotropic medium:

$$\nabla\times E=-i\omega B=-i\omega\mu H \tag{5.15}$$

$$\nabla\times H=(\sigma+i\omega\varepsilon)E \tag{5.16}$$

Eqs. (5.15) and (5.16) show the following relationship between the components of the E- and B-fields using curl operator:

$$\frac{\partial E_z}{\partial y}-\frac{\partial E_y}{\partial z}=-i\omega B_x=-i\omega\mu H_x \tag{5.17a}$$

$$\frac{\partial E_x}{\partial z}-\frac{\partial E_z}{\partial x}=-i\omega B_y=-i\omega\mu H_y \tag{5.17b}$$

$$\frac{\partial E_y}{\partial x}-\frac{\partial E_x}{\partial y}=-i\omega B_z=-i\omega\mu H_z \tag{5.17c}$$

$$\frac{\partial H_z}{\partial y}-\frac{\partial H_y}{\partial z}=(\sigma+i\omega\varepsilon)E_x \tag{5.18a}$$

$$\frac{\partial H_x}{\partial z}-\frac{\partial H_z}{\partial x}=(\sigma+i\omega\varepsilon)E_y \tag{5.18b}$$

$$\frac{\partial H_y}{\partial x} - \frac{\partial H_x}{\partial y} = (\sigma + i\omega\varepsilon)E_z \tag{5.18c}$$

Considering the EM waves to be vertically incident plane waves only, it follows that $\nabla_x = \nabla_y = 0$ and the EM field has only z-component. Since the H-field propagates vertically downward in the z-direction *while* the induced E-field does not have a z-component ($E_z = 0$). With these assumptions **Eqs.** (5.17a)–(5.17c) and (5.18a)–(5.18c) takes the form:

$$\frac{\partial E_y}{\partial z} = i\omega\mu H_x, \frac{\partial E_x}{\partial z} = -i\omega\mu H_y \tag{5.19}$$

$$-\frac{\partial H_y}{\partial z} = (\sigma + i\omega\varepsilon)E_x, \frac{\partial H_x}{\partial z} = (\sigma + i\omega\varepsilon)E_y \tag{5.20}$$

From **Eqs.** (5.12a) and (5.12b) we have

$$\frac{\partial E_x}{\partial z} = -ikE_x \tag{5.21}$$

Taking the ratio of **Eqs.** (5.19) and (5.21) we have

$$\frac{E_x}{H_y} = \frac{-i\omega\mu}{-ik} = \frac{\omega\mu}{k} = \frac{\omega\mu}{\sqrt{i\omega\mu\sigma}} = \sqrt{\frac{\omega\mu}{i\sigma}} \tag{5.22}$$

Again from **Eqs.** (5.12a) and (5.12b) we have

$$\frac{\partial E_y}{\partial z} = -ikE_y \tag{5.23}$$

Taking the ratio of **Eqs.** (5.19) and (5.23) we have

$$\frac{E_y}{H_x} = \frac{i\omega\mu}{-ik} = -\frac{\omega\mu}{k} \tag{5.24}$$

The magnetotelluric impedance Z (in m/s) is the ratio of the electric and magnetic fields measured at the surface and from **Eqs.** (5.22) and (5.24) we have that the orthogonal components of the field satisfy:

$$Z = \frac{E_x}{H_y} = -\frac{E_y}{H_x} \tag{5.25}$$

The relationship between the electric and magnetic fields is, therefore:

$$\begin{pmatrix} E_x \\ E_y \end{pmatrix} = \begin{pmatrix} 0 & Z \\ -Z & 0 \end{pmatrix} \begin{pmatrix} H_x \\ H_y \end{pmatrix} \tag{5.26}$$

A magnetic field lags behind the electric field by $\frac{\pi}{4}$ in a completely homogeneous and isotropic half-space with true resistivity equal to:

$$\rho = \frac{|Z|^2}{\omega\mu} = \frac{T}{2\pi\mu}|Z|^2$$

with $T = \frac{2\pi}{\omega}$ = time period in (s). This can be expressed as (Cagniard, 1953):

$$\rho = 0.2T\frac{|E_x|^2}{|H_y|^2} \tag{5.27}$$

where ρ = resistivity in Ωm, E = the horizontal electric field in mV/km, H = the orthogonal horizontal magnetic field in gamma, and T = period in s. When the earth resistivity is non-uniform, the right-hand sides of Eq. (5.27) provide apparent resistivities (instead of true resistivity), ρ_a, which are frequency (period) dependent, as is the case with 1D, 2D, or 3D situations. In the frequency domain, Z(ω) is complex and has an associated phase (φ). So phase of Z_{xy}

$$\varphi = \tan^{-1}\frac{\mathrm{Im}\left|\frac{E_x}{H_y}\right|}{\mathrm{Re}\left|\frac{E_x}{H_y}\right|} \tag{5.28}$$

The impedance phase gives additional information about the conductive structure. For a nonhomogeneous medium, phase $\neq \frac{\pi}{4}$.

5.6 Audio-magnetotelluric sounding

Natural source electromagnetic (EM) signals are generated in both the atmosphere and magnetosphere. Magnetotelluric (MT) signals, which typically operate at low frequencies (< 1 Hz), are generated by the interaction between the Earth's magnetosphere and solar wind, sunspot activity, and auroras. High-frequency signals (> 1 Hz) in the audio range (audio-frequency magnetotellurics, AMT) are generated primarily by thunderstorms and lightning worldwide.

Audio-frequency magnetotellurics (AMT) specifically uses the audio band, which spans from 0.1 Hz to 20 kHz, falling within the higher frequency range of the MT spectrum. AMT surveys tend to have shorter recording lengths compared to typical MT surveys and provide higher resolution of near-surface stratigraphy. Ground resistivity values are derived from the magnitude and ratios of the electric (Ex & Ey) and magnetic field (Hx & Hy) components, and these are subsequently mapped using inversion and modeling software. The resistivity values obtained correlate with geological features of the subsurface.

Two types of AMT data acquisition arrays are used: Scalar AMT and Tensor AMT. In Scalar AMT, data is collected from a series of electrode pairs measuring the electric field component Ex, with one magnetic field sensor measuring the Hy component along a survey line. In Tensor AMT, additional electric field (Ey) and magnetic field (Hx) components are measured, providing information on the impedance directionality at specific locations (Fig. 5.6). The Hz component, often recorded in lower frequency MT surveys, is typically not used in AMT surveys.

FIG. 5.6 Zonge's backpack portable AMT acquisition system allows for use in difficult terrain and relatively easy field logistics. *Source: Zonge International, Tucson, Arizona, USA.*

The depth of investigation for MT/AMT surveys ranges from tens of meters to several kilometers. AMT can image deeper subsurface features in more resistive ground by decreasing the floor frequency, although practical limitations related to the size of the features being resolved at depth do exist. Lateral survey resolution is generally estimated at 15% of the depth of investigation. This factor leads to a significantly different survey design for natural source AMT compared to traditional MT surveys.

In recent years, the AMT method has undergone significant improvements, allowing for the production of accurate 2D or 3D models for various mineral exploration applications, particularly those requiring moderate depth investigation and where resistivity variations are expected. For more detailed information on the application of AMT in mineral exploration, see the works of Strangway et al. (1973), Strangway and Koziar (1979), and Goldstein (1971).

5.7 Audio-frequency magnetic sounding

In the 1950s, ground and airborne electromagnetic (EM) surveys became the primary field method for geophysicists exploring for massive sulfides, particularly in Canada. However, a significant limitation of these methods was their depth of exploration. For airborne surveys, the depth was limited to only 20–60 m (Ward, 1959). The signal-to-noise ratio and signal strength were key factors that restricted the usefulness of this exploration method.

Slichter, 1955 proposed an innovative solution, suggesting the use of natural electromagnetic sources, such as lightning strikes associated with thunderstorms, to replace the artificially transmitted energy used in conventional EM methods. The aim was to overcome the depth limitations inherent in the traditional methods. Ward (1959) further explored this idea and concluded that audio-frequency magnetics (AFMAG) could be an effective exploration tool. He demonstrated AFMAG's ability to locate and identify buried conductive bodies, establishing a relationship that characterized AFMAG as a scalar quantity:

$$H_z = T_x H_x \quad (5.29)$$

where Hz is the vertical magnetic field, Hx is the horizontal magnetic field, and Tx is the transfer function (tipper in the horizontal direction). Despite these promising developments, AFMAG faced challenges such as poor-quality data and difficulties in replicating successful trials, which hindered its broader acceptance in the mineral exploration industry (Ward et al., 1966, 1968). Ward et al. (1966) recognized the potential of AFMAG as a geological mapping tool but noted several limitations, including:

- Significant seasonal and diurnal variations in the 150 Hz and 510 Hz frequency bands used in AFMAG, which hinder year-round application.
- Power line interference, which could be stronger than the natural fields, thus limiting the area of applicability.
- AFMAG's tendency to delineate faults, shears, and carbonaceous horizons makes it challenging to locate sulfide bodies due to their relatively small response.
- Some sulfide ore bodies failed to produce measurable responses, particularly in airborne AFMAG surveys.
- The labor-intensive nature of ground-based AFMAG surveys.

Despite these challenges, Ward et al. (1966) concluded that AFMAG was particularly suitable for deep exploration, fault mapping, locating sulfides buried beneath deep glacial ice, and exploration in rugged terrain. However, further field, experimental, and theoretical work was necessary to fully understand and optimize the method.

In the early 1970s, Vozoff (1972) corrected Ward's scalar estimation (Eq. (5.29)), redefining the relationship between the vertical and horizontal components of the magnetic field and the tipper tensor. Vozoff's revised formula is as follows:

$$H_z(f) = \{T_x(f)^* H_x(f)\} + \{T_y(f)^* H_y(f)\} \quad (5.30)$$

where f is the frequency, Hx, Hy, and Hz are the components of the magnetic field in three orthogonal directions, and Tx and Ty are the two horizontal components of the tipper tensor. This equation became the foundation for the AFMAG exploration method.

In the mid-1980s, Labson et al. (1985) made significant improvements to the AFMAG method, introducing a spectral, Fourier-domain derived tipper measurement that was repeatable throughout the year, regardless of the polarization state of the natural source field. They also introduced a remote referencing technique, which improved data quality by allowing measurements during periods of low signal strength and enabling the use of portable sensors. By utilizing a broad frequency range, this new method proved effective for exploring depths ranging from several meters to several kilometers.

Despite these advancements, the airborne AFMAG method was once again overshadowed by inductive EM systems. However, in 2000, Geotech Ltd revitalized AFMAG development, applying the signal processing techniques proposed by Labson et al. (1985). Their efforts culminated in the creation of ZTEM (Z-axis Tipper Electromagnetics) in 2005. ZTEM became the first commercial airborne AFMAG system in 30 years when it was brought into operation in 2006 (Legault, 2012). This development marked a significant milestone in the use of AFMAG technology for airborne mineral exploration.

5.8 Controlled-source audio-magnetotelluric sounding

The controlled-source audio-frequency magnetotellurics (CSAMT) method has been shown to effectively reject ambient noises, leading to relatively high signal-to-noise ratios (Nabighian, 1988; Zonge and Hughes, 1991; Fu et al., 1993; Xu and Liu, 1995; Tang and He, 2005). With a powerful transmitter that can reach tens of kilowatts and with carefully chosen survey parameters, the method is capable of producing a significant depth of investigation. This feature makes CSAMT particularly useful for applications such as mineral exploration, geothermal studies, hydrocarbon exploration, deep tunnel detection, and groundwater contamination assessment (Wannamaker, 1997; Unsworth et al., 2000; Di et al., 2002, 2004; Takahashi, 2004; Wannamaker et al., 2004; Pellerin and Wannamaker, 2005; He et al., 2006; Miller et al., 2006, 2008; Troiano et al., 2009; An and Di, 2010; An et al., 2012). According to Nichols et al. (1994), Wu and Shi (1996), and Wu et al. (1996), the method has also been successfully applied to the study of deeply buried groundwater. The ratio of the orthogonal horizontal components, $\frac{E_x}{H_y}$, gives the Cagniard resistivity (Cagniard, 1953). For an electrical dipole source over a uniform half-space, the far-zone electrical and magnetic fields are expressed as follows (Goldstein, 1971; Goldstein and Strangway, 1975):

$$E_x = \frac{I.AB.\rho_1}{2\pi r^3}.(3\cos^2\theta - 2) \tag{5.31}$$

$$E_y = \frac{3.I.AB.\rho_1}{4\pi r^3}.\sin 2\theta \tag{5.32}$$

$$E_z = (i-1)\frac{I.AB.\rho_1}{2\pi r^2}.\sqrt{\frac{\mu_0\omega}{2\rho_1}}.\cos\theta \tag{5.33}$$

$$H_x = -(i+1)\frac{3I.AB}{4\pi r^3}.\sqrt{\frac{2\rho_1}{\mu_0\omega}}.\cos\theta.\sin\theta \tag{5.34}$$

$$H_y = (i+1)\frac{I.AB}{4\pi r^3}.\sqrt{\frac{2\rho_1}{\mu_0\omega}}.(3\cos^2\theta - 2) \tag{5.35}$$

$$H_z = i\frac{3I.AB.\rho_1}{2\pi\mu_0\omega r^4}.\sin\theta \tag{5.36}$$

It follows from **Eqs.** (5.31) and (5.35) that the ratio of $\frac{E_x}{H_y}$ can be written as

$$\rho_s = \frac{1}{\mu_0\omega}\left|\frac{E_x}{H_y}\right|^2 \tag{5.37}$$

The Cagniard resistivity is defined as

$$\rho_s = \frac{1}{5f}\frac{|E_x|^2}{|H_y|^2} \tag{5.38}$$

with the impedance phase given by

$$\varphi = E_{phase} - H_{phase} \tag{5.39}$$

Note that the Cagniard resistivity represents the true resistivity of the subsurface only when the ground is uniform. Otherwise, the Cagniard resistivity is considered an apparent resistivity of the subsurface (Cagniard, 1953). A CSAMT survey involves transmitting a controlled signal at a range of frequencies into the ground from one location (the transmitter site) and measuring the received electric and magnetic fields at the receiver site. The system can supply an output power of up to 30 kW to the electric dipole transmitter, with frequency ranges spanning from 0.25 Hz to 8192 Hz (Fig. 5.7). A remote, grounded dipole is employed for electromagnetic signals, with two grounded current electrodes spaced approximately 1.5 km apart (Fu et al., 2013). Grounded current electrodes are typically constructed using multiple galvanized steel stakes hammered into the ground or by constructing small, aluminum foil-lined mud pits. Salt water is generally used to ensure good electrical contact with the ground. For high-power surveys, the transmitter and generator are typically mounted on trucks. The receiver system consists of a computer-controlled digital receiver, magnetic coils, and nonpolarization electrode amplifiers. These components filter, process, and record the received signals at individual stations. The transmitted signal is detected using short grounded dipoles (electrode pairs) and magnetic field detectors or sensors. The receiver electrode spacing is typically 40 m. The multichannel receiver instrument can acquire electrical field data at seven stations and one

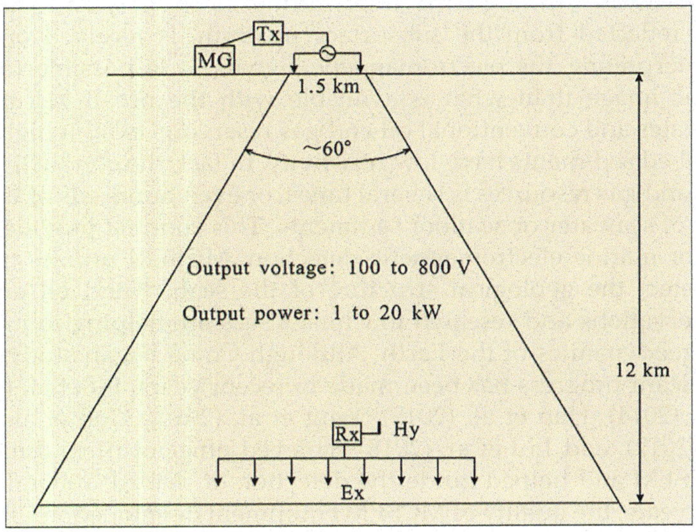

FIG. 5.7 Sketch of CSAMT observation system. The system is used in Metropolitan Beijing for groundwater exploration (Modified).

magnetic field simultaneously. Before conducting the survey, the receiver and coil calibrations are checked for consistency. The receiver dipoles are surface wires, grounded with small, porous ceramic "pots" buried about ½ inch into the soil. Magnetic field detectors are cylindrical coils of wire about three to four feet long, placed on the ground. Depending on logistical factors, one to seven stations can typically be collected in one setup. After acquiring data at one field setup, the crew moves the receiver, pots, and wires to the next set of stations. Lateral resolution is determined by the station spacing, which typically ranges from 10 to 200 m. The strength of the received signal is proportional to the station spacing.

Before geologic interpretation, the acquired data are corrected for noise, static, and terrain effects. The static effect on CSAMT survey results arises from near-surface conductive inhomogeneities, which can cause electric field distortions over a broad time window. If not properly addressed, this effect can obscure deep structure information in the measurements. Medium-to-strong static effects were anticipated in the survey area due to near-surface conductive inhomogeneities under the thick vegetation. Several methods for correcting this effect are available (Torres-Verdín and Bostick, 1992).

5.9 Marine controlled-source electromagnetic method

Various earth resources, such as oil, natural gas, and rare metals, are present under the seabed, and many of these resources are yet to be identified and exploited. Developing necessary tools for the marine controlled-source electromagnetic method (MCSEM) is crucial for the investigation and evaluation of oil resources and natural gas hydrates (Di et al., 2020). The MCSEM method involves using a towed cable to transmit signals near the sea surface, while an array of ocean-bottom receivers is used to receive the signals from the subsurface. A high-power electromagnetic pulse with a frequency range of 0.1–10 Hz is sent into the seawater through the transmitting electrodes, and the electromagnetic response signal, reflected from the subsurface medium, is received on the seafloor. By inverting and interpreting the electromagnetic signals, it is possible to obtain a clearer seafloor dielectric image than what is available with the ocean MT method (Fig. 5.8). Natural gas hydrates and conventional oil and gas reservoirs exhibit high resistivity, while seawater and seabed sediments have low resistivity. In fact, the conductivity of natural gas hydrates and oil and gas resources is several times, or even hundreds of times, smaller than the conductivity of seawater or seafloor sediments. This contrast provides favorable physical conditions for marine electromagnetic detection. MCSEM utilizes these conductivity differences to detect the geological structure of the seabed and oil and gas resources. Geophysical observations and research in China's seas are helping to increase the understanding of the geodynamics of the Earth. Although China began studying MCSEM relatively late, significant progress has been made in recent years. Fu et al. (2009), Chen et al. (2009), Yin et al. (2014), Han et al. (2015), Yang et al. (2015), Guo et al. (2016), Cai et al. (2015), Li et al. (2017), and Liu et al. (2018) achieved efficient three-dimensional forward modeling of MCSEM and helped guide the detection of seabed minerals and oil and gas targets. In recent years, the quality of MCSEM equipment developed in China has advanced rapidly. In 2017, Central South University completed the development of a deep-sea 6000 m towed transient electromagnetic system, which includes a deck control system, a 10,000 m

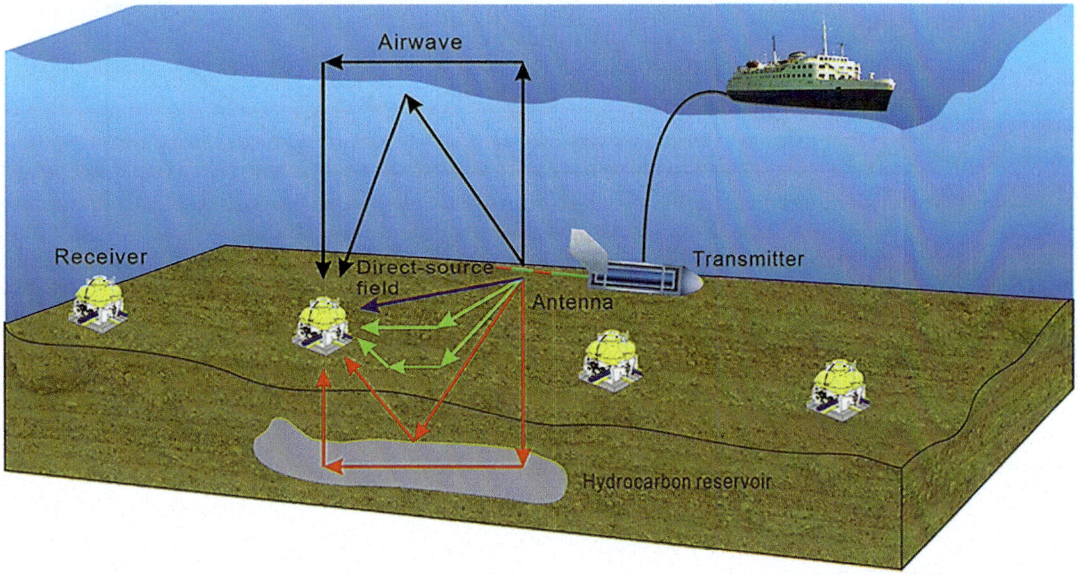

FIG. 5.8 The scheme of marine controlled-source electromagnetic (MCSEM) method. *Source: Li et al., 2022—Open Access.*

photoelectric composite cable, an instrument cabin tow, and an antenna trailer. A large electromagnetic anomaly was discovered in the southwest Indian Ocean Ridge hydrothermal zone during the 30th research cruise, confirming the system's effectiveness (Di et al., 2020; Luo and Li, 2015).

5.10 Surface-to-borehole electromagnetic method

The surface-to-borehole electromagnetic (EM) detection method plays a crucial role in baseline geoscience data for sustainable exploration and resource management. This method uses a dipole electrode grounded on the crustal surface to transmit an electromagnetic source into the subsurface, while the electromagnetic response signal is observed in a well, as shown in Fig. 5.9 (Chen et al., 2017, 2019; Di et al., 2020). By avoiding surface electromagnetic interference and low-resistance coatings, it enables the detection of subsurface anomalies with greater precision. Multiple sources arranged around the borehole are used to determine the spatial position and extent of the anomalous body, allowing for accurate positioning and tracking of underground target bodies.

This technique is vital for the sustainable exploration and management of natural resources, such as mineral deposits and groundwater reservoirs. By utilizing dipole electrodes placed on the crustal surface and detecting EM response signals from a borehole, the method provides high-resolution data on the spatial extent and location of subsurface anomalies. The ability to generate detailed subsurface images with minimal interference from surface noise is a significant advantage. The surface-to-borehole EM method supports efficient resource

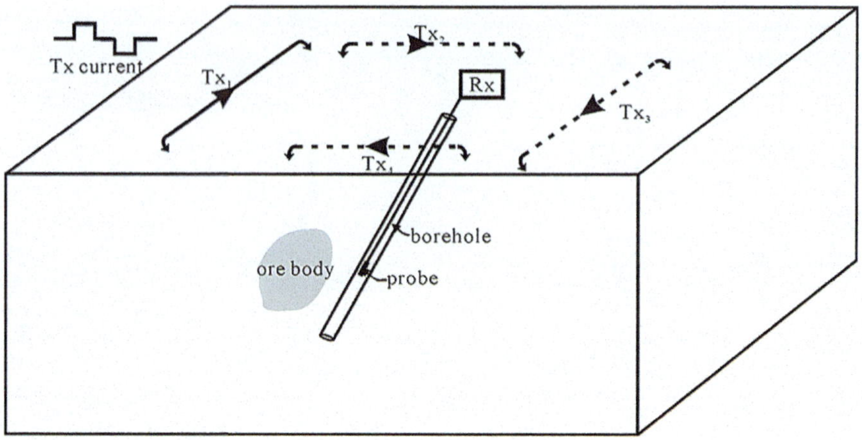

FIG. 5.9 The scheme of the well-ground electromagnetic (EM) method. *Source: Chen et al. (2017).*

management by improving the accuracy of subsurface exploration, reducing the need for invasive drilling, and ensuring the sustainable use of geoscience data for resource evaluation and management.

5.11 Time domain ground-based electromagnetic method

The transient electromagnetic (TEM or TDEM) method has undergone significant development since the mid-1980s. Initially, the inductive methods (TEM and FDEM) were designed primarily for mineral investigations. In the 1960s and 1970s, frequency-domain methods dominated, being highly sensitive to low-resistivity mineral deposits within high-resistivity host rocks (>10^5 Ωm). However, if the area was covered with a thick low-resistivity layer (up to 100 m), frequency-domain methods struggled to penetrate it due to the resistivity contrast. This limitation led to the development of the TEM method, which, at the time, only allowed for qualitative interpretations, providing indicative information about mineralizations. Key developments in the TEM method were made by Sidorov and Tickshaev (1969), Kaufman (1989), and Kaufman and Morozova (1970).

The TEM method measures the amplitude of the signal as a function of time, where a transmitter sends a pulse—typically a current that is switched off rapidly—and measurements are made after the primary fields have dissipated. The measurements are based on secondary fields induced by eddy currents, using an induction coil, which eliminates the geometric setup difficulties associated with frequency-domain methods.

The TEM field equations are derived from Maxwell's equations (Nabighian and Macnae, 1991; Strack, 1992), which can be solved independently in orthogonal directions using scalar potentials, denoted φ_E (electric) and φ_M (magnetic). These potentials do not generate vertical electric or magnetic field components and are referred to as the transverse electric (TE) and transverse magnetic (TM) modes in electromagnetic exploration (Strack, 1992).

For interpretation, the apparent resistivity (ρ) is introduced as a way to represent the resistivity of a homogeneous, isotropic ground that would yield the same voltage-current relationship as measured in the real earth. The apparent resistivity depends on the sounding method employed. The maximum depth from which useful information can be extracted from the TEM signal, also known as the time-domain skin-depth, is a complex function influenced by factors such as transmitter current, the geometry of the transmitter/receiver loop, the conductivity of the target area, and noise levels.

The diffusion depth (δ_t) for the time-domain signal at a time t (time elapsed after the initial transmitter switch-off) is given by Nabighian and Macnae, 1991 as:

$$\delta_t = \sqrt{\frac{2t}{\sigma\mu}} \tag{5.40}$$

where σ is the ground conductivity (S/m), and μ is the magnetic permeability (usually taken as μ_0 for free space).

Investigating the behavior of the transient magnetic field over a conductive half-space, Nabighian and Macnae, 1991 assumed that all magnetic field sources above the surface induce only horizontal electric fields and currents in a uniform or horizontally layered ground. This leads to a quasi-static approximation of electromagnetic field propagation in a conductive half-space, described as a diffusion process with current spreading downward and outward at approximately 30° to the surface (Nabighian, 1979; Nabighian and Macnae, 1991). The local current maximum in the horizontal plane propagates with diminishing amplitude and decreasing velocity (v), as described by:

$$v = \frac{1}{\sqrt{2\sigma\mu t}} \tag{5.41}$$

At zero time, the downward component of current density aligns with downward diffusion. Nabighian (1979) shows that the combined effect of all induced ground currents at or above the surface can be approximated by a single current filament, often visualized as a smoke-ring representation of the transient response. This current filament moves with a downward velocity (v) described by:

$$v = \frac{d}{dt}\left[z + \frac{4}{\sqrt{\pi}}\lambda\right] \tag{5.42}$$

where $\lambda = \sqrt{\frac{2}{\sigma\mu}}$, its horizontal dimensions increase in proportion to the diffusion depth $\sqrt{\frac{2t}{\sigma\mu}}$; its radius (r) (units m) thus expands as:

$$r = \sqrt{\frac{8Ct}{\sigma\mu}} = \sqrt{\frac{4.37t}{\sigma\mu}} \tag{5.43}$$

where C is a constant derived through the series expansion of the perturbation kernel for the magnetic Hertz potential of an arbitrary aperiodic loop over a homogeneous half-space (Nabighian, 1979; Arnason, 1989). At late times, the induced voltage in the receiver coil is approximately given by the equation:

$$V(t,r) \approx I_0 \frac{C(\mu_0 \sigma r^2)^{\frac{3}{2}}}{10\pi^{\frac{1}{2}} t^{\frac{5}{2}}} \quad (5.44a)$$

where $C = A_r n_r A_s n_s \frac{\mu_0}{2\pi^3}$, t = time elapsed after the transmitter current is turned off (s), A_r = cross-sectional area of the receiver loop (m²), A_s = cross-sectional area of the transmitter loop (m²), n_s = number of windings in the transmitter loop, n_r = number of windings in the receiver loop, r = radius of the transmitter loop, V (t, r) = transmitter voltage (V), I_0 = current in transmitter loop (A), and μ_0 = magnetic permeability in free space = $4\pi \times 10^{-7}$ (H/m).

As detailed by Spies (1989), by combining these decay rates, a late time approximation for the voltage $\left(V = \frac{dB}{dt}\right)$ response in an in-loop receiver over a homogeneous half-space can be equated by:

$$V(t) = \frac{I_0 \sigma^{\frac{3}{2}} \mu_0^{\frac{5}{2}} a^2}{30\sqrt{\pi t^{\frac{5}{2}}}} \quad (5.44b)$$

where I_0 is the current in the transmitter loop and a is its separation from the vertical axis (equivalently its radius). **Eq.** (5.44b) thus shows that the magnitude of the magnetic field response varies as $t^{-\frac{5}{2}}$ at late times—any variation indicating inhomogeneous ground, departing from a simple layered half-space to a complex, multidimensional terrain. The late time apparent resistivity (ρ_a) is defined by solving Eq. (5.44a), leading to Eq. (5.45):

$$\rho_a = \frac{\mu_0}{4\pi} \left[\frac{2I_0 \mu_0 A_r A_s n_r n_s}{5t^{\frac{5}{2}} V(t,r)} \right]^{\frac{2}{3}} \quad (5.45)$$

Finally, by combining the equation for diffusion depth (δ_t) with the late time voltage approximation, an expression can be derived for the depth of investigation (d) as a function of the ground conductivity, transmitter current, and subsequent voltage response for an in-loop receiver (Spies, 1989):

$$d \approx 0.55 \left(\frac{I_0 A}{\sigma V} \right)^{\frac{1}{5}} \quad (5.46)$$

where A is the equivalent loop area. This equation shows that the maximum depth of investigation is controlled by the size of the transmitter loop, the current used, and the sensitivity of the receiver. In practical field conditions, the effective depth of investigation based on time-dependent diffusion is limited by the point at which the signal diminishes to a level below the background noise (Macnae et al., 1984). Beyond this threshold, no further useful data can be obtained.

Over the years, various transient EM surveys have been developed for geophysical exploration. Some of the most promising systems include the SIROTEM, UTEM, LOTEM, and MTEM methods. The SIROTEM system, based on the Russian MPPO-1 unit, was redesigned and commercialized by CSIRO in the 1970s and is widely used in mineral exploration (Buselli and O'Neill, 1977; McCracken et al., 1986). The UTEM system, developed by the University of

Toronto, uses a fixed horizontal transmitter loop to send low-frequency current through the loop, with receivers measuring the magnetic and electric field components (Lamontagne et al., 1978). The UTEM system is primarily used for mineral exploration due to its limited depth range. The LOTEM system was designed for deep penetration in oil and gas exploration, using a grounded wire transmitter and electric field receivers. Finally, the MTEM system, an advancement of TDEM, incorporates multiple receivers and transmitters and combines parametric sounding and profiling for more successful mineral exploration (Wright et al., 2002; Hobbs et al., 2005; Ziolkowski et al., 2007).

5.12 Time domain long-offset electromagnetic method

To address exploration and production challenges that require deep penetration, a version of transient electromagnetics (EM) is often used in which the depth of investigation is less than or equal to the transmitter-receiver separation. This method is referred to as long-offset transient electromagnetics (LOTEM). Both magnetic and electric transients, or a combination of both, can be measured. The theoretical foundation for this method was provided by Eadic (1981) and Kaufman and Keller (1983), who demonstrated that resistive layers can be effectively identified in transient EM soundings by measuring the electric fields from a galvanic source. Electromagnetic methods using an inductive source primarily respond to conductive units in the geoelectric section (Verma and Mallik, 1979), which is also the case for passive EM methods. The technique is further explained by Keller et al. (1984), Strack, 1992, Router and Strack (1991), Nabighian (1979), Gunderson et al. (1986), and Strack et al. (1989) with respect to exploration challenges. In the LOTEM method, a carefully controlled electric current is transmitted through the Earth using a grounded wire transmitter. When the current is abruptly turned off or reversed, induction currents are generated in the subsurface. To understand the behavior of these induced currents, electromagnetic fields are calculated at various times after the current is turned off.

5.12.1 MTEM method and system development

The MTEM (Multitransient Electromagnetic) method was initially proposed by Ziolkowski and Wright (2001). In this method, all receivers are arranged in a stationary array, while the transmitter is moved along the survey line to measure the response at different offsets (Fig. 5.10A). The array configuration is illustrated in Fig. 5.10B, where R1–R16 represent the offset locations for the sounding points.

A distributed MTEM acquisition system was developed, as shown in Fig. 5.11A. This system utilizes a PRBS (Pseudorandom Binary Sequence) source waveform emitted by the transmitter, while the receivers simultaneously record the signal. A square wave current pulse, typically several tens of amperes, is injected into the subsurface, inducing eddy currents that diffuse outward and downward with time, forming a "smoke ring" system (Nabighian, 1979; Gunderson et al., 1986; Strack et al., 1989; Xue et al., 2013). The transient electric and magnetic fields observed at the receiver sites are recorded in a time-synchronous mode, similar to seismic recordings. This multichannel system is derived from a seismic system (Ruter and Strack, 1991). Figs. 5.11A and 5.11B display images of the developed

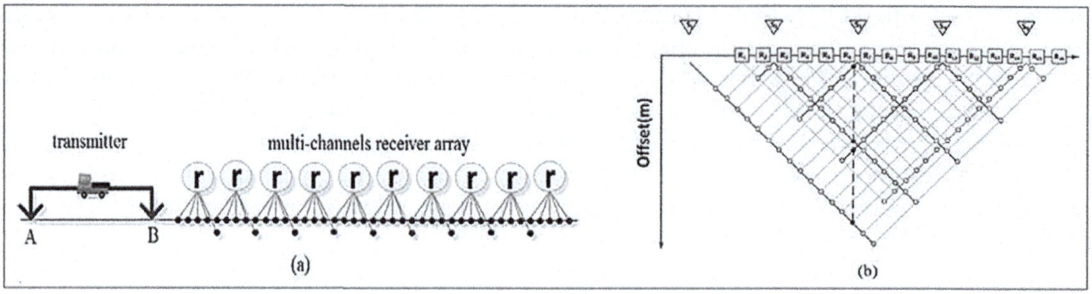

FIG. 5.10 The configuration for the MTEM. (A) Inline array configuration, (B) schematic of the multifold offset coverage. *Source: Di et al. (2018).*

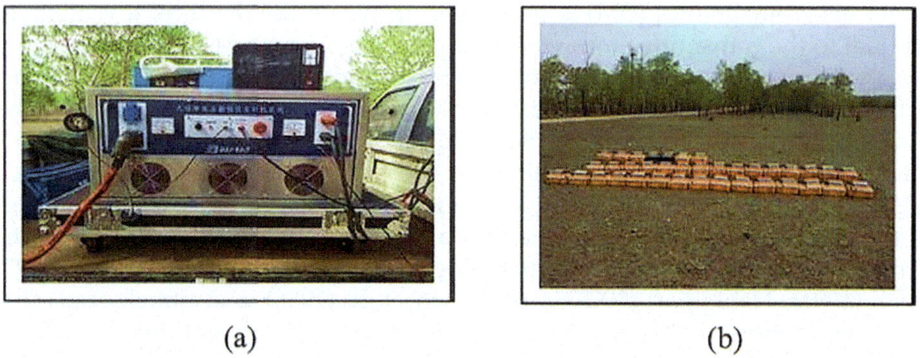

FIG. 5.11 MTEM system used in the field survey. (A) Transmitter, (B) receivers. Source: Di et al. (2018).

MTEM system, which uses a maximum PRBS order of 16, a maximum transmitting voltage of 1000 V, and a current of 50A. The system can support up to 200 receiver channels (Di et al., 2016).

5.12.2 MTEM data acquisition and results

MTEM data acquisition and processing were conducted along a 4.8 km profile over the Cao-si-yao molybdenum ore deposit in Inner Mongolia, North China (Fig. 5.12A). A strong earth impulse response curve was obtained even at long offsets. The peak times from the collected data were extracted and used to calculate the subsurface apparent resistivity. The resulting 2D inversion (Fig. 5.12B) revealed a resistive structure that aligned with known geological information about the ore body, as shown in Fig. 5.12C (Di et al., 2017).

5.13 Time domain short-offset electromagnetic method

The Short-Offset Grounded-Wire Transient Electromagnetic Method (SOTEM) was developed from the Long-Offset Transient Electromagnetic Method (LOTEM) (Xue et al., 2013;

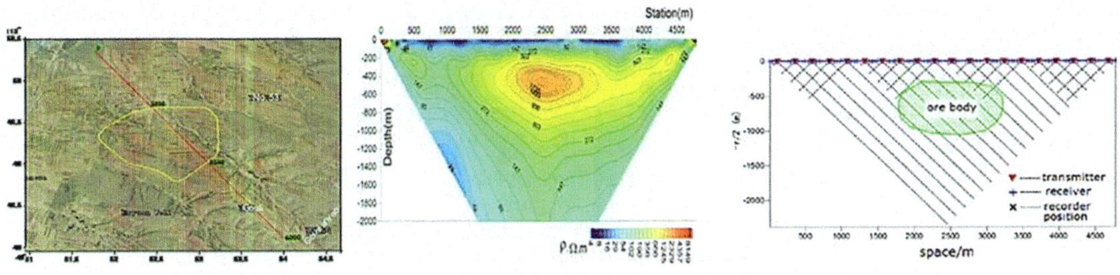

FIG. 5.12 MTEM field survey. (A) Survey line layout, (B) apparent resistivity section, (C) geological interpretation result. Source: Di et al. (2018).

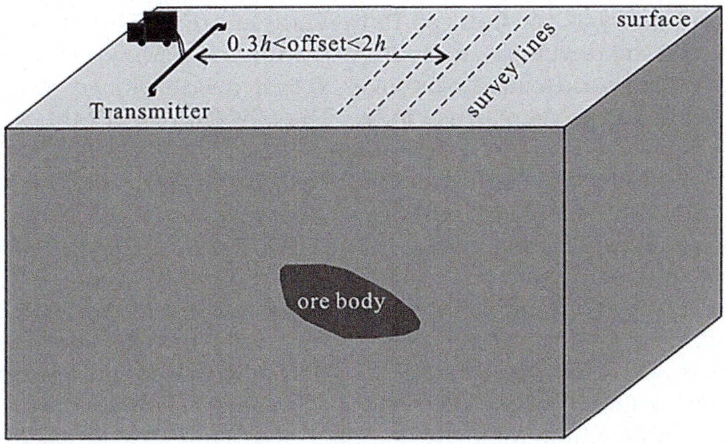

FIG. 5.13 Schematic of the SOTEM method, showing the grounded-wire source positioned parallel to the survey lines at an offset distance of approximately 0.3 to 2 times the maximum detection depth (h). *Source: Xue et al. (2022)—Open Access).*

Xue, 2018). The electromagnetic signal is generated by a grounded line source, and the response is measured in the near-source region along the east-west direction. The offset typically ranges from 0.5 to 1.5 times the detection depth, allowing for large detection depths with smaller offsets (Fig. 5.13). Data acquisition for SOTEM can be carried out using modern electromagnetic instruments such as the V8 and GDP-32. Chen et al. (2016) explored the response characteristics, workflow, and data interpretation methods for SOTEM based on several case studies. The SOTEM method has been successfully applied in the delineation of mined-out coalfield areas, mineral resource exploration, and engineering geological investigations, with targets located as deep as 1500 m underground (Chen et al., 2015; Zhou et al., 2015; Li et al., 2018). Although systems like the V8 and GDP-32 are commonly used for SOTEM data acquisition, the development of specialized equipment for SOTEM has become a priority.

The Short-Offset Transient Electromagnetic (SOTEM) method is a near-source electromagnetic technique. Unlike traditional far-source methods, the offset in SOTEM is significantly smaller, ranging from 0.3 to 2 times the maximum depth. As a result, SOTEM offers a higher signal-to-noise ratio and greater sensitivity. Its application at the Xiaoshan Mine in Henan Province, located in the southern margin of the North China Craton (NCC), demonstrated that the SOTEM method is highly effective for deep ore exploration, particularly in mountainous regions.

5.14 Time domain airborne electromagnetic method

The development of transient airborne electromagnetic (AEM) technology was driven by the need for efficient mineral exploration over large areas at a reasonable cost. The first attempts with airborne TEM systems in the 1950s were notably successful in base metal exploration in Canada, with >10 systems operational during that decade. The most successful system from this period was the Induced Pulse Transient (INPUT) system. Canada and the Nordic countries led the development and application of AEM systems, and by the 1970s, the method was in use worldwide. Of the over 30 systems developed since the inception of AEM, only a few are still in regular use today. The GEOTEM and MEGATEM systems are digital upgrades of the INPUT system. The concept of a transient helicopter system has only recently matured, with the development of systems such as AeroTEM, NEWTEM, Hoistem, VTEM, and SkyTEM. The AeroTEM, NEWTEM, Hoistem, and VTEM systems are primarily designed for mineral exploration, while the SkyTEM system is intended for mapping near-surface geological structures for groundwater and environmental studies, offering a faster alternative to ground-based TEM surveys. The transmitter and receiver coils, power supplies, laser altimeters, GPS, electronics, and data logger are carried as a sling load on the helicopter's cargo hook. An image of the SkyTEM system in operation is shown in Fig. 5.14.

In groundwater surveys, the SkyTEM system operates at a flying speed of 15–20 km/hr (4.1–5.5 m/s), providing a high moment stack size of around 1000 transients. To ensure vertical resolution and safety, the helicopter maintains an altitude of about 50 m, while the carrier frame stays at 15–20 m. In forested areas, the flying altitude is adjusted to the height of the trees.

5.15 ZTEM and airmt airborne electromagnetic systems

5.15.1 Introduction of the systems

To enhance mineral exploration success, an industry-wide consensus is necessary to expand the "discovery space" by exploring beneath cover and to greater depths. Over the past decade, airborne electromagnetic (AEM) systems have evolved with increasingly higher moments, and advancements in sensor calibration and postacquisition processing technologies have significantly improved data quality. As an alternative to conventional AEM, the Z-axis Tipper Electromagnetic (ZTEM) and Airborne Magnetic Tensor (AirMt) systems were developed to measure the transfer functions of audio-frequency natural electromagnetic sources from airborne platforms. The ZTEM system measures tipper

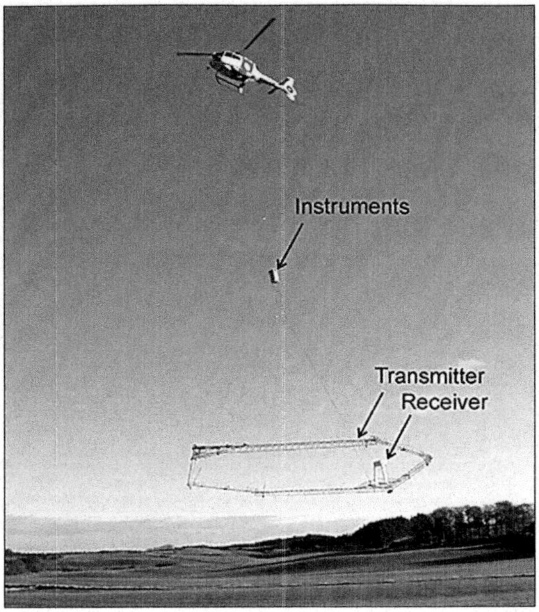

FIG. 5.14 The SkyTEM system in operation, with calibration conducted at the national test site in Denmark and routine data quality checks performed during helicopter refuelling and battery replacement at approximately 1.5 h intervals (Modified).

transfer functions, while the AirMt system measures the rotational invariant of the transfer functions. Ancillary data collected by both systems include radar altimeter, receiver altitude, GPS elevation, and total magnetic intensity. For both ZTEM and AirMt, data are typically recorded from 30 Hz to 720 Hz, enabling detection depths of up to 1 km or more, depending on terrain conductivity. This makes the method highly effective for mapping large-scale geological structures. **Eq.** (5.14), when applicable to ZTEM, suggests that electromagnetic techniques are less sensitive to conductivity anomalies deeper than 1 skin depth. This further implies that lower frequencies yield better resolution at greater depths (Fig. 5.15).

Both ZTEM and AirMt leverage Geotech's extensive logistical and technical expertise gained from its helicopter-borne Versatile Time Domain Electromagnetic (VTEM) system, which has been in commercial operation since 2002, with continuous generational improvements. The first commercial ZTEM surveys were launched in 2006, followed by the first commercial AirMt surveys in 2009. Currently, there are eight operational ZTEM systems and one AirMT system in use worldwide. These systems have been deployed for surveys across Australia, North America, South America, Africa, and the Middle East, targeting various mineralization types including Sedimentary Exhalative (Sedex) deposits, Volcanogenic Massive Sulfide (VMS) ore bodies, Iron Oxide Copper Gold (IOCG) deposits, Ni–Cu–PGE systems, porphyry, uranium, and precious metals, for a range of major and junior exploration companies.

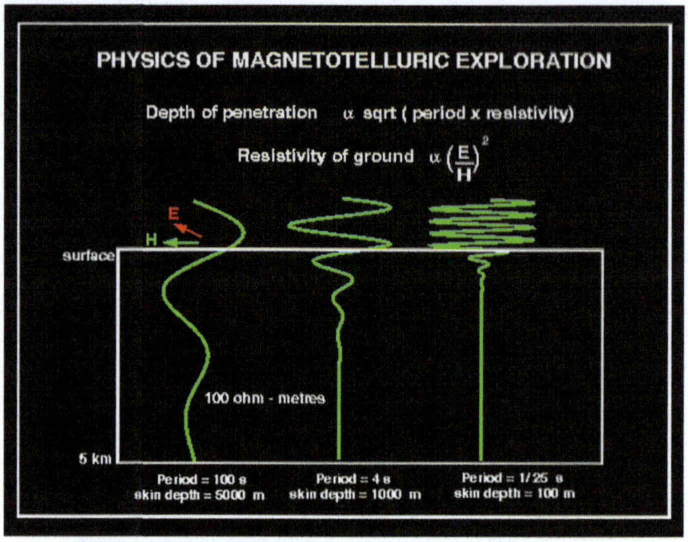

FIG. 5.15 Illustration demonstrating attenuation of amplitude with increasing frequency (Modified).

5.15.2 Background information along with theoretical consideration

5.15.2.1 Tensor impedance

1D impedance tensor: The conductivity of a 1D layered earth changes only with depth. The impedance tensor, Z, in this case is expressed as:

$$Z_{1D} = \begin{pmatrix} 0 & Z_{xy} \\ -Z_{xy} & 0 \end{pmatrix} \tag{5.47}$$

The apparent resistivity and phase in a layered earth are defined as:

$$\rho_a(\omega) = \frac{1}{\omega\mu_0}|Z(\omega)|^2 \text{ and } \varphi = \tan^{-1}\left(\frac{\mathrm{Im}Z_{xy}}{\mathrm{Re}Z_{xy}}\right) \tag{5.48}$$

2D impedance tensor: The conductivity for a 2D earth changes in one lateral direction and with depth but is constant in the strike direction:

$$Z_{2D} = \begin{pmatrix} Z_{xx} & Z_{xy} \\ Z_{yx} & Z_{yy} \end{pmatrix} \tag{5.49}$$

The 2D assumption requires that the strike length is significantly longer than the skin depth (Eq. 5.13) and all variations of the fields parallel to strike are zero ($\frac{\partial}{\partial x} = 0$). Furthermore, the EM fields are orthogonal and can be decoupled into a component with the E-field parallel to strike and the B-field parallel to strike. The two modes are referred to as E-polarization [transverse electric (TE) mode] and B-polarization [transverse magnetic (TM) mode], respectively.

The horizontal axes of the field layout are normally not aligned along the electromagnetic strike. The diagonal elements $Z_{xx} = —Z_{yy}$ are equal in amplitude but have opposite signs; the off-diagonal Z_{xy} and Z_{yx} are independent values. If the x direction is parallel to the strike direction, Z_{2D} impedance tensor becomes:

$$Z_{2D} = \begin{pmatrix} 0 & Z_{xy} \\ Z_{yx} & 0 \end{pmatrix} \tag{5.50}$$

The strike direction in a 2D earth is obtained by rotating the coordinate system to the direction minimizing the diagonal elements of the tensor, Z_{xx} and Z_{yy}. The 90° ambiguity in the electric strike is resolved by calculating the *Tipper* strike from H_z.

3D impedance tensor: The conductivity in a 3D earth varies in all directions (x, y, and z). The impedance tensor in this case is expressed as:

$$Z_{3D} = \begin{pmatrix} Z_{xx} & Z_{xy} \\ Z_{yx} & Z_{yy} \end{pmatrix} \tag{5.51}$$

All the elements of the 3D impedance tensor must be considered, because they are nonzero irrespective of rotation. According to Flóvenz et al., 2012, it is very essential in a 1D to invert for some rotational invariants defined in such a way that they average over directions. Then one does have to deal with the question of rotation. Three such invariants exist:

$$Z_B = \frac{Z_{xy} - Z_{yx}}{2} \tag{5.52}$$

$$Z_{det} = \sqrt{Z_{xx}Z_{yy} - Z_{xy}Z_{yx}} \tag{5.53}$$

$$Z_{gm} = \sqrt{-Z_{xy}Z_{yx}} \tag{5.54}$$

All these parameters give the same values for a 1D earth response. For 2D, Z_{det}(determinant) and (geometric mean) reduce to the same value, but Z_B(arithmetic mean) is different. For 3D responses, all these parameters are different (Flóvenz et al., 2012). For 3D responses, all these parameters are different (Flóvenz et al., 2012).

5.15.2.2 Tipper

For a non-1D earth, a part of the magnetic field tips into the vertical. The vertical magnetic component H_z is expressed as a linear combination of horizontal magnetic field components H_x and H_y (Everett and Hyndman, 1967), where:

$$H_z = AH_x + BH_y \tag{5.55}$$

form a complex vector often called the Wiese-Parkinson vector, induction vector, or tipper (Parkinson, 1959; Wiese, 1962). The Tipper magnitude is given by:

$$T = \sqrt{A^2 + B^2} \tag{5.56}$$

For a 1D earth, A=B = 0. For a 2D Earth, the structure can be rotated such that the x-axis is in the strike direction where A = 0, but $B \neq 0$. This is done by properly minimizing|A|. This strike direction is called the Tipper strike. The expression for calculating the tipper coefficients is:

$$A = \frac{H_z H_x^* H_y H_y^* - H_z H_y^* H_y H_x^*}{H_x H_x^* H_y H_y^* - H_x H_y^* H_y H_x^*} \tag{5.57}$$

$$B = \frac{H_z H_y^* H_x H_x^* - H_z H_x^* H_x H_y^*}{H_y H_y^* H_x H_x^* - H_y H_x^* H_x H_y^*} \tag{5.58}$$

In the context of magnetotelluric (MT) and audio-frequency magnetic (AFMAG) methods, let * denote the complex conjugate. The AFMAG method, first developed by Ward (1959), utilized two orthogonal coils towed behind an airborne platform to measure the tilt angle of the plane of polarization of natural magnetic fields within the 1 Hz to 20 kHz frequency band. These natural magnetic fields are primarily generated by atmospheric thunderstorm activity and can propagate over large distances with minimal attenuation due to the Earth-ionosphere waveguide. In a one-dimensional Earth model, where the tilt angle of the magnetic field is zero, the AFMAG method was highly effective for detecting the presence of conductors.

However, as pointed out by Ward et al. (1966), the direction and amplitude of the natural magnetic fields are highly variable both randomly over time and periodically with the seasons, which results in the AFMAG data being nonrepeatable. In response to this limitation, Labson et al. (1985) employed MT processing techniques for ground-based measurements of orthogonal horizontal magnetic fields. They demonstrated that repeatable "tipper" data could be recovered from these measurements, improving the robustness and consistency of the data collected in the presence of natural magnetic field variations.

The AFMAG method described by Labson et al. (1985) remained largely undeveloped for many years. However, with the recent commercialization of the ZTEM system (Legault, 2012; Pare and Legault, 2010) (Fig. 5.16) and the AirMt system (Kaminski et al., 2010) (Fig. 5.17) by Geotech, significant advances have been made in airborne magnetotelluric methods. These modern systems have enabled more efficient and effective exploration by utilizing the principles of AFMAG in combination with improved technological platforms for data acquisition and processing.

ZTEM measures the tipper components as the transfer function of a vertical magnetic field measured from an airborne receiver (r) to the horizontal components measured at a ground-based reference receiver (r_0):

$$H_z(r) = W_{zx}(r, r_0) H_x(r_0) + W_{zy}(r, r_0) H_y(r_0) \tag{5.59}$$

AirMt is a method used to measure the rotational invariant of the transfer function between magnetic field measurements taken by an airborne receiver and those taken by a reference ground-based receiver. The method generalizes the Weiss-Parkinson relationship, which is applied to model the linear relationship between the three components of a magnetic

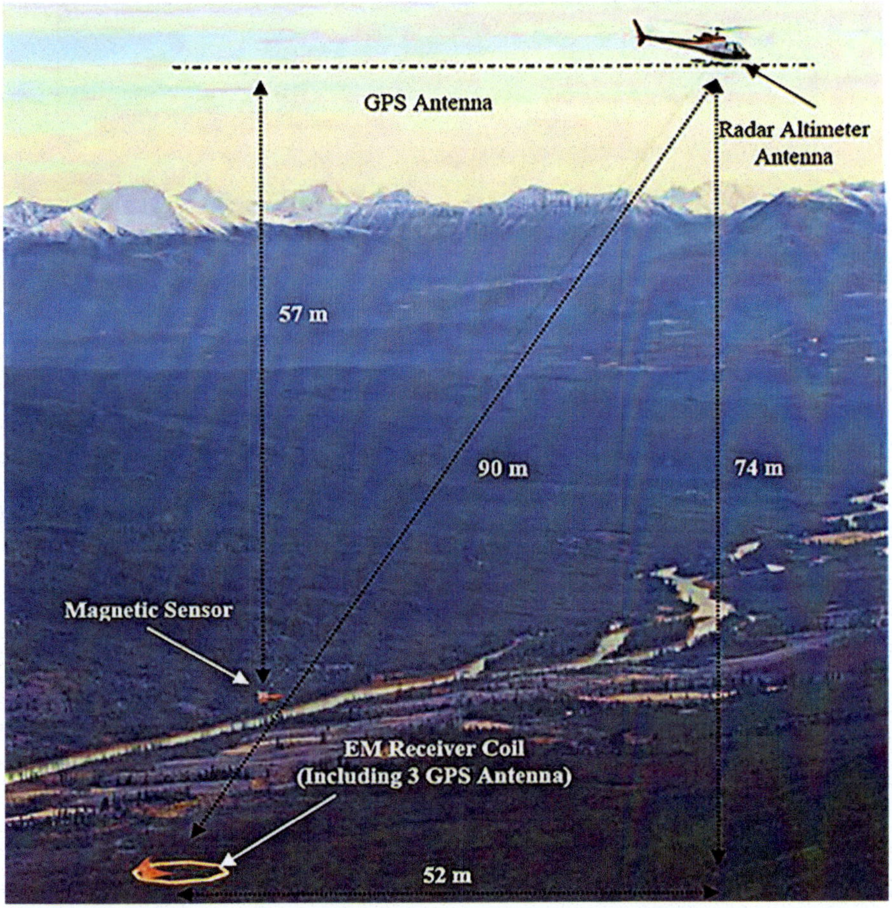

FIG. 5.16 ZTEM (z-axis Tipper electromagnetic) system configurations (Modified).

field at the airborne receiver and the horizontal magnetic field components measured at the reference ground receiver. In this context, the magnetic field components at the receiver r are linearly related to the horizontal components of the magnetic field at the reference receiver r_0. The relationship is captured through a transfer function that models the interaction between these fields, adjusting for rotational invariants that account for the orientation of both receivers.

$$\begin{bmatrix} H_x(r) \\ H_y(r) \\ H_z(r) \end{bmatrix} = \begin{bmatrix} H_{xx}(r,r_0) & H_{xy}(r,r_0) \\ H_{yx}(r,r_0) & H_{yy}(r,r_0) \\ H_{zx}(r,r_0) & H_{zy}(r,r_0) \end{bmatrix} \begin{bmatrix} H_x(r_0) \\ H_y(r_0) \end{bmatrix} \quad (5.60)$$

In the context of electromagnetic field measurements, let W_1 and W_2 represent the first and second columns of the transfer function matrix, respectively. These columns can be used to define a variable K, given by:

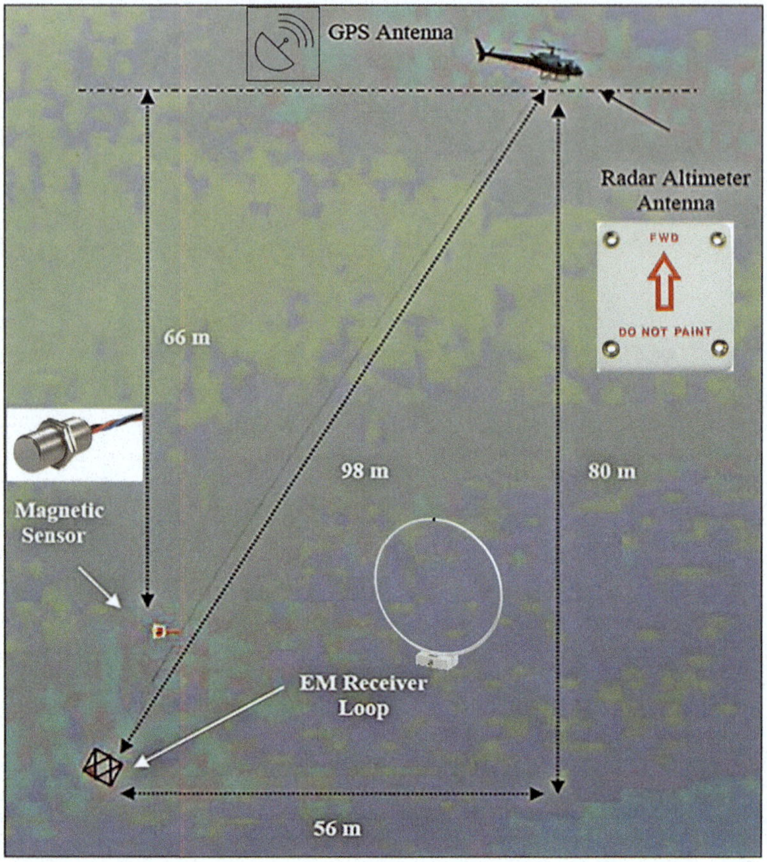

FIG. 5.17 AirMT (Airborne Magnetic Tensor) system configurations (Modified).

$$K = W_1 \times W_2$$

This results in the complex scalar K, which is further expressed as:

$$K = K.\frac{\mathrm{Re}(K)}{|\mathrm{Re}(K)|} \tag{5.61}$$

Here, K is referred to as the amplification parameter (AP), and it is rotationally invariant (Kuzmin et al., 2010). This means that the amplification parameter is independent of the sensor's orientation, which is an important property for accurate measurement in varying conditions. This property of rotational invariance ensures that the amplification parameter K remains consistent regardless of the sensor's alignment during measurement, which is a key aspect of systems like helicopter-borne or fixed-wing Zero-Temperature Electromagnetic (ZTEM) systems and AirMt systems.

Typically, the transfer functions in such systems are derived from measurements taken at several frequencies ranging from 20 Hz to 800 Hz. For these frequencies, the resulting skin depths typically range between 600 m and 2000 m, depending on the terrain's conductivity, which is important for subsurface imaging.

5.16 Ground penetrating radar or georadar

5.16.1 Introduction

Ground Penetrating Radar (GPR) is a relatively recent advancement in geophysical exploration, developed to investigate subsurface structures with high resolution. GPR utilizes high-frequency electromagnetic waves (typically ranging from 10 MHz to 1000 MHz) to acquire detailed subsurface information (Kirsch, 2006). The technique involves sending pulses of energy into the ground, which then propagate downward. These waves are reflected back to the surface by boundaries where there are contrasts in the electrical properties of the materials, such as changes in conductivity or permittivity. The reflected signals are recorded and analyzed to obtain insights into the subsurface (Fig. 5.18).

The data from GPR is typically presented in a format known as a "trace," which shows the signal's amplitude as a function of time. Each trace includes the transmitted pulse followed by subsequent reflections from subsurface objects or layers, revealing the presence of contrasts in material properties at various depths.

GPR is commonly applied in a range of fields, including environmental monitoring, engineering, archaeology, and shallow geophysical investigations. The Geological Survey of India (GSI), Polar Studies Division, has also procured a GPR system for glaciological studies and regularly conducts GPR surveys for polar research (Shrivastava et al., 2017). The survey over a sub-glacial lake at Schirmacher Oasis, Antarctica, was conducted by the Geological Survey of India (GSI) as part of the XXII Indian Scientific Expedition to Antarctica (ISEA), 2003. The survey was carried out using a RAM AC GPR system equipped with a 250 MHz shielded antenna. This GPR system was used to investigate the subsurface structure beneath the ice, focusing on the identification and characterization of the sub-glacial lake. The GPR data

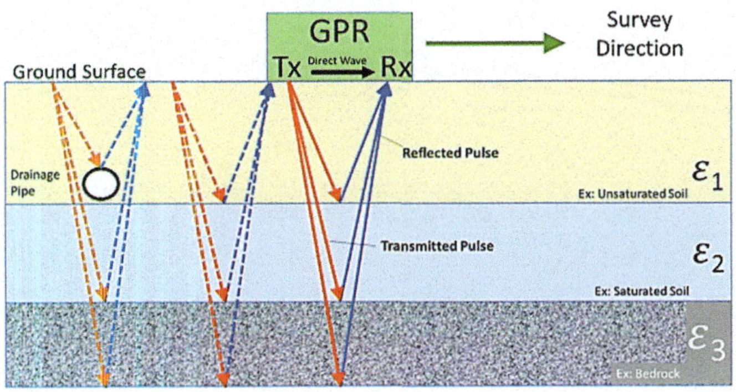

FIG. 5.18 Schematic of a Ground Penetrating Radar (GPR) survey showing the propagation of reflected waves through subsurface materials with varying dielectric permittivity (ε). Subsurface anomalies, such as a draining pipe, are detected, with dashed lines indicating the survey progression. *Source: US EPA.*

provided high-resolution images of the ice layers, offering insights into the depth, extent, and composition of the sub-glacial lake. This survey is a crucial component of glaciological studies, contributing to the understanding of the dynamics of sub-glacial environments, which are critical for ice sheet stability and the broader climate system (Fig. 5.19). The collected data supports ongoing research in glaciology and polar studies, further advancing scientific understanding of polar regions (Sinharay, 2019).

Its noninvasive nature allows for the mapping of subsurface features without the need for excavation. Like other geophysical methods, the interpretation of GPR results is nonunique and should be corroborated with direct physical evidence. Typical applications of GPR include:

- Mapping the depth to bedrock or the water table
- Determining the depth and thickness of soil or sediment layers on land or beneath freshwater bodies
- Locating subsurface cavities, fractures, or voids in bedrock
- Detecting buried objects such as pipes, drums, tanks, cables, and boulders
- Identifying the boundaries of landfills or trenches, mapping contaminants, and conducting archaeological surveys.

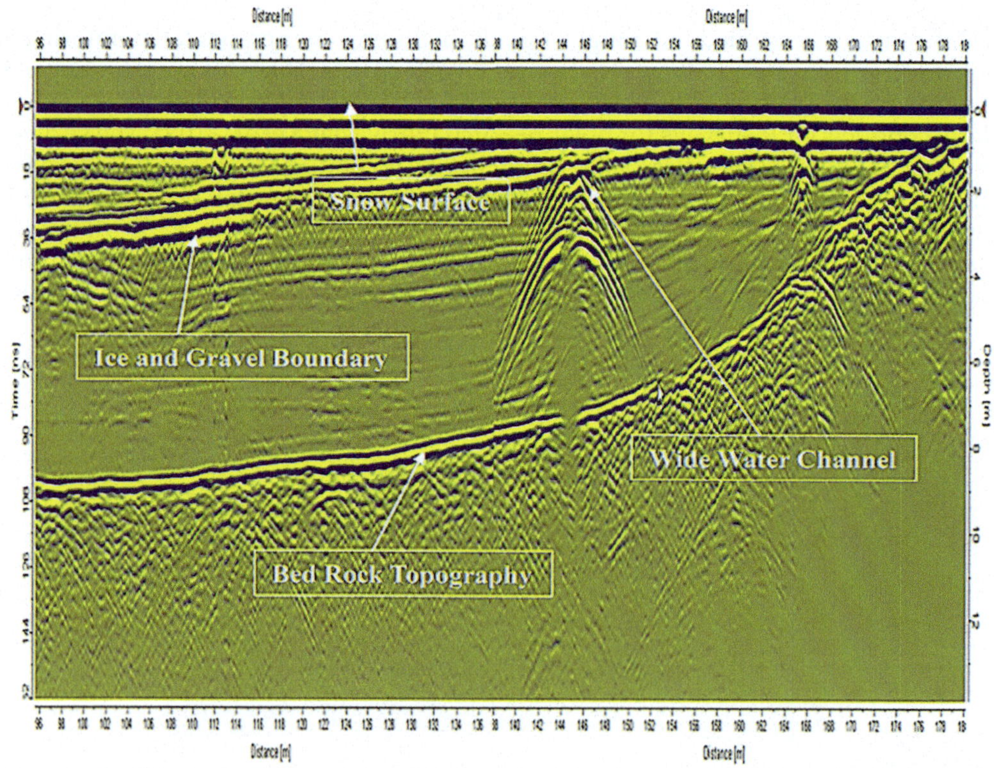

FIG. 5.19 GPR section of a sub-glacial lake at Schirmacher Oasis, Antarctica, acquired during XXII ISEA using RAM AC GPR with 250 MHz shielded antenna (Modified).

In addition, GPR is often integrated with other geophysical methods, such as seismic, resistivity, or electromagnetic techniques, to enhance subsurface characterization and reduce uncertainties in data interpretation. The integration of GPR with other methods improves the overall reliability and resolution of the investigation.

Currently, GPR is recognized as a standard technique for subsurface investigations. It is widely used in fields requiring shallow subsurface imaging, and several guidelines exist for its application. The American Society for Testing and Materials (ASTM) has established an approved standard guide for the use of GPR in subsurface investigations. Comprehensive overviews of GPR techniques can be found in textbooks such as Daniels (1996) and Reynolds (1997), which provide detailed discussions on the method's principles, applications, and technical considerations.

5.16.2 Theory of operation

The Ground Penetrating Radar (GPR) system emits a high-frequency electromagnetic pulse (radio wave) into the ground (Fig. 5.20). Unlike continuous signals, GPR transmits discrete pulses, each containing a set of electromagnetic waves oscillating at a specific frequency. As these radio waves travel through the subsurface, they undergo distortion due to the varying electromagnetic properties of the materials they encounter, such as electrical conductivity (σ), magnetic permeability (μ), and dielectric permittivity (ε).

At high frequencies, especially in poorly conductive media, the conduction term in the electromagnetic wave equations becomes negligible compared to the displacement current term. Under these conditions, the electric field equation simplifies as described by **Eqs.** (5.8a) and (5.8b).

$$\nabla^2 E = \mu_0 \varepsilon \frac{\partial^2 E}{\partial t^2} \tag{5.62}$$

with a similar equation for the magnetic field. The first term of **Eq.** (5.8b) is related to the familiar conduction of electricity in a conductor. Maxwell introduced the second term and called it the displacement current. **Eq.** (5.57) has the familiar form of the wave equation, which describes the propagation of an elastic disturbance [Eqs. u = Asin(kx—ωt) = Asink(x—ct) (3.55) & $k\frac{dx}{dt} - \omega = 0$ (3.56) in which the wave number (k), angular frequency (ω), and

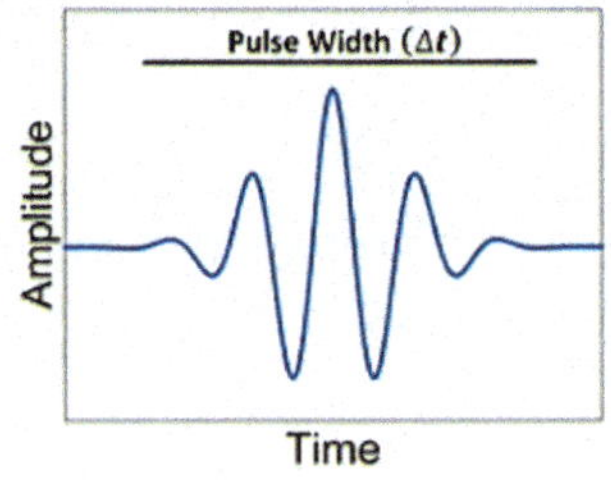

FIG. 5.20 Wavelet representing the pulse of EM waves within the GPR signal. *Source: GeoSci Developers, Version 0.0.1, 2017.*

velocity (c) are defined and related by $k = \frac{2\pi}{\lambda}$, $\omega = 2\pi f = \frac{2\pi}{T}$, $c = \lambda f = \frac{\omega}{k}$; see Lowrie (2006), page 133]. Analogously, **Eq.** (5.57) describes the propagation of the electric part of an electromagnetic wave. By comparing with the seismic wave equations, we see that the E and B fields in an electromagnetic wave have the same velocity v, where

$$v^2 = \frac{1}{\mu_0 \varepsilon} \tag{5.63}$$

The basic unit of electromagnetic wave travel time is the nanosecond (ns), where 1 ns = 10^{-9} s and the velocity of an electromagnetic wave in air is 0.3 m/ns. In a vacuum, the wave velocity is equal to the velocity of light c, given by $c^2 = \frac{1}{\mu_0 \varepsilon}$. Using the relationship of dielectric permittivity $\varepsilon = \kappa\varepsilon_0$, we get considering the dielectric constant (κ)

$$v^2 = \frac{c^2}{\kappa} \tag{5.64}$$

where $\varepsilon = \varepsilon_0\varepsilon_r$, ε_r is the relative permittivity of the material, and ε_0 is the permittivity constant, or permittivity of free space $\varepsilon_0 = 8.854 \times 10^{-12}\,C^2N^{-1}m^{-2}$. In constant electrical fields, ε_r is known as the dielectric constant, κ. κ is 5–20 in earth materials, and the velocity of an electromagnetic wave in the ground is found to be about 0.2c–0.6c. To simplify further discussion, suppose that the electromagnetic disturbance propagates along the z-axis

(i.e., $\frac{\partial}{\partial x} = \frac{\partial}{\partial y} = 0$), so that $\nabla^2 = \frac{\partial^2}{\partial z^2}$. For this one-dimensional case

$$\frac{\partial^2 E}{\partial z^2} = \mu_0 \varepsilon \frac{\partial^2 E}{\partial t^2} = \frac{1}{v^2}\frac{\partial^2 E}{\partial t^2} \tag{5.65}$$

If we compare this equation [with u = Bsink(x + ct) (3.58), Lowrie (2006), page 133] for a seismic wave, we see that the solution for a component E_i of the electric field is (Lowrie, 2007)

$$E_i = E_0 \sin 2\pi\left(\frac{z}{\lambda} - ft\right) \tag{5.66a}$$

or, according to Kirsch (2006), in exponential notation, Eq. (4.65a) can be written as

$$E = E_0 e^{i(\beta z - \omega t)} e^{-\alpha z} \tag{5.66b}$$

In the context of electromagnetic wave propagation in the ground, the relationship between the wavelength (λ), frequency (f), and wave velocity (v) is given by:

$$f\lambda = v$$

Additionally, the wave exhibits both phase and attenuation characteristics, which are described by the phase constant (β) and the attenuation constant (α), respectively. These parameters influence how the wave propagates and decays in the subsurface. The propagation of high-frequency electromagnetic waves in the ground behaves similarly to seismic waves. In this analogy, the electric field (E) and the magnetic field (B) are orthogonal to each other, as well as to the direction of wave propagation. The equation associated with the wave's behavior, **Eqs.** (5.61a) and (5.61b), accounts for both phase variations with respect to time and space, as well as the attenuation of the wave as it propagates through the medium. The first exponential term in the equation represents the phase variations of the wave in both time and space, with β being the phase constant. The second exponential term describes the attenuation of the wave, governed by the attenuation constant α\alphaα. Both β and α are frequency-dependent and are related to the dielectric permittivity (ε) of the medium (Kirsch, 2006).

$$\alpha = \frac{\omega}{c_0}\sqrt{\frac{\varepsilon_r'}{2}\left(\sqrt{1+\tan^2\delta}-1\right)}\,\text{Neper}/\text{m} \tag{5.67}$$

$$\beta = \frac{\omega}{c_0}\sqrt{\frac{\varepsilon_r'}{2}\left(\sqrt{1+\tan^2\delta}+1\right)}\,\text{radian}/\text{m} \tag{5.68}$$

where $c_0 = 3 \times 10^8$ m/s = speed of the light in vacuum (or air), ε_r'= the real part of dielectric permittivity, and tanδ = the loss tangent is the ratio of conduction to displacement currents and defined as (Kirsch, 2006)

$$\tan\delta = \frac{\sigma}{2\pi f\varepsilon_0\varepsilon_r} \tag{5.69}$$

for tanδ = 1 the conduction currents are equal to the displacement currents. Electromagnetic wave propagation utilized in GPR takes place for tanδ < 1, while for tanδ > 1 diffusive processes govern (the case of low frequency electromagnetic like RMT or AEM).

It is important to note that the attenuation constant α\alphaα is small when the loss tangent (tanδ) is much <1, which occurs when the conductivity (σ) is low, leading to minimal absorption. As the frequency increases, electromagnetic signals are attenuated more rapidly with depth. This attenuation is caused not only by the geometrical spreading of the signal (spherical divergence), which reduces the signal's intensity with distance but also by the absorption of the signal due to the conductivity of the ground materials.

At low-loss conditions, the velocity of electromagnetic wave propagation in the medium can be approximated by the following simplified equation:

$$v = \frac{3\times10^8}{\varepsilon^{\frac{1}{2}}} \tag{5.70}$$

where v is the velocity in meters per second (m/s) and ε is the dielectric permittivity of the material. This equation assumes minimal loss and provides an estimate for the wave's velocity based on the material's dielectric properties.

5.16.3 System components

Ground Penetrating Radar (GPR) operates based on the "echo principle," similar to reflection seismology. A brief radar pulse, typically lasting only a few nanoseconds (10^{-8} s), is emitted from a mobile antenna located on the ground surface. This pulse travels through the ground as a ray, encountering refractions, reflections, and diffractions when it crosses boundaries with varying dielectric constants. A second antenna, functioning as the receiver, is placed near the transmitter to detect near-vertical reflections from subsurface discontinuities, similar to seismic reflection methods. To minimize the impact of diffractions and other noise sources, signal processing techniques from seismic reflection seismology are also applied to GPR data. This allows for the creation of detailed images of the shallow subsurface structure (Fig. 5.21).

The depth of penetration of GPR depends on the conductivity of the material being investigated. In materials with low conductivity (e.g., a few milliSiemens/m), GPR can typically penetrate depths of over 30 m. In certain conditions, such as in thick polar ice or salt deposits, the penetration can extend up to 5000 m. However, in most soils and rocks, GPR penetration is usually <10 m. In highly conductive materials, such as smectite clays or rocks with conductive pore fluids, the depth of penetration can be limited to less than one meter.

Despite its advantages, GPR is sensitive to unwanted noise from both geological and cultural sources. Geological sources of noise include inhomogeneities like boulders, animal burrows, tree roots, and other subsurface features that can cause unwanted reflections or scattering. Cultural sources of noise arise from human-made objects such as vehicles, buildings, fences, power lines, and trees. While shielded antennas can help minimize these unwanted reflections, electromagnetic interference from devices like cellular phones, two-way radios, television, and microwave transmitters may still introduce noise into GPR data.

5.16.4 Mode of operation—reflection and refraction of plane waves

At a plane boundary between two media (1) and (2) with different dielectric properties (ε) and conductivities (σ), an incident electromagnetic wave is partially reflected in medium (1) and partially refracted into medium (2). The incident, reflected, and refracted waves all lie in the same plane. The angle of incidence (θ_1) is equal to the angle of reflection (θ_1) as illustrated in Fig. 5.22. The angle of refraction (θ_2) is related to the angle of incidence (θ_1) by Snell's Law, which for electromagnetic waves is expressed as:

$$\frac{\sin\theta_1}{\sin\theta_2} = n_{21} = \sqrt{\frac{\varepsilon_2}{\varepsilon_1}} = \frac{v_1}{v_2} \tag{5.71}$$

where v_1 and v_2 are the velocities of the wave in media (1) and (2), respectively, and n_{21} is the index of refraction for the transition from medium 1 to medium 2. If the velocity in medium 2 (v_2) is greater than the velocity in medium 1 (v_1), the wave refracts toward the horizontal direction, causing the wavefront to spread out. Conversely, if $v_2 < v_1$, the wave refracts toward the vertical, causing the wavefront to contract.

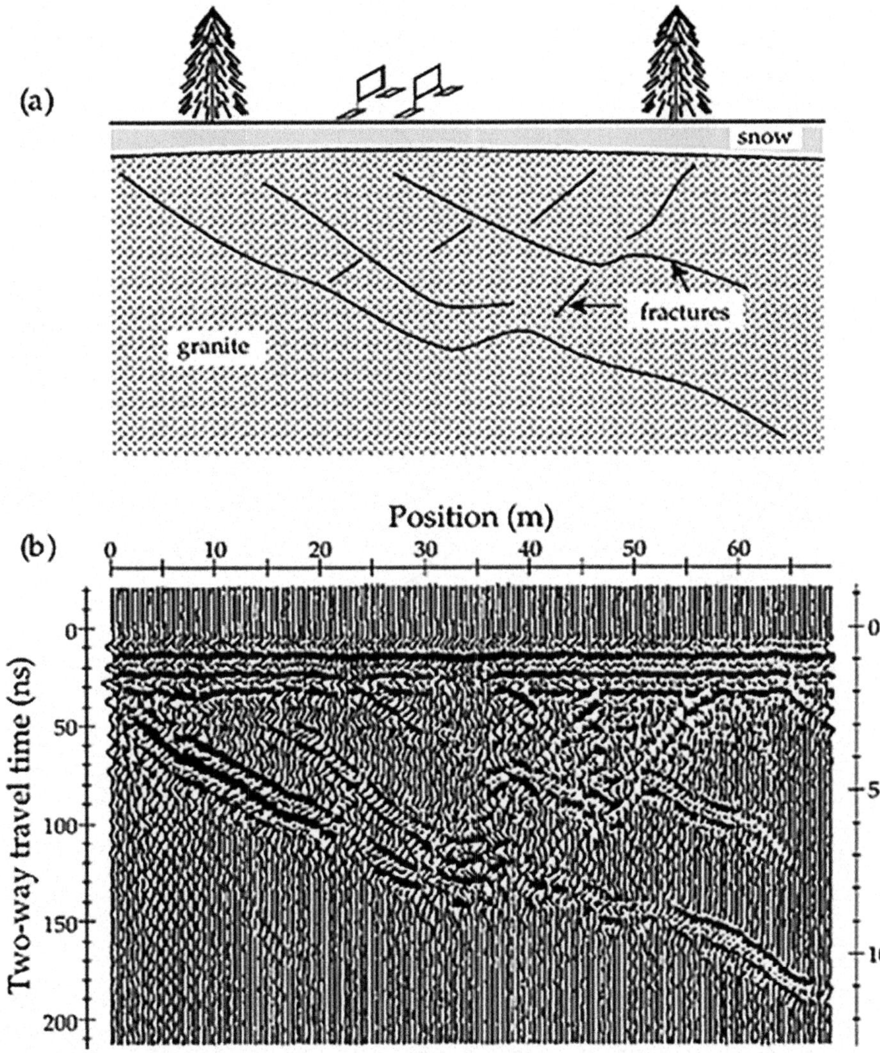

FIG. 5.21 Fracture pattern in granitic bedrock revealed by ground penetrating radar: (A) geological cross-section, (B) processed georadar reflection section. *Source: Lowrie (2007).*

Just like in refraction seismology, radio waves can experience critical refraction. This occurs when the incident angle (θ_1) is such that the refracted wave propagates along the interface at velocity v_2 and the angle of refraction (θ_r) reaches 90°, resulting in the formation of a head wave (Fig. 5.23). The critical angle (θ_c) is given by:

$$\sin\theta_c = \frac{v_1}{v_2} \tag{5.72}$$

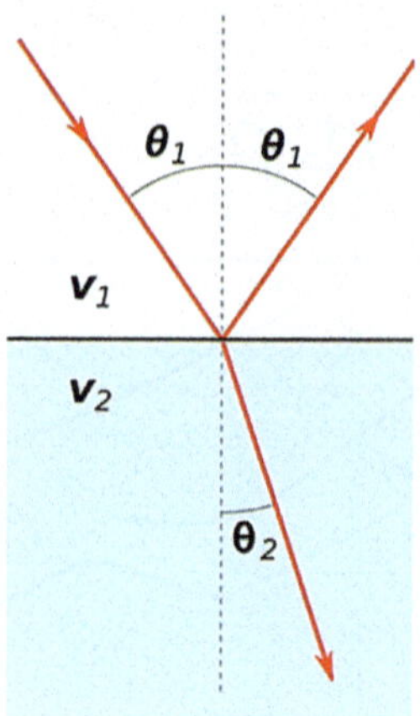

FIG. 5.22 Reflection and refraction of an incoming radio wave. *Source: GeoSci Developers, Version 0.0.1, 2017.*

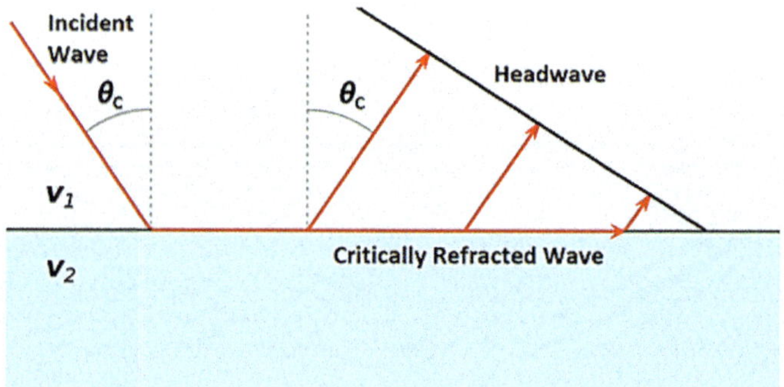

FIG. 5.23 Critical refraction at interface and the resulting head wave. *Source: GeoSci Developers, Version 0.0.1, 2017.*

Critical refraction occurs only when $v_1 < v_2$. For all angles of incidence $\theta_1 > \theta_c$, total reflection occurs. In the far field of a transmitting antenna, electromagnetic waves are typically approximated as plane waves. In the case of normal incidence ($\theta_1 = 0°$), only the dielectric contrast (ε) is considered, as described in the more general Fresnel equations (von Hippel, 1954a).

The reflection coefficient, which represents the reflected part of the incident unit amplitude, is given by:

$$r = \frac{v_2 - v_1}{v_2 + v_1} = \frac{\sqrt{\varepsilon_1} - \sqrt{\varepsilon_2}}{\sqrt{\varepsilon_1} + \sqrt{\varepsilon_2}} \tag{5.73}$$

The refracted (or transmitted) part of the wave is given by:

$$t = \frac{2v_2}{v_2 + v_1} \tag{5.74}$$

The reflection coefficient r can be either positive or negative, with values between $-1 < r < 1$. The magnitude of r determines how much of the incident wave is reflected. It is important to note that these equations are valid for ideal dielectric materials. In the case of conductive (lossy) materials, even for normal incidence, the reflection coefficient can become complex, leading to a deformation of the reflected wave.

Multiple reflections, which occur between the surface and a strong reflector, are typically not significant for GPR, as there is usually considerable absorption in the soil and rock that reduces the intensity of the reflected waves.

5.16.5 GPR ray geometry and travel times

The propagation of electromagnetic waves in Ground Penetrating Radar (GPR) can be represented by ray theory, similar to optics and reflection seismics. In a simple two-layer model with a reflector at depth h, the wave paths and travel time curves can be described by four different paths. Fig. 5.24 illustrates the ray path scheme for GPR.

At time t = 0 s, the transmitter (Tx) generates a pulse of radio waves. As depicted in Fig. 5.24, there are multiple paths that the radio waves can take to reach the receiver (Rx). Two direct waves with different phase velocities and amplitudes travel along the ground surface: the air wave and the ground wave (Banos, 1966). Each ray path can be explained as follows:

Path 1: Direct air wave

This wave travels through the air in a direct line from the transmitter to the receiver. Since radio waves in the air propagate at the speed of light ($c = 3 \times 10^8 \text{m/s}$), the direct air wave is always the first signal detected by the receiver. The time taken for the air wave to reach the receiver is given by:

$$t_{air} = \frac{x}{c} \quad (5.75)$$

where x is the distance between the transmitter and receiver. The velocity of the air wave can be derived from the slope of the direct air wave on the radargram (Fig. 5.25) (Kirsch, 2006).

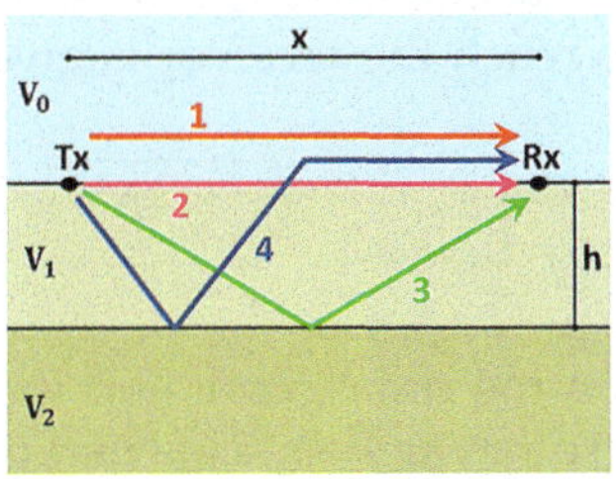

FIG. 5.24 Radio wave signals measured by a receiver for a horizontal 2-layer earth. *Source: GeoSci Developers, Version 0.0.1, 2017.*

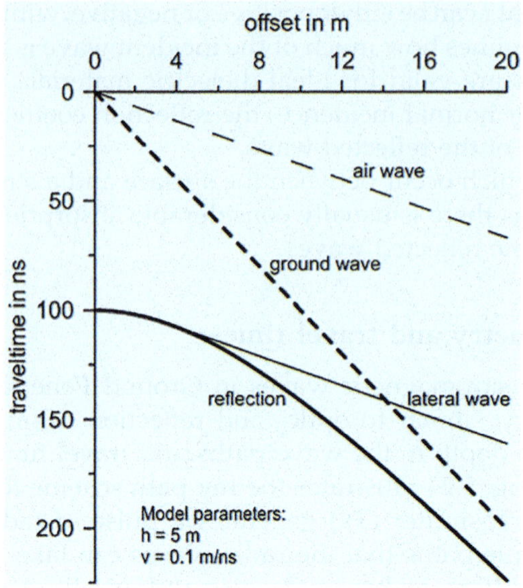

FIG. 5.25 GPR travel time diagram (radargram) for the horizontal two-layer earth (Modified).

Path 2: Direct ground wave

- The ground wave travels along the surface interface at velocity v_1. Since $v_1 < c$, the ground wave arrives later than the air wave. The time it takes for the ground wave to reach the receiver is given by:

$$t_{ground} = \frac{x}{v_1} \tag{5.76}$$

The direct ground wave is also visible on the radargram. Like the air wave, the velocity of the direct ground wave can be derived from the slope of its corresponding line.

Path 3: Reflected wave

- The reflected wave travels through medium 1 at velocity v_1. Since the reflected wave follows a longer path compared to the direct ground wave, it arrives later. The time taken for the reflected wave to reach the receiver is given by:

$$t_{reflected} = \frac{\sqrt{x^2 + 4h^2}}{v_1} \tag{5.77}$$

The reflected wave is shown on the radargram, and its arrival time follows a hyperbolic curve, which distinguishes it from other signals. This part of the curve is useful for estimating the velocity of the top-most layer. Notably, the slopes of the direct ground wave and the reflected wave are parallel.

Path 4: Critically refracted at surface

- The critically refracted wave, denoted in blue, occurs because $v_1 < v_0$, causing the reflected waves to undergo critical refraction at the surface. This wave propagates along the surface interface and has a velocity close to the speed of light. The time for this wave to reach the receiver is given by:

$$t_{critical} = \frac{x}{c} + \text{constant} \tag{5.78}$$

The arrival time of the critically refracted wave follows a linear relationship. Although the critically refracted wave might not be easily visible in the radargram, its existence is still significant and may be inferred from the data.

5.16.6 GPR amplitudes

Understanding amplitude decay is crucial for interpreting Ground Penetrating Radar (GPR) measurements and estimating penetration depth. The amplitude of the signal decreases with distance due to two main factors: geometric spreading and absorption. The amplitude of a wave reflected at depth h, with a reflectivity r, can be expressed using the following equation:

$$\frac{E_r}{E_1} = \frac{r}{2h} e^{-2\alpha h} \tag{5.79}$$

where E_r = amplitude of the reflected wave at depth h, E_1 = amplitude of the wave at unit depth, r = reflectivity coefficient of the interface (material property describing the fraction of energy reflected back), h = depth of the reflecting layer (in meters), α = attenuation coefficient (in m^{-1}), defined as $\alpha = \frac{\alpha'}{8.686}$, where α' is in decibels per meter (dB/m), and $e^{-2\alpha h}$ = exponential term representing signal absorption during its two-way travel (down to the reflector and back).

The first factor influencing amplitude decay is geometric spreading. As the wave propagates, its amplitude decreases inversely with the distance 2 h. This is represented by the term $\frac{1}{2h}$, which reflects the spreading of energy over a larger area. This phenomenon is a natural consequence of wave propagation in a three-dimensional medium. The second factor is attenuation, which is captured by the exponential term $e^{-2\alpha h}$. This term quantifies the energy lost due to absorption as the wave travels through the medium and returns to the surface. The extent of attenuation depends on the material properties and the distance traveled by the wave. Finally, the reflectivity coefficient r plays a significant role in determining the proportion of energy reflected back to the source. This coefficient is influenced by the contrast in material properties at the interface. Higher reflectivity indicates a greater fraction of energy being reflected, which affects the amplitude of the reflected signal. Together, these factors—geometric spreading, attenuation, and reflectivity—govern the behavior of signal amplitude in GPR measurements, providing critical insights into subsurface properties and layer interfaces.

For waves diffracted back toward the source, such as by small objects or features, the radar equation provides an alternative to **Eq.** (5.74). The equation is expressed as:

$$\frac{P_r}{P_t} = q_s \frac{G^2\lambda^2}{(4\pi)^3 h^4} e^{-4\alpha h} \tag{5.80}$$

where P_t = transmitted power (in watts), P_r = received power (in watts), q_s = scattering cross-section of the diffractor (in m^2), G = antenna gain, relative to a spherical dipole (unitless), λ = wavelength at the center frequency (in meters), h = distance between the antenna and the diffractor (in meters), α = attenuation coefficient (in m^{-1}), and $e^{-4\alpha h}$ = exponential term accounting for signal absorption over the four-way travel path (to the diffractor and back for both transmitted and scattered waves).

The first feature of **Eq.** (5.75) is the power decay with distance. The received power decreases with the fourth power of the distance h, as represented by the term $\frac{1}{h^4}$. This significant decay reflects energy spreading and the fact that diffractors are often much smaller than the wavelength of the signal, leading to weaker returns. The antenna gain, G^2, is another important factor. It quantifies how effectively the antenna focuses energy in a specific direction. Higher antenna gains result in stronger received signals, enhancing the sensitivity and effectiveness of the GPR system. The dependence on wavelength, λ^2, also plays a role. Lower-frequency signals with longer wavelengths generally result in stronger scattering, as the interaction with diffractors is more pronounced at these wavelengths. This characteristic is crucial in selecting the appropriate radar frequency for specific subsurface investigations. The exponential term $e^{-4\alpha h}$ represents the attenuation of the signal due to the medium's absorptive properties. It accounts for the four-way travel path, incorporating the energy loss experienced by both the transmitted and scattered signals as they traverse the medium.

Additional observations on signal decay complement the understanding of GPR measurements. The direct wave's field strength decreases with the square of the distance from the source in the far field, as described by Banos (1966). Similarly, the ground wave is further reduced due to absorption and scattering. Lateral waves also experience decay approximately proportional to the square of the distance, as noted by Brekovskikh (1980). Practically, **Eq.** (5.74) is ideal for estimating the amplitude of reflected waves and is particularly useful for mapping subsurface layer interfaces. In contrast, **Eq.** (5.75) is applicable to diffracted waves and is effective for detecting small objects such as buried pipes or voids. This equation also aids in optimizing GPR system design by considering transmitted power, antenna gain, and wavelength. This explanation provides a structured understanding of GPR amplitudes and their implications, helping in the planning and interpretation of subsurface investigations.

5.16.7 Estimation of exploration depth in GPR (ground penetrating radar)

Estimating the exploration depth in Ground Penetrating Radar (GPR) surveys is critical for selecting appropriate equipment and interpreting results. The depth of exploration depends on factors like the electromagnetic properties of the subsurface, the GPR system frequency, and antenna characteristics. The frequency of the GPR system affects both resolution and depth penetration. Higher frequencies (1000–2500 MHz) offer better resolution but are limited to shallow depths, ideal for locating small objects or characterizing near-surface features. Lower frequencies (50–250 MHz) provide deeper penetration at the cost of resolution, making them suitable for investigating larger geological structures or deeper subsurface layers.

Ground Penetrating Radar (GPR) is widely used for mapping near-surface structures, and in a study by the CSIR-National Geophysical Research Institute (NGRI), a GPR survey was conducted using the Terra Sirch-3000 m with 2D data collection (Sonkamble and Chandra, 2021). This survey targeted various applications, including civil engineering, environmental, archaeological, geological features, aquifer contamination, and seawater intrusion studies in different regions of India (Fig. 5.26). The study utilized 200 MHz and 400 MHz frequency antennae, enabling depth identification up to 3–8 m (Sonkamble and Chandra, 2021; Sonkamble et al., 2013).

GPR data was collected using Geologic Scan, Utility Scan, Structure Scan, and Terra SIRch modes, processed with RADAN (version 5.0) software for clear anomaly detection. The data was validated through excavation, electrical resistivity tomography, and soil and water chemistry. Key findings include the identification of caving in a trenchless MS pipe, discovery of an ancient temple (Fig. 5.27), detection of a dolerite dyke, clandestine pipes, fresh groundwater zones, and seawater intrusion (Sonkamble and Chandra, 2021).

This study highlights GPR as an effective tool for rapid and precise mapping of shallow subsurface complexities, proving valuable for a range of environmental and earth management applications. The results offer insights for researchers, government agencies, industries, and consultants in making quick decisions on field sites. GPR has been widely used by CSIR-NGRI for applications in civil engineering, archaeology, geology, and environmental management across India.

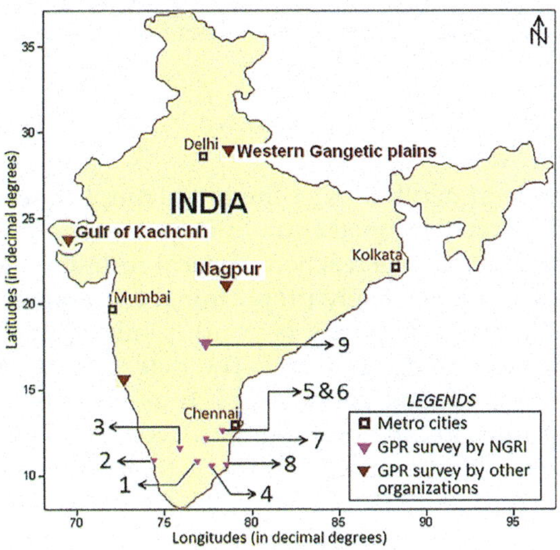

FIG. 5.26 Locations of GPR surveys in India for various applications (Modified).

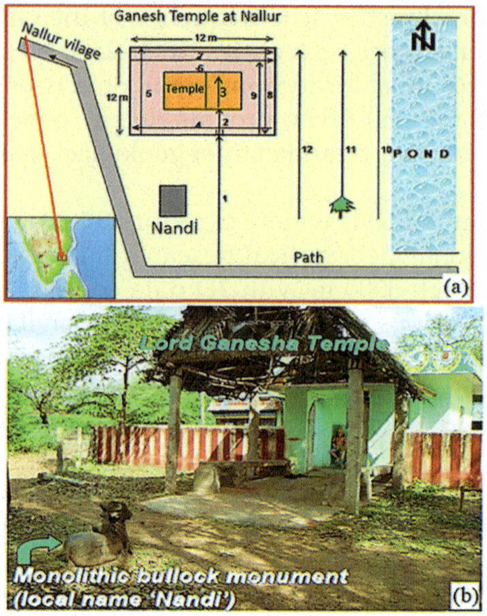

FIG. 5.27 Figure showing (A) locations of GPR profiles, and (B) field snapshot of Lord Ganesha temple in Tanjavur District, Tamil Nadu (Modified).

Dielectric permittivity significantly influences depth penetration. Materials with low permittivity (e.g., dry sand, gravel) allow deeper radar signal penetration, while materials with high permittivity (e.g., clays, wet soils) cause rapid signal attenuation. Attenuation, related to dielectric properties and material conductivity, limits usable signal depth. Signal attenuation occurs due to scattering, absorption, and spreading. High-conductivity materials like wet clay or saline water absorb signals more, reducing exploration depth. Low-conductivity materials like dry sand or granite permit deeper signal travel. Attenuation can be calculated using

$$A(x) = A_0 e^{-\alpha x} \tag{5.81}$$

where $A(x)$ = the amplitude at depth x, A_0 = the initial amplitude, α = the attenuation coefficient, and x = the depth. When the signal amplitude becomes too weak to be distinguishable from noise, the depth of exploration is reached. Vertical resolution determines the minimum detectable layer thickness and is inversely proportional to frequency. High frequencies offer finer resolution for shallow investigations (1–5 m), while lower frequencies, though less precise, enable deeper exploration (over 10 m). The detection of reflectors depends on contrasts in dielectric properties. Strong contrasts, such as between rock and water, produce clearer reflections and improve depth detection. Weak contrasts, like similar soil layers, limit detection. The maximum exploration depth can be estimated using

$$D = \frac{c}{2f\sqrt{\varepsilon_r}} \tag{5.82}$$

where D = the maximum depth of exploration, c = the speed of light in a vacuum, f = the frequency of the GPR system, and ε_r = the relative dielectric permittivity of the material. For multi-layer models, depth estimation involves calculating signal travel times and velocities for each layer. In Ground Penetrating Radar (GPR) surveys, estimating the exploration depth is essential for selecting appropriate equipment and interpreting results. For multi-layer models, the total depth can be calculated by considering the thickness and signal velocity in each layer. The total depth D is the sum of the thicknesses of all layers, given by

$$\mathbf{D} = \sum_{n=1}^{N} \mathbf{d_n} \text{ and } \mathbf{T} = \sum_{n=1}^{N} \mathbf{t_{2n}} = \sum_{n=1}^{N} \frac{2\mathbf{d_n}}{\mathbf{v_n}} \tag{5.83}$$

where D = total depth to the deepest layer, T = total two-way travel time for all layers, N = total number of layers, T_{2n} = two-way travel time for the nth layer, d_n = thickness of the nth layer, and v_n = velocity of the radar signal in the nth layer. Reflections between layer interfaces must also be accounted for when interpreting GPR data.

The central frequency required for exploration can be estimated using

$$f_m \geq \frac{36000}{\rho\varepsilon} \tag{5.84}$$

where f_m = the central frequency in MHz, ρ = the DC resistivity in Ωm (inverse of conductivity σ), and ε = the dielectric permittivity. The radar wave velocity v_n in each layer depends on the dielectric permittivity of that layer. The relationship is:

$$v_n = \frac{c}{\sqrt{\varepsilon_n}} \tag{5.85}$$

where v_n = wave velocity in nth layer, c = speed of light in vacuum ($\approx 3\times10^8$ m/s), and ε_n = the dielectric permittivity of the nth layer. For each layer, the dielectric permittivity ε_n can be related to resistivity ρ_n and frequency f_m using the relationship:

$$\varepsilon_n \approx \frac{1}{f_m \rho_n} \tag{5.86}$$

Signal attenuation is another key factor in determining exploration depth. The attenuation coefficient α\alphaα in dB/m is given by

$$\alpha = 1640\frac{\sigma}{\sqrt{\varepsilon}} \tag{5.87}$$

where α = attenuation coefficient in dB/m, σ = DC conductivity in S/m (Siemens per meter), and ε = the dielectric permittivity of the material. The **attenuation** increases with higher conductivity (σ) and with lower dielectric permittivity ε. This means materials that conduct electricity more (like wet soil or salt-rich environments) will absorb radar signals more quickly, reducing penetration depth. **Permittivity** ε, which is related to the material's ability to store electrical energy, also plays a role in signal attenuation. A lower permittivity leads to higher attenuation.

The maximum exploration depth h_{max} is constrained by the total absorption, which must not exceed 60 dB. Using

$$A = \alpha \cdot 2h_{max} \tag{5.88}$$

where A = total absorption in dB, and $2h_{max}$ = the total absorption over the path (down and back). For the GPR system to function effectively, we want the two-way absorption to be limited to a maximum of 60 dB. This is the rule of thumb that ensures the signal remains detectable. Thus, we impose the condition:

$$2h_{max} \cdot \alpha \le 60(\text{dB})$$

Substituting $\alpha = 1640\frac{\sigma}{\sqrt{\varepsilon}}$ into this equation:

$$2h_{max} \cdot 1640\frac{\sigma}{\sqrt{\varepsilon}} \le 60$$

Now solve for h_{max}:

$$\mathbf{h_{max}} \le \frac{60\sqrt{\varepsilon}}{2.1640\sigma}$$

Simplifying:

$$\mathbf{h_{max}} \le \frac{60\sqrt{\varepsilon}}{3280\sigma}$$

We know that $\sigma = \frac{1}{\rho}$ is the **resistivity** in Ωm (Ohm-meters). Therefore:

$$\mathbf{h_{max}} \le \frac{60\sqrt{\varepsilon}}{3280.\frac{1}{\rho}} = \frac{60\rho\sqrt{\varepsilon}}{3280}$$

Simplifying further:

$$\mathbf{h_{max}} \le \frac{0.018\rho\sqrt{\varepsilon}}{1}$$

Finally, we obtain the formula for the maximum exploration depth h_{max}:

$$\mathbf{h_{max}} \le 0.018\rho\sqrt{\varepsilon} \tag{5.89}$$

This formula gives an **upper bound for the exploration depth** for GPR, where the signal loss due to absorption does not exceed 60 dB. **Resistivity** (ρ) and **dielectric permittivity** (ε) are key material properties that directly impact the depth to which GPR can be applied. For materials with higher resistivity (low conductivity) and lower permittivity, the depth of penetration increases, as there is less attenuation of the signal. If $\rho = 100\ \Omega m$ (resistivity of dry sand) and $\varepsilon = 5$ (typical for dry sand):

$$h_{max} \leq 0.018.100.\sqrt{5} \approx 0.018.100.2.236 = 4.03m$$

This suggests that, with these material properties, the maximum depth h_{max} for which GPR can be used effectively (with <60 dB of absorption) is approximately 4 m. The formula $h_{max} \leq 0.018\rho\sqrt{\varepsilon}$ provides a useful guideline for estimating the **maximum depth** to which GPR can be successfully applied without excessive signal attenuation. This depth depends on the resistivity and permittivity of the subsurface materials, and helps in selecting appropriate GPR operating frequencies and planning GPR surveys for different ground conditions.

5.17 Conclusions

The chapter titled *"Electromagnetic Surveying Techniques"* in the proposed Elsevier publication *"Baseline Geoscience Data for Sustainable Exploration and Resource Management"* provides a comprehensive understanding of electromagnetic (EM) methods as crucial tools for subsurface exploration. It emphasizes the versatility of EM techniques in mapping geological structures, identifying mineral resources, and assessing groundwater systems, while ensuring sustainable exploration practices.

The discussion highlights that the effectiveness of EM surveying depends on several interrelated factors, including subsurface conductivity, frequency ranges, and data acquisition methods. Low-frequency methods are suitable for deeper investigations, while high-frequency techniques offer finer resolution for shallow targets. The chapter underscores the role of resistivity and permittivity in governing the depth of exploration and the resolution of subsurface features, emphasizing the need for precise calibration of these parameters for accurate interpretation.

In conclusion, the chapter posits that electromagnetic surveying techniques, when integrated with complementary geoscience methods and tailored to specific geological conditions, can significantly enhance resource management and exploration efficiency. It advocates for the adoption of sustainable and minimally invasive practices, ensuring that resource exploration aligns with environmental stewardship and long-term ecological balance. This foundational knowledge serves as a guide for researchers and practitioners in optimizing the use of EM surveys for resource and environmental management.

CHAPTER

6

Remote sensing techniques

6.1 Geospatial technology-based applications in mineral exploration

Geospatial technology has become a crucial tool in mineral exploration, especially in light of the increasing demand for mineral resources due to industrialization. The development of geospatial technologies has a rich history that began in the 19th century with cartography, evolving over the years with the advent of aerial photography, satellite imagery, and computer systems. Today, space technology plays a significant role in advancing human capabilities, enabling more precise mapping and management of natural resources. As populations grow and resources become stretched, geospatial techniques are now essential for efficient land resource management.

Geospatial technology refers to the tools and methods that deal with spatial information about geographic features, collected in real time, to aid in analysis and decision-making. The technology integrates software for mapping geographic locations and assessing the impact of human activities on the environment. This includes tools like Geographic Information Systems (GIS), Global Positioning Systems (GPS), remote sensing (RS), and geofencing. GIS is particularly important because it combines datasets with spatially referenced maps, offering detailed, layered analyses of complex data, which can be crucial for many sectors, including mineral exploration.

In the context of mining, the demand for mineral resources is rapidly increasing to meet the needs of various industries such as healthcare, infrastructure, transportation, and communication. However, many mineral reserves are becoming depleted, necessitating more efficient techniques for discovering new resources. Traditional methods of mineral exploration, which relied on manual processes, were often time-consuming, costly, and prone to errors, leading to significant financial losses. This has prompted the development of advanced technologies like geospatial technology, which reduces mining risks, improves efficiency, and lowers operational costs.

Geospatial technologies have revolutionized mineral exploration in several ways. They are used in mineral reserve estimation, mine design, material handling optimization, and enforcing safety measures in mining operations. The application of these technologies has not only improved the accuracy of mineral exploration but also enhanced decision-making

Baseline Geoscience Data Generation for Sustainable Exploration of Natural Resources
https://doi.org/10.1016/B978-0-443-45221-5.00004-7

processes, making it easier to map deposits, monitor mining activities, and predict future resource availability. With the increasing integration of artificial intelligence (AI) and robotics, geospatial technology will continue to advance the mining industry, driving down costs while improving safety and operational efficiency. This evolution ensures that geospatial technology will remain a cornerstone in the future of mining exploration, resource management, and sustainability (Said et al., 2023).

6.2 Introduction to remote sensing

Conventional mapping and exploration techniques may not be sufficient in difficult and remote areas, creating the need for modern technologies like remote sensing to identify potential mineral deposits. Remote sensing involves measuring the properties of objects on the Earth's surface using data collected from aircraft and satellites. Various sensors collect this data, which is then analyzed to obtain information about the objects, areas, or phenomena being studied. Initially, remote sensing was adopted as a technique to gather information about distant objects without physical contact (Fisher, 1975). In practical terms, remote sensing captures electromagnetic or acoustic signals (Gupta, 2018).

The earliest aerial photographs were taken in 1858 using balloons, but since the 1930s, aerial photography with aircraft has been widely used for resource surveys. More recently, unmanned aerial vehicles (UAVs) have also been employed. Satellite imagery became available to geologists in the 1960s (Roonwal, 2017), and the development of satellite remote sensing began with the launch of Landsat-1 in 1972 for civilian applications (Simonetti et al., 2014). In 1979, Seasat-1 became the first radar imaging satellite (Ouchi, 2013), marking the start of a new era in remote sensing. Since then, many advancements have occurred, such as the development of higher spatial resolution optical and radar systems, hyperspectral sensors, and products like digital elevation models (DEMs), along with the integration of machine learning techniques in processing remote sensing data (Ali et al., 2015). Today, satellite series like Sentinel and Planet Labs are transforming the sector by providing high spatial, spectral, and temporal resolution data at low or no cost.

The data collected through remote sensing can take many forms, such as variations in gravity, acoustic wave distributions, or electromagnetic energy distributions. For instance, gravity meters collect data on variations in gravitational force, while sonar systems, like a bat's echolocation, measure variations in acoustic waves. Our eyes also acquire data by detecting variations in electromagnetic energy.

Mineral resource mapping is a crucial aspect of geological exploration and often involves a range of studies, including spectral analysis (Longhi et al., 2001), geological mapping (Harris, 1991), structural mapping (Liu et al., 2000), identifying hydrothermal alteration zones (Podwysocki et al., 1983), and mineral alteration mapping (Tangestani and Moore, 2002). Remote sensing has proven valuable in these areas, helping to identify ferric oxide and oxyhydroxide minerals (Farrand, 1997), gold exploration (Spatz, 1997), and hyperspectral imagery analysis (Neville et al., 2003), as well as facilitating integration with Geographic Information Systems (GIS) and Global Positioning Systems (GPS) (Akhavi et al., 2001).

As most of the easily accessible surface and near-surface mineral deposits have already been found, there is now a greater focus on locating deposits deep beneath the Earth's surface or in hard-to-reach regions. Geophysical methods are generally required for deep exploration, but remote sensing can still provide valuable information about potential areas by analyzing surface features seen in aerial photographs and satellite images. For example, in Australia, which is largely covered by sediments, comparing known gold deposits with areas of weathered basement rocks (where gold and other minerals are likely to be found) highlights the potential of remote sensing to locate new deposits beneath sedimentary layers.

Despite its utility, remote sensing data has an important limitation when it comes to mineral exploration: its penetration depth. Data collected through remote sensing typically reaches only a few micrometers in the near-infrared region, a few centimeters in the thermal infrared, and possibly a few meters in the microwave region (especially in hyper-arid regions). Therefore, remote sensing interpreters must often rely on indirect clues, such as the geological setting, alteration zones, associated rocks, structure, and vegetation anomalies, as it is rare to pinpoint the exact location and mineralogy of a deposit solely based on remote sensing data. In this context, both multispectral and hyperspectral sensors, which can detect specific mineral types, play a significant role in identifying ore minerals or their pathfinders. A multispectral image can be processed using image filtering algorithms to decompose high and low-frequency components, which correspond to spatial details and background information, respectively (Wang et al., 2021). The results from these techniques are increasingly integrated into operational exploration models using GIS technology, which plays a critical role in modern mineral exploration (Bonham-Carter, 1994; Memmi and Pride, 1997).

Remote sensing (RS) is the science of acquiring data from remote objects or areas (usually from aircraft or satellites), while GIS is a system for capturing, storing, evaluating, populating, aligning, managing, and displaying geographical data. Combining these methods addresses many of the shortcomings of traditional exploration techniques. India has launched several remote sensing satellites, beginning with IRS-1A in 1988 and continuing with more recent launches, including the Aditya-L1 satellite on September 2, 2023. The India Space Research Organization (ISRO) uses advanced satellite and airborne sensors to revolutionize geo-exploration by capturing high-resolution spectral, 3D, and radar data. These cutting-edge tools—including hyperspectral imagers, synthetic aperture radar (SAR), and light detection and ranging (LiDAR) that help geoscientists to calculate precise distances of the object and to map hidden mineral reserves, besides analysis of structural lineaments, and locating deep-seated groundwater zones. Several key advanced remote sensing tools are deployed for geo-exploration of natural resources, which include hyperspectral imaging (HySIS) that meant for acquiring hundreds of narrow, contiguous wavelength bands across the electromagnetic spectrum (visible to shortwave infrared). Geo-exploration utility is very useful for generating essential specific "spectral signatures" that help directly identify and map surface minerals, rare earth elements, and alteration zones related to ore bodies. Additionally, geological structures like faults, folds, and shear zones are mapped with the application of InSAR (Interferometric SAR) to detect subtle ground deformation or aquifer compaction. Geo-exploration utility provides high-resolution Digital Elevation Models (DEMs). It is crucial for characterizing terrain, mapping geohazards, and uncovering hidden geological landforms beneath dense forests or beneath the inaccessible regions with hostile terrain. Multi-spectral Imaging with Cartosat & RESOURCESAT series) helps in collects data in discrete, wide spectral bands with ultra-high spatial resolution (down to 30 cm). This facility allows comprehensive geomorphological mapping, identification of geomorphic proxies, such as alluvial fans or paleochannels), and structural analysis. Geospatial Portals & Processing. The Inia space research organization (ISRO) not only captures data but also provides free, centralized platforms to access and analyze it for research and exploration. The most effective portal launched by ISRO-India is Bhuvan, which is a flagship web-geoportal of India that provides open access to OGC-compliant geoscientific layers, including lineament maps, geomorphology, and glacial lakes. Indian institute of Remote Sensing (IIRS) and National Remote Sensing Centre (NRSC) of ISRO have their own outreach programs where these institutes actively conducts training courses and capacity-building programs on utilizing advanced remote sensing for geological analysis of the remote sensing based data and information. Similarly, Geological survey of India (GSI) has its own data bank under the name of Bhukosh for conducting detailed analysis for geo-exploration of natural resources.

Field-observed reference data is typically obtained using a Global Navigation Satellite System (GNSS). GNSS technology is also crucial for other remote sensing tasks, such as guiding aircraft during sensor data collection and correcting and referencing raw image data. The first GNSS, the U.S. Global Positioning System (GPS), was originally developed for military purposes but is now widely used in civilian applications worldwide, including vehicle navigation, surveying, and location-based services on mobile phones. Other GNSS constellations, such as Galileo (Europe), GLONASS (Russia), Bideu/compass (China), Gagan (India) are being developed and will significantly improve accuracy and reliability for users. The U.S. GPS system currently includes at least 24 satellites in precise orbits, providing global coverage for navigation and positioning, while the European Galileo system, scheduled for completion in 2020, will include 30 satellites, offering enhanced compatibility and accuracy with U.S. GPS signals (Lillesand et al., 2015). The detection and delineation of active sub-surface faults have now become possible by estimating total electron contents through continuous monitoring of GNSS system at local and regional scale.

6.3 Imaging spectrometry for Earth remote sensing

Imaging of the Earth's surface from aircraft and spacecraft without the use of imaging spectrometry is limited by low spectral resolution and a limited number of spectral bands, typically fewer than 10 bands, each having a width of 100 to 200 nm. Hyperspectral sensors, or imaging spectrometers, have revolutionized remote sensing by acquiring image data in many narrow spectral bands (usually <40 nm wide) that are contiguous. The ultimate aim of hyperspectral imaging is to produce detailed spectral reflectance curves for each pixel (picture element) in the image. These curves provide a much higher level of detail compared to conventional imaging systems such as Landsat scanners, which use broadband, low-resolution bands (Freden and Gordon, 1983). The reflectance spectrum in the range of 0.4 to 2.5 μm can be used to identify a wide range of surface materials that cannot be identified by systems with broader spectral bands.

The term "hyperspectral" refers to the multidimensional nature of the spectral data set, capturing a broad spectrum of wavelengths. Many minerals and rocks exhibit unique spectral signatures with absorption features that are typically 20 to 40 nm wide. Hyperspectral sensors can capture these narrow spectral features, enabling precise mapping of surface mineralogy based on spectral characterization (Van der Meer, 1998). Unlike broader band sensors like Landsat's Thematic Mapper (TM), hyperspectral sensors can directly identify minerals based on their diagnostic spectral features (Rajesh, 2004).

Most hyperspectral remote sensing continues to be carried out using airborne systems such as the Compact Airborne Spectrographic Imager (CASI), the Airborne Visible and Infrared Imaging Spectrometer (AVIRIS), the SWIR Full Spectrum Imager (SFSI), the Hyperspectral Digital Imagery Collection Experiment (HYDICE), Probe-1, and HyMap. Examples of satellite sensors include Hyperion, the Fourier Transform Hyperspectral Imager (FTHSI), and the Australian Resource Information and Environment Satellite (ARIES-1). The reflectance spectra of minerals measured by various spectroradiometers with different spectral resolutions are stored in spectral libraries in digital format (Grove et al., 1992; Clark et al., 1993), and these libraries are used for correlating and classifying the hyperspectral data.

Several methods have been developed to extract information from high-spectral resolution image data, often processing spectral information on a pixel-by-pixel basis. These methods include binary encoding (Goetz et al., 1982), spectral unmixing (Adams et al., 1986), relative absorption band depth mapping (Crowley et al., 1989), waveform characterization (Okada and Iwashita, 1992), classification techniques (Cetin et al., 1993), spectral angle mapping (SAM; Kruse et al., 1993), decorrelation stretching (Abrams and Hook, 1995), probability density function analysis (Nedeljkovic and Pendock, 1996), constrained energy minimization (CEM; Farrand and Harsanyi, 1997), cross-correlogram spectral matching (Van der Meer and Bakker, 1997), back-propagation neural networks (BPN; Yang et al., 1999), and geophysical inversion (Van der Meer, 2000).

For instance, Beraton et al. (1997) demonstrated that AVIRIS could be effectively used for mapping potassium metasomatism in well-exposed geological areas. The Cuprite mining district in Nevada, a well-known test site for remote sensing due to its exposed altered bedrock, minimal vegetation cover, and alluvial deposits, has been the subject of numerous studies. AVIRIS-derived mineral distribution maps of this region have contributed to advancing remote sensing techniques (Abrams et al., 1977; Kruse et al., 1990; Rast et al., 1991; Clark et al., 2003).

In particular, Resmini et al. (1997) illustrated the use of CEM for mineral mapping in the Cuprite area, a technique that requires only the spectrum of the mineral to be mapped and no prior knowledge of the background materials. This method has proven to be a rapid and effective approach for mineral mapping. Yang et al. (1999) demonstrated the high classification accuracy of back-propagation neural networks (BPN) over spectral angle mapping (SAM) in the Cuprite area. Furthermore, Neville et al. (2003) showed the ability of constrained linear unmixing procedures to create mineral fraction maps using data from SFSI and AVIRIS over the same area. These mineral abundance maps are often integrated with other exploration data, such as digital elevation models (DEMs), to provide a more comprehensive perspective on the geology of the area.

6.4 An overview of the electromagnetic remote sensing process

This chapter discusses electromagnetic energy sensors used in airborne and spaceborne platforms to assist in inventorying, mapping, and monitoring Earth's resources. These sensors gather data on how different surface features emit and reflect electromagnetic energy. The data is then analyzed to provide information about the resources being studied. Fig. 6.1 shows the

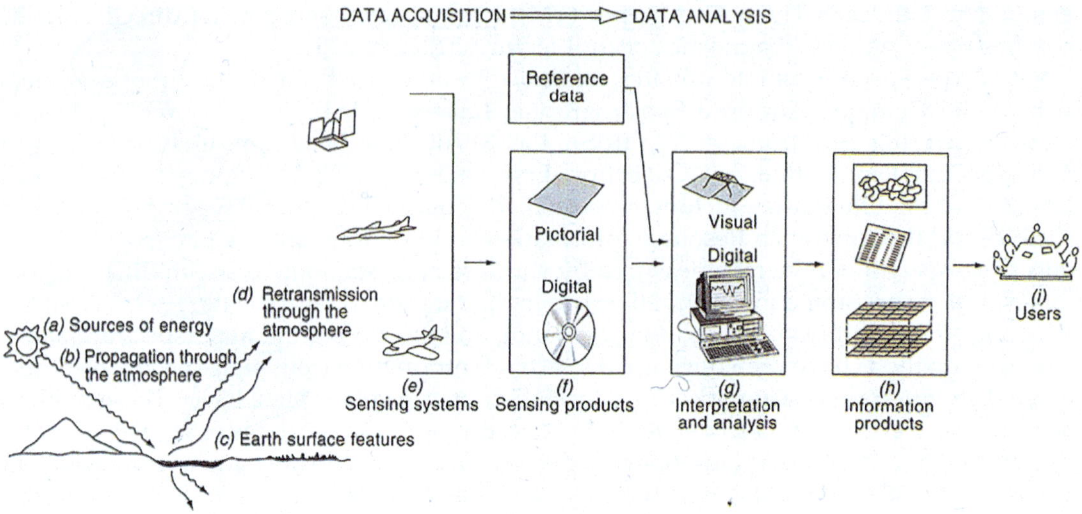

FIG. 6.1 Electromagnetic (EM) remote sensing of Earth resources (Modified).

generalized processes and elements involved in electromagnetic remote sensing of Earth resources. The two main processes in this field are data acquisition and data analysis (Lillesand et al., 2015; Ballhorn, 2012).

In the data acquisition process, sensors record variations in how Earth's surface features reflect and emit electromagnetic energy. Electromagnetic energy comes in various forms, including visible light, radio waves, heat, ultraviolet rays, and X-rays, all of which travel according to the wave theory. According to this theory, electromagnetic energy travels in a sinusoidal, harmonic manner at the speed of light. Electromagnetic energy consists of both an electric wave (E) and a magnetic wave (M), which are at right angles to each other and perpendicular to the direction of propagation (Fig. 6.2) (Lillesand et al., 2008; Ballhorn, 2012).

Remote sensing often categorizes electromagnetic waves based on their wavelength within the electromagnetic spectrum, as shown in Fig. 6.3. The most common unit for measuring wavelength is the micrometer (μm), where 1 μm equals 1×10^{-6} meters. Although regions of the electromagnetic spectrum are named (such as "ultraviolet" and "microwave") for convenience, there is no exact boundary between one spectral region and the next (Art et al., 1994; Lillesand et al., 2008).

The human eye can only detect visible light within the wavelength range of 0.4 to 0.7 μm. However, remote sensing technology can detect electromagnetic energy outside of this range, including reflective and radiant energy from Earth's surface and atmosphere, such as the reflective qualities of clouds. To display this data visually, false color assignments (red, green, and blue, or RGB) are used to represent various electromagnetic bands, and the combination of these colors helps express the true physical reflective qualities of objects in the image (Lillesand et al., 2004, 2015; Lillesand and Kiefer, 1979).

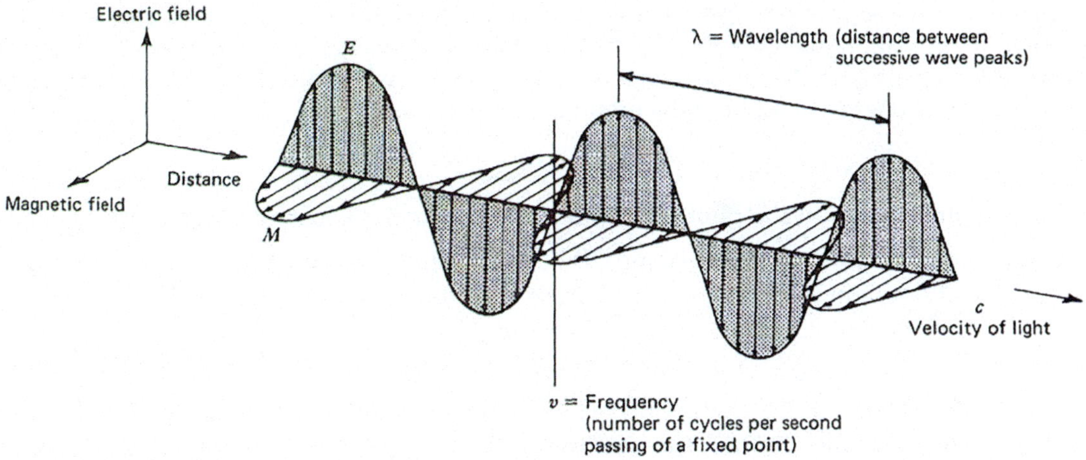

FIG. 6.2 Components of an electromagnetic wave. This diagram shows a sinusoidal electric wave (E) and a similar magnetic wave (M), both at right angles to each other and perpendicular to the direction of propagation (Modified).

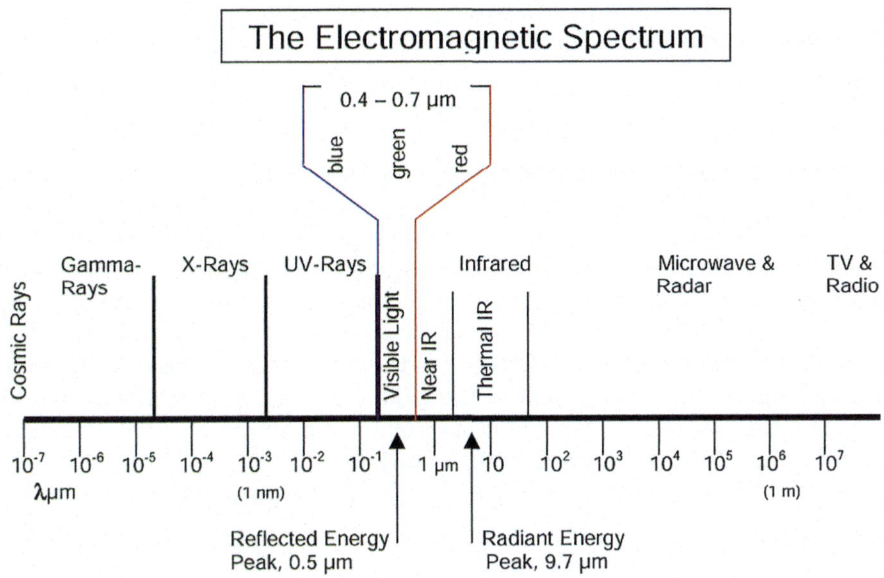

FIG. 6.3 Electromagnetic (EM) spectrum. The range of EM bands that can be transmitted through Earth's atmosphere is often referred to as atmospheric windows. *Source: IJERT.*

While the wave theory explains many characteristics of electromagnetic radiation, the particle theory offers insights into how electromagnetic energy interacts with matter. The particle theory suggests that electromagnetic radiation consists of discrete units called photons or quanta. The energy of a photon is given by the equation:

$$E = h\nu \tag{6.1}$$

where E is the energy of the photon (in Joules), h is Planck's constant (6.626×10^{-34} Js), and υ is the frequency (in cycles per second). By combining the wave equation ($\upsilon = \frac{c}{\lambda}$), where c is the velocity of light and λ is the wavelength, the energy equation becomes:

$$E = \frac{hc}{\lambda} \tag{6.2}$$

This shows that the energy of a quantum is inversely proportional to its wavelength. Longer wavelengths have lower energy content, which is important in remote sensing. For instance, long-wavelength radiation, such as microwave emission, is more difficult to detect than shorter-wavelength radiation, like thermal infrared (IR) energy. This lower energy content in long-wavelength radiation means that systems using such wavelengths must observe large areas to capture detectable signals.

The sun is the most prominent source of electromagnetic radiation for remote sensing. However, all matter with a temperature above absolute zero (0 K or −273°C) emits electromagnetic radiation. The amount of energy an object radiates depends on its surface temperature, as described by the Stefan-Boltzmann law:

$$M = \sigma T^4 \tag{6.3}$$

where M is the total radiant exitance from the surface (W/m^2), σ is the Stefan-Boltzmann constant (5.6697×10^{-8} $W/m^{-2}K^{-4}$), and T is the absolute temperature (°K). This law shows that the total energy emitted by an object increases rapidly as its temperature rises. However, it's important to note that this law is idealized for blackbody radiation, a theoretical material that absorbs and re-emits all energy. Real-world objects only approximate this ideal.

Fig. 6.4 illustrates the energy distribution of blackbodies at temperatures ranging from 200 to 6000 K. The area under each curve represents the total radiant exitance M, and the curves also show that higher temperatures cause the peak wavelength of radiation to shift toward shorter wavelengths. Wien's displacement law relates the temperature of a blackbody to the wavelength at which the maximum energy is emitted:

$$\lambda_m = \frac{A}{T} \tag{6.4}$$

where λ_m is the wavelength of maximum emission (in μm), A is a constant (2898 μm.K), and T is the temperature (in°K). For the sun, with a temperature of about 6000 K, the peak radiation occurs at around 0.5 μm, while Earth's surface temperature of about 300 K results in

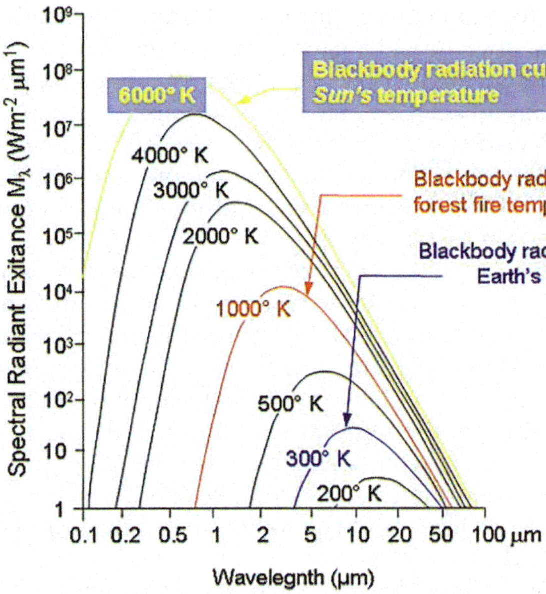

FIG. 6.4 Spectral distribution of energy from blackbodies at various temperatures. The spectral radiant exitance M represents energy emitted per unit wavelength, with total radiant exitance given by the area under the curves. *Source: Intech Open—Open Access.*

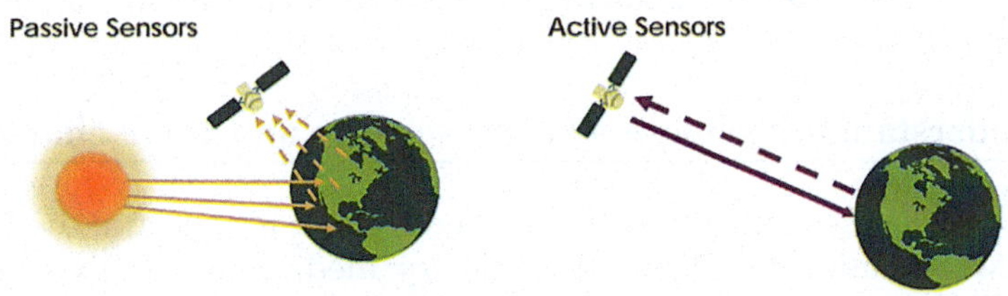

FIG. 6.5 Diagram of a passive sensor vs. an active sensor. *Source: NASA.*

peak radiation at around 9.7 µm, which is in the thermal infrared range. Thermal infrared radiation is emitted by Earth's surface features but cannot be seen with the naked eye. However, it can be detected by radiometers and scanners.

The general dividing line between reflected and emitted IR energy is around 3 µm. Below this, energy is primarily reflected, while above this wavelength, emitted energy predominates. Remote sensing systems can be categorized as either "active" or "passive." Active systems, like radar, lidar, and sounders, emit their own energy to observe objects, while passive systems, such as spectrometers and radiometers, rely on naturally available energy (Fig. 6.5).

For remote sensing images to be integrated into Geographic Information Systems (GIS) and accurately represent reality, they must be georeferenced, meaning aligned with a reference coordinate system. Georeferencing involves defining ground control points with precise geographic coordinates. These points can be extracted from an image, an already corrected aerial photo, or field data collected via GPS (Longley et al., 2005). Image rectification and restoration processes, which correct geometric distortions and noise, are also essential to ensure accurate representation (Lillesand et al., 2008). Orthorectification is a specific type of rectification that corrects for terrain displacement.

Remote sensing plays a crucial role in the study of mineralized areas, complementing field, lithologic, and structural mapping, as well as GIS. In mineral resource mapping, remote sensing helps detect spectral anomalies caused by mineralization processes. Lithologic mapping is the first step, followed by structural and alteration mapping, as mineral deposits often align with geological structures and hydrothermal alteration zones. Hyperspectral remote sensing, which identifies specific mineral absorption features, is also valuable in mineral exploration. Finally, integrating remote sensing data with other geoscience data, such as geochemical and geophysical information, is essential for identifying potential mineral deposits. GIS is an ideal tool for this integration, enabling the analysis of large datasets with different attributes and formats (Rajesh, 2004).

Satellite imagery is commonly produced using electro-optical sensors that measure reflectance in the visible, near-infrared (VNIR), shortwave-infrared (SWIR), and thermal infrared (TIR) parts of the electromagnetic spectrum. Examples include Landsat Multispectral Scanner (MSS), Landsat Thematic Mapper (TM), and Advanced Spaceborne Thermal Emission and Reflection Radiometer (ASTER). Other satellites, such as RADARSAT, ERS, and JERS, use synthetic aperture radar (SAR) to detect microwave energy (Rajesh, 2004).

6.5 Understanding remote sensing sensors and their characteristics

Remote sensing (RS) sensors are instruments designed to measure the properties of electromagnetic (EM) radiation that leave a surface or medium, typically due to scattering or emission. The most commonly measured property is radiance, which is a function of wavelength. However, these sensors can also measure other parameters, such as polarization, depending on the sensor type and application.

The technology behind active and passive sensors differs significantly. Active sensors emit their own energy, such as microwaves or lasers, and measure the reflected energy from the target. In contrast, passive sensors detect natural radiation, such as sunlight or thermal energy, reflected or emitted from the Earth's surface. These differences in energy emission and detection principles define the capabilities and applications of each sensor type (Figs. 6.6A and 6.6B).

When designing remote sensing sensors, several characteristics play a critical role in determining their suitability for specific tasks. These include spatial resolution, spectral resolution, temporal resolution, and radiometric resolution. Understanding these characteristics and the trade-offs between them is crucial for selecting the appropriate sensor for a particular application. For instance, sensors with high spatial resolution can detect finer details but may require more data processing or result in a smaller field of view.

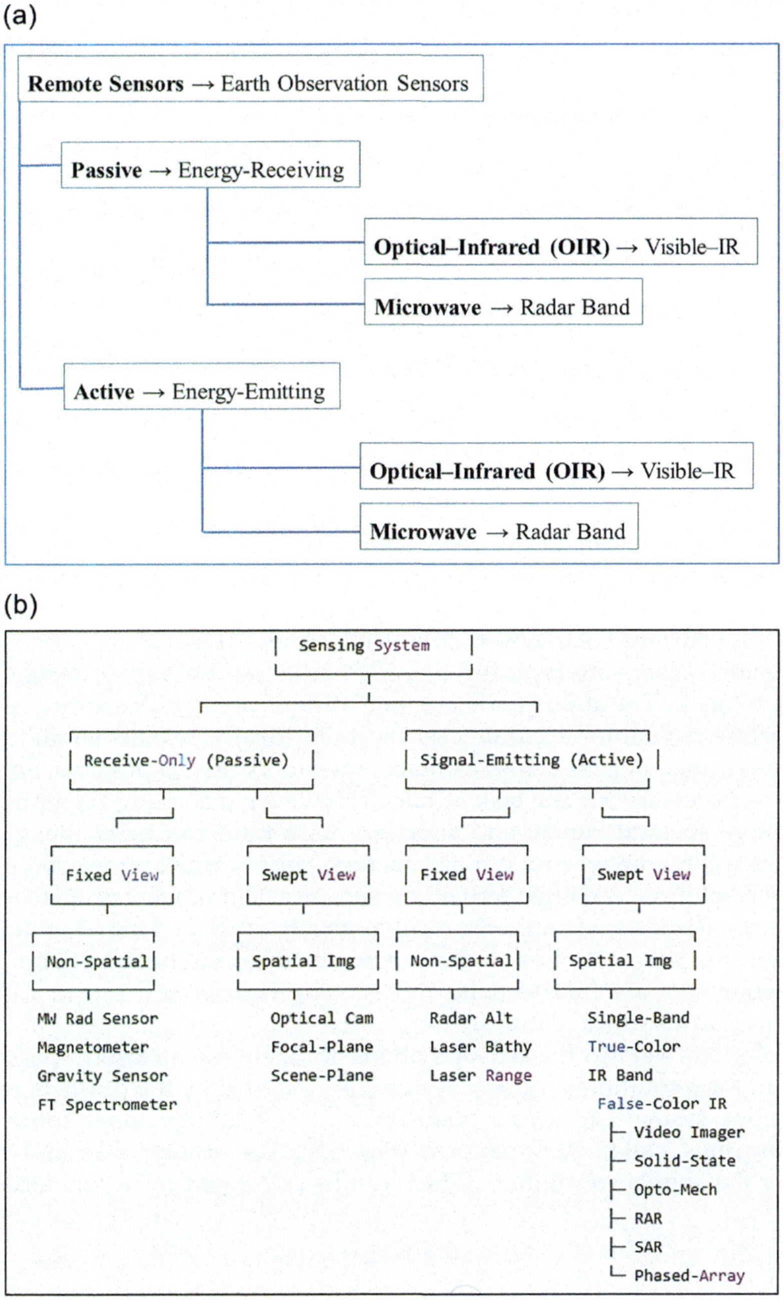

FIG. 6.6 (A) Comparison of active and passive remote sensing sensors. (B) Classification of remote sensing sensors: Active vs. passive systems. *Source: GNR401 Dr A. Bhattacharya.*

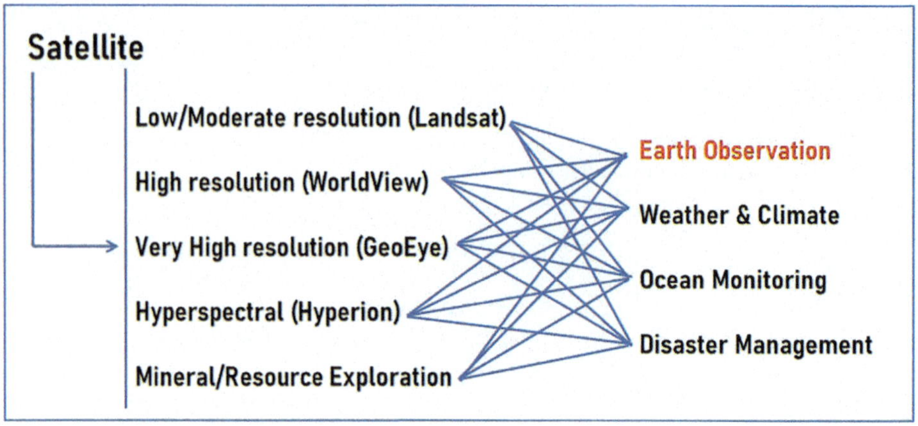

FIG. 6.7 Comparison of remote sensing resolutions in satellite systems. *Source: GNR401 Dr A. Bhattacharya.*

The Landsat and SPOT satellite systems were among the first global monitoring systems to systematically acquire moderate-resolution data for Earth observation. These systems set the foundation for modern satellite-based environmental monitoring and continue to serve as benchmarks for current Earth observation technologies (Fig. 6.7).

Remote sensing sensors are typically classified based on both their design and mapping accuracy requirements. The ability to detect small differences in reflectance or emittance from the Earth's surface is vital for accurate classification. Ideally, sensors should be able to distinguish between small objects across as many spectral bands as possible, and acquire data as frequently as necessary for the task at hand. However, achieving the optimal balance of spatial resolution, spectral bands, and signal-to-noise ratio can be challenging due to the strong interdependence between these parameters. For example, improving one parameter, such as spatial resolution, can often reduce the sensor's field of view or increase the required processing power. Remote sensing sensors are typically evaluated based on four key parameters: spatial resolution, spectral resolution, radiometric resolution, and temporal resolution. These four parameters are fundamental in assessing how well a sensor can capture and analyze various Earth surface features.

Spatial resolution refers to the sensor's ability to distinguish closely spaced objects. The theoretical limit for resolution in optical systems is governed by the diffraction of EM waves. When an imaging system captures a point object, the resulting image forms a bright disc surrounded by rings, called Airy patterns (Fig. 6.8). The ability to resolve two objects is determined by the angular resolution, which can be calculated using the formula:

$$\theta = 1.22\frac{\lambda}{D} \tag{6.5}$$

where λ is the wavelength and D is the diameter of the lens aperture (Fig. 6.9). In remote sensing systems that use discrete detectors, such as Landsat MSS or IRS LISS, spatial resolution is typically expressed by the Instantaneous Field of View (IFOV). The IFOV represents the ground area captured by each detector element. For example, a Landsat ETM + scene has

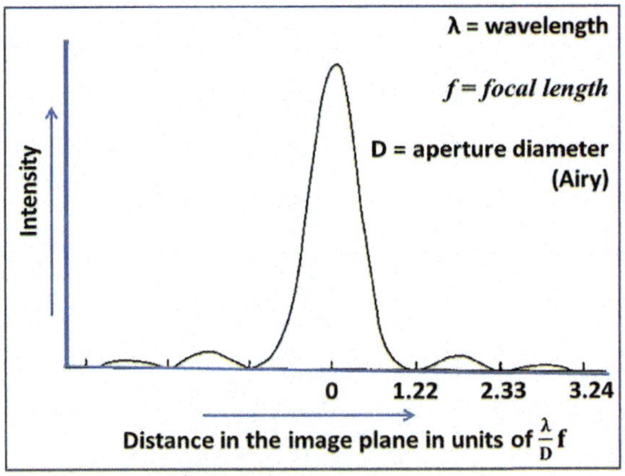

FIG. 6.8 Distribution of energy in the Airy pattern due to circular aperture. The diameter of the first minima is given by 2.44(λ/D)f, where f is the focal length. *Source: GNR401 Dr A. Bhattacharya.*

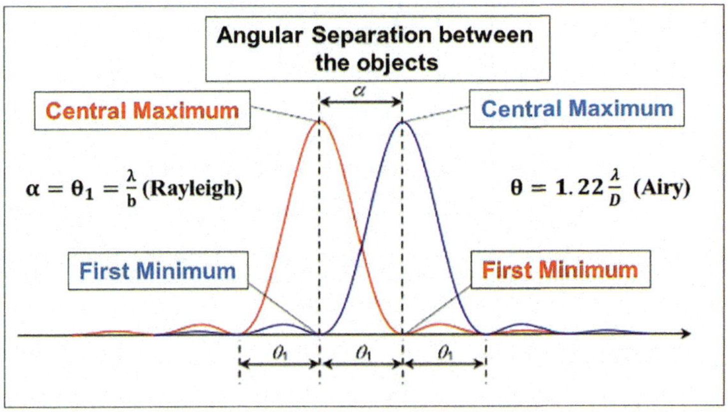

FIG. 6.9 Theoretical limit for resolving two objects in terms of angular resolution. Two objects can be resolved if the peak of the Airy pattern of one object falls on the first minima of the other, separated by 1.22(λ/D)f. *Source: GNR401 Dr A. Bhattacharya.*

an IFOV of 30 m in bands 1–5 and 7, 60 m in band 6, and 15 m in band 8. The IFOV is particularly useful because it describes a sensor's resolution independent of the altitude at which it operates (Fig. 6.10).

Spectral resolution refers to the sensor's ability to measure the reflected or emitted radiation across specific wavelengths. Multispectral remote sensing systems use spectral bands to distinguish various surface features. The spectral bandwidth (Δλ) defines the range of wavelengths captured by each band. A smaller bandwidth results in higher spectral resolution (Fig. 6.11). For example, the first spectral band of the IRS LISS system captures wavelengths from 0.45 μm to 0.52 μm. While ideal spectral systems should have a perfect response within a given bandwidth, in practice, the spectral response is not always ideal. The response is often defined as the 50% of the peak value on either side, and methods such as the method of moments are used to calculate accurate spectral band values.

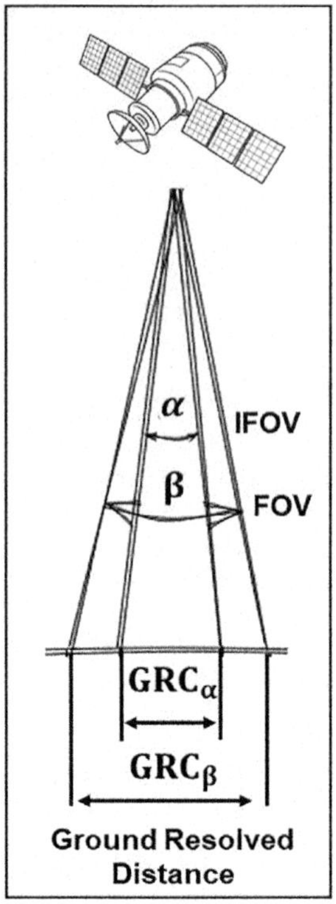

FIG. 6.10 Spatial resolution, IFOV (=d/f in radians, where d is the dimension of the detector and f is the focal length of the sensor), FOV, and swath in remote sensing sensors. *Source: GNR401 Dr A. Bhattacharya.*

Equation for spectral band range:

$$\lambda_{min} = \lambda_0 - \frac{\Delta\lambda}{2} \tag{6.6}$$

$$\lambda_{max} = \lambda_0 + \frac{\Delta\lambda}{2} \tag{6.7}$$

This allows the measurement of reflected or emitted radiation in specific regions of the electromagnetic spectrum, facilitating the identification of various surface features based on their spectral signatures.

Temporal resolution refers to the time it takes for a satellite to complete one full orbit cycle, which is also known as the **revisit time (Sensor: Landsat, MODIS, Commercial Orb View, Revisit time: 16 days, 2 days, 1–2 days)**. This interval is influenced by several factors, including the satellite's **sensor capabilities**, the extent of **swath overlaps**, and the **latitude** at which the satellite is operating. Satellites with higher temporal resolution are capable of capturing images of the same location more frequently. This can be achieved through the ability to **maneuver their sensors** or by experiencing **increased swath overlap at higher latitudes**, enabling more frequent observations of specific areas.

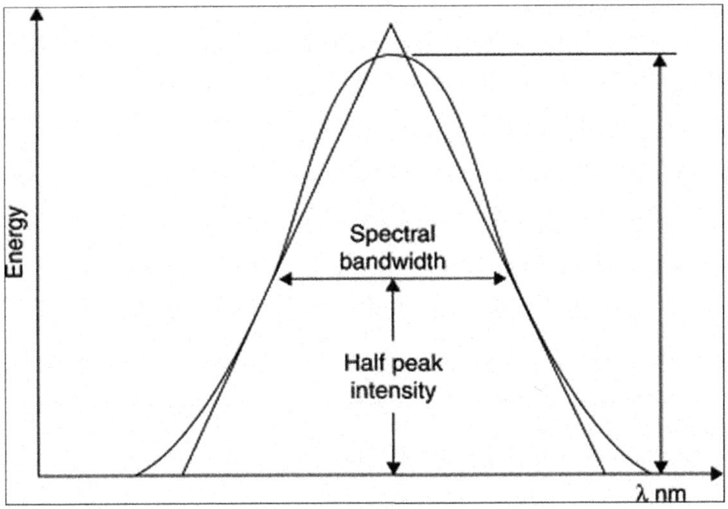

FIG. 6.11 Spectral band definition in multi-spectral remote sensing. The x-axis represents wavelength (λ), typically in nanometers or micrometers, and the y-axis represents spectral radiance (W/m^2). Each spectral band is shown as a rectangular region with a central wavelength (λ_0) and a bandwidth (Δλ). *Source: GNR401 Dr A. Bhattacharya.*

Radiometric resolution is the ability of a sensor to detect small differences in radiance, which is crucial for distinguishing between objects in an image. The contrast between different objects can be quantified using the Contrast Ratio (CR) and Contrast Modulation (CM) equations, which compare the radiance values of adjacent areas. The Modulation Transfer Function (MTF) describes how much the imaging system degrades the contrast between objects. It is a measure of a sensor's ability to preserve image detail across different spatial frequencies. In applications like computer classification, the relative differences in radiance between pixels are often more important than the absolute values. However, for tasks such as multi-date data analysis or data fusion, having accurate absolute radiance values becomes essential.

In conclusion, selecting the optimal remote sensing sensor requires a deep understanding of its design characteristics and the trade-offs between spatial, spectral, radiometric, and temporal resolutions. By carefully balancing these parameters, sensors can be tailored to meet the specific needs of various Earth observation tasks, from environmental monitoring to land use classification.

6.6 Satellite data processing levels

Satellite data processing involves several stages that transform raw satellite data into usable information. At Level 0, the raw data is collected directly from the satellite sensors, often including radiometric measurements, but it may contain errors such as noise or distortions. No corrections or calibrations have been applied at this stage, and the data is typically not usable for analysis. At Level 1, the raw data is calibrated to correct for sensor biases, radiometric distortions, and geometric issues. The data is then converted into a more standard format, such as reflectance or radiance, and may include geometric corrections and orthorectification to align the data with geographic coordinates.

Level 2 processes the data to extract geophysical information, such as surface temperature, vegetation indices, or atmospheric properties. Atmospheric corrections are applied, and spectral indices, such as NDVI, are derived to provide meaningful environmental variables tailored for specific analysis. At Level 3, data is aggregated over time and space to create products for trend analysis and long-term monitoring. This involves temporal compositing, spatial resampling, and data fusion from multiple sources, resulting in gridded maps or patterns used for environmental monitoring and resource management. Level 4 involves generating high-level geophysical models, forecasts, or simulations by integrating satellite observations with other data sources. The outputs from this level include forecasting models or long-term environmental simulations, such as climate predictions or agricultural yield forecasts. Each level of processing refines the satellite data, transforming it from raw sensor output to actionable insights that support scientific research, policy-making, and resource management.

6.7 Projections and coordinate systems

In remote sensing and geospatial data analysis, understanding projections and coordinate systems is essential for accurately representing the Earth's surface and interpreting spatial data. Projections and coordinate systems are used to transform the three-dimensional surface of the Earth onto a two-dimensional map or image, allowing for precise location and measurement of features.

A coordinate system is a framework used to define the positions of points in space. The most common coordinate systems are the Geographic Coordinate System (GCS) and the Projected Coordinate System (PCS). The GCS uses latitude and longitude to specify locations on the Earth's surface, with latitude (North-South) and longitude (East-West) as angular measurements, typically expressed in degrees. The origin of the GCS is the intersection of the equator and the prime meridian (0° latitude, 0° longitude), located in the Gulf of Guinea, off the coast of West Africa. The PCS, in contrast, transforms the Earth's surface onto a flat plane, using x and y coordinates (easting and northing) measured in linear units such as meters or feet. This transformation is done through a mathematical process called projection, which distorts the Earth's surface to a flat plane in a way that can be useful for mapping and analysis.

Projections are mathematical methods used to represent the Earth's curved surface on a flat map. Since the Earth is a sphere (or more accurately, an oblate spheroid), it is impossible to represent it perfectly on a two-dimensional surface without some form of distortion. Different projections are designed to minimize distortions for specific purposes. A cylindrical projection involves projecting the Earth onto a cylinder, and the map is formed by unrolling the cylinder. An example of this is the Mercator projection, which is widely used for navigation and shows straight lines as constant bearing angles. In a conic projection, the Earth is projected onto a cone, which is then unrolled into a flat surface. This type of projection is often used for mapping smaller regions, particularly those in mid-latitudes, with the Albers

Equal Area Conic projection as an example. Azimuthal (or planar) projection involves projecting the Earth onto a flat plane, often used for polar maps, such as the Lambert Azimuthal Equal Area projection, which preserves area relationships around a central point. Equal area projections, like the Mollweide projection, preserve the relative size of areas, meaning the areas on the map are proportional to those on the Earth's surface.

One of the most famous cylindrical projections is the Mercator projection, which preserves angles but distorts size, particularly at higher latitudes. It is commonly used for navigation and world maps. The Universal Transverse Mercator (UTM) is a global map projection system that divides the world into a series of 6° longitudinal zones, each with its own coordinate system, minimizing distortion within each zone. UTM is widely used for large-scale mapping, particularly for topographic and cadastral maps.

When working with geospatial data, it is essential to choose an appropriate coordinate system and projection for the intended analysis. The choice of projection depends on the region of interest and the specific characteristics, such as area, distance, or shape, that need to be preserved. In summary, projections transform the Earth's curved surface into a flat map, while coordinate systems define how spatial locations are represented and measured. Understanding both concepts is crucial for accurate geospatial analysis and ensuring that data from different sources can be correctly integrated and compared.

6.8 Application of remote sensing in mineral resource mapping

6.8.1 Spectral signature of minerals and rocks

In remote sensing, objects are identified through three spectral responses: selective absorption, selective reflection, and selective emission (Anbazhagan and Arivazhagan, 2010; Paramasivam, 2016). Since Hunt and Ashley (1979), many studies have compiled comprehensive mineral and rock spectra, analyzing absorption features across wavelengths from visible to thermal infrared, providing a foundation for interpreting remotely sensed spectroscopic data (Gaffey, 1985; Clark et al., 1990; Salisbury et al., 1991; Christensen et al., 2000). Spectroscopic criteria are crucial in hyperspectral image analysis for mineral and alteration mapping, helping researchers understand the relationship between unaltered rocks and alteration products for reconnaissance and exploration (Van der Meer et al., 1997; Mazzarini et al., 2001; Ferreira et al., 2002).

Spectroscopy studies the interaction between matter and electromagnetic radiation, analyzing light emitted, reflected, or scattered by solids, liquids, or gases (Clark, 1999; Gupta, 2005). Spectra, representing the response against wavelength or frequency, are obtained from rocks and minerals in various forms (Gaffey, 1986). Absorption characteristics reveal chemical information based on reflected or emitted light, and modern spectrometers provide greater precision over broader wavelengths (Clark, 1999). Spectral measurements, using spectroradiometers in the field and laboratory, aid in diagnosing minerals, rocks, and surface features (Clark, 1999; Chung and Fabbri, 2020). Laboratory studies include rocks like peridotite, dunite, and magnesite formations (Ranganathan and Siddan, 2021; Geological Survey of India (GSI), 2006; Satish Kumar et al., 2011; Chetty, 1996; He et al., 2015; Friend and Janardhan, 1984).

Van der Meer and De Jong (2003) noted that spectral reflectance* in the visible and near-infrared (VNIR; 0.4–1.0 μm) regions is a cost-effective method for determining mineralogy and chemical composition. Distinctive spectral features of wall rocks, altered rocks, and ores help in mineral exploration using remote sensing data (Bailin and Xingli, 1991). Spectral features in the VNIR region are influenced by electronic transitions in metals like Fe, Mn, Cu, Ni, and Cr. Fig. 6.12A shows the reflectance spectra of iron-bearing minerals, with features from ferrous (Fe^{2+}) ion transitions. Iron oxide spectra range from 0.35 to 1.5 μm, with absorption anomalies indicating hematite (shorter than 0.9 μm) and jarosite or goethite (≥0.9 μm) (Zadeh et al., 2014; Hunt and Ashley, 1979). The migration of electrons from iron to oxygen results in a broad absorption band at wavelengths shorter than 0.55 μm, notably affecting iron oxides and hydroxides (Fig. 6.12B), which accounts for their yellow, orange, red, and brown appearance.

*Spectral reflectance is the ratio of upwelling to downwelling radiant flux for a surface, accounting for illumination, observation, and sensor properties. Remote sensing reflectance (R_{rs}) is given by $R_{rs}\lambda = \frac{L_w \lambda \theta \varphi}{E_d^{0+} \lambda}$, where L_w is water-leaving radiance, the fraction of upwelling radiance passing the air-water interface in zenith direction θ and azimuth direction φ and E_d^{0+} is plane downwelling irradiance. Reflectance is commonly used in remote sensing after compensating for atmospheric and interface effects, and can be estimated through radiative transfer models (Mobley, 1994).

Vibrational transitions in the near-infrared (1.1–2.5 μm) region provide detailed mineralogical information, with key transitions related to OH^- ions or water molecules in the shortwave-infrared (SWIR) region (Gupta, 2005; Mustard and Pieters, 1987). Absorption features at 1.9, 1.4, 1.14, and 0.94 μm indicate molecular water in minerals (Fig. 6.12A). The bending of Al-OH and Mg-OH bonds causes distinct absorption features in aluminous micas and clay minerals (Fig. 6.12C), common in magnesium-rich hydroxylated minerals like talc, chlorites, and serpentines. Carbonates exhibit absorption near 2.3 μm in the SWIR. The mid-infrared region shows high reflectance anomalies in rocks and minerals, including a notable absorption band at 2.2 μm (Hunt and Salisbury, 1970b; Hunt and Salisbury, 1971; Hunt, 1979). Quartz exhibits a trough between 8 μm and 9 μm due to Si–O bond stretching (Fig. 6.12D), visible in both silicate and nonsilicate minerals. Silicate and nonsilicate minerals, such as carbonates and iron oxides, significantly alter the spectra of silicate rocks (Hunt and Salisbury, 1970a; Kahle et al., 1996). Rock spectra, which are mixtures of individual mineral spectra, can be distinguished in the thermal emission region under ideal conditions (Lahren et al., 1988; Sabine et al., 1994).

Representative spectra of sedimentary, igneous, and metamorphic rocks reveal distinct absorption features related to their mineral compositions (Rajesh, 2004; Salisbury and Hunt, 1974). In sedimentary rocks, carbonate-related features dominate, with absorption bands between 1.9 and 2.3 μm. These rocks typically show water absorption bands at 1.4 μm and 1.9 μm, and clay-shales present an additional absorption between 2.1 μm and 2.3 μm. Limestones and calcareous rocks exhibit carbonate absorption bands at 1.9 μm and a stronger one at 2.35 μm, while dolomites show ferrous ion absorption at 1.0 μm due to Mg^{2+} substitution by Fe^{2+}. In igneous and metamorphic rocks, the SiO_2 content influences the Si–O bond

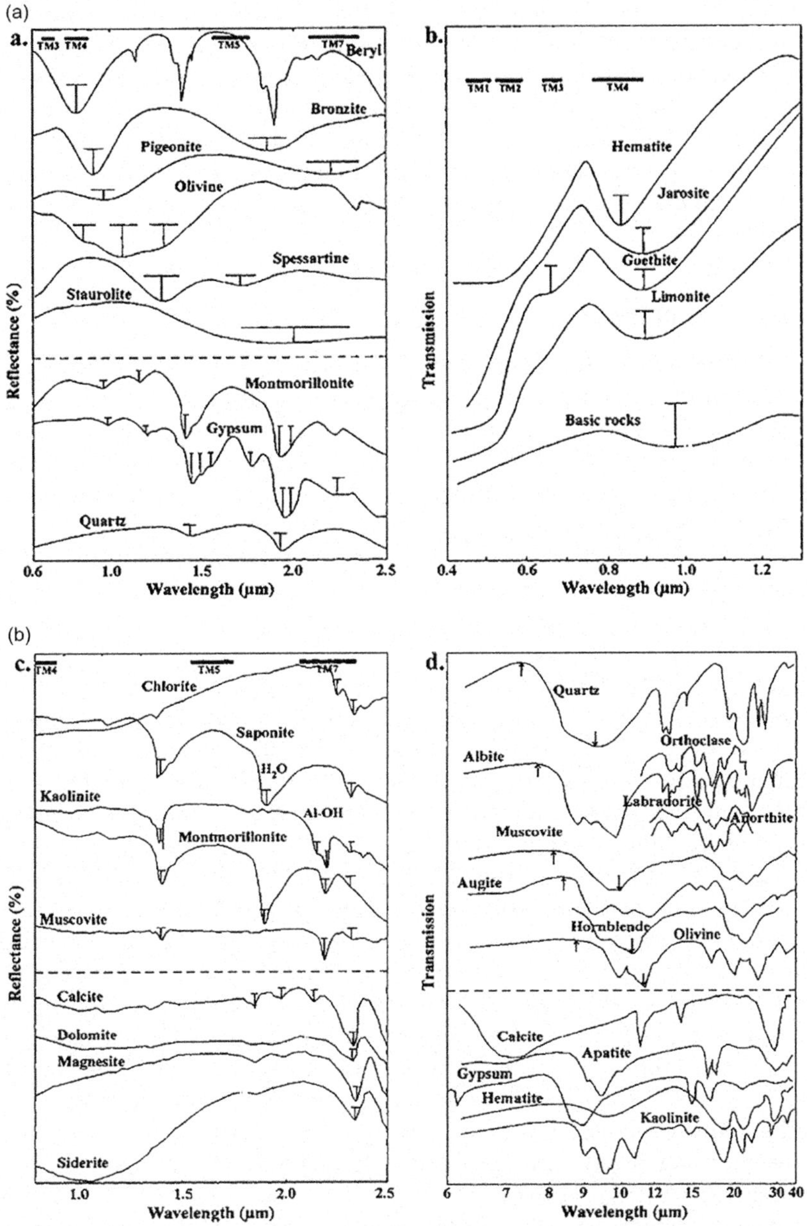

FIG. 6.12 (A), (B): Spectra of iron-bearing minerals (above the dashed line) and minerals with chemically bound water (below the dashed line). Absorption features are marked by T-shaped symbols. Panel (A) shows iron-bearing minerals, while panel (B) shows iron oxides and hydroxides. (C), (D): (C) Spectra of clay minerals and micas (above the dashed line) and carbonate minerals (below the dashed line), with absorption features indicated by T-shaped symbols. (D) Spectra of silicates (above the dashed line) and non-silicates (below the dashed line) in the mid- or thermal-infrared, showing the shift of the short-wavelength peak and main trough as minerals transition from felsic to mafic compositions. Spectra are offset for clarity, with thick horizontal bands indicating Landsat TM spectral widths. *Source: Rajesh (2004).*

stretching absorption, with the feature shifting to longer wavelengths as SiO_2 content decreases (Vickers and Lyon, 1967). For rocks like granites, a composite signature is used to represent the entire class. Metamorphic rocks such as termolite schists show broad absorption from ferrous ions. Marbles exhibit strong carbonate absorptions at 1.9 μm and 2.3 m, and high reflectivity in the near infrared. Additionally, water and hydroxyl bands are observed in schists, marbles, and quartzites (Hunt and Salisbury, 1976). Laboratory experiments confirm that rocks can be classified using airborne or satellite sensor data when sufficient spectral detail is available.

6.8.2 Lithologic mapping

Lithologic mapping is crucial in mineral resource exploration, identifying geological features through remote sensing. Key lithological information is derived from factors such as geological setting, weathering, landforms, drainage, structural features, soil, vegetation, and spectral characteristics. In sedimentary rocks, bedding patterns, visible on hillsides or faulted areas, are strong indicators. Linear features form ridges and valleys, with drainage density and vegetation patterns providing additional clues. In arid regions, sandstone outcrops display distinct colors based on cementing materials, with silica- and carbonate-cemented sandstones appearing pale, while iron-rich ones are yellow, orange, or red. Mudstones and siltstones exhibit closely spaced, dendritic drainage due to poor drainage.

Intrusive igneous rocks, due to their massive, isotropic, and homogeneous nature, are easily identified by their shape (e.g., batholiths, dykes) and distinct image tones in remote sensing. Extrusive igneous rocks like cones and craters are identified by landforms, image tone, and drainage patterns, especially if young. Metamorphic rocks are more challenging to interpret due to complex structures and erosion resistance. Interpretation depends on the balance between stable silicates (like quartz) and unstable ferromagnesian silicates.

Topographic features, including drainage, often reflect geological structure and lithology. Synthetic Aperture Radar (SAR) enhances topographic and structural mapping, while Landsat is effective for mapping surface cover types such as vegetation. SAR can aid lithologic mapping by detecting variations in outcrop patterns and rock features (Blom and Daily, 1982; Rebillard and Evans, 1983). Landsat TM band 4 (0.76–0.90 μm) often shows dark tones for low-lying vegetation on marble. Combining Landsat with higher-resolution data like SPOT or IRS improves lithological discrimination (Fraser et al., 1997; Mickus and Johnson, 2001; Chatterjee et al., 2003). SAR combined with geophysical data enables better correlation of surface and subsurface features (Pedroso et al., 2001). Stern et al. (2002) showed that the MOMS-2P satellite images were more effective than Landsat for mapping Neoproterozoic sedimentary rocks in northern Ethiopia.

Remotely sensed images serve as effective reconnaissance geologic maps for mineral resource mapping. Studies have shown that remote sensing data can enhance existing geological maps (Rothery, 1987; Abrams et al., 1988). Krishnamurthy (1997) used IRS LISS-I and II data to update geological maps in Karnataka, India, refining lithological boundaries, identifying unmapped rock units, and mapping lineaments. A flowchart for creating modified geological maps from remote sensing data is shown in Fig. 6.13.

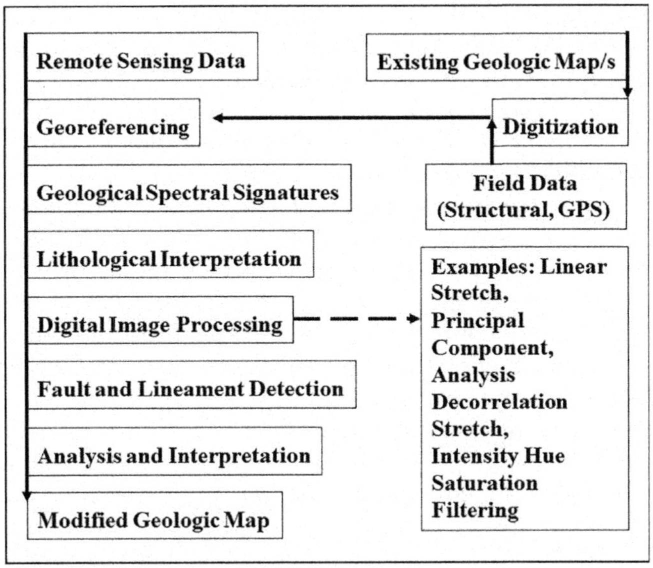

FIG. 6.13 Flowchart depicting the general processing steps to obtain a modified geologic map using remote sensing data. *Source: Rajesh (2004).*

Classified images offer geometric accuracy for precise sample site location, aiding mineral resource studies. Various methods, including geological, geophysical, geochemical, and remote sensing data, are employed for geological mapping and mineral deposit discovery. These methods include traditional field surveys, statistical processes, and machine learning techniques like random forests, support vector machines, and logistic regression (Harris and Grunsky, 2015; Othman and Gloaguen, 2017; Pal et al., 2020; Shirmard et al., 2021; Wang et al., 2021). Wang et al. (2021) developed a hybrid method combining random forests and metric learning to map the distribution of Himalayan leucogranites, integrating remote sensing and geochemical data for improved accuracy.

6.8.3 Structural mapping

Mineral deposits are often associated with deformation of the lithosphere, and many ore formation theories incorporate tectonic or deformational concepts (O'Leary et al., 1976). Lineament analysis, particularly digital lineament studies, has proven essential in locating major mineral deposits, showing a strong correlation between lineaments and mineral deposits (Kutina, 1969; Katz, 1982; Liu et al., 2000; Rein and Kaufmann, 2003). Short linear features (up to 10 km) are discernible in aerial photos but are less visible in satellite imagery, marking individual geological features like folds, faults, joints, and veins. These features reflect near-surface conditions but are not reliable for locating concealed deposits. Longer features (10–200 km) suggest the regional geometry of folds, faults, and structures, while features greater than 200 km are best studied using Landsat mosaics, displaying nearly orthogonal patterns globally. Liu et al. (2000) demonstrated lineament analysis from satellite imagery for mineral exploration in northeast Brazil.

Aerial photographs offer evidence of bedding through ridge occurrences and tonal differences in beds, with several studies using stereoscopic SPOT images to determine the attitude of faults and lithological units (Berger et al., 1992; Bilotti et al., 2000). Fu et al. (2004) applied Landsat TM/ETM stereoscopic images and high-resolution IRS-IC PAN images to delineate Quaternary deformation along the Tian Shan belt, China. Remote sensing excels in delineating vertical and high-angle faults, with mineralization often occurring along fault lines. However, low-angle faults are challenging to interpret due to the planar nature of images.

Radar imagery (e.g., RADARSAT) and Landsat data are valuable for studying structurally complex terrains, particularly when combined with fieldwork. RADARSAT images reflect topographic features like erosional escarpments due to lithological unit variations (Evans et al., 1986). Radar imagery also proves useful in forested areas, where it provides clear topographic data without interference from ground cover. Its stereoscopic capability aids in lineament identification (Sharma et al., 1999).

Various algorithms for extracting lineaments from digital images and aerial photos have been proposed (Budkewitsch et al., 1994; Raghavan et al., 1995; Koike et al., 2001; Costa & Starkey, 2001). Principal Component Analysis (PCA), applied to SAR data from Seasat-SAR and Shuttle Imaging Radar (SIR-B), has enhanced topographic data for structural and lineament mapping (Masouka et al., 1988). Integrating lineament data from satellite imagery (Landsat TM, SAR) with GIS and geophysical data has proven effective in mineral exploration (Akhavi et al., 2001). Zakir et al. (1999) found that lineament intersection density was nearly double in altered zones compared to unaltered ones. Multi-technique approaches combining remote sensing, geophysical, and geological methods enhance the effectiveness of lineament intersection analysis in mineral exploration (Chernicoff et al., 2002).

6.8.4 Alteration mapping

Mineral deposits are often linked to hydrothermal alteration of surrounding rocks, with the style and extent of alteration reflecting the type of deposit (de la Vega-Marquez Fernandez et al., 2001). Host rocks of hydrothermal deposits show chemical interactions with fluids, forming alteration halos that provide larger exploration targets than the deposits themselves (Pirajno, 1992). Remote sensing has proven effective in detecting these altered areas, as certain minerals—such as iron-bearing minerals (hematite, goethite), hydroxyl-bearing minerals (clays, micas), and hydrated sulphates (gypsum, alunite)—show distinct spectral features (Hunt and Ashley, 1979; Prost, 1980; Podwysocki et al., 1983; Gladwell et al., 1983; Townsend, 1987; Clark et al., 1990; Fraser, 1991). These minerals, although not directly detectable, can serve as indicators of hydrothermal alteration associated with gold (Vincent, 1997). Hydroxyl-bearing minerals like clays have diagnostic features in the infrared spectrum, aiding identification of hydrothermal alteration (Ruiz-Armenta and Prol-Ledesma, 1998).

Weathering can produce similar minerals to hydrothermal alteration, complicating remote sensing analysis, which is why field verification is essential (Buckingham and Sober, 1983). Iron oxides often give red, brown, orange, or yellow hues, while clay minerals produce pale colors like yellow, violet, green, and beige.

Landsat imagery is particularly useful for hydrothermal mineral identification due to its mid-infrared bands, which capture spectral signatures of hydrothermal minerals (Rajesh,

2004; Ruiz-Armenta and Prol-Ledesma, 1998). The Landsat Operational Land Imager (OLI) sensor, covering wavelengths from 0.4 μm to 2.3 μm, detects hydroxyl and iron oxide mineral assemblages (Osinowo et al., 2021). Merging Landsat data with high-resolution panchromatic imagery (e.g., IRS) via techniques like intensity-hue-saturation (IHS) transform can improve spatial and spectral resolution (Chavez et al., 1991). ASTER data, known for its high spatial resolution, is also effective in creating accurate alteration maps. Son et al. (2021) applied WorldView-3 (WV-3) data combined with field shortwave-infrared (SWIR) spectroscopy for alteration mapping in Mongolia, supported by conventional methods like X-ray diffraction and fluorescence.

Techniques such as Band Math (BM), Decorrelation Stretch (DS), and Principal Component Analysis (PCA) have long been used for highlighting alteration zones (Mars and Rowan, 2006; Sabins, 1999; Van der Meer et al., 2012; Mars, 2018). Advanced methods like Spectral Angle Mapper (SAM), Matched Filtering (MF), Mixed Tuned Matched Filter (MTMF), and Adaptive Coherence Estimator (ACE) have been applied for mapping hydrothermal alteration minerals (Ducart et al., 2006; Bedini, 2011, 2019; Son et al., 2019; Kruse et al., 2015; Salehi and Tangestani, 2018). Machine learning algorithms, such as Artificial Neural Networks (ANN), Support Vector Machines (SVMs), and Extreme Learning Machines (ELM), are increasingly used for mineral identification in remote sensing, especially for hyperspectral data, though their use in multispectral imagery remains limited (Thompson et al., 2001; Yousefi et al., 2018, 2020; Hu et al., 2018).

6.9 Types of remote sensing data acquisition techniques

Remote sensing utilizes various imaging techniques across the visible, infrared, and thermal infrared regions of the electromagnetic spectrum. Satellite programs like Landsat, SPOT, and IRS, as well as high-resolution satellites, multispectral and hyperspectral techniques, and microwave remote sensing, are essential for Earth observation. Radar and laser systems, along with drones, are also used for surface observation.

India's space program began in the 1960s under Dr. Vikram Sarabhai's guidance, with the launch of IRS-1A in 1988 marking the start of India's remote sensing efforts. Since then, India has launched several satellites, including Cartosat for high spatial resolution and RISAT for radar imaging. The Polar Satellite Launch Vehicle (PSLV) has been highly successful, with a 95% success rate as of 2023. These satellites serve both India and its neighboring countries, supporting resource management, disaster mitigation, and climate studies. India's achievements extend beyond Earth observation, as the country has successfully completed missions to the Moon (Chandrayaan) and Mars (Mangalyaan) on the first attempt (Adimurthy, 2015; Chauhan et al., 2012; Menon, 2008).

Mishra et al. (2017) and Roy et al. (2017) reviewed India's remote sensing systems, highlighting the country's extensive constellation of satellites. The Indian space agency has launched a series of Earth observation satellites, from Indian Space Research Organisation (ISRO) (1979) to Cartosat-2E (2017), with both optical and microwave sensors providing data for land, ocean, and atmospheric applications. For a summary of major remote sensing satellite systems, see the overview table (Tables 6.1 and 6.2).

TABLE 6.1 Remote sensing satellite and data overview.

Sensor (Mission)	Provider	Orbital height	Spatial resolution	Swath width
LISS-4 (Resourcesat-2) *Linear imaging self-scanning sensor*	*ISRO, India*	*817 km* *Sun synchronous*	*5.8 m*	*70 km*
Data use: Integrated land and water resources management at the micro level with enhanced multi-spectral and spatial coverage. To further carry out studies in advanced areas of user applications like improved crop discrimination, crop yield, crop stress, pest/disease surveillance, disaster management, and urban management. (i) Launch period: April 20, 2011. (ii) Radiometric resolution 10-bit. (iii) Temporal resolution: 24 days. (iv) Spectral resolution: 0.52–0.86 μm. (v) Spectral bands: 3 (visible and near-infrared; VNIR).				
LISS-3 (Resourcesat-2)	*ISRO, India*	*817 km* *Sun synchronous*	*23.5 m*	*141 km*
(i) Launch period: April 20, 2011. (ii) Radiometric resolution: 10-bit. (iii) Temporal resolution: 24 days. (iv) Spectral resolution: 0.52–1.70 μm. (v) Spectral bands: 4 (3 spectral bands in VNIR and 1 in SWIR).				
AWiFS (Resourcesat-2) *Advanced wide field sensor*	*ISRO, India*	*817 km* *Sun synchronous*	*56 m*	*740 km*
(i) Launch period: April 20, 2011. (ii) Radiometric resolution 12-bit. (iii) Temporal resolution: 24 days. (iv) Spectral resolution: 0.52–1.70 μm. (v) Spectral bands: 4 (3 spectral bands in VNIR and 1 in SWIR).				
PSLV-C36 (Polar satellite launch vehicle) (Resourcesat-2A; Three-tier imaging system; AWiFS, LISS-3 and LISS-4 sensors)	*ISRO, India*	*817 km* *Sun synchronous*	*5.8 m, 23.5 m, 56 m*	*740 km*
Data use: Resourcesat-2A is intended to continue the remote sensing data services to global users provided by Resourcesat-1 (Mission 2003) and Resourcesat-2 (Mission 2011). (i) Launch period: December 07, 2016. (ii) Spectral resolution: (0.52–0.59, 0.62–0.68, 0.77–0.86; 0.52–0.59, 0.62–0.68, 0.77–0.86, 1.55–1.70; 0.52–0.59, 0.62–0.68, 0.77–0.86, 1.55–1.70) μm. (iii) Spectral band: VNIR and SWIR.				
Cartosat-2 F (PSLV-C40 Mission)	*ISRO, India*	*505 km* *Sun synchronous*	*65 cm (PAN), 2 m (MS)*	*10 km*
Data use: Cartostat-2 F is the eighth satellite in the Cartosat-2 series. The data from the satellite can be used for detailed mapping and other cartographic applications at the cadastral level, urban and rural infrastructure development and management, as well as applications in Land Information System (LIS) and Geographical Information System (GIS). (i) Operation period: January 12, 2018. (ii) Radiometric resolution: 10-bit. (iii) Temporal resolution: 4 days. (iv) Spectral resolution: 0.50–0.85 μm (PAN), 0.43–0.90 μm (MS). (v) Spectral bands: 4 + PAN.				

Sensor (Mission)	Provider	Orbital height	Spatial resolution	Swath width
RISAT-2BR1 (PSLV-C48) (SAR)	*ISRO, India*	*576 km Sun synchronous*	*1 × 0.5 m 0.5 × 0.3 m*	*5–10 km*
Data use: It is a radar imaging Earth observation satellite. Its function is to help in agriculture, forestry, and disaster management support. (i) Launch period: December 11, 2019.				
EOS-1 (RISAT-2BR2) (PSLV-C49) (SAR)	*ISRO, India*	*575 km Sun synchronous*	*8–20 m*	*100 km*
Data use: The Earth Observation Satellite-01 (EOS-01) is a satellite imaging mission for surface observations including agriculture, forestry, flood monitoring, and also military surveillance. (i) Launch period: November 7, 2020. (ii) Temporal resolution: 3–4 days.				
EOS-3 (GISAT-1) (GSLV-F10)	*ISRO, India*	*35,786 km Sun synchronous*	*42–1180 m Multispectral (VNIR and LWIR), Hyperspectral (VNIR and SWIR)*	*160–470 km*
Data use: It would provide near real-time imaging of the large area of the Indian subcontinent at frequent intervals. This satellite will be helpful for quick monitoring of natural disasters or any short-term events/episodes. It would help obtain spectral signatures of agriculture, forestry, mineralogy, etc. It would also help in disaster management by issuing regular warnings, cloud properties, snow and glacier, and oceanography. GISAT is the first of its kind and unique geostationary imaging satellite for simultaneous and integrated observations of land, ocean, and atmosphere at relatively high resolution using the hyperspectral technique. (i) Launch period: August 12, 2021. (ii) Spectral resolution: 0.45–0.875 μm (MS, VNIR), 0.38–1.0 μm (Hyperspectral, VNIR), 0.90–2.5 μm (Hyperspectral, SWIR), 7.1–13.5 μm (MS, LWIR). (iii) Spectral bands: 6 (MS, VNIR), 158 (Hyperspectral, VNIR), 256 (hyperspectral, SWIR), 6 (MS, LWIR).				
EOS-4 (RISAT-1A) (PSLV-C52) (SAR)	*ISRO, India*	*529 km Sun synchronous*	*1–50 m*	*10–225 km*
Data use: EOS-04 or Earth Observation Satellite-04 (formerly known as RISAT-1A) is an Indian Space Research Organisation Radar Imaging Satellite designed to provide high-quality images under all weather conditions for applications such as agriculture, forestry and plantations, soil moisture and hydrology, and flood mapping. (i) Launch period: February 14, 2022. (ii) Spectral resolution: Radar. (iii) Spectral bands: N/A.				
MSS (Landsat 1–5)	*NASA, USA*	*705 km Sun synchronous*	*80 (MS), 240 (TIR) m*	*185 km*
Data use: The objective of MSS was to provide repetitive daytime acquisition of high-resolution, multispectral data of the Earth's surface on a global basis and to demonstrate that remote sensing from space is a feasible and practical approach to efficient management of the Earth's resources. (i) Launch period: 1972–1992. (ii) Radiometric resolution: 8-bit. (iii) Temporal resolution: 16–18 days. (iv) Spectral resolution: 0.5–1.1 μm, 10.4–12.6 μm. (v) Spectral bands: 4–5.				

Continued

TABLE 6.1 Remote sensing satellite and data overview—cont'd

Sensor (Mission)	Provider	Orbital height	Spatial resolution	Swath width
Landsat 4, 5 TM	*NASA, USA*	*705 km* *Sun synchronous*	*30 m, 120 m*	*185 km*
Data use: Oceanography, aerosols, bathymetry, vegetation types, peak vegetation, biomass content analysis, moisture analysis, thermal mapping, mineral deposit identification. (i) Launch period: July 16, 1982 and March 01, 1984, respectively.				
Landsat 7 *ETM +*	*NASA, USA*	*705 km* *Sun synchronous*	*15 m, 30 m,* *60 m*	*183 km*
Data use: Oceanography, aerosols, bathymetry, vegetation types, peak vegetation, biomass content analysis, moisture analysis, thermal mapping, mineral deposit identification. (i) Launch period: April 15, 1999.				
Landsat 8 *OLI (B1–B9)* *TIRS (B10-B11)*	*NASA, USA*	*705 km* *Sun synchronous*	*15 m, 30 m,* *60 m, 100 m*	*185 km*
Data use: Oceanography, aerosols, bathymetry, vegetation types, peak vegetation, biomass content analysis, moisture analysis, cloud cover analysis, thermal mapping, soil moisture estimation. (i) Launch period: February 11, 2013.				
*Landsat 9 (see) *OLI (B1-B9)* *TIR (B10-B11)*	*NASA, USA*	*705 km* *Sun synchronous*	*15 m, 30 m,* *100 m*	*185 km*
Data use: Oceanography, aerosols, bathymetry, vegetation types, peak vegetation, biomass content analysis, moisture analysis, cloud cover analysis, thermal mapping, soil moisture estimation. (i) Operation period: September 27, 2021.				
I-M Imager (GOES 11–15)	*NESDIS, USA*	*35,790 km* *Sun synchronous*	*1 km (VNIR),* *4 km (SWIR),* *8 km* *(Moisture),* *4 km (TIR)*	*8 km*
Data use: The imager will monitor changing weather patterns over the United States and provide imagery throughout the 24 h day to assist weather forecasters. (i) Launch period: October 16, 1975. (ii) Radiometric resolution: 10-bit. (iii) Temporal resolution: 0.25–2 h. (iv) Spectral resolution: 0.55–12.5 μm. (v) Spectral bands: 5.				
IKONOS	*GeoEye, USA*	*681 km* *Sun synchronous*	*80 cm B/W* *3.2 m RBGiR*	*11.3 km*
Data use: Mapping, change detection, planning (engineering, natural resources, urban, infrastructure), land-use, EIA, tourism, military, crop management, environmental monitoring. (i) Launch period: September 24, 1999. (ii) Radiometric resolution: 11-bit. (iii) Temporal resolution: 3–5 days. (iv) Spectral resolution: 0.526–0.929 μm (PAN), 0.445–0.853 μm. (v) Spectral bands: 4 + PAN.				

Sensor (Mission)	Provider	Orbital height	Spatial resolution	Swath width
SAR (RADARSAT-1)	*CSA, Canada*	*798 km* *Sun synchronous*	*8–100 m*	*45–500 km*
Data use: Environmental monitoring, ice monitoring, marine surveillance, disaster management, hydrology, mapping, geology, agriculture, and forestry. (i) Launch period: November 4, 1995. (ii) Temporal resolution: 24 days. (iii) Spectral resolution: Radar. (iv) Spectral bands: N/A.				
SAR (RADARSAT-2)	*CSA, Canada*	*798 km* *Sun synchronous*	*3–100 m*	*20–500 km*
Data use: Ice and environmental monitoring, marine surveillance, disaster management, resource management, and mapping. (i) Launch period: December 14, 2007. (ii) Temporal resolution: 24 days. (iii) Spectral resolution: Radar. (iv) Spectral bands: N/A.				
RADARSAT Constellation	*CSA, Canada*	*600 km* *Sun synchronous*	*1 × 3 m*	*300 km*
Data use: Provides complete coverage of Canada's land and oceans. RADARSAT Constellation Mission (RCM) provides services to Canadians in areas such as maritime surveillance, disaster management, and ecosystem monitoring, e.g., floods, agriculture, ice, biodiversity, wetlands, and change detection. (i) Launch period: June 12, 2019.				
AVHRR/3	*NOAA, USA*	*830–870 km* *Sun synchronous*	*1090 m*	*833 km*
Data use: Cloud and surface mapping, land-water bounds, recognition of snow and ice, sea surface temperatures, and cloud cover analysis at night. (i) Launch period: May 13, 1998.				
HRVIR (SPOT 4, 5)	*SPOT Image, France*	*832 km* *Sun synchronous*	*10 (PAN), 20 (MS) m*	*60 km*
Data use: The HRVIR instrument, along with the Vegetation Monitoring Instrument (VMI), produces multi-purpose imagery of both the land and the ocean, specifically data concerning albedo and reflectance, vegetation, and landscape topography. (i) Launch period: March 24, 1998 (SPOT 4) and May 4, 2002 (SPOT 5). (ii) Radiometric resolution: 8-bit. (iii) Temporal resolution: 26 days. (iv) Spectral resolution: 0.61–0.68 (PAN), 0.50–1.75 μm. (v) Spectral bands: 4 + PAN.				

Continued

TABLE 6.1 Remote sensing satellite and data overview—cont'd

Sensor (Mission)	Provider	Orbital height	Spatial resolution	Swath width
VEGETATION (SPOT 4, 5)	*SPOT Image, France*	*832 km Sun synchronous*	*1150 m*	*2250 km*
Data use: The most important advance is the addition of the "Vegetation" instrument, with four spectral bands to allow continuous, worldwide crop monitoring. The data may be used for **crop forecasts** and environmental studies. (i) Launch period: March 24, 1998 (SPOT 4) and May 4, 2002 (SPOT 5). (ii) Radiometric resolution: 10-bit. (iii) Temporal resolution: 26 days. (iv) Spectral resolution: 0.43–1.75 μm. (v) Spectral bands: 4.				
HRG (SPOT 5)	*SPOT Image, France*	*832 km Sun synchronous*	*2.5–5 (PAN), 10 (VNIR), 20 (SWIR)*	*60 km*
Data use: This was designed to improve the knowledge and management of Earth's resources, detecting and forecasting phenomena involving climatology and oceanography, and monitoring human activities and natural phenomena. (i) Launch period: May 4, 2002 (SPOT 5). (ii) Radiometric resolution: 8-bit. (iii) Temporal resolution: 26 days. (iv) Spectral resolution: 0.48–0.71 (PAN), 0.50–1.75 μm. (v) Spectral bands: 4 + PAN.				
SPOT 6	*SPOT Image, France*	*832 km Sun synchronous*	*1.5 m (PAN), 6 m (MS)*	*60 km*
Data use: SPOT 6, like all the SPOT missions, was aimed at supplying high-resolution, wide-area optical imagery. This was designed to improve the knowledge and management of Earth's responses, detecting and forecasting phenomena involving climatology and oceanography, and monitoring human activities and natural phenomena. (i) Launch period: September 9, 2012.				
SPOT 7	*SPOT Image, France*	*832 km Sun synchronous*	*1.5 m (PAN), 6 m (MS)*	*60 km*
Data use: SPOT 7, like all the SPOT missions, was aimed at supplying high-resolution, wide-area optical imagery. This was designed to improve the knowledge and management of Earth's resources, detecting and forecasting phenomena involving climatology and oceanography, and monitoring human activities and natural phenomena. The final satellites in the SPOT series, SPOT 6 and SPOT 7, were identical satellites and assured continuity of high-resolution, wide-swath data. (i) Launch period: June 1, 2014.				
MODIS	*NASA, USA*	*705 km Sun synchronous*	*250 m (B1-B2) 500 m (B3-B7) 1000 m (B8-B36)*	*2330 km*

Sensor (Mission)	Provider	Orbital height	Spatial resolution	Swath width
Data use: Aerosols, land and cloud boundaries and properties, ocean biology, biogeochemistry, atm. water vapour, sea surface and atmospheric temperature, and cloud analysis. (i) Launch period: Board the Terra (EOS AM) in December 1999 and Board the Aqua (EOS PM) in May 4, 2002.				
ASTER	*NASA, USA, and METI, Government of Japan*	*705 km* *Sun synchronous*	*15–90 m*	*60 km*
Data use: Vegetation, ecosystem dynamics, hazard and disaster monitoring, change detection, Earth science, land cover analysis. (i) Operation period: ASTER is an imaging instrument on the **Terra satellite** launched in December 1999.				
ALI (EO-1)	*NASA, USA*	*705 km* *Sun synchronous*	*10 (PAN), 30 (MS) m*	*37 km*
Data use: The most important goal of EO-1 was to **validate the Advanced Land Imager** (ALI) instrument for future Earth-Observing satellites. Built and tested at the MIT Lincoln Laboratory, ALI is a hybrid multisensor system for the measurement of Earth's surface reflectance. (i) Launch period: November 21, 2000. (ii) Radiometric resolution: 12-bit. (iii) Temporal resolution: 16 days. (iv) Spectral resolution: 0.48–0.69 μm (PAN), 0.433–2.35 μm. (v) Spectral band: 9 + PAN.				
Hyperion (EO-1)	*NASA, USA*	*705 km* *Sun synchronous*	*30 m*	*7.5 km*
Data use: EO-1 also validated the scientific value of orbital imaging spectroscopy with Hyperion, the first spaceborne hyperspectral land imaging instrument, and demonstrated that a moderate-spatial/high-spectral resolution imager could be used in self-correction of atmospheric effects—to retrieve the apparent surface reflectance from top of atmosphere (TOA) radiances—and systematic errors. (i) Launch period: November 23, 2000. (ii) Radiometric resolution: 12-bit. (iii) Temporal resolution: 16 days. (iv) Spectral resolution: 0.353–2.577 μm. (v) Spectral band: 220.				
QuickBird	*Digital Globe, USA*	*482 km* *Sun synchronous*	*65 cm B/W* *2.62 m RGBiR* *61 cm B/W* *2.44 m RGBiR*	*16.8–18 km*
Data use: Mapping, change detection, planning (engineering, natural resources, urban, infrastructure), land-use, EIA, tourism, military, crop management, environmental monitoring. (i) Launch period: October 18, 2001.				

Continued

TABLE 6.1 Remote sensing satellite and data overview—cont'd

Sensor (Mission)	Provider	Orbital height	Spatial resolution	Swath width
MERIS (Envisat-1)	*ESA*	*790 ± 10 km Sun synchronous*	*300 m (land), 1200 m (ocean)*	*1150 km*
Data use: The Medium Resolution Imaging Spectrometer (MERIS) was a programmable spectrometer on board the Envisat mission, operating in the solar reflective spectral range. Although primarily dedicated to ocean colour observations, MERIS broadened its scope of objectives to atmospheric and land surface-related studies. MERIS had a high spectral and radiometric resolution and a dual spatial resolution, within a global mission covering open ocean and coastal zone waters and a regional mission covering land surfaces. (i) Launch period: March 01, 2002. (ii) Radiometric solution: 12-bit. (iii) Temporal resolution: 35 days. (iv) Spectral resolution: 0.39–1.04 μm. (v) Spectral bands: Up to 15.				
AATSR (Envisat-1)	*ESA*	*800 km Sun synchronous*	*1000 m*	*500 km*
Data use: It was designed primarily to measure Sea Surface Temperature (SST). The primary scientific objective of the instrument was to establish continuity with its predecessors, ATSR-1 and ATSR-2. The second scientific objective was to ensure the production of measurements over land, such as Land Surface Temperature (LST). Land and cloud measurement objectives were met through the use of an additional visible focal plane assembly, which can lead to indications of: vegetation biomass, vegetation moisture, and vegetation health and growth stage. The visible channels can also contribute to the measurement of cloud parameters, like water/ice discrimination and particle size distribution. (i) Launch period: March 01, 2002. (ii) Radiometric solution: 12-bit. (iii) Temporal resolution: 35 days. (iv) Spectral resolution: 0.55–12 μm. (v) Spectral bands: 7.				
ASAR (Envisat-1)	*ESA*	*800 km Sun synchronous*	*30–1000 m*	*100–400 km*
Data use: The ASAR instrument aboard Envisat mainly supported ocean (wave characteristics, ocean fronts, coastal dynamics, oil slicks, and ship traffic), land (global vegetation monitoring, forestry, geology and topography, agriculture, natural hazards, flooding, hydrology and water management, and urban studies), and snow and ice land (ice concentration determination, sea ice mapping and classification, ice feature identification, sea ice motion monitoring, iceberg tracking, ship routing through sea ice, ice sheet dynamics, and snow cover) applications. The instrument was also used to monitor natural disasters (volcanic activity, earthquakes, land subsidence, and flooding). (i) Launch period: March, 2002. (ii) Temporal resolution: 35 days. (iii) Spectral resolution: Radar. (iv) Spectral bands: N/A.				

Sensor (Mission)	Provider	Orbital height	Spatial resolution	Swath width
Sentinel-2A, Sentinel-2B, Sentinel-2C	*ESA*	*786 km Sun synchronous*	*10–60 m (VNIR and SWIR)*	*290 km*
Data use: The mission supports a broad range of services and applications such as agricultural monitoring, emergency management, land cover classification, and water quality. (i) Launch period: June 23, 2015; March 7, 2017; 2024. (ii) Temporal resolution: 10 days. (iii) Radiometric resolution: 12-bit. (iv) Spectral resolution: 0.015–0.185 μm. (iii) Spectral bands: 13.				
CALIOP (CALIPSO)	*NASA, USA and CNES, France*	*705 km Sun synchronous*	*333 m*	*0.1 km*
Data use: The global, multiyear dataset obtained from CALIOP provides a new view of the Earth's atmosphere and will lead to an improved understanding of the role of aerosols and clouds in the climate system. (i) Launch period: April 28, 2006. (ii) Radiometric resolution: 22-bit. (iii) Temporal resolution: <16 days. (iv) Spectral resolution: 0.532–1.604 μm. (v) Spectral band: 2.				
HRC (CBERS-2B)	*CBERS, China/ Brazil*	*780 km Sun synchronous*	*2.7 m*	*27 km*
Data use: The data is used to **monitor water resources, agriculture, urban growth, and land use, with a large focus on controlling the deforestation and burning of the Amazon rainforest.** The CBERS-2B spacecraft suffered a power system failure on May 10, 2010, leaving it unable to continue operations. It remains in orbit. (i) Launch period: September 19, 2007. (ii) Radiometric solution: 8-bit. (iii) Temporal resolution: 26 days. (iv) Spectral resolution: 0.45–0.85 μm. (v) Spectral bands: 1.				
CCD (CBERS-2B)	*CBERS, China/ Brazil*	*778 km Sun synchronous*	*20 m*	*113*
Data use: Coherent Change detection, also known as CCD, is an intelligence gathering process in which two time-lapsed high-resolution satellite radar images of one geographic area on the surface of Earth, or other planets, are compared in order to detect and measure changes to that area. Coherent Change Detection can provide very useful and otherwise unavailable insights for a variety of missions—military C4ISR, climate science, space exploration, civil engineering, agriculture, forestry, asset monitoring, disaster management, and more. (i) Launch period: September 19, 2007. (ii) Radiometric solution: 8-bit. (iii) Temporal resolution: 26 days. (iv) Spectral resolution: 0.45–0.73 μm. (v) Spectral bands: 5.				

Continued

TABLE 6.1 Remote sensing satellite and data overview—cont'd

Sensor (Mission)	Provider	Orbital height	Spatial resolution	Swath width
GeoEye-1	*GeoEye, USA*	*681 km* *Sun synchronous*	*0.41 (PAN),* *1.65 (MS) m*	*15.2 km*
Data use: The satellite imagery is used for a variety of purposes, including but not limited to defence, disaster response, air and marine transportation, oil and gas exploration, mining production and exploration, mapping of remote regions, location-based services, insurance and risk management, agricultural crop management, and more. (i) Launch period: September 06, 2008. (ii) Radiometric resolution: 11-bit. (iii) Temporal resolution: <3 days. (iv) Spectral resolution: 0.45–0.80 (PAN) μm, 0.45–0.92 μm. (v) Spectral band: 4 + PAN.				
TANSO-FTS (GOSAT/Ibuki)	*JAXA, Japan*	*666 km* *Sun synchronous*	*10.5 m*	*160 km*
Data use: The Greenhouse Gases Observing Satellite (GOSAT), also known as Ibuki, is an Earth Observation Satellite and the world's first satellite dedicated to greenhouse gas monitoring. (i) Launch period: January 23, 2009. (ii) Radiometric solution: 8-bit. (iii) Temporal resolution: 3 days. (iv) Spectral resolution: 0.758–14.3 μm. (v) Spectral bands: 4.				
WorldView-2 (WV 2)	*DigitalGlobe, USA*	*770 km* *Sun synchronous*	*0.46 (PAN),* *1.85 (MS) m*	*16.4 km*
Data use: Creation and update of topographic maps and plans of scales up to 1:5000, engineering survey for construction works of oil and gas facilities, inventory of forests and forest management, monitoring of the crop state, yield forecasting, creation of maps for urban planning purposes, emergency monitoring, and research. (i) Launch period: October 08, 2009. (ii) Radiometric resolution: 11-bit. (iii) Temporal resolution: 1.1–3.7 days. (iv) Spectral resolution: 0.45–0.80 μm (PAN) μm, 0.45–1.04 μm. (v) Spectral band: 8 + PAN.				

Note: *AATSR*, Advanced Along-Track Scanning Radiometer; *ALI*, Advanced Land Imager; *ASAR*, Advanced Synthetic Aperture Radar; *ASTER*, Advanced Spaceborne Thermal Emission and Reflectance Radiometer; *AVHRR*, Advanced Very High Resolution Radiometer; *CALIOP*, Cloud-Aerosol Lidar with Orthogonal Polarization; *CALIPSO*, Cloud-Aerosol Lidar and Infrared Pathfinder Satellite Observation; *CBERS*, China-Brazil Earth Resources Satellite; *CCD*, Coherent Change Detection; *CNES*, Centre National d'Etudes Spatiales; *EO*, Earth Observing; *ESA*, European Space Agency; *ETM+*, Enhanced Thematic Mapper Plus; *GISAT*, Geo Imaging Satellite; *GOES*, Geostationary Operational Environmental Satellite; *GSLV*, Geosynchronous Satellite Launch Vehicle; *HRC*, High Resolution Camera; *IKONOS*, Commercial Earth Observation Satellite comes from the Greek word eikon for "image"; *ISRO*, Indian Space Research Organization; *LISS*, Linear Imaging Self Scanner; *MERIS*, Medium Resolution Imaging Spectrometer; *METI*, Ministry of Economy, Trade, and Industry; *MS*, Multispectral; *MSS*, Multispectral Scanner; *NASA*, National Aeronautics and Space Administration; *NESDIS*, National Environmental Satellite Data and Information Service [was created by the National Oceanic and Atmospheric Administration (NOAA)]; *NOAA*, National Oceanic and Atmospheric Administration; *OLI*, Operational Land Imager; *PAN*, Panchromatic; *RGB*, red, green, and blue; *RISAT*, Radar Imaging Satellite; *SPOT*, *Satellite pour l'Observation de la Terre*; *SWIR*, Shortwave Infrared; *TANSO-FTS*, Thermal And Near-infrared Sensor for carbon Observation (TANSO)-Fourier Transform Spectrometer (FTS); *TIR*, Thermal Infrared; *TM*, Thematic Mapper; *USA*, United States of America; *VNIR*, Visible and Near-Infrared; *ZAXA*, Japanese Aerospace Exploration Agency.

Rogan and Chen (2004); Khorram et al. (2012).

TABLE 6.2 Landsat 9 Operational Land Imager and thermal infrared sensor spectral bands.[a]

Spectral band	Wavelength (μm)	Resolution (m)
Operational Land Imager		
Band 1—ultra blue (coastal/aerosol)	0.435–0.451	30
Band 2—blue	0.452–0.512	30
Band 3—green	0.533–0.590	30
Band 4—red	0.636–0.673	30
Band 5—near-infrared (NIR)	0.851–0.879	30
Band 6—shortwave-infrared (SWIR) 1	1.566–1.651	30
Band 7—shortwave-infrared (SWIR) 2	2.107–2.294	30
Band 8—panchromatic (black and white)	0.503–0.676	15
Band 9—cirrus	1.363–1.384	30
Thermal-infrared sensor		
Band 10—thermal-infrared (TIR) 1	10.60–11.19	100
Band 3—thermal-infrared (TIR) 2	11.50–12.51	100
Thermal bands are acquired at 100 m resolution but are resampled to 30 m.		

[a]*Landsat 9 is the latest satellite in the Landsat series—it continues Landsat's irreplaceable record of Earth's land surface. It launched from Vandenberg Space Force Base on September 27, 2021. Landsat 9 largely replicates its predecessor, Landsat 8 (Masek et al., 2020). TIRS-2 measures thermal-infrared radiation, or heat (brightness temperature), of Earth's surfaces. The instrument was built at NASA's Goddard Space Flight Center in Greenbelt, Maryland. TIRS-2 provides an upgraded version of the Landsat 8 TIRS instrument by improving reliability and correcting known issues with stray light (Montanaro et al., 2022).*
Source: USGS.

Several countries outside North America and Europe have established successful satellite programs. The Indian Space Research Organization (ISRO) initiated its Earth Observing Satellite (EOS) program with the launch of IRS-1A in 1988, marking a new era in satellite-based resource monitoring, coastal management, disaster mitigation, and climate change studies (Ramana Murthy et al., 2022). By July 30, 2023, ISRO's Polar Satellite Launch Vehicle (PSLV) had completed 59 launches, 56 of which successfully reached their intended orbits, resulting in a 95% success rate (97% including one partial failure). All launches were conducted from the Satish Dhawan Space Centre (formerly Sriharikota Range, SHAR). Tables 6.3–6.5 list ISRO's PSLV launches (PSLV-The Travel).

India has successfully executed both a Moon mission (Chandrayaan) and a Mars mission (Mangalyaan) on its first attempts (Adimurthy, 2015; Chauhan et al., 2012; Menon, 2008). The country has also made significant strides in human resource development in space technology (Deekshatulu and Roy, 2000). Chandrayaan-3, launched on July 14, 2023, by ISRO from the Satish Dhawan Space Centre, marked India as the fourth country to land on the Moon. This mission, consisting of a lander (Vikram) and rover (Pragyan), focused on the lunar South Pole region, known for its ice deposits that could provide insights into the history of the Moon, Earth, and the solar system, and serve as a resource for future crewed missions (Hari

TABLE 6.3 Launch system status.

Variant	Launches	Successes	Failures	Partial failures
PSLV-G (Standard)	12	10	1	1
PSLV-CA (Core Alone)	17	17	0	0
PSLV-XL (Extended)	25	24	1	0
PSLV-DL	3	3	0	0
PSLV-QL	2	2	0	0
Total as of September 2023	59	56	2	1

TABLE 6.4 Decade-wise summary of PSLV launches.

Decade	Successful	Partial success	Failure	Total
1990s	3	1	1	5
2000s	11	0	0	11
2010s	33	0	1	34
2020s	9	0	0	9
Total	56	1	2	59

TABLE 6.5 Indian Remote Sensing (IRS) launch log.

Serial no.	Satellite	Date of launch	Launch vehicle	Status
1	IRS-1A	17 March 1988	Vostok, USSR	Mission completed
2	IRS-1B	29 August 1991	Vostok, USSR	Mission completed
3	IRS-P1 (also 1E)	20 September 1993	PSLV-D1	Crashed, due to launch failure of PSLV
4	IRS-P2	15 October 1994	PSLV-D2	Mission completed
5	IRS-1C	28 December 1995	Molniya, Russia	Mission completed
6	IRS-P3	21 March 1996	PSLV-D3	Mission completed
7	IRS-1D	29 September 1997	PSLV-C1	Mission completed
8	IRS-P4 (Oceansat-1)	27 May 1999	PSLV-C2	Mission completed
9	Technology Experiment Satellite (TES)	22 October 2001	PSLC-C3	Mission completed
10	IRS-P6 (Resourcesat-1)	17 October 2003	PSLV-C5	Mission completed
11	IRS-P5 (Cartosat-1)	5 May 2005	PSLV-C6	Mission completed

Serial no.	Satellite	Date of launch	Launch vehicle	Status
12	IRS-P7 (Cartosat-2)	10 January 2007	PSLV-C7	Mission completed
13	Cartosat-2A	28 April 2008	PSLV-C9	In service
14	IMS-1	28 April 2008	PSLV-C9	Mission completed
15	RISAT-2	20 April 2009	PSLV-C12	In service
16	Oceansat-2	23 September 2009	PSLV-C14	In service
17	Cartosat-2B	12 July 2010	PSLV-C15	In service
18	Resourcesat-2	20 April 2011	PSLV-C16	In service
19	Megha-Tropiques	12 October 2011	PSLV-C18	Mission completed
20	RISAT-1	26 April 2012	PSLV-C19	Mission completed
21	SARAL	25 February 2013	PSLV-C20	In service
22	Cartosat-2C	22 June 2016	PSLV-C34	In service
23	Scatsat-1	26 September 2016	PSLV-C35	In service
24	Resourcesat-2A	7 December 2016	PSLV-C36	In service
25	Cartosat-2D	15 February 2017	PSLV-C37	In service
26	Cartosat-2E	23 June 2017	PSLV-C38	In service
27	Cartosat-2 F	12 January 2018	PSLV-C40	In service
28	RISAT-2B	22 May 2019	PSLV-C46	In service
29	Cartosat-3	27 November 2019	PSLV-C47	In service
30	RISAT-2BR1	11 December 2019	PSLV-C48	In service
31	EOS-1 (RISAT-2BR2)	7 November 2020	PSLV-C49	In service
32	EOS-3 (GISAT-1)	12 August 2021	GSLV-F10	Crashed, due to launch failure of GSLV
33	EOS-4 (RISAT-1A)	14 February 2022	PSLV-C52	In service
34	DS-EO	30 June 2022	PSLV-C53	In service
35	EOS-D2	7 August 2022	SSLV-D1	In service
36	OneWeb India-1	23 October 2022	LVM3-M2	In service
37	EOS-06	26 November 2022	PSLV-C54	In service
38	EOS-07	10 February 2023	SSLV-D2	In service
39	OneWeb India-2	26 March 2023	LVM3-M3	In service
40	TeLEOS-2	22 April 2023	PSLV-C55	In service

Continued

TABLE 6.5 Indian Remote Sensing (IRS) launch log—cont'd

Serial no.	Satellite	Date of launch	Launch vehicle	Status
41	NVS-01	29 May 2023	GSLV-F12	In service
42	Chandrayaan-3	14 July 2023	LVM3-M4	In service
43	DS-SAR	30 July 2023	PSLV-C56	In service
44	Aditya-L1	2 September 2023	PSLV-C57	In service

SSLV, Small Satellite Launch Vehicle; *LVM*, Landslide Vulnerability Mapping; *DS*, Distributed Scatterer; *NVS*, Navigation Satellite; *SARAL*, Satellite with ARgos and ALtiKa.

Note: The initial versions are composed of the 1 (A, B, C, D). The later versions are named based on their area of application, including Oceansat, Cartosat, and Resourcesat. Some of the satellites have alternate designations based on the launch number and vehicle (P series for PSLV). From 2020, the name was changed to the generic EOS, which stands for Earth Observation Satellite (Jones, 2020).

et al., 2023). The China-Brazil Earth Resources Satellite (CBERS) program, established in 1988, launched CBERS-2B in 2007. This satellite, with its medium-resolution (20 m) multispectral CCD camera, has been a low-cost alternative to other sensors like Landsat ETM + (Ponzoni and Albuquerque, 2008). A study by Redo (2012) utilized data from multiple Landsat missions and CBERS satellites to analyze deforestation in Bolivia. The Japan Aerospace Exploration Agency (JAXA), formed in 2003, launched the Greenhouse Gases Observing Satellite (GOSAT) in 2009. GOSAT is designed to measure atmospheric concentrations of CO_2 and CH4 near Earth's surface (Morino et al., 2010(; Kuze et al., 2009).

6.10 Characteristics of remote sensing satellite imagery

Remote sensing images are characterized by spectral, spatial, radiometric, and temporal resolutions, which depend on the satellite orbit configuration and sensor design. Spectral resolution refers to the bandwidth and sampling rate of the sensor as it collects information about a scene. High spectral resolution is defined by a narrow bandwidth (e.g., 10 nm, where 1 nm = 0.001 micrometre). Multispectral imagery typically includes 3 to 10 bands, whereas hyperspectral imagery captures hundreds or thousands of narrower bands, offering higher spectral resolution (Fig. 6.14). Panchromatic imagery consists of a single broad band covering a wide range of wavelengths.

Spatial resolution refers to the smallest features in an image that can be separated, while radiometric resolution measures the dynamic range or the number of discrete signal strengths a sensor can record. A higher radiometric resolution allows for more detail and can distinguish both high and low contrast objects. For example, the Landsat 7 sensor has an 8-bit

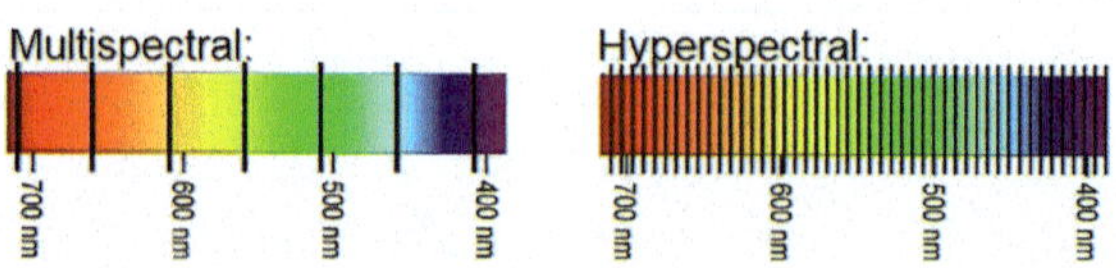

FIG. 6.14 Remote sensing spectral resolution. *Source: NASA.*

resolution (256 gray values), while Ikonos-2 has an 11-bit resolution (2048 gray values). Temporal resolution is the time interval between consecutive images of the same location, which aids in monitoring land use changes (Pradhan and Lee, 2007). Due to trade-offs in data volume and signal-to-noise ratio (SNR), remote sensing systems typically have either high spatial resolution or low spectral resolution, but not both simultaneously.

6.11 Remote sensing software

Remote sensing software is essential for extracting, analyzing, and visualizing data collected through satellite and airborne sensors. These software tools offer powerful capabilities for mapping, classification, and modeling, allowing users to interpret geographic data for various applications, such as environmental monitoring, urban planning, and disaster management. Below are some of the top remote sensing software packages:

6.11.1 ENVI (Environment for Visualizing Images)

- **Overview**: ENVI is one of the most widely used software programs for processing and analyzing remote sensing data. It offers advanced tools for image analysis, including machine learning-based classification, change detection, and spectral analysis.
- **Key Features**:
  - Image enhancement, classification, and filtering
  - 3D visualization and geospatial analysis
  - Support for a wide range of file formats
  - Advanced algorithms for analysis of hyperspectral and multispectral data

6.11.2 Arcgis (arcmap, arcgis pro)

- **Overview**: ArcGIS is a comprehensive geographic information system (GIS) that supports remote sensing tasks. It integrates remote sensing data with GIS data for spatial analysis and visualization.
- **Key Features**:
  - Integrated tools for remote sensing image analysis
  - Powerful spatial analysis and map creation tools
  - Raster analysis and image classification
  - Customizable Python scripts for automation

6.11.3 QGIs (Quantum GIS)

- **Overview**: QGIS is an open-source GIS platform that also includes several features for processing remote sensing data. While it may not be as powerful as some commercial

software, it is highly customizable and suitable for users who need a flexible, cost-effective solution.
- **Key Features**:
  - Multispectral and hyperspectral image processing
  - Support for a wide variety of remote sensing data formats
  - Raster and vector analysis
  - Large library of plugins for remote sensing tasks

6.11.4 ERDAS IMAGINE

- **Overview**: ERDAS IMAGINE is a high-performance software suite used for remote sensing data analysis. It is known for its robust capabilities in image processing, 3D modeling, and photogrammetry.
- **Key Features**:
  - Multi-dimensional image processing
  - Advanced geospatial analytics, including land use/land cover classification
  - 3D visualization and model generation
  - Remote sensing data fusion and enhancement tools

6.11.5 PCI Geomatics

- **Overview**: PCI Geomatics offers powerful tools for remote sensing image processing, geospatial data analysis, and automated feature extraction. It is often used in industries such as agriculture, forestry, and environmental monitoring.
- **Key Features**:
  - Automated and manual image classification
  - Image fusion, correction, and enhancement
  - Geospatial data analytics and model building
  - Advanced radar and LiDAR data processing

6.11.6 Google Earth Engine

- **Overview**: Google Earth Engine is a cloud-based platform for geospatial data analysis, designed to handle large-scale environmental monitoring and analysis. It offers access to vast repositories of satellite imagery and geospatial data.
- **Key Features**:
  - Large archive of remote sensing data
  - Cloud computing for fast data processing
  - Time-series analysis for environmental change detection
  - Integration with various satellite data sources like Landsat and Sentinel

6.11.7 MATLAB with image processing toolbox

- **Overview**: While not a traditional remote sensing software, MATLAB is often used for custom analysis of remote sensing data. The Image Processing Toolbox and other specialized toolboxes allow for advanced data analysis, processing, and visualization.
- **Key Features**:
  - Customizable analysis tools for remote sensing data
  - Advanced image processing algorithms
  - Integration with other software like ArcGIS and ENVI
  - Visualization of 3D data and large datasets

6.11.8 GDAL (Geospatial Data Abstraction Library)

- **Overview**: GDAL is an open-source library for reading and writing geospatial raster data formats. Although it is more of a library than standalone software, many remote sensing applications use GDAL for processing and transforming satellite imagery and other remote sensing data.
- **Key Features**:
  - Supports over 200 raster formats
  - Tools for reprojection, transforming, and analyzing raster data
  - Ability to convert between different file formats

6.11.9 Sentinel Application Platform (SNAP)

- **Overview**: SNAP is a free, open-source software package developed by the European Space Agency (ESA) for processing data from the Sentinel satellites. It supports various remote sensing techniques, particularly those for Earth observation data.
- **Key Features**:
  - Specializes in processing Sentinel-1, Sentinel-2, and other EO data
  - Preprocessing, calibration, and atmospheric correction tools
  - Tools for SAR (Synthetic Aperture Radar) and optical imagery analysis
  - Integration with other geospatial analysis tools

6.11.10 Agisoft Metashape

- **Overview**: Agisoft Metashape is a photogrammetry software that specializes in creating 3D models from satellite, aerial, and ground-based imagery. It is often used in drone-based remote sensing applications.
- **Key Features**:
  - 3D reconstruction and modeling
  - Georeferencing and orthophoto generation
  - Point cloud processing
  - Supports both optical and LiDAR data

These remote sensing software options vary in functionality, from advanced geospatial analysis to 3D modeling and cloud-based processing. Some are designed for professional and academic users, while others are more accessible and customizable for smaller projects or open-source implementations. Selecting the best software depends on the specific needs of the user, whether it's advanced image processing, large-scale data analysis, or custom modeling.

6.12 Indian remote sensing (IRS) data applications

Indian Remote Sensing (IRS) data is utilized for a wide range of applications, particularly under the National Natural Resources Management System (NNRMS), which aims to support sustainable resource management. Below is a list of key applications of IRS satellite data:

6.12.1 Agriculture

- **Crop Monitoring**: IRS data is used for monitoring crop health, detecting diseases, and estimating crop yield.
- **Crop Classification**: Satellite imagery helps in classifying crops and assessing crop types across large regions.
- **Irrigation Management**: Remote sensing helps in evaluating irrigation coverage and detecting areas of water scarcity.
- **Soil Moisture Estimation**: IRS data is used to estimate soil moisture levels, which are critical for agricultural planning (Latwat et al., 2019; Lillesand and Kiefer, 2000; Sekhar et al., 2022; Safdary et al., 2022; Prasanth Kumar et al., 2021; de la Torre et al., 2021; UN, 2021).

6.12.2 Forestry

- **Forest Cover Mapping**: IRS satellites are used to map forest cover, track changes, and estimate biomass.
- **Deforestation and Reforestation Monitoring**: IRS data helps in monitoring deforestation rates and assessing reforestation programs.
- **Forest Health and Fire Risk**: Remote sensing can identify areas affected by pests, diseases, and potential fire hazards (Baldo et al., 2023; GodGodson et al., 2021).

6.12.3 Water resources management

- **Water Bodies Monitoring**: IRS data is used to map lakes, rivers, reservoirs, and wetlands, and monitor their water levels (Siva Sankar et al., 2021; Dilip Kumar et al., 2019; Chowdary et al., 2009).

- **Groundwater Assessment**: Satellite data aids in assessing groundwater potential and distribution by mapping surface features and changes (Sharma, 2018; Rajesh et al., 2021; Mohammed Aslam et al., 2003; Arulbalaji et al., 2019; Bagari et al., 2013).
- **Flood and Drought Monitoring**: IRS data helps in monitoring flood-prone areas and drought conditions by analyzing water availability (Panigrahy, 2017).

6.12.4 Land use and land cover mapping

- **Urban and Rural Development**: Satellite data helps track urbanization, land use changes, and the development of infrastructure like roads and buildings (Kumar et al., 2014).
- **Zoning and Planning**: IRS data is used for land-use planning and assessing the suitability of land for various purposes (e.g., residential, agricultural, industrial) (Rawal et al., 2020).

6.12.5 Disaster management

- **Flood Monitoring**: IRS data is instrumental in monitoring flood extent, forecasting flood events, and assessing damage (Vengala et al., 2021; Zhang et al., 2002; Ok et al., 2012; Farhadi and Najafzadeh, 2021).
- **Cyclone and Storm Tracking**: IRS satellites are used for tracking cyclones, storms, and assessing their impact on coastal areas.
- **Drought Assessment**: Satellite data helps in monitoring drought conditions, water scarcity, and assessing the impact on crops and vegetation (Ray et al., 2022; Dutta et al., 2015; Seiler et al., 1998, 2000; Domenikiotis et al., 2004; Quiring and Ganseh, 2010).

6.12.6 Environmental monitoring

- **Pollution Monitoring**: Remote sensing is used to monitor air quality, water pollution, and the impact of industrial activities on the environment.
- **Ecosystem Health**: IRS data is used for monitoring ecosystems, including wetlands, forests, and coastal areas, to assess their health and biodiversity (Subhas Ashutosh and Roy, 2021).

6.12.7 Coastal and marine applications

- **Coastal Zone Management**: IRS data is used to map and monitor coastal areas, including shoreline erosion, mangrove forests, and coral reefs (Ramana Murthy et al., 2022).
- **Marine Resource Management**: Satellite data is used for monitoring marine resources, including fisheries, and assessing oceanographic conditions.

6.12.8 Geological studies and mineral exploration

- **Mapping of Geological Features**: IRS data aids in the mapping of geological formations, including identifying faults, mineral deposits, and tectonic features.
- **Mineral Exploration**: Remote sensing helps in identifying mineral-rich areas, including assessing the presence of precious metals and other natural resources. Advanced remote sensing techniques, including Landsat and ASTER data, were used to study the mineralogy of beach sediments and hydrothermal alteration zones in India, such as the Chavara deposits in Kerala and the Chitradurga Schist Belt in Karnataka (Rejith et al., 2021; GSI, 2022).

6.12.9 Urban planning and infrastructure development

- **Urban Sprawl Monitoring**: IRS satellites help track the expansion of urban areas, helping in planning for infrastructure and resource distribution (Roy and Kasemi, 2021; Parveen and Ilahi, 2022).
- **Transportation and Road Network Mapping**: Satellite data is used to map existing and planned road networks, railways, and airports for better planning and development (Nuis, 2006, 2015; Rao et al., 2005).

6.12.10 Climate change and environmental impact assessment

- **Carbon Sequestration Monitoring**: IRS data is used to assess carbon sequestration potential in forests and other ecosystems.
- **Climate Change Monitoring**: Remote sensing is used to track changes in land surface temperature, vegetation cover, and other climate indicators to assess climate change impacts.

6.12.11 Disaster risk reduction

- **Vulnerability Assessment**: IRS data is utilized to assess the vulnerability of regions to natural hazards such as earthquakes, landslides, and tsunamis (Raju et al., 2016; Nageswara Rao, 1991; Kharol et al., 2008).
- **Post-Disaster Damage Assessment**: After disasters, satellite data helps in evaluating the extent of damage and aiding in recovery efforts.

6.12.12 Biodiversity conservation

- **Habitat Mapping**: IRS data assists in mapping and monitoring wildlife habitats, crucial for biodiversity conservation.

- **Protected Area Management**: Remote sensing helps in monitoring protected areas, such as national parks and wildlife sanctuaries, to assess changes and manage resources effectively.

These diverse applications of IRS data play a critical role in enhancing India's capacity for resource management, disaster preparedness, environmental monitoring, and sustainable development.

6.13 Earth observations: Credible data for diverse uses

Remote sensing satellites orbit Earth, providing near real-time data that allows for the analysis of both current and historical conditions. This technology supports scientific research and various practical applications (Ridwan et al., 2018). For mineral resource mapping, remote sensing offers unique advantages. Traditional geologic maps can be inaccurate and general, and fieldwork can be challenging in areas with poor accessibility or dense vegetation. Remote sensing enables the detection of geologic structures and mineral alteration patterns that might be missed on the ground. Image interpretation, especially through false-color renditions, enhances the visibility of subtle features, aiding in the identification of mineral-rich areas and structural features not apparent through fieldwork alone. This approach can also extend known mineral areas by identifying lineaments beyond established boundaries.

Traditional mineral exploration methods in Alwar Basin, Rajasthan, India, have relied heavily on physical rock characterization and extensive fieldwork (Heron, 1953; GSI, 2011). However, remote sensing technologies have emerged as a valuable tool in mineral exploration. Chattoraj et al., 2020 utilized Advanced Spaceborne Thermal Emission and Reflection Radiometer (ASTER) data, specifically the visible near-infrared (VNIR) and shortwave-infrared (SWIR) bands, to detect surface mineralization in the Mundiyawas-Khera area. Using the Spectral Angle Mapper (SAM) method, they identified mineralization signatures, including galena, chalcopyrite, and malachite. The accuracy of the spectral anomalies was validated using the Pixel Purity Index and X-ray diffraction data. The spectral anomaly maps were then integrated with other geospatial data, such as lineament density and geochemical anomalies, through the weight of evidence method. This study demonstrates the potential of remote sensing in identifying unexplored, mineral-rich areas, using synoptic, multispectral, and multitemporal data for rapid exploration and delineation of metallogenic provinces.

The integration of various remote sensing data types, along with ancillary data from multiple sources, has been a driving force behind numerous scientific investigations, from estimating forest biomass to mapping the surface of Mars in search of minerals (Anderson, 2010; Behera et al., 2015; Roy and Ravan, 1996). Remote sensing data is also enhancing the development of improved geographical information systems (GIS), which in turn can be applied to education, land management, natural resources management, environmental monitoring, and aeronautical applications. Additionally, government organizations are increasingly providing access to data portals, making it available to the general public. In India, remote sensing data has been extensively applied across several operational themes, including agriculture, forestry, biodiversity, groundwater targeting, water resources, land use, urban sprawl, geology, environmental monitoring, coastal zone management, marine resources, snow and glacier studies, disaster monitoring and mitigation, and infrastructure development (Behera, 1999).

6.14 Contribution of RS in prospecting exploration studies

The coastal tracts of India, particularly those along Kerala's Chavara and Tamil Nadu's Manavalakurichi, contain significant concentrations of strategic heavy minerals like sulfur, lead, petroleum, zinc, mercury, platinum, nickel, graphite, and others. Remote sensing has played a vital role in studying the mineralogy of these coastal placer deposits (Ilmenite, rutile, garnet, monazite, zircon, and sillimanite) using Landsat multispectral satellite data (Rejith et al., 2021). Ilmenite is the predominant mineral, followed by monazite, sillimanite, rutile, zircon, garnet, leucoxene, and kyanite. Advanced hyperspectral techniques, including minimum noise fraction transformation, pixel purity index, and n-dimensional visualizer, followed by spectral angle mapper (SAM) classification, were employed to study the distribution of these minerals. SAM algorithms successfully identified and mapped the end members of ilmenite and quartz from Landsat imagery, enabling eco-friendly exploration of these valuable minerals.

Recent advancements in remote sensing have improved the mapping of mineral occurrences, though quantitative estimation of mineral abundances remains challenging. Rejith et al. (2022) used EO-1 Hyperion data, spectral angle mapper (SAM), continuum removed band depth analysis, and random forest (RF) regression to map and estimate beach placer minerals along the Cuddalore coastal stretch in Tamil Nadu, India. The SAM method showed an 81.06% accuracy in detecting placer mineral deposits like zircon. The RF regression model, based on the correlation between geometric parameters of mineral spectra and in-situ mineral concentration, successfully predicted placer mineral concentrations with low RMSE values (0.082 for reference spectra and 0.156 for image spectra), demonstrating the potential of EO-1 Hyperion data for quantifying mineral resources.

Additionally, Wan et al., 2021 used hyperspectral aerial imagery, ground spectrum tests, and sampling analysis for mineral exploration in the Huaniushan area of Gansu Province, China. The study applied hyperspectral remote sensing techniques, including SAM, minimum noise fraction transform, and mixed tuned matched filtering, to classify and quantitatively analyze mineral and rock mass occurrences. The study achieved an 80% classification accuracy, with the recognition rate for the main ore body—iron-manganese cap lead-zinc oxide ore—reaching 81%. These findings underline the effectiveness of hyperspectral remote sensing in enhancing mineral exploration, supporting the creation of more accurate mineral and geological maps, and guiding future exploration efforts.

6.15 NISAR observation system

The NASA–ISRO Synthetic Aperture Radar (NISAR) mission is a landmark Earth-observation initiative jointly developed by the Indian Space Research Organisation (ISRO) and the National Aeronautics and Space Administration (NASA). It employs a dual-frequency microwave imaging system that integrates both L-band and S-band Synthetic Aperture Radar (SAR) instruments. Together, these sensors provide full-polarimetric and interferometric capabilities, enabling precise, high-resolution observations of the Earth's land and ice-covered surfaces. NISAR relies on a SweepSAR architecture that allows it to capture wide-swath imagery at regular intervals. The satellite revisits most land and ice regions, including remote islands, major ice sheets, sea-ice zones, and selected ocean areas, every twelve days, regardless of atmospheric conditions or lighting (Source: https://www.isro.gov.in/Mission_GSLVF16_NISAR_Home.html).

The mission is designed to generate consistent, repeatable global monitoring of the Earth's dynamic surface. By exploiting both the L-band radar contributed by NASA and the S-band radar developed by ISRO, NISAR provides complementary imaging capabilities that capture variations in vegetation, soil moisture, surface roughness, and ground deformation. The combination of two radar frequencies from a single satellite platform is unprecedented and greatly enhances the spatial and temporal richness of the observations.

NISAR's scientific objectives focus on improving our understanding of key Earth system processes. Central to the mission is the measurement of global woody biomass and the detection of changes in forest structure, which are essential for carbon-cycle assessments. The mission also aims to monitor the growth and spatial extent of active crops, track the seasonal behavior of wetlands, and observe the evolution of glaciers, sea ice, and the large ice sheets of Greenland and Antarctica. A major component of the mission is the use of interferometric SAR measurements to characterize land-surface deformation. This includes detecting movements associated with earthquakes, volcanic activity, landslides, and anthropogenic processes such as groundwater withdrawal, hydrocarbon extraction, and subsurface compaction.

NISAR carries two primary instruments: The S-band SAR system developed by ISRO and the L-band SAR system provided by NASA. ISRO is responsible for building the satellite bus, the S-band radar, the high-speed data-downlink system, the launch vehicle, and overall mission operations. NASA contributes the L-band radar, the onboard solid-state recorder, the high-capacity downlink subsystem, the GPS navigation receiver, and the long boom assembly that deploys the 12-meter reflector. Both agencies operate ground-station networks to receive and process the data, ensuring continuous global coverage. All processed data products will be distributed openly, supporting a wide range of scientific and applied uses.

The availability of dual-frequency SAR data from a single orbiting platform makes NISAR an exceptionally powerful tool for Earth system science. L-band radar penetrates vegetation canopies and dry soil more effectively than higher-frequency sensors, while S-band radar provides enhanced sensitivity to surface roughness and structural changes. Together, they enable advanced analyses of topography, vegetation conditions, ice dynamics, and temporal changes to the Earth's surface. This integrated capability supports long-term studies of land deformation, environmental hazards, climate-related processes, and ecosystem change, while also contributing to rapid-response assessments following natural disasters.

Although NISAR is not specifically designed to detect mineral deposits, it provides several types of information that can indirectly support mineral exploration. Interferometric SAR data allow precise monitoring of ground deformation, revealing subsidence or uplift associated with groundwater withdrawal, fluid migration, or mining-induced stress changes. Such deformation patterns also help infer subsurface structures and fault zones that may guide geologic interpretation. The combined L-band and S-band observations improve geological mapping in regions where vegetation cover or remote terrain limits field accessibility. Structural features such as lineaments, fractures, and bedrock patterns become more apparent in radar imagery, aiding exploration teams in targeting prospective areas. The mission also supports environmental monitoring by documenting land-use changes, vegetation loss, and surface disturbances caused by mining operations, thereby contributing to sustainable resource management and regulatory compliance.

A defining feature of the NISAR mission is its open-data policy. All data products, including calibrated radar images and derived interferometric measurements, will be freely accessible to the global community. This open distribution model will create a long-term, consistent archive of dual-frequency SAR observations, supporting both scientific research and operational applications across many disciplines, including geology, hydrology, climate science, agriculture, environmental monitoring, and natural hazard assessment.

NISAR represents a major advancement in global radar-based Earth observation. Its integration of L-band and S-band SAR systems, combined with its frequent revisit capability and open-data policy, makes it an exceptionally powerful platform for monitoring environmental processes and assessing natural hazards. The mission supports a broad spectrum of scientific and applied investigations by providing consistent, high-quality radar observations across multiple temporal and spatial scales.

Although NISAR is not specifically designed for direct mineral detection, it contributes meaningfully to mineral exploration through its ability to monitor ground deformation, delineate geological structures, and capture environmental conditions in areas of interest. These datasets strengthen geological interpretation in regions where ground access is limited or where vegetation obscures surface features. By delivering continuous and reliable measurements of the Earth's dynamic surface, NISAR enhances our capacity to understand surface processes, assess environmental change, and support the sustainable management and exploration of natural resources.

6.16 Conclusions

In the context of geoscience exploration and resource management, remote sensing techniques play a pivotal role in obtaining critical data for sustainable decision-making. Among the various remote sensing approaches, hyperspectral and multispectral remote sensing have emerged as key tools for detailed Earth observation. These techniques provide valuable insights into the Earth's surface, enabling the monitoring of land use, environmental changes, and resource distribution with high spatial, spectral, and temporal precision.

Hyperspectral remote sensing uses a large number of narrow, contiguous spectral bands, providing detailed spectral information that can reveal subtle differences in material properties. This allows for more accurate identification and classification of surface materials, offering significant advantages in areas such as mineral exploration, vegetation monitoring, and water quality assessment. The ability to capture continuous spectra across a wide range of wavelengths enables hyperspectral sensors to detect specific spectral signatures that are often invisible to traditional multispectral sensors.

On the other hand, multispectral remote sensing involves measurements in a smaller number of broader spectral bands. While less detailed than hyperspectral sensing, multispectral data are still extremely valuable for broader, large-scale applications such as land cover classification, agriculture monitoring, and forest management. The selected spectral bands in multispectral systems are designed to capture information from key regions of the electromagnetic spectrum, such as visible, near-infrared, and shortwave infrared wavelengths, which are particularly useful in distinguishing different surface features based on their spectral signatures.

The spectral signature of a material represents its unique reflectance or emissivity characteristics across specific wavelengths of the electromagnetic spectrum. By analyzing these spectral signatures, remote sensing techniques enable the identification of surface materials, helping geoscientists assess the composition of soils, rocks, vegetation, and water bodies, as well as track changes over time. The power of spectral signature analysis is especially evident in applications related to resource exploration, where distinguishing between similar-looking materials can make a substantial difference in the discovery of valuable resources.

Overall, remote sensing technologies—both hyperspectral and multispectral—are indispensable tools for geoscience and resource management, offering a noninvasive, efficient, and cost-effective means of collecting baseline data. As the field continues to evolve, advancements in sensor technology, data processing techniques, and analytical models will further enhance our ability to monitor and manage the Earth's resources in a sustainable manner, aligning with global goals of environmental preservation and sustainable exploration.

CHAPTER

7

Geospatial surveying techniques: GIS and GPS

7.1 Introduction to geospatial surveying

7.1.1 Overview of Geographical Information System (GIS) and Global Positioning System (GPS)

Geographic Information Systems (GIS) are computer-based platforms designed to collect, manage, analyze, and visualize geospatial data, enabling professionals to derive actionable insights from spatially referenced information. In mineral exploration, GIS plays a pivotal role by integrating multisource and multiscale exploration data, offering superior efficiency over traditional methods in analyzing large volumes of spatial data for target generation. A key application of GIS in this domain is mineral prospectivity mapping (MPM), which helps in identifying potential resource locations by leveraging its advanced mapping and spatial analysis capabilities (Bonham-Carter, 1994; Ford et al., 2016; Sun et al., 2019). However, GIS-based MPM faces challenges in modeling the complex multiscale ore-forming geological processes essential to understanding mineral deposit formation. These limitations often restrict its effectiveness when applied to traditional deposit-scale forensic models (Porwal and Kreuzer, 2010; Yousefi et al., 2019).

The effectiveness of GIS is inherently dependent on the accuracy, comprehensiveness, and currency of the underlying spatial and thematic data, such as administrative boundaries and land cover types. Traditionally derived from topographic or land-use maps, these datasets may be generalized or outdated, thereby limiting their utility. The integration of GIS with satellite-based technologies like Remote Sensing (RS) and the Global Positioning System (GPS) addresses this gap by providing high-precision, up-to-date geospatial information that supports more reliable analysis and decision-making.

The Global Positioning System (GPS) is a satellite-based navigation and positioning system made possible by advancements in satellite technology, including high-frequency signal transmission and precise atomic clocks. GPS determines an observer's three-dimensional position—latitude, longitude, and height—as well as Universal Coordinated Time (UTC) and velocity vectors, using signals from a constellation of at least four orbiting satellites

Baseline Geoscience Data Generation for Sustainable Exploration of Natural Resources
https://doi.org/10.1016/B978-0-443-45221-5.00009-6

(McNamara, 2008). GPS operates in real time and through postprocessing modes, offering flexibility in diverse surveying and mapping applications. It is widely used across land, air, and sea navigation, as well as in cadastral surveys, urban planning, and environmental monitoring. The ability to accurately determine the coordinates of geographical features makes GPS indispensable for data collection in mapping and for integration into GIS platforms. GPS has also been integrated with RS techniques like aerial scanning and photogrammetry to enhance data accuracy in georesource mapping (Ahmed El-Rabbany, 2002).

The GPS consists of 24 satellites orbiting the Earth at an altitude of approximately 20,200 km, providing continuous global coverage (Fig. 7.1). Since the early 1990s, the establishment of permanent International Global Navigation Satellite Systems (GNSS) Service Interpersonal Guilt Scale (IGS) stations in India has contributed to large-scale crustal deformation studies, which have expanded considerably over the past decades (Tiwari et al., 2014). GPS surveying is crucial in numerous fields, particularly in the exploration and management of natural resources. The ability to measure precise geographic locations supports accurate mapping, monitoring, and spatial analysis—key elements in decision-making for resource extraction, land-use planning, and environmental assessment.

Advanced GPS surveying techniques such as Differential GPS (DGPS) and Real-Time Kinematic (RTK) positioning are widely used to ensure high positional accuracy, even in complex environments. DGPS uses a network of ground-based reference stations to compare received signals to known positions and transmit correction data to nearby receivers,

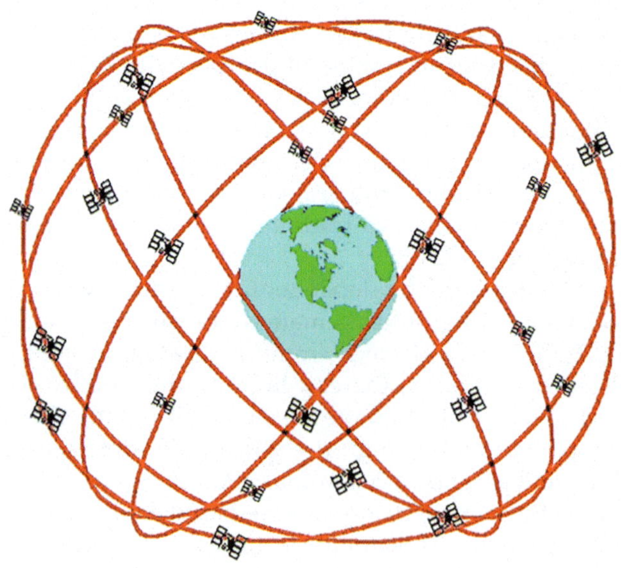

FIG. 7.1 Nominal GPS constellation of 24 satellites in six orbital planes (equally spaced 60 degrees apart), four satellites per plane, providing between five and eight visible satellites from any point on Earth. *Source: Y. E.-S. Fouda, 2004.*

significantly improving positional accuracy. RTK GPS, on the other hand, enables centimeter-level precision by transmitting correction data from a fixed base station to a mobile rover in real time. These advanced techniques enhance data collection in applications requiring high accuracy, including infrastructure development, natural resource management, and topographic surveys.

Despite its accuracy, GPS surveying is not without challenges. Signal obstructions caused by tall buildings, rugged terrain, or dense vegetation can hinder signal reception, resulting in positional inaccuracies. Multipath errors—caused by signal reflections—also contribute to potential inaccuracies. Moreover, atmospheric disturbances such as ionospheric and tropospheric delays can affect signal quality. These issues are commonly mitigated through advanced correction techniques like DGPS and RTK, along with improvements in satellite technology and signal processing.

The integration of GPS with GIS and RS enhances geospatial capabilities by combining real-time data collection, spatial modeling, and large-scale image analysis. This synergy supports a wide array of applications, including geological mapping, environmental monitoring, emergency response, and sustainable resource development. By combining GPS-derived positional data with thematic and remotely sensed datasets, it is possible to construct detailed and dynamic geospatial models that provide valuable insights into resource availability and land-use changes. For instance, in mineral and water resource exploration, such integration facilitates high-resolution geological mapping that aids in more effective and environmentally conscious decision-making.

As technology continues to evolve, the precision and versatility of GPS surveying are expected to improve further. Innovations in satellite signal correction, geospatial data integration, and automated data processing will enhance the functionality and reliability of GPS. This will further expand its application in geoscience, land management, and sustainable development, making GPS an increasingly essential tool in modern geospatial practices. In conclusion, GPS surveying, when combined with GIS and RS, supports accurate mapping, monitoring, and analysis in natural resource exploration and management. Despite challenges such as atmospheric interference and environmental obstructions, techniques like DGPS and RTK, along with continuous technological advancements, ensure their ongoing effectiveness and relevance. GPS will continue to be a cornerstone in the sustainable exploration and management of Earth's natural resources.

7.1.2 Deposit models in mineral exploration

Deposit models describe the geological characteristics and regional settings of mineral deposits (Carranza, 2017). Although mineral deposits can be categorized into genetic types, each deposit retains uniqueness, often due to the specific geological conditions at the site. To make **mineral prospectivity mapping (MPM)** more geologically relevant, an in-depth understanding of **ore-forming geological processes** across various scales is necessary (Kreuzer et al., 2020). These processes are integral to improving MPM models and their applications, but translating these geological processes into practical exploration criteria has proven challenging (McCuaig et al., 2010; Hagemann et al., 2016). This complexity often arises from incomplete geological data and the evolving nature of mineral systems theory.

7.1.3 Understanding ore-forming geological processes

Mineral deposits are the result of complex Earth processes that operate across a range of spatial and temporal scales. These processes, often referred to as **mineral systems**, form the basis of modern exploration targeting (Wyborn et al., 1994; McCuaig et al., 2010). Mineral systems theory, increasingly adopted by academic institutions, industries, and governmental agencies, faces challenges in translating these processes into practical models that can be applied in the field (Huston et al., 2006; Hronsky, 2004, 2011). The complexity of these processes, coupled with gaps in geological knowledge, presents significant uncertainty in effective mineral exploration (Hagemann et al., 2016a; Hronsky, 2011).

7.1.4 Challenges in mineral exploration targeting

Several challenges complicate mineral exploration, including the **lack of reliable datasets**, the absence of standardized methodologies for integrating geological knowledge, and the need for advanced tools like geophysical surveying and drilling (Harris, 1990; Knox-Robinson and Wyborn, 1997). Additionally, as mineral exploration data grows in size, managing and interpreting this data becomes increasingly complex. The integration of geological, geophysical, and geochemical data is critical for efficient exploration, but data availability and quality continue to pose significant challenges (Porwal and Kreuzer, 2010; Hronsky and Kreuzer, 2019). Overcoming these challenges will enable the development of more accurate and effective models for mineral exploration.

7.1.5 Integration of GPS and GIS for resource mapping

The integration of GPS and GIS enables precise data collection and real-time tracking for applications ranging from georesource mapping to environmental monitoring. The role of GPS is pivotal in ensuring the accurate location of features during surveys and field studies. High-precision GPS receivers are used for specific tasks like cadastral mapping, topographic surveys, and linear mapping, making them an essential tool in natural resource exploration. Furthermore, GPS's real-time capabilities reduce the time and labor typically required by traditional surveying methods, making it especially useful for large-scale mapping operations.

GPS technology also enhances remote sensing techniques by providing georeferencing for aerial surveys, ensuring that aerial imagery and sensor data are accurately aligned with geographic coordinates. This capability has transformed industries ranging from urban planning to disaster management by providing accurate, timely data.

7.2 GIS and GPS: motivation, inspiration, and objectives

7.2.1 Motivation for integrating GIS and GPS in geospatial surveying

The increasing volume and variety of data in modern surveying, particularly in **mineral exploration**, have made it challenging to efficiently manage and organize this data within a

Geographical Information System (GIS). In mineral exploration, complex interrelationships exist among **ore-forming processes** that govern the location, size, and grade of mineral deposits. This complexity necessitates an advanced **information system** that simplifies and organizes these dynamic relationships, making it easier to identify and characterize the similarities and differences among various components of the system (Gill and Chew, 2018). Through this process of simplification, an information system enhances the effective management and understanding of these complex systems.

In this context, GIS plays a crucial role by enabling practitioners to select and prioritize the most informative features from various datasets, such as mineralization-related geological features. GIS facilitates data analysis, modeling, and decision-making, offering the tools necessary to handle the large and intricate datasets involved in mineral exploration. As noted by Lang et al. (2018), these systems allow for a more comprehensive understanding of mineralization, helping to expand knowledge of complex natural phenomena and ultimately driving the success of mineral exploration.

Furthermore, the integration of GIS with other data collection and survey methods, including GPS technology, enhances the accuracy and efficiency of geospatial data acquisition. By incorporating GPS coordinates, the spatial resolution of exploration data improves, allowing for precise mapping of geological features critical for targeting mineral resources.

7.2.2 Key objectives in the use of GIS and GPS for modern surveying techniques

The primary objective of integrating GIS and GPS in modern surveying is to create an efficient, accurate, and comprehensive system for managing and analyzing geospatial data. This integration allows surveyors to capture precise geographical coordinates of natural and artificial features and seamlessly incorporate them into GIS platforms for further spatial analysis.

GIS and GPS together enable real-time data collection, highly accurate positioning, and seamless data integration, enhancing the ability to conduct topographic surveys, resource mapping, and environmental monitoring. For example, the combination of GPS's positioning accuracy and GIS's spatial analysis capabilities allows for better-targeted resource mapping, including the identification of mineral deposits and the assessment of natural hazards like landslides, forest fires, and oil spills.

Additionally, the use of GIS and GPS in geospatial surveying improves data accuracy and efficiency in resource management. By integrating these technologies, the reliability of geospatial datasets is significantly enhanced, which is particularly valuable in mineral exploration, where even slight errors in location or measurement can lead to substantial financial losses. Fig. 7.2 illustrates the interaction between an Exploration Information System (EIS) and GIS in the context of mineral systems (Yousefi et al., 2021).

7.2.3 Benefits of integrating GIS and GPS

Integrating GIS and GPS provides several key benefits that have transformed modern surveying and exploration techniques. The most significant benefits include:

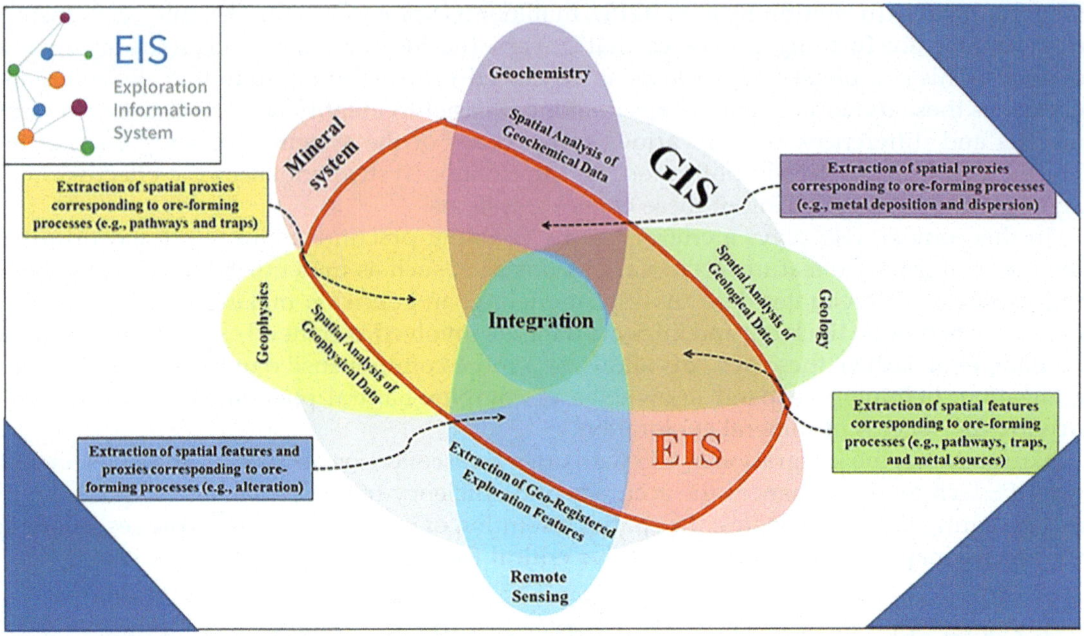

FIG. 7.2 Influence area of an Exploration Information System (EIS) in conjunction with Geographic Information System (GIS) and mineral systems. *Source: Modified.*

Improved Data Accuracy:

GPS allows for highly precise geospatial positioning, and when this data is integrated into GIS, it enables accurate mapping of large geographical areas, reducing errors in the data collection process.

Spatial Analysis and Modeling:

GIS enables the analysis of spatial patterns and relationships, which is crucial for mineral exploration targeting. By utilizing both technologies, practitioners can better understand geological features and processes that influence mineral deposit formation, such as geochemical signatures or geophysical anomalies.

Enhanced Decision-Making:

The integration of GIS and GPS allows for better-informed decision-making by providing a unified, real-time view of geospatial data. This capability supports exploration efforts by offering insights into potential mineralization and guiding surveyor activities more effectively, leading to cost savings and increased exploration efficiency.

Time and Cost Efficiency:

The use of GPS reduces the time and labor required for field data collection, particularly when surveying large, remote, or difficult-to-access areas. GPS also enhances the efficiency of aerial surveys and photogrammetry in capturing high-quality geospatial data.

Data Integration Across Multiple Scales:

By combining GPS data with GIS tools, surveyors can integrate geospatial information from various sources, creating a comprehensive dataset that covers multiple spatial scales.

This integrated approach helps to address complex challenges in mineral exploration, where both large-scale geological processes and small-scale features must be considered in the exploration process.

7.2.4 Exploring the role of GIS in mineral systems

The use of information systems for mineral exploration has proven essential in handling the complex datasets that arise from the study of ore-forming processes. These systems provide an effective framework for structuring data and facilitating decision-making tasks. As noted by McCuaig et al. (2010) and Hagemann et al. (2016a), exploring the interrelationships between geological features and mineral deposits requires systems that can handle both spatial and nonspatial data.

The Exploration Information System (EIS) can be integrated with GIS to enhance the understanding of mineral systems. Yousefi et al. (2019) categorized ore-forming processes into three subsystems, each with its own set of features that can be mapped using GIS technology. By progressively converting raw data into information, knowledge, and ultimately insight, GIS systems play a pivotal role in mineral exploration targeting.

Through this integration, an EIS can help practitioners identify spatially mappable criteria for mineral exploration. The ability to map these processes across multiple scales allows exploration efforts to focus on the most promising targets, improving the efficiency of exploration and reducing uncertainty in decision-making. Fig. 7.2 illustrates the influence area of an EIS in conjunction with GIS and mineral systems, highlighting how GIS can transform raw exploration data into actionable insights.

By integrating GIS and GPS, the benefits in data accuracy, spatial analysis, and decision-making are clearly evident. This combination of technologies has redefined modern surveying and exploration, offering significant advantages in terms of time, cost, and overall exploration efficiency.

7.3 Applications of GIS-based techniques in surveying and mapping

7.3.1 Role of GIS in mapping, geospatial analysis, and visualization

Mineral exploration targeting is a highly complex decision-making process, influenced by various factors such as the quality of exploration data and the robustness of the underlying conceptual targeting model. Two key risk factors—the quality of exploration data and the effectiveness of the targeting model—significantly impact the decision-making process. Geographic Information Systems (GIS) can be utilized not only for compiling, integrating, and interpreting diverse exploration data but also for targeting through the application of powerful mathematical algorithms. This approach, known as mineral potential modeling (MPM) or mineral prospectivity mapping (MPM), is a widely used technique in modern mineral exploration (Yousefi et al., 2021).

GIS-based techniques have become particularly prominent in mineral prospecting, where the identification of exploration targets typically involves the review and integration of various datasets, such as remote sensing data, geological mapping, geochemical and geophysical surveys, and drilling data (Fig. 7.3) (Zuo, 2020; Marjoribanks, 2010; Moon et al., 2006). More

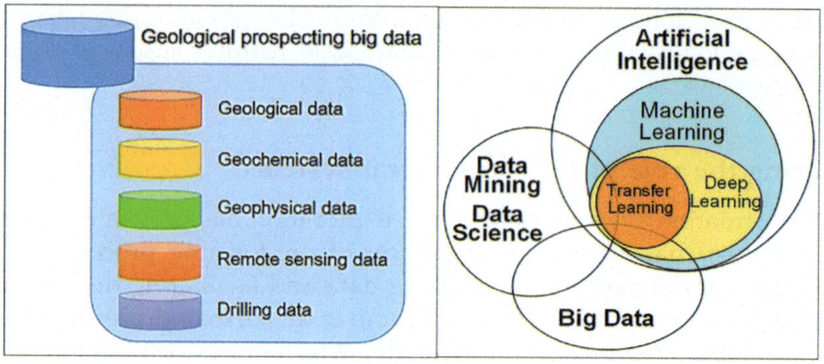

FIG. 7.3 Geological prospecting big data for Mineral Prospectivity Mapping (MPM). *Source: Modified.*

recently, GIS-based mapping of **mineral prospectivity** has gained significant attention due to its ability to offer objective and reproducible results, making it a valuable tool for prioritizing exploration targets and assessing various types of geological evidence (Behnia, 2007; Boleneus et al., 2001; Carranza and Sadeghi, 2010; Ford and Hart, 2013; Gonzalez-Álvarez et al., 2010; Nykanen and Ojala, 2007; Porwal et al., 2010; Raines, 1999). GIS-based methods allow exploration teams to systematically assess and justify the need for further investigations (Andrada de Palomera et al., 2014).

7.3.2 GIS techniques in mineral exploration and prospectivity mapping

There are various GIS-based methods employed for mapping mineral prospectively (Zuo, 2020) (Fig. 7.4), including logistic regression (Harris and Pan, 1999), evidential belief functions (Carranza and Hale, 2003), fuzzy logic (Carranza and Hale, 2001; Porwal et al., 2003), machine learning algorithms such as neural network techniques (Singer and Kouda, 1999), support vector machines (Zuo and Carranza, 2011), random forests (RF) (Rodriguez-Galiano et al., 2015; Carranza and Laborte, 2015a, b; Harris et al., 2015), extreme learning (Chen and Wu, 2017), deep learning algorithms such as deep autoencoder networks (Chen, 2015; Xiong et al., 2018), convolutional neural networks (CNN) (Li et al., 2019; Sun et al., 2020), weights-of-evidence modeling (Bonham-Carter et al., 1989), and Bayesian network classifiers (Porwal et al., 2006).

A **function of MPM** can be expressed for each location x_i as follows:

$$p(x_i) = f(m_1, m_2, \ldots\ldots m_j) \tag{7.1}$$

where i = 1, 2 ..., n_i = 1, 2 ..., n, and n is the total number of grid cells in a given area; p(xi) represents the probability or possibility of locating a mineral deposit at xi; mj denotes an evidence layer, and j represents the total number of evidence layers. The function f (·) refers to the method used in MPM for integrating multiple evidence layers into a mineral potential map.

Among these various methods, the Weights-of-Evidence (WofE) technique is often preferred due to its transparency and ease of interpretation, making it more suitable compared

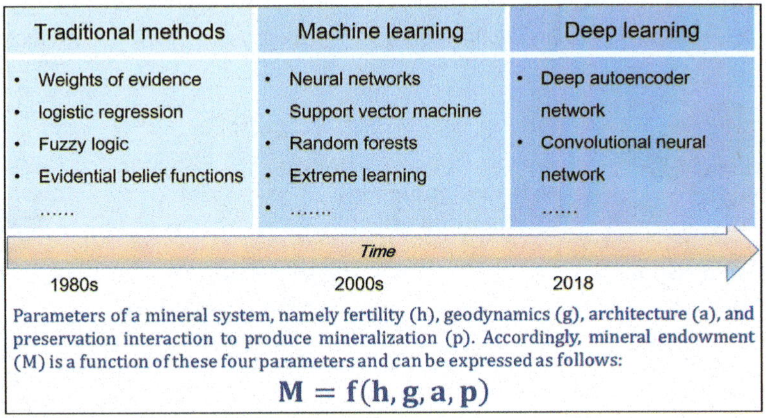

FIG. 7.4 Mathematical functions for integrating evidence layers for MPM.

to methods such as neural networks, neuro-fuzzy systems, and Bayesian network classifiers (Agterberg, 2011; Agterberg and Cheng, 2002; Bonham-Carter et al., 1989; Ford and Hart, 2013; Porwal et al., 2010). Though logistic regression can yield high-quality results, it is computationally intensive, and its regression coefficients are often challenging to interpret (Deng, 2009). Conversely, weighted logistic regression produces unbiased probabilities, but its coefficients can exhibit relatively large variances (Agterberg, 2011). The fuzzy logic method, while effective in accommodating multiclass variables, depends significantly on the reliability and correctness of the underlying exploration models.

7.3.3 Benefits of the Weights-of-Evidence (WofE) method

The WofE method is a quantitative, data-driven approach that employs a log-linear form of the Bayesian probability model (Bonham-Carter, 1994). This method is particularly valuable because it provides conservative estimates and is highly suitable for areas that have been only partially studied (de Quadros et al., 2006). Additionally, WofE permits the use of zero weights for missing or unknown data, preventing the exclusion of some layers of evidence that may occur in other methods like logistic regression (Agterberg, 2011; Bonham-Carter et al., 1989; Deng, 2009). Through WofE, the relative importance of each individual evidence layer can be estimated statistically, minimizing subjective biases when quantifying spatial associations between evidence layers and mineral deposits (Zuo, 2020) (Fig. 7.5).

The WofE method can calculate weights for different classes within each evidence layer and produce a probability map indicating the likelihood of mineral deposit occurrence. This can be achieved even with a limited number of known mineral deposits (Carranza, 2004). However, it is important to note that the WofE method assumes conditional independence (CI) between layers of evidence. When this assumption is violated, biased estimates of probabilities may arise, leading to overestimation of probabilities. The performance degradation generally increases as the number of predictor variables grows (Singer and Kouda, 1999).

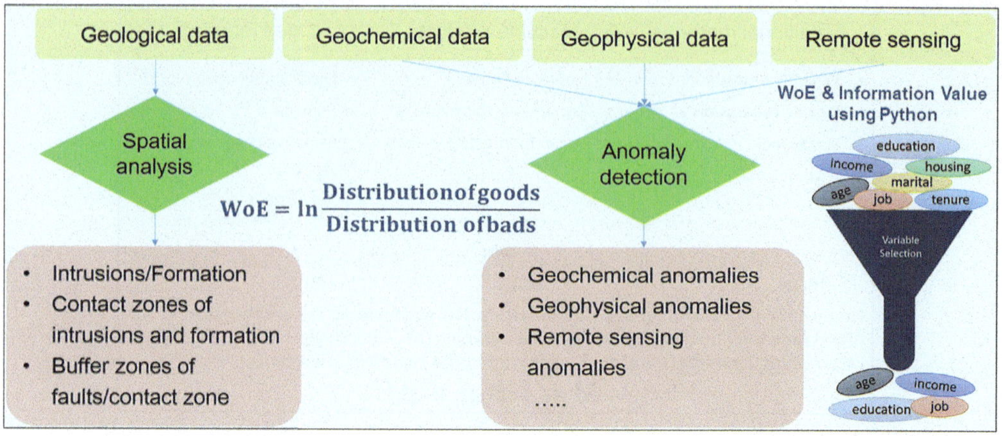

FIG. 7.5 Evidence layers created by spatial analysis and anomaly detection methods. *Source: Modified.*

Therefore, it is essential to test for conditional independence before applying the WofE modeling approach (Bonham-Carter et al., 1989). The WofE calculations and CI tests were conducted using the Spatial Data Modeller (SDM) module in ArcGIS software (Sawatzky et al., 2009).

7.4 Exploring GPS surveying techniques

7.4.1 Importance of satellite-based positioning for accurate field data collection

The use of satellite-based GPS in surveying is critical for obtaining highly accurate location data. GPS receivers measure the time it takes for signals from multiple satellites to reach the receiver, calculating precise locations with high accuracy. This is essential in fields such as resource exploration, where accurate location data is crucial for identifying potential resource-rich areas, and in land management, where GPS ensures that boundaries are defined accurately for legal and development purposes.

Despite the advancements in GPS technology, challenges such as signal obstructions and multipath errors can still affect positioning accuracy, especially in dense urban environments or rugged terrains (Fig. 7.6A & Fig. 7.6B).

These challenges occur when the GPS signal is reflected off nearby objects or blocked by buildings, trees, or mountains, causing errors in the measurement process. To mitigate these errors, GPS receivers often rely on correctional techniques, such as DGPS or RTK, which enhance the accuracy of field data collection by correcting these errors in real time (Fig. 7.7A & Fig. 7.7B).

In addition to signal-related errors, atmospheric conditions—specifically ionospheric and tropospheric delays—can introduce minor inaccuracies into GPS measurements. These delays affect the speed at which signals travel through the atmosphere and can cause slight shifts in the calculated position. Ongoing research and technological innovations aim to reduce these atmospheric errors, ensuring even greater precision in GPS-based surveying methods.

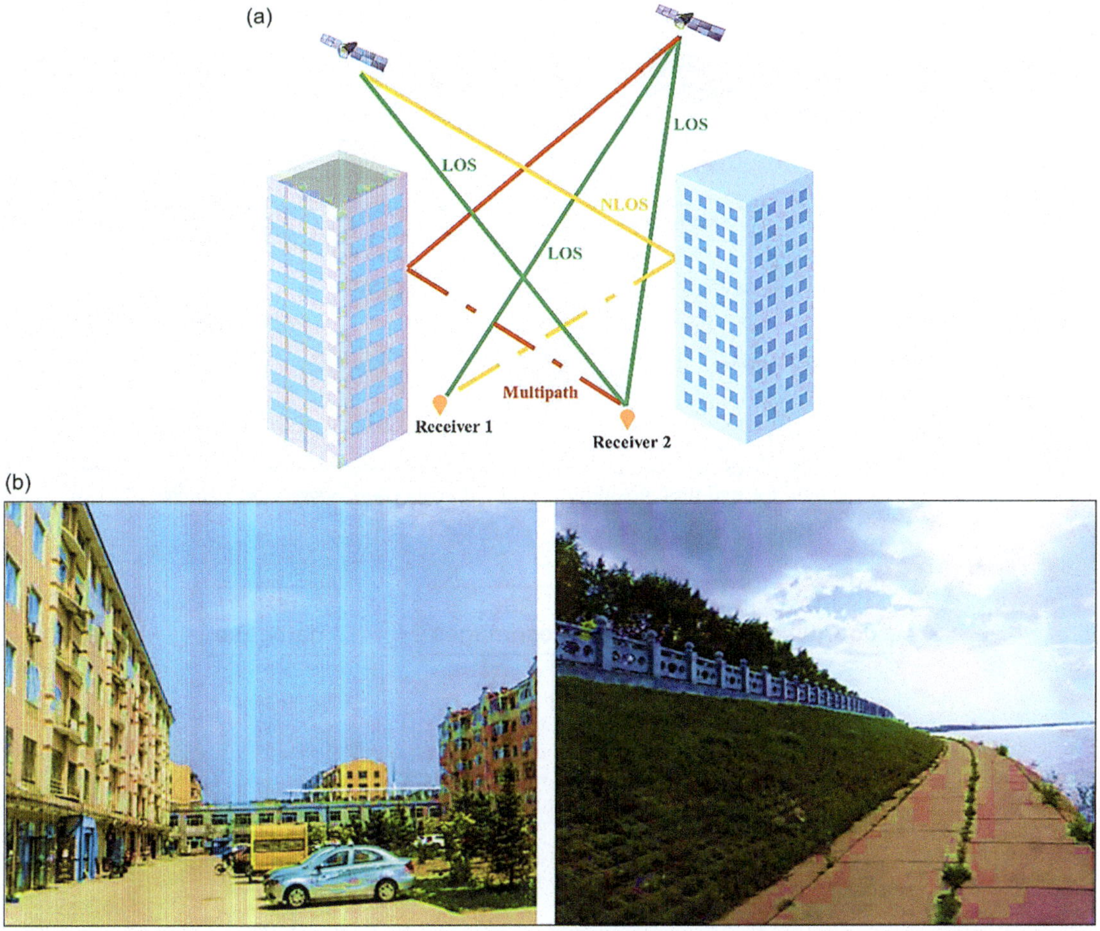

FIG. 7.6 (A) Global Positioning System (GPS) signal interference and multipath effects in urban environments. (B) GPS surveying: Data collection for relative positioning test in (a) Dense urban environment and (b) Riverside environment. *Source: Yuan et al., 2020.*

7.4.2 GPS applications in geospatial surveying

The combination of GPS with other geospatial tools, such as GIS and RS, enables the development of sophisticated geospatial models. These integrated models are particularly valuable in the context of resource exploration. For instance, combining GPS data with remote sensing imagery allows for the identification of areas with high resource potential, providing valuable insights for targeted exploration. Through spatial analysis and data integration, these models help optimize resource extraction strategies and minimize environmental impact by providing an accurate representation of the terrain and natural features (Fig. 7.8).

In environmental monitoring, GPS surveying supports the precise tracking of natural features such as water bodies, vegetation, and landforms. Accurate mapping of environmental conditions allows for better planning and management of ecosystems, as well as the

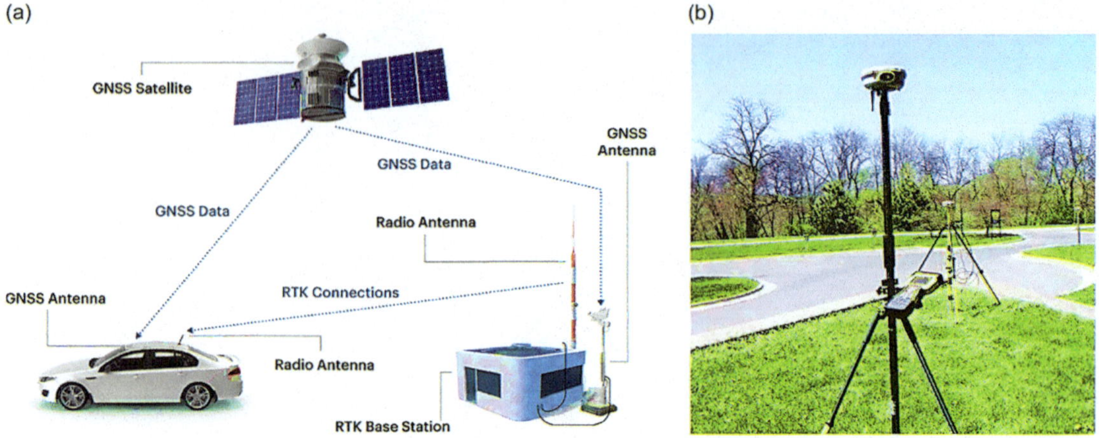

FIG. 7.7 (A) Real-Time Kinematic (RTK) base station setup for high-precision surveying and positioning. (B) RTK surveyor rover setup in the field for real-time positioning and data collection. *Source: Cavli Hubble.*

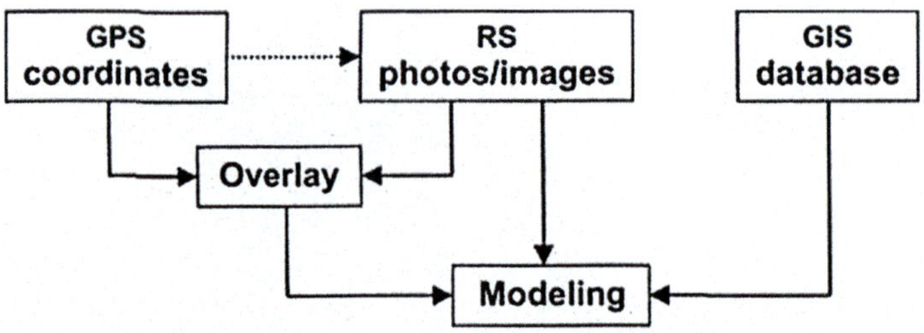

FIG. 7.8 The hierarchical model of integration for GPS and remote sensing in resource exploration. *Source: J. Gao, American Society for Photogrammetry and Remote Sensing, 2002.*

assessment of potential impacts from development projects, such as mining or construction. By utilizing GPS in combination with remote sensing data, environmental changes can be monitored over time, allowing for more informed decision-making and supporting sustainability efforts.

Another key application of GPS surveying is in land-use planning, where it assists in the accurate delineation of boundaries and the management of land resources. Whether for urban planning or agricultural zoning, the high precision of GPS surveying allows planners to create accurate maps, assess land features, and determine the best use of the land in a sustainable manner.

7.4.3 The future of GPS surveying: Innovations and advancements

As the field of GPS surveying continues to evolve, innovations in satellite technology, signal processing, and data integration are expected to drive further improvements in accuracy and efficiency. These advancements will enable GPS to provide even more precise measurements, particularly in challenging environments, and will enhance the integration of GPS data with other technologies for comprehensive geospatial analysis.

The integration of GPS with other data sources, such as GIS and RS, will continue to expand the capabilities of geospatial surveying. These technologies work together to create detailed and accurate models of the landscape, allowing for more effective decision-making in resource management, environmental protection, and land development. As the technology advances, GPS will remain a critical tool in sustainable natural resource management, helping to ensure that resources are extracted and managed in a way that minimizes environmental impact and supports long-term sustainability.

In conclusion, GPS surveying is an essential tool in modern geoscience and resource exploration. The ability to collect precise location data is vital for mapping, monitoring, and managing natural resources. Despite challenges posed by environmental obstacles and signal interference, the application of advanced correctional techniques, such as DGPS and RTK, ensures that GPS remains an indispensable tool in geospatial surveying. As technology continues to improve, GPS surveying will play an increasingly important role in the sustainable exploration and management of natural resources, providing valuable insights for decision-making and supporting efforts toward environmental sustainability.

7.5 GIS, GPS, and Remote Sensing (RS): Roles, integration, and applications

7.5.1 Role of RS and GPS in overcoming limitations

Remote Sensing (RS) and GPS are key technologies that address these limitations. Aerial photographs and satellite imagery provide original, up-to-date data that is more accurate and spatially comprehensive than traditional maps. GPS offers an efficient method for real-time field data collection, which is especially useful in dynamic environments where rapid changes must be captured.

The integration of RS, GPS, and GIS technologies has made it possible to overcome the data limitations inherent in traditional mapping methods. The combined use of these technologies enables more precise, timely, and reliable data collection, providing a more accurate understanding of the spatial characteristics of the environment (Fig. 7.9).

7.5.2 Integration of GIS, GPS, and RS technologies: Convergence and evolution

While the development of RS, GPS, and GIS technologies occurred somewhat independently over time, they have gradually converged into a unified framework. Today, these technologies are deeply intertwined, with data collected through GPS, satellite imagery, and aerial photogrammetry being processed and visualized within GIS platforms. In fact, most of the foundational data used in GIS applications today is obtained through these methods, which include cultural features, roads, buildings, water features, vegetation, topography, soils, geology, and more. Gao (2002) and Merchant & Narumalani (2009) have noted how the

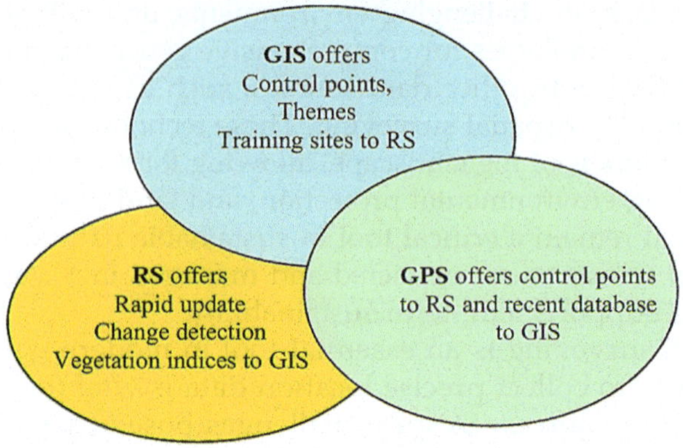

FIG. 7.9 Integrated framework of GIS, GPS, and remote sensing for geospatial data analysis. *Source: Baral, 2004.*

integration of GIS, RS, and GPS has revolutionized spatial data analysis, enabling a more comprehensive understanding of complex geographic and environmental processes.

7.5.3 Applications across various sectors

The integration of GIS, RS, and GPS technologies has proven invaluable across a variety of sectors, particularly those that require real-time, georeferenced data. For example, resource management, environmental monitoring, emergency response, mobile mapping, and logistics all benefit greatly from the integrated use of these technologies. As Najibullah Hasan Zurmotai (2020) suggests, these systems support real-time decision-making by providing accurate spatial data for informed responses in sectors such as research, conservation, and education.

Case studies that exemplify these applications can be seen in areas like mineral exploration, where GIS, GPS, and RS work together to optimize exploration strategies by providing detailed, accurate data about the terrain, mineral deposits, and environmental considerations (Fig. 7.10). Similarly, agricultural monitoring benefits from this integration, where farmers use these technologies for precision farming, tracking soil health, crop conditions, and water use in real time (Table 7.1).

7.5.4 Proposed models of integration

The integration of GIS, RS, and GPS can be categorized into four proposed models: linear, interactive, hierarchical, and complex. These models reflect the increasing sophistication and utility of integrated systems. The linear model represents basic data collection with a minimal level of interaction, whereas the interactive model allows for real-time data exchange between the systems. The hierarchical model involves a more structured approach where one system (e.g., GIS) acts as the backbone for processing and managing data collected by others (e.g., GPS, RS). The complex model reflects advanced integration, where all systems work seamlessly together in a comprehensive spatial analysis framework, suitable for complex decision-making in fields like environmental management, urban planning, and disaster response.

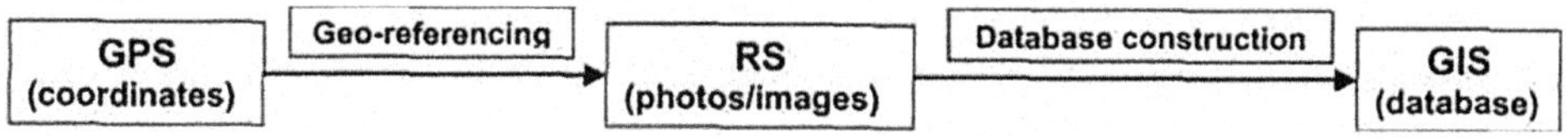

FIG. 7.10 The linear model of integration of GIS, GPS, and remote sensing in mineral exploration. *Source: Gao, 2002.*

TABLE 7.1 Agricultural monitoring using Geographic Information System (GIS), Global Positioning System (GPS), and Remote Sensing (RS).

Step	Action	Technology involved
1. Data collection	Collect data using RS for large-scale coverage (satellite/Unmanned Aerial Vehicle [UAV] imagery), GPS for location tagging, and ground truth data (soil, crop, weather).	RS, GPS
2. Data integration	Integrate RS data (imagery) into GIS to analyze vegetation, soil conditions, and land use. Use GPS to map locations.	GIS, RS, GPS
3. Data analysis	Perform spatial analysis in GIS, using RS for vegetation index (e.g., Normalized Difference Vegetation Index [NDVI]), water stress, and GPS for field monitoring.	GIS, RS, GPS
4. Decision-making	Make informed decisions on fertilization, irrigation, pest control, crop rotation, and harvest, based on data insights.	GIS, RS, GPS
5. Monitoring and feedback	Continuously monitor crops with RS and update GIS systems with new data for ongoing agricultural management.	RS, GIS
6. Optimized management	Achieve optimized agricultural practices by using GIS, GPS, and RS data to allocate resources, increase yields, and improve crop management.	GIS, GPS, RS

7.5.5 Challenges and the digital divide

While the integration of GIS, RS, and GPS provides new tools for professionals across various sectors, it also raises challenges, particularly related to the digital divide. Despite the rapid advancements in technology, there remains a gap in the understanding and accessibility of these systems. Some professionals or regions lack the resources or training to fully utilize the potential of these technologies. Ensuring that GIS, RS, and GPS technologies are accessible to a broader audience and providing adequate education and training will be essential to maximizing their potential in solving complex geographic problems.

7.6 Techniques for GPS surveying in natural resource exploration

GPS surveying plays a crucial role in natural resource exploration, offering precise location measurements that are essential for tasks such as mineral exploration, water resource management, and land resource mapping. Several techniques are used to achieve high accuracy in these applications, each with its own strengths and limitations, making them suitable for different types of surveys and terrain conditions. The primary techniques include static surveying, kinematic surveying, and RTK surveying, each contributing to precise data collection for various resource exploration needs.

7.6.1 Static surveying

One of the most common methods used in GPS surveying is static surveying, where the GPS receiver is stationary at a fixed point for an extended period. During this time, the receiver collects signals from multiple satellites, and the data is averaged to produce highly accurate location measurements. Static surveys are typically used for geodetic surveys and boundary mapping, where high precision is essential. The extended data collection period allows for signal averaging, which minimizes errors caused by atmospheric conditions, multipath interference, and other environmental factors, making this method ideal for projects requiring high accuracy (Fig. 8.1).

7.6.2 Kinematic surveying

In contrast to static surveying, kinematic surveying involves a moving GPS receiver. This technique is typically used when continuous data collection is needed, such as in topographic mapping or land surveying over large areas. The GPS receiver is placed on a moving platform—such as a vehicle, drone, or boat—allowing for real-time position tracking as the platform moves across the terrain. While kinematic surveying provides less precision than static surveying, it offers greater flexibility and efficiency, especially for dynamic or large-scale surveying projects where real-time data is necessary (Fig. 8.2).

Kinematic surveying is particularly useful in scenarios such as highway surveys, construction projects, and large-scale land use studies where it is impractical to stop and set up a stationary receiver at every point of interest. However, the reduced accuracy due to the receiver's movement is often compensated for by the ability to cover larger areas in less time.

7.6.3 Real-Time Kinematic (RTK) surveying

The RTK technique offers even higher precision than traditional GPS methods. RTK surveying uses carrier-phase measurements in addition to standard GPS signals, allowing for centimeter-level accuracy in real time. This technique involves a base station with a known location that continuously broadcasts correction data to the rover receiver. The rover, placed at the point of interest, uses this correction data to adjust its position in real time, ensuring highly accurate location measurements throughout the surveying process.

RTK is particularly beneficial in applications requiring real-time data, such as in construction site management, mining operations, and resource exploration, where precise positioning is crucial. The real-time data corrections allow RTK to significantly reduce errors caused by factors like atmospheric conditions, multipath interference, and satellite geometry, making it one of the most accurate and widely used methods in modern GPS surveying (Fig. 8.3).

In addition to its accuracy, RTK surveying enhances operational efficiency by reducing the time and effort required to obtain high-precision data. This is especially useful in industries like mining and land development, where accurate positioning is critical for decision-making processes related to resource extraction and land-use planning.

7.6.4 Integration of GPS with GIS and RS

The techniques discussed above are often integrated with other technologies, such as GIS and RS, to enhance the overall effectiveness of data collection in natural resource exploration. For example, satellite imagery and aerial photography (obtained through RS) can provide detailed spatial context for the GPS data, enabling a more comprehensive analysis of the surveyed area. When combined with GIS, which can process and analyze the spatial data from GPS and RS, these integrated systems allow for the creation of detailed resource maps, 3D models, and predictive models for natural resource exploration.

This integration improves the accuracy of data collection and provides additional layers of information that can aid in decision-making. For instance, in mineral exploration, GIS can be used to visualize the spatial relationships between geological features, mineral deposits, and other factors influencing resource extraction. Similarly, in water resource management, the combination of GPS data and RS can be used to monitor and model water bodies, facilitating better management and conservation strategies.

7.6.5 Advantages and limitations

Each of these techniques—static, kinematic, and RTK—has distinct advantages and limitations. Static surveys are ideal for highly accurate and long-duration measurements but are limited by the need for the receiver to remain stationary for extended periods, making them unsuitable for dynamic or large-scale surveying tasks. Kinematic surveys offer greater flexibility and efficiency for large or dynamic surveys, but at the cost of some accuracy. RTK surveying, while providing the highest accuracy, requires a base station and can be affected by certain limitations, such as signal obstructions or the need for clear line-of-sight to satellites.

The choice of technique depends on various factors, including the required accuracy, the terrain being surveyed, the duration of the survey, and the resources available. By selecting the appropriate method, surveyors can ensure that the data collected meets the needs of the project and supports effective decision-making for natural resource exploration.

7.7 Utilizing GIS for model-driven surveying and exploration

This section introduces the classical approach to GIS-based applications in Mineral Prospectivity Mapping (MPM). The first step in this approach is to define a conceptual model for the targeted mineral deposit. This model can be based on data-driven models, knowledge-driven models, hybrid models, or empirical functions, such as logistic-based modeling approaches. The conceptual model dictates the selection of spatial geoscience datasets that will be used in mineral exploration targeting through MPM (Carranza, 2008, 2017).

Once the conceptual model is defined, GIS-based analysis of spatial geoscience datasets results in the creation of predictor maps, which depict the spatial relationships between exploration data and the targeted mineral deposit. These maps are integral for guiding the exploration process. Fig. 7.11 illustrates the hierarchical relationship between data, information, knowledge, and insight in the context of GIS-based MPM. It shows how raw geospatial

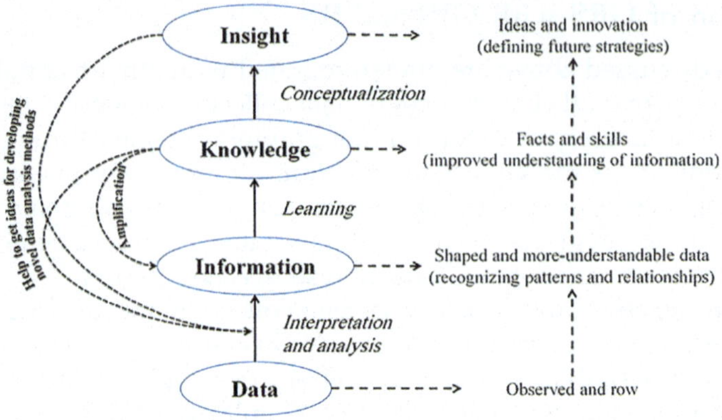

FIG. 7.11 Conceptual framework for GIS-based MPM. *Source: Yousefi et al., 2021.*

and geoscience data are processed and analyzed through GIS techniques to produce actionable insights for mineral exploration targeting.

The development of conceptual models must account for local geological contexts, as the characteristics of geological settings vary widely. Based on the defined conceptual model and the geological characteristics of an area, the next essential step is to generate predictor maps and proxies derived from relevant data sets, such as geochemical, geophysical, and geological data (McCuaig et al., 2010; Yousefi and Carranza, 2015b, 2016).

Data-driven approaches to MPM are particularly well-suited for brownfield areas, which are well-explored and data-rich. In these areas, multiple known mineral occurrences (such as deposits, prospects, and showings) can be used to quantify spatial and genetic relationships between exploration indicators and mineral occurrences (Bonham-Carter et al., 1989; Bonham-Carter, 1994; Singer and Kouda, 1999; Pan and Harris, 2000; Mihalasky and Bonham-Carter, 2001; Carranza and Hale, 2002; Carranza and Hale, 2001a; Carranza, 2008, 2009a; Porwal et al., 2003a, 2003c, 2004, 2006b; Nykanen, 2008; Harris et al., 2003; Chung and Fabbri, 1993; Carranza et al., 2005; Zuo and Carranza, 2011; Abedi et al., 2012; Harris and Grunsky, 2015; Carranza and Laborte, 2016; Zuo and Wang, 2020). However, data-driven methods can be biased by the locations of known mineral occurrences used as training sites, leading to exploration biases and uncertainties (Porwal et al., 2015; Coolbaugh et al., 2007). These biases are inherent to data-driven targeting models and present challenges, especially when the known mineral occurrences may not fully represent the range of mineralization (Yousefi and Nykanen, 2016).

On the other hand, knowledge-driven methods of MPM are often applied in poorly explored, greenfield environments, where both exploration data and mineral occurrences are scarce (Bonham-Carter, 1994; Carranza and Hale, 2001b; Porwal et al., 2003c; Carranza, 2008; Gonzalez-Alvarez et al., 2010; Markwitz et al., 2010; Lusty et al., 2012; Abedi et al., 2013; Lisitsin et al., 2013; Elyasi et al., 2019). These methods can be influenced by the subjective judgment of analysts, which may lead to potential biases (Yousefi et al., 2012, 2014; Yousefi and Carranza, 2015a, 2015b, 2015c). Joly et al. (2012) proposed that knowledge-driven methods should be carried out first, prior to data-driven methods, to mitigate this type of bias. A significant challenge in knowledge-driven MPM is the difficulty in assigning robust weights to the feature classes that indicate the presence of the mineral deposit

(Knox-Robinson, 2000; Porwal et al., 2003c, 2006b; Nykanen et al., 2008a, 2008b; Lusty et al., 2012; McKay and Harris, 2016).

Hybrid methods, which combine both data-driven and knowledge-driven approaches, represent a fusion of data- and knowledge-driven strategies for mineral exploration targeting (Porwal et al., 2004, 2006b; Chung and Fabbri, 1993; Roy et al., 2006). These methods incorporate both known deposit sites and expert opinions for the assignment of weights. Researchers have proposed empirical functions combined with expert judgment to improve the assignment of weights to deposit evidence features in data and knowledge-driven methods of mineral exploration targeting (Chung and Fabbri, 1993; Luo and Dimitrakopoulos, 2003; Carranza, 2009b). These hybrid methods use expert knowledge to assign weights indirectly.

A noteworthy technique that has been applied to both greenfield and brownfield settings is the use of logistic sigmoid (S-shaped) functions for weight allocation in mineral exploration targeting. This approach does not require known deposit sites, categorization of continuous spatial exploration data, or expert judgment, making it more flexible and less prone to biases. Logistic functions are particularly useful in areas with limited or no known mineral occurrences. Studies have shown that logistic sigmoid functions can effectively assign continuous weights to exploration spatial data, yielding better results than traditional categorized methods (Cox and Snell, 1989; Tsoukalas and Uhrig, 1997). Researchers, including Yousefi and Carranza (2015a, 2015b, 2016, 2017), Yousefi and Nykanen (2016, 2017), and Almasi et al. (2017), have demonstrated that continuous weighting methods outperform categorized approaches in terms of exploration accuracy.

In conclusion, GIS modeling techniques and spatial analysis play a critical role in improving surveying outcomes in mineral exploration. By employing model-driven approaches, exploration efforts can be optimized, minimizing biases and uncertainties while enhancing the accuracy of targeting and resource discovery in both brownfield and greenfield environments.

7.8 Leveraging GIS for managing geophysical surveying applications

7.8.1 GIS in managing and analyzing data collected during geophysical surveys

The evolution of information technology has profoundly transformed the role of Geographic Information Systems (GIS) in spatial data management, analysis, processing, and presentation. Initially rooted in conventional cartography and image processing, GIS has progressed significantly, embracing advanced tools like 3D visualization and dynamic graphic features. This shift has enabled GIS to find applications in a wide array of fields. Geophysical exploration projects, which often involve data acquisition from hundreds of spatial locations, generate extensive datasets that require substantial time and effort for management during data processing and interpretation. GIS plays a pivotal role by offering a graphical interface for georeferenced data, thus enhancing the management and analysis of these large datasets.

GIS is generally divided into two major types: raster-based GIS and vector-based GIS. Raster-based GIS relies on image processing techniques for handling satellite imagery and remote sensing data. In contrast, vector-based GIS utilizes relational database management systems (RDBMS) to store data, including geographic coordinates and attributes, which are then represented as vector graphics in the GIS interface. Modern systems tend to merge both methods, enabling the overlay of satellite imagery with multiple layers of vector graphic data (Khan et al., 2011).

The application of GIS extends across various domains, including geological mapping (Dey and Ghosh, 2008; Colman-Sadd et al., 1997), environmental monitoring (Viger, 2008), waste management (Muttiah et al., 1996), meteorology (Massons et al., 1996), defense, urban growth analysis (Acevedo and Masuoka, 1997), traffic flow analysis (Jiang and Liu, 2009), health, planning, ecology, and animal habitat studies (McNoleg, 1996), agriculture, forestry (Wulder, 1998), soil erosion management (Molnar and Julien, 1998), earthquake analysis (Louvari and Kiratzi, 1997), hazard management (Audisio et al., 2009), tunnel analysis (Choi et al., 2009), digital elevation models (Skidmore, 1989; Kienzle, 2004), soil analysis (Farewell and Farewell, 2010), mineral resource exploration (Zhou et al., 2007), geothermal resource exploration (Noorollahi et al., 2007), glaciology (Knoll et al., 2009), hydrology, and groundwater exploitation (Abdella and Alfredsen, 2010; Merkel et al., 2008; Portoghese et al., 2005). GIS has also demonstrated substantial utility in new areas, particularly in the development of GIS-based project management software that is utilized in geophysical applications to oversee project datasets.

7.8.2 Integration of geophysical data with GIS for improved understanding of geological structures

To ensure comprehensive GIS solutions, it is essential to integrate a GIS graphical front-end, an RDBMS, and associated processing applications (PA) (Khan, 2000). Two main approaches to implementing PAs with GIS are commonly recognized: loose-coupled and tight-coupled approaches (Karimi and Houston, 1996). The selection of the appropriate approach depends on factors such as the complexity of geophysical PAs, data volume, memory requirements, and the level of interaction needed between the GIS and the PA. The tightly coupled approach is favored for implementing a GIS-based project manager. This method presents two integration options: extended toolboxes and embedded GIS. Given that geophysical problems are often complex and computationally intensive, it is not feasible to implement them within GIS as toolboxes. Consequently, the embedded GIS concept is adopted to develop a geophysical application component, thereby integrating GIS functionality within processing applications.

The GIS-based project manager functionality is implemented as an ActiveX (Palmer, 1997) software component, which can be utilized by geophysical processing applications. This component is object-oriented, reusable, and features a set of properties, methods, and events that allow customization and control over data access. It includes a prebuilt GIS and a project explorer, both of which can be tailored by adjusting component properties to meet the specific needs of a given geophysical application. The component generates and maintains the project database in Microsoft Access format and is compatible with any geophysical software developed in a programming language that supports the Component Object Model (COM) (Rogerson, 1996).

The GIS project manager is made up of three core subcomponents: the project database, GIS as the core technology, and the project explorer. The project database is an RDBMS where, for any new project, a database is created and updated using component methods to assign processing parameters and store the filenames of all project datasets along with their respective data types. The GIS component acts as the project's front-end, displaying a project map that shows the geographic location of all datasets and their processing status. Datasets are represented graphically in the GIS interface as various symbols (e.g., boxes, rectangles, polygons, circles) with different colors indicating their current processing status. The GIS also provides direct access to the associated datasets.

7.8.3 Efficient data management and interaction in large geophysical projects

The project explorer is another front-end component that organizes project datasets into a data tree, categorized by their respective data types. Like the GIS interface, the project explorer allows direct access to each dataset listed in the tree (Project Explorer). A functional diagram of the GIS project manager component is shown in Fig. 7.12.

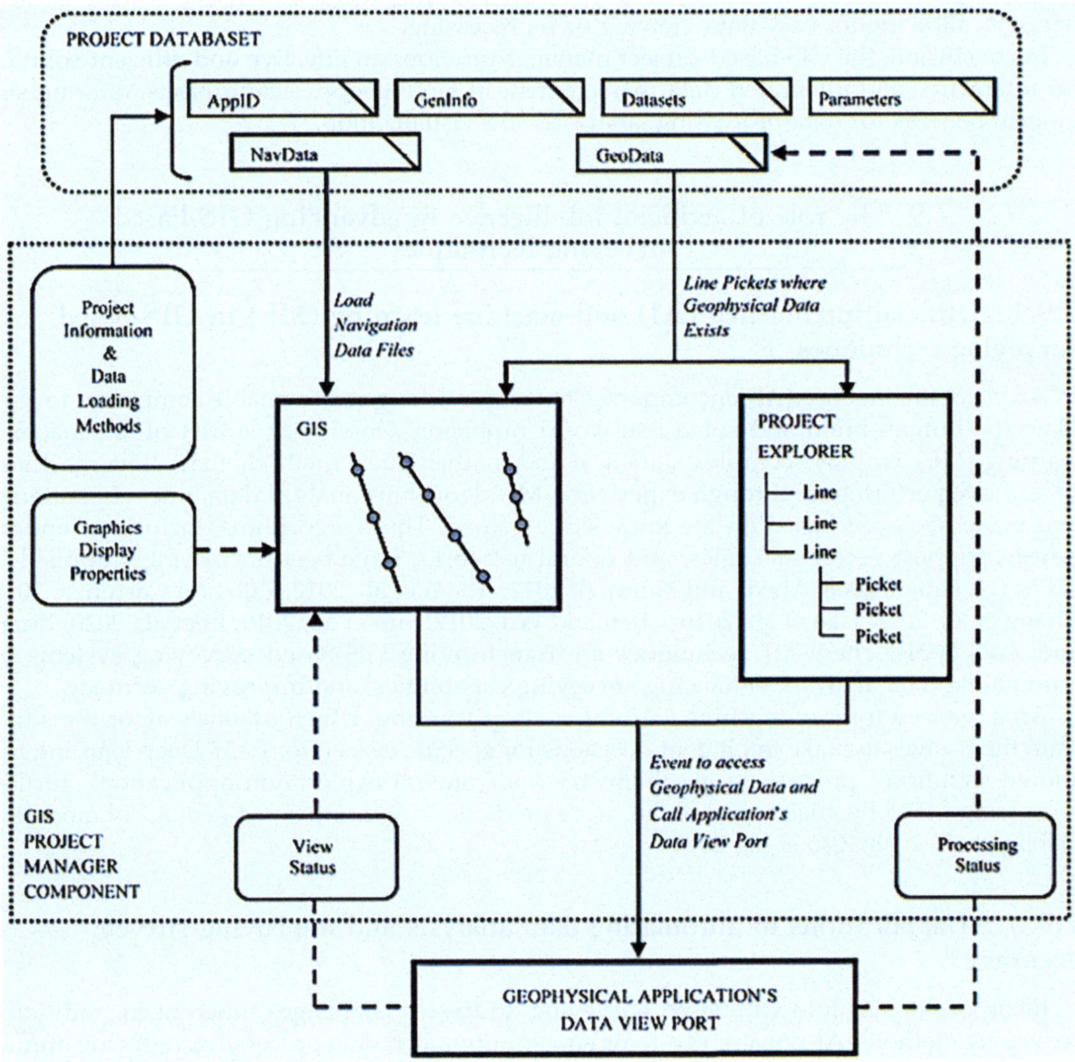

FIG. 7.12 The complete functional diagram of the GIS project manager component, along with the project database and application's data view port. *Source: Khan et al., 2011.*

The GIS project manager is an interactive and efficient tool for accessing, viewing, and processing geophysical datasets in large projects. This component has been effectively implemented in seismic refraction data processing software, which operates in three stages: picking arrival times, computing refractor models, and calculating statics. When a new project is initiated, all datasets and processing parameters are defined in the database. Either the GIS interface or the project explorer can be used to open and process datasets according to the predefined job sequence. This approach enables the project manager to serve as an efficient tool for data management in seismic exploration projects. One of the key advantages of the GIS-based project manager is that once datasets are defined in the database, there is no need to reload them each time. Loading the project database grants immediate access to all defined datasets, allowing for easy data viewing or reprocessing.

In conclusion, the GIS-based project manager provides an effective and efficient solution for interactive and integrated data management in large geophysical projects, offering significant benefits for data processing, analysis, and visualization.

7.9 The role of artificial intelligence in advancing GIS-based surveying techniques

7.9.1 Artificial intelligence (AI) and machine learning (ML) in GIS-based surveying techniques

Artificial intelligence (AI) encompasses techniques designed to enable computers to simulate the human brain and solve real-world problems. One major subset of AI, machine learning (ML), employs complex statistical and mathematical methods that allow machines to improve performance through experience. ML algorithms analyze data, uncover patterns, and make decisions based on the knowledge gained. These algorithms, including random forests, support vector machines, and neural networks, have been increasingly applied in MPM (Nykanen, 2008; Abedi and Norouzi, 2012; Abedi et al., 2012; Zuo and Carranza, 2011; Zhang et al., 2016; Gao et al., 2016; Chen and Wu, 2017; Sun et al., 2019; Li et al., 2020; Xiong and Zuo, 2021). These ML techniques are transforming GIS-based surveying systems by automating data analysis, enhancing surveying capabilities, and improving accuracy.

An advanced form of machine learning is deep learning, which involves algorithms that train themselves to make intelligent decisions for specific tasks (Fig. 7.13). Deep learning has shown significant promise in geochemistry and mineral exploration applications, further advancing MPM by enabling more accurate predictions and improved geological modeling (Xiong et al., 2018; Zuo et al., 2019).

7.9.2 AI applications in automating data analysis and improving survey accuracy

Incorporating AI into GIS-based surveying addresses challenges inherent in traditional surveying methods. AI-powered systems enable automated data processing, reducing human error and eliminating biases introduced during manual interpretation. The integration of machine learning algorithms with GIS enhances surveying capabilities by efficiently

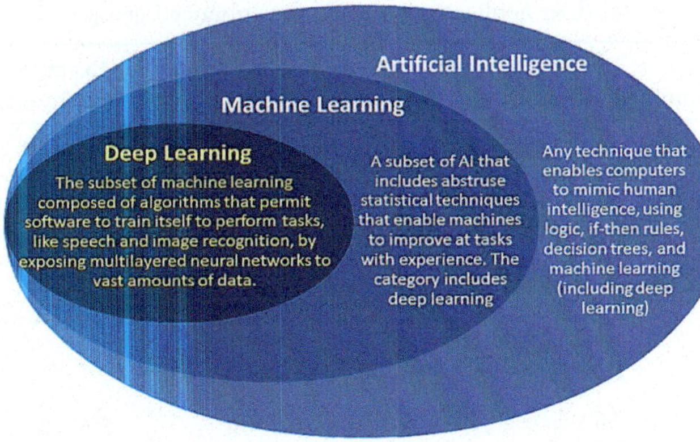

FIG. 7.13 Comparison of Artificial Intelligence (AI), Machine Learning (ML), and Deep Learning (DL) technologies in GIS-based surveying and mineral exploration. *Source: Geospatial world.*

assessing large datasets, identifying complex patterns, and improving the accuracy of geological models.

The use of AI to automate data analysis in GIS-based surveying can significantly reduce time and effort compared to traditional methods. This has the potential to optimize geospatial models, leading to more precise predictions of mineral deposits and a better understanding of geological processes. AI also plays a vital role in minimizing human bias, as the algorithms work independently of human interpretation, ensuring more objective results.

7.9.3 Challenges and future directions in AI for GIS-based surveying

While AI offers numerous benefits, its practical implementation in MPM and other GIS-based surveying applications faces several challenges. These include generating sufficient training samples, addressing imbalanced training data (Xiong and Zuo, 2017; Prado et al., 2020; Zuo and Wang, 2020; Rahimi et al., 2021), and integrating data science into AI methodologies (Zuo, 2020). Overcoming these challenges is essential for the successful application of AI in mineral exploration and geological modeling.

Further questions remain about the use of AI in MPM, such as:

- What achievements do we expect to gain by applying artificial intelligence to MPM?
- What problems have been solved through AI that could not be addressed by traditional methods?
- How can AI methods improve the understanding of complex geological processes related to ore formation?
- How can AI methods be applied to translate ore-forming processes into mappable exploration criteria?
- Can AI methods reduce user bias compared to traditional methods?

As AI continues to evolve, its integration with GIS will play an increasingly important role in improving the accuracy and reliability of geological surveys and mineral exploration (Tables 7.2 and Table 7.3). In conclusion, AI is playing a crucial role in advancing GIS-based

TABLE 7.2 AI Techniques and their applications in GIS for enhanced geospatial analysis.

AI Technique	GIS Application	Description	Examples/Tools Machine Learning (ML)
Machine Learning (ML)	Spatial data analysis	ML algorithms, like supervised and unsupervised learning, help classify data, detect patterns, and predict spatial outcomes.	Support Vector Machines (SVM), k-means clustering, regression models.
Deep Learning (DL)	Geospatial image analysis	DL models, such as Convolutional Neural Networks (CNN) and Recurrent Neural Networks (RNN), improve classification and detection in large-scale geospatial datasets.	CNN for object detection, RNN for time-series analysis.
Natural Language Processing (NLP)	Text and social media geotagging	NLP is used to extract insights from geotagged text data and improve geospatial information systems by analyzing unstructured data.	Geotagging social media, sentiment analysis, location-based data mining.
Geographic Object-Based Image Analysis (GEOBIA)	Remote sensing	GEOBIA integrates object identification, image processing, and machine learning to extract geographic data from remote sensing imagery.	Object-based classification using satellite images for land-use/cover mapping.
Predictive modeling	Disaster and risk assessment	AI-driven predictive models are used to anticipate disasters, disease outbreaks, and other geospatial events for better preparedness and response strategies.	Disease outbreak prediction, landslide risk models, flood prediction.
Image classification & object detection	Land use and infrastructure mapping	AI is used for detecting objects in satellite/aerial images and classifying land cover, infrastructure, or vegetation for urban planning and environmental monitoring.	Convolutional Neural Networks (CNN), Object detection for roads, buildings, vegetation.
Data fusion	Enhanced GIS database	AI combines data from multiple sources, such as satellite images, GPS, and sensor networks, to create richer, more accurate GIS datasets.	Integration of social media, GPS, IoT devices, and satellite data for comprehensive GIS analysis.
Real-time data analysis	Emergency response and urban planning	AI algorithms analyze real-time geospatial data for instant decision-making, such as during disasters or urban traffic management.	Real-time GIS analysis for disaster response, traffic management systems.
Geospatial data quality control	Ensuring accurate GIS data	AI techniques are used for data validation and error detection in geospatial datasets to improve overall quality and consistency.	Anomaly detection in GIS data, automated error correction.
Route optimization	Logistics and transportation	AI-powered GIS helps identify optimal routes for transportation by considering traffic, weather, and other variables.	Logistic planning, smart transportation networks.
Automated feature extraction	Reducing manual input	AI automates the extraction of features such as roads, buildings, and vegetation from satellite images, reducing manual effort.	Remote sensing analysis, feature extraction using AI models.

AI Technique	GIS Application	Description	Examples/Tools Machine Learning (ML)
Environmental monitoring	Climate and pollution	Monitoring	AI helps in continuous monitoring of environmental variables such as air quality, water quality, and deforestation.
Wildlife conservation	Biodiversity and habitat management	AI is applied to monitor wildlife movement and behavior patterns for effective conservation strategies.	Wildlife tracking, habitat monitoring using AI.

TABLE 7.3 Benefits of integrating AI with GIS for improved decision-making and efficiency.

Benefit	Description
Enhanced decision support	AI improves spatial analysis, providing decision makers with better tools for interpreting complex geospatial data.
Improved data quality	AI techniques help in cleaning and validating geospatial data, ensuring better accuracy and consistency.
Automation	Many routine GIS tasks, such as data processing and map generation, can be automated using AI, increasing efficiency.
Predictive insights	AI helps in creating predictive models for disasters, climate change, or urbanization trends.
Real-time analysis	AI enables the analysis of live data, allowing for timely interventions in fields like disaster response and urban planning.
Cost efficiency	AI optimizes resource allocation, reducing operational costs in areas like logistics, transportation, and urban development.

surveying techniques, particularly in MPM. Through machine learning and deep learning approaches, AI is helping automate data analysis, improve survey accuracy, and optimize geospatial models. As AI technology continues to evolve, it will offer new opportunities to interpret and model geological data more accurately, paving the way for more efficient and sustainable resource exploration.

7.10 Applications of GIS in mining and resource surveying

With global concerns regarding the exhaustion of high-quality ore deposits near the surface, the risks associated with mine development are increasing in terms of economic, technical, and environmental considerations (Carvalho, 2017). The decline in both mineral production and profits for mining companies has led to the adoption of large, heavy machinery and advanced mining technologies to extract lower-quality ore deposits from greater depths. Furthermore, as mine-induced disasters and hazards become more difficult to properly monitor and control, the negative impact on work safety and the surrounding environment has become evident. As a result, there is a growing need for systematic and strategic mine planning, operation, and environmental management based on optimization techniques or decision-making systems to enhance mineral productivity, operational efficiency, and stability in the mine environment (Choi et al., 2020).

To achieve these objectives, Geographic Information System (GIS)-based methodologies have been successfully employed to design and optimize the mine development process. GIS is a comprehensive framework used to capture, store, query, analyze, and display spatial data across various applications (Chang, 2010). In the field of geoscience, numerous applications have been proposed to manage and control natural phenomena using GIS (Lewis, 1997; Lukasheh et al., 2001; Wang et al., 2005; Gajos and Sierka, 2012; Hou et al., 2017; Wang and Xie, 2018). However, in the mining sector, although GIS has been widely applied for mine development, limited attention has been given to evaluating GIS-based methods and applications specifically in mine planning, operation, or environmental management. These applications can be organized into several key areas.

7.10.1 Case studies and applications of GIS in mining surveying

GIS-based techniques have been employed in various stages of mining surveying, including exploration, resource estimation, and environmental management. Case studies demonstrate how GIS technology has significantly improved efficiency in mineral exploration by enabling more accurate mapping of ore bodies, mineral prospectivity, and geological features. By integrating spatial data, GIS has enhanced the process of identifying and assessing mineral deposits, allowing for better resource estimation and risk management during exploration. Additionally, GIS applications are used to assess environmental impacts, particularly in terms of habitat mapping, water resource management, and tracking the effects of mining on surrounding ecosystems.

7.10.2 GIS in mining: A focus on mineral prospectivity and geological mapping

GIS technology plays a crucial role in mineral prospectivity and geological mapping, providing a systematic approach to the identification of potential mining sites. Through spatial analysis, GIS integrates data from various sources, such as geological surveys, RS, and geophysical measurements, to assess the likelihood of mineral deposits. This approach not only enhances the accuracy of geological maps but also helps in identifying areas with the highest potential for resource extraction. Additionally, GIS helps in the visualization of geological structures and ore bodies, which assists in improving mineral exploration and mining operations.

7.10.3 GIS-based methods and applications for mine planning

GIS is instrumental in developing comprehensive mine plans by analyzing topography, geotechnical properties, and ore distribution. Through spatial data integration, GIS facilitates efficient route planning for material handling, infrastructure development, and optimizing mine design. The system helps to determine the most feasible extraction methods, material flow, and logistical operations, improving the overall operational efficiency. GIS-based mine planning also supports real-time updates, allowing mine planners to quickly adjust plans based on new information or changing conditions in the mining environment.

7.10.4 Literature studies on various mine operation systems using GIS

Several studies have highlighted how GIS enhances operational efficiency in mining operations. Through real-time monitoring and control systems, GIS helps track equipment, personnel, and materials, which contributes to better management of mine activities. Additionally, GIS aids in logistics management by optimizing routes for the transport of ore and waste, reducing costs and time. Studies have also shown how GIS applications can be integrated with other technologies, such as GPS, to improve mine operations and ensure that resources are being utilized efficiently and safely.

7.10.5 Review of GIS-based methods for mine environmental management and reclamation planning

GIS plays a critical role in environmental management and land reclamation during and after mining operations. By creating detailed maps of habitats, water resources, and potential hazards, GIS supports efforts to minimize the environmental impact of mining activities. It allows for better management of waste disposal, contamination, and the monitoring of biodiversity in areas surrounding the mine. Furthermore, GIS assists in reclamation planning by tracking the changes in land use over time, ensuring that postmining reclamation efforts are well-coordinated and effective. The system also supports the restoration of disturbed land by simulating various reclamation strategies and their environmental impacts.

7.10.6 Advantages and disadvantages of using GIS in mining

GIS offers significant advantages in terms of improved data integration, spatial analysis, and decision-making. It enhances the efficiency of mine planning, operations, and environmental management by providing a comprehensive and visually intuitive platform for managing complex datasets. However, the implementation of GIS in mining comes with challenges. High setup and operational costs, as well as the need for specialized training, can be barriers to adoption. Additionally, GIS data accuracy and integration remain critical issues, particularly when combining data from various sources or managing large-scale mining operations with complex environmental considerations.

7.10.7 Suggestions for future directions for GIS in mining

The future of GIS in mining lies in the integration of emerging technologies such as artificial intelligence (AI), machine learning, and real-time data collection through the Internet of Things (IoT). These technologies could significantly enhance the predictive capabilities of GIS, allowing for automated mine operations and improved safety measures. GIS could also evolve through the integration of 3D modeling, which would provide more dynamic representations of mine sites and operations. Moreover, user-friendly interfaces and advancements in mobile GIS technologies could make the system more accessible to mine operators, enabling onsite decision-making and data collection in real time. Future advancements may also include incorporating virtual reality (VR) and augmented reality (AR) for mine visualization, offering more immersive and interactive methods for mine planning and monitoring.

7.11 GIS in prospecting and exploration studies

The integration of Geographic Information Systems (GIS) into the field of prospecting and exploration has revolutionized the way mineral deposits are discovered and assessed. GIS applications provide a powerful platform for storing, analyzing, and visualizing complex geological data, which allows for more efficient and accurate mineral exploration. The use of GIS in prospecting and exploration has enabled geologists and mining companies to better understand the spatial distribution of mineral resources, improving the decision-making process in resource identification and extraction.

7.11.1 GIS applications in mineral exploration and resource identification

Mineral exploration is a critical phase in the mining industry, where the primary objective is to identify and delineate new ore bodies or mineral resources. GIS has become an essential tool in this process due to its ability to integrate a wide variety of spatial and nonspatial data, including geological, geophysical, geochemical, and RS data. GIS aids in mineral exploration by providing:

Geospatial Data Integration: GIS allows geologists to integrate data from various sources, such as geological maps, satellite imagery, remote sensing, and field survey data. This helps in creating detailed geological models that represent the distribution of mineral deposits and potential resource targets.

Spatial Analysis: GIS tools enable geospatial analysis to identify mineralization patterns and relationships between geological features such as faults, rock types, and mineral occurrences. By using spatial statistics and modeling, GIS helps to define areas of high mineral prospectivity, thereby optimizing exploration efforts.

Remote Sensing: Remote sensing data, including satellite imagery and airborne geophysical surveys, can be analyzed within GIS to detect surface anomalies indicative of underlying mineral deposits. GIS helps in processing and interpreting remote sensing data, improving the accuracy and efficiency of exploration activities.

Geochemical Analysis: Geochemical data, including soil, water, and rock samples, can be analyzed using GIS to identify geochemical anomalies and potential mineralization zones. By mapping the distribution of geochemical elements, GIS helps to pinpoint areas where further exploration is warranted.

Risk Assessment and Resource Estimation: GIS plays a key role in resource estimation and risk assessment by providing spatially explicit data on the distribution of ore bodies. It helps in determining the potential economic value of a deposit by analyzing factors such as ore grade, volume, and accessibility. GIS also allows for the assessment of environmental and logistical risks associated with mineral exploration.

Visualization: One of the key benefits of GIS is its ability to create visual representations of complex data. 3D visualization of geological formations, ore bodies, and mineralization patterns enables exploration teams to better understand subsurface conditions and make informed decisions on where to drill or invest further resources.

7.11.2 Specific case studies of GIS-based surveying in global regions

GIS has been widely used in mineral exploration across various global regions. Several case studies demonstrate the effectiveness of GIS-based surveying techniques in identifying mineral resources and enhancing exploration efforts. Some of these case studies include:

Case Study 1: Exploration in Western Australia

In Western Australia, GIS-based mineral exploration has been successfully applied to discover gold and other precious metal deposits. The integration of geological, geophysical, and geochemical data within a GIS platform allowed geologists to identify mineralization patterns and refine drill targets. The use of GIS to integrate RS data, such as hyperspectral imagery, enabled the identification of alteration zones, which are indicative of gold mineralization. This GIS-driven exploration approach has led to more efficient resource identification and a significant reduction in exploration costs.

Case Study 2: Gold Exploration in Ghana

In Ghana, GIS has been used extensively in gold exploration, particularly in the Ashanti Belt, one of the most significant gold-producing regions in West Africa. By integrating geological mapping, geophysical data, and soil geochemistry into a GIS platform, exploration companies have been able to identify prospective gold deposits with greater precision. GIS was used to generate 3D models of ore bodies and assess the viability of different mining methods. This helped improve decision-making during the exploration phase and has contributed to Ghana's status as a top global gold producer.

Case Study 3: Copper Exploration in Chile

Chile, known for its vast copper reserves, has employed GIS in mineral exploration for decades. GIS has been utilized to map copper-bearing zones, integrate geological and geophysical data, and analyze mineralization patterns at the regional and local scales. The combination of geochemical data, remote sensing, and geological surveys within GIS platforms has enabled more accurate targeting of copper exploration sites, reducing the amount of trial and error in the field. GIS technology has also been used to optimize drilling programs, improve resource estimation, and manage exploration data more effectively.

Case Study 4: Diamond Exploration in Canada

In Canada's Northwest Territories, GIS has been applied in the exploration of diamond deposits. The application of GIS in this case included the integration of remote sensing data, geological mapping, and geophysical surveys to analyze the distribution of kimberlite pipes, which are the primary host for diamonds. GIS-based modeling has enabled exploration teams to refine the identification of kimberlite clusters and target areas for further drilling. The success of GIS in this exploration process has contributed to the discovery of several significant diamond deposits in the region.

Case Study 5: Iron Ore Exploration in Brazil

In Brazil, GIS has played an integral role in iron ore exploration, particularly in the Carajas Mineral Province, which is one of the world's largest iron ore mining regions. GIS-based methods have been used to map the geological formations and geophysical anomalies associated with iron ore deposits. By integrating topographic data, remote sensing imagery, and geochemical surveys, GIS has helped identify new iron ore resources and optimize exploration programs. GIS has also been used to create detailed resource models, aiding in the estimation of ore volumes and helping companies better manage their exploration projects.

7.11.3 Advantages and limitations of GIS in mineral exploration

GIS provides numerous advantages in mineral exploration, including:

Improved Decision-Making: GIS facilitates data-driven decision-making by integrating various types of spatial and nonspatial data, allowing exploration teams to prioritize high-prospect areas.

Increased Exploration Efficiency: GIS accelerates the exploration process by helping to identify areas with the highest potential, reducing the amount of time and resources spent on unproductive sites.

Better Risk Management: GIS helps assess environmental and operational risks by providing a spatial context for the location of resources, infrastructure, and sensitive environmental zones.

However, there are limitations to GIS in mineral exploration:

Data Accuracy: The accuracy of GIS analysis is highly dependent on the quality of the data being used. Inaccurate or outdated data can lead to flawed exploration models and decision-making.

High Costs: The implementation of GIS technology and the acquisition of high-quality spatial data can be expensive, particularly in remote or undeveloped areas.

Complexity: GIS systems can be complex and require specialized training and expertise, making it challenging for some exploration teams to fully utilize the technology.

7.11.4 Future directions for GIS in exploration

The future of GIS in mineral exploration looks promising, with emerging technologies such as machine learning, artificial intelligence (AI), and 3D modeling likely to enhance its capabilities. These technologies will allow for more advanced predictive modeling, automated data analysis, and real-time exploration monitoring. Additionally, the continued development of remote sensing technologies, such as Unmanned Aerial Vehicles (UAVs) and hyperspectral imaging, will further improve the accuracy and efficiency of GIS-based exploration techniques.

In conclusion, GIS has proven to be an invaluable tool in mineral exploration, providing exploration teams with powerful capabilities for data integration, spatial analysis, and resource identification. With ongoing technological advancements, the role of GIS in prospecting and exploration will only continue to expand, offering even more precise, cost-effective, and sustainable methods for identifying mineral resources globally.

7.12 Case studies on GIS-based surveying techniques

7.12.1 GIS in mineral exploration: A case study of northeastern Algeria

The integration and processing of diverse geoscientific data, including geological, depositional, geophysical, and geochemical datasets, within a Geographic Information System (GIS) played a pivotal role in accurately determining the spatial distribution and characteristics of Pb–Zn–Fe–Ba (Cu, Hg, F, Sr) mineralization in the Setifien/Hodna region and the eastern Saharan Atlas of northeastern Algeria. The use of GIS, particularly its spatial modules, provided significant insights into the spatial relationships between mineralization and tectonic lineaments or geological trends in the region.

One of the most important aspects of this case study was the integration of gravimetric and aeromagnetic data, which proved essential in identifying key tectonic lineaments. These lineaments, which represent significant geological features, were found to be strongly correlated with mineralization zones. The presence of mineral deposits, especially Pb, Zn, Cu, Fe, and other trace elements like Hg, Sr, and F, was consistently found along these tectono-lineaments and at the boundaries of sedimentary basin margins. This correlation provides strong evidence for the hypothesis that tectonic lineaments play a crucial role in the localization of mineralization (Saadia et al., 2019).

The spatial distribution of mineralization was further analyzed through the application of ordinary kriging interpolation to field-measured concentrations of Pb–Zn–Cu–Fe (%). The results of this analysis revealed distinct linear mineral zoning that closely aligned with the directional trends of the tectonic lineaments. These directional trends are thought to reflect deformation corridors, which may have significantly influenced the emplacement of the mineralized deposits. This suggests that these corridors could represent key targets for future exploration efforts, highlighting the potential of GIS to identify and refine prospective mineral exploration sites.

In addition to the spatial analysis provided by GIS, geostatistical methods, including multivariate analysis of geological, depositional, topographical, geochemical, and geophysical data, were employed to further refine and specify the distribution properties of mineralization in the Setifien/Hodna and eastern Saharan Atlas regions. These advanced GIS techniques allowed for a detailed understanding of the spatial relationships between various geological features, enabling more efficient and targeted exploration.

By utilizing GIS in combination with advanced geostatistical tools, this study not only improved the efficiency of mineral exploration but also enhanced the accuracy of identifying and delineating mineral deposits in northeastern Algeria. The use of GIS provided a comprehensive framework for analyzing complex geological and geochemical datasets, offering valuable insights into the spatial distribution of mineral resources and guiding future exploration efforts.

7.12.2 GIS in oil and gas exploration: A case study from Ethiopia

Geographical Information Systems (GIS) have been instrumental in the oil and gas exploration efforts conducted by PETRONAS (National Petroleum Limited, a Malaysian global energy group) in Ethiopia. The project involved the integration and analysis of a wide range of geological datasets, all of which contributed to optimizing the planning and decision-making processes for exploration.

The exploration team utilized GIS to develop integrated geological maps, incorporating a variety of data types, including topographical maps, Digital Elevation Models (DEM), lithological data, strike and dip orientations, fold structures, fault lines, satellite imagery, cultural data, and leads and prospects maps. By combining these diverse datasets, GIS allowed the team to identify the most promising areas for potential oil and gas deposits, significantly enhancing the precision of their exploration efforts.

One of the primary applications of GIS in this project was determining the optimal layout of seismic survey lines. Using ArcGIS as the primary mapping tool for seismic survey planning, the team was able to streamline the process, reducing both time and costs. This approach improved the efficiency of seismic surveys, which are essential for evaluating the subsurface geology and identifying potential hydrocarbon reservoirs (Latef and Kahar, 2008).

A major advantage of GIS in this context was its ability to compare and integrate various datasets, either individually or in combination. This capability enabled exploration geoscientists to extract the maximum amount of information from the study area, including detailed surface topography, fault structures, and dip-strike patterns. By analyzing these geological features, the team improved their understanding of the region's subsurface characteristics, ultimately helping to identify the most promising prospects for oil and gas exploration.

In summary, GIS played a critical role in enhancing the efficiency, accuracy, and effectiveness of the oil and gas exploration process in Ethiopia. By leveraging the power of spatial data integration and analysis, PETRONAS was able to optimize its exploration strategy and identify key areas for further exploration, contributing to the success of the project.

7.12.3 Geospatial data infrastructure: A case study with GSI and BRGM

The Geological Survey of India (GSI), in collaboration with the Bureau of Geological and Mining Research (Bureau de Recherche Géologique et Minières; BRGM) of France, conducted an important multivariate analysis study aimed at base metal exploration in the Tons Valley, located in the northwestern region of the State of Uttarakhand. This study focused on the integration of various geospatial datasets to assess the region's base metal potential, particularly for minerals like copper, lead, zinc, nickel, cobalt, and silver.

A variety of datasets were employed during this study, including topographic contours, geochemical data, geophysical data, land use/geology, and lineament information. Notably, geochemical data for copper, which exhibited an anomalous distribution pattern, were used as a key layer for further analysis. Additionally, geophysical data, such as chargeability (induced polarization), self-potential, and resistivity, were integrated into the study to identify areas of mineralization.

To ensure the targeted exploration was as effective as possible, weightages were assigned to each dataset through consultations with the mineral exploration team working on the project. This collaborative approach led to the preparation of map layers, which were overlaid in a GIS environment. These map layers were then draped onto the Digital Elevation Model (DEM) of the area to aid in visualization and further analysis.

The GIS analysis revealed two significant anomalies: the first coincided with an already known zone of mineralization, while the second anomaly, located at depth, was deemed a priority for drilling. This indicated the potential for discovering a new mineralized zone. This case study effectively demonstrated the power of multivariate analysis in mineral targeting using GIS technology. The integration of diverse geospatial datasets facilitated a more accurate and informed decision-making process, highlighting the indispensable role of GIS in modern geological surveys and base metal exploration.

7.12.4 GIS-based groundwater assessment: A case study of the Krishna River basin

Groundwater occurrence and movement within a flow system are governed by a variety of natural factors, including topography, geology, geomorphology, lineament structures, soil, drainage networks, and land use/land cover (LULC) (Das et al., 1997). Given the complexity of natural geological and hydrogeological systems, effective and systematic planning is essential for optimal groundwater exploitation. A key aspect of groundwater management involves the characterization of aquifer systems and the identification of groundwater potential zones in various geological settings.

This study applied an integrated GIS-based approach, using weighted index overlay analysis (WIOA), to assess groundwater potential zones within the Krishna River Basin in India. The GIS-based results were validated by comparing them with actual groundwater levels observed in the field (Harini et al., 2018). The study utilized spatial analysis tools in ArcGIS 10.1 to integrate various thematic layers, including geology, geomorphology, soil, slope, LULC, drainage density, lineament density, and annual rainfall distribution.

Data for the thematic layers were sourced from a range of reliable datasets, including Geological Survey of India maps, the European Digital Archive of Soil Maps, Bhuvan (ISRO's Indian Geo-Platform), and 30-meter resolution global land cover data. These layers were combined using WIOA to identify groundwater potential zones across the region. The final output was a map that classified the Krishna River Basin into five categories of groundwater potential: very good, good, moderate, poor, and very poor.

The GIS analysis results were found to closely align with the actual groundwater conditions observed during field investigations, confirming the effectiveness of the GIS-based assessment in mapping groundwater potential. The study underscores the capability of GIS and remote sensing techniques to accurately identify and classify groundwater potential zones on a regional scale. This case study highlights the significant advantages of using GIS and remote sensing for groundwater assessment and management. These indirect methods offer substantial cost and time savings, providing valuable tools for sustainable groundwater resource planning and exploitation.

7.12.5 GIS-based groundwater potential survey in the Deccan volcanic province

Kumar et al. (2020) conducted a study to identify groundwater potential zones (GWPZs) using Remote Sensing (RS), Geographic Information System (GIS), and the Analytic Hierarchy Process (AHP) in the Chandrabhaga River Basin, located in the Nagpur district of Maharashtra, India, within the Deccan Volcanic Province (DVP). The study integrated various thematic layers to assess and map the groundwater potential of the region.

The thematic layers included geology, geomorphology, land use/land cover (LULC), slope, drainage density, lineament density, and soil maps. These layers were prepared using a combination of conventional data sources, topographic maps, and satellite imagery, specifically Landsat-8 (Thematic Mapper (TM) and Operational Land Imager [OLI]) images acquired in November 2015. Additionally, the Shuttle Radar Topography Mission (SRTM) Digital Elevation Model (DEM) data with a 30 m resolution was used to generate slope and drainage density maps.

The AHP technique was employed to assign weightages to each thematic layer, based on their relative importance in determining groundwater potential. These layers were reclassified and integrated into the GIS environment using ArcGIS 10.4. The resulting integrated data produced a groundwater potential map for the study area, classifying it into five distinct groundwater potential zones: very poor: 11.77%, poor: 21.73%, moderate: 30.13%, good: 25.34%, and very good: 11.02%. These zones were validated by comparing the GIS-based results with actual field data from existing borewell yields, showing a strong correlation. A cross-correlation coefficient (R = 88%) was calculated, demonstrating a robust relationship between the mapped groundwater potential zones and the observed groundwater conditions in the field. The formula for the cross-correlation coefficient is:

$$R = \frac{\sum PD}{n\sigma_p\sigma_d} \tag{7.2}$$

where P represents the deviation from the mean of rainfall, p (mean rainfall estimation)—p′ (mean monthly rainfall), D represents the deviation from the mean of depth to groundwater level, d (mean depth to groundwater)—d′ (observed depth to groundwater), σp and σd are the standard deviations of the rainfall and depth series, respectively, and n is the number of data points. Since groundwater levels are influenced by rainfall recharge (Todd, 1980), the deviations in the depth to the groundwater table can be attributed to changes in rainfall patterns, which were effectively captured in this analysis.

The study demonstrated the successful identification and classification of groundwater potential zones within the Chandrabhaga River Basin using GIS, RS, and AHP methodologies.

The generated groundwater potential map provides a valuable tool for water resource management and sustainable groundwater exploitation in the region. By leveraging remote sensing data and spatial analysis techniques, the study presents a cost-effective, efficient, and accurate methodology for large-scale groundwater potential assessments.

7.12.6 GIS-based prospectivity of epithermal gold-silver deposits in the Deseado Massif, Argentina

Andrada de Palomera et al. (2014) observed that earlier prospectivity models for epithermal Au–Ag deposits in the Deseado Massif, southern Argentina, had produced regional-scale prospectivity maps. However, these maps offered limited guidance for exploration at smaller district scales due to their relatively low level of detail. As numerous districts within the Deseado Massif remain underexplored, the authors stressed the need for higher-resolution prospectivity maps that could provide more practical insights for exploration at the district level. The study aimed to develop enhanced, higher-resolution prospectivity models for low- and intermediate-sulphidation epithermal deposits (LISEDs) at both regional (1000 m spatial resolution) and district (100 m spatial resolution) scales.

To achieve this, the authors employed two key geological evidence datasets, which included geological maps, alteration data, and structural features, among others. The weights-of-evidence (WofE) method was used to assess the relationships between these geological features and the occurrence of LISEDs. In addition, the leave-one-out cross-validation (LOOCV) technique was applied to evaluate the reliability and accuracy of the models. The prospectivity models were generated using different data sources for the regional and district-scale models—Advanced Spaceborne Thermal Emission and Reflection Radiometer (ASTER) data for the district scale and Landsat Enhanced Thematic Mapper Plus (ETM) + data for the regional scale.

The results of the district-scale prospectivity model were notably more effective, reliable, and useful than those of the regional-scale model, primarily for two key reasons: first, the incorporation of higher-resolution evidential layers at the district scale resulted in more detailed and accurate geological maps; second, the use of ASTER data, as opposed to Landsat (ETM+) data, provided better resolution and more accurate detection of mineralization-related features, such as alteration zones and structural lineaments.

Consequently, the district-scale model offered a significantly more reliable foundation for identifying areas with high prospectivity for LISEDs within the Deseado Massif. The findings suggested that the integration of higher-resolution data layers and improved spatial accuracy enhanced the understanding of geological processes associated with the formation of epithermal deposits. The study also identified several directions for further refining district-scale prospectivity mapping:

More Accurate Determination of Mineralization Age: A better understanding of the temporal relationship between mineralization and lithological units could enhance the precision of prospectivity models.

Understanding the Relationship between LISEDs and Geological Evidence: Further exploration of the connection between hydrothermal clay alteration and the presence of LISEDs could improve model accuracy. This includes refining techniques for detecting clay minerals associated with epithermal deposits.

Incorporation of Additional Data Layers: Adding more geochemical and geophysical data would enrich the geological understanding and improve the predictive capabilities of the models.

Andrada de Palomera et al. (2014) demonstrated that district-scale prospectivity mapping for low- and intermediate-sulphidation epithermal deposits is far more effective for guiding exploration activities in the Deseado Massif compared to regional-scale models. The higher resolution of the evidential data layers, along with more accurate geological maps and remote sensing data, significantly enhanced the effectiveness and practical utility of the prospectivity models. This approach not only improves the targeting of exploration activities in the Deseado Massif but also offers a valuable methodology for other regions with similar geological characteristics.

7.12.7 GIS-based surveying of diamondiferous kimberlite potential in Namibia

Namibia, located on the southwestern coast of Africa, has a robust economy largely driven by the extraction of its extensive fluvial diamond placer deposits along the Orange River, as well as alluvial and eluvial placers along its coastline. In addition to these primary sources of diamonds, kimberlite pipes—which are potential sources of diamonds—have been discovered in neighboring countries like Botswana and South Africa, where they are actively being mined. As a result, the exploration for diamond-bearing kimberlite pipes has expanded into Namibia, with the hope of discovering similar deposits within its borders. The search for diamondiferous kimberlite pipes is guided by several criteria:

Identification of Archean Cratonic Areas: These regions, believed to possess stable, ancient continental crust, are often potential hosts for kimberlite pipes.

Zones of Crustal Weakness: These areas, often marked by intrusive bodies (such as dikes, sills, or kimberlite pipes) or geophysical data showing weaknesses in the Earth's crust, are potential locations where diamond-bearing kimberlites may form.

Geothermal Gradients: The rate at which temperature increases with depth is a key factor, as kimberlite pipes are typically associated with specific geothermal conditions.

Chemistry of Indicator Minerals: Minerals such as diamond inclusions and chrome-rich garnets, which are often found in kimberlite pipes, can serve as indirect indicators of diamond-bearing formations.

To support the search for diamonds in Namibia, these geological and geophysical features have been integrated into a GIS platform, utilizing ArcInfo software. GIS tools allow the integration of spatial data with geological features, significantly enhancing the ability to identify high-potential areas for further exploration. By compiling a variety of geological data layers, such as geophysical data, geological maps, thermal gradients, and chemistry of indicator minerals, the GIS-based prospecting tool provides a spatially detailed analysis of potential diamond-bearing areas.

This GIS-based approach has proven invaluable in systematically identifying regions in Namibia that are most likely to host kimberlite pipes, similar to those found in Botswana and South Africa. The integration of diverse geological and geophysical datasets within a GIS environment improves the accuracy of exploration efforts by facilitating better decision-making and targeting the most promising exploration zones.

Looking forward, the continued application and development of GIS technologies in diamond exploration in Namibia are expected to refine the precision of identifying high-potential areas for diamond discoveries. Additionally, the integration of advanced geophysical surveys and RS data may further enhance these models, providing new insights into the spatial distribution of kimberlite pipes.

7.12.8 GIS-based geological mapping and mineral prospectivity in Hamissana, northeast Sudan

In recent years, geological activities and various mineral prospecting and exploration programs have intensified along the Red Sea Hills in parts of Hamissana, located in northeast Sudan, to enhance geological mapping and evaluate mineral potential (Almakki Mohamed et al., 2021). This study explores the application of remote sensing and Geographic Information System (GIS) techniques for geological mapping and gold mineralization prospecting in the region. To differentiate lithological units and identify wall rock alteration zones—key indicators for gold mineralization—various digital image processing techniques were applied to Landsat-8 Operational Land Imager (OLI) images. The following steps were utilized:

Image Sharpening: This technique was employed to enhance the spatial resolution of the images, thereby providing more detailed information for geological interpretation.

Contrast Stretching: Applied to improve the image contrast, this process made the images more interpretable and enhanced the differentiation between geological features.

Principal Component Analysis (PCA): PCA was used to transform the image data, resulting in enhanced images that were more interpretable than the original dataset.

Ratio Imaging: Multiple ratio images were prepared, combined, and displayed as RGB color composite ratio images. This composite allowed for a more precise identification of alteration zones within the study area.

The enhanced satellite images clearly revealed alteration zones extending from the northeast to the southwest, primarily within areas composed of acid metavolcanics and silica-barite rocks. These alteration zones are considered potential targets for gold mineralization. Using the enhanced satellite images in the GIS environment, a geological map of the area at a scale of 1:400,000 was generated, facilitating the efficient and accurate delineation of geological features.

To validate the presence of gold mineralization, X-ray fluorescence (XRF) analyses were performed on selected samples taken from the wall rock alteration zones. The results confirmed the presence of gold-bearing minerals within the identified alteration zones. A flowchart diagram (Fig. 7.14) outlines the methodology used in this study for geological mapping and gold mineralization zone delineation. This flowchart summarizes the steps from remote sensing data acquisition and processing to the final geological map production and gold mineralization zone identification.

This case study demonstrates the effective use of remote sensing and GIS technologies in geological mapping and gold mineralization prospecting in the Red Sea Hills region of northeast Sudan. The integration of satellite imagery, image processing techniques, and geochemical analyses provides an efficient and accurate method for identifying potential gold-bearing zones. The results suggest that GIS-based methodologies can

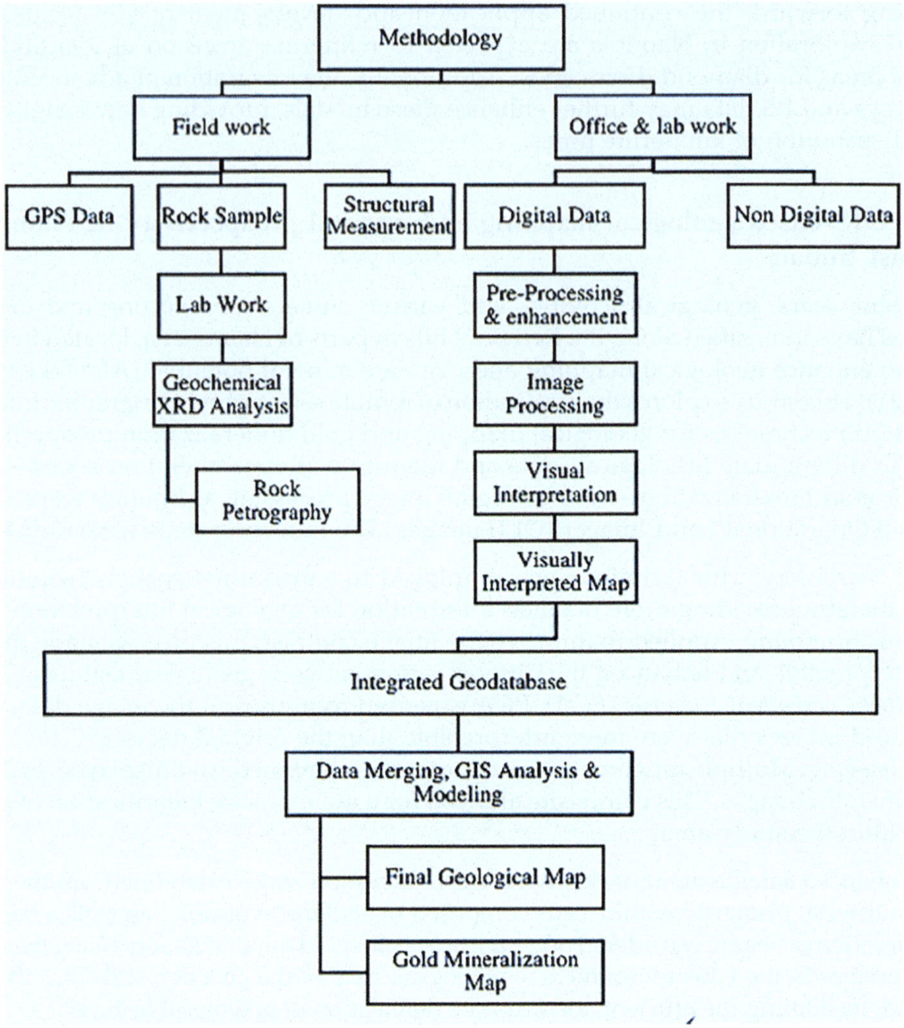

FIG. 7.14 The flowchart of the adopted methodology of geospatial data. *Source: Almakki Mohamed et al., 2021.*

significantly enhance mineral exploration efforts, improving decision-making and resource management.

7.12.9 Geospatial tools in the exploration of radioactive mineralization: A case study from Wadi Dahab, southeastern Sinai, Egypt

The exploration and mapping of mineral resources are critical to the sustainable development of industrialized nations, particularly those involved in power generation and urban development. Modern geospatial tools, such as remote sensing (RS) and GIS, have become essential in delineating lithological features, structural lineaments, hydrothermal alterations,

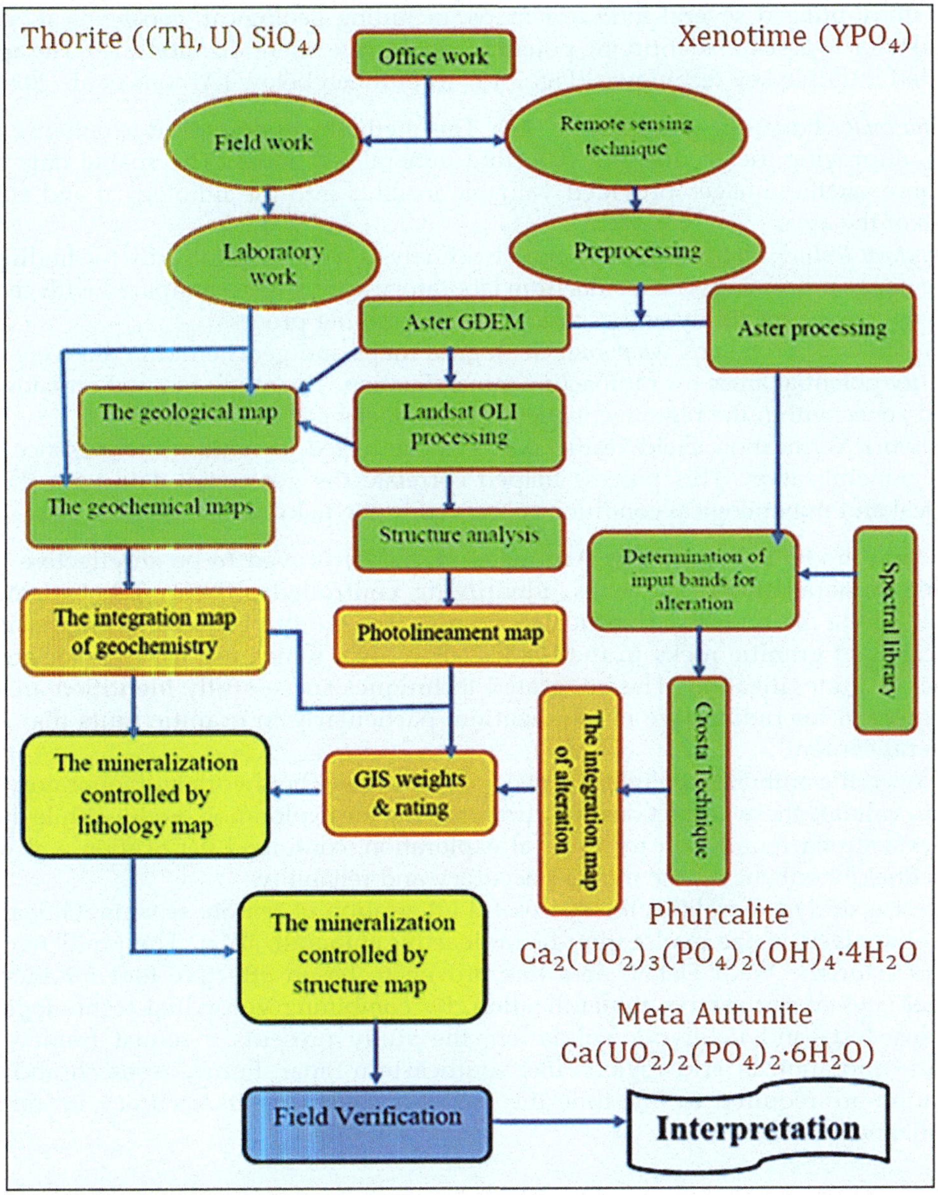

FIG. 7.15 A flow chart simplifying the data description and the followed methodology steps. *Source: Modified.*

radiometric anomalies, and geochemical element distributions in mineralized zones. Arnous et al. (2021) conducted a study on radioactive mineralization in the Wadi Dahab (WD) area of southeastern Sinai, Egypt, with the goal of creating a predictive model for exploring radioactive mineralization sites.

The study utilized several indicator maps, including geological, geochemical, and geophysical data, to assist in identifying potential radioactive mineralization sites. The approach integrated multiple key techniques (Fig. 7.15), as outlined below (Arnous et al., 2021):

Digital Image Processing of Spaceborne Data: This method was employed to enhance satellite images, improving the accuracy of detecting mineralized zones. The spatial data derived from these satellite images provided valuable insights into the lithological and structural features of the area.

Laboratory Testing: Fieldwork and laboratory analysis were used to verify the findings from satellite image processing. The results from laboratory testing were compared with geospatial data to ensure the reliability and accuracy of the mapping process.

Geological and Geochemical Mapping: Geological maps and geochemical data were utilized to identify potential areas for radioactive mineralization. The study focused on known mineralized zones within the region, enhancing the model's predictive power.

Fieldwork Verification: Field verification was conducted to confirm the presence of radioactive mineralization. This process helped correlate the geospatial data with the actual geological and mineralogical conditions observed in the field.

The final model developed by Arnous et al. (2021) proved to be an effective tool for mapping mineral alteration zones, identifying controlled structural lineaments, and delineating the main radioactive-rich rock units. The study particularly emphasized the significance of granitic rocks in the Wadi Dahab area, which are likely associated with radioactive mineralization. The integrated techniques successfully identified areas with high potential for radioactive mineralization, particularly in granitic units that are rich in mineral content.

However, the authors highlight the need for additional fieldwork and laboratory examination to validate these predictions and further refine the exploration model. While the study provides a strong foundation for mineral exploration, continued verification and research are essential for enhancing the model's accuracy and reliability.

This case study exemplifies the successful integration of remote sensing, GIS, and geochemical analysis in the exploration of radioactive mineralization. The predictive model developed for the Wadi Dahab area has proven to be an effective tool for identifying potential sites of radioactive mineralization. By combining geospatial technologies with laboratory tests and fieldwork validation, the study presents a robust framework for mineral exploration in arid regions like southeastern Sinai. Future research and further verification are required to fine-tune the model and confirm its accuracy in identifying mineralization zones.

7.12.10 GIS-based geological mapping and hydrothermal alteration in central Jebilets, Morocco

The Jebilets massif, located in the western Meseta of Morocco, forms part of the Paleozoic Hercynian chain and is home to several polymetallic mines, including well-known sites such as Draa Sfar, Koudiat Aicha, Laachach, Bir Enhass, and Kettara. These mines are renowned for their rich deposits of copper, lead, and zinc, which are primarily attributed to volcanic massive sulphide (VMS) deposits, with pyrrhotite being the dominant mineral. The mineralization in these deposits is often closely associated with hydrothermal alteration zones, characterized by distinct zonation of alteration minerals such as chlorite, sericite, illite, and iron oxides.

To gain a deeper understanding of the spatial distribution of these hydrothermal alteration minerals, geological mapping, lineament extraction, and hydrothermal alteration mapping were carried out in the central Jebilets region. Abdelouhed et al. (2021) employed GIS and remote sensing (RS) techniques, using ArcGIS, ENVI, and PC Geomatica 2016 software for data analysis and processing.

Remote Sensing for Hydrothermal Alteration Mapping: ENVI software was utilized to process satellite imagery and extract data on hydrothermal alteration minerals, with a focus on the silicate group of minerals (e.g., chlorite, sericite, illite). The band rationing method was applied to enhance the spectral signatures of alteration minerals, facilitating the precise identification of alteration zones.

Geological Mapping and Lineament Extraction: ArcGIS was used to georeference various datasets (including satellite images and geological surveys) and integrate them into a unified platform for analysis. The GIS tools enabled the extraction of structural lineaments and geological features, which were essential for understanding the mineralization control mechanisms and identifying spatial relationships between alteration zones and ore bodies.

Geospatial Integration and Visualization: The processed data from both ENVI and ArcGIS were integrated and overlaid, allowing the creation of detailed geological maps, lineament maps, and hydrothermal alteration mineral maps. These maps were essential for highlighting areas with potential for mineralization and for guiding further exploration efforts in the Jebilets region.

The GIS and RS-based approach significantly enhanced the understanding of mineralization and alteration patterns in the central Jebilets area. The alteration mineral mapping effectively delineated the distribution of key alteration minerals, which are indicative of hydrothermal processes associated with VMS-type deposits. Lineament extraction further helped identify structural controls such as faults and fractures, which likely played a role in the localization of the mineral deposits.

The geospatial data generated from this study not only contributed to the exploitation of the central Jebilets deposits but also offered valuable insights into metallotects that could be useful for exploring similar deposits within the broader Jebilets Guemassa metallogenic province.

This case study underscores the vital role that GIS and remote sensing technologies play in enhancing the efficiency and accuracy of geological mapping, lineament extraction, and hydrothermal alteration mineral mapping. By integrating satellite imagery with geospatial analysis, the study provides valuable insights into the structural and mineralogical characteristics of the central Jebilets region, making it a powerful tool for ongoing exploration efforts. These findings can also guide future exploration of similar mineralization systems within the Jebilets Guemassa metallogenic province. The integration of these technologies presents a comprehensive framework for effective resource management and exploration in geologically complex regions.

7.12.11 GIS-based mineral detection and mapping in Cuddapah district, Andhra Pradesh, India

The use of Remote Sensing (RS) and Geographic Information Systems (GIS) in mineral exploration has become an invaluable tool for mapping and detecting minerals in regions with significant geological interest. In this study, Ramachandra et al. (2020) applied digital image processing techniques and Landsat Enhanced Thematic Mapper (ETM)+) imagery to map structural features and detect minerals in the Cuddapah district of Andhra Pradesh, India. This region is of growing importance due to its mineral exploration potential and increasing industrial investment. The study specifically focused on band rationing as a method for detecting specific minerals and surface features.

Data source

Landsat (ETM+) Imagery: The study utilized medium-resolution satellite imagery from Landsat (ETM+), which includes seven bands and provides crucial spectral information for mineral detection. The Cuddapah district was identified as having high mineral exploration potential, and the goal was to map and detect minerals using remote sensing techniques.

Band rationing technique

Band Rationing is a technique where one band of a satellite image is divided by another to highlight specific features based on their spectral characteristics. This method is particularly useful for detecting minerals with distinct spectral signatures. The following band combinations were used in the study: Clay Minerals: Band 5/Band 7 of the Landsat (ETM+) imagery; Iron Oxide Minerals: Band 3/Band 1 of the Landsat (ETM+) imagery; Ferrous Minerals: Band 5/Band 4 of the Landsat (ETM+) imagery. These combinations were applied to create individual layers representing different minerals.

Data integration and layer stacking

The three individual layers representing clay minerals, iron oxide minerals, and ferrous minerals were stacked together to form a composite band-rationing map for the study area. This final map enabled the visualization and interpretation of mineral distribution across the Cuddapah district.

Field and RS integration

To validate and refine the results, field surveys were conducted to compare the remote sensing findings with on-the-ground observations. The study emphasizes the importance of integrating field data with remote sensing techniques, which not only replace outdated geological maps but also provide more accurate and up-to-date information on mineral distribution.

Results and application

The band-rationing maps effectively highlighted areas rich in clay, iron oxide, and ferrous minerals within the Cuddapah district. These minerals are important indicators of certain geological formations and processes associated with mineralization. The study demonstrated the effectiveness of using Landsat imagery and digital image processing techniques to enhance mineral detection, especially in regions like Cuddapah, where traditional geological surveys may have limitations in scope or resolution.

The application of band rationing in this study has proven to be a highly effective tool for mineral exploration and mapping in the Cuddapah district. By utilizing Landsat (ETM+) imagery and applying specific band ratios, the study successfully identified mineral-rich areas and their spatial distribution. The integration of remote sensing with field data has further enhanced the accuracy and reliability of geological maps, making this approach a valuable resource for both ongoing and future mineral exploration efforts.

This study highlights the potential of remote sensing and GIS techniques as a cost-effective and efficient alternative to traditional geological mapping methods. By using advanced remote sensing data, this approach offers a more detailed and timely means of exploring and identifying mineral resources in regions like Cuddapah.

7.13 Fieldwork applications: Best practices and methodologies

When conducting field GPS surveys for natural resource exploration, following best practices and methodologies is critical to ensure accurate data collection and reliable analysis. The process begins with thorough site preparation: ensuring that the GPS receiver is properly calibrated and that the survey area is free from obstructions such as tall buildings, dense vegetation, or other factors that might impede satellite signals. Choosing the appropriate GPS equipment is equally important; for surveys requiring high accuracy, high-precision receivers such as RTK or Differential GPS (DGPS) systems should be used.

Satellite visibility plays a significant role in the accuracy of GPS surveys. Field teams should position themselves in open areas with minimal overhead obstruction to maximize satellite visibility. It is important to maintain a stable receiver setup, allowing the GPS receiver to acquire enough satellite signals for reliable positioning. The duration of the survey will depend on the technique being employed—static surveys require a longer data collection period for more precise measurements, while kinematic surveys can be completed more quickly but may sacrifice some accuracy.

A critical element in fieldwork applications is real-time data monitoring. When using advanced systems like RTK or DGPS, it is essential to continuously monitor the data being collected. Real-time feedback enables field crews to adjust their positions or surveying techniques as necessary, ensuring data quality throughout the survey process. In addition, redundant measurements or resurveys at key points can help confirm data consistency and correct any potential errors.

Once field data is collected, postprocessing is a crucial step. Specialized software can apply correction factors, remove erroneous data points, and enhance the quality of the data. Additionally, integrating GPS data with GIS and other geospatial tools improves the spatial accuracy of the collected data, enhancing the quality of analysis and decision-making.

By adhering to these best practices—proper equipment calibration, maintaining satellite visibility, utilizing real-time feedback, performing redundant measurements, and integrating GPS data with GIS—surveyors can ensure the accuracy and reliability of their GPS data. This will ultimately facilitate more efficient and effective exploration and management of natural resources.

7.14 Future trends in GPS surveying for sustainable resource exploration

The future of GPS surveying for sustainable resource exploration is being shaped by emerging technologies and innovations that promise to enhance the accuracy, efficiency, and environmental sustainability of resource management. As new advancements continue to develop, the GPS surveying landscape will evolve, focusing on greater precision, improved efficiency, and eco-friendly practices.

7.14.1 Advancements in satellite technology and GNSS integration

The launch of new, more precise GPS constellations will significantly enhance positioning accuracy, especially in challenging environments where traditional systems may struggle. This will make it easier to conduct surveys in remote and rugged locations, such as forests or mountains, which are critical for natural resource exploration. The integration of GNSS, including Galileo, BeiDou, and Global Navigation Satellite System (GLONASS), alongside traditional GPS, will provide multifrequency signals for improved accuracy and redundancy. This will help ensure more reliable positioning worldwide, increasing the versatility of GPS in a variety of terrains (Liu and Zhang, 2023).

7.14.2 Integration with other geospatial technologies

The integration of GPS with GIS, Remote Sensing (RS), and machine learning is poised to revolutionize resource exploration. Real-time processing and analysis of data will allow surveyors and geologists to identify potential resource-rich areas more efficiently. Machine learning algorithms will play a key role in analyzing large datasets from GPS and remote sensing, automatically detecting resource patterns, and predicting potential mineral or energy sources. These technologies will also enhance resource monitoring, improving the ability to track environmental changes and the impact of resource extraction on ecosystems.

7.14.3 Drone and UAV integration

Drones and unmanned aerial vehicles (UAVs) equipped with high-precision GPS are transforming the way resource exploration is conducted. Drones can cover vast areas quickly and gather topographic data and high-resolution imagery. This allows for dynamic, real-time surveying, reducing the need for extensive ground-based surveys and increasing the speed and efficiency of data collection. Drones also provide the advantage of reaching difficult-to-access locations, such as hazardous or remote sites, improving safety and reducing the environmental footprint of traditional surveying methods (Singh and Gupta, 2022).

7.14.4 Real-time high-precision GPS technologies

Real-Time Kinematic (RTK) and Precise Point Positioning (PPP) technologies are becoming more affordable and widely available, providing centimeter-level accuracy in real time. This is crucial for tasks such as detailed resource mapping and monitoring land use changes during exploration activities. The use of RTK and PPP ensures that even the smallest variations in landforms or ecosystems are accurately captured, leading to more precise decision-making. These technologies will also be pivotal in providing immediate insights into environmental conditions, enabling real-time feedback and adjustments during the exploration process.

7.14.5 Sustainability and environmental monitoring

Sustainability will continue to be a primary focus of GPS advancements in resource exploration. New GPS technologies will improve the monitoring of environmental impacts associated with resource exploration and extraction. These tools will allow for real-time tracking of disruptions to habitats, ecosystems, and water sources, enabling companies to address environmental concerns immediately.

By integrating GPS data with environmental sensors and GIS, exploration companies and regulatory bodies can track landform changes, monitor water quality, and measure vegetation health. This real-time monitoring can ensure that exploration activities comply with sustainable practices, reducing the overall environmental impact and helping mitigate potential harm to ecosystems.

In conclusion, the future of GPS surveying in sustainable resource exploration is closely linked to the continuous integration of advanced technologies. These innovations, such as multiconstellation GNSS, drones, real-time GPS, and machine learning, will significantly enhance surveying precision, speed, and efficiency. Furthermore, these advancements will play a key role in supporting sustainable practices by improving environmental monitoring, enabling ecofriendly resource management, and ensuring long-term sustainability in resource extraction.

7.15 Case studies on the integration of GPS, GIS, and remote sensing for natural resource exploration

7.15.1 Western Australia: Integrating GPS, GIS, and remote sensing for mineral exploration

In Western Australia, a study by McMahon and Mulligan (2015) demonstrates the effective use of GPS, GIS, and remote sensing technologies for mineral exploration. The approach combined GPS-based ground surveys with satellite imagery and GIS to identify potential mineral deposits in the region. The integration of high-precision GPS data with remote sensing and GIS allowed for the creation of accurate geospatial maps of the terrain, which were then analyzed to target areas with high mineral potential.

Using high-accuracy GPS receivers, the team collected point data that was then integrated into a GIS platform. This data was overlaid with satellite imagery, which provided critical information about vegetation, soil moisture, and geological features indicative of mineral-rich areas. Remote sensing data further enhanced the ability to detect subtle geological patterns that could suggest the presence of valuable minerals.

The combination of these technologies allowed for more targeted and cost-effective exploration. The integration of GPS, GIS, and remote sensing not only improved the accuracy of the geological maps but also reduced the time and resources traditionally spent on extensive ground surveys. This approach significantly increased the likelihood of discovering mineral deposits, thereby optimizing exploration efforts and resource allocation. The case study illustrates the substantial benefits of combining GPS with GIS and remote sensing in modern mineral exploration, showcasing how these technologies can work synergistically to enhance exploration efficiency and precision.

7.15.2 GPS and remote sensing for forest resource mapping in Canada

In Canada, Global Positioning System (GPS) and remote sensing technologies have been effectively utilized for mapping forest resources in remote areas where traditional surveying methods are difficult to implement. A study by Gauthier and Drapeau (2006) employed GPS for ground-truthing and satellite imagery for a broader view of forest cover, species distribution, and biomass estimation.

Researchers used GPS receivers to collect ground data, including tree species, height, and health, which were then integrated into a Geographic Information System (GIS). Satellite images, particularly those from Landsat, were used to identify forest types, density, and distribution across large areas. By combining GPS data with remote sensing, the team was able to develop a comprehensive model of the forest resources in the region.

This integration of GPS, GIS, and remote sensing significantly enhanced the accuracy of forest resource mapping and enabled more efficient management of timber harvesting. Additionally, it supported conservation efforts by identifying ecologically sensitive areas in need of protection due to biodiversity concerns. The study highlights the value of combining ground-based GPS data with satellite imagery to improve forest management, resource allocation, and environmental protection strategies.

7.15.3 GPS and GIS in coastal resource management in the United States

In the United States, GPS and GIS technologies have played a vital role in coastal resource management, particularly in the Florida coastal wetland, where efforts focus on wetland restoration and monitoring saltwater intrusion caused by rising sea levels (McKinney and Pappas, 2017).

GPS data was employed to survey coastal elevations and monitor changes in wetland boundaries over time. This data, combined with remote sensing technologies such as high-resolution satellite imagery and aerial photography, provided critical information about vegetation health, water levels, and the overall state of the wetland ecosystems. The integration of this data into a GIS platform enabled the development of models to predict potential environmental impacts and track restoration progress.

By combining GPS, GIS, and remote sensing, the project facilitated real-time monitoring of coastal wetland conditions, including saltwater intrusion and vegetation changes. This approach proved essential in developing more effective management strategies for the restoration of the wetland ecosystem, ensuring the long-term health and resilience of the region's coastal resources. The integrated use of these technologies allowed for informed decision-making and adaptive management in the face of environmental challenges.

7.15.4 GPS-GIS integration in oil and gas exploration in the Middle East

In the Middle East, the integration of GPS and GIS technologies has significantly enhanced oil and gas exploration, transforming how geologists and engineers approach site identification and resource evaluation (Al-Khayyat and Nasr, 2010). The use of GPS for precise positioning combined with GIS for advanced spatial analysis allows for more efficient exploration and site selection.

GPS data was crucial in accurately marking drilling locations and defining the boundaries of exploration areas, ensuring the correct placement of equipment and reducing the risk of errors. GIS tools were then utilized to analyze a range of geophysical data, including seismic information, well logs, and subsurface measurements, enabling a detailed understanding of geological formations. This integration helped assess the viability of potential drilling sites by providing insights into subsurface conditions.

By merging GPS with GIS, the oil and gas industry in the region was able to streamline the exploration process, reducing time and costs associated with inaccurate site identification. This combination also improved the precision of resource estimation and enhanced environmental monitoring, making the exploration process both more efficient and environmentally conscious. The ability to analyze large-scale geological features in real time has further advanced exploration efforts, increasing the success rate of drilling and resource extraction.

7.15.5 Sustainable mining and resource monitoring with GPS, GIS, and remote sensing in South America

In South America, a significant project in the Andes Mountains leveraged GPS, GIS, and remote sensing technologies to monitor and manage the environmental impacts of mining activities, with a particular focus on deforestation and soil erosion (Sanchez and Rojas, 2014). This integration of advanced geospatial technologies provided crucial insights into the environmental consequences of mining, supporting more sustainable practices.

GPS technology played a key role in tracking the precise location of mining operations and monitoring the resulting changes in terrain and vegetation. By continuously recording coordinates, GPS allowed for accurate tracking of mining site expansion and its impact on the surrounding environment.

Remote sensing technologies, including satellite imagery, were utilized to detect deforestation and land use changes caused by mining activities. These satellite images offered a broader perspective, allowing the team to assess large-scale environmental alterations over time and monitor forest cover reduction or soil degradation.

GIS was employed to analyze and model the spatial distribution of environmental impacts. By combining data from GPS and remote sensing, GIS tools enabled researchers to predict future environmental changes based on ongoing mining activities. This predictive modeling helped in the development of more effective strategies for mitigating deforestation, soil erosion, and other ecological concerns.

The integration of these technologies enhanced the sustainability of mining operations by providing accurate, real-time data. This allowed mining companies and environmental authorities to implement better management and mitigation strategies, ensuring that the environmental impact of mining activities was minimized. The ability to monitor and predict changes also facilitated adaptive management practices, making it easier to respond to emerging issues promptly.

These case studies illustrate the power and versatility of integrating GPS, GIS, and remote sensing technologies in natural resource exploration and management. By providing precise, actionable data, these technologies play an essential role in promoting sustainability, improving decision-making, and minimizing environmental harm in resource extraction processes.

7.15.6 Integrating GPS, GIS, and remote sensing for forest resource mapping and management in Madhya Pradesh, India

In Madhya Pradesh, forest management authorities have successfully integrated GPS, GIS, and remote sensing technologies to enhance the monitoring of forest cover, biodiversity, and deforestation trends (Mishra and Sahu, 2020). This integration has proven to be invaluable in tracking and managing the region's forest resources more effectively.

Remote sensing data from satellites like Landsat and Sentinel have provided high-resolution imagery to assess changes in forest density, vegetation health, and land use patterns over time. By utilizing these satellite-based tools, authorities can gain a broader perspective of the region, identifying areas that are experiencing significant deforestation or degradation.

Ground-based GPS data was used to collect precise location information on key forest features, such as boundary lines, tree species, and biodiversity hotspots. This GPS data was then integrated into GIS platforms, allowing for the mapping and analysis of spatial patterns of forest cover and its changes.

The combination of GPS, GIS, and remote sensing technologies has greatly improved forest management in Madhya Pradesh. It has allowed for more accurate monitoring of forest health and biodiversity, enabling authorities to detect illegal logging activities quickly. Furthermore, these technologies have supported biodiversity conservation efforts by identifying critical habitats that need protection.

In conclusion, the integration of GPS, GIS, and remote sensing in Madhya Pradesh has significantly enhanced the ability to monitor and manage forest resources, improving the efficiency of conservation efforts and helping authorities address environmental threats such as deforestation and biodiversity loss. These technologies have become essential tools for sustainable forest management in the region.

7.15.7 Water resource management in Rajasthan, India

In the arid region of Rajasthan, GPS and remote sensing technologies have been effectively integrated to monitor groundwater levels and manage irrigation systems, addressing the challenges posed by water scarcity (Patel and Kumar, 2019). The combination of satellite imagery and GPS data collected through field surveys has provided crucial insights into land use patterns, crop water requirements, and areas most vulnerable to water stress.

Satellite-based remote sensing data, such as images from Landsat and Sentinel, have been used to analyze vegetation cover, water bodies, and soil moisture, which are vital indicators for monitoring agricultural water use and groundwater levels. GPS data, collected during field surveys, has been used to precisely locate water sources, irrigation channels, and wells, and to map areas of land affected by over-extraction or insufficient irrigation.

This integration of GPS and remote sensing has significantly improved water resource management in Rajasthan. It has facilitated the planning and optimization of irrigation systems, ensuring more efficient use of water. Additionally, the technology has helped identify recharge zones for groundwater and monitor their sustainability, promoting the management of this limited resource.

By enabling real-time data collection and spatial analysis, these technologies have empowered decision makers to design more effective water conservation strategies and mitigate the impacts of water scarcity in Rajasthan's agricultural regions.

In conclusion, the integration of GPS and remote sensing in water resource management has proven to be a powerful tool for enhancing irrigation practices, improving groundwater recharge, and ensuring the sustainable use of water in the arid landscapes of Rajasthan. These technologies play a crucial role in addressing the challenges of water scarcity and ensuring the resilience of the region's agricultural systems.

7.15.8 Coastal zone management in Kerala, India

In Kerala, the Kerala State Disaster Management Authority (KSDMA) utilizes GPS, GIS, and remote sensing technologies to monitor coastal erosion and safeguard vulnerable areas along the Arabian Sea. The integration of these technologies plays a crucial role in managing the state's coastal zones, especially in light of the increasing risks posed by erosion, flooding, and the effects of climate change.

Remote sensing data, particularly from satellites, provides a detailed and comprehensive view of shoreline changes, enabling the monitoring of coastal erosion, sediment movement, and sea-level rise over time. High-resolution satellite imagery allows for the tracking of changes in landforms and vegetation along the coast, which are essential for understanding the dynamics of coastal erosion.

GPS technology is used to accurately mark the boundaries of coastal areas, ensuring precise demarcation of vulnerable zones. It helps in mapping the extent of erosion and identifying critical areas that require immediate intervention, such as those prone to flooding or damage from storm surges.

The integration of GPS, GIS, and remote sensing has led to improved coastal zone management by facilitating better data collection, analysis, and decision-making. This integration supports more effective disaster preparedness, enabling authorities to plan mitigation measures, design coastal protection infrastructure, and create early warning systems.

Additionally, this technological synergy aids in developing policies and strategies for mitigating erosion and flood risks, enhancing the resilience of coastal communities, and safeguarding vital ecosystems such as mangroves and coral reefs.

In conclusion, the use of GPS, GIS, and remote sensing technologies in coastal zone management in Kerala has provided valuable tools for monitoring and protecting the state's coastal regions. By integrating these technologies, KSDMA has been able to strengthen disaster preparedness and develop more effective measures to address coastal erosion, thereby reducing risks and protecting coastal infrastructure and livelihoods.

7.15.9 Mineral exploration in Jharkhand, India

In Jharkhand, a state known for its vast reserves of coal, iron ore, and bauxite, GPS, GIS, and remote sensing technologies are being effectively utilized to enhance mineral exploration and ensure sustainable mining practices (Kumar and Singh, 2022). These technologies play a pivotal role in mapping and monitoring mining activities, providing critical data for resource management and environmental protection.

Remote sensing data, primarily from satellites, is used to gather high-resolution imagery of the region, which helps in identifying mineral-rich areas and tracking land-use changes due to mining. By analyzing satellite imagery, the government can monitor environmental changes, such as deforestation, soil erosion, and water contamination, which are associated with mining activities.

GPS technology plays a crucial role in precisely locating mining sites and monitoring excavation activities. It ensures accurate positioning for exploration and allows for the continuous tracking of mining operations, thereby improving resource estimation and mapping. This is particularly important for determining the extent and quality of mineral deposits.

GIS is then used to integrate GPS and remote sensing data, enabling the overlay of mineral deposit maps with environmental datasets. This integration helps identify areas with the highest resource potential while also factoring in environmental constraints, such as protected areas or zones with high biodiversity. The combination of these technologies helps authorities and mining companies make informed decisions, balancing the need for resource extraction with environmental conservation efforts.

Through this integration, the state has streamlined its mineral exploration process, improving resource estimation and environmental monitoring. Furthermore, the technologies support adherence to sustainable mining practices, ensuring that mining activities have minimal environmental impact and contribute to the region's long-term economic growth.

In conclusion, the use of GPS, GIS, and remote sensing in Jharkhand's mineral exploration efforts has greatly enhanced the efficiency and sustainability of mining operations. By providing accurate, real-time data for resource mapping, monitoring, and environmental protection, these technologies help ensure that the state's mineral wealth is extracted responsibly and sustainably.

7.15.10 Agricultural land use mapping in Punjab, India

In Punjab, India's leading agricultural state, GPS, GIS, and remote sensing technologies are increasingly being employed to enhance agricultural productivity and sustainability (Gupta and Mehta, 2021). These integrated technologies enable the monitoring of crop health, soil fertility, and land use patterns, helping farmers make informed decisions to improve crop yield and optimize resource use.

Remote sensing through satellite imagery provides valuable insights into crop health by detecting early signs of stress, pest infestations, or disease. This data allows farmers to take proactive measures to address potential issues, thereby improving crop yield and reducing the need for chemical inputs. By analyzing changes in vegetation cover over time, satellite images also help in assessing the overall health and productivity of different crops across the region.

GPS technology is used for ground verification, ensuring the accuracy of remote sensing data by enabling the precise positioning of field boundaries and monitoring crop conditions in real time. Furthermore, GPS-guided machinery, such as tractors and harvesters, helps farmers optimize their operations, reducing fuel consumption, minimizing input waste, and increasing efficiency in planting, irrigating, and harvesting crops.

GIS integrates the data collected from GPS and remote sensing, providing farmers and agricultural authorities with detailed maps that reveal land use patterns, soil fertility levels, and areas prone to crop diseases or water stress. These insights are crucial for planning and optimizing the use of fertilizers, irrigation, and other resources, promoting sustainable farming practices.

The integration of these technologies has enabled significant improvements in crop yield predictions, allowing farmers to plan better and adjust their practices for maximum productivity. Moreover, it supports sustainable agricultural practices by optimizing the use of fertilizers and water, reducing environmental impacts, and improving the overall sustainability of farming in Punjab.

In conclusion, the application of GPS, GIS, and remote sensing in Punjab has revolutionized agricultural land use mapping, enabling farmers to adopt precision agriculture techniques that improve productivity while promoting sustainability. This approach has the potential to significantly enhance food security and resource management in the region, ensuring that agricultural practices can meet future challenges.

7.15.11 Remote sensing and GIS for bauxite and limestone exploration in Odisha, India

In Odisha, a mineral-rich state in India, GPS, GIS, and remote sensing technologies have been effectively integrated for bauxite and limestone exploration (Sahoo and Pandit, 2020). These advanced technologies are used to map and assess mineral deposits, identify potential mining sites, and monitor the environmental impact of mining activities in the region.

Satellite imagery from sources like Landsat and Sentinel is employed to capture detailed views of the terrain, enabling the identification of geological features associated with bauxite and limestone deposits. This remote sensing data, when combined with GPS-based field surveys, helps to accurately delineate areas containing valuable minerals. GPS is used for precise positioning during field surveys, ensuring that data collected onsite is georeferenced, which is crucial for creating reliable mineral maps.

GIS serves as a powerful tool for integrating diverse data sources, such as geological, environmental, and land-use information, allowing for comprehensive analysis and visualization of mineral deposits and their surrounding environments. By combining these datasets, GIS enables the creation of detailed mineral maps that aid in decision-making for exploration and mining activities. This approach also helps in identifying environmentally sensitive areas that should be protected or avoided during mining operations.

The integration of GPS, GIS, and remote sensing in bauxite and limestone exploration in Odisha has led to several benefits, including more efficient resource mapping and reduced exploration costs. These technologies allow for the identification of mineral-rich areas with greater accuracy and less reliance on costly and time-consuming ground surveys. Moreover, the ability to assess environmental impacts early in the exploration process ensures that sustainable practices are incorporated into the planning and management of mining activities.

In conclusion, the use of GPS, GIS, and remote sensing in Odisha has significantly enhanced the efficiency and sustainability of bauxite and limestone exploration. These integrated technologies not only help streamline the exploration process but also contribute to more responsible and environmentally conscious mineral extraction practices.

7.15.12 GPS-GIS integration for coal exploration in Chhattisgarh, India

In Chhattisgarh, one of India's leading coal-producing states, the integration of GPS, GIS, and remote sensing technologies is playing a pivotal role in the exploration and management of coal resources (Verma and Kumar, 2021). This integration enhances the accuracy of coal exploration by combining precise field data collection with spatial analysis, leading to more efficient resource identification and management.

In the exploration process, GIS is employed to analyze satellite data, integrating geological maps and remote sensing imagery to identify areas with a higher probability of containing coal deposits. This analysis provides a comprehensive spatial understanding of coal-rich zones, helping geologists and resource managers pinpoint promising sites for further investigation. GPS technology is then used in the field to mark exact locations of mineral resources, ensuring that the data gathered during field surveys is georeferenced and accurately tied to real-world coordinates.

Remote sensing, on the other hand, plays a critical role in monitoring the environmental impacts of coal mining. By using satellite imagery and aerial surveys, it is possible to track land use changes, such as vegetation loss, soil erosion, and habitat disturbance, caused by mining operations. This environmental monitoring helps to ensure that mining activities are carried out with minimal impact on the surrounding ecosystems. The integration of GPS, GIS, and remote sensing technologies in Chhattisgarh has led to several key benefits:

Enhanced Coal Resource Identification: The combination of field data, satellite imagery, and spatial analysis helps in more accurate mapping of coal deposits, reducing the risk of misidentification and improving exploration efficiency.

Improved Exploration Accuracy: The use of GPS ensures precise positioning of mineral resource locations, eliminating errors typically associated with traditional methods of exploration and enabling more effective decision-making.

Environmental Monitoring: Remote sensing data allows for continuous monitoring of environmental changes due to mining activities, helping stakeholders assess and mitigate negative impacts on the surrounding ecosystems.

Sustainable Mining Practices: By providing comprehensive data on both mineral deposits and environmental conditions, this integration supports the adoption of sustainable mining practices. It ensures that coal extraction is carried out responsibly, with due consideration for the surrounding environment and communities.

These benefits highlight how the integration of GPS, GIS, and remote sensing technologies enhances both the efficiency and sustainability of coal exploration in Chhattisgarh. By streamlining the exploration process and providing valuable environmental insights, this integrated approach helps ensure more responsible resource extraction and better land use management.

The integration of GPS, GIS, and remote sensing in coal exploration is revolutionizing the way natural resources are managed and extracted in India. In Chhattisgarh, these technologies are improving the accuracy of coal exploration, enhancing environmental monitoring, and promoting sustainable mining practices. This case study exemplifies how geospatial tools can be used to improve resource management and ensure that mineral extraction is carried out in an environmentally responsible manner.

This section reflects the integration of GPS, GIS, and remote sensing in natural resource management, not only enhancing exploration efficiency but also promoting environmental sustainability. It ties into other case studies from India, demonstrating how these technologies are being applied across different sectors such as forestry, water management, agriculture, and mining to improve both the efficiency and sustainability of resource use.

7.15.13 Case studies on the integration of GPS, GIS, and remote sensing for geological and resource exploration by the Geological Survey of India (GSI)

The Geological Survey of India (GSI) has been a pioneer in integrating GPS, GIS, and Remote Sensing (RS) technologies for geological exploration and resource management. Below are some notable case studies where GSI has successfully applied these technologies in various geological and resource exploration activities:

7.15.13.1 Geological mapping and resource exploration in the eastern Himalayas

The GSI employed GIS, GPS, and remote sensing to map geological features and mineral resources in the Eastern Himalayas (Geological Survey of India, 2017). Remote sensing data from satellites, including Landsat and Indian Remote Sensing (IRS) satellites, was used to capture geological structures such as faults, folds, and rock formations. GPS technology was used to accurately locate geological samples during fieldwork, and GIS integrated this data to create detailed geological maps.

This integrated approach was pivotal in identifying valuable mineral deposits, including graphite, coal, and iron ore, in the region. The combination of these technologies also aided in better understanding the region's geological features, streamlining resource exploration, and enhancing hazard mapping in seismic zones.

7.15.13.2 Groundwater resource mapping in Rajasthan

In Rajasthan, GSI utilized GPS and remote sensing technologies to assess groundwater resources and their sustainability (Geological Survey of India, 2019). Satellite imagery from the IRS series was employed to analyze land use patterns and vegetation indices, highlighting areas facing groundwater depletion. Field surveys using GPS technology were conducted to accurately locate boreholes and wells, which were integrated into a GIS platform alongside satellite data.

This integration allowed for the creation of a comprehensive groundwater resource management system, improving water management practices in the region. It supported the identification of groundwater potential zones, informing water conservation projects and ensuring sustainable groundwater use in the arid region of Rajasthan.

7.15.13.3 Coastal erosion monitoring and coastal zone management

GSI utilized GPS, GIS, and remote sensing technologies to monitor coastal erosion and land changes along India's eastern and western coasts (Geological Survey of India, 2020). Remote sensing data, including imagery from IRS and Radio Detection and Ranging (uses microwave signals) (RADARSAT) satellites, was used to track shoreline changes over time, while GPS data provided ground-truth information for more precise location-based assessments.

GIS played a crucial role in integrating these data sets, creating coastal zone management maps that delineated erosion-prone areas and supported disaster preparedness efforts. This project enabled a better understanding of coastal erosion patterns and allowed for the implementation of protective measures to safeguard vulnerable coastal areas, infrastructure, and ecosystems.

7.15.13.4 Earthquake hazard mapping and seismic risk assessment in northeast India

In northeast India, GSI employed GPS, GIS, and remote sensing technologies to study seismic hazards and map earthquake-prone areas (Geological Survey of India, 2021). GPS data was used to monitor crustal movements and tectonic activities, while remote sensing from satellites provided insights into surface deformation and fault lines.

The integration of these technologies in GIS allowed the creation of seismic hazard maps, highlighting areas most at risk for earthquakes and other natural disasters. This information significantly contributed to enhancing disaster preparedness in these high-risk seismic zones, assisting in the development of early warning systems and mitigation strategies.

7.15.13.5 Mineral exploration in the Aravalli range

The GSI conducted a study in the Aravalli Range of Rajasthan to explore mineral resources, such as limestone, marble, and copper (Geological Survey of India, 2018). The integration of GIS, GPS, and remote sensing facilitated the efficient mapping of mineral-rich areas. Satellite images were used to detect surface mineral alterations, and GPS technology pinpointed the locations of mineral outcrops and boreholes for ground-truthing.

GIS was used to create comprehensive geological maps that integrated mineral data with environmental considerations and land use. This study led to the identification of high-potential mineral zones and promoted the sustainable management of these mineral resources, while minimizing the environmental impacts of mining operations.

These case studies from the Geological Survey of India (GSI) highlight the transformative impact of integrating GPS, GIS, and remote sensing technologies in geological and resource exploration. They demonstrate how these technologies enhance the accuracy of geological mapping, improve resource management practices, and contribute to the sustainable use of natural resources in India. These applications showcase the potential of geospatial technologies to not only improve exploration and resource management but also to play a crucial role in environmental protection and disaster risk management. By leveraging these technologies, GSI has been able to streamline the exploration process, enhance decision-making, and ensure the sustainability of India's natural resource sectors.

7.16 The complementary roles of both technologies in modern surveying practices

GPS and GIS serve complementary roles in modern surveying practices, each bringing its unique capabilities to the table. GPS offers high precision for spatial positioning, while GIS provides the ability to analyze, interpret, and visualize the data in various ways. Together, these technologies enable comprehensive mapping of geographical features, resource management, and environmental monitoring.

For instance, in mineral exploration, GPS allows for accurate location marking of drilling sites, while GIS integrates this data with geological and environmental datasets, leading to a more sustainable approach to resource extraction. In forest management, GPS is used for collecting precise data on tree health and biodiversity, and GIS enables the creation of maps to monitor forest cover and identify areas of illegal logging. In agriculture, GPS-driven machinery helps optimize land use and crop yields, while GIS helps in making informed decisions about irrigation, fertilization, and pest control.

Furthermore, remote sensing technologies add another layer to this synergy by providing aerial and satellite data, which can be integrated with GPS and GIS for more comprehensive analysis. This combination helps monitor environmental changes, including deforestation, water levels, and coastal erosion, ensuring more sustainable practices across various sectors.

7.17 Future directions and technological advancements in geospatial surveying

Looking ahead, the future of geospatial surveying with GPS and GIS technologies is full of exciting advancements. Emerging technologies such as real-time GPS (e.g., RTK and PPP) offer even higher levels of accuracy, enabling centimeter-level precision in surveying. Advanced satellite systems (e.g., multiconstellation GNSS and Galileo) will provide more robust, reliable positioning even in challenging environments, enhancing data quality for geospatial surveys.

Drones and unmanned aerial vehicles (UAVs), equipped with high-precision GPS, are also becoming an integral part of geospatial surveying. These technologies enable the rapid collection of data over large and difficult-to-reach areas, enhancing both the speed and efficiency of surveys. The integration of machine learning and artificial intelligence (AI) with GIS and GPS will further improve data processing, allowing for the automated interpretation of large datasets, which will significantly reduce time and labor costs.

Moreover, there will be a continued focus on sustainability in resource exploration and monitoring. As the demand for natural resources increases, advanced GPS and GIS technologies will provide better tools for monitoring the environmental impacts of resource extraction. This will be critical for balancing resource use with environmental conservation, supporting the global shift toward more sustainable and eco-friendly practices in industries such as mining, forestry, and agriculture.

Data integration and real-time monitoring will be at the forefront of future developments, ensuring that stakeholders have access to up-to-date, accurate information for making informed, timely decisions. The convergence of GPS, GIS, and remote sensing will continue to drive innovations that push the boundaries of what is achievable in natural resource management and geospatial surveying. This conclusion effectively summarizes the content and highlights the significant progress and future potential of integrating GPS and GIS for geospatial surveying.

7.18 Conclusions

Chapter 7 provides a comprehensive overview of how Geospatial Surveying Techniques, particularly the integration of Geographical Information Systems (GIS) and the Global Positioning System (GPS), are transforming sustainable natural resource exploration. The chapter emphasizes that modern mineral and resource exploration relies on accurate spatial data, efficient mapping, and advanced analytical tools to understand complex geological processes.

The discussion highlights how GIS supports spatial data management, modeling, and visualization, while GPS ensures precise geolocation and field data acquisition. When integrated, these technologies enhance exploration efficiency, accuracy, and decision-making. Their applications extend from mineral and groundwater prospecting to mining operations, environmental monitoring, and land-use management. The inclusion of Remote Sensing (RS) further strengthens data interpretation and broad-scale monitoring capabilities.

Emerging innovations—such as Artificial Intelligence (AI), Machine Learning (ML), Real-Time Kinematic (RTK) GPS, UAV-based surveying, and GNSS integration—are paving the way for automation, improved precision, and sustainable geospatial practices. Numerous global and regional case studies demonstrate the effectiveness of GIS-GPS integration in diverse geological and resource management contexts.

In conclusion, GIS and GPS form the backbone of modern geospatial surveying. Their convergence with remote sensing and intelligent data analytics promotes sustainable, data-driven exploration and management of natural resources. Future advancements will continue to enhance their interoperability, accuracy, and applicability, reinforcing their critical role in achieving efficient and environmentally responsible exploration practices.

CHAPTER

8

Drone-based geoscientific surveying techniques

8.1 Introduction

The rising demand for raw materials, such as rare earth elements (REE) (set of 17 metallic elements) and lithium, makes the exploration and extraction of mineral deposits critical. Identification of Earth's hidden treasures is becoming increasingly difficult, owing to the continued depletion of easily accessible deposits and the social stigma surrounding mining activities. Efficient methods that minimize invasive and costly drilling are key for the discovery of potentially profitable mineral resources. However, a gap in observation scales remains. Traditional ground-based surveys (such as rock and soil sampling), although detailed, can only cover some 15–30 km per day, and large-scale mapping via helicopter, plane, and satellite fails to provide sufficient resolution to efficiently map small-scale (<1 km^2) geological features. The use of unmanned aerial vehicles (UAVs) for industrial applications goes back to the 19th century. In 1860, balloons were used to take pictures for remote sensing purposes (Moore, 2010). In 1903, pigeons carrying a breast-mounted aerial camera were used for photography (Rambat, 2011). Around the beginning of World War I, aerial torpedoes, which are known as the origin of drones, were developed (Keane and Carr, 2013; Kindervater, 2016). In recent years, attention to research and development of unmanned aerial vehicles has been growing by academic and industrial communities worldwide (Cai et al., 2010; Hassanalian and Abdelkefi, 2017). Drones have been increasingly used in mineral exploration (e.g., Booysen et al., 2020; Jackisch et al., 2020). Difficult terrain or outcrops unreachable on foot or by vehicles can be rapidly surveyed with UAVs from a safe distance with minimal personnel on site, ensuring safety, efficiency, and speed16,1. Depending on the defined mission, drones are generally classified widely based upon their configurations (Alvarado, 2018; DJI, 2019; McLeod et al., 2013; Shahmoradi et al., 2020; Freire and Cota, 2017; Hennage et al., 2019; Mining, 2019) (Figs. 8.1 and 8.2). Drones can be grouped into nine categories, such as fixed wing, flapping wing, rotary wing, tilt rotor, ducted fan, helicopter, ornithopter, and unconventional types.

Baseline Geoscience Data Generation for Sustainable Exploration of Natural Resources
https://doi.org/10.1016/B978-0-443-45221-5.00019-9

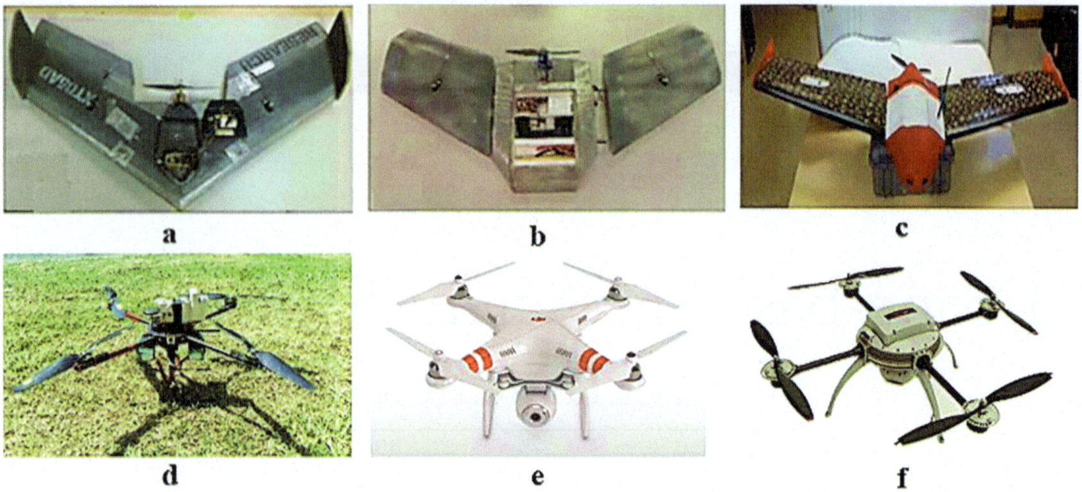

FIG. 8.1 Views of some utilized drones in surface mining (A) Teklite, (B) GoSurv, (C) Swamp Fox, (D) Quadcopter, (E) Phantom 2 Vision +, (F) Aeryon Scout. *Source: Shahmoradi et al. (2020).*

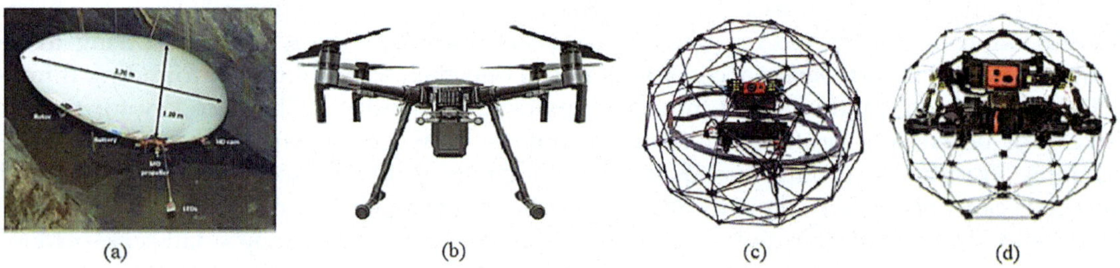

FIG. 8.2 Commonly used drones in underground mines (A) Zeppelin, (B) DJI M210, (C) ELIOS 1, (D) ELIOS 2. *Source: Shahmoradi et al. (2020).*

Drones have a variety of capabilities for both military and civilian utilization (Hassanalian and Abdelkefi, 2017; Hassanalian et al., 2015; Hassanalian et al., 2018). These capabilities, along with the demand for unmanned technologies, have resulted in the integration of drones into civil practices (Finn and Wright, 2016). Toward this end, new unmanned aerial vehicles are being developed that can perform various missions in a variety of environments (Finn and Wright, 2016; Hassanalian et al., 2018). For example, drones are utilized in a vast range of civilian applications such as search and rescue, surveillance, firefighting; weather monitoring, surveying (Rouse et al., 2019), power infrastructure monitoring (Karjalainen et al., 2016), and urban planning and management (Zlatanova et al., 2016). Drones have also been used for building environment monitoring (Rouse et al., 2019) and urban traffic monitoring (Leitloff et al., 2014; Barmpounakis and Golias, 2016), ecological and environmental monitoring (Lee and Choi, 2016), species distribution modeling (Hodgson et al., 2016), population ecology (Hodgson et al., 2016), and ecological monitoring and conservation (Hardin and Jensen, 2011). Archaeology and cultural heritage (Fernandez-Hernandez et al., 2015), human and social understanding (Rohrbach et al., 2016; Xiang et al., 2018), personal and business drones for photography and videography, and even delivery services (Rouse et al., 2019) are other applications of drones. In addition, the unmanned aerial systems have been successfully used in different industries, such as agriculture (Elijah et al., 2018), oil and gas (Whipple et al., 2019), construction (Li and Liu, 2019), environmental protection (Vergouw et al., 2016), mining (Lee and Choi, 2016), etc. Recently, the mining industry has shown increased interest in the use of drones for routine operations in surface and underground mines (Green, 2013; McLeod et al., 2013; Ranjan et al., 2018; Rupprecht and Pieters, 2018; Dunnington and Nakagawa, 2017; Hoffmann and McAllister, 2018). There are two main advantages of using drones in mining operations (Green, 2013). First, drones equipped with different types of sensors can conduct a quick inspection of an area, either in an emergency situation or for hazard identification. Second, inspection and unblocking of blocked box-holes and ore-passes can be done using drones. Drones can also be used for blockage inspection, explosive, and package delivery. Lee and Choi categorized the applications of drones in the mining industry, including surface, underground, and abandoned mines, as demonstrated in Table 8.1.

Drones present the perfect trade-off between coverage and scale of observation, and are essential when ground access is impossible, dangerous, or logistically costly. Drones can carry lightweight sensors that, for example, capture changes in the earth's magnetic field as well as a continuous spectrum of reflected sunlight in the visible and near-infrared (VNIR) regions (hyperspectral imaging). Subtle variations in the measured properties can be used to determine the abundance and composition of key minerals at Earth's surface. For example, specific spectral absorption features and magnetic anomalies could indicate the presence of iron alteration related to mineralization. The hyperspectral and magnetic data collected by the drones can be used alongside high-resolution true colour cameras or laser scanning sensors to place mineral deposits into a 3D geographical context. Therefore, economic mineral deposits can be identified down to depths of several hundred meters. Drones are becoming important tools for mineral exploration by contributing to the safe, efficient, and sustainable provision of the high-tech metals that are required by modern society. For example, drone-based hyperspectral imaging has been used to rapidly map rare-earth-element-rich minerals in Namibia (country in Southern Africa). In addition, in Greenland (North American

TABLE 8.1 The applications of drone-based surveying in mining.

Surface mine	Underground mine	Abandoned mine
Mine operation	Geotechnical characterization	Subsidence monitoring
3D mapping	Rock size distribution	Recultivation
Slope stability	Gas detection	Landscape mapping
Mine safety	Mine rescue mission	Gas storage detection
Construction monitoring		Acid drainage monitoring
Facility management		

Lee and Choi (2016).

autonomous territory **in the Kingdom of Denmark**), drone-based magnetic surveys were deployed to identify subsurface ore potential at a fraction of the cost of traditional surveys. Drones have the potential to provide non-invasive and eco-friendly platforms from which the environmental impact of exploration and mining activities (such as soil erosion, acid mine drainage, and vegetation stress) can be assessed (Jackisch, 2020).

8.2 Mineral exploration employing drones, contemporary geological satellite rs and GIS procedures

Modern remote sensing using satellites replaced film cameras and balloons used as platforms in aerial photography. Sikakwe (2023) used multispectral and hyperspectral remote sensors and geographic information system (GIS) in mineral exploration to evaluate mineral potential. Landsat, Advanced Spaceborne Thermal Reflection and Emission Radiometer (ASTER), System Pour l'Observation de la Terre (SPOT), Indian Remote Sensing (IRS), Sentinel-2, and Airborne Visible Infrared Imaging Spectrometer (AVIRIS) sensor image data were used in the analyses. Image data processing by minimum noise fraction (MNF), principal component analysis (PCA), band rationing (BR), and independent component analysis (ICA) improves data for easy minerals recognition. Earth Resources Data Analysis System (ERDAS) Imagine software is for numerical image processing, while geological, geophysical, and geochemical survey data are integrated with image data using ArcGIS software to delineate mineral potential zones. The mineral potential map is classified into low, moderate, good, and excellent zones. Innovations such as cloud computing, big data analytics, drones, and machine learning algorithms have been introduced in mineral exploration due to depletion in available mineral reserves. These technologies are useful in mineral exploration in hard-to-reach areas. The review shows that the ASTER sensor has a higher capability to detect hydrothermally altered minerals than the Landsat, SPOT, and Sentinel-2 sensors. Hyperspectral AVIRIS can identify minerals in hydrothermally altered volcanic areas better than all multispectral sensors due to its narrow band widths. The review shows that a combination of more than one remote sensor in mineral search gives more credible results than using an individual sensor. The application of satellite remote sensing in geological mapping covers a wider area than aerial photography. However, the remote sensing method has a limitation of depth, because the electromagnetic spectrum (EMS) measures properties of surface geological features, not from the subsurface. Mineral exploration using satellite remote sensing and GIS can be applied in different geological settings.

8.3 Drone technology for natural resource exploration

In surface mines, drones are used in mine operations, 3D mapping, slope stability analysis, and related mine operations. In underground mines, drones are used in geotechnical characterization, rock size distribution, gas detection, and mine rescue missions. In abandoned mines, drones are used in subsidence monitoring, cultivation, landscape mapping, gas storage, and acid drainage monitoring (Shahmoradi et al., 2020). Drones are suitable for strategic planning of surface mines to achieve pit and leach pad design, road design, and mineral exploration. Sensors on drones correspond with the drone size and the purpose. Based on the aim of the aerial investigation, existing drones include RGB sensors, ultrasonic sensors, infrared sensors (IR), stereo camera, laser range finders (LRFs), and ultra-wide band radar (UWB) (Shahmoradi et al., 2020). For drones to be used in mineral exploration, the data acquired by drones must be versatile, accurate, and adapted to the task. In addition, the platforms should be robust and of low maintenance to ensure operational use in remote locations (Booysen et al., 2021). The use of small unmanned airborne systems (UAS) to obtain geophysical data will definitely alter the manner in which geophysical surveys are carried out. Drones are used at mining sites for geological and structural analysis through remote sensing, aerial geophysical (Park and Choi, 2020). Shahmoradi et al., (2020) used drone-driven data to evaluate a marble quarry in Iran. The study convinced that drone-based imagery in the initial step of dimension stone exploration conserves the time and cost of exploration.

In geological structural analysis through remote sensing, UAVs are used mainly to first acquire visible infrared, multispectral, and hyperspectral data (Park and Choi, 2020). High-resolution UAS-based multispectral orthomosaics are used to identify surficial iron staining, which serves as a proxy for outcropping sulphide mineralization (Jackish et al., 2021). Shrestha et al. (2022) used UAV drone technology for image capturing for surveying and mapping of open-cast mines in surface areas of Nepal (South Asia). The study showed that UAV drone technology in mine survey is fast, safe, reliable, more accurate, and easy for calculating volume, analysis, and interpretations. Porras et al. (2021) applied a drone magnetic survey for mineral exploration in Don Jacobo in copper-cobalt-nickel strata-bound mineralization, iron mined in Beltic Cordillera, Spain. A vapour magnetometer was installed on a Matrice 600 Pro Hexacopter. Twenty-four parallel survey lines flown at 5 m/s orthogonal to the regional strike of the geologic structure and mineralization, with 50 m separation and 20 m flight height over the studied area. Data interpretation modeled two high magnetic susceptibility bodies with residual magnetization close to the existing mines and surface mineral expressions. These bodies could be associated with potential unmined zone connected with normal faults in the southern portion of the study area. The study improved the geological and mining potential of the area to design prospective surveys. Drone-based data are used in the initial phase of mineral exploration (Shahriari et al., 2022). A high-resolution imaging done on marble quarry, Kerman Province, Iran showed lithological and geological structural mapping using a pixel of 3 cm/pix at an altitude of 70 m, ground control point made to attain the desired accuracy. Agisoft PhotoScan was used for photogrammetric processing. Faults and fractures were delineated from geology map using high-resolution orthophoto and hillshade made in ArcGIS software. The density map was produced, and the deposit subdivided into five structural regions. The priority areas of exploration for development activities and suitable methods were defined for each region. It was concluded that drone-based imagery in the first phase of dimension stone exploration should be considered to reduce the cost of exploration.

Sustainable exploration and mining of REEs are required to implement the energy transition, e-mobility, and enable the economic growth of society. Independent and ethical supply of critical raw materials has been strongly encouraged in Europe1 as well as in the USA2. Despite several exploration initiatives all around the world, only a few REE projects (e.g.,

Browns Range, Australia) reached the extraction stage during the last years. Within this context, REE deposits can be divided into two general categories. Primary deposits are associated with magmatic processes and hydrothermal fluid mobilization and precipitation, whereas secondary deposits originate from the movement of REE minerals from their place of origin, such as through sedimentary concentration and weathering5. Five main types of REE deposits are currently exploited: (1) carbonatites, (2) magmatic alkaline deposits, (3) ionic clay deposits, (4) laterite deposits, and (5) placer deposits. Currently, the most important source of REEs is carbonatite deposits and their associated alkaline igneous rocks6. Booysen et al. (2020) recently proposed a novel approach for the exploration of REEs applying remote sensing techniques. Booysen et al. (2020) employ drones with hyperspectral sensors to detect REEs at the earth's surface and thus contribute to a rapidly evolving field at the cutting edge of exploration technologies. They showcase for the first time the direct mapping of REEs with lightweight hyperspectral UAV platforms. Their solution has the advantage of quick turnaround times (<1 d), low detection limits [<200 ppm for neodymium (Nd)], and is ideally suited to support exploration campaigns. This procedure was successfully tested and validated in two areas: Marinkas Quellen, Namibia, and Siilinjarvi, Finland. This strategy should invigorate the use of drones in exploration and for the monitoring of mining activities.

NGRI develops drone-based geophysical survey system. The indigenous system will be useful for mineral exploration, mapping of basement topography, and for aquifer mapping (Source: Hyderabad, The Hindu). National Mineral Development Corporation Ltd (NMDC), Country's largest iron ore producer, a Central Public Sector Enterprises (CPSE) under Ministry of Steel signed a MoU on March 2022 [Source: Press Information Bureau (PIB) Delhi] with IIT Kharagpur for drone-based mineral exploration (Fig. 8.3). NMDC and IIT Kharagpur would develop spectral products, methods and algorithms for exploration using the Drone (UAV) for mining. The collaboration between NMDC and IIT Kharagpur would further lead to the development of software spectral tools for mineral excavation and capacity-building programs on mining technology. Data from a recent survey shows that the drone industry, over the next five to seven years, could potentially generate about 100,000 direct and indirect job opportunities. The National Institute of Technology (NIT) in Andhra Pradesh announced

FIG. 8.3 Drone-based mineral exploration. *Source: Press Information Bureau, Delhi, March 20, 2022—Open Access.*

it would offer a minor degree in drone studies (Source: Open Gov March 2022—MoU for drone-based mineral exploration in India).

Satellite remote sensing data and UAV/Drone images have been conjunctively used for identification and mapping of probable indicators of economic mineral deposits in Girar area of Lalitpur district, UP, India (Uniyal et al., 2022). Area of interest for UAV/Drone survey was selected from the digitally enhanced IRS P6 LISS-IV images. Anomalous orientation, disposition, and abrupt changes in the shape and pattern of landforms, particularly the ridges at some places, have been referred to as geomorphic anomalies. Lineaments with a few hundreds of meters of length have been delineated on spatially enhanced images and DEMs generated from UAV/Drone data. Sudden change in slope and relatively steeper valley walls are distinguishable on UAV data-based Digital Elevation Model (DEM), thereby vindicating the existence of lineaments. Intersections of ENE–WSW and NW-SE trending lineaments in Girar ridge area have been marked as structural anomalies. UAV/Drone images have not only yielded precise information about the spatial extent of anomalies (earlier been identified on IRS P6 LISS-IV satellite images), but have also helped in the identification of smaller geomorphic and structural anomalies, which were otherwise difficult to identify on satellite images. Further, the field investigations have revealed the rock outcrops of Banded Haematite Quartzite (BHQ) within these geomorphic and structural anomalies, with BHQ considered to be potential host rocks for iron oxides (Fe_2O_3). The structural and geomorphic anomalies demarcated on digitally processed UAV/Drone images and DEMs of this area seem to be the probable indicators for iron deposits.

8.4 Marut Drones develops India's first mineral exploration drone for NMDC

Dronetech start-up Marut Drones has delivered India's first mineral exploration drones to NMDC Ltd. The drones, equipped with advanced sensors and a coaxial (X8) design, enhance precision in mineral exploration and mapping, thereby contributing to the promotion of sustainable mining practices. This milestone marks a breakthrough in the mining and exploration space, revolutionizing prospecting for critical minerals like lithium, copper, gold, diamond, rare earth elements, and other essential minerals in India. Dronetech start-up Marut Drones in February 2024 said it has delivered India's first mineral exploration drones to state-run iron ore miner NMDC Ltd. The Hyderabad-based start-up has supplied two customized Octacopter drones to the Navratna PSU (Fig. 8.4). The mineral exploration drones are equipped with cutting-edge features, including advanced sensors such as magnetometer, hyperspectral, and lidar, that enhance their capabilities for accurate exploration and mapping. Its innovative coaxial design not only ensures stability but also facilitates precise manoeuvrability, a crucial aspect for effective mineral exploration, Marut added.

Pointing out that drone-based hyperspectral magnetic exploration has been initiated for the first time by NMDC, Marut said that the spacing and altitude for the flight are determined through mission planning, followed with hyperspectral survey, which ensures a 30% overlap of the captured land during the drone flight. The acquired data is later processed using Hyperspec III and ENVI software by NMDC. Similarly, in drone-based magnetic surveying, the drones are calibrated, and the spacing and altitude are decided before planning the mission. Geometrics Magarrow Magnetometer is used for magnetic data acquisition, and

FIG. 8.4 Marut Drones, a start-up specializing in drone technology, has announced the delivery of India's inaugural mineral exploration drones to the government-owned iron ore mining company NMDC Ltd. The start-up provided NMDC with two specially tailored octacopter drones as part of this collaboration. *Source: NMDC—Open Access.*

Geometrics and Geosoft Oasis Montage software are employed for data processing by NMDC. Every aspect of this drone, including the motors and the propeller systems, has been made in India, and it is believed that this is one of the biggest breakthroughs in the mining and exploration space.

8.5 Integration of drone-based surveying with other geophysical methods

Drone-based surveying can be effectively integrated with other geophysical methods such as ground-penetrating radar (GPR), magnetic surveys, and seismic imaging. This section explores how the combination of drone technology and these traditional geophysical methods can provide a more comprehensive data set, enhancing the accuracy and efficiency of mineral exploration. The integration allows for better subsurface mapping, more precise identification of mineral deposits, and greater exploration in challenging terrains. The section also discusses the potential benefits, such as cost-effectiveness, time savings, and improved accessibility, as well as some challenges that come with integrating these technologies.

8.5.1 The role of drones in geophysical surveys

This subsection discusses the role drones play in complementing traditional geophysical methods. Drones equipped with high-resolution cameras and sensors, such as LiDAR and multispectral sensors, can gather detailed surface data, enhancing the interpretation of subsurface information obtained through geophysical methods. Their flexibility and accessibility allow drones to reach difficult terrains, such as mountainous or remote regions, where traditional survey methods may be challenging or dangerous. Drones are also capable of rapidly covering large areas, capturing surface data that can be processed to identify areas of interest for more detailed geophysical analysis. Additionally, the use of drones reduces the overall cost and time involved in data collection, making it possible to gather more data in less time compared to ground-based surveys.

8.5.2 Synergy between drones and magnetic surveys

Magnetic surveys are effective in detecting subsurface mineral deposits, particularly ferromagnetic materials. By combining drone-captured high-resolution aerial imagery with magnetic data, it becomes possible to correlate surface features with magnetic anomalies, offering a more comprehensive understanding of the subsurface environment. Drones equipped with magnetometers can perform high-resolution, high-coverage magnetic surveys quickly and efficiently. The data collected from these surveys can be integrated with satellite imagery and GIS to enhance the accuracy of anomaly detection. Case studies from regions where drones have been used in conjunction with magnetic surveys for mineral exploration demonstrate improved resolution and detection capabilities, leading to more effective and precise exploration efforts.

8.5.2.1 Case study: Drone-based magnetic surveys for mineral exploration and mapping

The rising demand for raw materials, particularly rare earth elements (REE) and lithium, highlights the need for efficient mineral exploration and extraction methods. As easily accessible deposits continue to deplete, there is a growing need for innovative and non-invasive exploration techniques. Traditional ground-based surveys, though detailed, are limited in scale, typically covering only 15–30 km per day. On the other hand, large-scale aerial methods such as satellite, helicopter, or plane-based surveys lack the resolution needed to effectively map small-scale geological features (<1 km^2). Drone-based surveys offer an ideal solution, providing a balance between coverage and observation scale, especially in areas where ground access is difficult or dangerous.

Drones equipped with lightweight sensors are transforming mineral exploration by capturing variations in the Earth's magnetic field, providing a cost-effective and detailed method for detecting subsurface ore potential. These drone-based magnetic surveys allow for high-resolution mapping of magnetic anomalies, which serve as crucial indicators of mineralization. Such anomalies often signal the presence of minerals like iron alteration, commonly linked to mineral deposits. Additionally, hyperspectral imaging—using drones to capture a continuous spectrum of reflected sunlight in both the visible and near-infrared regions—helps detect subtle variations in the Earth's surface composition, making it easier to identify specific mineral types. When combined, magnetic and hyperspectral data offer a comprehensive understanding of mineral deposits, extending to depths of several hundred meters (Jackisch, 2023).

In Namibia, drone-based hyperspectral imaging was utilized to rapidly map rare-earth-element-rich minerals. By combining this imaging with drone-based magnetic surveys, the exploration team was able to identify potentially profitable mineral resources efficiently and non-invasively. This case underscores how drones can map and assess mineral deposits without the need for costly drilling, providing a more sustainable and effective approach to mineral exploration.

In Greenland, drones equipped with magnetic sensors have been used to explore areas with challenging terrain. These drone-based surveys have successfully identified subsurface ore bodies that would have been difficult or impossible to detect using traditional methods. The integration of magnetic data with high-resolution imagery from drones has proven to be essential in identifying and mapping mineral deposits in regions that are difficult to access.

Drone-based surveys offer significant advantages in non-invasive exploration, allowing for eco-friendly surveys with minimal environmental disruption. This approach is less invasive than traditional drilling, reducing risks such as soil erosion, acid mine drainage, and vegetation stress. Additionally, drone surveys are much more cost-effective than traditional methods conducted by helicopter or plane, enabling broader exploration of potential mineral deposits, especially in remote or hard-to-reach areas. With the ability to capture high-resolution data from both magnetic and hyperspectral sensors, drones provide superior resolution and accuracy for detecting mineralization. The integration of data from drones with high-resolution true-color cameras or laser-scanning sensors also allows for the creation of a 3D geographical context, providing a more detailed view of a mineral deposit's spatial extent and depth.

Drone-based magnetic surveys are becoming increasingly important in mineral exploration due to their cost-effectiveness, resolution, and ability to minimize environmental impact. As global demand for critical minerals such as rare earth elements and lithium continues to rise, drones present a promising solution for the sustainable and efficient exploration of mineral deposits.

8.5.2.2 UAV-borne magnetic survey for diamond exploration: A case study over kimberlite-carbonatite intrusion at Khaderpet, South India

This case study presents the use of a drone (UAV)-borne magnetometer (MagArrow) survey in diamond exploration over a kimberlite pipe with carbonatite at Khaderpet, located within the Wajrakarur Kimberlite Field (WKF) in South India (Phani et al., 2025). The kimberlite rock at Khaderpet is a chloritized kimberlite-granite breccia, which shows characteristics of an ultramafic lamprophyre. Because the kimberlites in the WKF respond well to magnetic surveys, a magnetic sensor was chosen for this investigation. The survey included ten east-west (E-W) traverses, covering 15 km over an area of 1.8 square kilometers. This was completed in just 35 min, while a traditional ground magnetic survey would have taken about a week.

The data revealed a strong magnetic "low" of about—130 nT in the reduced to the pole (RTP) image, where carbonatite is located among granite-kimberlite breccia. The magnetic "low" anomaly matches the outline of the kimberlite rock as interpreted by earlier studies. The results from this survey were confirmed with a conventional ground magnetic survey, which also showed the same anomaly (magnetic "low") representing the kimberlite-granite breccia-carbonatite. This was contrasted with the surrounding "high" magnetic area of granitic and gneissic country rocks. In this investigation, it was concluded that the diamond-bearing carbonatite hosted within chloritized kimberlite breccia extends to a depth between 7.3 m and 41 m from the surface, which matches the drilling results from previous exploration. This study demonstrates that drone-based magnetic surveys can effectively identify anomalies and causative bodies within the background of country rocks, helping to locate potential kimberlite and related rocks in other cratonic areas, provided that geological, geochemical, and indicator mineral evidence supports the findings.

8.5.2.3 Aeromagnetic gradiometry with UAV: A case study on the Kalakan iron ore deposit, Kurdistan, Iran

Unmanned Aerial Vehicles (UAVs) have evolved from military applications to widespread use in various geophysical surveys, including magnetometry. UAV-borne magnetometry offers benefits such as reduced operational costs, high precision, and improved safety (Porras et al., 2021). However, challenges such as electromagnetic interference from UAVs have been addressed through advanced sensor technologies (Walter et al., 2019). This study explores the use of lightweight magneto-inductive (MI) sensors mounted on a Hexacopter UAV for magnetic gradient measurements. The focus is on the Kalakan iron ore deposit near Qaleh Rutaleh village, Iran, where previous terrestrial magnetometry revealed a northeast-trending anomaly (Shahsavani and Smith, 2024). The aim is to compare the results of two magnetometer setups: the finite-difference vertical gradient and the calculated vertical derivative.

The survey site, Kalakan iron ore deposit, located about 20 km from Divandarreh city in Kurdistan, Iran, was chosen for its favorable topography and the lack of significant electromagnetic interference from human-made infrastructure. The regional total magnetic intensity (TMI) at the site is 48,000 nT, with inclination and declination values of 54.78° and 5.33°, respectively. A Hexacopter drone with Pixhawk 2.4.8 controller was used, carrying two lightweight MI sensors. The flight was conducted at a height of 6 m with a speed of 5 m/s. The survey covered profiles planned perpendicular to the iron ore deposit's strike, ensuring optimal data coverage. The acquired magnetic data was processed using Oasis Montaj software, and a minimum curvature grid with a 10 m cell size was created. The TMI data for the lower and upper sensors is shown in Fig. 8.5. The vertical gradient was computed by subtracting the measurements of the two sensors, and results are shown in Fig. 8.6.

Aeromagnetic surveying using drones for gradiometry presents significant challenges. The use of two sensors on a drone can reduce flight endurance and increase operational costs. However, the MI sensors used in this study are lightweight and offer reasonable resolution, enabling the use of two sensors without significantly increasing the drone's weight. The finite-difference gradient, calculated by subtracting the TMI map of the upper sensor from the lower sensor's map and dividing by their distance, is compared with the calculated vertical derivative map. The results show that for most positive anomalies, the finite-difference gradients are greater than the calculated derivatives. For many negative anomalies, the finite-difference gradients are less than the derivatives. Additionally, the calculated derivative exhibits high-frequency noise (around 0.35 Hz), while the finite-difference gradient is smoother and contains less high-frequency content. The comparison along four profiles demonstrates the capability of the proposed method.

Table 8.2 summarizes the differences, showing that the finite-difference gradient is higher by up to 10 nT/m for positive anomalies. Despite the advantages, installing two sensors increases the drone's weight, which impacts flight endurance. Also, at high elevations or in magnetically quiet areas, the sensitivity of the MI sensors becomes insufficient, making precise gradient measurements difficult. The development of drones for aeromagnetic surveys has driven the advancement of lightweight magnetometer sensors. However, for gradiometry measurements, two sensors are still necessary to achieve optimal results. In this study, Shahsavani and Smith, 2024 applied two lightweight MI sensors to a Hexacopter drone, suspending them with three-meter cables to mitigate electromagnetic noise from the drone's motors. The survey was conducted over a small iron ore deposit across six profiles, each around 300 m long. The gradient results obtained by subtracting the lower sensor from the upper sensor were compared with calculated vertical derivatives for validation. The finite-difference gradient results showed an increase of up to 10 nT/m compared to the derivative results, demonstrating the method's effectiveness.

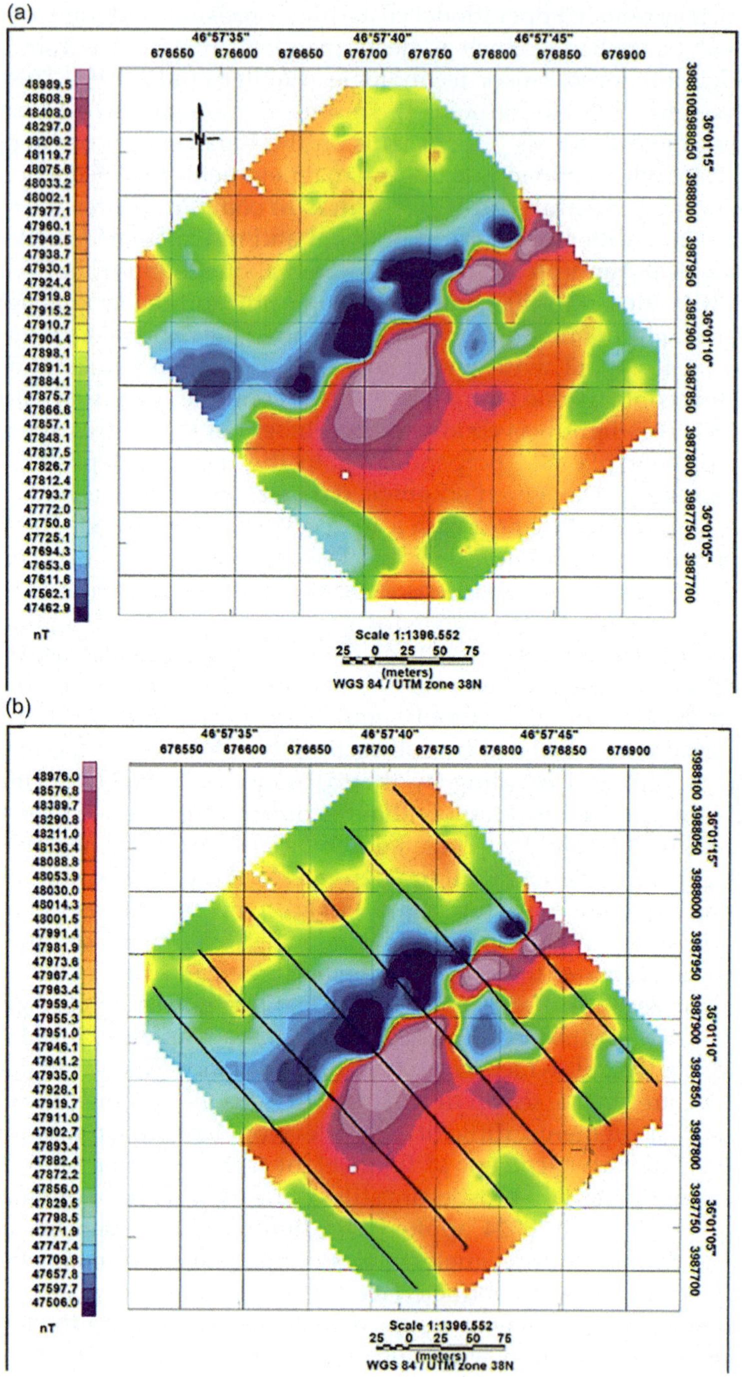

FIG. 8.5 Total magnetic intensity (A) measured by the lower sensor and (B) measured by the upper sensor in Kalakan iron ore deposit. *Source: Shahsavani and Smith, 2024—Open Access.*

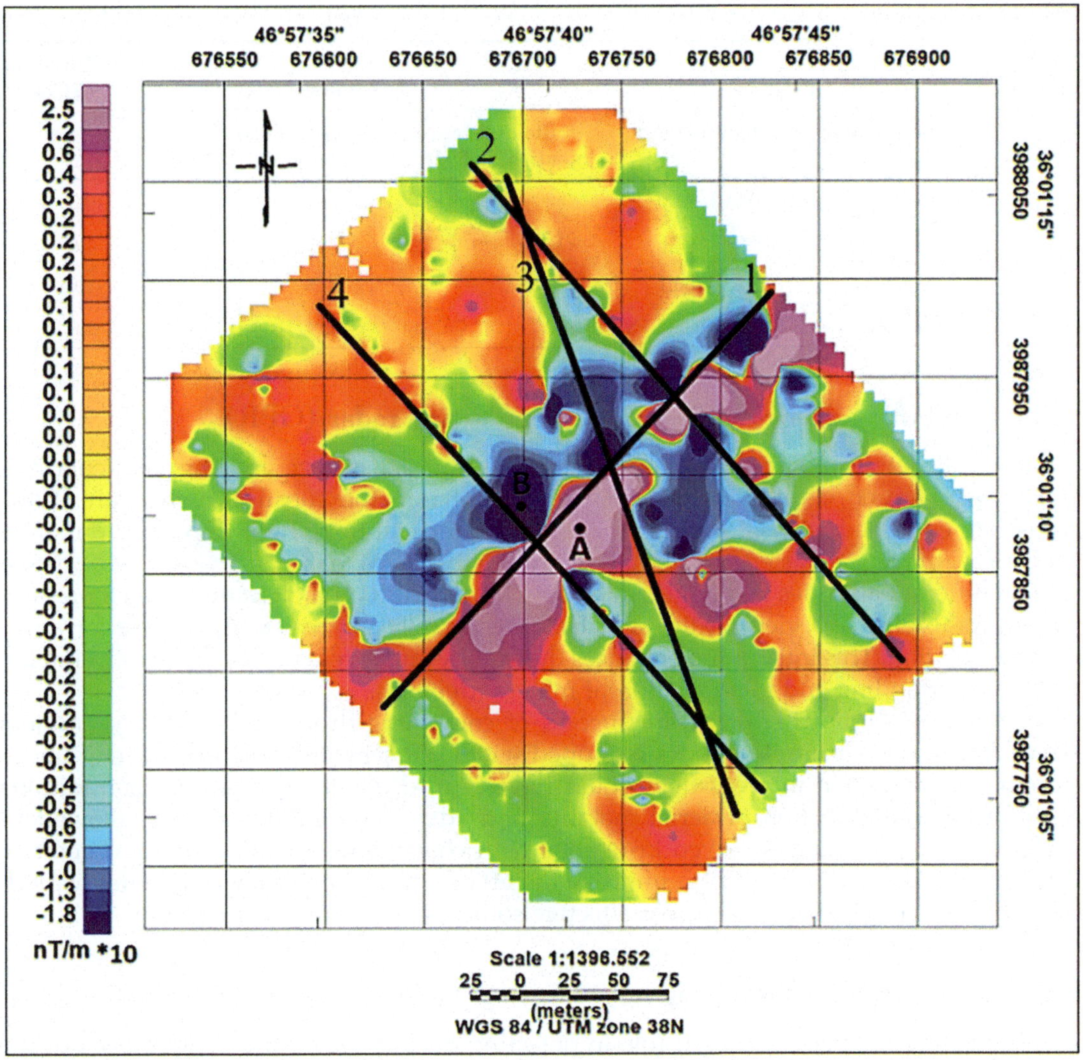

FIG. 8.6 The vertical derivative calculated from the lower sensor measurements in the Kalakan iron ore deposit. The black lines from 1 to 4 show some random profiles. Points A and B indicate two poles of the anomaly. *Source: Shahsavani and Smith, 2024—Open Access.*

Although the sensors are light, installing two sensors decreases the drone's flight time compared to using a single sensor. This method is more suitable for **detailed investigations** or **high-resolution surveys**, particularly where shallow information is required. However, at higher elevations, the **magnetic gradient** becomes weaker, making it difficult for low-resolution sensors to measure the gradient accurately. Thus, the drone should fly as **close to the Earth's surface** as possible for best results.

TABLE 8.2 Maximum gradient comparisons across profiles in the Kalakan iron ore deposit, Kurdistan, Iran.

Profile number	Position		Maximum		Difference (nT/m)
	Longitude (m)	Latitude (m)	Finite-difference gradient (nT/m)	Calculated derivative (nT/m)	
1	676,780.3	3,987,937.0	33.7	24.4	9.3
2	676,754.5	3,987,897.6	8.1	3.5	4.6
3	676,713.7	3,987,860.1	34	23.6	10.4
4	676,781.7	3,987,933.7	26.2	22.4	3.8

8.5.3 Combining drones with ground penetrating radar (GPR)

Ground Penetrating Radar (GPR) is widely used for subsurface imaging, particularly in mineral exploration, as it provides detailed insight into the subsurface structure. When drones are equipped with GPR sensors, they can offer real-time, high-resolution data over large areas, making it easier to map and analyze subsurface features. The ability of drones to fly at different altitudes and angles allows for varied GPR data collection, which provides a more comprehensive view of the subsurface, especially in areas with diverse terrain or dense vegetation. This integration enhances the quality and extent of data gathered, enabling better understanding and identification of mineral deposits.

Several case studies highlight the success of combining drones with GPR for mineral exploration. In one example, drones equipped with GPR sensors were used to map underground mineralization in areas that would have been difficult to access using traditional methods. The combination of aerial mobility and advanced GPR technology allowed for detailed subsurface imaging, even in regions with challenging terrain or thick vegetation. These integrations have proven to improve the efficiency and accuracy of mineral exploration, offering a non-invasive and cost-effective solution to subsurface mapping.

8.5.4 Drones and seismic imaging

Seismic imaging plays a crucial role in detecting deeper subsurface formations, but it is often costly and time-consuming. Drones offer a valuable solution by being integrated into seismic surveys, where they can transport lightweight geophysical equipment to remote or challenging sites, reducing the logistical challenges that are typically associated with traditional seismic methods. Drones can deploy geophones or other seismic sensors at precise locations, allowing for the collection of data that complements the seismic waves generated by ground-based methods. This capability improves the efficiency of seismic surveys by streamlining the process and reducing the time required for equipment setup.

An example of the successful application of drones in seismic imaging is seen in environments where access is difficult, such as mountainous or forested areas. Drones have been used to transport and deploy seismic equipment in these challenging environments, significantly improving the speed and accuracy of subsurface mapping. The integration of drone technology in seismic imaging enables faster data collection, more precise placement of sensors, and enhanced overall survey results, making it a more efficient and cost-effective solution for subsurface exploration.

8.5.5 Seismic data acquisition utilizing a group of unmanned aerial vehicles in desert environments

8.5.5.1 *Introduction*

Seismic exploration plays a key role in mineral and hydrocarbon prospecting, as well as in engineering surveys. However, conducting these surveys in desert and mountainous regions presents significant challenges due to extreme weather conditions and the lack of infrastructure. Furthermore, shallow karst formations can create dangerous conditions during oil well drilling. To overcome these challenges, unmanned aerial vehicles (UAVs) are being explored to automate seismic data collection. UAVs are highly mobile, making them ideal for use in difficult terrain, and they are already being used in applications such as magnetic surveys, geohazard mapping, and seismic data acquisition. The Autonomous Seismic Acquisition Device (ASAD), developed since 2021, uses UAV swarms to automate seismic exploration. The ASAD system has been successfully tested in desert conditions, such as in Saudi Arabia (Timoshenko et al., 2025) (Figs. 8.7A and 8.7B).

8.5.5.2 *Background and previous studies*

Seismic exploration works by recording the reflected seismic waves from subsurface layers using ground-based sensors. These waves are typically generated by seismic sources like explosives or vibrators. Modern seismic systems have expanded to include large sensor networks, often involving hundreds of thousands of nodes. UAVs have been introduced to seismic operations for sensor deployment and data collection. Initial UAV systems focused on deploying individual seismic nodes, but more recent developments use UAV swarms controlled centrally to improve operational efficiency. Path planning algorithms, such as A* and artificial potential fields (APF), enable real-time obstacle avoidance and dynamic swarm coordination in 3D environments, allowing UAVs to work together more effectively. By automating seismic data collection, UAVs address logistical challenges such as extreme weather, difficult terrain, and the absence of infrastructure, reducing the need for large teams and improving the overall speed and efficiency of seismic exploration.

The ASAD system is made up of three main components: The Swarm Control Station (SCS), a group of ASADs, and the Hit Recording Module (HRM). The system operates on four control levels: user, high (mission planner), middle (control firmware), and low (executive subsystems) (Fig. 8.8). The SCS communicates with the ASADs using a Wi–Fi module connected to a router, with an amplifier and a LoRa communication module enhancing communication range. The Flight Planning and Control System (FPCS) employs multi-threaded architecture to manage the ASADs by handling task management, path planning, and scheduling (Fig. 8.9). The A* algorithm is used for efficient flight routing. Each ASAD consists of a quadrotor platform, a geophone sensor package, and a sensor deployment system (SDS) for ground coupling during data collection (Fig. 8.10). All ASADs are controlled by an onboard Raspberry Pi 4. The software architecture uses the Robot Operating System (ROS) to organize multiple nodes responsible for different system functions, including

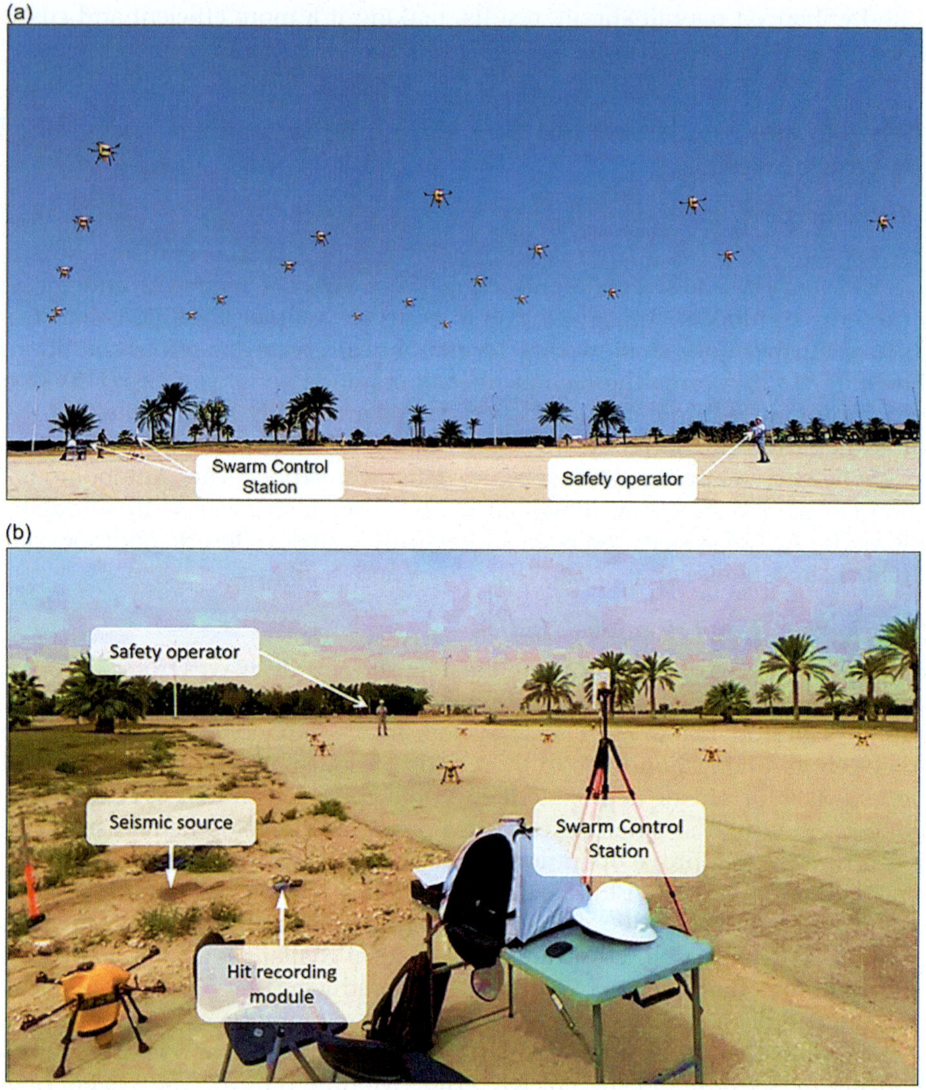

FIG. 8.7 Autonomous mission execution by a group of 20 ASADs controlled by the Flight Planning and Control System (FPCS). (B) System setup for testing with 9 ASADs. **Panel A**, *Source: Timoshenko et al. (2025)—Drones—Open Access;* **Panel B**, *Source: Timoshenko et al. (2025)—Open Access. Materials and Methods*

communication, autopilot, motor control, and diagnostics. The Hit Recording Module (HRM) follows a similar architecture for data acquisition, ensuring that all sensors are synchronized and functioning properly during the mission (Fig. 8.11).

8.5.5.3 Results

During field testing, ASADs were deployed in a 3 × 3 formation along a 500 m trace, with each ASAD positioned near a geophone array. The mission was divided into three stages,

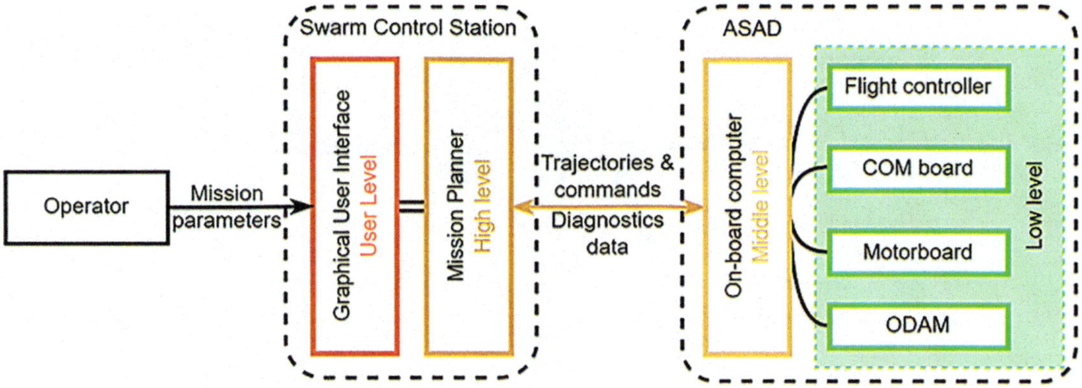

FIG. 8.8 Overview of the system architecture.

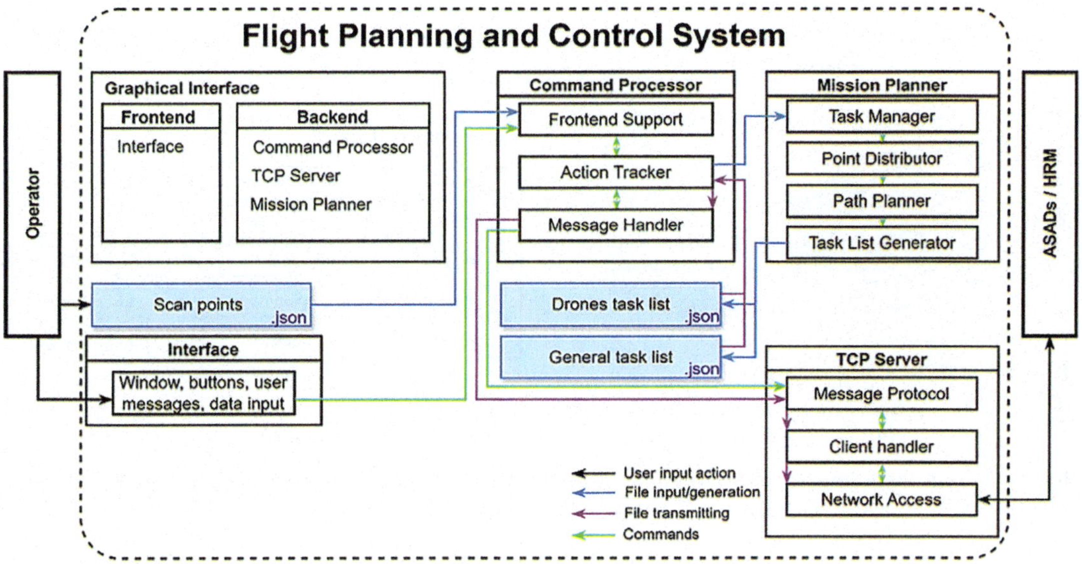

FIG. 8.9 FPCS (Flight Planning and Control System) architecture with detailed components. *Source: Timoshenko et al. (2025)—Open Access.*

due to the need for battery replacements, with 18 flights used to record data across 500 m, with a 25 m distance between each recording point (Figs. 8.12A and 8.12B).

The data collected by the ASADs demonstrated a high correlation with the traditional geophone system, showing that the UAV-based system could collect high-quality seismic data. In terms of frequency, the amplitude spectrum from the ASAD sensors showed a broader range than that of conventional geophone arrays (Fig. 8.13). A Signal-to-Noise Ratio (SNR) map indicated variations in data quality based on surface conditions, with improved SNR in specific areas. The average GNSS positioning error was found to be 1.09 m, with a range from 0.47 to 1.87 m.

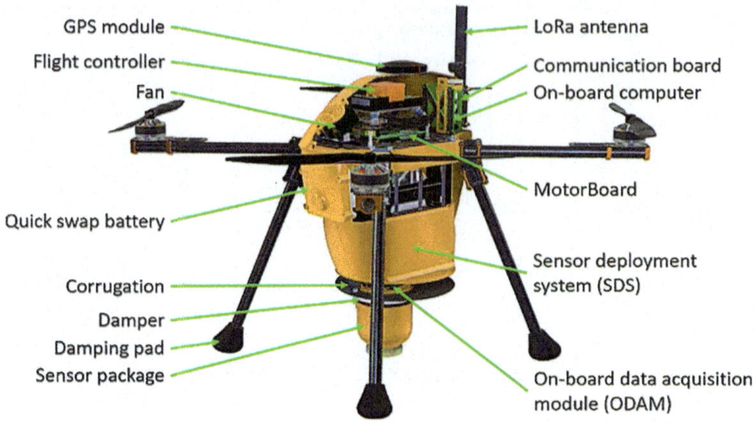

FIG. 8.10 Layout and configuration of the ASAD (Autonomous System for Advanced Deployment). *Source: Timoshenko et al. (2025)—Open Access.*

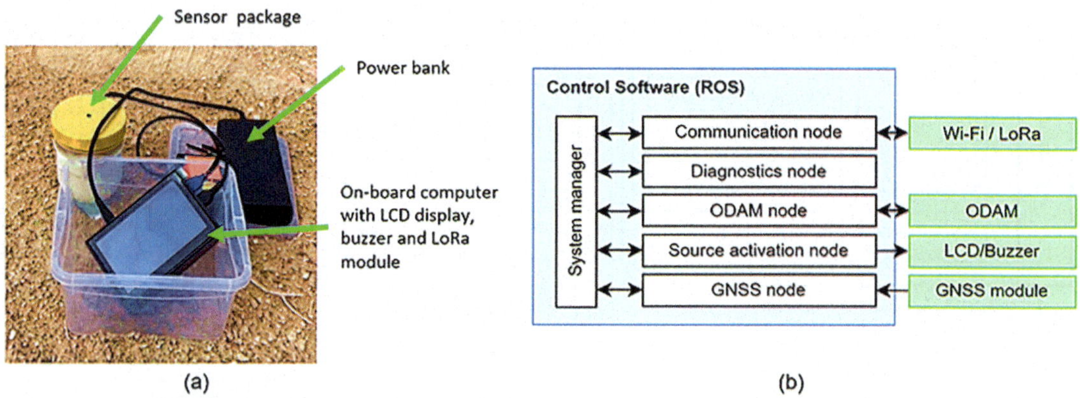

FIG. 8.11 Hit Recording Module: (A) Photo of the module; (B) Software layout and functionality of the Hit Recording Module. *Source: Timoshenko et al. (2025)—Open Access.*

8.5.5.4 Discussion

The ASAD system offers numerous advantages for seismic operations, including the ability to cover large areas more efficiently with fewer UAVs and to automate tasks like sensor deployment and transportation. This is in contrast to traditional nodal systems, which require extensive manpower and resources. The technology can be applied to various seismic tasks, such as engineering surveys, near-surface characterization, and shallow cavity detection. For large-scale engineering surveys, the system would need to support larger UAV groups, which would be determined by logistical constraints such as communication, power, and battery capacity. Multiple teams could manage swarms of up to 100 ASADs, with swarms launched from different areas and coordinated by a service crew. As the system scales, robust communication becomes more critical. This could be achieved through the use of high-altitude UAVs for swarm control or by having the ASADs function as radio re-translators in a distributed

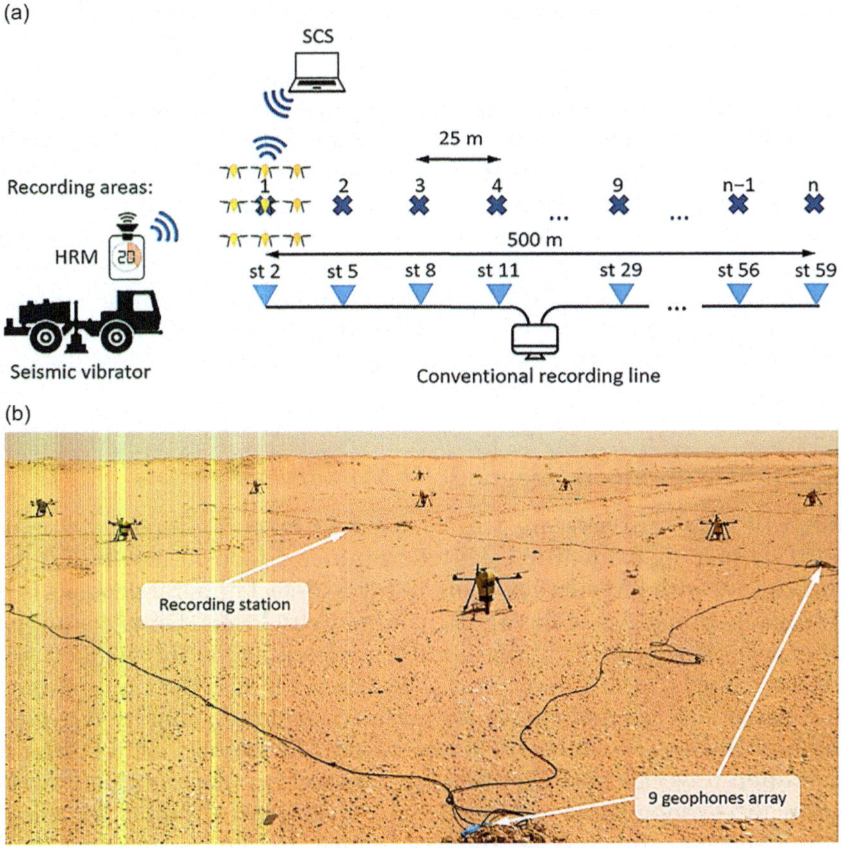

FIG. 8.12 Field experiment setup: 3 ASAD formation scheme. (B) Placement of 9 ASADs at station #2, positioned above an array of 9 geophones. *Source: Timoshenko et al. (2025)—Open Access.*

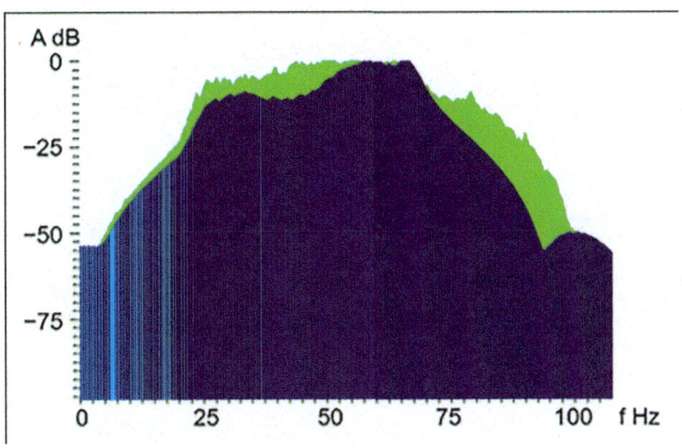

FIG. 8.13 Amplitude spectra of ASAD (green) and geophone (blue) traces. *Source: Timoshenko et al. (2025)—Open Access.*

network. To further expand the system, improvements in communication infrastructure, including long-range radio systems, are necessary.

In transitioning to an industrial-scale system, additional design modifications are needed to optimize weight, size, and sensor coupling to the ground. Field tests revealed issues with weak Signal-to-Noise Ratio (SNR), which were caused by swaying of the UAVs in the wind and improper coupling of the sensors to uneven surfaces. Solutions include improving the SDS mechanism, using landing gear to stabilize the robots on uneven surfaces, and employing more sensitive seismic sensors to reduce noise. Future work also involves implementing obstacle avoidance strategies using maps generated by aerial photography, optimizing swarm navigation patterns, and enhancing the Flight Planning and Control System (FPCS) to improve trajectory planning and mission optimization. This would improve overall operational efficiency, including reducing deployment time, data collection time, and battery replacement time.

8.5.6 Integration with electromagnetic (EM) surveys

Electromagnetic (EM) surveys are widely used to map conductivity variations in the Earth's subsurface, helping to identify areas where mineral deposits may be located (Fig. 8.14). By integrating drones with EM surveys, their ability to detect mineral deposits is significantly enhanced, especially in regions with varying conductivity levels. Drones equipped with EM sensors can collect data from multiple altitudes, providing more detailed coverage and improving the accuracy of the survey. This integration reduces the time required for traditional ground-based surveys and allows for more efficient data collection over large areas.

An example of the successful use of drone-based EM surveys can be seen in a specific mineral exploration project where drones were deployed to map conductivity variations in a remote region. By carrying EM sensors and flying at different altitudes, the drones provided a more comprehensive and precise dataset, which led to more accurate localization of mineral deposits. This case demonstrates how drone-based EM surveys can improve exploration efficiency and the precision of mineral discovery, making them a valuable tool in modern geophysical methods.

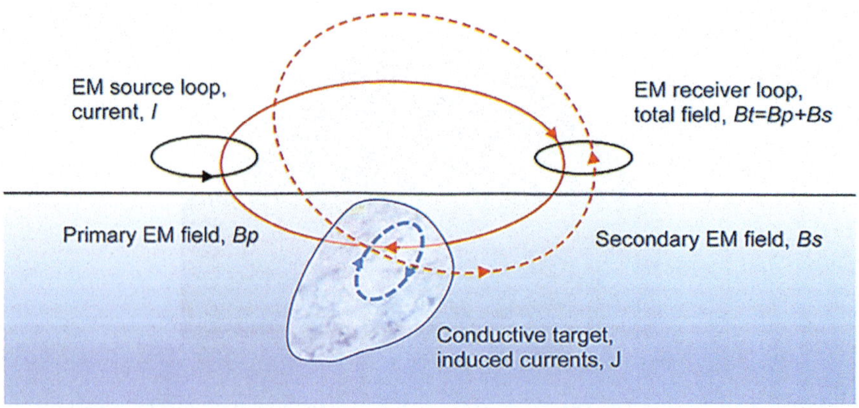

FIG. 8.14 Principles of geophysical inductive electromagnetic (EM) methods: Representation of primary field, induced currents, and secondary EM field. *Source: Pirttijarvi (2022)—Open Access.*

8.5.7 Drone-based electromagnetic survey system for geophysical applications: Case studies from the Lintumaansuo and Ervastinranta test sites, Finland

Geophysical electromagnetic (EM) methods are commonly used for geological mapping, mineral exploration, groundwater studies, and geotechnical investigations. Airborne EM methods are particularly useful because they can avoid obstacles like lakes, rivers, swamps, and ravines. While manned aircraft are traditionally used for these surveys, drones, or unmanned aerial vehicles (UAVs), offer several advantages. Drones are versatile, quick to deploy, cost-effective, and environmentally friendly. At present, magnetic surveys are the only type of geophysical survey typically conducted with drones. However, the limited payload capacity of drones restricts the use of other methods, such as EM and radiometric surveys.

Radai Ltd, a Finnish company, has been working on a solution to this challenge. They have developed Louhi, a drone-based frequency-domain EM survey system, as part of the EU-funded Horizon 2020 project NEXT-New Exploration Technologies. The Louhi system offers two operation modes: one uses a large loop on the ground as the EM source, while the other uses a small, portable EM transmitter loop. Both systems feature a lightweight, stand-alone three-component EM receiver that can be easily towed by a drone (Figs. 8.15A and 8.15B). This article discusses the technology developed by Radai Ltd, the current version of their EM device, and results from field and flight tests. These tests demonstrate the potential of this drone-based EM system for geophysical surveys (Pirttijarvi, 2022).

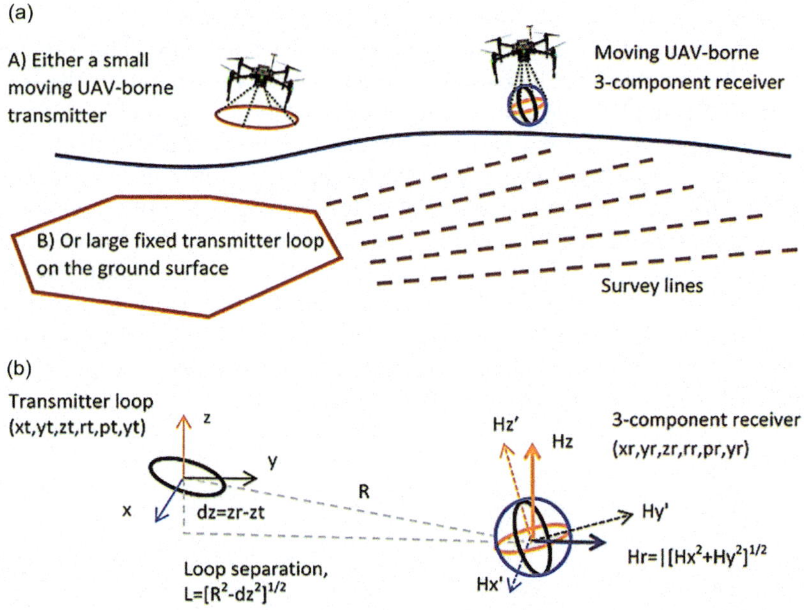

FIG. 8.15 Concept of unmanned aerial vehicle (UAV)-based electromagnetic (EM) survey method: Primary field generation by drone-carried current loop or ground-based large loop, and EM response measurement with 3-component receiver. (B) The parameters of a two-loop electromagnetic system. (xt, yt, zt) refer to XYZ coordinates of the transmitter and (xr, yr, zr) refer to those of the receiver. (rt, pt, yt) refer to roll, pitch, and yaw angles of the transmitter and (rr, pr, yr) to those of the receiver. (Hx′, Hy′, Hz′) are the three measured components of the magnetic field in device coordinates and (Hx, Hy, Hz) are the (away from transmitter) components of the magnetic field in world coordinates. *Source: Pirttijarvi et al. (2022)—Open Access.*

Case Study 1: Lintumaansuo Test Site, Finland

The Lintumaansuo test site, located in Kiiminki, 20 km north of Oulu, Finland, features a SE-NW-directed zone of black schists (pyrrhotite) that causes both electromagnetic conductivity and magnetic susceptibility anomalies (Honkamo, 1988). Along the Lintumaansuontie road, where the Louhi EM tests were conducted, a magnetic anomaly shows two closely located magnetized bodies. Previous VLF measurements (Mursu, 1991) did not detect separate conductors. In October 2020, Radai performed field tests at Lintumaansuo using the Louhi prototype 3 receiver system. Both small (1 m diameter) moving loop and large (120 m × 120 m) fixed loop systems were used. The separation between the small loop transmitter and receiver was about 25 m. Original data, including digital maps and terrain models, are available in the underlying data (Pirttijarvi et al., 2021). The orientation-corrected sensor data (Bu=Bz, Bn=Bx, Be=By) and the B_z/B_r ratio from the small loop measurement are shown in Fig. 8.16.

The radial component (B_r) represents the horizontal field, and B_z is the vertical component. The B_z/B_r ratio is presented as both real and imaginary components, along with amplitude and phase (a tan (B_z/B_r)). In the small loop measurement, the up component (B_u) corresponds to the Horizontal Loop Electromagnetic (HLEM) or Slingram response. A negative peak at X = 170 m indicates the NE location of the conductor. A weaker minimum at X = 210 m suggests either a second conductor or the SW side of a thick conductor. The data were interpreted using 1D inversion with Lempo1D software (Pirttijarvi, 2014), which minimizes the fit between measured data and modeled response. Lempo1D is private, but similar open-source software like Em1dinv (Auken et al., 2005) and Aarhusinv (Auken et al., 2015) can be used for similar interpretations.

The results from the Lempo1D inversion and the fit between measured and modeled responses are shown in Fig. 8.16 for the small loop system. Due to short loop separation, the small loop data was interpreted with a single-layer model. The pseudo-section in Fig. 8.16 is similar to apparent resistivity. The fixed loop data used a two-layer model, but the accuracy of the layer thickness is uncertain since only one frequency was used. Despite high noise from inaccurate real-time orientation correction, both datasets were successfully interpreted. The pseudo-sections illustrating conductivity/resistivity along the profiles provide meaningful geological information. Both the moving loop data (150–250 m) and fixed loop data (300–400 m) show a conductive zone approximately 100 m wide, with different resistivity values on opposite sides. The fixed loop data resolves two separate conductors. At the start of the fixed loop profile, resistivity values are overestimated due to the receiver being close to the loop. This spiking effect is evident at X = 50 m. The results from the fixed loop system are shown in Fig. 8.17.

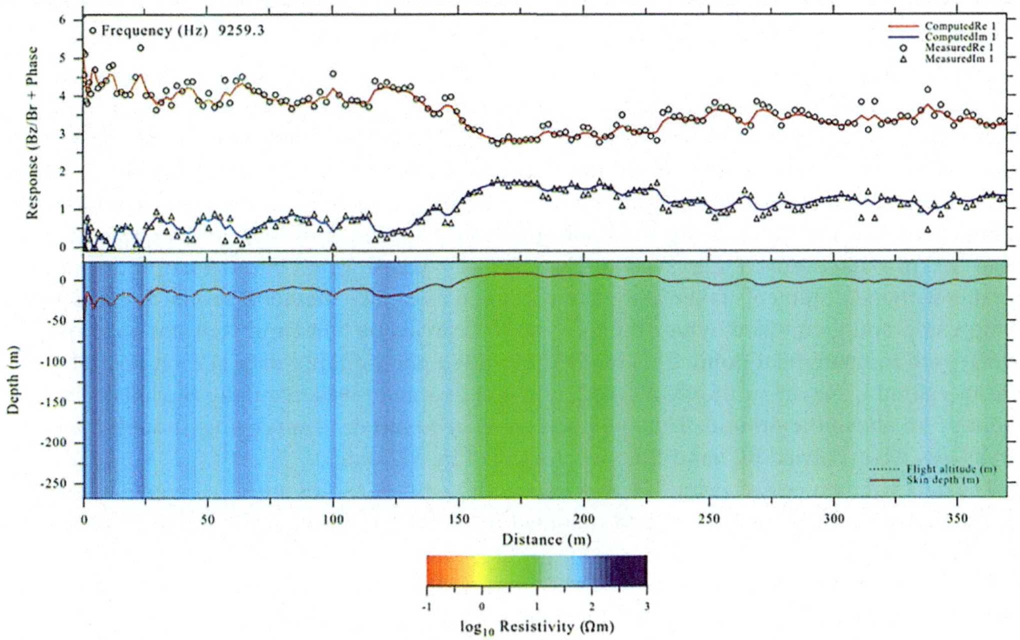

FIG. 8.16 Lempo1D inversion results for small moving loop data from Lintumaansuo. The top graph shows the fit between measured data (symbols) and modeled response (solid lines). The bottom graph depicts the resistivity pseudo-section along the profile. *Source: Pirttijarvi et al. (2022)—Open Access.*

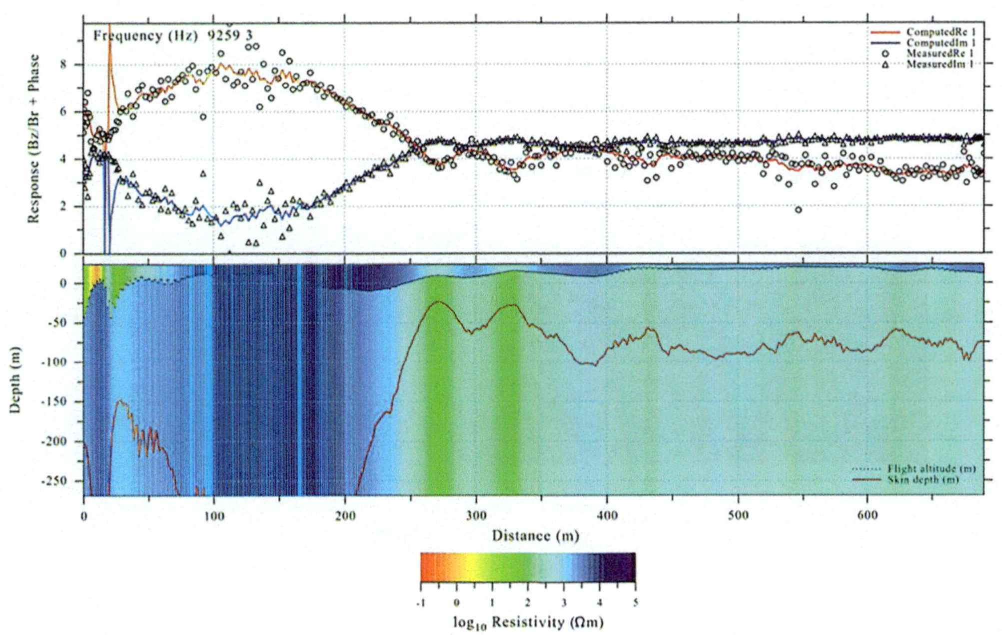

FIG. 8.17 Lempo1D inversion results for fixed loop data from Lintumaansuo. *Source: Pirttijarvi et al. (2022)—Open Access.*

Case Study 2: Ervastinranta Test Site, Finland

The Ervastinranta test site is located in Haukipudas, about 20 km north of Oulu, Finland. The site features an east-west (EW) directed contact between conductive sandstones (20 Ωm) of the Muhos formation to the south and resistive mica schists of the Kiiminki formation to the north (Honkamo, 1988). Previous studies by the University of Oulu indicate that the contact is nearly vertical and the thickness of the sandstones is around 100 m close to the contact. In May 2020, Radai conducted field tests using the Louhi prototype 3 receiver system in Ervastinranta with a large (100 m × 100 m) fixed loop system. The data were transformed into the amplitude of the B_r/B_z ratio, with measurements taken at the same location using two methods. Profile #1 was measured using a stop-and-go method, where the receiver was carried by hand and kept stationary for about 10 s at each measurement point. This method provided good-quality data at each point. Profile #2 was measured using an on-the-fly method, where the receiver was manually carried at a height of about 0.5 m above the ground. This method is similar to airborne surveying, though the receiver sways less when carried by hand than when carried by a drone.

The data were processed using 1D inversion with Lempo1D software. The inversion results and the fit between the measured data and modeled response are shown in Figs. 8.18 (Profile #1) and 8.19 (Profile #2).

Fig. 8.18 shows the results from Profile #1 (stop-and-go method), which has less noise, and the data points are more clustered around the true values. Fig. 8.19 shows the results from Profile #2 (on-the-fly method), which has a higher noise level, but the interpreted model is very similar to the one from Profile #1. Both profiles were interpreted using a two-layer model, where the top

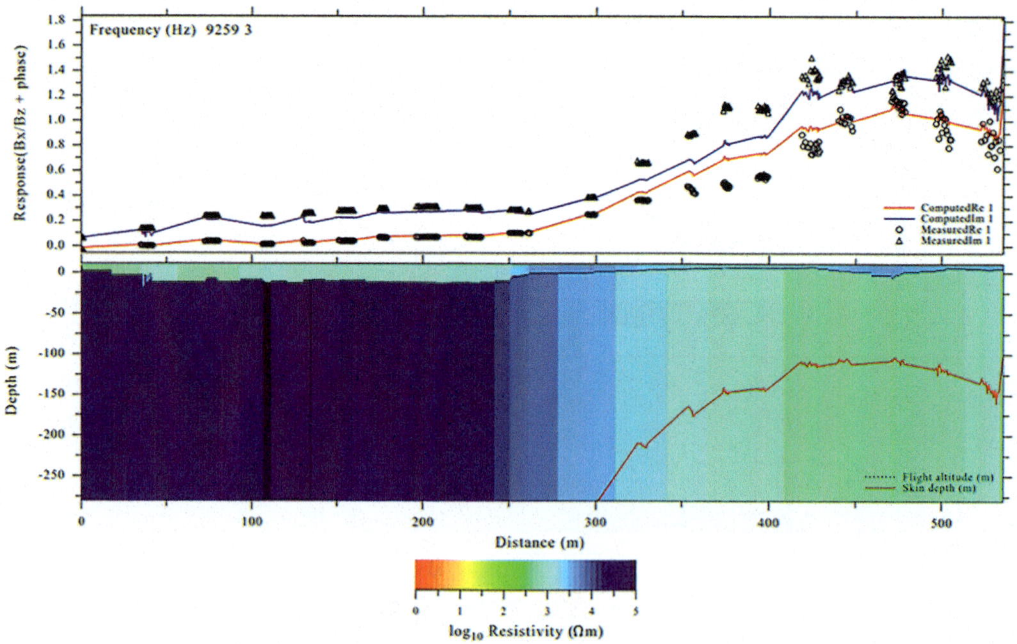

FIG. 8.18 Lempo1D inversion of fixed loop data from Ervastinranta profile #1 (stop-and-go method). *Source: Pirttijarvi et al. (2022)—Open Access.*

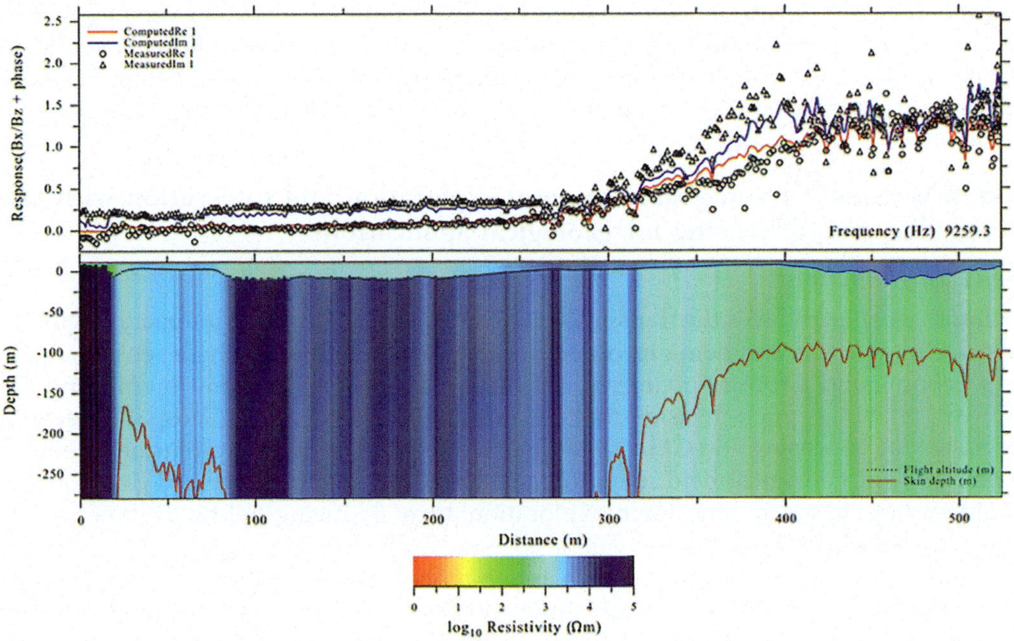

FIG. 8.19 Lempo1D inversion of fixed loop data from Ervastinranta profile #2 (on-the-fly method). *Source: Pirttijarvi et al. (2022)—Open Access.*

layer represents the soil overburden and the bottom layer represents the basement rocks. Despite the noise in Profile #2, both inversions showed the same geological contact between the resistive Kiiminki mica schists (north) and the conductive Muhos sandstones (south) at X = 320–350 m. The Louhi EM system proved to be effective for geological mapping, providing consistent results across both measurement methods.

8.5.8 Leveraging drone data in geospatial platforms (GIS)

Drone-collected data, such as high-resolution imagery and 3D point clouds, can be seamlessly integrated with other geophysical data in Geographic Information Systems (GIS) platforms. This integration enables geoscientists to perform spatial analysis, allowing them to visualize the relationships between surface features and subsurface anomalies. By combining drone data with geophysical methods, the resolution and accuracy of geological mapping are significantly enhanced, aiding in the identification of potential exploration sites. This approach allows for a more comprehensive understanding of the area being studied, improving decision-making and exploration planning.

An example of the effective use of GIS platforms in mineral exploration involves the integration of drone imagery, magnetic survey data, and seismic data. By combining these diverse data sources within a GIS environment, exploration teams can gain a more detailed and accurate view of the subsurface, resulting in a more efficient workflow. This integration not only improves the quality of geological mapping but also streamlines the exploration process, leading to faster identification of areas with high potential for mineral deposits.

8.5.9 Case study: Drone-based mineral exploration and integration with GIS and satellite remote sensing for geological applications

8.5.9.1 Introduction

Remote sensing involves the transmission of energy through space, interacting with matter at the Earth's surface, and being recorded by sensors on platforms. These sensors, deployed on spaceborne and airborne platforms, are essential for surveying and monitoring Earth's resources, enabling quick decision-making in mineral exploration. Geospatial techniques, especially remote sensing and GIS, play a significant role in mineral exploration by enhancing the efficiency of geological field mapping and facilitating rapid, informed decisions. These techniques help in narrowing down exploration areas, reducing fieldwork time, and minimizing overall exploration costs.

Various remote sensing methods, such as aerial photographs, Landsat, hyperspectral, and thermal imagery, are used for distinguishing and classifying geological features and mineral deposits. GIS integrates various data layers—geological maps, satellite images, and geochemical and geophysical data—to create mineral prospect maps. Despite the advantages of these methods, their application in mineral exploration remains underutilized due to misconceptions about data accuracy, resolution, and costs. Overcoming these barriers is essential for advancing geospatial methods in the geosciences.

Notable studies integrating Landsat, Sentinel-2, and ASTER data for mineral exploration, such as those by KWang et al., 2014 and Alarifi et al., 2022, demonstrate how remote sensing and GIS can be used to identify potential mineral zones. These studies show the benefits of using geospatial techniques, including cost-effectiveness and time efficiency, compared to traditional geophysical methods.

8.5.9.2 Overview

This review draws from a wide range of literature sourced from platforms like Google Scholar, Wiley Online Library, Academia, Elsevier, and ResearchGate. The research included both open-access articles accessed through subscriptions and full-text articles requested via ResearchGate (Sikakwe, 2023). The literature covered satellite remote sensing sensors and studies from various global regions, highlighting the broad applicability of remote sensing in geological and mineral exploration. Key search terms for the review included geological satellite sensors, multispectral and hyperspectral sensors, GIS in mineral exploration, big data analytics, drones, and machine learning.

Remote sensing began with early platforms such as balloons and kites for aerial photography, later advancing with the introduction of airplanes, which provided more stable platforms for aerial photography and laid the groundwork for modern remote sensing technologies. The launch of satellites like Landsat-1 marked a significant milestone, revolutionizing the field by allowing the use of electromagnetic radiation to collect data from the Earth's surface for geological mapping and mineral exploration.

Remote sensing sensors, particularly optical sensors, are passive in nature, capturing data across the ultraviolet, visible, near-infrared, and thermal regions of the electromagnetic spectrum. These sensors are used for mineral classification, soil mapping, and the detection of geological features. Advances in sensor technology, such as hyperspectral sensors, have provided higher resolution data that allows for more detailed geological mapping, including the identification of rock types and mineral compositions.

Microwave sensors, including Synthetic Aperture Radar (SAR) systems, are active sensors that operate independently of sunlight, enabling them to penetrate through soil and vegetation. This capability makes them ideal for mapping geological structures and identifying mineralization zones.

The choice of remote sensing platform depends largely on the mineral being explored. Spaceborne satellites are typically used for large-scale surveys, while airborne platforms such as aircraft and drones are often employed for high-resolution data collection in more targeted areas. The growing capabilities of these platforms, combined with advancements in data processing and analysis tools, have furthered the potential of remote sensing in mineral exploration and geological studies.

The increasing demand for minerals, combined with the decreasing discovery of new deposits, has led to the development of efficient and innovative techniques in mineral exploration. One key advancement is cloud computing, which addresses the challenges of managing vast geological datasets by enabling data sharing and analysis across different platforms. The Geological Cloud 1.0, launched in 2017, facilitates the integration of geological data, improving resource management and enhancing the overall exploration process.

Another significant development is big data analytics. With its high volume, variety, and velocity, big data analytics allows for the analysis of complex geological datasets. By integrating artificial intelligence (AI) and machine learning, big data analytics helps improve exploration efficiency, reduce operational costs, and identify patterns in mineral deposits. Studies have shown the potential of using deep learning algorithms to identify new mineralization zones, improving the accuracy and effectiveness of exploration efforts.

Drone technology has also made a significant impact on mineral exploration. UAVs (unmanned aerial vehicles) are used for geological analysis, 3D mapping, and structural assessments in both surface and underground mines. Drones equipped with various sensors, such as infrared and multispectral cameras, provide high-resolution, cost-effective, and accurate data, making exploration in remote areas more efficient. Several studies have demonstrated the effectiveness of drones in identifying iron mineralization and improving exploration productivity.

8.5.9.3 Machine learning in mineral exploration

Machine learning (ML) and deep learning have significantly advanced mineral exploration by processing vast remote sensing datasets and identifying relationships within mineral deposits. ML, a branch of artificial intelligence (AI), allows computers to recognize patterns and make predictions based on large, complex datasets. ML algorithms such as Support Vector Machine (SVM), Artificial Neural Networks (ANN), and Random Forest (RF) are commonly used to map mineral prospectivity, classify mineral deposits, and enhance exploration techniques.

ML has been instrumental in predicting mineral deposits by analyzing historical and modern geochemical data. For example, ML models have been used to predict elemental concentrations in geochemical exploration (Zhang et al., 2022) and to identify prospective areas for tungsten deposits (Liu and Yonghang, 2022). These models integrate geological, geochemical, and geophysical datasets to produce predictive maps for exploration.

ML is also valuable in classifying mineral deposits by identifying lithological units, alteration types, and indicator minerals using remote sensing data. Techniques like Decision Tree (DT), SVM, and Random Forest (RF) help generate mineral potential maps and aid in the classification of ore bodies based on various input data. Studies have shown that ML is effective in mapping mineral deposits such as zinc, copper, and gold, and in evaluating mineral potential in large-scale exploration areas (Gao et al., 2016; Singh et al., 2022). In conclusion, machine learning has become an essential tool in modern mineral exploration, offering improved prediction accuracy, data processing efficiency, and resource identification.

8.5.9.4 Classification of satellite remote sensors and their role in mineral mapping

Satellite remote sensors are classified based on their spatial resolution, which determines their suitability for different applications, including mineral exploration. Coarse resolution sensors, with a spatial resolution greater than 250 m, such as NOAA, AVHRR, Terra MODIS, and Landsat MSS/TM 4–5, are used for large-scale monitoring of extensive areas or cosmic scale applications. Medium resolution sensors, ranging from 10 m to 250 m, include instruments like Landsat ETM + and IRS satellites, providing a balance between spatial detail and area coverage, ideal for broader geological surveys. High-resolution sensors, such as Ikonos, Quick Bird, SPOT-5, and WorldView, offer detailed imagery with resolutions <10 m, suitable for smaller-scale mapping and mineral exploration, although they require greater data storage and processing capabilities.

Multispectral and hyperspectral sensors are particularly useful in mineral mapping due to their ability to detect a wide range of wavelengths. Multispectral sensors, with 3–12 spectral bands, are effective for mapping larger areas, while hyperspectral sensors, which contain hundreds of spectral bands with finer spectral resolution, are highly valuable for precise mineral identification. For instance, AVIRIS hyperspectral scanners can effectively identify mineral compositions in rock outcrops. The use of multispectral and hyperspectral sensors, such as those from Landsat, ASTER, and SPOT, has proven successful in identifying lithologies, geological structures, and hydrothermally altered zones associated with mineral deposits. These sensors have been applied in mapping gold, copper, and other mineral deposits, as well as detecting mineralogical changes in regions like South Africa, Sudan, Oman, and Australia.

In mineral exploration, the classification of satellite sensors by spatial resolution enables geologists to choose the appropriate sensor for specific tasks, from large-scale geological monitoring to detailed mineral identification, ensuring effective and efficient exploration.

Hyperspectral remote sensing is a powerful tool in mineral exploration, especially in challenging environments like the Arctic. Unlike traditional methods, hyperspectral sensors identify a wide range of surface minerals, such as hydroxyl-bearing minerals, clays, sulphates, and silicates, helping to locate potential ore-bearing hydrothermal zones. These sensors offer high spatial resolution and cover a wide spectrum, from visible light to near-infrared wavelengths, enabling precise mineral identification.

China pioneered hyperspectral remote sensing in the 1980s, using sensors like the Field Imaging Spectrometer System (FISS) and Modular Airborne Imaging Spectrometer (MAIS). These tools have been used in regions such as the Qilian Mountains and Xinjiang Province for mineral and oil exploration. Hyperspectral techniques have proven effective in identifying minerals in magmatic, hydrothermal, and sedimentary deposits, with successful applications in kimberlite, porphyry, and SEDEX Pb–Zn–Ag deposits.

The technology has also been used in areas with difficult access, such as Canada's Arctic regions, to map indicator minerals like nickel, copper, and platinum. Advances in deep learning have further enhanced the ability to extract information from hyperspectral data, improving mineral discrimination. For example, studies using sensors like AVIRIS-NG have successfully mapped hydrothermal alterations in minerals like talc and kaolinite. Additionally, hyperspectral imaging integrated with other techniques, such as petrography and geochemical data, has been used to study lithology and ore grades, offering valuable insights for mineral exploration.

Image improvement techniques enhance remote sensing data to better identify features and improve analysis. These methods include contrast stretching, linear combination, principal component analysis (PCA), spatial filtering, and band combination. Contrast stretching increases the visibility of features, while PCA reduces correlated data into uncorrelated components, making the data easier to interpret.

PCA has been used in studies like mapping lithology and structural features using ASTER and Landsat 8 OLI data. It helps condense waveband data into principal components that are more interpretable. PCA is also used in combination with other techniques like band ratio (BR) and Minimum Noise Fraction (MNF) to extract mineralogical information, such as hydrothermal alteration minerals.

Other methods, like Independent Component Analysis (ICA), uncover hidden features in hyperspectral data that PCA and Factor Analysis (FA) might miss. For instance, ICA has been used alongside PCA and MNF for lithological separation in mineral exploration. Tools like Spectral Angle Mapper (SAM) and Adaptive Coherence Estimator (ACE) are also effective in identifying minerals by analyzing spectral data. These techniques have been applied successfully in regions with complex geological formations, such as in Iran for mapping Pb–Zn mineral deposits.

8.5.9.5 Spectral bands for mineral deposit distribution

Shortwave infrared (SWIR) bands are essential for identifying minerals such as ferric/ferrous, clays, carbonates, sulphates, and mica. These minerals have unique spectral characteristics, particularly in the SWIR range. Hyperspectral imaging in the VNIR (Visible and Near-Infrared) and SWIR (Shortwave Infrared) range (0.4–2.5 μm) is effective for detecting minerals and rock types, as different iron-containing minerals show distinct absorption in the VNIR, while hydroxyl-bearing minerals (e.g., clays, sulphates, carbonates) have unique signatures in the SWIR. Landsat OLI, enhanced with edge improvement and filtering, shows that bands 4 and 5 are optimal for isolating structural features.

Geographical Information Systems (GIS) integrate spatial and non-spatial data to support decision-making and spatial analysis. In mineral exploration, GIS combines geological, geochemical, and geophysical data to generate mineral potential maps, which focus future investigation efforts on specific areas. GIS tools like ArcGIS and QGIS help manage vast datasets and facilitate faster processing, allowing for the identification of mineral occurrences and reducing exploration costs. GIS also enables the creation of 3D models and the use of multi-source map overlays to generate accurate predictions for mineral deposits (Fig. 8.20) (Sikakwe, 2023).

Regression analysis is often used in mineral exploration to establish relationships between variables and predict mineral potential. For example, multivariate regression analysis of remote sensing data has been applied to delineate mineral resource potential (Monsouri et al., 2018). The regression models, such as those developed from ASTER satellite data, are used to create mineral potential maps based on the relationships between spectral bands and mineral deposits. The adjusted R-squared (R^2_{adj}) value is a key indicator of model accuracy, with values close to 1 suggesting a well-fitted model for predicting mineral locations (Granian et al., 2015).

Remote sensing has evolved from early platforms like pigeons, kites, and balloons to modern satellites, offering broader coverage and more sophisticated data collection compared to traditional aerial photography. Satellite remote sensing provides advantages over

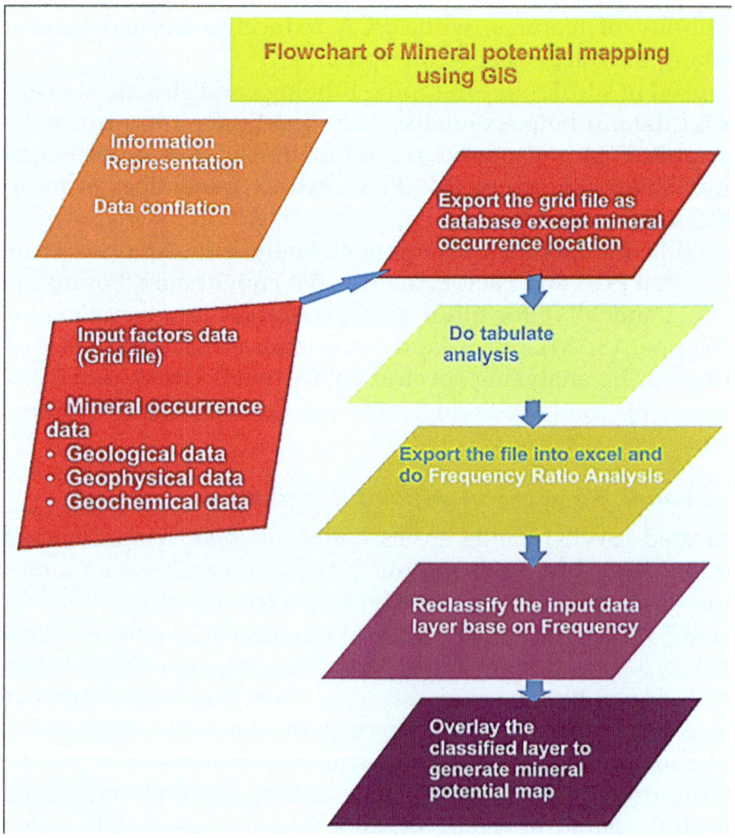

FIG. 8.20 Flowchart of GIS application in mineral potential mapping. *Source: Self designed.*

aerial photography, such as larger coverage areas and the ability to gather data continuously, unaffected by weather conditions. While passive remote sensors rely on external light sources, active sensors have their own energy source, allowing them to operate day and night and penetrate obstacles like clouds and smoke.

Satellite remote sensing plays a significant role in mineral exploration, particularly in combination with GIS, for processing geological, geophysical, and geochemical data. This approach allows for efficient exploration in challenging terrains, with lower costs and reduced time compared to traditional fieldwork and geophysical surveys. However, the method has limitations, primarily due to its inability to directly sense subsurface features, requiring complementary geophysical and geochemical investigations.

While multispectral and hyperspectral satellite sensors are effective for identifying mineral zones based on surface reflectance, their limitations in depth and difficulty in distinguishing similar materials remain challenges. Advances in technologies like cloud computing, drones, machine learning, and big data analytics are enhancing mineral exploration, particularly in remote areas. These technologies offer more precise, three-dimensional data than traditional methods, opening new avenues for discovering mineral deposits.

Despite these advances, there is a shortage of geoscientists with expertise in geospatial technologies, which hinders the full potential of remote sensing in geoscience. To address this, greater emphasis on geospatial literacy and specialized training is needed to support the growing demand for advanced geospatial analysis in geoscience, ultimately improving mineral exploration efforts.

8.5.10 Advantages of integrating drone-based surveying with other geophysical methods

Integrating drone technology with traditional geophysical methods offers several key benefits that significantly enhance the effectiveness of mineral exploration. One of the primary advantages is increased accuracy. Drone data provides more detailed and accurate surface-level insights, which improves the interpretation of geophysical data, allowing for better decision-making in the exploration process. This enhanced precision also leads to improved exploration efficiency. By reducing the time required for field surveys, drones enable faster data acquisition, which leads to more efficient resource allocation and quicker identification of potential mineral deposits.

The combination of drones with other geophysical methods also results in cost-effectiveness, particularly in remote or difficult-to-reach areas. Traditional methods often require significant manpower, equipment, and time, whereas drones can reduce these costs by quickly covering large areas and reducing the need for ground-based surveys. Additionally, drones enable high-resolution imaging and sensor data collection, enhancing data resolution. This high level of detail allows for the detection of subtle geological features that may otherwise be missed by traditional methods.

Lastly, drones help minimize the environmental impact of exploration activities. Their ability to perform non-invasive surveys reduces disruption to the surrounding environment compared to traditional exploration methods that may involve more intrusive techniques like drilling. This makes drone-based surveying a more sustainable and eco-friendly option for mineral exploration.

8.5.11 Developing multi-sensor drones for geological mapping and mineral exploration: Setup and first results from the MULSEDRO project

The case study focuses on the use of Unmanned Aerial Systems (UAS) for geological applications, specifically within the EU-funded MULSEDRO project (MULti-SEnsor DROnes) (Heincke et al., 2019). This project develops lightweight UAS systems capable of collecting both magnetic and hyperspectral data for mineral exploration. The case study highlights the use of both multi-copter and fixed-wing UAS, each offering unique advantages. Multi-copters are used for high-resolution data collection over rugged terrain at low altitudes, while fixed-wing UAS cover larger areas quickly due to their higher speed and longer endurance. Additionally, an integrated positioning system (IPS) was developed to provide accurate data in areas with poor GPS reception, such as mining tunnels and narrow valleys. The systems were designed to operate within regulatory constraints, focusing on take-off weights under 5 kg for easy deployment during geological field campaigns. The general concept of the MULSEDRO project is summarized in Fig. 8.21.

The case study presents the use of multi-sensor Unmanned Aerial Systems (UAS) for geological exploration, specifically in the context of the EU-funded MULSEDRO project. Two UAS systems were developed for this project: a fixed-wing system by Radai Oy, Finland (Fig. 8.22A), and a multi-copter system by the Helmholtz Institute Freiberg for Resource Technology (HZDR-HIF) (Fig. 8.22B). The fixed-wing UAS offers long flight times (up to three

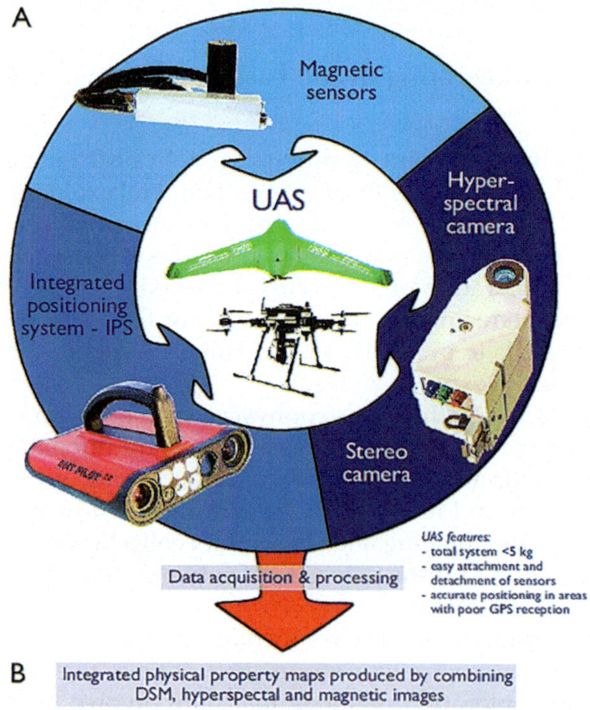

FIG. 8.21 (A) UAS multi-sensor systems for magnetic and hyperspectral data collection. (B) Data integration with digital surface models (DSMs) to map physical properties in the Siilinjarvi carbonatite-glimmerite complex, Central Finland. *Source: Heincke et al., 2019—Open Access.*

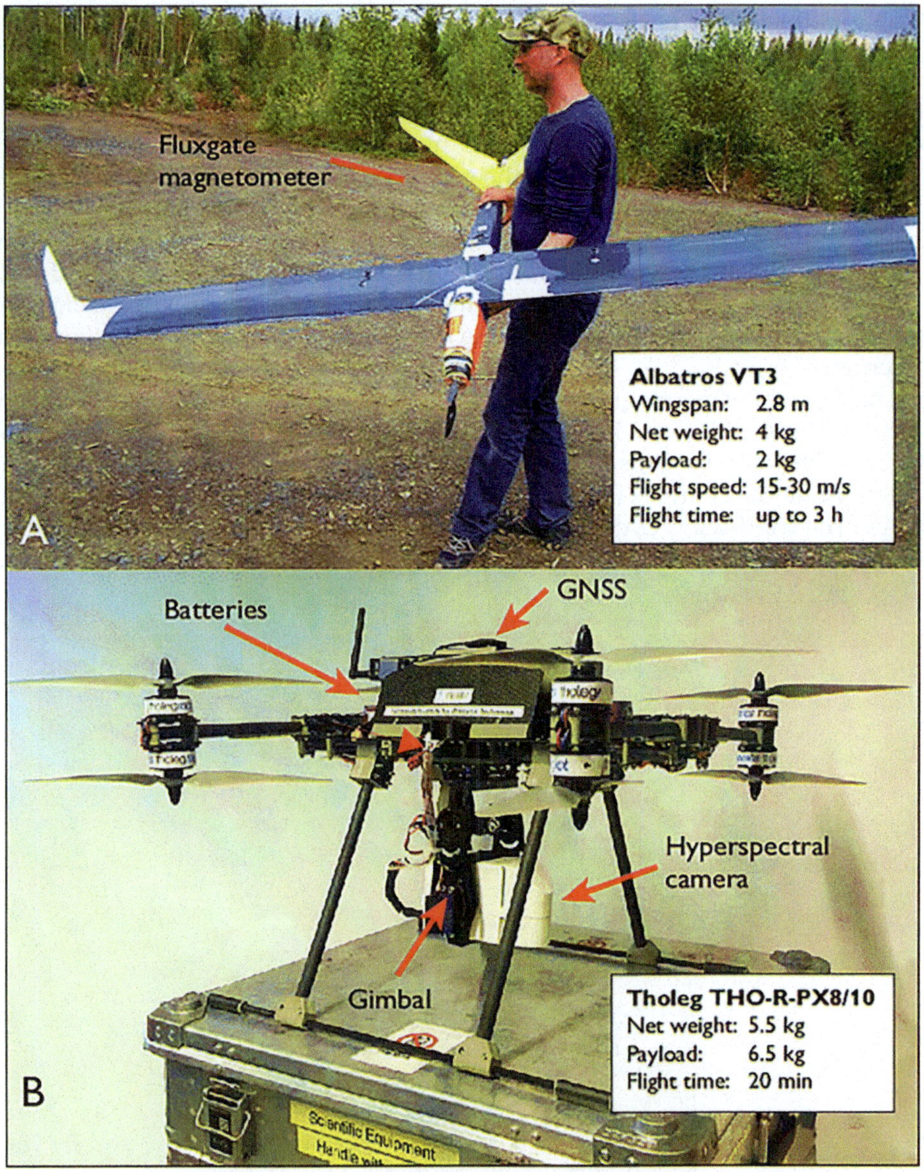

FIG. 8.22 The UAS systems used within the project. (A) Custom-built fixed-wing UAS from Radai Oy equipped with a magnetometer. (B) Customized multi-copter from HZDR-HIF equipped with a hyperspectral frame camera, tested in the Siilinjarvi carbonatite-glimmerite complex, Central Finland. *Source: Heincke et al., 2019—Open Access.*

hours) and smooth flight trajectories, ideal for acquiring high-quality magnetic and multispectral data. The multi-copter system is flexible, allowing quick attachment and detachment of various sensors for efficient surveying.

Both UAS platforms are equipped with three-component fluxgate magnetometers for magnetic surveys and Global Navigation Satellite Systems (GNSS) receivers and Inertial

Measurement Units (IMUs) to measure the UAS's position and orientation. The magnetic data is corrected for diurnal variations using base stations, and flight paths are autonomous along predefined lines. The multi-copter allows for surveys with higher resolution, achieving magnetic maps comparable to ground-based studies with line spacing as narrow as 7 m.

Hyperspectral data are collected with the multi-copter using the Rikola Hyperspectral Imager camera, which captures images in the visible and near-infrared spectrum (504–900 nm). To ensure consistent data, the multi-copter maintains a steady position during data acquisition. The hyperspectral system is particularly effective for mapping iron-bearing minerals and vegetation features associated with mineralization. Additionally, a high-resolution digital surface model (DSM) is created through a photogrammetry survey using a commercial fixed-wing UAS equipped with an RGB camera (Fig. 8.23A) for accurate draping of the hyperspectral data onto the terrain.

A key challenge of using magnetic sensors on multi-copters is electromagnetic noise from the motors. However, this noise is filtered out during data processing, ensuring high-quality magnetic data. The UAS systems were tested in Central Finland, specifically in the Siilinjarvi carbonatite-glimmerite complex (Fig. 8.24A), where they conducted magnetic and hyperspectral surveys. The fixed-wing UAS mapped large areas with magnetic anomalies (Figs. 8.24B and 8.24C), while the multi-copter gathered high-resolution data over outcrops (Fig. 8.24D).

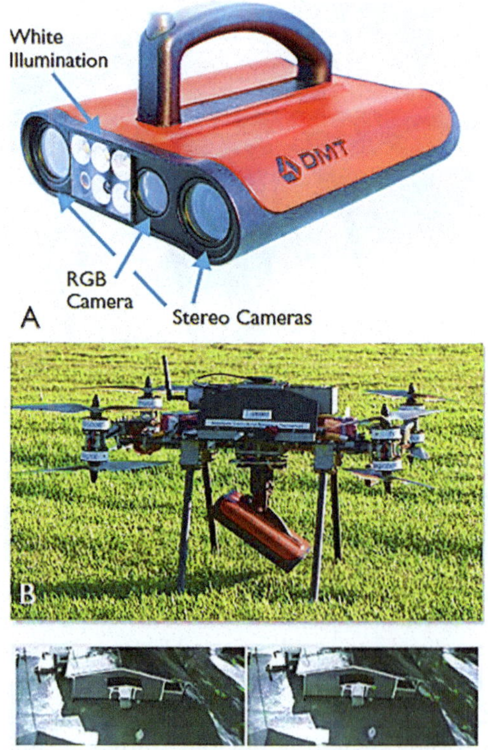

FIG. 8.23 The integrated positioning system PILOT3D. (A) Handheld device. (B) Mounted on a UAS. (C) Images obtained by the stereo camera during a UAS test flight in the Siilinjarvi carbonatite-glimmerite complex, Central Finland. *Source: Heincke et al., 2019—Open Access.*

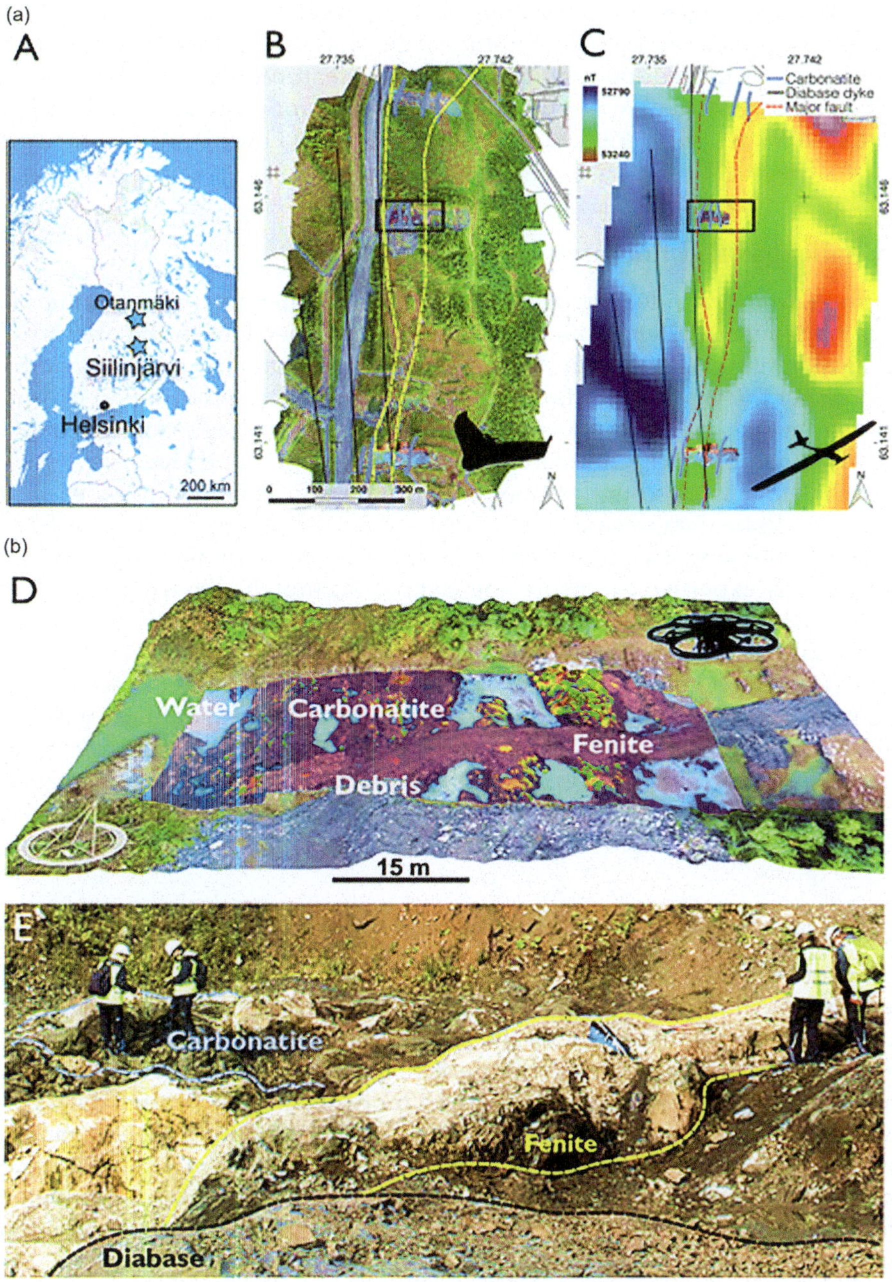

FIG. 8.24 (A) Location of field test sites in Finland, with results presented only from the Siilinjärvi site. (B) Orthophoto from photogrammetry. (C) Total magnetic intensity maps from fixed-wing UAS surveys of Siilinjärvi. (D) Processed hyperspectral data draped on the digital surface model, showing fenite-hosting rocks (green shading). (E) Ground sampling at one of the outcrops covered by hyperspectral surveying, conducted in the Siilinjarvi carbonatite-glimmerite complex, Central Finland. *Source: Heincke et al., 2019—Open Access.*

Ground-based investigations supplemented the UAS data, including handheld XRF, spectroradiometric, and rock samples for laboratory analysis. The combined datasets revealed a major fault zone and helped identify key lithological features. In the future, the integration of multi-spectral sensors on fixed-wing UAS is planned for large-area exploration, and the UAS-based systems will be deployed in remote areas like Greenland for further geological studies.

8.5.12 Challenges and limitations of integration

While integrating drone technology with geophysical methods offers numerous advantages, there are several challenges and limitations to consider. One of the primary challenges is the technical limitations of drones. Drones have limited payload capacity and battery life, which can restrict the types of equipment they can carry and the duration of surveys. This limitation can impact the ability to conduct long-duration or high-capacity data collection over large areas, requiring careful planning and optimization of drone use.

Another challenge is the complexity of data processing. The integration of multiple data sources, such as drone data, seismic data, and magnetic data, requires advanced processing and analytical techniques. This can be complex and time-consuming, often requiring specialized software and expertise to ensure the data is accurately interpreted and integrated. Handling large datasets from different sources can also pose logistical difficulties, particularly when managing and storing the information.

Lastly, regulatory and safety concerns are important when using drones for geophysical surveys. Drone operations must comply with local aviation regulations, which may vary depending on the region. Additionally, safety considerations must be carefully accounted for, especially when flying drones over sensitive exploration sites. Ensuring that drones are operated safely and within the legal frameworks is crucial to avoid accidents and ensure the success of the survey.

8.5.13 Future directions and potential developments

The future of integrating drones with geophysical methods holds significant promise, with various advancements on the horizon that could further enhance the efficiency and accuracy of natural resource exploration. One of the most notable areas of development lies in sensor technology. As sensor capabilities continue to evolve, drones will be able to carry more advanced instruments capable of capturing even more detailed and varied data. For instance, sensors with greater sensitivity and wider detection ranges will improve the ability to identify subtle geological features, which can be crucial for pinpointing mineral deposits. This progression in sensor technology will also enable drones to operate in a broader range of environments, including those with more complex geological or climatic conditions.

Another transformative advancement is the application of Artificial Intelligence (AI) in data analysis. As drones collect vast amounts of data during geophysical surveys, AI-powered algorithms will play an increasingly important role in processing and analyzing this information. AI can enhance the accuracy and speed of data interpretation by automatically identifying patterns, correlations, and anomalies in the collected data. Machine learning algorithms, in particular, can be trained to recognize the geophysical signatures of specific mineral types, significantly reducing the time needed for human analysis. This integration of AI not only improves the precision of mineral detection but also makes the overall survey process more efficient, enabling geoscientists to make faster, data-driven decisions.

In addition to advancements in sensor technology and AI, the development of automated drone fleets represents another exciting frontier for large-scale exploration projects. The use of coordinated fleets of drones, capable of working together autonomously, has the potential to revolutionize how large areas are surveyed. These fleets could be deployed to cover vast terrains much faster than individual drones, all while collecting high-resolution data across multiple geophysical methods simultaneously. The ability to manage fleets of drones efficiently will allow for real-time adjustments and adaptive survey strategies, optimizing data collection and reducing costs. Moreover, with advancements in drone automation and remote operation capabilities, these fleets can operate in challenging or hazardous environments, minimizing the risk to human survey teams.

As drone technology continues to evolve, we can also expect improvements in battery life, payload capacity, and overall flight endurance. These advancements will increase the operational capabilities of drones, allowing them to conduct longer surveys and carry heavier, more powerful equipment, thus expanding the scope of exploration activities. Additionally, the growing adoption of autonomous drones means that they will be able to conduct surveys with minimal human intervention, leading to safer, more efficient operations in remote or difficult-to-reach locations.

The continued integration of drones with geophysical methods will lead to significant strides in the efficiency, cost-effectiveness, and precision of natural resource exploration. The combined advancements in sensor technology, AI, and autonomous drone fleets will likely propel the industry forward, offering unprecedented opportunities for discovering and mapping mineral deposits in a more sustainable and environmentally friendly manner. As these technologies evolve, the potential for drones to transform geophysical surveys in the coming years seems limitless, opening up new possibilities for faster, more accurate, and cost-effective resource exploration.

8.6 Conclusions

The integration of drone-based surveying techniques into the geoscience domain has ushered in a new era of precision, efficiency, and accessibility in natural resource exploration. This chapter has demonstrated how drones, when combined with advanced satellite remote sensing (RS), GIS, and various geophysical methods, are transforming mineral exploration and geological mapping. From magnetic and electromagnetic surveys to ground penetrating radar (GPR) and seismic imaging, drones offer a flexible and cost-effective platform for acquiring high-resolution data in challenging and remote terrains.

The case studies presented—spanning India, Iran, Finland, and other regions—underscore the growing global reliance on unmanned aerial vehicles (UAVs) for site-specific exploration, particularly in mineral-rich but logistically complex environments. Notably, projects such as Marut Drones collaboration with NMDC and the MULSEDRO initiative highlight the momentum toward multi-sensor integration and the development of specialized UAV systems tailored for geological applications.

While the synergy between drone technologies and traditional geophysical methods has unlocked significant advantages—including improved spatial resolution, reduced human risk, and real-time data acquisition—it is not without challenges. Technical limitations related to payload capacity, data processing complexity, regulatory constraints, and the need for interdisciplinary expertise remain critical barriers to widespread adoption.

Looking forward, advancements in AI-driven data analytics, autonomous UAV operations, and sensor miniaturization are expected to enhance the capability and reliability of drone-based surveying. Continued innovation and cross-disciplinary collaboration will be essential for realizing the full potential of UAVs in sustainable and responsible natural resource exploration.

In summary, drone-based surveying techniques have become indispensable tools in modern geoscience, offering a scalable and integrative approach to understanding Earth's subsurface and surface features. As technologies evolve, drones will continue to play a pivotal role in shaping the future of exploration, mapping, and environmental stewardship.

CHAPTER

9

Seismic surveying techniques

9.1 Introduction

Seismic surveying has several impressive classical studies to explore subsurface natural resources associated with shallow and deeper formations at different depths (Mishra and Naskar, 2024). The near source exploration of the targeted mineral deposits and renewable subsurface energy sources are associated with different causative formations at comparatively shallow layers, which are reliably explorable using seismic refraction and reflection methods in combination of other geophysical methods of exploration having reliable probing potential at varying depths. This is so because the origin of mineral resources is closely related to the deeper subsurface processes in the crustal and subcrustal layers at varying depths (Mishra and Naskar, 2025, 2026). Recently, it has been documented that the seismic imaging using 3D seismic tomographic techniques of various scales of resolution of the imaged structures has found its diverse applicability in the exploration of natural resources and in providing engineering solutions to the subsurface-related developmental projects (Zhao, 2015; Mishra, 2025). The deeper hidden layers embedded with varying structural heterogeneities having appreciable acoustic impedance contrasts (velocity x density) are very much decipherable with higher resolution using both controlled or active source (e.g., hitting the iron plate by sledge hammer; vibriosis; weight drops from height; varying charges of explosive blasts; Betsy gun and airgun for marine surveys) tomography and uncontrolled or passive source (e.g., earthquakes; ambient seismic noise) tomography. There are several classical studies made by different researchers using seismic tomography to explore the nature and contents of the different subsurface layers in the diverse tectonic settings with high resolvability of structures at the deep crustal to subcrustal depths (Aki et al., 1977; Zhao, 1992; Zhao, 2001; Zhao et al., 2002; Mishra and Zhao, 2003, 2004; Singh et al., 2011; Singh and Mishra, 2015; Mishra et al., 2023). Most of mineral, oil, gas, and engineering solution provider-based companies deploy controlled source 3-Seismic tomography to get information about the causative bodies and formational details lying at few hundred meters (e.g., Seguin et al., 1994) while detailed crustal and subcrustal processes are studied by deploying uncontrolled or passive source 3D seismic tomography to image structures of few tens of kilometers (e.g., Zhao, 2015; Mishra, 2025).

Seismic surveys are generally an extension of the originally understood earthquake seismology (Kearey et al, 2002) providing information about the Earth's layers as the waves propagate through them. These layers, in turn, give information about their physical

Baseline Geoscience Data Generation for Sustainable Exploration of Natural Resources
https://doi.org/10.1016/B978-0-443-45221-5.00011-4

properties and composition. A seismic survey is a method of investigating subterranean structures, particularly as related to the exploration of petroleum, natural gas, and mineral deposits. These methods utilize the propagation of elastic waves in the Earth, which are reflected and refracted when they intersect a boundary between rocks of different acoustic impedance. The detection of the reflected or refracted components allows the boundaries to be mapped. Seismic methods are becoming increasingly important tools to help unravel structures hosting mineral deposits at great depths for mine planning and exploration. The technique is based on determining the time interval that elapses between the initiation of a seismic wave at a selected shot point (the location where an explosion generates seismic waves) and the arrival of reflected or refracted impulses at one or more seismic detectors (geophones). Seismic air guns are commonly used to initiate the seismic waves. This technique has largely replaced the practice of exploding dynamite underground. Electric vibrators or falling weights (thumpers) may also be employed at sites where an underground explosion might cause damage—for example, where caverns are present. Upon arrival at the detectors, the amplitude and timing of waves are recorded to produce a seismogram (a record of ground vibrations).

Application after the discovery of X-ray in the Medical Sciences revolutionized the concept of seismic imaging of the subsurface layers by deploying advanced tools of 3D seismic tomography that is based on unique concept of deciphering layers of near-surface and deeper sub-surface associated with varying amount of structural heterogeneities within the layer as well as outside at the boundary of two different layers by developing a versatile algorithm of ray trace of pseudo bending combined with Snell's law, for the first time in the history of 3D seismic inversion (Zhao, 1992; Zhao, 2015). This classical concept of Zhao (1990) found its wide-scale applicability all over the globe, including the exploration of minerals and finding engineering solutions for the critical structures even for the areas of complicated seismotectonic settings wherein active (controlled as for example explosive/vibroseis for shallow information) and passive (uncontrolled as per example earthquake for deeper information) source can be used to generate seismic wave within the subsurface layers for conspicuous and judicious imaging of underneath strata (Mishra, 2025).

Generally, the density of rocks near the surface of the Earth increases with depth. Seismic waves initiated at a shot point at or near the surface may reach the receiving point by reflection, refraction, or both. When the shot point is close to the receiving point, reflected waves usually reach the receiving point first. At greater distances, however, the seismic pulse travels faster by the refraction path because its velocity is greater along the top of the lower, denser layer than it is through the upper layer; in this case, the refracted wave arrives first. Interpretation of the depths and media reached by seismic waves thus depends on the distance between shot points and receiving points and the densities of the strata. The results of a seismic survey may be presented in the form of a cross-sectional drawing of the subsurface structures as if cut by a plane through the shot point, the detector, and Earth's center. Such drawings are called seismic profiles.

Seismology is a scientific discipline concerned with the study of earthquakes and the propagation of seismic waves within the Earth. Exploration seismology has been established since 1923 and has been used extensively in natural resource exploration (Sheriff and Geldart, 1995; Dentith and Mudge, 2014). As a branch of geophysics, it has provided significant information about the composition and state of the planet's interior. The goals of seismological investigations may be local or regional, such as determining subsurface faults and other structures in petroleum (a natural resource and complex mixture of hydrocarbons that occur in the Earth in liquid form; the liquid and gaseous phases of petroleum constitute the most important of the primary fossil fuels) or mineral exploration. Alternatively, they may be of global significance, such as attempts to determine structural discontinuities in the Earth's interior, the geophysical characteristics of island arcs, oceanic trenches, or mid-oceanic ridges, or the elastic properties of Earth materials generally. Artificially generated seismic waves recorded during seismic surveys are used to collect data in oil and gas prospecting, mineral exploration, and engineering.

However, mineral prospecting using seismic methods is not straightforward and still requires improvements in all parts and aspects of the workflow. Economic metallic deposits usually have a strong seismic contrast (Eaton et al., 2003; Malehmir et al., 2012), a product of velocity and density (elastic impedance), with their host rocks. Therefore, in the presence of favorable geometry, size, and signal-to-noise ratio, they should be detectable using seismic methods. While seismic methods have considerably better resolution and penetration depths than other geophysical methods, the high acquisition and processing costs pose a constraint in using them routinely for mineral exploration purposes. There is, therefore, a high demand to reduce this cost in order for the method to be established in the mineral exploration sector, similar to a variety of other geophysical methods. A decrease in receiver costs and higher-quality sensors has allowed some contributions, but on the source side, this is still an issue. Within an ongoing project involving petrophysical, geological, and geophysical studies, Malehmir et al. (2017) have examined the potential of a newly developed micro-electro-mechanical system (MEMS)-based seismic land streamer (Brodic et al., 2015) designed for urban environments, for cost-effective mine planning and mineral exploration at two sites in Finland (Malehmir et al., 2017) and Sweden (Malehmir et al., 2017) respectively.

To date, tens of 2D and 3D surface seismic surveys have been acquired in Canada, Europe, Australia, and South Africa (Malehmir et al., 2012) to help in targeting mineral deposits at depth or for designing deep mines. The steadily increasing usage of reflection seismic methods demonstrates that they are finally becoming recognized and established within the mining sector. This brings new opportunities for geophysicists, but also new challenges. India has 26 sedimentary basins covering an area of 3.36 million sq km spread over onland, shallow water and deep water. As a base to launch future exploration and production (E & P) activities in the country, appraisal of all unappraised areas by acquiring geoscientific data with state-of-the-art technology has been embarked on. The generated data is instrumental in identifying the prospective areas, carving out the blocks in Open Acreage Licensing Policy etc [Source: Directorate General of Hydrocarbons (DGH), Ministry of Petroleum & Natural Gas, Government of India, October 2023]. As a step towards Prime Minister's vision of reducing India's energy imports, Ministry of Petroleum & Natural Gas has launched the Central Sector Scheme christened as National Seismic Program (NSP) and approved by Cabinet Committee of Economic Affairs for appraisal of onland part of Indian sedimentary basins where no/scanty data is available. Appraisal of unappraised offshore areas up to exclusive economic zone and drilling of parametric wells are also being taken up separately. The NSP project commenced in September 2016 and total ~46960 line kilometer (Lkm) 2D seismic data acquisition, processing and interpretation (API) thereof have been completed covering Category-I, II and III basins (Table 9.1).

TABLE 9.1 Details of areas taken up under the NSP campaign.

Category-I	
Assam Shelf & Assam Arakan	5246 line kilometer (Lkm)
Cambay	1469 Lkm
Krishna-Godavari	981 Lkm
Rajasthan	2752 Lkm
Category-II	
Andaman	264 Lkm
Kutch	1809 Lkm
Mahanadi	2481 Lkm
Saurashtra	2306 Lkm
Vindhyan	5373 Lkm
Category-III	
Himalayan Foreland	1564 Lkm
Ganga-Punjab	3858 Lkm
Bengal	911 Lkm
Satpura-South Rewa-Damodar	1124 Lkm
Narmada	1618 Lkm
Deccan Syncline	10,436 Lkm
Bhima	612 Lkm
Kaladgi	257 Lkm
Cuddapah	1387 Lkm
Pranhita-Godavari	735 Lkm
Chhattisgarh	1778 Lkm

Source: Directorate General of Hydrocarbons (DGH), Ministry of Petroleum & Natural Gas, Government of India, October 2023.

The data generated under the NSP campaign using long spreads (12–18 km) and state-of-the-art technology has enabled the ascertainment of the presence of sub-trappean Mesozoic sediments in some areas. The NSP data has provided significant leads toward the extension of existing producing basins, as the information about the tectonic framework has improved. So far, 11 blocks have been awarded from areas appraised under NSP (7 in the Rajasthan Basin, 2 in the Mahanadi Basin, and 2 in the Assam Shelf) (Fig. 9.1).

The interested exploration and production stakeholders can view geoscientific data from anywhere in the world and form an opinion regarding the prospectivity of the blocks prior to bidding for them. Further details of NSP data can be accessed through the National Data Repository website (https://www.ndrdgh.gov.in/NDR/).

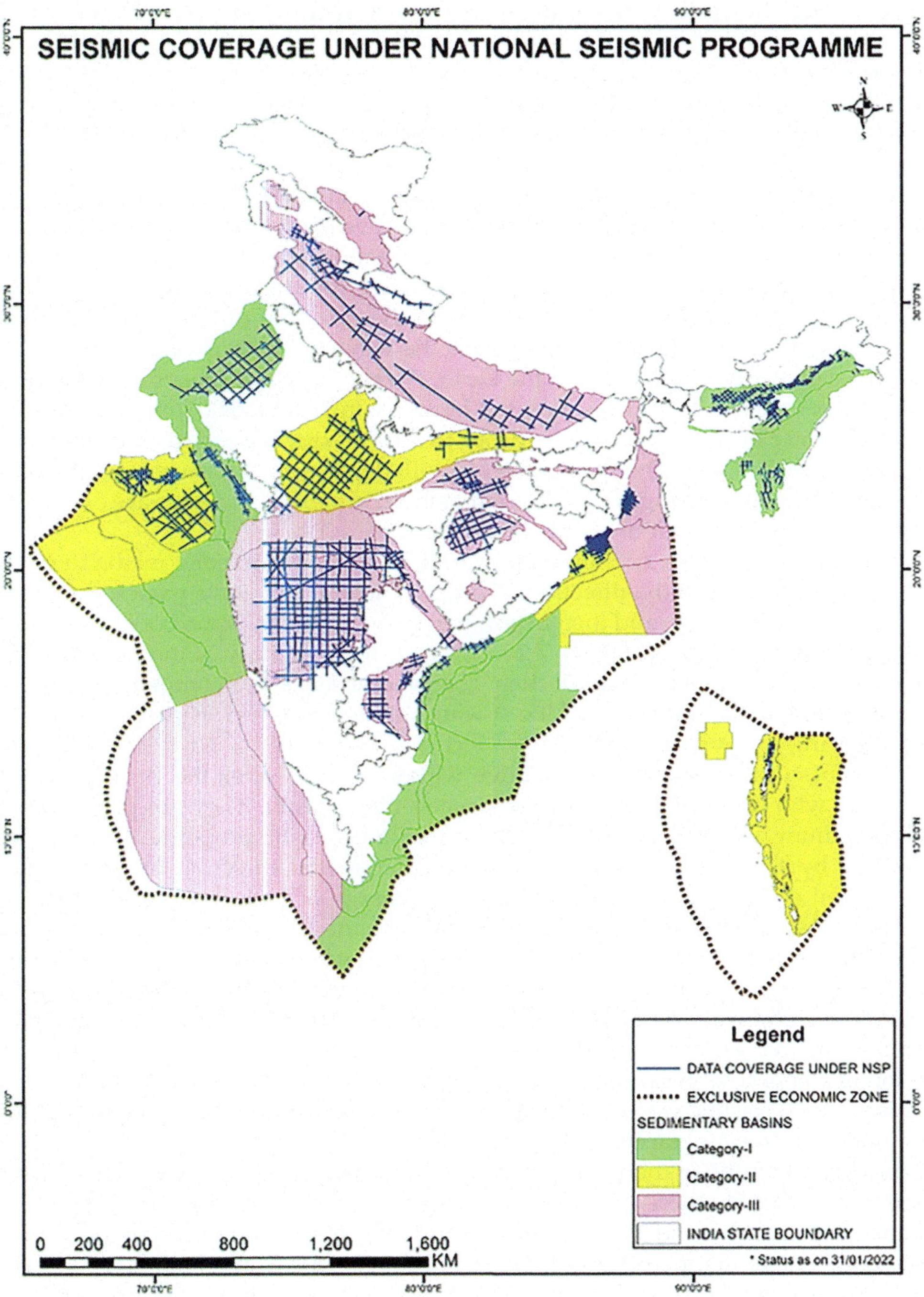

FIG. 9.1 Seismic coverage under National Seismic Programme (NSP). *Source: Directorate General of Hydrocarbons (DGH), Ministry of Petroleum & Natural Gas, Government of India, October 2023. https://www.dghindia.gov.in/index.php/page?pageId=200&name=Activities*

9.2 Seismic wave and mineral exploration—an overview

Seismic waves are vibrations generated by an earthquake, explosion, or similar energetic source and propagated within the Earth or along its surface. Earthquakes generate four principal types of elastic waves; two, known as body waves, travel within the Earth, whereas the other two, called surface waves, travel along its surface. Seismographs record the amplitude and frequency of seismic waves and yield information about the Earth and its subsurface structure. Of the body waves, the primary, or *P*, wave has the higher speed of propagation and thus reaches a seismic recording station faster than the secondary, or *S*, wave. *P* waves, also called compressional or longitudinal waves, give the transmitting medium—whether liquid, solid, or gas—a back-and-forth motion in the direction of the path of propagation, thus stretching or compressing the medium as the wave passes any one point in a manner similar to that of sound waves in air. In the Earth, *P* waves travel at speeds from about 6 km/s in surface rock to about 10.4 km/s near the Earth's core, some 2900 km below the surface. As the waves enter the core, the velocity drops to about 8 km/s. It increases to about 11 km/s near the center of the Earth. The speed increase with depth results from increased hydrostatic pressure as well as from changes in rock composition; in general, the increase causes *P* waves to travel in curved paths that are concave upward.

S waves, also called shear or transverse waves, cause points in solid media to move back and forth perpendicular to the direction of propagation; as the wave passes, the medium is sheared first in one direction and then in another. In the Earth, the speed of *S* waves increases from about 3.4 km/s at the surface to 7.2 km/s near the boundary of the core, which, being liquid, cannot transmit them; indeed, their observed absence is a compelling argument for the liquid nature of the outer core. Like *P* waves, *S* waves travel in curved paths that are concave upward.

Of the two surface seismic waves, Love waves—named after the British seismologist A.E.H. Love, who first predicted their existence—travel faster. They are propagated when the solid medium near the surface has varying vertical elastic properties. Displacement of the medium by the wave is entirely perpendicular to the direction of propagation and has no vertical or longitudinal components. The energy of Love waves, like that of other surface waves, spreads from the source in two directions rather than in three, and so these waves produce a strong record at seismic stations even when originating from distant earthquakes. The other principal surface waves are called Rayleigh waves, after the British physicist Lord Rayleigh, who first mathematically demonstrated their existence. Rayleigh waves travel along the free surface of an elastic solid such as the Earth. Their motion is a combination of longitudinal compression and dilation that results in an elliptical motion of points on the surface. Of all seismic waves, Rayleigh waves spread out most in time, producing long wave duration on seismographs.

Seismic methods utilize the propagation of elastic waves in the Earth. These waves are reflected and refracted when they intersect a boundary between rocks of different acoustic impedance in the Earth, and the detection of the reflected or refracted components allows the boundaries to be mapped (Wright, 1981). Although seismic methods have found broad and sophisticated applications in petroleum exploration, they have contributed far less to metals exploration. The reasons are twofold: the geologic environments containing most metal deposits are commonly more complex than the petroleum environments, and whereas the petroleum industry has invested a great deal of money in optimizing the seismic method for the petroleum environment, the mining community has not made a corresponding investment. Thus, seismic exploration has a reputation for being expensive and useful in only a few special cases. Geophysicists generally agree, however, that research and development could substantially improve the method for mining applications. Seismic methods are effective at target depths beyond those appropriate for the electrical methods that now dominate mineral exploration. In addition, seismic methods can often detect smaller targets. Consequently, the cost-effectiveness of seismic techniques will increase as targets become smaller and/or deeper and as optimization of these techniques for the mining environment is achieved through research and development.

9.3 Seismic reflection method in mineral exploration and crustal imaging

The seismic reflection method has been used extensively in mineral exploration and for imaging crustal structures within hard rock environments (Ahmadi, 2015). There are generally different methods in exploration seismology that can be implemented to address specific problems. One of the most commonly used methods is the seismic reflection method (Sheriff and Geldart, 1995; Yilmaz, 2001) in which reflected or diffracted waves are studied (Telford et al., 1990; Sheriff and Geldart, 1995). The geometry of the subsurface structures and their accurate delineation are of great interest to exploration geoscientists. Knowledge about geological structures may have different applications in various projects, such as geotechnical investigations, mine planning, reservoir estimation, and oil, gas, and mineral exploration (Malehmir et al., 2012; Telford et al., 1990; Denti th and Mudge, 2014). In mineral exploration, the seismic reflection method has demonstrated proven capability for almost six decades (Schmidt, 1959; Wu et al., 1995; Eaton et al., 1996; Milkereit et al., 2000; Adam et al., 2003; Bohlen et al., 2003; Malehmir et al., 2012). The strength of the method lies in its good signal penetration, which makes it capable of imaging deep targets and structures (Malehmir et al., 2012, and references therein). This benefit led the mining industry to take advantage of the method and seek solutions to their prospecting problems using the seismic reflection method (Malehmir et al., 2012). Apart from its application in natural resource exploration, the seismic reflection method may also be used to image crustal structures (Lie and Andersson, 1998; Juhojuntti et al., 2001; Hobbs et al., 2006; Martinez Poyatos et al., 2012). Understanding the geometry of crustal structures is important and may have applications in the risk assessment of natural hazards such as earthquakes, studies of geodynamical processes, and detecting mineral deposits associated with geological structures (Williams et al., 1995; Hurich, 1996; Larkin et al., 1996; Haugen and Schoenberg, 2000; Juhlin et al., 2000; Juhojuntti et al., 2001; Hobbs et al., 2006; Goleby et al., 2006; Ji and Long, 2006; Juhlin et al., 2010; Campbell et al., 2010; Juhlin and Lund, 2011; Cheraghi et al., 2011; Lundberg et al., 2012; Ehsan et al., 2014; Jackson et al., 2015).

The seismic reflection method is sensitive to the seismic acoustic impedance (acoustic impedance = -seismic velocity × density) contrast of the rocks through which the waves propagate. In general, a reflection from a structure may be observed if there is a significant impedance contrast between the structure and the surrounding media. The reflection seismic method has resolution at depth superior to the other geophysical techniques used in mineral exploration, has excellent depth penetration, and offers the opportunity to characterise hitherto unused rock properties in situ. Furthermore, metallic mineralisation and altered rock types can exhibit strong acoustic impedance contrasts with country rock. However, hard rock environments, typically those with steeply dipping (75°-90°) irregular boundaries and disruptive structures, can be far more complex and disordered than the layered earth environments commonly encountered in soft rock exploration. It is these steeply dipping features which pose particular challenges to the seismic method. There appear to be two aspects to the problem of steeply dipping features—imaging and recognition.

9.3.1 Imaging

Vertical layers and boundaries cannot be imaged directly by conventional, straight wave path reflection seismics. With the energy source and the geophone receivers all located on a level ground surface, it's just not possible—it is simply a matter of geometry (Fig. 9.2). Unless survey procedures and arrays are specially adapted, this nonimaging problem effectively applies to steeply dipping layers and boundaries too (Harvey, 2022). The dip of layers and boundaries that can be directly detected can be expanded by displacing the geophone array in the down dip direction, but there are practical limitations to how far this can be taken. Also, as dip angles increase, azimuthal direction effects increase, implying that 3D arrays are essential for complex hard rock environments.

Oil and gas exploration seismic section patterns typically comprise pervasive sub-horizontal reflectors mapping out the layered, relatively flat lying stratigraphy. Steeply dipping structures are then inferred by tracing out the disruptions, dislocations and truncations of these sub-horizontal reflectors. When most of these reflectors are absent, as is often the case in hard rock environments, recognition and tracing of sub-vertical structures is not possible. The end result of these factors is that sub-vertical reflectors will not normally be present in the seismic results. And in a complex hard rock, nonlayered environment, in the absence of extensive sub-horizontal reflectors, inferring the presence of any sub-vertical structures will be that much more difficult. This disparity can bias the interpretation.

Normally, the seismic reflection method is carried out in two phases; seismic data acquisition, and then seismic data processing. Enhancing a clear reflection originating from a geological structure, requires sorting source gathers into a set of gathers in which

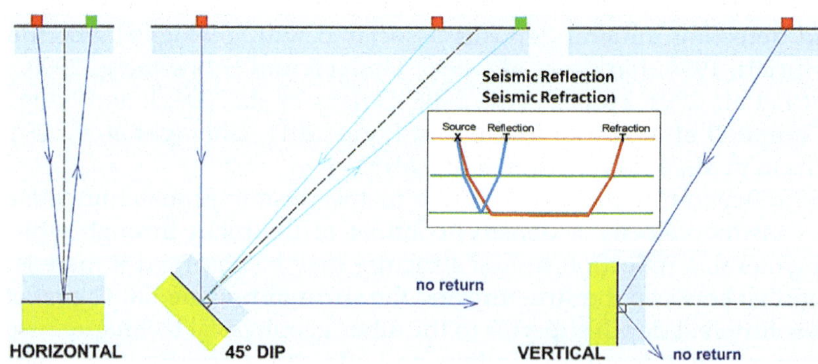

FIG. 9.2 Sample wave paths for horizontal, dipping and vertical boundaries Minerals geophysics (Modified).

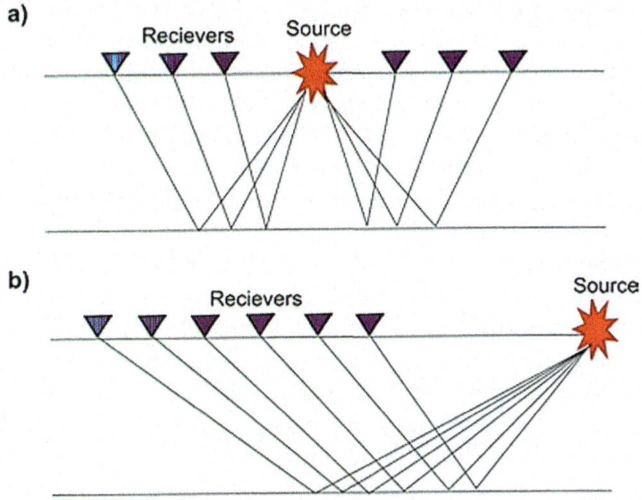

FIG. 9.3 Two types of seismic spreads: a) split-spread and b) end-on spread. *Source: Ahmadi, 2015. https://uu.diva-portal.org/smash/record.jsf?pid=diva2%3A844513&dswid=5671*

their seismic traces have a common midpoint (CMP). These gathers are named CMP or CDP (common depth point) gathers and the number of traces within each CDP gather equals the fold of that gather. In order to achieve enough fold coverage for imaging, each reflecting point must be recorded more than once (Sheriff and Geldart, 1995). In CMP or CDP recording a seismic spread should move continuously down the seismic line to record and obtain the desired fold coverage. There are a number of seismic spread types (fixed-spread, split-spread and end-on spread) that can be used (Sheriff and Geldart, 1995).

In the fixed-spread the receiver line is fixed and the source moves along the seismic profile. This type of spread is suitable when the length of the profile is relatively short. Split-spread is a common survey type where the source is located in the center of the receiver line (Fig. 9.3A). Receivers are spaced regularly depending upon the desired lateral resolution. Another common seismic survey type is the end-on spread. In the end-on configuration, the source is located at one end of the receiver line (Sheriff and Geldart, 1995) and receivers are spaced regularly (Fig. 9.3B).

9.4 Seismic methods in mineral exploration with state-of-the-art technology

Across the globe, the mineral industry is seeking technology to improve exploration efficiency at depth and to help design safer and more productive mines. Seismic methods are increasingly used by this industry for a wide range of commodities, including base metals, uranium, diamonds, and precious metals. Seismic methods provide high-resolution images of geologic structures hosting mineral deposits and, in a few cases, can be used for direct targeting of mineral deposits. Applications are not limited to only surface seismic surveys but also include borehole seismic methods such as vertical seismic profile (VSP) and crosshole imaging. In many parts of the world, exploration for mineral deposits is moving progressively but persistently to greater depths, relying on knowledge gained from previous exploration campaigns and on new exploration tools and techniques used to guide deep and costly boreholes to the most suitable targets. As they maintain high resolution with depth, seismic methods are particularly effective at detecting and delineating deep geological units and structures (Bellefleur et al., 2019). This section contains references, from petrophysical studies to data acquisition, processing, and imaging, as well as 2D and 3D seismic modeling of mineral deposits and their host rock structures, which illustrate the importance of seismic methods not only in the hydrocarbon industry but also in the mineral industry.

To be fully embraced in mineral exploration, seismic data need to be acquired quickly, more cheaply, and with minimal environmental impact, while also addressing the often brownfield, highly noisy environment using seismic land streamers (Malehmir et al., 2017). Seismic land streamers are not new and have been used since the 1980s, primarily for urban or, in general, near-surface (>100 m) applications (Eiken et al., 1989; Pugin et al., 2009; Miller et al., 1986; Dehgahnnejad et al., 2017; Malehmir et al., 2015; Inzaki, 2004; Krawczyk, et al., 2012). Malehmir et al. (2017) used a digital-based, 240 m long seismic land streamer (Inazaki, 2004; Malehmir et al., 2015) and a drop hammer to develop cost-effective seismic mineral exploration methods down to approximately 850 m depth, targeting iron-oxide (predominantly magnetite, though there are horizons where hematite is rich or present) mineralization in the Bergslagen mineral district of Central Sweden as a pilot project. A combined streamer (100–3C-MEMS, 2–4 m spacing) and 75 wireless recorders (mixed 10 Hz and MEMS, 10 m spacing) were used. A Bobcat-mounted drop hammer, weighing 500 kg, was used to generate the seismic signal. Within four days, approximately 3.5 km of seismic data were acquired using 2–10 m source and receiver spacing. Reflection data processing results clearly image the mineralization as a set of strong, high-amplitude reflections, likely extending slightly beyond the known 850 m depth. This is encouraging and suggests that such a cost-effective (50% cheaper than conventional methods) exploration method can be used in the area and elsewhere to delineate similar depth-range iron-oxide deposits. Downhole-logging investigations, including full-waveform sonic and laboratory density measurements, were conducted (Maries et al., 2016) prior to this study, suggesting that iron-oxide deposits in the study area should be seismically detectable.

The benefit of state-of-the-art data acquisition and novel technology is demonstrated in Naghizadeh et al. (2019). They show clear improvement in reflector detection and definition achieved during the acquisition and processing of deep seismic reflection profiles acquired in Canada by the Metal Earth project. Metal Earth is dedicated to understanding the processes responsible for the differential metal endowment in Archean greenstone provinces and uses seismic methods as one of the tools to identify key geological, geochemical, and geophysical attributes of metal sources, transport pathways, and economic concentrations at the crust-mantle scale. Seismic improvements are demonstrated by a comparative analysis with earlier Lithoprobe surveys and are attributed to the increased bandwidth of the signal, better estimates of refraction and reflection velocities used in processing, and especially the prestack time migration of the data. Reidel et al. (2018) present a case study of an underground VSP experiment, which was designed to image a semimassive sulfide deposit located in the Kylylahti polymetallic mine, Finland. They utilized and compared data measured with a conventional VSP system with three-component geophones and a novel fiber-optic distributed acoustic sensing (DAS) system. Both systems, deployed in boreholes located near the sulfide deposit, successfully recorded seismic reflections from the deposit and nearby geological contacts. In particular, they emphasize the potential benefit of using fiber-optic DAS systems, which provide reflection data with sufficient quality, with less logistical effort and a higher acquisition rate.

Seismic imaging, commonly known as migration, has a significant impact on results. Methods relying on depth imaging, because they can take into account significant lateral and vertical velocity variations, typically provide the best results. However, depth imaging methods were used only recently for mineral exploration (Urosevic et al., 2017) because they require accurate velocity models that are particularly difficult to obtain with the low signal-to-noise ratio seismic data typical of mining environments. Singh et al. (2019) compare and analyze results from time and depth imaging approaches, including a specialized Kirchhoff prestack depth migration referred to as coherency-based Fresnel volume migration (FVM). The comparative study is performed with 3D seismic data acquired over the Kylylahti mine area, Finland. The 3D survey is characterized by an irregular shooting geometry and relatively large receiver intervals (50 m), resulting from the utilization of a receiver spread initially designed for passive seismic measurements and shot points from a 2D seismic survey, VSP, and randomly distributed shot points. The coherency-based FVM resulted in a cleaner image of steeply dipping reflections located at a shallow depth (<1500 m). In comparison, standard time-domain processing and imaging failed to produce a convincing image of the top ~1500 m of the subsurface, which was the primary interest of the survey. Heinonen et al. (2019) show that an advanced prestack depth-imaging algorithm based on FVM recovers meaningful and robust seismic images from 2D seismic data acquired with sparse shot points. This suggests that 2D seismic reflection profiling with sparse geometry can be used as a reconnaissance tool for mineral exploration in areas with limited access for active seismic surveys. Heinonen et al. (2019) obtained these results by varying the number of shot points used in the depth migration algorithm on two approximately 6 km long seismic reflection profiles acquired with 20 m source spacing at the polymetallic Kylylahti massive sulfide mine site, Finland. Steep reflectors in the shallow (<500 m) part of the seismic sections are imaged especially well in the FVM migration result and correlate with black schist interlayers with mica schist observed in the geological map. Cichostepski et al. (2019) present results of the inversion of prestack seismic data to evaluate the sulfur content within a carbonate reservoir. The strata-bound sulfur deposits occur at depths of about 300 m, with sulfur content reaching up to 30%. The sulfur-bearing strata and host rocks were imaged on five 2D high-resolution shallow seismic profiles processed to preserve relative amplitude relationships. Cichostepski et al. (2019) used a simultaneous inversion procedure to recover porosity changes and sulfur content and demonstrate that sulfur content is dependent on the carbonate reservoir's porosity.

Seismic inversion techniques have been frequently employed in the petroleum sector to find hydrocarbon bearing layers in the subsurface from seismic data (Bosch et al., 2010; Maurya and Singh, 2018). Seismic reflection data from the Krishna–Godavari (KG) basin in India is utilized to estimate subsurface acoustic impedance using maximum likelihood seismic inversion (MLSI) and model-based seismic inversion (MBSI) technique (Richa et al., 2022). The seismic frequency band is limited to about 15–80 Hz, and hence the low and high frequency components are often missing for the inversion which is one concern during inversion. One way to reduce these uncertainties uses some additional data which is derived from well log data. This well log data supply low as well as higher frequency components to the inversion. As a result, the final inversion findings are dependent on both the seismic data and the supplementary information acquired from the well log data (Mavko and Mukerji, 1998). Using well log data, the user may be able to determine shale volume (V_{sh}), water saturation (S_w), porosity (φ), permeability (k), elasticity [AI (acoustic impedance), SI (shear impedance)], reflectivity coefficient (R), and other data that the user requires (Crain, 2006). The present study focused on the analysis of well log data as well as seismic data of the KG basin to find the prospective zone. Three main hydrocarbonbearing zones are identifed with an average gamma ray value of 50 API (American Petroleum Institute) units at the depth range of (1918–1960 m), 58 API units (2116–2136 m), and 66 API units (2221–2245 m). The average resistivity is found to be 17 Ohm-m, 10 Ohm-m, and 12 Ohm-m and average porosity is 15%, 15%, and 14% of zone 1, zone 2, and zone 3 respectively. The analysis of petrophysical parameters and diferent cross plots showed that the reservoir rock is of sandstone with shale as a seal rock. On the other hand, two types of seismic inversion namely MLSI and MBSI are used to estimate subsurface acoustic impedance. The inverted section is interpreted as two anomalous zones with very low impedance ranging from 1800 m/s*g/cc to 6000 m/s*g/cc which is quite low and indicates the presence of loose formation.

Shale gas, shale oil, tight sand, heavy oil, gas hydrates, and coal bed methane (CBM) are the most important unconventional hydrocarbon resources of the 21st century. Tight sand and shale plays are characterized by low permeability; to exploit hydrocarbons economically, their permeability is increased artificially by creating fractures, a process termed fracking. The ever-widening gap between hydrocarbon imports and domestic production in India mandates the efficient exploitation of unconventional hydrocarbons. The prevailing hydrocarbon market does not support the acquisition of specialized seismic data and specialist software for unconventional hydrocarbon exploitation. It is imperative to devise cost-effective ways of exploiting tight sand, shale oil, and gas. Natural fractures in tight sand and shale plays are helpful in production, and their characterization is important. Since seismic amplitude varies along and across swarms of fractures, Multiazimuth seismic data can be analyzed for fracture characterization. Onshore seismic data, which are acquired and processed for conventional seismic exploration, can be used in unconventional exploration and exploitation after 5D interpolation of the seismic data in inline, cross line, offset, azimuth, and frequency. Yadav et al. (2019) analyzed seismic amplitude variation with offset and azimuth using conventional seismic data from the Cambay Basin in India for fracture characterization by making the data rich in azimuth through 5D interpolation.

Several deposit types with major economic importance are inherently associated with the large volume of rocks infiltrated by hydrothermal fluids that, through precipitation and trapping processes, lead to the formation of deposits. Identification and recognition of such hydrothermal alteration provide important evidence that an area is favorable for the occurrence of deposits and can guide exploration in areas with the highest discovery potential. Thus, understanding and characterizing the signature of hydrothermal alteration on seismic data is of primary importance for exploration. Schetselaar et al. (2019) used an integrated analysis of seismic rock properties, lithogeochemical data, and mineral compositional data, estimated via scanning electron microscopy–energy dispersive X-ray spectroscopy, to provide insights into the effects of hydrothermal alteration on seismic reflectivity in the footwall of the Lalor volcanogenic massive sulfide (VMS) deposit, Canada. Footwall alteration of the

Lalor VMS deposit is associated with an enhanced seismic reflectivity that can be tracked laterally and at depth on a 3D seismic volume and is caused by the high abundance of garnet, anthophyllite, and, particularly, cordierite in rocks with mafic protolith. These minerals are constituents of amphibolite facies assemblages that are diagnostic of the metamorphosed and intensely altered footwall of volcanogenic massive sulfide ore lenses.

Junno et al. (2019) use self-organizing map (SOM) analysis to predict missing seismic velocity values from other available borehole data at the Kevitsa Ni–Cu–PGE deposit, Finland. SOM is used to predict missing seismic velocities from other geophysical, geochemical, geological, and geotechnical data to support the interpretation of a series of reflections observed beneath the Kevitsa resource area. Predicted reflectivity logs show that some contacts between olivine pyroxenite/olivine websterite dominant host rocks of disseminated sulfide mineralization and altered olivine pyroxenite variants (metaperidotite) are reflective. This indicates that alteration can potentially cause significant reflectivity within the Kevitsa intrusion.

9.5 Land-based seismic methods in mineral exploration—case study

Seismic methods were not used for hard rock mineral exploration until relatively recently (Best, 2015). However, they were applied to mineral exploration in sedimentary environments because of the similarities to petroleum exploration [e.g., uranium exploration in the Athabasca Basin (O'Dowd et al., 2006) and potash exploration in Saskatchewan (Gendzwill, 1969; Gendzwill and Stead, 1992)]. Advances in hard rock acquisition and processing procedures, along with more and better measurements of density and velocity of igneous and metamorphic rocks and minerals, provided the impetus for the development of 2D and 3D seismic methods for hard rock mining applications. For example, the number of geophone channels has significantly increased, so that now 10,000 or more channels can be deployed. This leads to better spatial sampling, larger offset sampling, and the collection of multicomponent seismic information. Acquisition systems can now be deployed without using cables (wireless acquisition systems), making the collection of data in difficult areas easier. See the special section entitled "Seismic Methods in Mineral Exploration" in (Malehmir et al., 2012) (issue 5, volume 77) and (Eaton et al., 2003) for recent information on seismic applications to mineral exploration. Reflection seismic methods are also used in conjunction with other information for mine planning (Malehmir et al., 2012). Reflection seismology, seismic refraction, seismic tomography, and 2D and 3D seismic surveys for deposits from 500 to 2000 m depth—below the reach of conventional geophysical techniques—are expensive but are the right tools for areas with mining infrastructure (Gandhi and Sarkar, 2016). The refraction technique uses a long array of geophones to sense refracted waves, while the reflection technique uses a condensed array near the source to sense reflected waves (Fig. 9.4). Reflection surveys are widely used to map the upper crust for oil and gas exploration. First utilized in the early 1920s, the reflection seismic method, an aspect of the seismic geophysical method, has been associated with some of the most profound structural observations (Kearey et al., 2002). Refraction surveys are commonly used for shallow mapping in groundwater supply and contamination studies.

The following provides an example of 3D seismic imaging of VMS (pyrite versus pyrrhotite rich ore) deposits in the Flin Flon mining camp in Manitoba, Canada (Bellefleur et al., 2015; Malinowski and White, 2011; Malinowski et al., 2008, 2012; White et al., 2011; White

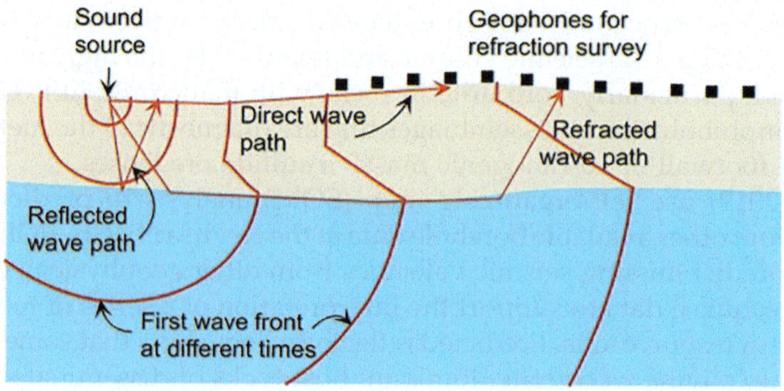

FIG. 9.4 Schematic of a seismic refraction survey. The lower (darker) layer transmits waves at higher velocity than the upper layer. *Source: Fitts, 2024. https://s100.copyright.com/CustomerAdmin/PLF.jsp?ref=cddd7b86-affd-4cc1-9c98-71e3c8d932ce*

et al., 2012). The Flin Flon mining camp in Manitoba is located within the Paleoproterozoic Flin Flon metavolcanics belt and is close to the town of Flin Flon, Manitoba, near the Saskatchewan border at ~55° north.

The Main, 777, and Callinan deposits lay along the west limb of the hidden lake synform (White et al., 2011, 2012). The Flin Flon main deposit closed in 1990 but the 777 (14.2 million tons) and Callinan (8.4 million tons) deposits are still being mined. However, the Callinan deposit was nearly mined out when the 3D survey was carried out in 2007. The 777 deposits are generally more massive than the Callinan deposits and contain more pyrrhotite than pyrite relative to the Callinan deposits (White et al., 2011, 2012). The deposits tend to be elongated in the dip direction and have thicknesses between 10 and 30 m (Fig. 9.5). The 3D seismic survey covered an area of ~17 km^2 with over 900 shot locations and >3300 receiver stations (Best, 2015). Refraction and residual static corrections were carried out along with a conventional dip move out (DMO) time processing flow. Poststack migration, poststack DMO migration, and prestack time migration were carried out on the data volume. The poststack DMO migration visually provided the best results. White et al. (2012) (Fig. 9.6A) shows the migrated data for a vertical section over the Callinan and 777 ore zones. The black lines are the locations of the ore lenses. Note that the correlation with reflectors is better for the Callinan zones (Fig. 9.6B). More diffraction occurs at depth for the 777 ore lenses (Best, 2015). Further details on the processing and interpretation of these data can be found in White et al. (2012).

CBM is a natural gas produced from coal seams. CBM is a mixture of hydrocarbons, predominantly methane, with minor amounts of other gases such as carbon dioxide and nitrogen. CBM is generated either through a biological process resulting from microbial action or through a thermal process caused by increasing heat with the depth of the coal. High-rank coals buried at great depths (>300–1200 m) are suitable for the exploitation of CBM. Identification of faults, thickness, and depth of coal seams is essential in CBM exploration, as these factors are responsible for the formation of methane. A 2D high-resolution seismic survey (HRSS) plays a unique role in mapping coal seams and their fine structure, including small-scale faults. In this technique, the property of acoustic impedance (the product of the sound velocity in a material and the density of the material) is of great importance in identifying different rock formations associated with coal, and it has been successfully

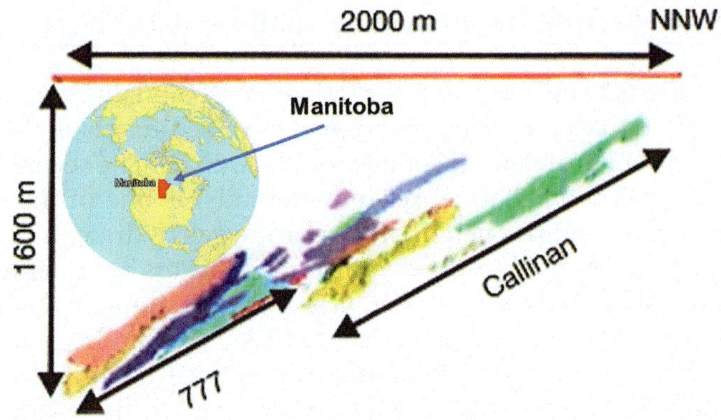

FIG. 9.5 Cross-section of the Callinan and 777 deposits (individual deposits in different colors) showing them plunging to the SSE in the Flin Flon mining camp in Manitoba, Canada (modified).

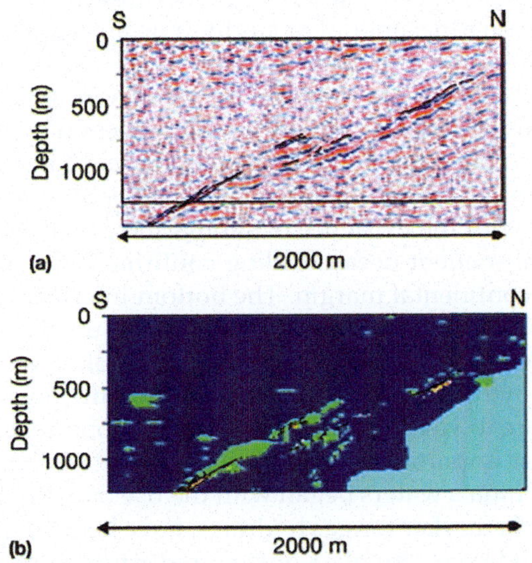

FIG. 9.6 (A) Vertical slice through the 3D migrated volume. The black rectangle outlines the area covered by the model in (b). (B) Corresponding vertical slice from the 3D lithologic model based on drillhole information and used for seismic modeling (Malinowski et al., 2012 for more details on the modeling). Dark blue, basalt; light blue, sandstone; green, rhyolite; and orange, ore. The ore lenses are shown in black in the two plots (Modified).

implemented. The study area is situated in the central part of the Sohagpur coalfield, which forms the southern part of the Rewa master Gondwana basin in the large Son-Narmada valley of Central India (Seshunarayana et al., 2008). In the Sohagpur coalfields, Barakar is the main coal-bearing formation. The Barakar formations are traversed by a number of basic dykes and sills. The HRSS method was applied in a 1 sq km area of the Sohagpur coalfield along 10 profiles covering 9.20 Lkm, and coal seams as well as structural features were identified at a depth range of 300–450 m.

9.6 Seismic technique in submarine shallow natural gas—case study

Exploration for crude oil and natural gas originated in the latter part of the 19th century when geologists began to map land features to search for sediments that had a high potential to contain crude oil and/or natural gas (Monteleone et al., 2022). The seismic method (including single-channel seismic and multichannel seismic methods) is one of the most effective techniques for the identification and prediction of submarine gas hydrate. It plays a key role in the discovery of submarine gas hydrate. The identification markers are mainly the BSRs, velocity, amplitude anomaly, and variation with offset. Gas hydrates are a crystalline solid formed of water and gas. They look and act much like ice, but they contain huge amounts of methane; they are known to occur on every continent and exist in vast quantities in marine sediments in a layer several hundred meters thick directly below the sea floor and in association with permafrost in the Arctic. Gas hydrates are not stable at normal sea-level pressures and temperatures, which is the primary reason that they are challenging to study. Remote seismic methods, which measure the compressional wave (P-wave) velocity (V_P) and shear wave (S-wave) velocity (Vs), can be used to assess the distribution and concentration of marine gas hydrates in situ (Priest et al., 2005). Estimating the amount of gas hydrate and free gas is difficult in deep seas, even with scientific coring and downhole measurements. Well data may be incompatible between holes of a site as well as with depth in the same hole (Ghosh et al., 2010).

Markl et al. (1970) identified an abnormally strong reflection on the single-channel seismic profile, which is parallel to the seabed and oblique to some weak reflection layers. Following the 11th voyage of deep-sea drilling, this anomalous reflector was named the BSR and was thought to be associated with the existence of submarine gas hydrate (Hollister et al., 1972). Deep-sea drilling and subsequent ocean drilling confirmed the existence of gas hydrate in seabed sediments at the continental margin. The bottom interface of the BSR, corresponding to the gas hydrate stability zone, also achieved consensus (Pecher and Holbrook, 2000; Holbrook et al., 1996; Song et al., 2003). However, Holbrook et al. (1996) explained that the BSR is primarily caused by the underlying gas and not by the overlying gas hydrate, and its strength is highly sensitive to variations in gas concentration present in the pore spaces. The BSR has become the most important geophysical evidence for identifying gas hydrate, and gas hydrate research in China's waters began with the use of the BSR (Deng et al., 2004; Deng et al., 2008). BSRs, the main marker for gas hydrates, have been identified on seismic sections in the KG, Mahanadi, Andaman, Kerala-Konkan, and Saurashtra regions (Sain and Gupta, 2012). The BSR is weak and patchy on the seismic section in the Krishna-Godavari basin, located on the eastern Indian margin, where massive gas hydrates have been recovered at Site 10 of the Indian National Gas Hydrates Program Expedition 01 (NGHP-01). Large quantities of gas hydrates were recovered through coring at this site (Collett et al., 2008). The depth of the BSR near this site is approximately 160 m below the sea floor (mbsf) (Figure 1a, Wang et al., 2014). Kumar et al. (2019) studied the presence of gas hydrates in the Mahanadi offshore basin, located on the eastern margin of India, and identified a BSR on seismic sections based on its characteristic features. The presence of gas hydrates is characterized by low amplitude, high frequency, and low sweetness attributes, whereas free gas zones are characterized by bright amplitudes associated with low frequency and high sweetness below the BSR. Multichannel seismic data from offshore Andaman Islands, India, show reflection characteristics consistent with the presence of gas hydrate and free gas. A detailed study of these data reveals the presence of a prominent BSR. This BSR is associated with seismic blanking above and strong, bright reflections below, which are interpreted as free gas at the base of the hydrate stability zone (Satyavani et al., 2008).

In the submarine stratum containing gas hydrate, the sediment pore is filled and cemented by gas hydrate, and its seismic velocity increases, while the underlying stratum without hydrate often contains free gas and exhibits smaller wave velocity. Consequently, the stratum containing hydrate and the underlying stratum without hydrate often cause a negative strong reflection. At the same time, within the stratum containing hydrate, because the sediment pore is filled and cemented by hydrate, there is a homogeneous response in the seismic and acoustic wave characteristics; this appears as a blank reflection band on the seismic profile. Namely, the BSR is often associated with the blank reflection band on it. The BSR has obvious characteristics that are easy to identify (Wang et al., 2006):

BSR is generally almost parallel to the modern seabed, and in many cases, it intersects with the submarine sediment strata reflections; the BSR exhibits strong characteristics of reflection amplitude and polarity reversal compared to seabed reflection.

Bottom-simulating reflector shows a bright spot belt on the seismic profile, and there are often reflection blank areas on it and beneath it.

Bottom-simulating reflector is often distributed beneath seabed terrain heights or appears on the continental slope.

BSRs have different scales; the small ones span only a few thousand meters, while the larger ones can extend tens of kilometers. Some seismic profiles collected in Chinese waters reflect these characteristics of BSRs (Wang et al., 2011a & b; Liu et al., 2011; Song et al., 2007).

Ye, (2017) showed that the BSR is influenced by tectonic action, sedimentation action, sediment carbon content, hydrate content, and many other factors. The distribution of gas hydrate does not have a one-to-one correspondence relationship with the BSR; when using the BSR to identify submarine gas hydrate, one must pay attention to its combination with other markers (Wang et al., 2006). Su and Chen (2007), Su et al. (2012), Liu et al. (2021), and Chen et al. (2006) studied and concluded that in the two kinds of gas hydrate systems—diffusion system and venting system with a developed marine environment (also known as the strata diffusion type and structural leakage type, Wu et al., 2008; Wu et al., 2010)—the hydrate in the diffusion system generally has an obvious BSR. However, in the leakage system, due to the presence of free gas in the hydrate deposit layers, there is no obvious acoustic velocity and impedance difference between the hydrate layer and the underlying sedimentary layer on the seismic reflection records, and the BSR is generally not developed. Therefore, we cannot use the BSR to identify hydrate, but we can use related methods for identifying submarine shallow gas to identify it.

9.7 Seismic technique in shale gas exploration—case study

Seismic methods play a key role in identifying shale formations (clay, quartz, calcite minerals) structures, their extent, discontinuities, and other properties that may control their quality. Gas in shale deposits is predominantly methane (CH_4), derived from organic materials through biogenic and thermogenic processes under suitable conditions, where shale acts as both the source rock and reservoir (Curtis, 2002; Jarvie et al., 2007; Varma et al., 2014; Mishra et al., 2016; Mendhe et al., 2018b). Shale gas reservoirs are known for their large extent, low porosity, and permeability, as well as their high anisotropy, which distinguish them from conventional sand and carbonate reservoirs. As these properties inevitably influence the seismic responses of gas-saturated shales, seismic data targeting shale formations must receive special care in acquisition, processing, and interpretation (Farfour et al., 2021). To economically develop shale gas [shale gas is an unconventional natural gas found trapped within shale formations (U.S. Energy Information Administration, eia.gov, retrieved 6 August 2013)], operators strive to maximize borehole exposure to the reservoir and create a complex fracture network to increase permeability. Since the 1990s, a combination of horizontal drilling and hydraulic fracturing has made large volumes of shale gas more economical to produce, and some analysts expect that shale gas will greatly expand the worldwide energy supply (Krauss, 2009). Moreover, most shale operators use seismic data to map the thickness of the shale reservoir unit and identify possible hazards, such as faults, karst zones, and collapsed chimneys, helping drilling engineers stay in zone and better geosteer long laterals. Thus, seismic surveys are preferably conducted with wide azimuth angles, small trace intervals, high folds, uniform vertical and horizontal coverage, and long spreads to meet the needs of shale gas exploration in terms of structural and lithological interpretations as well as fracture prediction.

In fact, seismic data with wide azimuth angles and long offsets contain relatively complete information about the lithological and physical properties of shale layers, providing a data basis for lithological exploration. The seismic data with high folds, small trace intervals, and long offsets contain relatively complete features of the dynamic and kinematic characteristics of effective reflected waves and interference waves, which is favorable for suppressing interference waves, utilizing effective waves, and increasing the signal-to-noise ratio, the fidelity of the effective waves, and the imaging resolution of seismic data. With the improvement in the understanding of shale plays (a shale play is composed of **discovered [proved], undiscovered [probable], or possible gas and oil reserves that exhibit similar geological characteristics, and these plays are within geological depressions)**, in the processing stage of seismic data, the acquired data are subject to thorough processing operations to remove noise and improve signals from shales.

Seismic characterization is becoming an emerging technique for extending source rock properties beyond well locations and an effective tool for mapping their lateral and vertical heterogeneities. The Raghavapuram Formation of Early Cretaceous age in the KG basin, India, an organic-rich source rock, was studied for source facies characterization using seismic data (Omer et al., 2020). Poststack 3D seismic volume, prestack angle gather, and eight wells with a complete set of wireline logs and core data were used for total organic carbon (TOC) volume estimation in the Gudivada graben of the KG basin. Prestack simultaneous angle-dependent seismic inversion and amplitude versus offset (AVO) analysis were carried out together with a rock physics model of the formation, which generated a quantitative estimate of TOC content in the study area and helps to find the sweet spots for shale gas exploration. Impedance shows an inverse linear relationship with TOC in the study area. AVO analysis showed a Class IV anomaly for TOC-rich zones mainly present in the lower part of the Raghavapuram shale (high gamma, high resistivity formation). The present study tries to find a relationship between seismic attributes and TOC that can be applied without the need for well information and is thus a feasible method for investigating the source rock properties in unexplored basins.

A statistical review shows that global primary energy consumption grew at a rate of 2.9% in 2018, almost double its 10 year average of 1.5% per year, and the fastest growth since 2010. The paucity of conventional energy resources compels India to import energy, which weakens its energy security and fiscal health. The need of the hour is to aim toward better energy self-reliance. Organizations such as the Oil and Natural Gas Corporation of India, the DGH, the Geological Survey of India, Oil India Limited, the Central Mine Planning and Design Institute, the United States Geological Survey, and GAIL (India), to name a few, are working on this. Several geologic provinces are being evaluated for shale gas resource potential. Moitra et al. (2022) review the current state of shale gas exploration in India and discuss some of the prospects and challenges. Proterozoic basins in India,

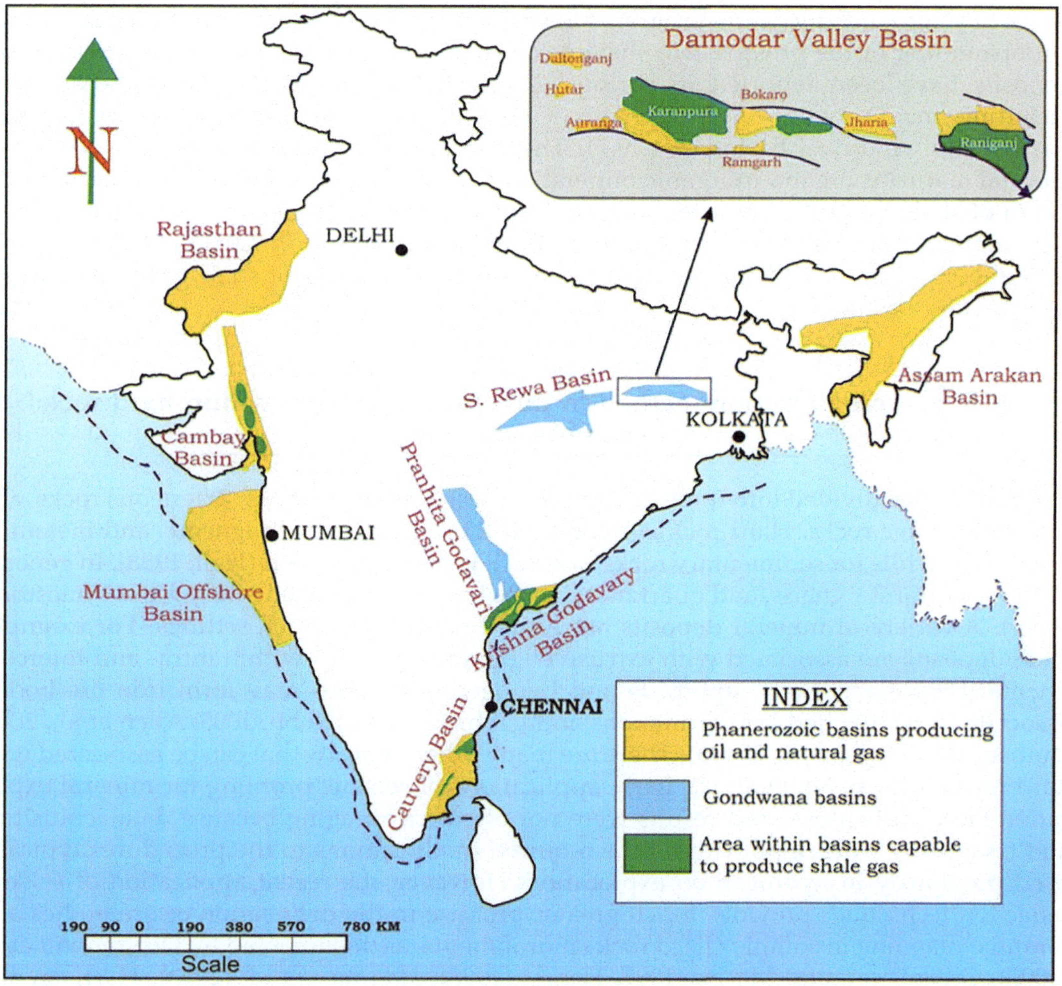

FIG. 9.7 Potential shale gas basins of India. *Source: Hasan et al., 2013; EIA (Energy Information Administration), 2011, 2017; Sain et al., 2014; Kumar, 2021. http://www.sapub.org/journal/contactus.aspx*

which are associated with prominent carbonaceous shale stratigraphic units, should also be brought into focus. Anjirwala and Bhatia, (2016) compare shale gas scenarios between India and the USA, including favorable conditions and challenges with the logistics of shale gas technology implementation. Padhy et al. (2016) describe various new opportunities for shale oil exploration from Paleocene-Eocene sequences of the Cambay Basin in the western part of India. New opportunities and favorable conditions for shale gas prospects are provided in Sain et al. (2014).

Encouraging results of shale gas production in different countries, viz. the United States of America (EIA, 2016), China, and Australia, have inspired India to explore shale gas resources, which may not only support meeting the growing energy demand but also assist in reducing the import of hydrocarbons (Varma et al., 2014; Mishra et al., 2016; Mendhe et al., 2018b; DGH, 2012; DGH, 2018). Kumar (2021) aims to review the current status of shale gas exploration programs in various sedimentary basins in India for a better understanding of the prospects of shale gas. Significant thicknesses of organic-rich shale deposits have been reported in various Gondwana and Tertiary sedimentary basins, including the Damodar Valley, Krishna-Godavari, Cauvery, Cambay, Assam-Arakan, Ganga, and Vindhyan basins (Fig. 9.7). The geochemical properties, such as TOC and thermal maturity, organo-inorganic mineral compositions, pore-network systems, and the depth of shale deposits, favor the potential prospects of shale gas in India. Several agencies, namely EIA (EIA (Energy Information Administration), 2011), ONGC, USGS, Schlumberger, and CMPDI, have reported enormous resources of shale gas in its various sedimentary basins.

9.8 Seismic reflection method in mineral exploration within hard rock terrain—case study

Rocks can be divided into three main groups:1) sedimentary rocks, 2) igneous rocks, and 3) metamorphic rocks. Hard rock is a common term used for both igneous and metamorphic rocks, while for sedimentary rocks, soft rock is frequently used (Hall, 1988). In general, hard rock (granite, gneiss, and quartzite) may be harder than sedimentary rocks (limestone, shale). A variety of mineral deposits are hosted within hard rock settings. For example, VMS deposits are associated with extrusive igneous processes within intra- and intercontinental formations. Both supergene and hypogene processes may form iron ore bodies associated with banded iron formations and gabbroic rocks (Robb, 2005; Allen et al., 2008; Duuring and Hagemann, 2013a). There are many other deposits that can be associated with hard rock (Au, Cu–Ni, PGE, U). Early applications of seismic profiling for mineral exploration faced difficulties, and results were not often encouraging because data acquisition and processing were done without the required modifications to the procedures typically used previously in hydrocarbon exploration. However, the recent application of seismic reflection techniques provides much greater promise in the delineation of ore bodies and for mine planning in complex hard rock environments, as documented by Drummond et al. (2000), Greenhalgh and Mason, 1997, Hearst (1998), Hammer et al. (2003), Harrison and Urosevic (2012), Hatherly et al. (1994), Juhlin and Palm, 2003, Malehmir et al. (2012), Milkereit et al. (2000), Perron and Calver (1998), Roberts et al., 2003a, Salisbury et al. (1997, 2000), Snyder et al. (2009), Spencer et al. (1993), and Urosevic et al. (1992, 2005). From an acquisition view point, the survey array needs to be expanded into the third dimension, with some geophones or shot points located at depth below the ground surface, and with the survey procedures and processing regimes modified accordingly. Given that many hard rock seismic surveys are carried out at mine sites, with the option to position geophone strings in boreholes or mine openings, this is not as far-fetched as it sounds. Specialist seismic survey companies are already working on this.

A 2D high-resolution seismic profile was acquired in Garpenberg, central Sweden, at an active mine area (Ahmadi, 2015). The processing strategies were successful in eliminating noise and boosting the signal in the data. A bright spot anomaly is interpreted to originate from near the top of the ore body, with the location of this anomaly consistent with the working geological model of the Lappberget ore body. Furthermore, DMO processing improved the image quality of the subsurface. In order to constrain the seismic interpretations, 3D ray-based modeling was carried out using the true acquisition geometry of the seismic profile and the geological model of the ore body based on the mine data and drilling. Given the geometry of the ore body, seismic modeling results suggested that imaging the top and bottom of the ore body is possible. Migration of the synthetic data and comparison with the real data indicate that the bright reflection does not come from the top of the modeled ore body but rather from within the ore volume. This study demonstrates that imaging the sulfide deposits at the site is possible, but their delineation is challenged with only one 2D survey and the absence of additional geological constraints.

A deep 2D seismic section was studied, and detailed 3D and 2D interpretations were made for the subsurface of the Cue-Weld Range area in Western Australia (Ahmadi, 2015). The seismic section displays a key characteristic of a superimposed folding system. Interpretation of the geological map and the repetition of greenstone units in roughly two perpendicular directions indicate the occurrence of two generations of deformation in the area. Furthermore, 3D constant velocity ray-tracing allowed the correlation of shallow reflections visible in prestack data with mafic volcanic rocks of the Glen Group and basaltic rocks of the Polelle Group in the area. In order to test the model, a 2D finite difference (FD) method was performed, and a comparison between the synthetic and real data confirmed the proposed structural model. The study suggested that using the seismic reflection method to help locate mineral deposits associated with the folding in the area is plausible. Acquiring more 2D data or 3D seismic data in the area may reveal more information about the complicated structures and deformation in the Cue-Weld Range area.

Since the discovery of the world-class Olympic Dam iron oxide copper-gold (IOCG) deposit in 1975, the Gawler Craton in South Australia has been subjected to scrutiny by various explorers using a multidisciplinary exploration approach (geology, geophysics, and tectonic analysis) in search of similar deposits. Potential field geophysical methods are traditionally used for IOCG exploration. The seismic reflection method provides the possibility for delineation of very complex geological structures, and this method might be effective for detecting the presence of IOCG deposits (Onojasun, 2016). Despite many technically superior attributes, no arguments for regional seismic exploration have been proposed, probably because a cost-benefit analysis has never been conducted at such a scale. In this study, Onojasun (2016) analyzes such a case by modeling a hill side IOCG deposit scenario where 2D seismic with relatively sparse source-receiver geometry is used to detect the presence of a possible intrusive package near a deep fault. The modeling results show that seismic reflection using 40 m geophones and 80 m shot spacing as an exploration tool is feasible, and that with the spacing halved, we can definitely recover reasonable images of the upper parts of the mineralization. The presence of such intrusives is clearly detectable, and with the seismic method, they are detectable from 100 to 1000 m deep. Thus, it is proposed that using 2D seismic is viable for IOCG exploration as it can detect mineralized intrusive structures along known favorable corridors or structures.

9.9 Building 3D rock models from seismic at the Tropicana gold mine, Western Australia

Seismic imaging technology involves the propagation of sound waves into the earth with the objective of imaging changes in the subsurface. The sound waves are sensitive to changes in rock impedance, which is the product of velocity and density. The final images represent the boundaries between different rocks, with the amplitude related to the contrast in the impedance above and below the boundary. There is often both random and coherent noise in the data, which further complicates how the seismic image relates to the subsurface. Understanding the seismic data and what it reveals about the subsurface can be challenging. The conversion of the seismic data into models of rock makes the results accessible to a larger audience. The key to making the conversion is to understand which rocks can be separated and to apply technology that can exploit all available data to build rock models. The technology chosen is geostatistical inversion, which is essentially geostatistical modeling with the addition of seismic data as an additional constraint. The geostatistical inversion exploits the detection limit of the seismic data to build detailed rock models. Data from the Tropicana gold mine were selected for testing because it consists of a large set of deep drillhole data, several sonically logged drillholes, and good-quality 3D seismic reflection data (Fig. 9.8) (Neroni and Hunt, 2022).

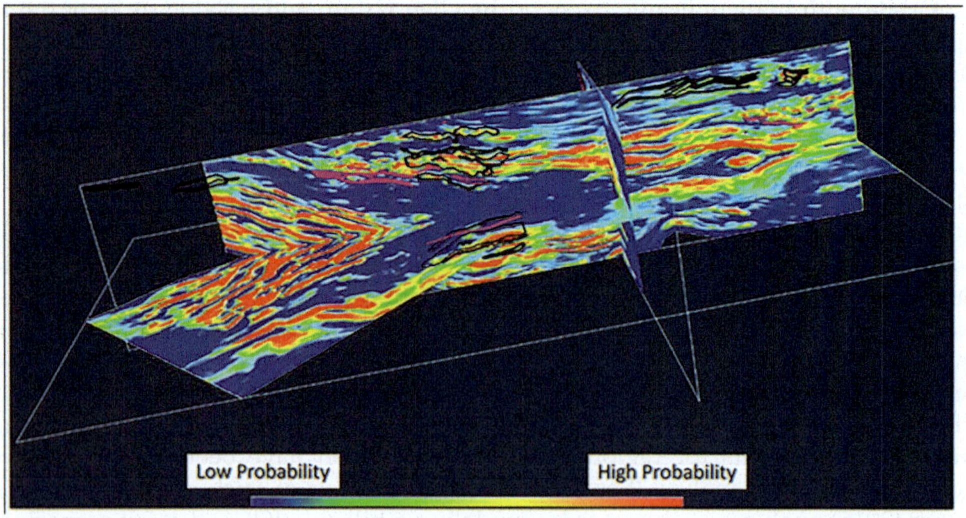

FIG. 9.8 Garnet gneiss probability volume derived from 3D seismic data over the Tropicana deposit. Garnet gneiss distribution from drilling is shown in black, while mineralization from drilling is depicted in pink. *Source: Neroni and Hunt, 2022; Jarvis et al., 2022. steveh@velseis.com*

CHAPTER

10

Marine mineral resource exploration

10.1 Introduction

The ocean is a body of salt water that covers approximately 71% of the Earth and contains 97% of Earth's water. The term ocean also refers to any of the large bodies of water into which the world ocean is conventionally divided. Distinct names are used to identify five different areas of the ocean: Pacific, Atlantic, Indian, Arctic, and Antarctic/Southern. Beginning in the twentieth century, some oceanographers labeled the seas around Antarctica the Southern Ocean, and in 2021, National Geographic officially recognized this fifth ocean. Seawater covers approximately 361,000,000 km^2 of the planet. The ocean is the primary component of the Earth's hydrosphere, and thus essential to life on Earth. The ocean influences climate and weather patterns, the carbon cycle, and the water cycle by acting as a huge heat reservoir. The oceans are an extensive source of living and nonliving resources. Despite its size and impact on the lives of every organism on Earth, the ocean remains a mystery. More than 80% of the ocean has never been mapped, explored, or even seen by humans. A far greater percentage of the surfaces of the moon and the planet Mars has been mapped and studied than of our own ocean floor.

Twelve of the largest urban areas in the world are located within 160 km of the coastline, demonstrating our increasing relationship with the oceans under different aspects, including fisheries, transport, leisure, security, and mineral exploitation. The salt in the seawater is naturally a mineral resource and source of important chemical elements, such as Chlorine (Cl), Sodium (Na), Magnesium (Mg), Potassium (K), Bromine (Br), Strontium (Sr), and Boron (B). But, only a few elements are commercially extracted from the seawater on a large scale, such as Na, Cl, Mg, and Br. Although the seawater is rich in elements with economic value, to a large extent, the mineral resources found in the oceans are related to specific geological environments, which reveals the interaction between the seawater and sediment deposition, biological activity, and magmatism. This shows that different processes contribute to enriching the mineral deposits and ore formation, such as precipitation, sedimentation, biological metabolism, diagenesis, and volcanic activity. Despite three decades of ocean research, the economic potential of the marine mineral resources is less than that expected in the 70's. The oil exploration is still very important as a major source of energy, but only a few other marine

Baseline Geoscience Data Generation for Sustainable Exploration of Natural Resources
https://doi.org/10.1016/B978-0-443-45221-5.00012-6

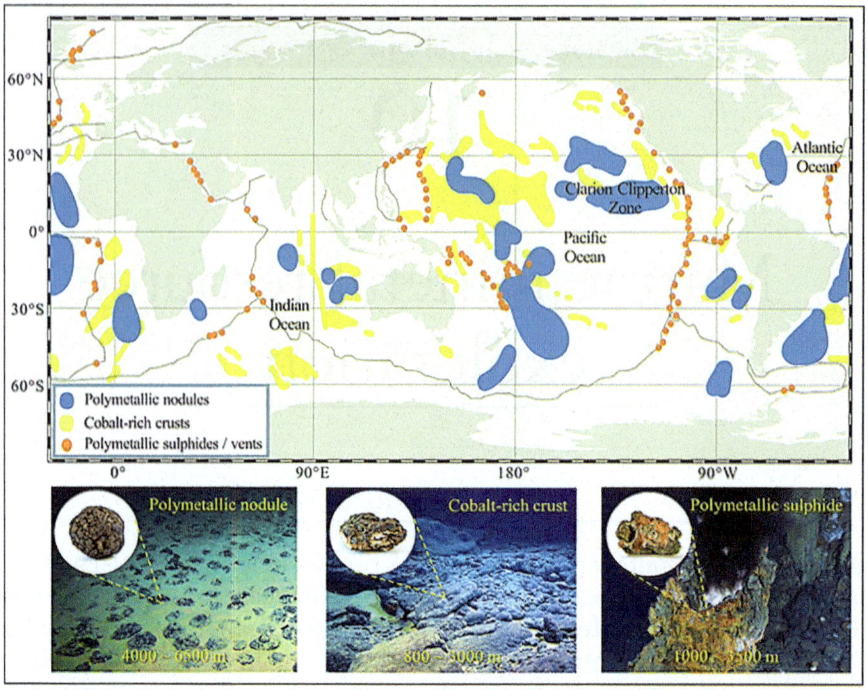

FIG. 10.1 Global distribution, occurrence state, and representative images of three major deep-seabed mineral resources. *Source: Guo et al., 2023, https://link.springer.com/article/10.1007/s40789-023-00580-x*

deposits are economically mined: marine placers, gravel and sand, shales, evaporites, and phosphorite. However, marine mineral resources (including the deep-sea deposits—polymetallic sulfides and nodules) may well represent an important resource in the medium and long term, depending on the changes in the international scenario. Therefore, such deposits, which may represent a strategic reserve, should be thoroughly investigated (Miller et al., 2018; Cuyvers et al., 2018). Fig. 10.1 shows the occurrence of different marine minerals in different environments in the oceans.

It is evident from the fact that the magnetotelluric (MT) method of exploration is a passive source-based geophysical exploration technique, which found its wider applications in the different tectonic environ to map sub-surface resources of different dimensions using land-based and sea-borne MT techniques supported by the advanced tools of data processing, which has been discussed in detail by Mishra and Naskar (2024).

10.2 Potential areas for deep-sea mining

Ferromanganese nodules in Clarion Clipperton Zone (CCZ), Peru Basin, Penrhyn Basin-Cook Islands exclusive economic zones (EEZ), and the Central Indian Ocean Basin; seafloor massive sulfide deposits in the EEZ of Papua New Guinea, Japan, and New Zealand as well as the Mid-Atlantic Ridge, and the Indian Ocean Ridge System (IORS); ferromanganese crusts in the Pacific Prime Crust Zone, and the Canary Islands Seamounts and the Rio Grande Rise in the Atlantic Ocean, and the REE rich deep sea muds around Minamitorishima Island in the western North Pacific are the major targets of economic interest for seabed mining (Sakellariadou et al., 2022). The development of technologies to collect manganese nodules first began in 1970. Later on, in 2021, a 25-ton robotic mining machine was developed by a Belgian company to collect manganese nodules. In 2022, an international mining company known as The Metals Company (TMC) announced a successful deepwater test of a manganese nodule collector vehicle at depths of 2470 m in the Atlantic Ocean. In the same year, a study was undertaken to understand the dynamics of sediment plumes generated when a prototype seabed manganese nodule collector vehicle was used at a depth of 4500 m in the CCZ in the Pacific Ocean. Such experiments shed light on the subsequent ambient sediment plumes generated during nodule collection on the seabed, the turbidity current dynamics, and the possible environmental effects of seabed mining, enabling the regulatory authorities to make decisions on future seabed mining (Munoz-Royo et al., 2022). Gonzalez et al. (Gonzalez et al., 2023) summarized a compilation work on seabed mineral deposit types with 692 occurrences comprising 1194 samples from different deposits such as ferromanganese crusts, manganese nodules, phosphorite, marine placer deposits, volcanogenic massive sulfides, and hydrothermal mineralization in pan-European seas based on extensive studies carried out utilizing geophysical surveys, sampling stations, underwater photography, and remotely operated vehicle (ROV) surveys, and mineralogical, geochemical and isotopic studies. Such studies help in identifying the potential areas for responsible future resource exploration and mining, and the standards to be used. Table 10.1 lists mining trials of deep seabed minerals and operational tests for mining equipment from 1978 to 2019.

10.2.1 A case study from Papua New Guinea, Pacific Ocean

Papua New Guinea's first offshore project at the international level "Deep Sea Mining of Submarine Hydrothermal Deposits—Solwara 1" was granted a 20-year license for mining the hydrothermal vents for copper (Cu), gold (Au), and silver (Ag) around a mile (1.6 km) below the ocean's surface (0.1 square km of the seafloor) and given to an underwater mineral exploration company known as Nautilus Minerals Inc., but failed before the extraction phase (Leal Filho et al., 2021). The maritime communities from the Bismarck and Solomon Seas have resisted the Nautilus Minerals Inc. experimental project since 2008 and appealed to their government to not only revoke the project but also to put a total ban on further seabed mining projects in their customary waters. The mining activity was experimental; there were no examples from anywhere in the world, and Papua New Guinea has no regulatory framework for such activities. Moreover, environmental concerns have also been raised by these indigenous communities, suggesting that mining will cause irreversible damage, disrupt their cultural practices, and affect food sources. The company stopped its activities in 2018, and by 2019, Nautilus, the first company to ever receive a deep-sea mining license, had gone bankrupt before extracting any minerals, and the Papua New Guinea government, which had invested in the project, was left with millions of dollars in debt (Chin and Hari, 2023). However, in 2022, some areas in the CCZ in the Central Pacific Ocean were licensed for mining by the International Seabed Authority (ISA), although several groups, including

TABLE 10.1 Mining trials of deep-seabed minerals and tests of some mining systems.

Time (year)	Country/Company	Water depth (m)	Project content
1978	Ocean Management Inc (OMI)	5500	A
1978	Ocean Mining Association (OMA)	4570	A
1979	Ocean Mining Corporation (OMCO)	5000	A
1989	Germany/the former Ministry of Science and Technology of the Federal Republic of Germany	4140–4180	B
1990	Russia/Moscow Institute of Geological Exploration	79	C
1996	India/National Institute of Ocean Technology (NIOT); Germany/University of Siegen	500	D & E
1997	Japan/Polymetallic nodules Mining system research and development project	2200	F
1998	Japan/the Geological Survey of Japan, National Institute of Advanced Industrial Science and Technology (GSJ-AIST)	1600	G
2002	Japan/Japan Oil, Gas and Metals National Corporation (JOGMEC)	1600	D
2006	India/NIOT	450	F
2009	Korea/Korea Institute of Geoscience and Mineral Resources (KIGAM)	100	H
2012	Japan/JOGMEC	1600	E
2013	Korea/Korea Institute of Ocean Science and Technology (KIOST)	1370	F
2015	Korea/KIOST; Korea/Korea Research Institute of Ships and Ocean Engineering (KRISO)	1200	C
2016	China/Changsha Research Institute of Mining and Metallurgy	300	H
2017	Japan/JOGMEC	1600	C and E
2018	China/Changsha Research Institute of Mining and Metallurgy	500	E
2018	Netherland/Royal IHC	300	D
2019	China/Institute of Deep-sea Science and Engineering, Chinese Academy of Sciences (IDSSE, CAS)	2500	F

Notes: (A) Mining trial for polymetallic nodules; (B) Disturbance and recolonization experiment for deep seabed sediments; (C) Test of the hydraulic lifting system; (D) Test of the travel performance of the mining vehicle; (E) Test of the acquisition performance of the mining vehicle; (F) Test of the mining vehicle; (G) Economic and technical feasibility study of polymetallic sulfide mining; and (H) Test of the delivery system of the mining vehicle.
Source: Guo et al., 2023; Thiel and Schriever, 1990; Ding et al., 2009; Yang et al., 2020; Su, 2020.

Greenpeace, have been very strongly opposed to deep-sea mining as it might cause irreversible damage to the marine environment.

10.3 Deep-sea mineral deposits

The ocean contains a complex combination of processes leading to the formation of a wide variety of mineral categories. In addition to the well-known oil, natural gas, and gas hydrates, the deep oceans are also bestowed with massive metal deposits. These deep-sea mineral deposits can be divided into five types: manganese nodules, seafloor massive sulfides, ferromanganese crusts (also known as cobalt crusts because of the significant concentration of cobalt), REE rich marine mud, and marine phosphorite (Fig. 10.2). Deep ocean mineral deposits also provide valuable information on the evolution of the Earth, seawater, and insights into the past global climate history (Aitken and Foulc, 2019).

For the whole world to deliver net zero by 2050, large scale mining is more critical for metals such as lithium, cobalt, platinum, palladium, REE, gallium, tungsten, tellurium, and indium as these metals are essential for green technology applications such as making wind turbines, solar panels, fuel cells, electric vehicles, and data storage systems required to transition to a low carbon economy. Since land-based mineral deposits are depleting fast, seabed resources are seen as a new resource frontier for mineral exploration and extraction. They include mainly deep ocean mineral deposits, such as massive sulfides, manganese nodules, ferromanganese crusts, phosphorite, and REE-rich marine muds. Manganese nodules contain mainly manganese and iron, but also valuable metals like nickel, cobalt, and copper, as well as REE and platinum, which are used in making several high-technology and green-technology products. For example, deep-sea mud enriched in REE (>2000 μg/g) was found in the western North Pacific Ocean. High concentrations of REE range from 1727 to 2511 μg/g in the crust samples collected from the Afanasy Nikitin Seamount (ANS) in the Indian Ocean. However, these deposits usually have lower REE grades than land-based REE deposits, such

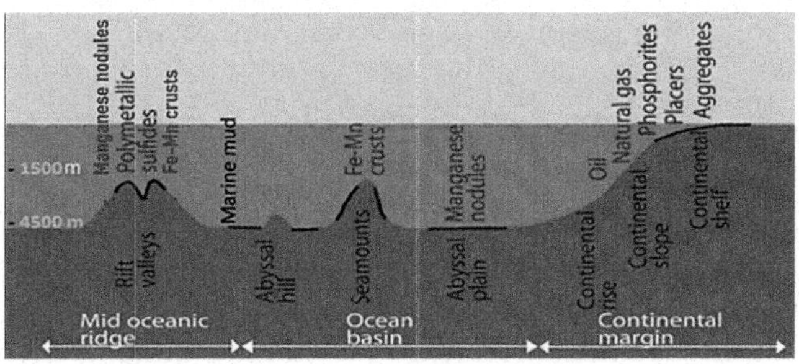

FIG. 10.2 Occurrence of manganese nodules, ferromanganese crusts, massive sulfide deposits, rare earth element (REE)-rich marine mud, and phosphorite at the ocean basins, mid-oceanic ridges, seamounts, and continental margins in the oceans. *Source: Vysetti, 2023, https://www.oaepublish.com/articles/mmm.2022.12*

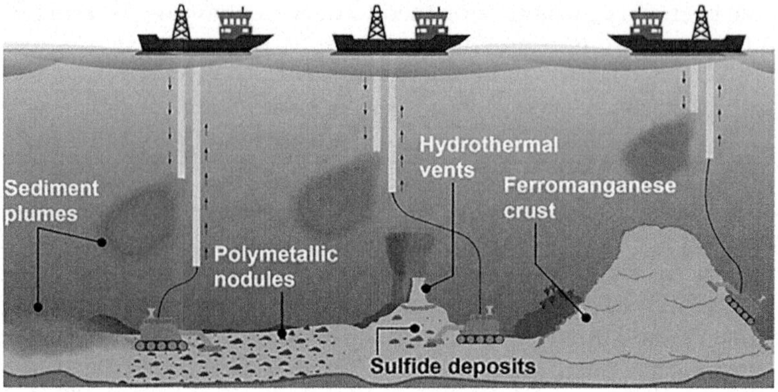

FIG. 10.3 Concept image with examples of using remotely operated vessels (ROVs) from a vessel on the surface for the extraction of different mineral materials during deep-sea mining.

as carbonatite-hosted deposits, but form greater potential volumes. Though the mining companies and their sponsoring countries are in the process of developing the required technologies to mine the three deep sea environments: abyssal plains, seamounts, and hydrothermal vents, due to severe concerns about the possible environmental damages, the International Seabed Authority (ISA) has not granted any mining permissions so far, although deep sea mining becomes inevitable in the future green energy revolution. The concept of deep-sea mining and the process of retrieving mineral deposits from the deep seabed are explained in Fig. 10.3.

Seabed minerals are mostly located at depths of 1–6 km and comprise three main types (Petterson et al., 2021) (Table 10.2). Diamonds are mined from the seabed by De Beers and others. Cobalt-rich ferromanganese formations are found at various depths between 400 and 7000 m. These formations are a type of manganese crust deposit. The substrates consist of layered iron and manganese (Fe–Mn oxyhydroxide deposits) that host mineralization (Maciag et al., 2019).

An overview of different types of major deep-sea minerals, which are described in a more detailed way in the following (Figure 1, Page 5, Balaram, 2023).

10.3.1 Ferromanganese (polymetallic) (Fe–Mn) nodules

The occurrence of seabed nodules has been known since the mid-nineteenth century (Murray and Renard, 1891). There are extensive deposits of high-grade polymetallic nodules resting on the seabed within the Clarion Clipperton Zone (CCZ) in the Pacific Ocean and the Indian Ocean Nodule Field (IONF) [International Seabed Authority (ISA), 2003; Mukhopadhyay et al., 2018]. These marine sedimentary rocks form in the oxidizing environment of the sediment-covered abyssal plains of the Indian, Pacific, and Atlantic Oceans at depths of 3000 to 6000 m. They contain huge quantities of iron and manganese (Table 10.3) and also contain significant quantities of several other metals such as copper, cobalt, nickel, REE, and platinum (Balaram, 1999).

TABLE 10.2 Minerals and related depths.

Type of mineral deposit	Average depth (m)	Resources found
Polymetallic nodules (Manganese nodules)	4000–6000	Nickel, copper, cobalt, and manganese
Manganese crusts	800–2400	Mainly cobalt, some vanadium, molybdenum, and platinum
Sulfide deposits	1400–3700	Copper, lead, and zinc, some gold and silver

Source: Ahnert and Borowski, 2000.

TABLE 10.3 Average concentrations of some strategic metals in different types of marine sediments from different oceans.

Ocean	Type of sediment	REE range (μg/g)	Fe %	Mn %	Ni %	Co %	Cu %
East Siberian Arctic Shelf (Sattarova et al., 2023)	Bottom sediments	104–220	–	–	–	–	–
Central North Pacific Ocean (Sa et al., 2018)	Siliceous sediments	810.4	–	–	–	–	–
Afanasy Niktin Seamount (ANS) in the eastern equatorial Indian Ocean (Balaram et al., 2013; Balaram et al., 2023)	Cobalt crust	1727–2511	15.75	19.61	0.40	0.53	0.10
Mid-Pacific seamount (Cui et al., 2009)	Cobalt-rich crusts	2085	16.76	24.38	–	0.71	–
Indian Ocean (Nath et al., 1992; Balaram et al., 1995)	Ferromanganese crusts/nodules	928–1570	13.94	15.70	–	–	–
Scotia Sea (Gonzalez et al., 2023)	Ferromanganese crusts	3400	20.36	22.85	0.28	0.42	<0.03
Seafloor massive sulfide deposits off Papua New Guinea, Pacific Ocean (Hoagland et al., 2010)	–	–	–	–	–	–	6.6
Seafloor massive sulfide deposit SW Pacific (Paropkari et al., 2010)	Pedestal slab	7.6	8.02	0.02	<0.01	<0.01	19.84
	Peripheral chimney	0.72	2.01	0.04	<0.01	<0.01	1.25
	Precipitate	1.77	7.92	16.15	<0.01	<0.01	0.01
Eastern South Pacific (Kato et al., 2011)	Deep-sea mud	1000–2230	–	–	–	–	–

Continued

TABLE 10.3 Average concentrations of some strategic metals in different types of marine sediments from different oceans—cont'd

Ocean	Type of sediment	REE range (µg/g)	Fe %	Mn %	Ni %	Co %	Cu %
North Pacific (east and west of Hawaiian Islands)	Deep-sea mud	400–1000	–	–	–	–	–
Minamitorishima Island in the Western North Pacific (Yamazaki et al., 2021)	REE-rich mud	500–1500 REE + Y	–	–	–	–	–
South China Sea (Zhong et al., 2018)	Ferromanganese nodule deposits	1460 (avg)	–	–	–	–	–
Indian Ocean	REE-rich mud	>400	–	–	–	–	–
Andaman Sea, Indian Ocean (Surya Prakash et al., 2012)	Ferromanganese crust, the summit of the southern seamount	1139	0.60	31.3	0.10	<0.01	<0.01
	Ferromanganese crust within the two peaks of the same seamount	2285	13.83	14.2	0.07	0.05	<0.01
Lakshadweep Sea, Indian Ocean (Charlu and Kalluraya, 2023)	Ferromanganese crust	La (200) & Y (150)	–	–	–	–	–
West Sewell Ridge, Andaman Sea	Ferromanganese crust	1600	–	–	–	–	–
Indian Ocean (Dinesh et al., 2020)	Manganese nodules	1186	–	–	–	–	–
Clarion-Clipperton Fracture Zone, Northeastern Pacific Ocean (Kim et al., 2022)	Deep-sea sediments	>700	–	–	–	–	–
Peru Basin (Toro et al., 2020)	Fe–Mn nodules	403	6.12	34.2	1.3	0.05	0.60
CCZ, Pacific Ocean (SPC, 2023)	Fe–Mn nodules	813	6.16	28.4	1.3	0.21	1.07
			5.9	28.1	–	–	–
Cook Island	Fe–Mn nodules	–	15.9	17.6	0.45	0.41	0.23
Cook Islands (Toro et al., 2020)	Fe–Mn nodules	1707	16.1	15.9	0.38	0.41	0.23
West Clarion-Clipperton Zone, Pacific Ocean (Qiu et al., 2022)	Marine sediments	454.7 (REE + Y)	–	<1	–	–	–
North Pacific Ocean near Minamitorishima Island, Japan (Takaya et al., 2018)	Deep sea mud	>5000 (REE + Y)	–	–	–	–	–

Ocean	Type of sediment	REE range (μg/g)	Fe %	Mn %	Ni %	Co %	Cu %
Mid-Pacific Ocean (Bu et al., 2003)	Fe–Mn nodules	1178–1434	15.94	24.38	0.38	0.56	0.12
Canary Islands, Atlantic Ocean (Marino et al., 2019)	Fe–Mn crusts	335–3485	28.4	13.79	0.25	0.44	0.06
Central Pacific (Halbach et al., 2017)	Fe–Mn crusts	678–2649	6.0-18	13-27	0.17–0.73	0.25–1.18	0.02–0.22
Pacific Ocean (Halbach and Marbler, 2009)	Fe–Mn crusts	1060	–	22.5	0.49	0.71	0.04
Atlantic Ocean (Hein et al., 2017)	Fe–Mn crusts	2221	–	14.5	0.26	0.36	0.09
Indian Ocean	Fe–Mn crusts	2363	–	17.0	0.26	0.33	0.11
Indian Ocean	Fe-Mn nodules	1039	7.14	24.4	1.10	0.11	1.04
Pacific Ocean (Piper, 1974)	Deep nodules	1326	–	–	–	–	–
	Shallow nodules	1398	–	–	–	–	–

Note: – Not Available

Source: Balaram, 2023.

They are mainly composed of tectomanganeate such as vernadite, birnessite, buserite, and todorokite minerals (Kuhn et al., 2017). These manganese nodules support life in the deep-sea ecosystem by providing critical habitat for a large number of unique, largely unknown, and understudied species. Exploration of seabed polymetallic nodules identifies the Clarion-Clipperton Zone (a 4.5 million km^2 stretch of the seafloor between Hawaii and Mexico that extends 4000 to 5500 m deep) and the Indian Ocean Nodule Field to be of economic interest (Menard and Shipek, 1958; Sakellariadou et al., 2022). The first known published estimate of the size of the CCZ deposit was by Mero (1965), who estimated 1656 Bt for the entire Pacific from a set of 29 photographs, 10 dredges, and 62 box cores. Nodules can be sampled in a meaningful manner using mechanical devices, although with minor issues of bias. Grade and moisture content are measured using the established methodology for land-based minerals. Tonnage of resource is determined via the abundance of nodules in kilograms per square meter of seabed. Recently, Li rich ferromanganese oxide deposits have also reported from the South Andaman Sea, Northeastern Indian Ocean, with Li concentrations up to 781 μg/g (Rajani et al., 2022). High concentrations of Li up to 311 μg/g in manganese nodules from the Peru Basin were reported by Hein et al. (2013). Li rich ferromanganese deposits (up to 738 μg/g) have also been reported in the Cocos-Nazca spreading center, Pacific Ocean (Zawadzki et al., 2022).

The geology of polymetallic nodule deposits appears to be relatively well understood, and multiple lines of evidence, such as physical samples, seabed photographs, and multibeam echosounder surveys, have led to a common view of the deposit's characteristics. Recent interest in developing these resources has resulted in efforts to advance questions of mineral rights (Watzel et al., 2020; Toro et al., 2020; Hein et al., 2020). These deposits are located on the seabed within international waters, and ownership of the minerals cannot be directly attributed to any particular nation (as is typically the case on land) (Parianos et al., 2021). Most notably, the United States has issued its own seabed licenses for polymetallic nodules within the CCZ, albeit so far without any overlap with the ISA contracts or reserved areas (National Oceanic and Atmospheric Administration, 2016; Lipton et al., 2016). The sampling methods and scales of mineralization are unique to this type of deposit but are not a hindrance to the estimation or reporting of mineral resources (Parianos et al., 2021).

10.3.2 Seafloor massive sulfide (SMS)

Seafloor hydrothermal systems often have accompanying sulfide ore deposits (Hannington et al., 1995), which have been regarded as submarine mineral resources to be explored and mined (Boschen et al., 2013). Hydrothermal deposits are large-scale geochemical anomalies; sulfur and metallic sulfide minerals in hydrothermal deposits are the main economic minerals in the crust. Seafloor massive sulfide (SMS) mineralization typically occurs at marine plate boundaries, such as mid-ocean ridges, volcanic arcs, and back-arc basins, where hydrothermal fluids erupt from deep in the crust, mix with cold seawater, and precipitate minerals on or below the seafloor (Hannington et al., 2005). SMS is the primary source of the basemetal Copper (Cu), Zinc (Zn), and Lead (Pb) and a large number of lower concentration metals including silver (Ag), Arsenic (As), Gold (Au), Bismuth (Bi), Cadmium (Cd), Cobalt (Co), Gallium (Ga), Germanium (Ge), Indium (In), Mercury (Hg), Molybdenum (Mo), Nickel (Ni), Rhenium (Re), Antimony (Sb), Selenium (Se), Tin (Sn), Tellurium (Te), and Thallium (Tl), and these metals typically form their sulfides or appear as trace elements in other sulfides and sulfosalts (Fontbote et al., 2017). The targets for mining include ore deposits associated with both active and inactive hydrothermal systems. Ore deposits associated with active systems are visible above the seafloor. Those associated with inactive systems are, however, likely to be buried below the seafloor, where they are protected from weathering (Ridley, 2010). Exploring methods that are applicable not only to visible but also to buried ore deposits should choose out-of-line survey or mapping methods from the viewpoint of efficiency. Although such exploration methods are established in on-land environments, efficient methods for marine environments are not necessarily the same as those on land.

SMS deposits are of growing economic interest as they contain Cu, Sn, Zn, and potentially traces of Ag and Au (Galley et al., 2007). Within the active zones of global mid-ocean ridges, and along submarine volcanic arc and back arc spreading centers, the total amount of Cu and Zn in accessible SMS deposits is estimated on the order of 3×10^{10} kg (Hannington et al., 2011). However, a clear quantification is poorly constrained, as off-axis occurrences, which are no longer connected to hydrothermal activity and potentially covered by sediments, are not detectable using the presently available technologies and may significantly increase this estimate (Andersen et al., 2017). To find and delineate possibly unknown SMS occurrences, new geophysical technologies are currently being developed to detect and quantify inactive, off-axis SMS sites to improve the global estimates of mineral deposits on the seafloor. Here, we focus on two systems: first, a towed, fixed-offset three-axis electric field controlled source electromagnetic (CSEM) system and secondly a coincident-loop transient electromagnetic method (TEM) system that was used successfully on a research cruise.

10.3.3 Ferromanganese crusts

Ferromanganese crusts or Fe–Mn crusts are formed on seamounts (Fig. 10.2) by the precipitation of colloidal components such as iron-manganese oxides from seawater and also, in some cases, from hydrothermal inputs, on volcanic and biogenic substrate rocks. Mineralogical and geochemical studies on the ferromanganese crusts collected from two seamounts in the Andaman Sea revealed two types of origin, namely hydrothermal and hydrogenous (Surya Prakash et al., 2012; Kuhn et al., 1998). These crusts contain high concentrations of elements such as cobalt (Co), rare earth elements (REE), niobium (Nb), platinum (Pt), tungsten (W), bismuth (Bi), nickel (Ni), manganese (Mn), tellurium (Te), and titanium (Ti). It was also observed that the majority of REE, including yttrium (Y), were found to be associated with Mn-oxide and Fe-oxyhydroxide phases (Koschinsky and Hein, 2003; Marino et al., 2017). Balaram et al. (2023) reported very high concentrations of ΣREE ranging from 1737 to 2520 μg/g in the ferromanganese crust samples collected from the Afanasy Nikitin Seamount (ANS) in the Eastern Equatorial Indian Ocean in 2012. There is also considerable enrichment of platinum in these marine sediments, as the bulk concentration varies between 0.14 and 1.02 μg/g and averages at 0.51 ± 0.24 μg/g (Halbach et al., 1989; Balaram et al., 2023). Several studies (Ren et al., 2021; Sujith et al., 2017; Hein and Koschinsky, 2014) indicated that the elemental enrichment in ferromanganese crusts is the result of comprehensive factors, including low sedimentation rate, the high metal content of the bottom seawater, a noncarbonate depositional environment, oxidizing conditions, microbial activity, certain bottom current conditions, and cosmic spherules. Table 10.3 presents concentrations of ΣREE and also some critical metals in different marine sediments, including ferromanganese crust samples.

10.3.4 REE-rich marine mud

Recent studies suggested that pelagic sediments (**settling of particles to the floor of the open ocean, far from land**) may contain large amounts of REE and have received global attention as an important reservoir of the global REE budget and a potential resource for future use. The first report by Kato et al. (2011) indicated that pelagic sediments have the potential to become important REE resources. They reported very high concentrations of ΣREE (up to 2230 μg/g) in marine mud on the deep-sea floor in the Pacific Ocean off Hawaii and Tahiti. Subsequently, deep-sea mud containing over 5000 μg/g REE content was discovered in the western North Pacific Ocean near Minamitorishima Island, Japan, by Takaya et al. (2018), Iijima et al. (2016), and Fujinaga et al. (2016). In general, Ca-phosphate (e.g., bioapatite fossils) and Fe–Mn oxyhydroxide have been considered important REE carriers in deep-sea sediments (Liao et al., 2022). Deep-sea mud enriched in REE (ΣREE > 2000 μg/g) was found in the western North Pacific Ocean, and the sediment cores collected near topographic highs in the Central Pacific Ocean yielded ΣREE 4489 μg/g (Ohta et al., 2021). Recently, Fujinaga et al. (2022) reported marine mud markedly enriched with REE (up to ΣREE 1120 μg/g) from the Aki umber deposit in the Northern Shimanto Belt, central Shikoku, southwest Japan. The geochemical features suggested that the Aki umber represents a hydrothermal iron oxyhydroxide type REY-rich mud deposited originally in a pelagic deep-sea setting during the Middle Cretaceous (113.2 to 100.5 mA).

Rare earth elements and yttrium (REY) are critical materials to many leading-edge technologies due to their unique physical and chemical properties. Applications of REY span a wide range, including hybrid vehicles, rechargeable batteries, wind turbines, light-emitting diodes, compact fluorescent lamps, screen display panels, and many medical and military technologies [(US Department of Energy (DOE), 2011); Wall, 2014]. The industrial utility of REY, especially in renewable energy technologies and electronics, has driven up the demand for the essential metals in recent years (US Department of the Interior and US Geological Survey, 2018; Alonso et al., 2012; Massari and Ruberti, 2013; Dutta et al., 2016). Yasukawa et al. (2014) studied the geochemistry and mineralogy of REY-rich mud in the eastern Indian Ocean. These sediments consist mainly of siliceous ooze, with subordinate zeolitic clay that contains relatively high REY concentrations. The maximum and average ΣREY contents of this material are 1113 and 629 μg/g, respectively, which are comparable to those reported from the Pacific Ocean. Takaya et al. (2018) estimated in detail the resource potential of REY-rich mud using Geographical Information System (GIS) software and proposed appropriate mineral processing techniques that can effectively enhance its economic value. Moreover, using a hydrocyclone separator enabled the selective recovery of biogenic calcium phosphate grains, which have high REY content (up to 22,000 ppm) and constitute the coarser domain in the grain size distribution. The enormous resource amount and the effectiveness of the mineral processing are strong indicators that this new REY resource could be exploited in the near future.

10.3.5 Phosphorite deposits

Phosphorites are phosphorus-rich sedimentary rocks, usually composed of carbonate hydroxyl fluorapatite, $Ca_5(PO_4, CO_3)_3F$. They occur in the global ocean in three general environments: (i) as nodules and crusts originally formed in oceanic environments: continental margins—shelf, slope, banks, and plateaus, (ii) seamounts, especially the old (Cretaceous) seamounts, for example, in the NW Pacific Ocean, and, (iii) lagoon/insular deposits (Hein et al., 2016). Marine phosphorites primarily occur along continental margins where upwelling large volumes of phosphate-rich cold water rise from great depths to the surface. In the warmer surface waters, phosphate precipitates out of the solution and then sinks to the seafloor, forming deposits of phosphorite (Kolodny, 1969; Wang et al., 2014). Phosphorite deposits are widespread on the seafloor of continental shelves and slopes along the western continental margins of both the Pacific and Atlantic Oceans. Some of the areas where these deposits form include the Peru-Chile margin, on plateaus such as Chatham Rise offshore New Zealand, the Blake-Bahamas Plateau off the southeast United States, offshore Baja California, and on the shelf off Namibia. Significant deposits of phosphorite are found in the Xisha Islands (Ritterbush, 2023). Gonzalez et al. (2016) reported extensive phosphorite deposits in upwelling areas in the Atlantic Iberian margins. They can also form on seamounts where ferromanganese crusts grow. Some of the phosphorite deposits contain ΣREE up to 2000 μg/g (Emsbo et al., 2015), although the composition of these rocks and REE concentrations mostly depend on their type and origin, and they also contain up to 4% fluorine (USGS, 2023). Seamount phosphorites contain higher concentrations of REE than continental margin phosphorites (Hein et al., 2016). In fact, land-based phosphorite deposits also offer similar potential for metals such as REE and phosphorus, which drive the exploration, mining, and extraction studies in the future. Fig. 10.1 shows the occurrence of different marine minerals in different environments in the oceans.

10.4 Deep-seabed mining: Frontiers in engineering geology and environment

10.4.1 Bibliometric analysis of ocean mineral resource development and literature trends

Through bibliometric analysis, the development trends of some key information (literature numbers, authors, and keywords) from research literature can be assessed in a more objective and reliable way using mathematical and statistical theories (Chen, 2004, 2006). The research literature data were retrieved from the Web of Science (WOS) database and the China National Knowledge Infrastructure (CNKI) database. Original data, including titles, authors, keywords, abstracts, etc., were output from these databases. The subject words used in the retrieval were "Ocean (or marine, undersea, underwater) + Mineral resources (or mining)." As a result, a total of 5838 effective literature records were obtained, including 4986 literature from the WOS database and 852 literature from the CNKI database, in a range of publication years from 1957 to 2022. The time of data retrieval was July 2022. The information from that 5838 literature was imported into the tool CiteSpace-6.1.R2 as basic data for bibliometric analysis (Guo et al., 2023).

The number of literature published in a certain time can reveal development trends in a research field to a certain extent. The variation trends of 5838 relevant literature in the field of ocean mineral resources over 65 years from 1957 to 2022 are shown (Figure 2, Page 4, Guo et al., 2023). The time period of the analysis was divided into 12 time spans, with one span consisting of five years (except for the first and last spans). In Stage III, the number of relevant studies increased substantially, with the average number of literature in each time span reaching ~1256 for the WOS database and ~195 for the CNKI database. Those trends highlight how ocean mineral resource development has become a hot research topic in the past 10 years. In addition, the trends could herald that this research topic will remain popular for a long time in the future.

10.4.2 Degrees of national concern and international cooperation

The institutional information extracted from research literature can inform the concern and cooperation between countries or regions in a research field to a certain extent. According to the data extracted from 4986 literature in the WOS database, a network graph of national concern and cooperation in research on ocean mineral resources was constructed (Figure 3, Page 5, Guo et al., 2023). The clusters in the figure represent the countries of the institutions conducting the research, and the links between the clusters represent cooperation between institutes in different countries. Statistical analysis revealed 135 clusters and 163 links (not all clusters and links are displayed in Figure 3, Page 5, Guo et al., 2023, due to interface limitations). The three countries with the most relevant literature are the United States, China, and Germany, which show large clusters. Those highly concerned countries or regions often have longer shorelines and need sufficient research to support their ocean mineral resources

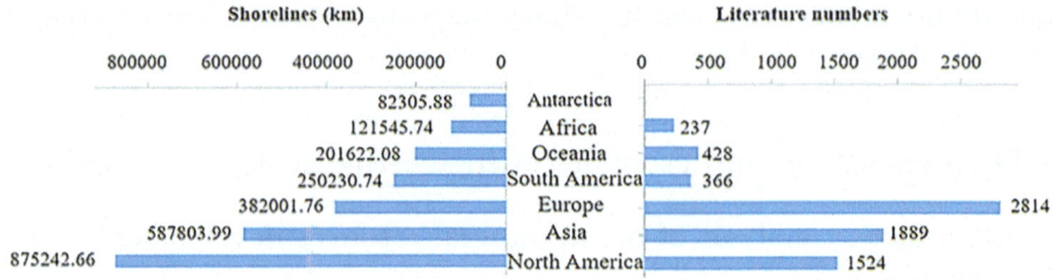

FIG. 10.4 Number of relevant literature versus shorelines by continent. *Source: Guo et al., 2023. https://link.springer.com/article/10.1007/s40789-023-00580-x*

development, as shown in Fig. 10.4 (here, the data from all countries were grouped into continents, and the number of published studies is plotted with shoreline length by continent; the shoreline length data are based on Liu et al. (2015).

As shown in (Figure 5a, Page 6, Guo et al., 2023), the five most common literature keywords from the WOS database were "heavy metal," "sediment," "trace element," "contamination," and "rare earth element" (sorted from more frequent to less frequent); the five most common literature keywords from the CNKI database were "polymetallic nodule," "marine sediment," "flooded mine," "mining vehicle," and "environmental impact" (Figure 5b, Page 6, Guo et al., 2023).

The popular research directions indicated by the keywords from the two databases are similar and include the mineral resources themselves (metals and nodules), resource development approaches, and the environmental effects of resource development. In addition, some keywords were often mentioned together, forming the keyword links in (Figure 5, Page 6, Guo et al., 2023). For example, "heavy metal" and "contamination" are linked in (Figure 5a, Page 6, Guo et al., 2023), while "polymetallic nodule" and "mining vehicle" are linked in (Figure 5b, Page 6, Guo et al., 2023).

10.5 Techniques for ocean mining and blue economy

In alignment with the principles of the Blue Economy—which emphasize sustainable use of ocean resources for economic growth, improved livelihoods, and ocean ecosystem health—ocean mining activities are typically structured into three main stages: exploration, extraction, and closure (China Association of Oceanic Engineering, 2021). These stages must be planned and executed in ways that minimize environmental disruption and ensure long-term benefits for coastal and marine communities.

The exploration stage occurs after an exploration permit has been granted but prior to the issuance of a mining permit. This phase focuses on comprehensive assessments of the marine environment, including geologic and hydrologic conditions, geotechnical properties of seafloor sediments, and the identification of potential natural and operational hazards. In accordance with sustainable development objectives, exploration activities should integrate environmental baseline studies and stakeholder consultations to inform responsible resource utilization strategies.

Once a mining permit is secured, the project advances to the extraction stage, where resource recovery operations are initiated. Extraction technologies must be selected and implemented with a focus on reducing ecological footprints, such as sediment plume control, biodiversity preservation, and energy efficiency. Techniques that support circular economic models, reduce waste, and encourage local economic participation are increasingly favored within blue economy frameworks.

Upon completion or suspension of extraction activities, the project transitions to the closure stage. This final phase involves the rehabilitation and restoration of the affected marine environment. Restoration practices should aim to reestablish ecosystem functions, promote biodiversity recovery, and monitor long-term environmental health. Strategic planning for closure not only addresses ecological concerns but also supports social and economic resilience in maritime regions.

The following sections provide a detailed overview of the specific techniques and technologies employed during each stage of ocean mining, evaluated through the lens of sustainable development and the blue economy.

10.5.1 Techniques for the exploration stage

10.5.1.1 Geophysical exploration techniques

A comprehensive exploration of an ocean mining project should cover the ocean life, meteorology, chemistry, and geology based on techniques from fields including engineering geology, geotechnical engineering, and biology. This study focuses on techniques related to geological exploration, which primarily include geophysical exploration, geochemical, and geotechnical exploration. Geophysical exploration can reveal the geological characteristics (geological background, stratigraphic substance composition, and geological structures) of the project area based on the local variation in geophysical fields. Geophysical exploration mainly considers the features of subsea stratigraphic substances (the density, elasticity, magnetic permeability, electrical conductivity, thermal conductivity, and radioactivity) to elucidate and explain the variation in geophysical fields. Thus, geophysical exploration is an indirect method. Geophysical techniques can provide efficient nonintrusive methods for detecting hidden minerals in terrestrial and marine environments. Active source methods like induced polarization (IP) (Han et al., 2013; Liu et al., 2014), electrical resistivity tomography (ERT) (Horo et al., 2020; Ishizu et al., 2019), the time domain transient electromagnetic method (TEM) (Cairns et al., 1996; Tao et al., 2013), and the controlled source frequency domain electromagnetic method (CSEM) (Gehrmann et al., 2019, 2020) have been applied in on-land and/or marine mineral exploration. TEM and CSEM are more popular because the sensors can be towed by a ship for continuous measurements in marine environments. Some surveys combine two (SP and ERT) (Horo et al., 2021) or more (SP, ERT, and IP) (Horo et al., 2020) of them for prospecting. Passive methods like gravity surveys (Araya et al., 2012) and the geomagnetic method (Szitkar et al., 2015; Honsho et al., 2016) are alternative techniques for exploring SMS; however, the magnetic method responds to any magnetized/less-magnetized body, thus this method requires geological information about the target area. The plume survey method is also used in marine mineral exploration (German et al., 2008). This method only responds to signals of plume fluids, while the fluid discharge location is sometimes different from the plume fluid location, which may lead to the wrong delineation

(Kawada and Kasaya, 2018). It is well known that more is known about the surface of the moon or Mars than about the surface of the Earth. This is because over 70% of the Earth is hidden under water that easily absorbs electromagnetic waves. While satellites and telescopes can look far into the universe, these tools cannot see to the bottom of our oceans. Therefore, we rely on sophisticated geophysical techniques to "see" the seafloor. These so-called "remote sensing" methods use electric and magnetic fields, variations in gravity, or sound waves (similar to echo-sounding techniques used by bats; bats are associated with the terrestrial environment, but have been observed over the ocean) to image the topography of the seafloor (GEOMAR—Research Center for Marine Geosciences)—(Leibniz Institute of Marine Sciences [IFM-GEOMAR] Archived 2012–03-27 at the Wayback Machine German Marine Research Consortium Retrieved 28 July 2011).

In on-land environments, the standard technique of exploring subsurface ore deposits is a remote sensing geophysical survey combined with geological information of the target area. The most effective methods for this purpose are electromagnetic methods with active sources (Meju, 2002). These methods induce electric currents into the ground and measure their secondary responses. Such methods can reveal the 2D or 3D distribution of subsurface electrical conductivity, which is sensitive to the occurrence of metal-containing materials such as sulfide minerals. These methods, which are being developed for use in marine environments (Wolfgram et al., 1986; Cairns et al., 1996; Jones, 1999; Kowalczyk, 2008), might also be used effectively in the ocean, but they are difficult to handle as tools for initial surveys without detailed information on the target area. Compared with the above techniques, the SP method has a direct response to ore bodies due to redox reactions as described by the geobattery model, thus being more suitable for SMS prospecting (Fig. 10.5) (Xie et al., 2023).

According to the seafloor features being detected, common geophysical exploration techniques include gravity exploration, seismic exploration, magnetic exploration, electromagnetic, and electrical exploration.

Gravity exploration

The application of **gravity survey** for **mineral and groundwater exploration primarily lies in delineating surface/shallow structures such as faults, fractures**, etc., which may be

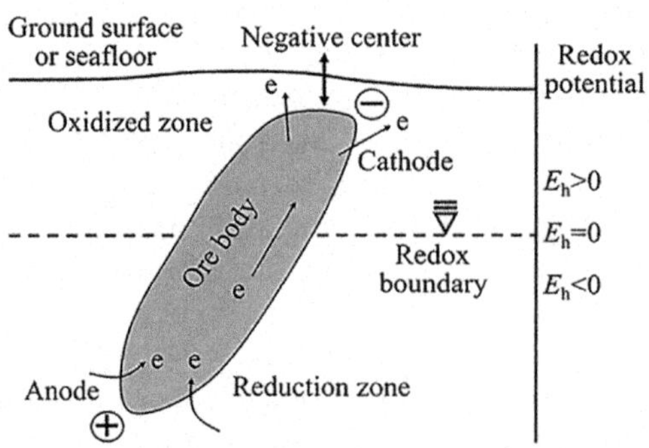

FIG. 10.5 A sketch of a natural geobattery in both on-land and marine environments, and E_h is the subsurface redox potential. *Source: Xie et al., 2023. https://www.sciencedirect.com/science/article/pii/S1003632623661771?via%3Dihub*

mineralized (Parasnis, 1997). The principle of gravity exploration is based on the local variation in the gravitational acceleration fields of the seafloor caused by the density differences of various rocks and ores, as shown in Figure 7 (Page 8, Guo et al., 2023). The local variation of the gravitational acceleration field (gravity anomaly) can be explored by gravity devices, mainly shipborne equipment (gravimeters and torsion balances). The plans and sections of the gravitational acceleration field can then be obtained and used to infer the position of mineral resources. Gravity exploration can be performed using fixed point (benchmark) surveys or regional continuous surveys (Essa et al., 2020).

The gravity exploration technique was developed in the early twentieth century. Baron Roland von Eotvos made the first gravity field measurement in 1901 in a subaquatic environment (Balaton Lake) using a self-developed torsion balance for measuring gravity variation (Vajk, 1949). Subsequently, the torsion balance was employed at sea by Hecker in 1903 (Torge, 1989). A gravity device was first attached to a submarine for exploration by Vening Meinesz in 1923 (Fujimoto et al., 2000), revealing changes in gravity in different oceanic regions; in particular, a large gravitational effect was found near the Indonesian Java Trench. Later, in the 1950s, gravity exploration approaches were widely applied in many countries, and the exploration efficiency was continuously improved. Some instruments used for gravity exploration include the pendulum ocean gravimeter of Graf-Askania and LaCoste & Romberg, and the vibration ocean gravimeter of the University of Tokyo and the Massachusetts Institute of Technology (Forsberg et al., 2015; Shinohara et al., 2015). Since the development of the Global Positioning System (GPS) based 3D positioning, the gravity exploration approaches have been greatly improved. The gravity anomalies caused by resource storage can be detected more accurately with the support of GPS. Recently, some undersea remotely operated vehicles (ROVs) or autonomous underwater vehicles with gravity exploration functions have been developed (Ballu et al., 1998; Gilbert et al., 2007; James et al., 2013).

Seismic exploration

The principle of seismic exploration is based on the propagation of seismic waves in strata with different elasticities. First, the seismic waves are artificially excited (by an explosive source, air gun, spark detonation gas, or resonator). The reflected or refracted waves at the interfaces between different elastic strata are received and recorded by hydrophones. Finally, the strata are identified by the different wave travel times and spectra. Seismic exploration is often used to explore oil and gas reservoir layers, coal fields, salt rock deposits, and some metallic mineral deposits. This technique can be used for continuous observation at sea and reach exploration depths of over a thousand kilometers, resulting in high efficiency.

In 1845, Mallet used artificially excited seismic waves to measure the propagation speed of elastic waves in the Earth's crust, which is considered the birth of seismic exploration (Graziano and Anita, 2005). Regarding the application of seismic exploration, a team of physicists and geologists (William P. Haseman, J. Clarence Karcher, Irving Perrine, and Daniel W. Ohern) first recorded clear reflection waves produced by an artificial seismic source in Oklahoma (United States) in 1921 (Schriever, 1952). Three oil storage fields were then discovered in the area through reflected seismic exploration in 1930 (Schriever, 1952). During the same period, many salt domes were discovered along the Gulf Coast using refracted seismic surveys (Sussman and Stallman, 1975). In the 1950s and 1960s, the spot photography technique of seismic exploration was replaced by analog tape recording, thereby improving the signal recording quality (Luo, 2009). Later in the 1970s, analog tape recording was replaced by digital tape recording, resulting in a complete system based on high-speed digital computers that combined digital recording, multiple overlay techniques, and seismic data processing. In the 1990s, a new seismic exploration technique, seismic imaging, was proposed. The method utilized the characteristics of no surface wave interference in water and forms color digital soil sections by means of densely displaying wave impedance (Mora and Tarantola, 1989). Recently, broadband seismic exploration techniques have been proposed for exploration in areas with large variations in water depth, rugged seabeds, and developed fractures (Moldoveanu et al., 2012).

From the perspective of testing dimension, seismic exploration approaches have developed from 1D to 4D (from the exploration of a point, to a line, to an area, and to multiple areas at different times). Time-varying 3D stereoscopic seismic images of the seabed layers can now be obtained using 4D seismic exploration approaches. In addition, hydrophone units have been placed on the seabed for long-term exploration, a technique known as ocean bottom seismometers (OBS) (Jocobson et al., 1991; Sha et al., 2015), as shown in (Fig. 10.6). The OBS exploration technique allows the seabed reception of seismic waves, thus avoiding the large amount of absorption and attenuation of seismic wave energy in seawater and improving the accuracy of the results (Zhang, 2015).

Magnetic exploration

The principle of magnetic exploration is based on the local anomalies in the geomagnetic fields caused by various magnetic rocks, ores, and other objects (Fig. 10.7). A magnetic field detector (proton precession magnetometer and gradiometer) is first fixed on a subsea drag body. The body is then dragged within a certain distance behind a moving ship, and the local magnetic anomalies of the stratum are recorded. This technique is suitable for exploring ore resources (iron ore, lead-zinc ore, and copper ore). However, this technique involves intermittent measurements, and the survey sensitivity is relatively low (Hsu et al., 2014).

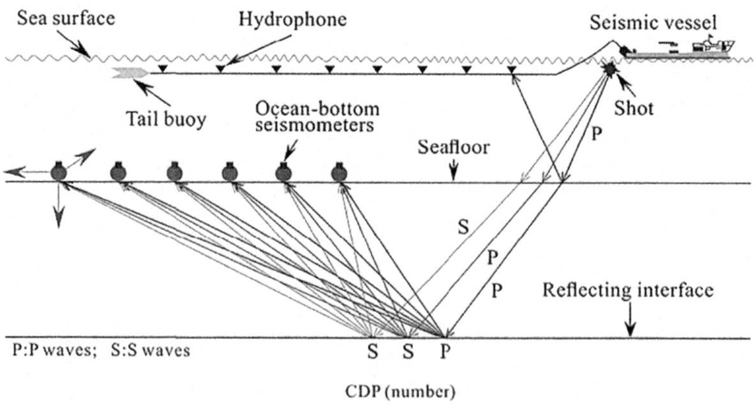

FIG. 10.6 Ocean bottom seismometer (OBS) principle. *Source: Guo et al., 2023. https://link.springer.com/article/10.1007/s40789-023-00580-x*

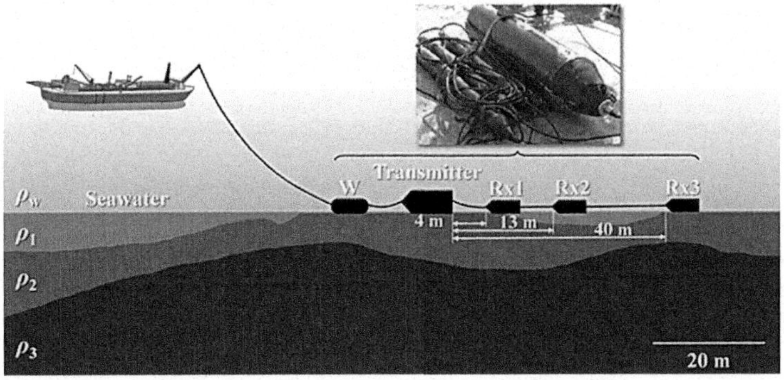

FIG. 10.7 Magnetic exploration working principle. The towed electromagnetic transmitter composed of three receivers, Rx1, Rx2, and Rx3, tuned to measure these magnetic fields different frequencies) and a transmitter (a horizontal magnetic dipole) forming an array with a total length of 40 m on the seafloor generates harmonic magnetic fields over a range of frequencies. *Source: Guo et al., 2023. https://link.springer.com/article/10.1007/s40789-023-00580-x*

Magnetic exploration is the earliest developed and most widely used among all geophysical exploration approaches, and some magnetite can be easily found using a magnetic compass. In 1870, Thaln and Tiberg constructed a universal magnetometer, and the magnetic exploration approach was formally proposed (Zhang, 2011). Schmidt invented a quartz blade magnetometer to find minerals other than iron ores in 1915. In the 1950s and early 1960s, magnetometers were installed on ships to conduct magnetic surveys of oceans (Howland-Rose et al., 1980). Later, a transient electromagnetic approach was developed with a deeper exploration depth and strong penetration in high resistance layers. In the 1980s, the rapid development of various computer technologies and high-precision magnetometers promoted the development of magnetic exploration based on high-precision and automatic interpretation (Chen, 2015). The inversion visualization technique for interpretation has been developed and realized (Zhang, 2011).

Electrical exploration

The principle of electrical exploration is based on the differences in the electrical properties (electrical conductivity, electromagnetic induction, and dielectricity) of rocks and ores (Fig. 10.8). Here, fields are of purely radial geometry and able to generate galvanic effects when they intersect subhorizontal, tabular bodies (Constable, 2010).

The implementation of this technique is similar to the implementation of magnetic exploration, except that the field sources in electrical exploration are electrical signals with different frequencies and waveforms. The spatial distributions of the electric and electromagnetic fields on the seafloor are then recorded and analyzed. Based on the current properties of the field source, the electrical exploration technique can be further divided into the direct current method and the alternating current method.

In the early twentieth century, Schlumberger conducted offshore measurements of direct current resistivity, marking the beginning of electrical exploration in the ocean (Allaud and Martin, 1977). Corwin (1976) discovered a 300 mV self-potential anomaly in Penobscot Bay, Myanmar, using the self-potential method (a type of direct current method) and found a

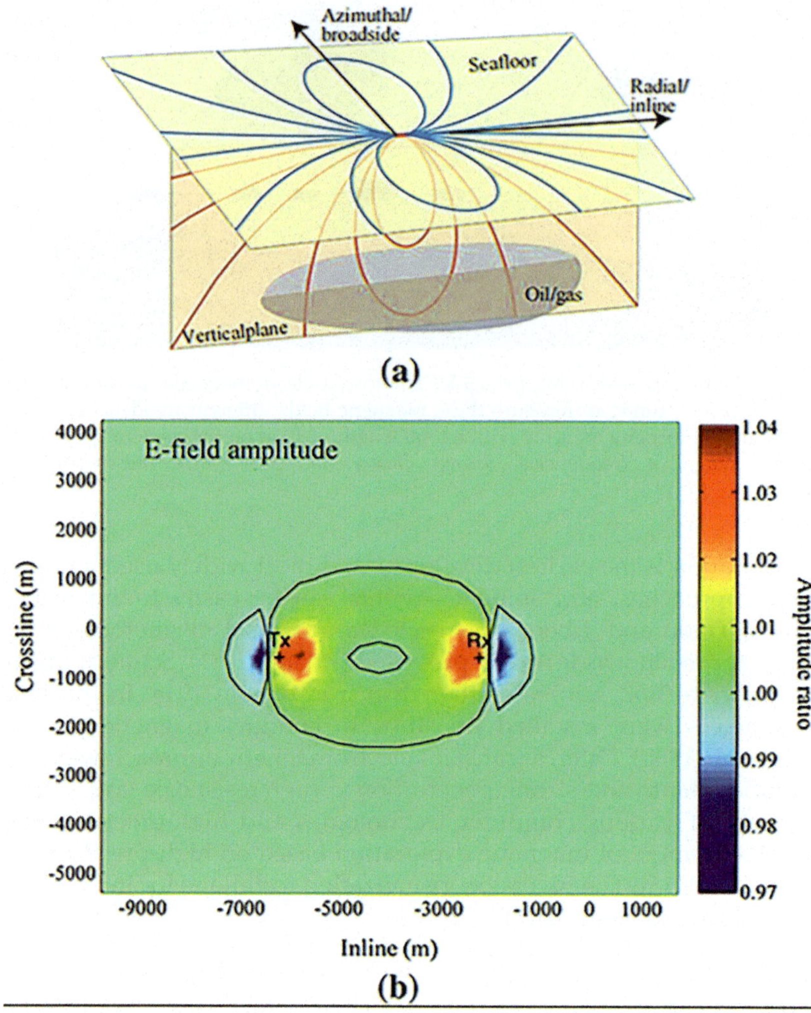

FIG. 10.8 Electrical exploration (A) Working principle (i.e., the dipole geometry of a near-seafloor transmitter, here the maximum vertical electric fields are below the transmitter in the inline direction, and fields are of purely radial geometry and able to generate galvanic effects when they intersect subhorizontal, tabular bodies such as oil and gas reservoirs) and (B) Example output, where Tx and Rx represent transmit and receive, respectively. *Source: https://link.springer.com and https://link.springer.com/article/10.1007/s40789-023-00580-x*

sulfide sedimentary deposit. Goto et al. (2007) developed Marine Autonomous Near-seafloor Tool for Analysis (MANTA), an ocean direct current resistivity measurement system for probing the top boundaries of methane hydrate fields; MANTA was successfully used to detect methane hydrate resources near Okinawa in Japan. Recently, using statically placed electrodes and drag electrode techniques, Simyrdanis et al. (2016) performed 2D and 3D direct current inversion to image the seafloor layers near the coast of northwestern Greece.

Electromagnetic exploration

Theoretically, mineral exploration using electromagnetic (EM) methods detectability is based on the different electrical resistivities of the ore body and its surrounding rocks or formations since the resistivity of the rocks can vary over several orders of magnitude (Fig. 10.9). The more conductive sulfide ores are very distinct from their wall rocks and that makes EM a powerful tool in mineral exploration (Guo et al., 2020).

Hence, EM methods are widely used in mineral exploration during airborne, onshore, offshore, and borehole surveys, respectively. The EM methods are continuously improved both by industry and academia. The academic community has a strong interest in developing new packages and software for the forward modeling and inversion of EM data. The efficiency of the EM equipment is also continuously improved due to academic research and development. Examples of EM instrumentation are V-8 multifunction receivers from Phoenix Geophysics Ltd. in Canada, GDP 32 II multifunction receivers from Zonge International in the United States, STRATAGEM EH4 systems from Geometrics Company in the United States, GMS-07E systems from Metronix in Germany, and TerraTEM from Monex Geosope in Australia. Most of the equipment above combines the frequency and time domain EM data acquisition methods, respectively. The mining industry contributes by providing practical experience and case studies. The future of successful mineral exploration using EM methods will be focused on concealed and deep target exploration, and possibly one day on resources on the seafloor. Artificial intelligence and deep learning techniques also provide

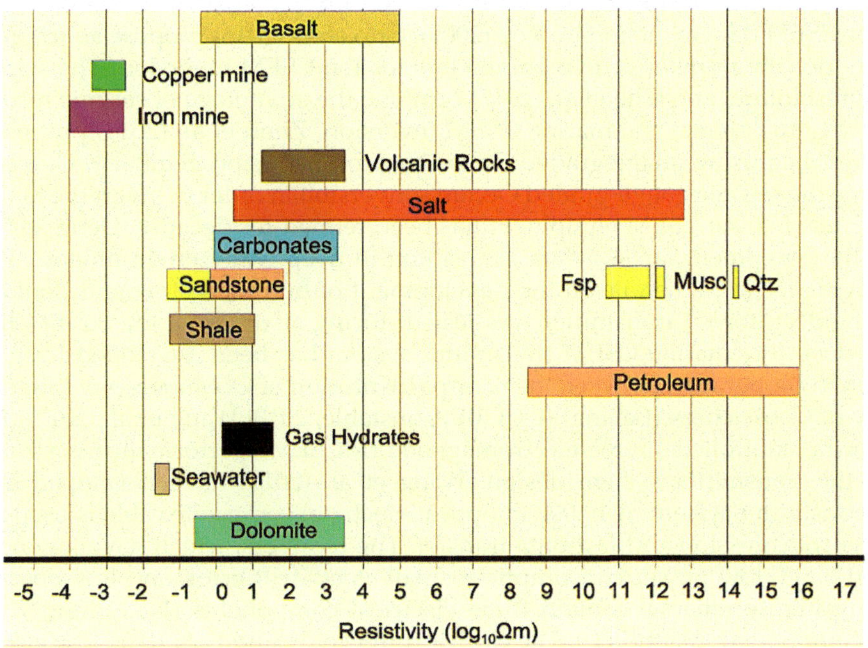

FIG. 10.9 Resistivity varies over several orders of magnitude in crustal materials. *Source: Modified.*

the potential to improve geophysical prospecting. It has already been successfully applied in seismic imaging and 3D interpretations in the oil exploration industry, and it will be increasingly used in mineral exploration.

Marine electromagnetic methods have also been rapidly developed during the last twenty years. Researchers designed the seafloor equipment for oil and gas exploration during the late 1990s. For offshore exploration, Statoil Hydrocarbon ASA (now Equinor), a Norwegian geophysical services company, successfully tested the marine controlled-source electromagnetic (CSEM) technique in Angola in November 2000 (Eidesmo et al., 2002). This new technique was also tested at the Troll area in Norway by EMGS ASA (Norwegian geophysical services Company) in 2003. The "seabed logging" (SBL) system locates the receivers on the seafloor, while the horizontal electric dipole (HED) source is towed over the seafloor for a distance of about 30 m. In order to increase the horizontal data density, the towed streamer electromagnetic (EM) method was presented by Petroleum Geo-Services (PGS) ASA (Norwegian geophysical services Company) using a similar marine seismic data acquisition system. The towed streamer EM method employed the HED source and a long streamer with EM receivers (Ziolkowski et al., 2010). It has been successfully tested in the North Sea (Linfoot et al., 2011; Zhdanov, 2012).

The controlled-source electromagnetic (CSEM) method has a high sensitivity for the relatively resistive hydrocarbon in offshore exploration. The success of the EM method's application for the search for oil and gas reservoirs is based on the fundamental fact that oil and gas-containing structures are characterized by very high resistivity, while the surrounding sea-bottom formations filled with salt water are very conductive. Therefore, a petroleum reservoir represents a clear target for EM methods. The classic inversion method was introduced as Occam's inversion by Constable et al. (1987). Occam's inversion was initially applied for CSEM data inversion by Key (2009). Guo et al. (2016) propose an irregular sparse mesh to reduce the number of free parameters for the CSEM inversion. The seismic image has a high resolution, and Guo et al. (2017) combine the advantage of seismic resolution and EM sensitivity to enhance the marine CSEM inversion. Zhou et al. (2014) propose to invert the electrical data using image-guided Gauss-Newton inversion. Similarly, Guo et al. (2017) could define the seismic image-guided Occam's inversion in order to invert the marine CSEM data. In China, marine CSEM equipment has been refined for decades. Deng and He (2013) designed the equipment for MT data acquisition in deep water environments. Wang et al. (2015) provide technical details on the development of both deep-tow and deployed transmitters. The deep-towed transmitter was tested during a survey in the South China Sea in 2014. The deep-tow marine CSEM transmitter system has been developed for gas hydrate exploration in cooperation between the Scripps Institution of Oceanography (SIO) and China University of Geosciences (Beijing) (CUGB) (Constable, 2013; Wang et al., 2017) (Fig. 10.10).

The marine controlled-source electromagnetic (MCSEM) performance is strongly influenced by the transmitter system design. Wang et al. (2017) have developed a deep-tow MCSEM transmitter system. A 10,000 m optical-electrical composite cable is used to support high-power transmission and fast data transfer. The receivers, which have low noise, broad acquisition band, low power consumption, small clock drift error, and low operating costs while being highly reliable, contain three electrical components (E_x, E_y, and E_z) and two magnetic components (H_x and H_y). The electric dipole of E_x and E_y is 10 m. The electric dipole of E_z is 4 m. The electrical and magnetic sensors are made of optimized Ag-AgCl electrodes

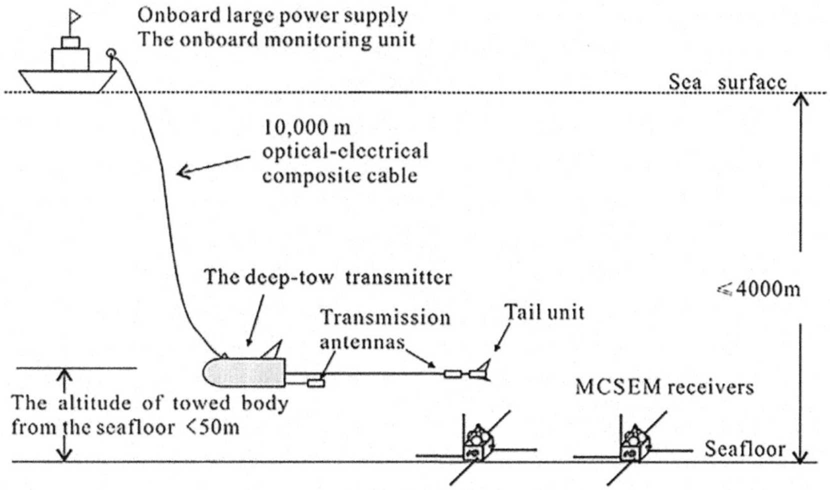

FIG. 10.10 Operation and geometry of marine controlled-source electromagnetic (MCSEM) experiment. Ultra-Short Base Line (USBL) is used to monitor the altitude of the towed body and tail unit. The real-time values could be displayed on a ship's lab screen. The operator could check the values and adjust the length of the winch cable to get the proper altitude of the towed body and tail unit. *Source: Modified*

and induction coils (Yan et al., 2013), respectively. The new induction coil sensor has a power consumption of 105 MW and a weight of 4.5 kg in air. An acoustic telemetry deck unit, releaser, modem, and release mechanism are used, which is called a self-pop-up system, to facilitate deployment and recovery operations for the recovery of the receiver.

Fig. 10.11 demonstrates the general survey configuration of the marine controlled-source electromagnetic (MCSEM) method (Zhdanov, 2018). A set of multicomponent electromagnetic receivers (typically 10 to 50) is deployed at the seafloor along the survey profile line(s). Usually, these receivers are dropped off from the survey vessel and fall down to the seafloor freely. The transmitting horizontal electric bipole typically has a length of several hundred meters and is towed by the survey vessel via an umbilical cable. This transmitter generates a low frequency (typically from 0.1 to 10 Hz) EM field, which propagates both upward in the seawater and downward within the sea bottom. The receivers measure the amplitude and the phase of the electric and/or magnetic fields generated by the transmitter. The recorded signal is formed both by the primary field from the transmitter and by the EM response from the geoelectrical structure beneath the seafloor. After the data acquisition is completed, the receivers float up to the surface by jettisoning away their weight, which previously held them to the seafloor, and are then retrieved by the survey vessel.

With many successful applications being reported, much attention has been paid to the research and development of this technology (Edwards, 2005; Smit et al., 2006; Hesthammer et al., 2010; Constable, 2010). The hardware development support is key to this technology. Scripps Institution of Oceanography (SIO) (Constable, 2006a, 2010, 2013), Electromagnetic Geoservices (EMGS) (Eidesmo et al., 2002), and the University of Cambridge (Sinha et al., 1990) have developed towed transmitters with multiple static receivers mainly for deep water (deeper than 500 m) applications. SIO has a long history of development of the MCSEM

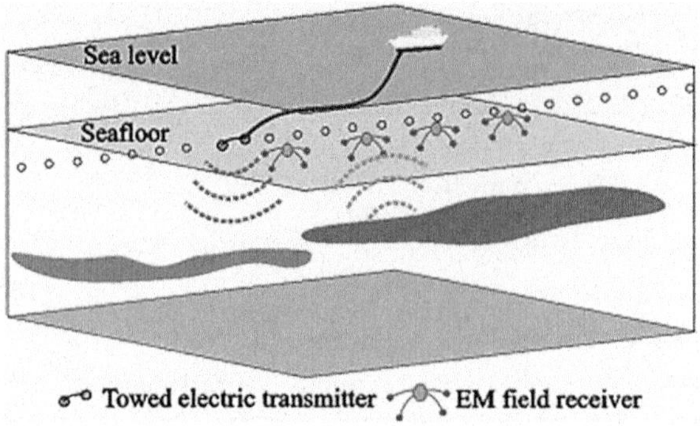

FIG. 10.11 General survey configuration of the MCSEM method. An electric bipole transmitter towed by the survey vessel generates an electromagnetic (EM) signal, while the fixed multicomponent EM field receivers located on the seafloor measure the EM response from the geoelectrical structure beneath the seafloor. *Source: Modified.*

transmitters (Cox et al., 1986; Webb et al., 1993; Constable, 2006a, 2013). SIO also developed the VULCAN system, which is a deep-towed CSEM receiver (Constable et al., 2012, 2016). PGS developed a cable-towing EM system with a single surface-towed transmitter and multiple synchronically towed receivers mainly for shallow water (McKay et al., 2015). A ship-towed, electric dipole-dipole transient electromagnetic system with a single transmitter and several towed receivers has been developed by the University of Toronto (Edwards, 1997; Yuan and Edwards, 2000). In all these systems, the key point is to develop a powerful and effective transmitter.

Electromagnetic methods represent an important tool in global mineral exploration. Outcropping mineral resources on land are increasingly exploited. Currently, geophysical prospecting has to focus on deep or concealed exploration targets. This will require the use of deep penetrating geophysical prospecting tools such as EM. More sophisticated equipment for EM prospecting will need to be developed to discover deep ore bodies under cover. Marine transient electromagnetic (TEM) techniques are already very powerful tools for offshore exploration. Airborne TEM techniques can be applied on land. However, these tools still need to be improved in their accuracy and efficiency, respectively. Marine TEM and CSEM methods have also been applied in the exploration of submarine metal deposits. However, the exploration of submarine mineral deposits is still not economic (Tao et al., 2013). The use of marine EM methods has, in recent years, focused on exploring submarine hydrothermal polymetallic sulfide deposits. Seafloor massive sulfide (SMS) exploration has been developed during the last decade (Franklin et al., 2005). The Trans-Atlantic Geotraverse (TAG) across the Mid-Atlantic Ridge has used a dipole-dipole electromagnetic system called Remotely Operated Vehicle (ROV) (Cairns et al., 1996). Recently, marine transient electromagnetic (TEM) and controlled-source electromagnetic (CSEM) surveys have been both used in SMS exploration. Marine TEM has been applied for the SMS detection and characterization (Nakayama and Saito, 2014). Marine TEM systems have been successfully tested over SMS mineral deposits. Tao et al. (2013) test a system consisting of coincident horizontal transmitter and receiver loops towed behind a ship at an altitude of 80 to 100 m above the seafloor and claim to have detected a known vent site at the South Atlantic Ridge. Marine CSEM systems have also successfully explored for SMS deposits (Cairns et al., 1996). In recent years, China has started to use SMS methods at the Southwest Indian Ridge (SWIR) (Tao et al., 2013). Li et al. (2016b) studied the data from hydrothermal sulfide deposits at the mid-Atlantic, mid-ocean ridge, and the southwest Indian Ocean. Guo et al., 2020 modeled these deposits in 3D using TEM forward modeling.

10.5.1.2 Geochemical exploration techniques

Many countries, such as the United States, Canada, China, Spain, the United Kingdom, Germany, and Japan, have begun extensive exploration studies in search of rich ocean floor deposits by geological, geophysical, geochemical, and mineralogical approaches (Vysetti, 2023). In this context, it is worth mentioning the application of portable analytical instruments such as the Raman spectrometer, laser-induced breakdown spectrometer (LIBS), near IR spectrometer, and a couple of nuclear techniques in geochemical exploration studies. When these miniature instruments were mounted in an ROV, they were successful in detecting metal-rich deposits in the deep oceans (Obhodas et al., 2018; Vysetti, 2022; Vysetti and Sawant, 2022). A new, more compact, and lighter underwater LIBS known as LIBSea II was developed for the investigation of the hydrothermal regions and polymetallic mineral areas. When the system was deployed on an ROV of Haima for a deep-sea trial, atomic lines of potassium (K), sodium (Na), calcium (Ca), and strong molecular bands of CaOH from a carbonate rock sample were obtained for the first time at depths of 1400 m (Liu et al., 2020). Such miniature analytical systems will have great potential to be used in deep-sea exploration studies in the future.

10.5.1.3 Geotechnical exploration techniques

Geotechnical exploration mainly serves to test and evaluate the geological conditions (composition, structure, formation age, and genetic type of each soil layer along with adverse geologic actions) and the geotechnical properties of the seafloor soils (particle size distribution, void ratio, liquid and plastic limits, compression modulus, cohesion, and internal friction angle) in the project area. Geotechnical exploration is a direct method. According to the test site, the techniques for geotechnical exploration include in-situ and laboratory testing.

In-situ testing

There are many types of in-situ testing techniques. According to current engineering survey reports of ocean wind farm projects and oil and gas pipeline projects, commonly used testing techniques include cone penetration test (CPT), standard penetration test, dynamic penetration test (DPT), vane shear test (VST), and full-flow penetrometer test (using T-bar and ball probes) (Guo et al., 2023). VST, the first technique to be used in ocean areas, was first applied over 100 years ago (Cadling and Odenstad, 1950; Briaud, 2013). The principle of VST is relatively simple, and few influencing factors are involved (Zhang et al., 2021a). While VST allows the shear strengths of seabed soils to be obtained quickly, it cannot achieve continuous testing along the depth direction of the soil layers. The most commonly used in-situ methods for determining the mechanical properties of seabed sediments include the CPT and cross-plate shear test (Wang et al., 2017; Chu et al., 2014; Tang, 2014). The CPT or the cone penetration test with pore pressure measurement (CPTu) can be used to evaluate the division of soil layers and soil character identification of marine thin soft soil. In the

1930s, the concept of CPT was developed, with the most representative CPT device being the Dutch cone (Stanton et al., 1939; Seed and De Alba, 1986). Begemann further developed CPT in the 1950s (Begemann, 1953) to obtain the frictional resistance of the penetration bar. CPT then began to be used for the engineering classification of soils. CPT was first used for underwater testing in the 1960s and extended to deep-sea testing in the 1980s (Hanzawa and Tanaka, 1992).

Based on the differences in detection methods, the CPT can be divided into the down-hole CPT and seabed CPT (Wang et al., 2017; Chu et al., 2014; Lu et al., 2004). A typical example of the down-hole CPT is Furgo's Wison-APB system, with a water depth of 3000 m (Chen et al., 2007). However, using the down-hole CPT requires drilling prior to detection, which inevitably disturbs the seabed soil; hence, its measurement result is not as accurate as that of the seabed CPT. The seabed CPT works differently from the down-hole CPT, involving continuous penetration of the probe into the soil from the seabed surface through the probe rod (Lu et al., 2004). In 2010, MARUM developed GOST, a commercial CPT system that functions in water depths up to 4000 m. Among the internationally known seabed CPTs are: Vandenberg DW-ROSON100, which has a working water depth of 4000 m and 50 m penetration depth (Lunne, 2010; Guo and Liu, 2015; Zhang et al., 2019); Fugro SEACALF (Bouwmeester et al., 2019), which has a 4000 m working water depth; NEPTUNE 5000 from Datem Company, the United Kingdom, which has a 3000 m working water depth and 20 m penetration depth; Geomil's MANTA-200, which has a 2000 m working water depth. Current CPT probes can carry numerous functional sensors, including pore pressure sensors, resistivity sensors, and wave transceivers (Lunne et al., 2009; Zou et al., 2020; Liu et al., 2021b).

Unlike CPT, DPT requires additional penetration power; thus, free-fall DPT penetrometers are mainly used for ocean soil testing. The first ocean free-fall DPT penetrometer was developed by the Memorial University of Newfoundland in 1973; this device had a limited testing water depth of 180 m and soil penetration depth of 4.5 m (Dayal and Allen, 1975). In 2006, MARUM also developed a DPT system, Deep Water Free-Fall Piezocone penetration Testing (DWFF-CPTU) (Lykousis et al., 2007). The maximum working water depth reached 4000 m, allowing the system to support deep-sea exploration. Full-flow penetrometer testing represents an improvement on CPT. The probes adopt T-bar and ball shapes, providing more accurate soil strength results. The full-flow penetrometer technique was first proposed and developed in the early 1990s by scholars at the Centre for Offshore Foundation Systems (Stewart and Randolph, 1991; Fan et al., 2020a). Subsequently, Low et al. (2008) proposed a new manually operated full-flow penetrometer system that is easy to use on board. Recently, Guo et al. (2022a) developed a comprehensive seabed soil testing system equipped with full flow penetrometers (ball and streamline probes), CPT, and VST probes. This system was used to evaluate the mechanical properties of sediments in the South China Sea (Guo et al., 2021a, 2021b, Guo et al., 2023a; Gu et al., 2022; Guo et al., 2023; Mayne et al., 2009).

Laboratory testing techniques

To eliminate interfering factors from the project site, representative soil samples from the project site can be tested in the laboratory (sometimes directly in the ship laboratory) using more precise and sensitive instruments than can be applied in the field. It should be noted that the quality of the soil samples directly affects the value of the test results. Subsea sampling is generally conducted by gravity sampling or drilling sampling. In gravity sampling, the samplers (surface grab samplers, which are generally used for sampling depths less than 1 m, and open-column samplers, which are generally used for sampling depths less than 6 m) freely drop to the seabed under the influence of gravity and are dragged back using a cable. In drilling sampling, holes are first drilled in the seabed, and a thin-walled sampler is then used to extract soil samples from the drill hole. The sampling depth for drilling sampling can reach ~60 m. Laboratory testing can then be used to obtain the geotechnical properties (physical, chemical, mechanical, and microcosmic indices) of the seabed soils based on soil mechanics theories (Mayne et al., 2009; Guo et al., 2023).

Most laboratory testing techniques for marine soils follow traditional soil tests. It is worth mentioning that the construction and development of submarine laboratories have accelerated in recent years. Submarine laboratories are designed according to the principle of saturation diving and can be moved or fixed. The working water depth of submarine laboratories has reached more than 300 m (Su, 2020). If submarine laboratories become commonplace, geotechnical exploration with laboratory testing techniques will become even easier. Regulations covering ocean engineering usually require geophysical exploration and geotechnical exploration to be conducted simultaneously so that comprehensive seafloor geological and geotechnical information can be obtained. Thus, there are many examples of applications that combine these two exploration techniques. For example, holes can be drilled first to obtain soil samples for laboratory testing, followed by geophysical exploration (electrical and magnetic tests) in the drill hole.

10.5.2 Techniques for the extraction stage

The extraction stage is the primary component of ocean mining and requires various devices and systems to work together, including the mining ship or platform, launch and recovery systems, resource collection systems, dewatering and subsea mining cargo handling systems, and subsea mining control and monitoring systems (American Bureau of Shipping, 2020). The techniques applied in the extraction phase often differ based on the physical form and depth of the resources being extracted. Overall, extraction techniques for ocean mineral resources can be divided into two types according to the systems used: mobile extraction techniques and fixed extraction techniques.

10.5.2.1 Mobile extraction techniques

Mobile collection techniques are generally used for seabed surface resource extraction (resources with granular or nodular shapes, like polymetallic nodules). These techniques rely on ship-dragged collection devices (chain buckets, crab buckets, and pipe-suction ends) or self-propelled collection devices (subsea mining vehicles and miniature submarines) for ocean resource mining (Bath, 1989). The collection devices are recovered after the temporary storage devices are filled. Some common mobile collection techniques are summarized in Fig. 10.12.

The development of mobile extraction techniques is related to the water depth of the mining operation. For shallow sea resource mining (placer), most extraction techniques are derived from land mining techniques and use mining vessels such as chain bucket vessels, crab bucket vessels, and pipe-suction end vessels. The cost of these extraction techniques increases as the water depth increases. The large-scale extraction of deep-seabed resources

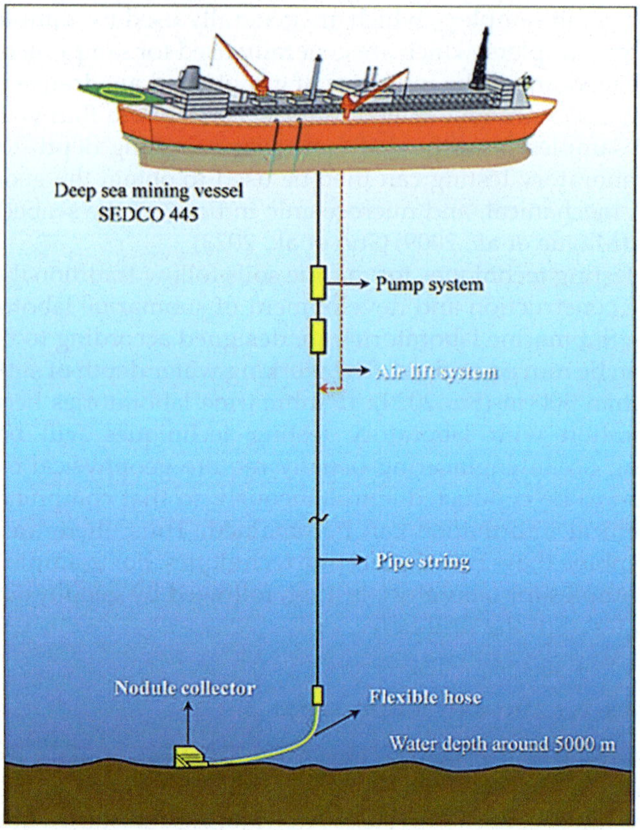

FIG. 10.12 Example mobile extraction techniques. *Source: Guo et al., 2023. https://link.springer.com/article/10.1007/s40789-023-00580-x*

(polymetallic nodules) began in the 1970s (Bath, 1989). In this system, the bucket is freely dropped from the vessel, and the worker on the vessel receives a signal when the bucket reaches the seabed. The bucket is then dragged using cables for collection. The continuous line bucket mining system was improved for use in deep-sea mining (Masuda et al., 1971). However, the system cannot adapt to the complex seabed topography, and the cables easily wind, resulting in low mining efficiency. In the 1980s, a shuttle vessel deep-sea mining system was developed by AFERNOD in France (Gauthier and Charles, 1979; Lenoble, 1981; Du et al., 2016). This system requires a submarine to dive into the seabed for collection and can work at seabed depths up to 6000 m, with the daily collection amount reaching 2000 t. However, the continuous power issues in this system were difficult to solve. During the same time period, a pipeline lift mining system developed in the United States began to be used in deep-sea environments (Chung, 2003), resulting in large-scale mining and high mining efficiency. Subsequently, Grebe (1997) proposed a cooperative mining system based on a flexible transfer pipe and an ore collector to adapt to the complex seabed terrain. More recently, subsea mining vehicles have emerged as a valuable tool for deep-seabed resource extraction

(Hong et al., 2019). These vehicles provide a flexible system with high mining efficiency and do not require a lifting system.

10.5.2.2 Fixed extraction techniques

Fixed extraction techniques are required for some resources in deep seafloor layers or bedrock (oil, natural gas, combustible ice, brimstone, and coal). Fixed collection techniques first drill through the seabed to the resource storage area and reinforce the drill-hole wall to form stable extraction wellheads. Working tunnels may be further dug in some projects. The extraction devices (centrifugal pumps, boring bits, and jet devices) then provide flow pressure, shaking force, or high temperature melting to release the resources, which are then transmitted to the platform or ship through conveyor belts, pipelines, or risers (Nian et al., 2018; Fan et al., 2022).

Fixed extraction techniques can be categorized based on their primary system and platform (Yu et al., 2015) (Fig. 10.13). The world's first offshore extraction well was drilled off the coast of California in the United States in 1897 (Silcox et al., 1987; Pan et al., 2006) for the extraction of offshore oil. A wooden drilling platform with a large size and shallow operating level was first employed. Steel offshore platforms began to be used in the 1950s and 1960s, resulting in the rapid development of offshore oil and gas resources (Nian et al., 2022). By the late 1960s, the operating water depth had exceeded 200 m. In 1973, the first concrete platform was built in the Ekofisk oilfield in the North Sea (Aitcin and Houle, 1986). In the 1980s, with the further development of deepwater oil and gas extraction, tension leg and guyed tower platforms were developed with working depths of 600 m. At present, semi-submersible platforms are mainly used for mining in deep water areas and can be used to extract multiple resources (oil, natural gas, combustible ice, brimstone, and coal). Representative systems include the Aker H-6 system from AKER Solutions in Norway, the DSS50 system from GustoMSC in the Netherlands, and the GVA7500M system from GVA in Sweden (Ted, 2014).

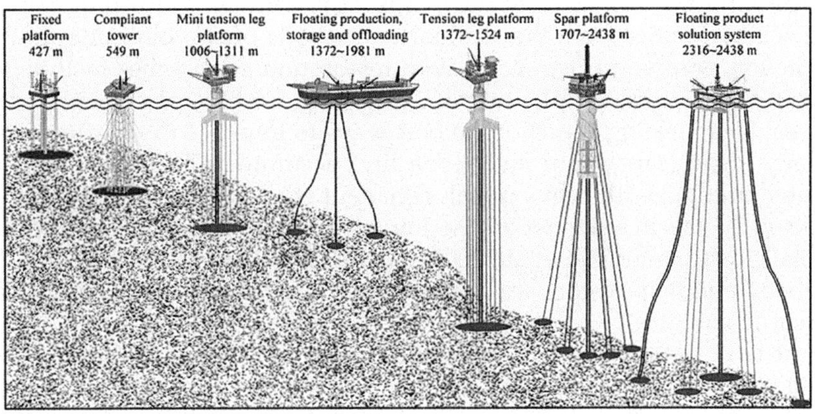

FIG. 10.13 Examples of resource extraction platforms. *Source: Guo et al., 2023. https://link.springer.com/article/10.1007/s40789-023-00580-x*

10.5.3 Techniques for the closure stage: Environmental concerns

Commercial mining activity in oceans can have widespread environmental consequences. For example, manganese nodules, which take millions of years to form, provide a critical habitat for an array of unique and largely understudied species, including deep-sea corals, sponges, sea urchins, starfish, jellyfish, octopus, squid, shrimp, and sea cucumbers. Deep-sea sediment plumes are created during mining activity, and the processed material that is discharged as wastewater by the surface support vessel can have devastating effects on marine biota (Gollner et al., 2017). Mining of manganese nodules is predicted to lead to a significant loss of biodiversity in the marine environment (Burns et al., 1978). Even in the earlier stages of understanding deep-sea ferromanganese nodules and their possible exploitation, several environmental studies, including small-scale in-situ artificial disturbances, were conducted (Ozturgut et al., 1978; Fukushima T. Overview, 2023; Sharma and Smith, 2019; Foell et al., 1990). For example, in order to monitor and understand the potential impacts on the recolonization of benthic biota by the disturbance that would be caused by commercial mining of manganese nodules, the first long term disturbance and recolonization experiment (DISCOL) was established in the Peru Basin during a German environmental impact study associated with manganesenodule mining in 1989 in the southeast Pacific Ocean in a circular area of the seabed measuring 10.8 km^2 at a depth of 4140 to 4,160 m (Gausepohl et al., 2020). In order to have more insight into the deep ocean, underwater robots are being extensively used as they are capable of tracking and recording high-resolution images of slow-moving and fragile zooplankton, gelatinous animals, and particles (Zhang et al., 2023).

To decrease the impacts on ocean ecological environments as much as possible, environmental restoration remains critical during the closure stage after the mining activities are finished. The regulations on ocean mining in many countries put forward clear requirements for closure activities (American Bureau of Shipping, 2020; China Association of Oceanic Engineering, 2021). In addition to controlling the mining scale, the environmental impacts of mining, and the carbon footprint of extraction, the mine tailings (ore waste resulting from mining, typically a mud like material) must be controlled and dealt with (by returning mine tailings back to the depth required by International Seabed Authority [ISA] or coastal states using the subsea mining downcomer and pumping system and returning tailings pipes for transporting seafloor mines from the seafloor to a mobile offshore mining unit).

After extraction, common ocean ecosystem restoration approaches include the construction of artificial algae farms, the planting of mangroves, and the placement of artificial reefs. Large-scale seaweed farming is one of the best ways to improve the marine ecological environment. Many algae (sargassum and stone lily) accumulate heavy metals and thus can purify seawater polluted with heavy metals (Phaneuf et al., 1999). Similarly, mangrove plants can absorb heavy metals in seawater and sediments through their roots, reducing the content of heavy metals in seawater (Ke et al., 2005). Artificial reefs can create a good environment for fish to inhabit and help restore sea areas with ecological imbalances and severe resource deficits (Fisher et al., 2015).

Still, now it is not known much about the impacts of deep-sea mining. Currently, the available data on the recovery of deep-sea biota following physical disturbances from potential mining is not enough to arrive at a firm decision. It also highlighted that there are many gaps in knowledge about deep-sea habitats and species, and how very little is known or understood about the risks of deep-sea mining. On the other hand, more than 80% of the seafloor remains unmapped at a resolution of 100 m or better by 2019, which is expected to be completed only by 2030 if everything works favorably (Showstack, 2019).

10.6 High Seas Agreement 2023

The world's oceans are facing a severe decline in biodiversity and degradation of ecosystems due to threats related to climate change and the expansion of human activities, such as shipping, overfishing, plastic pollution, and deep-sea mining (Tessnow-von et al., 2020; UN, 2023). Consequently, there is a pressing need for a more cohesive ocean governance framework, since the existing framework is too fragmented and incomplete to effectively secure conservation and sustainable use of marine biodiversity in areas beyond national jurisdiction. The high seas treaty aims to address the regulatory gaps (Gjerde et al., 2019).

The areas beyond national jurisdiction comprise the "high seas"' (water column) and the "area" (seabeds), making up about two-thirds of the ocean (Tiller et al., 2023; De Santo et al., 2019). The areas are currently regulated by different regional and sectoral agreements, such as Regional Fisheries Management Organisations (RFMOs) (Tessnow-von et al., 2020). However, they can only implement measures within their own respective mandate, and cooperation is lacking (Ardron et al., 2014). Additionally, only a few areas are covered, leaving the majority as good as unregulated. The remaining one-third of the ocean falls under national jurisdiction and is situated within the exclusive economic zones (EEZs). The exclusive economic zones extend 200 nautical miles (about 370 km) from the territorial sea baseline. The zones are established under the United Nations Convention on the Law of the Sea (UNCLOS), giving coastal states jurisdiction over the living and nonliving resources within the water and the seabeds (De Santo and Elizabeth, 2018).

In early 2023, more than 190 countries agreed on a long awaited pact to help safeguard the high seas, reverse biodiversity loss, and protect billions of marine species ranging from tiny plankton to giant whales from any threats such as pollution, overfishing, shipping, and deep sea mining, at a UN conference agreeing, on a common framework for establishing new protected areas in international waters. The United Nations high seas treaty, also known as the Biodiversity beyond National Jurisdiction (BBNJ) treaty, is a legally binding instrument for the conservation and sustainable use of marine biological diversity in areas beyond national jurisdiction (UN, 2023). It is an agreement under the United Nations Convention on the Law of the Sea (UNCLOS) (European Commission, 2023). The text was finalized during an intergovernmental conference at the United Nations on March 4, 2023, and adopted on June 19, 2023 (UN, 2023). Both states and regional economic integration organizations can become parties to the treaty (UN, 2023).

The treaty addresses four themes (Tessnow-von and Vadrot, 2020; UN, 2023): (1) Marine genetic resources (MGRs), including the fair and equitable sharing of benefits; (2) area-based management tools (ABMTs), including marine protected areas (MPAs); (3) environmental impact assessments (EIAs); and (4) capacity building (CB) and transfer of marine technology (TMT). The area-based management tools and environmental impact assessments relate mainly to conservation and sustainable use of marine biodiversity, while the marine genetic resources and capacity building and transfer of marine technology include issues of economic justice and equity (Tiller et al., 2023).

However, it is expected to take some time for the treaty to be formally adopted by member states and come into force. In fact, ISA is fully committed to protecting the marine environment and regulating economic, exploratory, and scientific activity in the deep sea and is currently negotiating the approval of a mining code, laying out the rules under which companies will be allowed to extract minerals from the seabed. In view of this, deep-sea mining is likely to be exempted in certain designated, relatively smaller areas, such as the CCZ, in the near future.

10.7 The blue planet: Earth's final frontier

The Earth's oceans form a continuous body of saltwater covering more than two-thirds of the planet and storing 97% of its water. With an average depth of about 3700 m (Charette and Smith, 2010), the oceans are widely considered to be Earth's final frontier. They control the global climate and weather and have provided humanity with many resources for millennia. Extending away from land, the oceans are divisible into three main regions: the continental shelf, where water depths are generally less than 200 m; the continental slope; and the flat or gently sloping abyssal plain, typically occurring at depths greater than 4000 m (Lusty et al., 2018).

Although a poorly defined term, the "deep" ocean may be considered to be the seafloor below 200 m, where, with little penetration of sunlight, photosynthesis is not possible. The deep seafloor covers about 60% of the Earth's surface and hosts a spectrum of geological settings, geomorphological features, and ecosystems. This diversity, and its long and dynamic history, results in the deep seafloor hosting mineral deposits that are both similar to those found on the continents as well as types unique to the oceans. Despite its fundamental importance, due to its sheer size, we lack basic information about many aspects of the deep ocean. The challenges of sensing what lies below the water surface in an alien environment for humans make access difficult, costly, and reliant on technology. After decades of ocean exploration, only about 10% of the ocean floor has been surveyed by ship-based sonar systems (Becker et al., 2009), and these provide only an average resolution of about 100 m squared.

10.8 Quantifying clay mineral sources in marine sediments by using end-member mixing analysis

Clay minerals in marine sediment are derived mainly from the land (Sionneau et al., 2008), and their components are likely controlled by the mixture of various detrital inputs (Singer, 1984). Detailed studies of the sources of clay minerals enable a better understanding of sediment transport and mixing processes. Clay minerals have the potential to trace sediment because of the specificity with which they can distinguish particular sources. The relative abundances of four major components (smectite, illite, chlorite, and kaolinite) differ greatly in different source areas as a result of diverse lithology, varying intensities of weathering, and different weathering mechanisms (Liu et al., 2016; Singer, 1984). Provenance studies of clay minerals have received a substantial amount of attention in recent years. It is generally accepted that clay minerals on the seafloor are compositionally similar to their terrestrial sources (Li et al., 2021). Trends in clay mineral distribution in the hydraulic environment are largely influenced by the strength and direction of currents (Singer, 1984). Numerous studies in the South Atlantic (Petschick et al., 1996), northeastern SCS (Liu et al., 2010), northern Yellow Sea (Li et al., 2014), Liaodong Bay (Bohai Sea) (Dou et al., 2014), Indonesian archipelago (Indian Ocean) (Gingele et al., 2001), Bay of Bengal (Indian Ocean) (Li et al., 2017), and Northern Gulf of Mexico (Sionneau et al., 2008) reveal that the spatial distributions of clay minerals fit well with the movement of ocean currents, and a plausible explanation can be established from the hydrodynamic viewpoint. Several mathematical models have been developed to evaluate the relative contributions of clay mineral sources. For example, Dou et al. (2010) established a set of linear equations of three components (smectite, chlorite, and kaolinite + illite). Li et al. (2017) established two linear equations comprising two components (smectite and kaolinite + illite). In these studies, linear extrapolation was used to calculate the source contributions. However, this approach requires the measured data to be completely unaffected by measurement errors (Woronow, 1991), which is rarely the case. Thus, a more accurate approach is critical for the quantification of clay mineral sources, for example, the unmixing technique.

The unmixing technique is applicable for quantitative analysis of natural tracers, for example, sediment grain sizes, chemicals, pollutants, detrital ages, stable isotopes, and magnetization (Derrien et al., 2018; Kruiver et al., 2001;Li et al., 2017; Sharman and Johnstone, 2017), but fewer studies have considered clay mineralogy. Two different methods have been developed: linear unmixing and bilinear unmixing (Sharman and Johnstone, 2017; Weltje, 1997). In linear unmixing analysis, end-members are obtained by field sampling, instead of statistical artefacts, which are used for bilinear unmixing. The real end-members are more geologically reasonable and more suitable than statistical artefacts, although their spatial and temporal variability are challenged (Uhlenbrook et al., 2002). End-member mixing analysis (EMMA), developed by Hooper et al. (1990) and Christophersen and Hooper (1992), has been widely and successfully applied in hydrological systems. By using this method, the measured concentrations are expressed by three potential end-members, and their relative contributions are evaluated by using a least-squares technique.

This study applied end-member mixing analysis (EMMA) to clay minerals in Taiwan Strait sediments to determine the proportional contributions from known riverine sources, which are called end-members. Three end-members were determined: the TW, ZM, and LZ end-members, and their proportional contributions were calculated by resolving inverse mixing problems. The TW end-member, representing the rivers of Taiwan (perhaps including the Yangtze River), was characterized by high illite and chlorite contents and contributed highly to the entire strait except for the western coastal area. The ZM end-member, representing the rivers of Zhu and Min, had the highest concentration of kaolinite and contributed mainly to the western coastal area. The LZ end-member, representing the rivers of Luzon (perhaps including the rivers of Zhejiang), was dominated by smectite and contributed to the southern region and the western coastal area. The clay mineral contents were well predicted by the linear mixing of the three end-members. The spatial trends of the end-member proportions are comparable with the ocean currents flowing through the Taiwan Strait, and thus the distribution patterns of clay minerals and source contributions can be explained from the hydrodynamic viewpoint (Li et al., 2021).

10.9 Marine dipole-dipole controlled-source EM and coincident-loop TEM experiments to detect SMS: Effects of 3D bathymetry

Seafloor massive sulfides (SMSs) are regarded as a potential future resource to satisfy the growing global demand for metals, including copper, zinc, and gold. Aside from mining and retrieving profitable amounts of massive sulfides from the seafloor, the present challenge is to detect and delineate significant SMS accumulations, which are generally located near mid-ocean ridges and along submarine volcanic arc and back arc spreading centers. Currently, several geophysical technologies are being developed to detect and quantify SMS occurrences that often exhibit measurable contrasts in their physical parameters compared to the surrounding host rock. Here, Haroon et al. (2018) use a short, fixed-offset controlled source electromagnetic (CSEM) system and a coincident-loop transient electromagnetic (TEM) system, which, in theory, allow the detection of SMS in the shallow seafloor due to a significant electrical conductivity contrast to their surroundings.

In 2016, CSEM and TEM experiments were carried out at several locations near the Trans-Atlantic Geotraverse hydrothermal field to investigate shallow occurrences of massive sulfides below the seafloor. Measurements were conducted in an area that contains distinct SMS sites located several kilometers off-axis from the Mid-Atlantic ridge, some of which are still connected to hydrothermal activity and others where hydrothermal activity has ceased. Based on the quality of the acquired data, both experiments were operationally successful. However, the data analysis indicates bias caused by 3D effects of the rough bathymetry in the study area; thus, data interpretation remains challenging. Therefore, the influence of 3D bathymetry for marine CSEM and TEM experiments was studied, focusing on shallow 3D conductors located beneath mound-like structures. The synthetic inversion models are analyzed for attributes associated with 3D distortions of CSEM and TEM data that are not sufficiently accounted for in conventional 1D (TEM) and 2D (CSEM) interpretation schemes. Before an adequate quantification of SMS in the region is feasible, these 3D effects need to be studied to avoid over/underestimation of SMS using the acquired EM data. The sensitivity of CSEM and TEM to bathymetry is investigated by means of 3D forward modeling, followed by 1D (TEM) and 2D (CSEM) inversion of the synthetic data using realistic error conditions. Subsequently, inversion models of the synthetic 3D data are analyzed and compared to models derived from the measured data to illustrate that 3D distortions are evident in the recorded data sets.

10.10 Application of self-potential (SP) in marine mineral exploration

The earliest and the most popular application of the SP method has been in mineral exploration since 1830, when Fox first observed negative SP anomalies related to sulfide deposits (Fox, 1830). In the early stage, the SP method is mainly applied to on-land mineral exploration (Epplebaum, 2019; Logn and Bolviken, 1974; Corry, 1985; Bhattacharya and Shalivahan, 2007;

Mendonca, 2008; Horo et al., 2020). Over the last nearly half a century, the SP method has seen increasing interest in detecting SMS deposits after Corwin's work (Corwin, 1976). Compared with on-land mineral exploration, marine SP signals generated by seabed mineralization can be detected using relatively simple instruments towed behind ships or autonomous underwater vehicles (AUVs). SP measurements in seawater have some advantages that the noise level is significantly reduced, the contact resistance of seawater to the electrode is low (usually less than 1 Ohm), the temperature and salinity of seawater are stable over time, meteorological and hydrogeological factors are reduced, and marine SP surveys are efficient and enable continuous measurements (Corwin, 1976; Heinson, 1999).

There have been lots of marine SP surveys aiming to detect the SMS deposits. For example, Brewitt-Taylor (1975) performed deep-sea trials of SP measurements at the Mid-Atlantic Ridge (MAR) to detect SMS deposits. Corwin (1973) began to develop the marine SP observation system in 1973 and observed an SP anomaly related to offshore extensions of the onshore deposits in 1976 (Corwin, 1976). Herzen et al. (1996) used an electrode array towed behind the submersible device Alvin and recorded negative potentials above the TAG hydrothermal mound. Heinson et al. (1999) conducted two marine SP measurements in the south of Eyre Peninsula, South Australia, Rose Canyon, San Diego, and California. Heinson et al. (2005) performed an SP measurement on the continental shelf of South Australia and suggested that the SP signals are probably due to the reduction-oxidation process across the conductive orebody below the seafloor because SP and magnetic data show limited spatial relevance. Beltenev et al. (2007, 2008, and 2009) conducted integrated geological-geophysical surveys to detect hydrothermal activity and sulfide mineralization within the MAR rift valley. Cherkashov et al. (2010, 2013) found a sulfide mineralization area in the inactive hydrothermal field at MAR. Recently, Safipour et al. (2017) made marine SP measurements over a known SMS deposit at the Palinuro Seamount in the Tyrrhenian Sea and observed SP signals at the site of SMS mineralization. Kawada and Kasaya (2017) observed negative SP anomalies over the SMS deposits in the Izena geothermal field. Wang et al., 2018 studied SP observation ways in detecting SMS deposits and provided a suggestion that gradient measurements are more suitable for offshore exploration. Kasaya et al. (2020, 2021) developed a system to monitor the sub-seafloor resistivity and SP concerning the physicochemical properties of ore deposits and the existence of hydrothermal fluid.

For data acquisition in the marine environment, in the early days, most SP studies typically used cables terminated with electrodes as the sensor and towed inline behind a ship (Corwin, 1976; Heinson et al., 1999, 2005; Beltenev et al., 2007, 2008, 2009; Kawada and Kasaya, 2017; Zhu et al., 2020). Towed arrays have many limitations (Xie et al., 2023). Therefore, AUVs equipped with SP instruments are introduced for marine SP measurements. For example, Kawada and Kasaya (2018) performed an SP survey using an AUV above the Sunrise deposit in the Myojin Knoll caldera of the Izu-Ogasawara arc. Constable et al. (2018) conducted three-component electric field observations using a commercial AUV. Kasaya et al. (2020) used multiple AUVs without a towing electrode cable for marine direct current resistivity and SP surveys. Zhu et al. (2021) built a new multicomponent electrical field observation system attached to an AUV. Due to the above successful cases of SP methods in marine mineral exploration, three field cases are given below to show the latest progress and application effect of SP surveys in marine prospecting (Xie et al., 2023).

10.10.1 Field case I—Myojin knoll, Izu-ogasawara arc, Southern Japan

The field case I is from Kawada and Kasaya (2018). The study area, Myojin Knoll, is located in the Izu-Ogasawara arc, southern Japan. The known Sunrise deposit has a size of about 400 m × 400 m (Iizasa et al., 1999). The survey was composed of two groups of survey lines (a total of 12 lines) using the AUV Jinbei (Hyakudome et al., 2012). The origin point was chosen near the center of the deposit. Integrating the electric field along the survey lines can produce a negative SP anomaly (about 300 m × 300 m) with an amplitude of a few millivolts, and there is no strong SP anomaly outside of the Sunrise deposit. The SP anomaly in the present survey agrees well with the results of an earlier survey by Iizasa et al. (1999). Kawada and Kasaya (2018) applied the probability tomography method proposed by Revil et al. (2001) and point source inversion to image a southward dipping dipole with a moment of approximately 1.4×10^3 Am and about 30 m below the southern part of the ore deposit, which indicates that the SP signals are probably contributed by the geobattery mechanism. The contribution of hydrothermal fluids to the SP anomalies is probably a secondary effect.

10.10.2 Field case II—Iheya minor ridge, Okinawa trough, Southern Japan

The field case II is from Constable et al. (2018). The study area is close to the Iheya minor ridge in the Iheya prospect of the Okinawa trough, where there is active hydrothermal circulation and known SMS deposits (Yeats et al., 2017). A three-axis electric field receiver (Ag-AgCl electrodes) was mounted to the hard lifting points of the AUV to record the cross-line, inline, and vertical electric fields. The SP distribution was reconstructed from the horizontal electric field data, where the largest anomaly is slightly over 25 mV and is located north of the center of the survey area, and a smaller 15 mV anomaly is located south of the area of the highest bathymetric relief, and the low SP extends along with the ridge structure to the northeast of the survey area. Marine SP anomalies are generally contributed to by either hydrothermal circulation (electrokinetic and/or thermoelectric phenomena) or the geobattery effect (electrochemical mechanisms). For the former, for example, Yamamoto et al. (2017) measured negative potentials related to hydrothermal fluids. For the latter, Corwin (1976) noted that measurements in marine sediments show lower oxidation potentials, which proves the assumption that SP anomalies generated by the geobattery mechanism are expected to be dipolar in nature. Safipour et al. (2017) and Kawada and Kasaya (2017) observed near-seafloor dipolar anomalies and interpreted them as caused by the geobattery mechanism (Sato and Mooney, 1960) over SMS deposits. In this case, although high conductivity suggests that the anomaly is associated with seafloor mineralization, Constable et al. (2018) thought that the source mechanism is more likely due to hydrothermal vents. If the source of the SP anomalies is SMS mineralization, there should be less localized and more dipolar SP signals.

10.10.3 Field case III—Yuhuang seamount, Southwest Indian Ridge (SWIR)

The field case III is from Zhu et al. (2020). The study area is located at the Yuhuang seamount on the ultraslow spreading Southwest Indian Ridge (SWIR). Two sulfide deposits with a diameter of about 500 m have been reported on the northwestern slope of the Yuhuang seamount (Liao et al., 2018). The hydrothermal field is likely inactive, which is good for testing the ability of the SP survey to characterize buried SMS deposits. In this case, the obvious negative SP anomaly is more than 20 mV in the middle of the two survey lines, which is located in the region of the known mineralization area. The SP anomalies measured at the Yuhuang seamount are more likely related to inactive SMS deposits instead of electrokinetic contributions of an active vent because there are no water plume (temperature/turbidity) anomalies associated with active vents, and SP signals pertaining to ore deposits are generally much stronger than electrokinetic contributions (Kawada and Kasaya, 2018; Revil et al., 2012). Besides, laboratory observations of SMS samples suggest that the outcrop sulfides primarily contain pyrite, chalcopyrite, and sphalerite, with ferric hydroxides identified on the surface of the samples. Similar pyrite samples also show high conductivity (Gehrmann et al., 2019), which means that the conductive minerals are physically connected rather than other electron transfer mechanisms (Gehrmann et al., 2019; Hordt et al., 2016). This study further strengthens the geobattery theory.

10.11 Marine mineral exploration with controlled source electromagnetics—case study

Gehrmann et al. (2019) reported the results of a geophysical study to estimate the distribution and size of seafloor ore deposits at the Trans-Atlantic Geotraverse (TAG) hydrothermal field, located south of the Azores, on the Mid-Atlantic Ridge. The Trans-Atlantic Geotraverse field hosts numerous deposits, all but one of which are hydrothermally inactive. Inactive deposits are economically more valuable but difficult to detect. Gehrmann et al. (2019) solution is a combination of high-resolution seafloor and sub-seafloor mapping techniques. This study presents new controlled source electromagnetic (CSEM), bathymetric, and magnetic results from the TAG field. Potential SMS targets were selected based on their surface expressions in high-resolution bathymetric data. High-resolution reduced-to-the-pole magnetic data show negative anomalies beneath and surrounding the SMS deposits, revealing large areas of hydrothermal alteration. The CSEM system presented here is comprised of the Deep-towed Active Source Instrument (DASI; Sinha et al., 1990) and two three-axis electric field receivers (VULCAN; Constable et al., 2016) towed at 350 m and 505 m behind DASI. The DASI-VULCAN array was towed between 50 m and 150 m above the seafloor at a constant speed of 1.852 km/hr. An alternating square waveform with a current amplitude of about 90 A and a base frequency of 1 Hz was transmitted through a 50 m long bipole antenna. Six multikilometer-long survey profiles were acquired across the active TAG mound and several inactive SMS deposits.

By focusing on exploration using ship-towed electromagnetic sensing and autonomous underwater vehicle-based mapping and characterization of their magnetic field, Gehrmann et al. (2019) exploit unique signatures identifying these deposits from their surrounding volcanic rock. These data enable us to delineate the extent of the deposits and provide a first estimate of their thickness. Gehrmann et al. (2019) integrate physical and chemical properties of ore and host rock samples, recovered by drilling, into the interpretation in order to estimate the full economic potential of the deposits. Controlled source electromagnetic data, sensitive to the electrical conductivity of SMS mineralization, further reveal a maximum thickness of up to 80 m and conductivities of up to 5 S/m. SMS samples have conductivities of up to a few thousand Siemens per meter, suggesting that remotely inferred conductivities represent an average of metal sulfide ores combined with silicified and altered host basalt that likely dominates at greater depths.

10.12 A case study of the minerals on the seabed of the West Coast of India

The west coast of India is a typical example of a passive continental margin. In the east, the margin is flanked by a narrow zone of onshore Tertiary sedimentary basin, formed as a consequence of down-faulting and graben formation. The regional fracture pattern, lineaments, and the rifting along the west coast produced the offshore sedimentary basins and associated volcanicity in the Laccadives area (Sinha Roy and Thomas Mathai, 1979). With the continuing exploitation of the reserves of natural resources from the Earth's land areas for the rapidly increasing industry, finite resources have diminished considerably. The need to win the same resources from the seabed is increasing day by day. India, with a 5700 km coastline and proven hydrocarbon resources within the Exclusive Economic Zone, has a tremendous stake in offshore exploration.

The occurrence and distribution of the mineral resources on the western continental margin of India, excluding hydrocarbons, are discussed by compiling all the available data (Rajendran, 1990). The margin is divided into four zones, viz: the northwest coast (Gulf of Kutch-Bombay), Konkan coast (Bombay-Mangalore), southwest coast (Mangalore-Cape Comorin), and the Laccadive area (Fig. 10.14). The exploration and assessment of the mineral resources are supported by geological and geophysical evidence (Rajendran, 1990).

As compared to the southwest coast, the heavy mineral deposits of the Konkan coast contain more rock-forming and unstable minerals like augite, hornblende, epidote, zoisite, etc., in addition to placer minerals like ilmenite, rutile, etc. The southwest and Konkan coasts are promising for heavy mineral exploitation; the Laccadives for calcareous sand, and the Vembanad estuary and the Gulf of Kutch for shale deposits. Offshore placer deposits of the Konkan coast are fairly high and widespread compared to their onshore deposits, whereas the offshore placer deposit potential of the southwest coast is yet to be studied. The offshore region of Bombay-Quilon is suggested for detailed exploration for phosphorite deposits. The terrigenous deposits of Kerala and Konkan coasts can be mined by wireline, bucket ladder, and hydraulic dredges, depending upon the location, and similarly, the biogenous deposits of Laccadives, Vembanad estuary, and the Gulf of Kutch by hydraulic suction dredges.

10.13 Marine CSEM data with a case study from the North Alex Mud Volcano, West Nile Delta, Egypt

Submarine mud volcanoes at the seafloor are surface expressions of fluid flow systems within the seafloor (Kopf, 2002). Active mud volcanoes show distinct geochemical signatures with changing salinity of pore fluids (Hensen et al., 2007). Also, large temperature gradients and occurrences of free gas (Feseker et al., 2009) can frequently be observed. Since the electrical resistivity of the seafloor is mainly determined by the amount and characteristics of fluids contained within the sediment's pore space (Fofonoff, 1985), electromagnetic methods

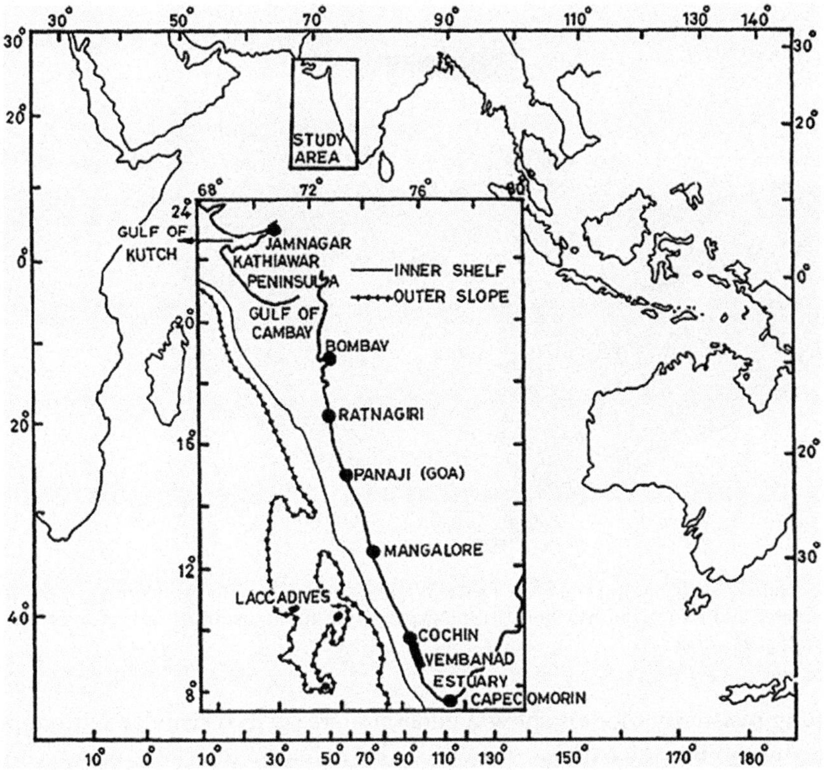

FIG. 10.14 The west coast of India. The width of the continental shelf varies from 300 km near Bombay to approximately 80 km near Porbandar, narrowing further toward 60 km near Cochin. *Source: Modified. https://oceanrep.geomar.de/id/eprint/41733/1/Thesis_msommer.pdf*

offer a promising approach to gain insight into a mud volcano's internal resistivity structure. To investigate this structure, Holz et al. (2015) conducted a controlled source electromagnetic experiment over the "North Alex" mud volcano, which is located about 50 km to the north of Alexandria (Egypt) in the West Nile Delta (WND). It was novel in the sense that the source was deployed and operated with a remotely operated vehicle (ROV), which allowed for a flexible placement of the transmitter dipole with two polarization directions at each transmitter location (Fig. 10.15) (Sommer et al., 2013).

For the interpretation of the experiment, the concept of rotational invariants from land-based electromagnetics is adapted to the marine case by considering the source normalized tensor of horizontal electric field components. The sensitivity of these rotational invariants is analyzed in terms of a 1D model and measurement geometries and associated measurement errors, which resemble the experiment at the mud volcano. The analysis shows that any combination of rotational invariants has an improved parameter resolution as compared to the sensitivity of the pure radial or azimuthal component alone. For the data set, which was acquired at the "North Alex" mud volcano, Holz et al. (2015) interpret rotational invariants in terms of 1D inversion on a common midpoint grid.

FIG. 10.15 Schematic sketch of the West Nile Delta (WND) experiment with the remotely operated vehicle (ROV) deployed transmitter and two nodal receivers. In the experiment, the transmitter and receivers are arbitrarily oriented. *Source: Holz et al., 2015.*

The resulting resistivity models show a general increase in resistivity with depth. The most prominent feature in the stitched 1D section is a lens-shaped interface, which can similarly be found in a section from seismic reflection data. Beneath this interface, bulk resistivities frequently fall in a range between 2.0 and 2.5 Ωm toward the maximum penetration depths. The lens-shaped interface is interpreted as the surface of a collapse structure, which was formed at the end of a phase of activity of an older mud volcano generation and subsequently refilled with new mud volcano sediments during a later stage of activity. Increased resistivities at depth cannot be explained by compaction alone, but instead require a combination of compaction and increased cementation of the older sediments, possibly in connection with trapped, cooled-down mud volcano fluids, which have a depleted chlorinity. At shallow depths (≤ 50 m), bulk resistivities generally decrease, and for locations around the mud volcanoes centre, 1D models show bulk resistivities in a range between 0.5 and 0.7 Ωm, which is interpreted in terms of gas saturation levels by means of Archie's law. After a detailed analysis of the material parameters contained in Archie's law, Holz et al. (2015) derived saturation levels between 0% and 25%, which is in accordance with observations of active degassing and a reflector with negative polarity in the seismic section just beneath the seafloor, which is indicative of free gas.

10.14 A case study of offshore marine placer gold at Nome, Alaska—GIS application

Miners, geologists, and engineers have been dealing with problems related to the analysis and manipulation of georeferenced information (i.e., spatial data) for many decades.

Conventionally, analog maps are used to store and display such information. A major component of many geological surveys consists of mapping the composition and structure of bedrock and surficial geology using traditional field methods as well as geochemical, geophysical, and remote sensing techniques. Integration of field survey data, maps, and other pertinent information for the purpose of mineral resource exploration and resource estimation is a very time-consuming task. However, geographic information systems (GIS) can accomplish such a task in a time-efficient and cost-effective manner. A GIS is a computer system capable of assembling, storing, manipulating, and displaying geographically referenced information, i.e., data identified according to their locations (USGS website). Practitioners also regard the total GIS as including operating personnel and the data that go into the system. With the increasing popularity and functional development of GIS in recent years, many mining companies started using GIS as the preferred tool for mine planning, analysis, and management. Moreover, the state and federal agencies involved in the mine permitting process are adopting the GIS format as the standard for communicating spatial data.

Abundant beach and offshore resources of heavy refractory minerals occur along the coast of Alaska. Nome, for example, is one of the most recently active areas of marine and beach placer mining in the State of Alaska (Koschmann and Bergendahl, 1968; Garnett, 2000). From as early as 1897 to 1962, the Nome area produced about 5 million ounces (oz) of gold (Koschmann and Bergendahl, 1968). Because of the huge amount of data in a variety of forms accumulated from placer gold exploration and production over the past century, the analysis and management of these data for future development of the Nome offshore gold resource is an enormous task. In the Nome Offshore Marine Mineral Resources Project, two relational geodatabases for storing various maps integrated with digital data sets have been created (Zhou et al., 2007). One geodatabase is known as the Integrated Geodatabase (IG), and another geodatabase, known as the Regularized 2.5D Geodatabase (R2.5DG), has been generated based on the IG. The IG served as a data warehouse and was used to manage various geological data, such as borehole, bedrock geology, surficial geology, and geochemical data. The R2.5DG could be used for a gold resource estimate at any given spatial domain. The GIS procedure developed in this project can be readily adapted for the analysis of offshore marine mineral data in other areas. This procedure also enables web-based GIS, which can be used by authorized users from remote computers.

Information on placer gold deposits can be updated, queried, visualized, and analyzed by making use of these geodatabases. Ore body boundaries, gold distribution, and the resource estimation at various cutoff grades can be calculated in a timely manner. The total placer gold resource near Nome is estimated from 113,767 oz (with a cutoff grade of 1000 mg/m^3) to 2,309,664 oz (with a cutoff grade of 0 mg/m^3). The average grade ranges from 0.233 g/m^3 (with a cutoff grade of 0 mg/m^3) to 1.929 g/m^3 (with a cutoff grade of 1000 mg/m^3).

10.15 Conclusions

Chapter 10 provides an in-depth exploration of the emerging field of marine mineral resource exploration and the technologies driving the Deep Ocean Mission. The chapter underscores the growing global interest in the ocean floor as a new frontier for sustainable mineral extraction, propelled by the increasing demand for critical and strategic metals essential for modern industries and clean energy technologies.

It discusses the major types of deep-sea mineral deposits—including ferromanganese nodules, seafloor massive sulfides (SMS), ferromanganese crusts, REE-rich marine muds, and phosphorite deposits—highlighting their geological significance, economic potential, and distribution across the world's oceans. The importance of engineering geology and environmental management in deep-sea mining is emphasized, especially in light of emerging policies such as the High Seas Agreement (2023), which promotes international cooperation and sustainable exploitation of marine resources.

Advanced exploration and extraction techniques—geophysical, geochemical, and geotechnical—along with innovative electromagnetic (CSEM) and self-potential (SP) methods, are described as key tools for identifying and characterizing subsea deposits. The chapter also presents several global case studies—from Papua New Guinea to Japan, India, Alaska, and Egypt—demonstrating practical applications of these technologies and the integration of GIS and modeling tools in marine exploration.

Ultimately, the chapter concludes that the deep ocean represents both a vast opportunity and a profound responsibility. Sustainable marine mineral exploration requires a balance between technological advancement, environmental protection, and international governance. As part of the blue economy vision, deep-sea exploration, guided by ethical and ecological principles, can contribute significantly to meeting future mineral demands while preserving the health of the planet's final frontier—the oceans.

CHAPTER

11

National geophysical mapping programs for mineral exploration of different countries: A global perspective

11.1 Overview of the importance of geophysical mapping for mineral exploration

Geophysical mapping plays a critical role in mineral exploration, serving as a noninvasive and cost-effective technique for identifying subsurface structures and mineral deposits. By utilizing a variety of methods, such as seismic, magnetic, electrical, and gravity surveys, geophysical mapping helps create detailed images of the Earth's subsurface, providing vital information for locating mineral resources. This is particularly valuable in regions where direct sampling may be limited due to environmental concerns or challenging terrain. Geophysical techniques are essential for reducing the exploration costs and risks associated with conventional drilling methods, as they allow exploration to focus on the most promising sites. As such, geophysical surveys are fundamental to the sustainable management of natural resources and the long-term viability of mining operations.

In recent years, the application of geophysical methods has expanded with the integration of cutting-edge technology, such as drones, satellite-based observations, and machine learning algorithms for data processing. These advancements have significantly improved the precision and efficiency of geophysical surveys, offering enhanced resolution and greater depth of exploration. For instance, the use of airborne electromagnetic surveys has become increasingly popular for exploring deep mineral deposits, particularly in areas where traditional ground-based methods face limitations. This technological leap allows for better resource targeting, minimizing the environmental footprint of exploration activities and enabling more sustainable practices in the industry. These innovations also contribute to more accurate resource estimation and improved decision-making for sustainable mining operations (O'Reilly et al., 2023).

Baseline Geoscience Data Generation for Sustainable Exploration of Natural Resources
https://doi.org/10.1016/B978-0-443-45221-5.00006-0

The importance of geophysical mapping is underscored by the growing need for sustainable resource exploration. As mineral resources become scarcer, the demand for new discoveries intensifies. Geophysical surveys provide an essential tool in exploring remote and unexplored regions, where traditional exploration methods may not be viable. Moreover, they are pivotal in advancing the goal of sustainable exploration by reducing the need for invasive techniques such as drilling, which can cause significant environmental impact. By improving the accuracy and effectiveness of resource exploration, geophysical mapping can help ensure that natural resources are extracted responsibly and efficiently, thereby contributing to the long-term sustainability of the mining industry (Zhao et al., 2022).

Globally, geophysical mapping has become an indispensable component of national geological surveys, contributing to the economic development of countries rich in mineral resources. Nations like Australia, Canada, and South Africa have long recognized the value of comprehensive geophysical data for advancing mineral exploration. Moreover, emerging economies are increasingly adopting these techniques to boost their resource exploration activities and attract foreign investment. As mineral exploration becomes more competitive and complex, the role of geophysical mapping continues to evolve, reinforcing its place as a cornerstone in the future of sustainable resource development (Xu and Zhang, 2024).

11.2 Efficacy of methodologies deployed for geophysical mapping

The effectiveness of methodologies used in geophysical mapping has evolved with advances in instrumentation, data processing, and integration techniques. Traditional methods such as ground-based magnetic and gravity surveys, seismic profiling, and resistivity measurements remain foundational, offering detailed subsurface information. However, their applicability is often limited by terrain accessibility, survey time, and cost. In contrast, airborne geophysical methods—such as magnetic, radiometric, and electromagnetic surveys—have proven highly efficient for regional-scale mapping, enabling rapid data collection across large, often remote areas. The efficacy of these methods is further enhanced when integrated with geological and geochemical datasets, enabling more accurate targeting of mineralized zones. The choice of methodology often depends on the exploration objectives, geological context, and available budget, but hybrid approaches that combine multiple geophysical techniques have shown the highest success rates in delineating potential mineral deposits.

11.3 Benefits and limitations of geophysical techniques in resource investigation

Geophysical techniques used by the Geological Survey of India (GSI) offer numerous benefits for resource investigation. Methods like magnetometry and resistivity surveys allow for the efficient identification of mineral deposits over large areas, providing valuable baseline data for exploration. These techniques help in detecting subsurface anomalies and geological structures that are difficult to observe through traditional methods, thus improving the accuracy of resource estimation and reducing exploration costs (GSI, 2020).

However, there are limitations. For example, magnetometry can be influenced by local magnetic interference, such as human-made structures, which may lead to false readings. Resistivity surveys may face challenges in areas with high groundwater conductivity, which can distort results. Similarly, seismic profiling requires high-quality equipment and extensive data processing, making it costly and time-consuming. Despite these limitations, these techniques remain indispensable for resource exploration, particularly when used in combination with other geophysical methods (Naskar et al., 2022).

11.4 Global approaches to geophysical mapping

Globally, countries adopt diverse strategies for geophysical mapping based on their geological frameworks, resource potential, and institutional capabilities. Developed countries like Canada, Australia, and the United States maintain comprehensive geophysical databases, generated through systematic national programs and often freely accessible to the public and industry. These programs typically employ high-resolution airborne geophysical surveys combined with ground truthing, resulting in reliable baseline datasets. Emerging economies are increasingly investing in national geophysical programs to attract foreign investment and promote sustainable resource development. Countries such as India, Brazil, and South Africa have initiated large-scale mapping projects using advanced technologies, often in collaboration with international agencies. The global trend emphasizes open data policies, cross-border collaborations, and integration of geophysical data with remote sensing and geological information systems to support sustainable mineral exploration.

11.5 Comparative analysis of geophysical mapping methodologies in various countries

Geophysical mapping methodologies vary across countries, influenced by local geological conditions, technological advancements, and resource exploration needs. In the United States, the United States Geological Survey (USGS) utilizes advanced techniques like airborne geophysics, magnetometry, and gravity surveys to map vast areas, enabling targeted mineral exploration. In Australia, the geophysical mapping approach integrates high-resolution magnetic surveys, seismic reflection, and geochemical analysis to assess mineral resources, with an emphasis on copper, gold, and rare earth elements. In Africa, countries like South Africa and Namibia have focused on geophysical techniques such as electrical resistivity and electromagnetic surveys to locate base metals and precious minerals in the largely unexplored terrains (Telford et al., 2015; Zana et al., 2019).

11.6 Contribution of global efforts to resource discovery and sustainable exploration

Global geophysical mapping initiatives significantly contribute to the discovery of new mineral resources and promote sustainable exploration. For example, in Australia, high-resolution aeromagnetic surveys have led to the identification of previously unknown mineral deposits, supporting the growth of the mining sector while minimizing environmental impact. Similarly, in Africa, geophysical surveys help identify critical resources, such as cobalt and gold, and guide exploration activities in ecologically sensitive areas, ensuring that exploration is carried out responsibly and sustainably (Telford et al., 2015). These international efforts also foster knowledge exchange, enabling countries to adopt best practices and enhance their exploration strategies.

11.7 The role of national geological surveys in supporting sustainable natural resource exploration

National geological surveys serve as cornerstone institutions in the sustainable exploration and management of natural resources, offering a scientific foundation for resource identification, evaluation, and planning. These organizations play a critical role in ensuring that mineral exploration and exploitation activities are environmentally responsible, socially acceptable, and economically viable. Globally, geological surveys such as the Geological Survey of India (GSI), the United States Geological Survey (USGS), Geoscience Australia, and the Geological Survey of Canada have been instrumental in generating and disseminating geoscientific data that support informed decision-making for sustainable resource development.

In the context of mineral exploration, national surveys like GSI contribute by collecting, analyzing, and publishing vital geophysical, geological, and geochemical data. This information forms the backbone of resource estimation and exploration targeting. For instance, GSI's comprehensive mapping programs assist in assessing regional geological potential and identifying areas favorable for mineral exploration (GSI, 2020). Employing techniques such as airborne surveys, gravity and magnetic mapping, and remote sensing, GSI enhances the ability to detect subsurface structures and mineralized zones—key elements in the pursuit of sustainable exploration strategies (Wadhawan et al., 2013).

Geological surveys also conduct environmental impact assessments and socio-economic evaluations as part of responsible exploration practices. These assessments are crucial for identifying ecologically sensitive areas and ensuring exploration does not lead to irreversible environmental harm. National surveys across the world are increasingly incorporating sustainability frameworks into their operational models. For example, GSI has promoted environmental considerations in its mineral assessment projects, advocating for a balance between economic development and ecological preservation (Boucher, 2014).

In terms of governance and policy, national geological surveys provide essential datasets that inform government decisions related to licensing, regulation, and environmental protection. These data support the design and implementation of resource conservation policies and land-use planning initiatives. In India, for example, GSI's continuous assessments have informed national policies on sustainable mining and mineral resource conservation. Similar roles are played by other geological surveys, such as the USGS in the United States and Geoscience Australia, which advise their respective governments on long-term mineral strategies and risk assessments (GSI, 2020).

Capacity-building is another key area where national surveys contribute significantly. These institutions organize training programs, technical workshops, and outreach initiatives to improve the skills of professionals in the mineral sector. Collaborations with universities and research institutions further facilitate the development of innovative technologies and methodologies for sustainable exploration. In this regard, GSI, along with its international counterparts, plays an essential role in advancing geoscience education and professional development (GSI, 2020).

On the international front, national geological surveys also participate in collaborative initiatives that support global data sharing and harmonization of exploration standards. Through platforms like the International Geoscience and Geoparks Programme (IGGP) and the United Nations Economic Commission for Europe (UNECE), these agencies contribute to a collective understanding of global mineral resource distribution and promote joint solutions to sustainability challenges. Such international cooperation enhances transparency, encourages responsible investment, and supports sustainable development across borders (UNEP, 2018).

In summary, national geological surveys—through their multifaceted roles in data generation, environmental stewardship, policy support, capacity-building, and international collaboration—are indispensable to the sustainable exploration of natural resources. While the Geological Survey of India exemplifies these contributions in the Indian context, similar roles are played by geological surveys worldwide. Together, these institutions help ensure that mineral exploration is scientifically robust, environmentally sound, and aligned with the principles of sustainability for the benefit of present and future generations.

11.8 Key technological advances in geophysical mapping

Recent technological advancements have significantly transformed geophysical mapping, enhancing data quality, accessibility, and interpretation capabilities (Anderson and McNeill, 2018). Drone-based geophysical surveys represent a major innovation, allowing rapid and cost-effective data collection in rugged or environmentally sensitive terrains. Equipped with lightweight magnetometers, gamma-ray spectrometers, and electromagnetic sensors, drones can acquire high-resolution geophysical data while minimizing the need for extensive ground crews. Satellite-based Earth observation technologies have also advanced, with missions like Gravity Recovery and Climate Experiment (GRACE) and Gravity Field and Steady-State Ocean Circulation Explorer (GOCE) contributing valuable data on gravity anomalies and magnetic fields. These satellite missions were launched by NASA (*National Aeronautics and Space Administration* [United States]) and ESA (*European Space Agency* [Europe]), respectively, to map Earth's gravity field with high precision. Additionally, innovations in data processing—including the use of 3D inversion modeling, real-time data acquisition systems, and cloud-based platforms—have streamlined workflows and improved the precision of subsurface imaging. These technological breakthroughs not only improve the efficiency of exploration but also reduce environmental footprints and operational risks.

11.9 Global geophysical mapping initiatives for mineral exploration

Numerous global geophysical mapping initiatives have been launched to support mineral exploration and build robust geoscientific infrastructure. Programs such as Australia's "Exploring for the Future," Canada's "Geo-mapping for Energy and Minerals" (GEM), and the European Union's "Minerals4EU" project exemplify coordinated national efforts to generate high-quality geophysical datasets. These initiatives often aim to map underexplored or frontier regions, using advanced airborne and satellite geophysical techniques to reveal subsurface structures and mineralization patterns. International collaborations, such as those under the Geological Survey of the African Union and initiatives by the International Union of Geological Sciences (IUGS), promote capacity building, data sharing, and standardization across borders. These global efforts not only facilitate mineral exploration but also support broader goals of sustainable development by promoting responsible resource management through science-based decision-making.

11.9.1 Future trends in global geophysical mapping

The landscape of global geophysical mapping is rapidly evolving with advancements in technology, data integration, and international cooperation. These trends promise to transform the ways in which mineral resources are discovered, explored, and managed. Below is a detailed discussion of the emerging trends in geophysical surveying and exploration, as well as the role of advanced technologies and international cooperation in shaping the future of mineral investigation.

11.9.2 Discussion of emerging trends in geophysical surveying and exploration

Geophysical surveying plays a crucial role in the discovery and management of mineral resources. As the demand for minerals increases globally, geophysical methods are being enhanced to meet the challenges of exploring increasingly complex and remote areas. Several emerging trends are shaping the future of geophysical surveying and exploration:

11.9.2.1 Increased use of remote sensing and satellite geophysics

One of the key trends in geophysical exploration is the growing reliance on remote sensing technologies, including satellite-based geophysical surveys. These technologies enable large-scale, high-resolution data collection across vast and often inaccessible areas, such as deep oceans, remote deserts, and polar regions. Satellite geophysics offers advantages in mapping geological features over large areas without the need for ground-based surveys.

Global Coverage: Remote sensing allows for the collection of data from large-scale regions, providing geophysicists with comprehensive views of subsurface geology that would have been difficult to obtain with traditional methods (Toth et al., 2020).

Cost and Time Efficiency: Satellite-based geophysics can significantly reduce the cost and time involved in conducting large-area geophysical surveys. This efficiency is particularly beneficial for countries with vast territories and limited resources (Mayer, 2018).

Applications in Exploration: Satellite data is increasingly being used for the detection of mineral anomalies, mapping geological structures, and identifying regions with potential for further exploration (Leber et al., 2017). This is especially beneficial in exploring areas where traditional ground surveys may be impractical.

11.9.2.2 Advancements in electromagnetic (EM) and resistivity surveys

Electromagnetic (EM) methods and resistivity surveys are gaining popularity due to their ability to detect hidden mineral deposits, particularly in areas with challenging terrain, such as covered terrains or deep bedrock. These methods work by measuring the response of the ground to electromagnetic fields or electrical currents.

Improved Sensitivity: Advances in EM and resistivity techniques have led to greater sensitivity, allowing geophysicists to detect smaller anomalies and mineral deposits at greater depths (Schmidt et al., 2019).

Enhanced Imaging and Mapping: The development of more sophisticated equipment, such as helicopter-borne EM systems and 3D resistivity imaging, has revolutionized the ability to map subsurface structures with high resolution and accuracy (Palacky, 2015). This enables a better understanding of complex geology and mineral deposit distributions.

11.9.2.3 Integration of geophysical and geological data with AI and machine learning

The use of artificial intelligence (AI) and machine learning (ML) in geophysical data analysis is a transformative trend in mineral exploration. By integrating these technologies, geophysicists can automate the interpretation of complex geophysical data sets, identify patterns, and predict mineral deposit locations with higher accuracy.

Data Integration: AI and ML algorithms can analyze large, multi-source data sets from geophysical surveys, geological reports, satellite imagery, and environmental data, providing a more holistic understanding of the subsurface (Mendenhall et al., 2020).

Predictive Modeling: Machine learning techniques allow for the development of predictive models that can identify areas with high mineral exploration potential based on historical data and patterns found in existing geophysical surveys (Chen et al., 2021).

Automation of Data Processing: The use of AI for automating data processing and interpretation reduces human error, speeds up analysis, and allows geophysicists to focus on refining exploration strategies based on the insights generated by the models (Cheng et al., 2023).

11.9.2.4 Enhanced seismic methods for deep and complex subsurface exploration

Seismic techniques have always been essential for mineral exploration, but new advancements are allowing for deeper and more complex subsurface exploration. Advanced seismic methods, such as 3D and 4D seismic imaging, provide high-resolution data of subsurface features, even at extreme depths or under complex geological conditions.

Deep Exploration: Modern seismic techniques allow geophysicists to explore deeper into the Earth's crust than ever before, enabling the discovery of deep mineral deposits that were previously inaccessible (Zhou et al., 2020).

Real-Time Imaging: 4D seismic surveys allow for real-time monitoring of subsurface conditions, which can improve the accuracy of exploration and allow geophysicists to adapt exploration strategies as new data becomes available (Zhang et al., 2018).

11.9.3 National geophysical mapping program of India with examples

India's mineral wealth is extensive, diverse, and spatially distributed across the country's major geological provinces. The **National Geophysical Mapping (NGPM) Program** implemented by the **Geological Survey of India (GSI)**, under the Ministry of Mines, plays a pivotal role in systematically delineating the spatial patterns of important mineral resources. This program includes gravity, magnetic, and radiometric surveys integrated with geological, geochemical, and remote sensing data to support sustainable mineral exploration. The following mineral-wise account demonstrates how national geophysical mapping has contributed to understanding their spatial distribution across India.

11.9.3.1 Ferrous minerals

IronOre

Iron ore is fundamental to India's metallurgical sector, with GSI's airborne magnetic and gravity surveys helping to identify major deposits. India holds one of Asia's largest iron ore reserves, primarily of **hematite and magnetite** types. These are found in proximity to coalfields, offering logistical advantages for steel production. The **Odisha-Jharkhand-Chhattisgarh belt** is the most prominent, mapped extensively by GSI. In Odisha, deposits in **Sundergarh, Mayurbhanj, and Kendujhar** (e.g., **Gurumahisani, Sulaipet, Kiruburu, and Bonai**) are key examples. Jharkhand's **Noamundi and Gua** mines, located in the **Singhbhum craton**, represent some of the country's oldest mapped deposits. Further extensions into **Chhattisgarh** include **Dalli-Rajhara and Bailadila** (Dantewara). In southern India, deposits in **Sandur-Hospet (Karnataka), Kudremukh (Chikkamagaluru)**, and **Shivamogga** are significant. Other mapped locations include **Maharashtra, Telangana, Andhra Pradesh, and Tamil Nadu**, and coastal **Goa** has also emerged as a prominent iron ore region.

Manganese

Manganese, vital for iron smelting and ferroalloy production, occurs mainly in the **Dharwar Supergroup**, and has been mapped by GSI using aerogeophysical surveys. Odisha leads in production, particularly in **Bonai, Kendujhar, Sundergarh, Gangpur, and Koraput**, which also overlap with iron ore belts. Karnataka contributes with deposits in **Ballari, Belagavi, Dharwar, Shivamogga, and Tumkur**. Maharashtra's **Nagpur and Bhandara**, Madhya Pradesh's **Balaghat-Chhindwara-Jhabua** belt, and minor producers in **Goa, Telangana, and Jharkhand** have also been identified through detailed field surveys and satellite interpretations.

11.9.3.2 *Nonferrous minerals*

Bauxite

Bauxite, the raw material for aluminum, occurs in **lateritic** and **tertiary deposits**, often located on plateau tops and hill ranges. Odisha ranks first in production, with mapped deposits in **Kalahandi, Sambalpur, Koraput, and Bolangir**. GSI's geomorphological and lithological mapping has also highlighted the **Lohardaga belt in Jharkhand, Bhavnagar and Jamnagar in Gujarat**, **Amarkantak Plateau (Chhattisgarh)**, and **Katni-Balaghat belt (Madhya Pradesh)**. Maharashtra's **Kolhapur and Ratnagiri**, and smaller pockets in **Tamil Nadu, Karnataka**, and **Goa**, have also been systematically studied.

Copper

India's copper reserves are mainly found in **Singhbhum (Jharkhand), Balaghat (Madhya Pradesh)**, and **Jhunjhunu-Alwar (Rajasthan)**. GSI's integrated geophysical mapping—including resistivity, magnetic, and induced polarization surveys—has aided in defining ore zones and structural controls. Minor copper belts in **Agnigundala (Andhra Pradesh)**, **Chitradurga (Karnataka)**, and **South Arcot (Tamil Nadu)** have also been delineated through such programs.

11.9.3.3 *Nonmetallic minerals*

Mica

Mica is an essential nonmetallic mineral, mainly used in the electronics industry. India's major mica belts have been mapped using remote sensing and lithological surveys. These include the **Hazaribagh-Giridih belt (Jharkhand), Nellore district (Andhra Pradesh)**, and **Jaipur-Bhilwara belt (Rajasthan)**. Secondary deposits have also been reported from **Karnataka, Tamil Nadu, Kerala, Maharashtra**, and **West Bengal**.

11.9.3.4 *Energy resources*

Coal

Coal is India's dominant fossil fuel and has been thoroughly mapped through stratigraphic drilling and geophysical profiling. Around 80% of deposits are of **bituminous Gondwana coal**, occurring in **Jharkhand-Bengal coalfields** (e.g., **Jharia, Raniganj, Bokaro, and Karanpura**). Other significant mapped coalfields include **Singrauli (Madhya Pradesh/Uttar Pradesh), Korba (Chhattisgarh), Talcher (Odisha)**, and **Wardha Valley (Maharashtra)**. **Tertiary coal** in **Meghalaya, Assam, Arunachal Pradesh**, and **Nagaland** has also been surveyed. **Lignite** deposits in **Tamil Nadu, Gujarat**, and **Jammu & Kashmir** have been mapped using geoelectric and seismic methods.

Petroleum

Petroleum resources, found in **Tertiary sedimentary basins**, have been explored through deep seismic sounding, gravity, and magnetic surveys. Early production was limited to **Digboi (Assam)**, but later expanded to **Ankaleshwar, Kalol, Mehsana (Gujarat)**, and the offshore **Mumbai High** basin. Other sedimentary basins under exploration include the **Krishna-Godavari**, **Cauvery**, and **Assam-Arakan** basins. Mapping efforts have also guided refinery locations—**field-based** (e.g., **Digboi**) and **market-based** (e.g., **Barauni**)—resulting in a network of **21 refineries as of 2011**.

Natural gas

Natural gas, often found with crude oil, is a growing segment of India's energy economy. Besides **Assam and Gujarat**, independent reserves have been mapped in **Odisha, Andhra Pradesh, Tamil Nadu, Tripura, and Rajasthan**. Offshore wells in **Maharashtra and Gujarat** have added significantly to reserves. **Tight gas sands**, such as those in deeper basins, require advanced geophysical imaging and hydraulic fracturing techniques. India is also pursuing cross-border pipeline projects, including the **TAPI (Turkmenistan-Afghanistan-Pakistan-India)** and the proposed **IPI (Iran-Pakistan-India)** pipeline (Fig. 11.1). These efforts involve international geostrategic coordination supported by geophysical feasibility studies.

11.9.3.5 Nonconventional energy sources

National mapping programs also support the identification of renewable energy zones. These **nonconventional resources**—solar, wind, hydro, biomass, and geothermal—are environmentally sustainable and increasingly viable. GSI and other agencies use **aero-geophysical, remote sensing, and GIS tools** to locate optimal zones for solar and wind farms. Though these sources have higher upfront costs, they are sustainable and equitably distributed, and their development aligns with India's energy security and climate commitments.

In conclusion, the **National Geophysical Mapping Programme of India**, primarily executed by **GSI**, has played a transformative role in understanding the spatial distribution of mineral and energy resources. Through its systematic and multi-disciplinary approach—integrating airborne, ground-based, and satellite technologies—it provides foundational datasets essential for **sustainable exploration, policy-making, and resource management**.

FIG. 11.1 Proposed natural gas pipelines in South Asia: Challenges and prospects. *Source: Indian Mines and Mineral Wealth.*

11.9.4 Description of the geological survey of India (GSI) and its geophysical mapping initiatives

India, with its unique geological and tectonic setting, is endowed with substantial mineral resources, making it a key player in global mineral exploration. The country's mineral resources include abundant reserves of iron, chromite, manganese, bauxite, coal, limestone, dolomite, mica, zinc, and graphite, while it remains deficient in resources like nickel, cobalt, molybdenum, copper, sulfur, and gold (Kumar and Pandey, 2022). Despite the existence of vast Obvious Geological Potential (OGP) areas—estimated to be 1 lakh km^2 for gold, 3 lakh km^2 for diamonds, and 1.6 lakh km^2 for base metals—there is still significant potential for discovering new mineral deposits across India. To address this, the Geological Survey of India (GSI), under the Ministry of Mines, has undertaken several large-scale national mapping projects to boost mineral exploration and sustainable resource management.

The GSI's geophysical mapping initiatives play a pivotal role in identifying and assessing mineral resources across India. Initially launched in 2001, GSI's National Geophysical Mapping Program (NGMP) integrates geological, geochemical, and geophysical data for a comprehensive understanding of India's geoscientific landscape. This initiative has seen expansion, with the OGP area growing from 5.71 lakh km^2 to 6.88 lakh km^2 for 18 minerals (Ministry of Mines, Government of India, 2022). The National Geophysical Mapping programme, along with other initiatives like National Geochemical Mapping (NGCM) and National Aero-Geophysical Mapping (NAGMP), is central to GSI's mission to provide high-quality baseline geoscience data for future systematic mineral exploration.

GSI's geophysical surveys focus on covering the entire country, with particular emphasis on OGP areas. The GSI aims to generate critical gravity and magnetic data, creating essential maps, such as 1 mGal gravity and 50 nT magnetic contour maps. These maps will support future mineral exploration activities, allowing stakeholders to target the most promising sites (GSI Annual Report, 2008–2009). Additionally, offshore geoscientific studies in India's Exclusive Economic Zone (EEZ) and Territorial Waters (TW) are being conducted to understand the potential for mineral resources along the coasts.

The GSI also prioritizes the integration of advanced geophysical techniques, such as Magnetotelluric (MT) surveys and Deep Seismic Reflection Surveys (DSRS), to improve the precision of deep-seated mineral targeting. These sophisticated methods are expected to define the crustal architecture of regions with mineral potential, especially in areas where shallow data may not be as revealing (Ministry of Mines, Government of India, 2021).

Furthermore, GSI has embraced a multidisciplinary approach by structuring its operations into five key missions: Baseline Surveys, Mineral Resource Assessments, Geoinformatics, Multi-disciplinary Geosciences, and Training and Capacity Building. This restructuring allows for the development of a comprehensive geoscience database, which is essential for both national development and the advancement of the mining sector (GSI Annual Report, 2008–2009).

In addition to these efforts, GSI has launched the National Geoscience Data Repository (NGDR), a digital platform for collating and integrating geoscientific data collected by GSI and other national agencies. This initiative aims to create a comprehensive, user-friendly repository that supports collaboration between government organizations, the private sector, academia, and other stakeholders in geoscience and exploration (Ministry of Mines, Government of India, 2021).

As part of its ongoing commitment to boosting exploration and ensuring sustainable development, GSI has set ambitious goals to complete several major surveys by 2024, which include the National Geochemical Mapping (NGCM), National Geophysical Mapping (NGPM), and the National Aero-Geophysical Mapping Programme (NAGMP). These programs are poised to revolutionize India's mineral exploration landscape, contributing to the discovery of new resources and the sustainable management of existing ones.

11.9.5 Objectives and outcomes of GSI's geophysical surveys, particularly in mineral exploration

The **National Geophysical Mapping Programme (NGPM)** of the Geological Survey of India (GSI) has a central objective of generating geophysical baseline data to aid in the identification and delineation of subsurface geological structures. This data plays a crucial role in developing conceptual models of mineralized belts, which are essential for efficient mineral exploration. The program uses a combination of geological, geophysical, and geochemical mapping to identify potential target areas for mineral exploration. It also helps in planning and formulating strategies for structural, tectonic, geotechnical studies, environmental hazard mitigation, and urban planning. Through these surveys, GSI aims to enhance the understanding of India's crustal architecture and deep-seated minerals.

One of the key components of GSI's geophysical surveys is the intensification of exploration for low-volume, high-value minerals such as gold, diamond, base metals, and platinum group elements (PGEs). This aligns with the objectives of **Mission-II** of GSI, which focuses on augmenting the exploration and resource assessment of both metallic (e.g., iron ore, copper, gold, and nickel) and nonmetallic minerals (e.g., graphite, limestone, and gypsum). Additionally, the surveys are crucial in estimating and searching for India's mineral resources, specifically targeting areas with high geological potential.

The outcomes of these surveys are significant in mineral resource exploration. For example, GSI's systematic ground gravity and magnetic surveys have mapped a total area of 707,872 sq km (5.25 lakh sq km in the Obvious Geological Potential (OGP) area) to generate gravity-magnetic anomaly maps (Fig. 11.2). These maps are vital in identifying new mineral deposits and predicting potential zones for further exploration. Moreover, the surveys have resulted in the delineation of over 110 target areas for exploration, including regions rich in base metals, gold, and rare earth elements (REEs), particularly in Rajasthan, Gujarat, Madhya Pradesh, and Uttar Pradesh (GSI, 2020).

Remote sensing and aerial surveys have also enhanced mineral exploration by allowing for large-scale multi-sensor aero-geophysical surveys and hyperspectral mapping. These technologies, used under GSI's **National Aero-geophysical Mapping Programme (NAGMP)**, provide high-resolution data over mineral-rich regions, leading to the identification of promising mineralized areas. Notably, hyperspectral data has enabled GSI to develop mineral alteration maps and spectral libraries for rocks and minerals, further refining exploration targets (GSI, 2020). In terms of marine exploration, GSI has conducted extensive offshore

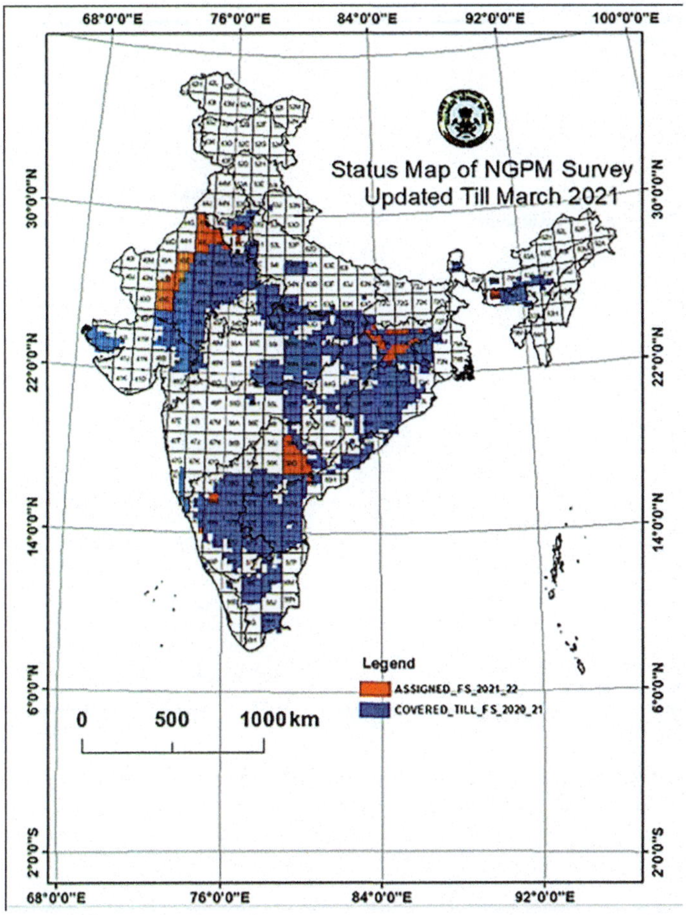

FIG. 11.2 Mission I baseline geoscience data generation. Status Map of National Geophysical Mapping Program (NGPM) in Geological Survey of India. *Source: GSI, 2021—dcport1gsigovin.*

surveys, particularly along the western and eastern coasts of India, through the **National High-Resolution Seabed Mapping and Natural Mineral Resource Evaluation** project. These surveys have resulted in the identification of areas rich in heavy minerals and construction sands, as well as phosphate-bearing sediments and Fe–Mn encrustations, contributing to India's mineral resource knowledge (Fig. 11.3) (GSI, 2020; Wadhawan et al., 2013).

In conclusion, GSI's geophysical surveys are an essential tool for locating and evaluating India's mineral resources, particularly in less explored or challenging regions. These surveys support ongoing and future exploration activities by providing a scientific basis for identifying and assessing the potential for mineral discoveries.

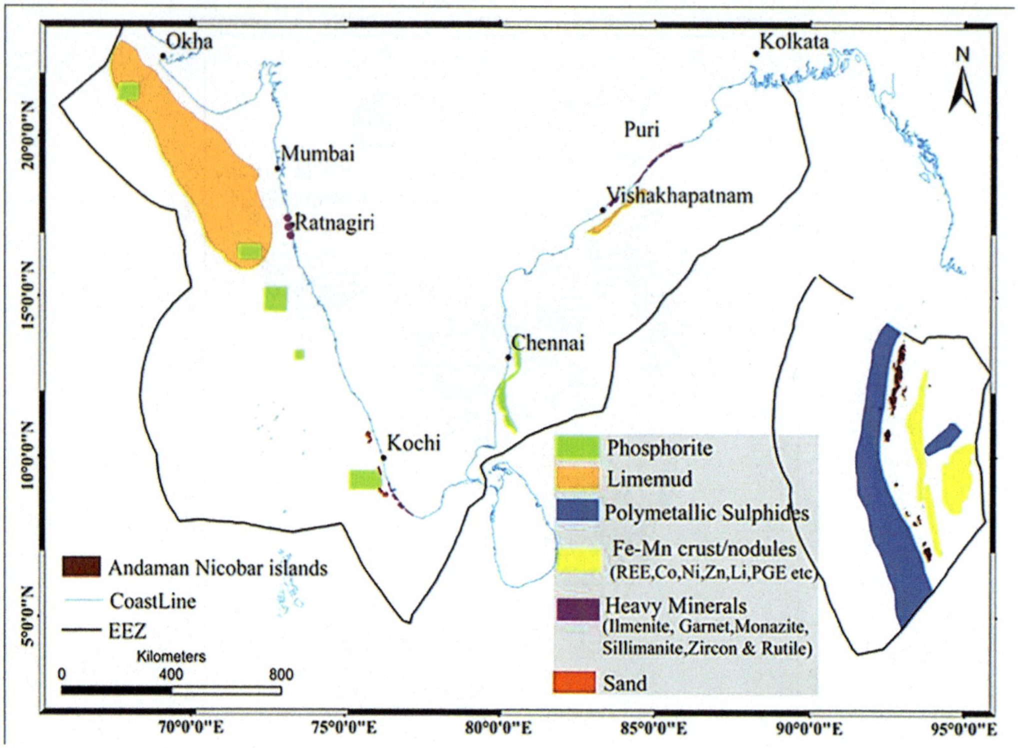

FIG. 11.3 Mission-I, baseline geoscience data generation. Occurrences/conceptualized areas for different mineral commodities within Indian EEZ. *Source: GSI, 2021—dcport1gsigovin.*

11.9.6 Crustal structure from gravity and magnetic data in parts of Nalgonda and Khammam districts, Telangana (India)

This case study focuses on **baseline geoscience data generation** through gravity and magnetic surveys, conducted in the northeastern part of the Cuddapah Basin (CB), specifically in the Palnadu sub-basin of Telangana, India (Naskar et al., 2020). The goal of the study was to characterize the subsurface features and provide insights into the region's crustal structure, essential for sustainable mineral exploration.

11.9.6.1 Gravity data interpretation

The **Bouguer gravity anomaly** survey reveals a **negative gravity anomaly** with amplitudes between −30 mGal and −85 mGal, indicating low-density subsurface features. The survey area shows high gravity anomalies in the north and east, with a medium to low gravity anomaly in the central and western parts, particularly near **Tummurukota** (Fig. 11.4). The high gravity anomaly is associated with a **basement high** beneath Kurnool Group sediments, which may be significant for uranium mineralization. The **gravity lineaments** (L6 to L8) near the eastern

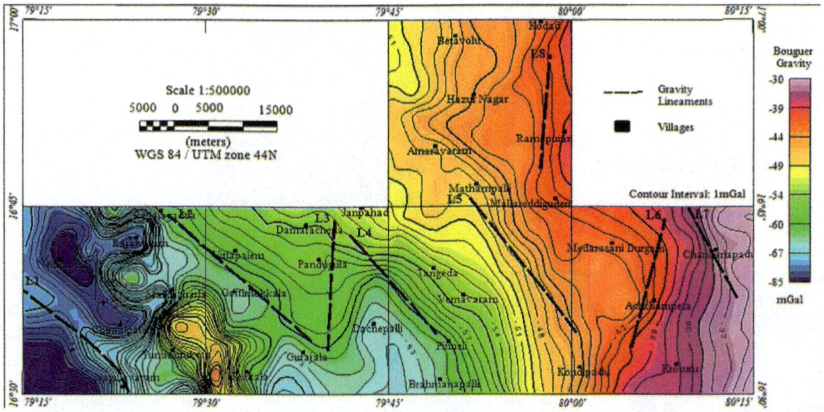

FIG. 11.4 Bouguer gravity anomaly contour map of parts of Khammam and Nalgonda districts, Telangana, India. *Source: Naskar et al., 2020—Personal.*

region reflect faulted contacts and crustal deformation due to tectonic movements, particularly the westward movement of the Eastern Ghats Mobile Belt (EGMB) against the foreland Cuddapah Basin (Fig. 11.5).

11.9.6.2 Magnetic data interpretation

The **magnetic anomaly** map shows bipolar anomalies ranging from −1449 nT to 1855 nT, with high magnetic anomalies primarily in the northern and eastern parts of the area, indicating the presence of **granite gneisses** and **Dharwar group rocks** (Fig. 11.6). These regions exhibit high magnetic susceptibility, especially in the **Peddavura Schist Belt**, reflected in both the ground and aeromagnetic data. The **aeromagnetic** map further confirms these anomalies, particularly in the northern and eastern areas, with lower magnetic anomalies observed in the central and western parts, likely due to the widespread presence of **Narji Limestone**.

11.9.6.3 Quantitative analysis

Spectral analysis of gravity and magnetic data indicates three distinct interfaces at depths of approximately 5.8 km, 1.8 km, and 1.1 km for gravity data, and two interfaces at 2.08 km and 1.2 km for magnetic data (Fig. 11.7A & b). These depths correlate with the **basement depth** and the volcanic-sedimentary sequence of the Cuddapah Supergroup and Kurnool Group. The **3D Euler deconvolution** of gravity data shows depth solutions ranging from <1500 m to over 2500 m, suggesting complex faulted contacts and geological structures, particularly along the N–S and NW-SE trending lineaments (Fig. 11.8).

Relevance to Chapter 11: National and International Geophysical Mapping for Sustainable Mineral Investigation

This case study contributes to **Chapter 11** by demonstrating the application of **geophysical mapping** in **sustainable mineral exploration**. The integration of gravity and magnetic data supports **data standardization**, helping establish baseline geoscience data that is essential for resource management and exploration. The study emphasizes the importance of using

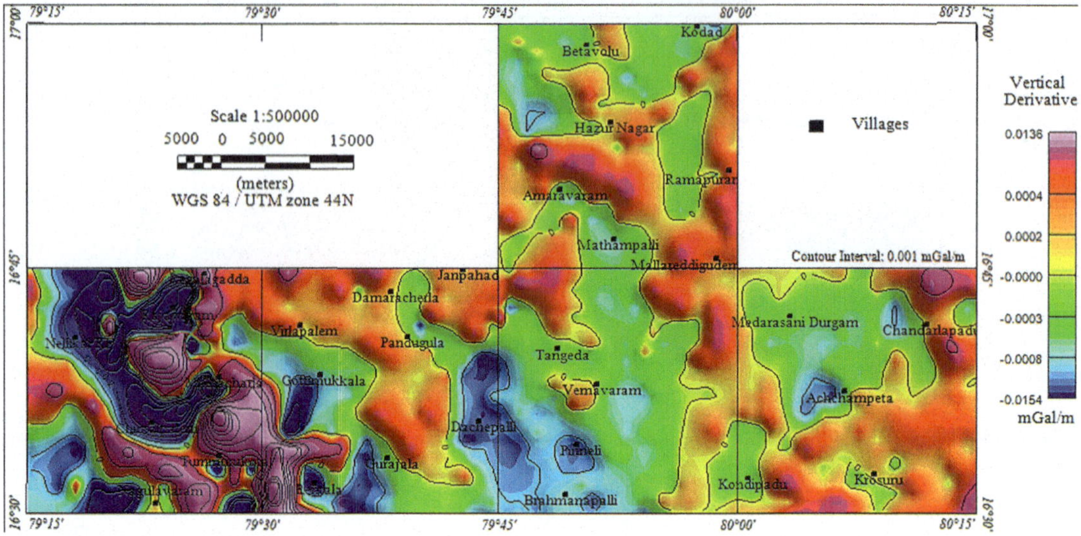

FIG. 11.5 First vertical derivative map of Bouguer gravity anomaly, parts of Khammam and Nalgonda districts, Telangana, India. *Source: Naskar et al., 2020—Personal.*

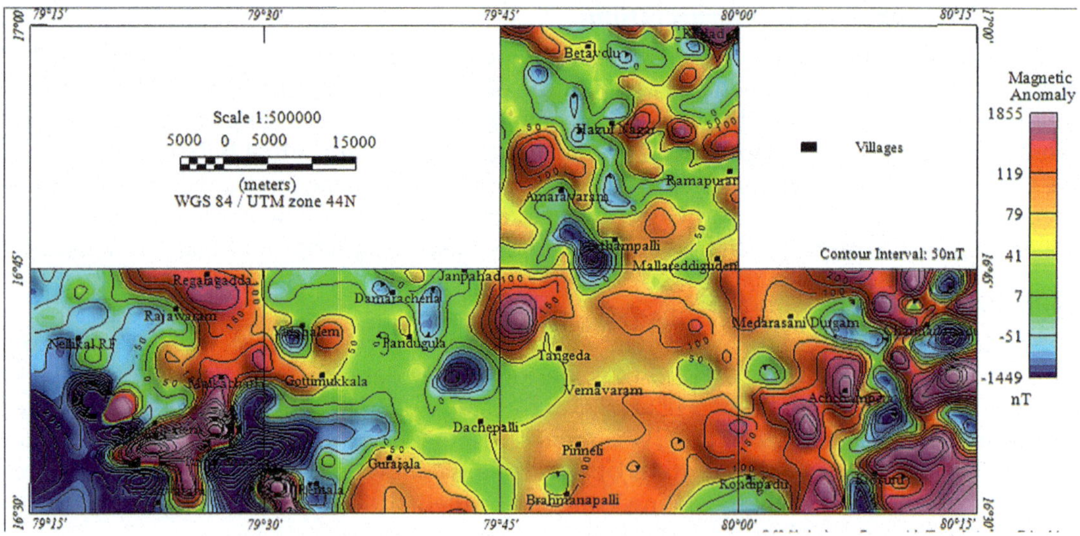

FIG. 11.6 Total field (TF) magnetic anomaly map of parts of Khammam and Nalgonda districts, Telangana, India. *Source: Naskar et al., 2020—Personal.*

high-resolution geophysical methods for identifying key geological features, such as **basement highs**, which are critical for targeting potential mineralization areas. Additionally, it highlights how **collaborative data sharing** and **data integration** across different geological regions can contribute to a **global geophysical mapping effort**, facilitating a more efficient and environmentally responsible exploration strategy for minerals.

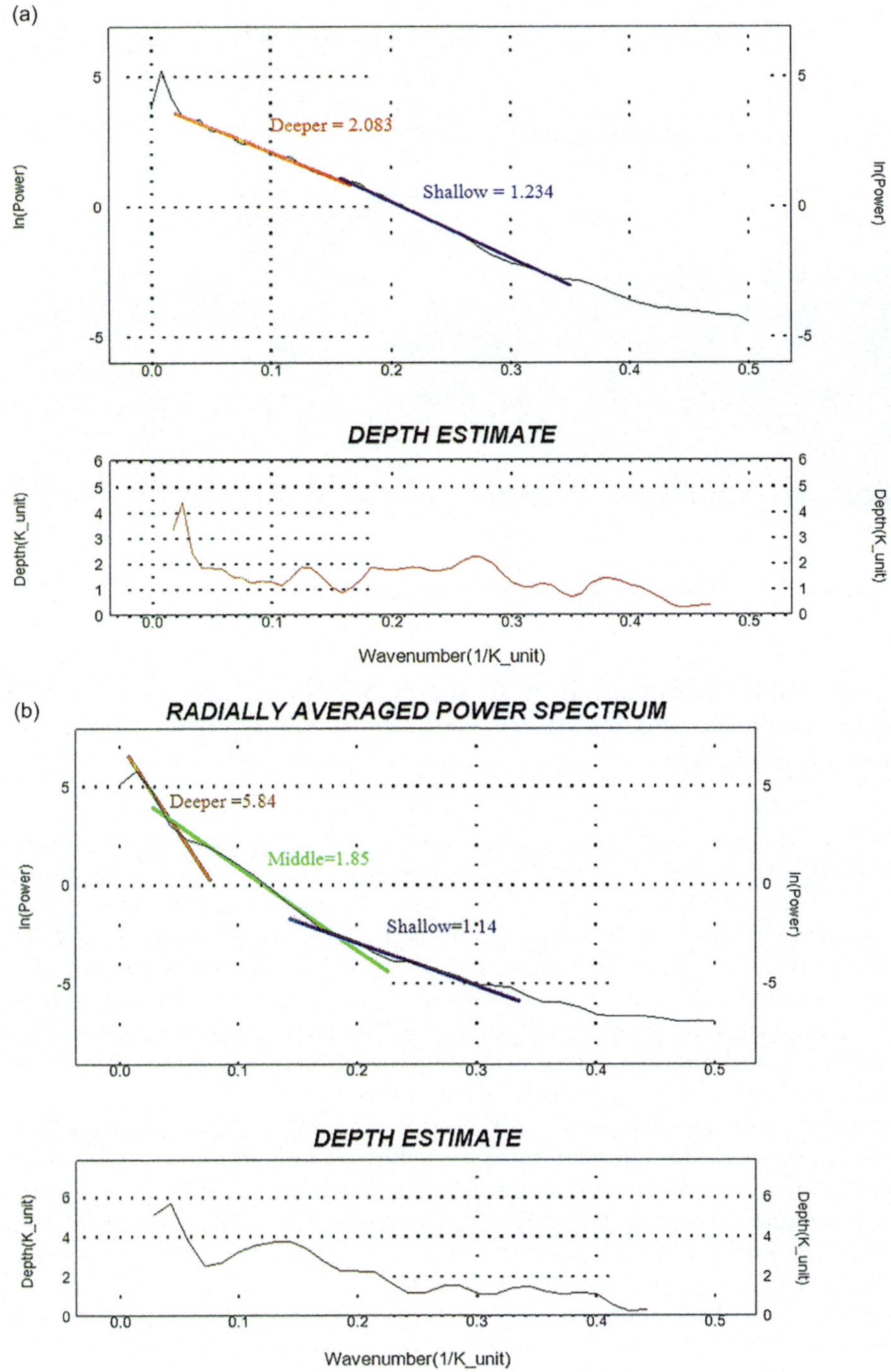

FIG. 11.7 (A) Radially averaged power spectrum of Bouguer gravity anomaly in Khammam and Nalgonda districts, Telangana, India. (B) Radially averaged power spectrum of total field (TF) magnetic anomaly in Khammam and Nalgonda districts, Telangana, India. *Naskar et al., 2020—Personal.*

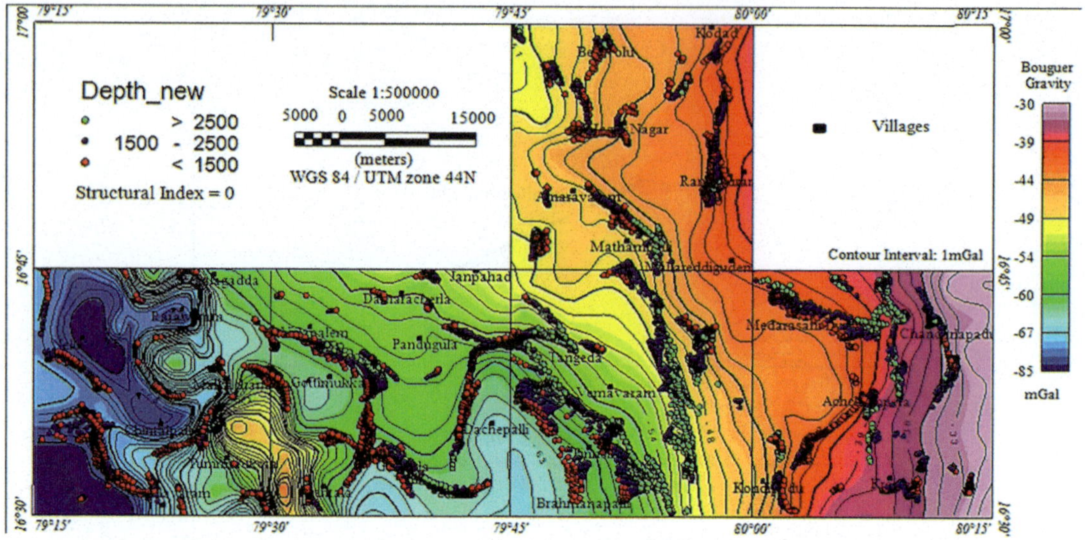

FIG. 11.8 Bouguer anomaly contour map with Euler depth solutions (SI Structural Index = 0) for parts of Khammam and Nalgonda districts, Telangana, India. *Source: Naskar et al., 2020—Personal.*

11.9.7 Case study: Delineation of geometry and spatial distribution of basic bodies using magnetic and resistivity methods in Sohagpur Coalfield Area, Madhya Pradesh, India

This case study focuses on the geophysical exploration conducted in the Sohagpur Coalfield Area, Madhya Pradesh, to delineate the geometry and distribution of basic bodies in the region (Naskar and Rai, 2018). Geophysical methods, including magnetic and resistivity surveys, were used to collect data, providing valuable insights for mineral exploration and resource extraction. The magnetic survey utilized a Scintrex Vertical Force (VF) Fluxgate Magnetometer (MF-2) with an accuracy of 1 to 10 nT. Surveys were conducted along N–S profiles at 1 km intervals, with observations at 200 m intervals. The magnetic data were corrected for diurnal and normal variations. Additionally, resistivity surveys were carried out using a Scintrex TSQ-3 resistivity unit, which measured potential differences between copper electrodes in a saturated copper sulfate solution.

The magnetic data collected along 29 traverses revealed distinct geological features and lithological boundaries. The Raniganj formation showed low amplitude magnetic anomalies due to its nonmagnetic nature, while basic bodies, such as exposed dykes, exhibited high amplitude anomalies. Based on the magnetic anomalies, three major anomaly zones (MZ-1, MZ-2, and MZ-3) were identified, indicating the lateral distribution of basic bodies. Anomaly Zone MZ-1, located in the northern part, was observed to extend across various villages, with a width of 2.5–3 km. MZ-2, south of MZ-1, was identified along the Son River, with a width of approximately 600–700 m, and showed variations in width due to interactions with MZ-1. MZ-3, in the southern part, extended eastward with a width ranging from 600 m to 1500 m.

These anomalies were clearly visible in the contour map and reflected the spatial disposition of the basic bodies (Fig. 11.9). The **Magnetic (VF) Contour Map** of the **Sohagpur Coalfield**, located in **Shahdol District, Madhya Pradesh, India**, represents the spatial

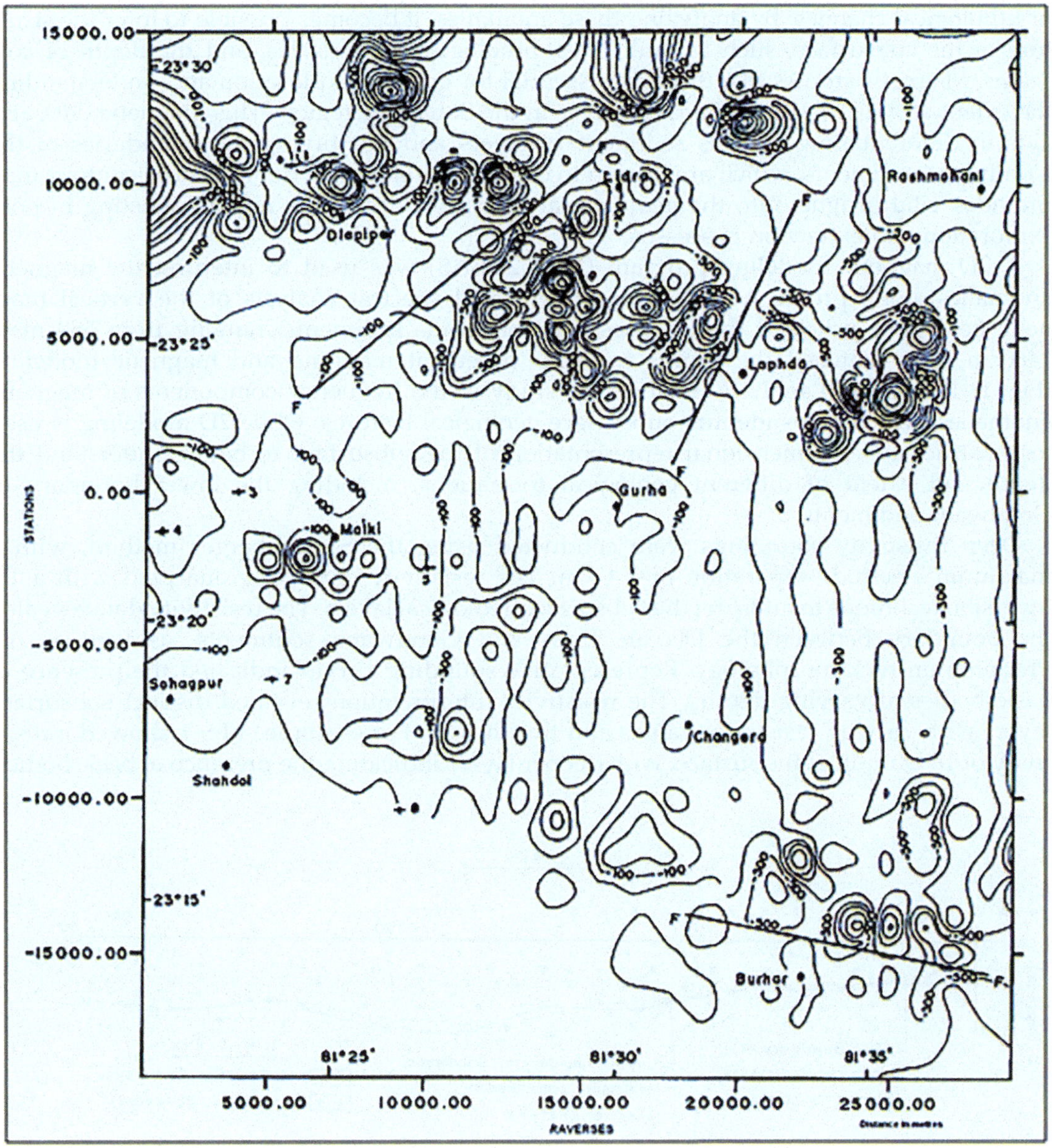

FIG. 11.9 Magnetic (VF) contour map of Sohagpur coal field, Shahdol district, Madhya Pradesh, India. *Source: Naskar and Rai, 2018—Personal.*

distribution of the earth's magnetic field variations over the coal field region. This map has been generated using **vector field (VF) data** to analyze subsurface magnetic anomalies caused by variations in geological formations, particularly the presence of coal deposits and associated strata. The contour lines on the map represent the magnetic intensity at different locations, with areas of higher intensity potentially indicating the presence of magnetic minerals or lithological changes. By analyzing these anomalies, it becomes possible to infer the structure of the subsurface, such as fault lines, lithological boundaries, and the depth of coal seams. Magnetic studies like these are essential for guiding exploration and understanding the extent and distribution of coal reserves in the Sohagpur region. This magnetic data also aids in differentiating various sedimentary layers and identifying the boundaries of the Gondwana sediments, which are known to contain valuable coal deposits. The contour map provides vital insights into the geophysical characteristics of the region, assisting in both exploration and extraction processes.

A 2D magnetic modeling program (MAG2DBAS) was used to interpret the magnetic anomalies along profile AO-OB. One-dimensional spectral analysis of the vertical magnetic field data indicated the thickness of Gondwana sediments, ranging from 680 m to 1460 m. This matched the results from 2D basement mapping and magnetic modeling (Fig. 11.10). Spectral analysis involves the study of the frequency components of magnetic anomalies, which helps identify subsurface geological features, while 2D modeling is used to construct a two-dimensional representation of the subsurface to better understand the depth and extent of different geological formations, including the lower boundary of Gondwana sediments.

Deep resistivity soundings were conducted using the Schlumberger method, with a maximum electrode separation of 2.4 km. The resistivity data were analyzed with a 1D inversion technique to interpret the subsurface geological layers. The resistivity data revealed the boundary between the Deccan Traps and Gondwana sediments, as well as the Precambrian rock morphology. Representative sounding curves indicated the presence of basic bodies at varying depths. The resistivity interpretation revealed distinct subsurface layers with varying resistivity values and thicknesses. For example, VES-1 showed a resistivity of 16 Ohm-m at the surface, with deeper layers indicating the presence of basic bodies.

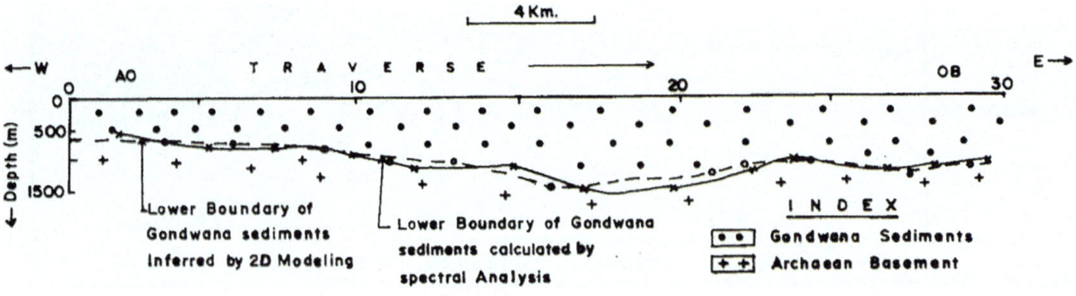

FIG. 11.10 Magnetic (VF) profile along AO-OB, Sohagpur coal field, Shahdol district, Madhya Pradesh, India. Lower boundary of Gondwana sediments inferred from 2D modeling and spectral analysis. *Source: Naskar and Rai, 2018—Personal.*

The resistivity data supported the magnetic survey findings, confirming the depth and extent of the basic bodies.

The vertical resistivity distribution and inferred lithology provided insights into the subsurface structure and nature of the basic bodies (Fig. 11.11). The resistivity profiles showed high resistivity values for the basic bodies, consistent with igneous rocks such as basalt or gabbro. The thickness of these bodies ranged from 0.7 m to 386.3 m, with variations indicating their heterogeneity. At greater depths, inhomogeneity in the basic bodies was observed, suggesting fractures or varying degrees of consolidation. The high resistivity basement rock could not be detected effectively due to the limitations of the resistivity method at such

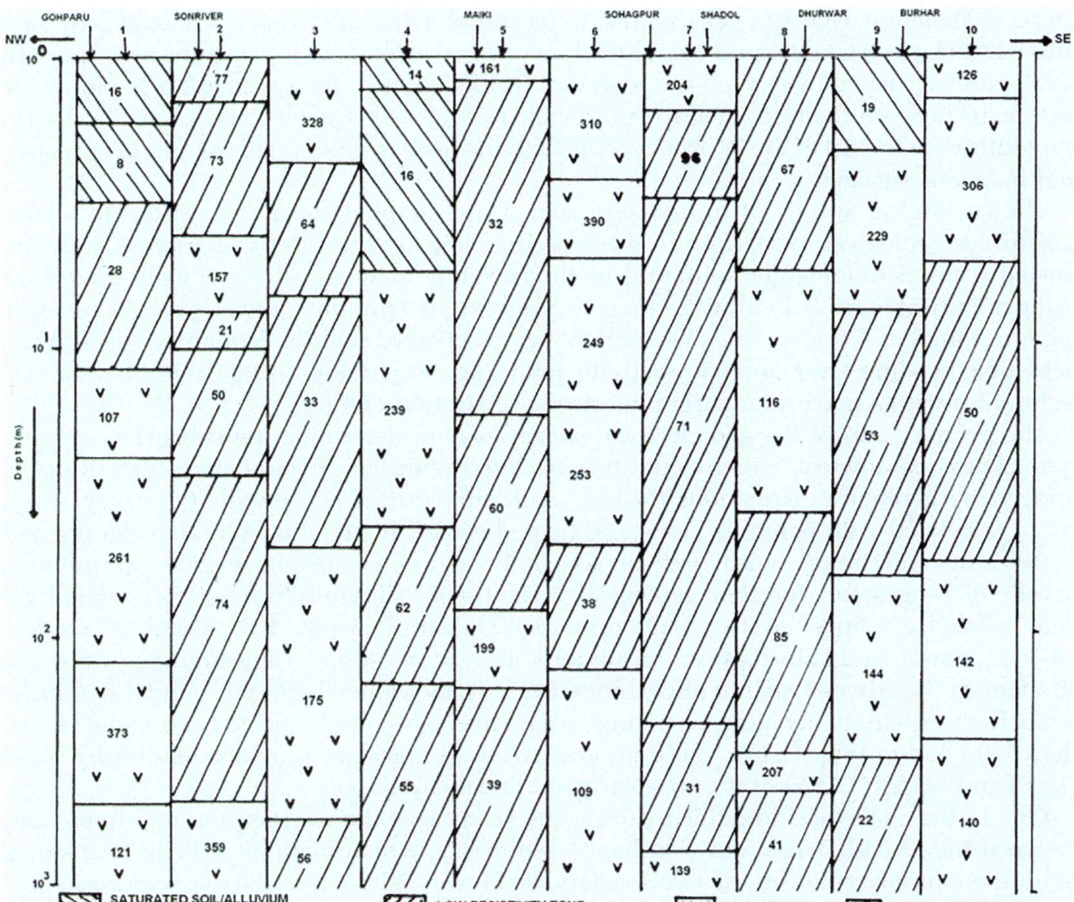

FIG. 11.11 The section represents the interpreted vertical resistivity distribution along the AO-OB profile, highlighting variations in subsurface resistivity correlated with different lithological layers. This analysis aids in understanding the geological composition and the depth of coal deposits within the Sohagpur coal field, Shahdol district, Madhya Pradesh, India. *Source: Naskar and Rai, 2018—Personal.*

depths. The integration of geo-electric interpretation with magnetic data enhanced the model of subsurface geology, where the magnetic anomalies corresponded to basic body distribution, and the resistivity data highlighted their electrical properties at different depths. By combining these findings, a more detailed and accurate subsurface model was constructed, crucial for resource assessment, exploration, and sustainable mining practices in the Sohagpur Coalfield Area.

11.9.8 Geophysical investigations for delineating Gondwana sediments below Deccan traps, Yeotmal and Wardha Districts, Maharashtra, India

The identification and delineation of hidden sediments beneath the Deccan Traps represent a challenging aspect of geoscientific exploration. Exploration for additional coal measures within known Gondwana basins and their covered extensions is of significant importance for enhancing the country's energy reserves. In this context, the delineation of Gondwana sediments beneath the Deccan Traps is crucial. To achieve these objectives, specific areas in the southwestern part of the Wardha valley, Maharashtra, were selected for detailed electrical and magnetic surveys.

The Geological Survey of India (GSI) carried out geophysical investigations to identify subsurface geological formations in these regions (Fig. 11.12). Electrical surveys were conducted using Schlumberger sounding methods, while magnetic surveys were undertaken using a Vertical Force (VF) MFD-2 Fluxgate Magnetometer in three distinct blocks across the region. A total of 439 magnetic observations were collected with N–S orientations, ensuring negligible heading error, and 95 resistivity soundings were taken using Schlumberger array techniques with a maximum current electrode separation of 4 km.

The primary aim of the geo-electrical survey was to determine the subsurface resistivity-thickness parameters, which could be correlated with the physical characteristics of the underlying geological formations (Naskar and Saha, 2015). The resistivity survey results, when correlated with borehole litho-logs, helped establish a relationship between the resistivity values and the geological units present in the region. The results revealed the following resistivity ranges for different geological formations: Alluvium/top soil/weathered trap had resistivity values up to 25 Ohm-m, the Deccan Traps had a resistivity range of 50–300 Ohm-m, and sub-trappean formations showed resistivity values ranging from 15 to 80 Ohm-m. Gondwana sediments exhibited resistivity values between 6 and 120 Ohm-m, while high-resistivity zones representing basement rocks were found to have values greater than 300 Ohm-m (Fig. 11.13). These findings were in line with previous studies by Nayak (1990) and Naskar (2005) in the Son-Narmada lineament area.

One of the challenges identified during the study was the overlapping resistivity range between the Deccan Traps and Gondwana sediments, which made it difficult to distinctly mark the interface between the two formations. However, the resistivity survey was able to successfully demarcate the Precambrian morphology due to a significant contrast in resistivity between the Deccan Traps/Gondwana sediments and Precambrian rocks. The field data obtained from the resistivity soundings were interpreted using a 1D inversion technique, employing the Resist software (Vander Velpen, 1988). This approach helped refine the model parameters, minimizing standard deviations and residual errors, to produce accurate

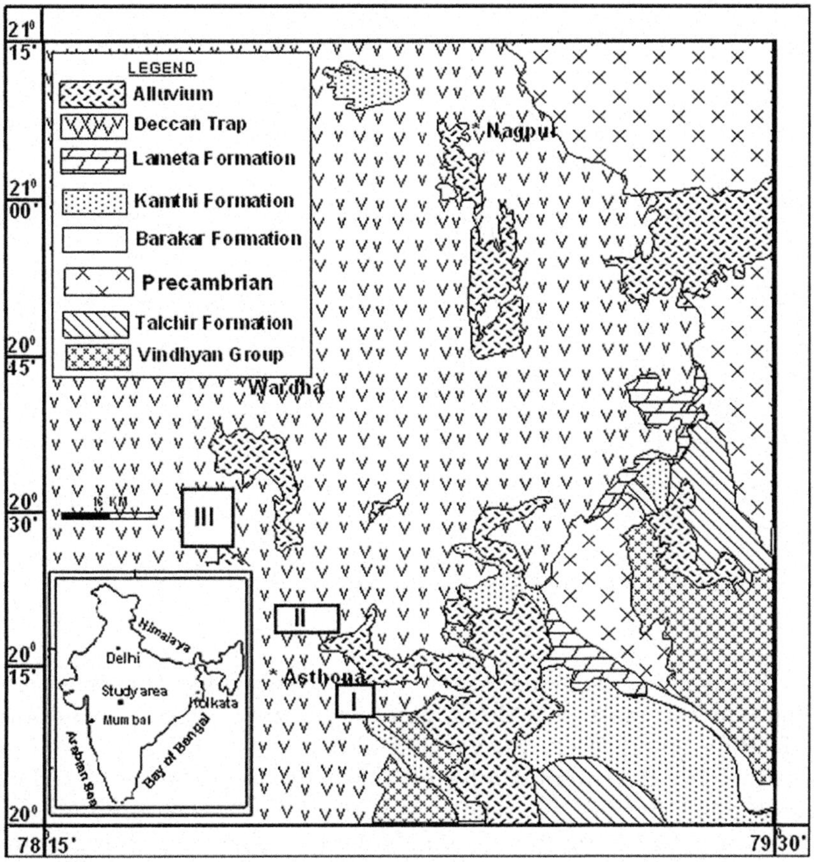

FIG. 11.12 Geological map of Wardha valley coalfields showing the areas of investigation, Yeotmal and Wardha districts, Maharashtra, India. *Source: Naskar and Saha, 2015.*

resistivity models of the subsurface. 1D inversion, while useful for initial interpretation, was supplemented by 2D and 3D inversion methods to provide a better approximation of the subsurface geological conditions. The data were also compared with gravity measurements to provide a more comprehensive understanding of the subsurface structure (Fig. 11.14).

To obtain a two-dimensional understanding of the subsurface configuration in the study area, two geo-electric sections—**AA′** and **BB′**—were selected based on the observed gravity response. Section **AA′** extends from **Kumbha in the northwest to Khairgaon**, while section **BB′**, oriented nearly orthogonal to AA′, runs from **Asthona to east of Mahagaon**.

Along section AA′, the uppermost layer is characterized by **low resistivity values ranging from 6 to 15 Ohm-m** and a thickness varying between 2 and 15 m (Fig. 11.15A). This layer corresponds to **alluvium/topsoil**. It is underlain by a higher resistivity layer, ranging from **25 to 325 Ohm-m** and **50 to 80 m thick**, which is interpreted as the **Deccan Traps**. Beneath this, a **low-resistivity zone** (ranging from **11 to 104 Ohm-m**) with a

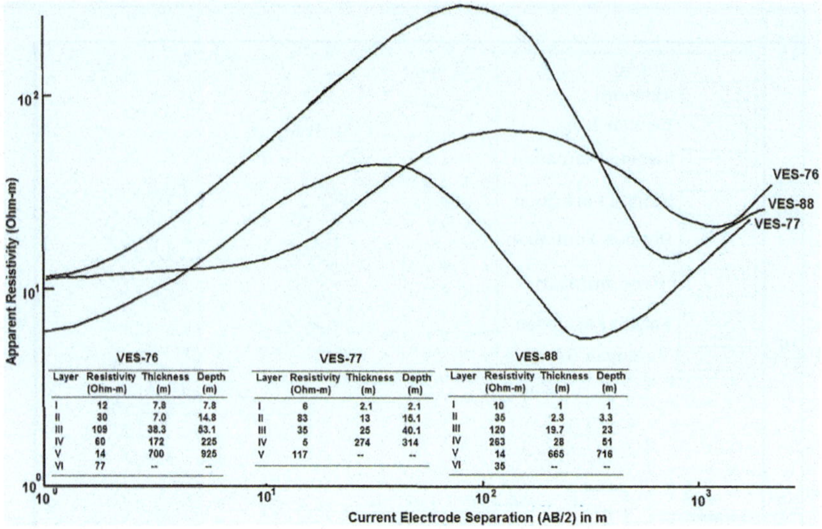

VES-76

Layer	Resistivity (Ohm-m)	Thickness (m)	Depth (m)
I	12	7.8	7.8
II	30	7.0	14.8
III	109	38.3	53.1
IV	60	172	225
V	14	700	925
VI	77	--	--

VES-77

Layer	Resistivity (Ohm-m)	Thickness (m)	Depth (m)
I	6	2.1	2.1
II	83	13	15.1
III	35	25	40.1
IV	5	274	314
V	117	--	--

VES-88

Layer	Resistivity (Ohm-m)	Thickness (m)	Depth (m)
I	10	1	1
II	35	2.3	3.3
III	120	19.7	23
IV	263	28	51
V	14	665	716
VI	35	--	--

FIG. 11.13 Representative resistivity sounding curves along with interpreted layer parameters in block-I: Ashtona-Kumbha area, Maharashtra, India. *Source: Naskar and Saha, 2015.*

thickness of 110–205 m is interpreted to represent the **sub-trappean formations**, including **Motur and Barakar formations**. A **thick, conductive zone** follows, exhibiting resistivity values of **6–18 Ohm-m** and **thickness varying between 340 and 700 m**, interpreted as the **Talcher Formation**. The **basement** is inferred at greater depths, represented by **high resistivity values (69–500 Ohm-m)**. The results of vertical electrical soundings (VES) correlate well with the gravity data along this section. Notably, a **basement depression** is identified in the central part of the **Bouguer anomaly map**, reinforcing the geoelectrical interpretation.

In section BB′, the uppermost layer shows resistivity values between 4 and 22 Ohm-m with thickness ranging from 2 to 10 m, representing alluvium/topsoil (Fig. 11.15B). Beneath this lies the Deccan Traps, having resistivity between 40 and 203 Ohm-m and a thickness of 26–79 m. Following the Deccan Traps is a low-resistivity zone (22–60 Ohm-m) with a thickness ranging from 24 to 143 m, attributed to sub-trappean formations, specifically the Lameta, Kamthi, and Motur formations. A thick, highly conductive zone lies below, with resistivity values ranging from 5 to 16 Ohm-m and thickness between 106 and 700 m, interpreted to be the Talcher Formation. The basement is identified beneath this conductive layer.

Interpretation of the resistivity sounding curves indicates that the high-resistivity basement occurs at a depth of 871 m in VES location Rs 76 (Mangli–Sindhi area), while it is shallower (313 m) in Rs 77, located to the south. The steep resistivity gradient suggests a possible fault zone in this area (Naskar and Das, 2009). However, no corresponding upliftment of the basement is observed in the gravity contours, suggesting that the inferred fault is more likely due to variations in depth rather than vertical displacement.

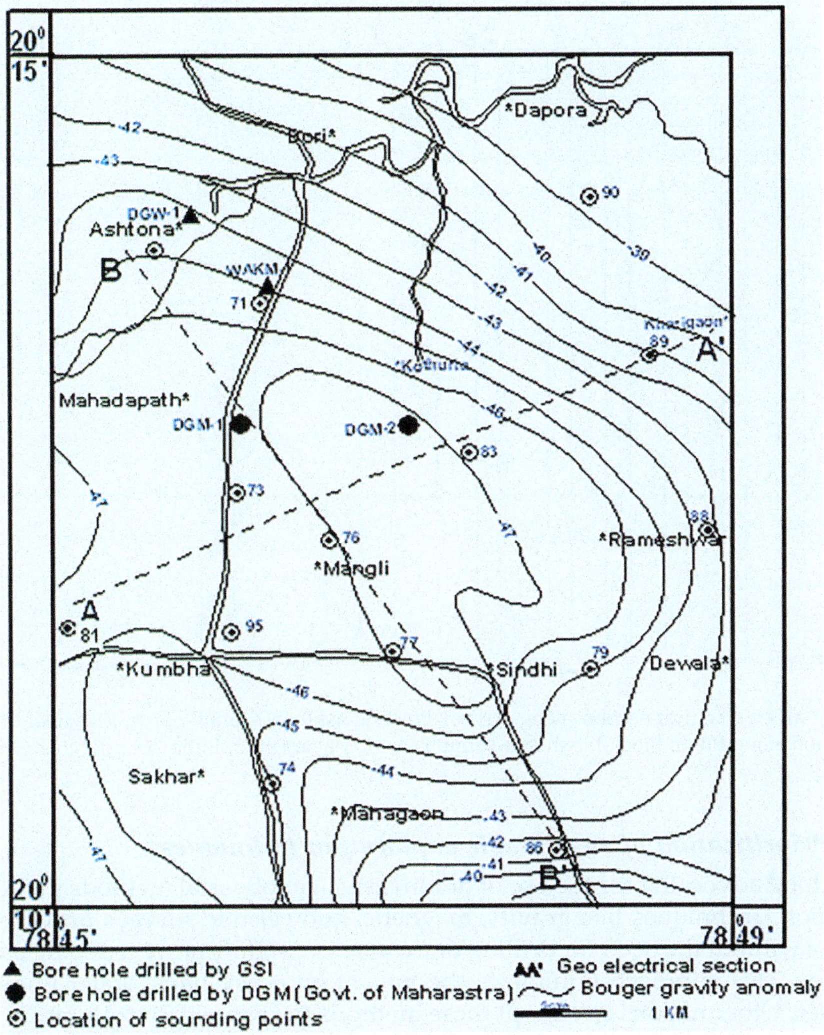

FIG. 11.14 Location map and Bouguer anomaly map (after Joga Rao et al., 1984) of Block-I: Kumbha area, Maharashtra, India. *Source: Naskar and Saha, 2015.*

11.9.9 Insights from GSI's experiences that can be applied to other nations, focusing on sustainable resource management and the use of geophysical data

The Geological Survey of India (GSI) has accumulated extensive experience in the sustainable management of natural resources, specifically through the integration of geophysical data in mineral exploration. Many of these insights can be applied to other nations aiming for sustainable resource management while effectively utilizing geophysical data. Below are some key lessons that can be adapted globally:

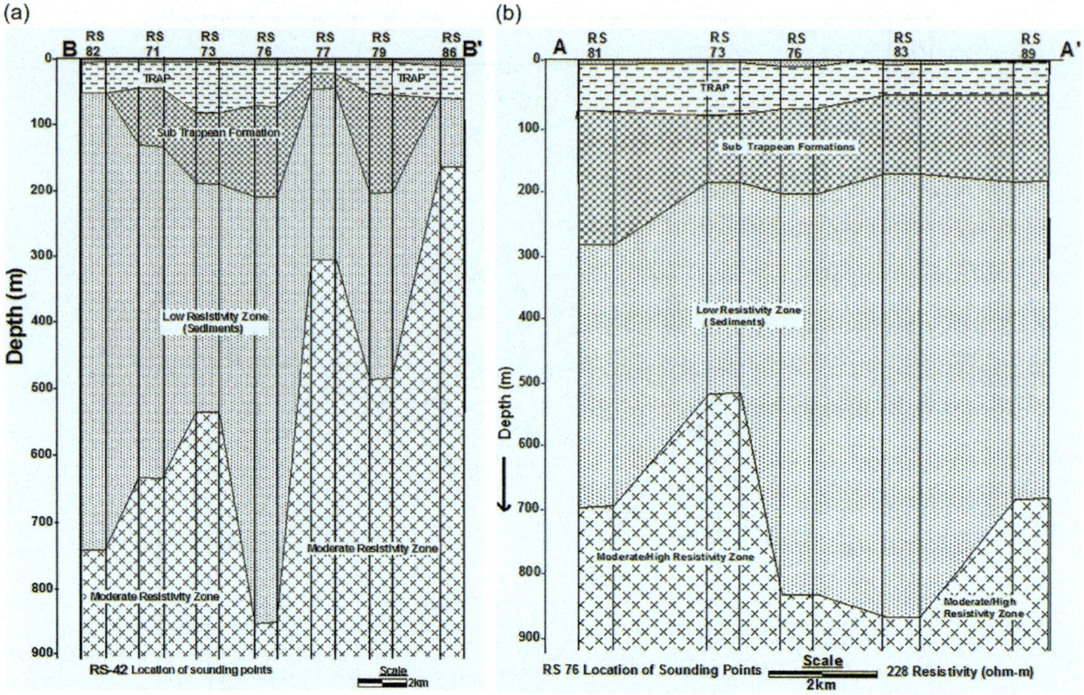

FIG. 11.15 a)Geo-electrical section along AA′ in Block-I: Ashtona-Kumbha area, Maharashtra, India. b)Geo-electrical section along BB′ in Block-I: Ashotna-Kumbha area, Maharashtra, India. *Source: Naskar and Saha, 2015.*

11.9.9.1 Prioritization of sustainable exploration techniques

GSI has long advocated for the use of noninvasive geophysical methods in the early stages of exploration. Techniques like gravity, magnetic, and seismic surveys provide critical subsurface data without the need for drilling or excavation, significantly reducing environmental disturbances. This approach minimizes the impact on ecosystems, water bodies, and local communities. Other nations can adopt these methods to ensure that exploration activities are less invasive, leading to sustainable resource management.

Global Application: Nations with significant mineral potential, like those in Africa, South America, or remote areas in Asia, can replicate GSI's focus on geophysical exploration techniques as the first line of action, reducing environmental degradation in sensitive areas.

11.9.9.2 Use of data integration for resource assessment and sustainability

GSI integrates various data sources—geophysical, geological, geochemical, and environmental—to assess mineral deposits comprehensively. This holistic approach helps understand not just the quantity of the resources but also the environmental implications and associated risks. For example, by combining geophysical data with environmental impact assessments, GSI can ensure that resource exploitation is balanced with ecological preservation.

Global Application: Other nations can build similar multidisciplinary teams that bring together geophysicists, geologists, environmental scientists, and policy experts. This approach allows for informed decision-making that accounts for both resource extraction potential and environmental sustainability.

11.9.9.3 Long-term monitoring and adaptive management

GSI employs continuous monitoring of geophysical data to track changes in mineral resources, water tables, and ground conditions. This long-term adaptive management model helps in the early identification of environmental impacts and ensures that exploration and extraction are adjusted accordingly.

Global Application: Countries with ongoing mining activities, such as in Australia, Canada, and Brazil, can incorporate long-term monitoring frameworks based on GSI's model. Periodic geophysical surveys can detect changes over time, such as groundwater contamination or land subsidence, enabling early interventions to preserve ecological balance.

11.9.9.4 Capacity building and technological advancement

GSI's approach involves constant upgrading of its methods and technologies. By investing in training and capacity-building programs for its workforce, GSI ensures that its personnel are equipped with the latest geophysical and environmental monitoring tools. This ongoing development of human resources helps in adapting to new challenges and technological advancements.

Global Application: Other nations should establish training programs in geophysical survey techniques and sustainable resource management to create a skilled workforce capable of handling modern challenges in resource exploration. This approach can be especially beneficial in emerging markets where the demand for skilled workers is high.

11.10 National geophysical mapping program of the USA with examples

11.10.1 National geophysical mapping program of Australia with examples

Australia's National Geophysical Mapping initiatives are globally recognized for their systematic and technologically advanced approach to mineral exploration. The country's vast, geologically diverse, and often remote terrain—particularly in regions such as the Outback—necessitates the use of cutting-edge geophysical tools and strategies to uncover subsurface mineral wealth.

The Geological Survey of Western Australia (GSWA) plays a central role in these efforts. It has extensively employed both airborne and ground-based geophysical methods, such as magnetometry, radiometry, gravity surveys, and electromagnetic and resistivity techniques, to map and assess subsurface structures critical for mineral exploration. These surveys provide valuable baseline datasets that serve as decision-support tools for both government and private-sector exploration activities.

A notable achievement of the GSWA is the successful discovery of new gold and nickel deposits in the Pilbara region. This was made possible through an integrated exploration strategy that combined high-resolution airborne geophysical surveys with detailed geological fieldwork and data modeling. The synergy between geophysics and geology significantly reduced exploration risk by identifying zones with strong mineralization potential before ground drilling commenced (Goleby et al., 2014).

Moreover, Australia's Exploration Incentive Scheme (EIS) and national programs like the Geophysical Archive Data Delivery System (GADDS) and AusGeoRef further enhance the availability and accessibility of geophysical data. These platforms provide open-access datasets covering large portions of the continent, supporting transparent and data-driven exploration decisions. Through such initiatives, Australia not only accelerates mineral discovery but also promotes sustainable development by minimizing unnecessary ground disturbance. In summary, the Australian geophysical mapping framework exemplifies best practices in national-scale resource exploration. Its success in regions like the Pilbara demonstrates how geophysical data, when strategically integrated with geological understanding, can unlock vast mineral potential even in the most challenging environments.

11.10.2 National geophysical mapping program of Africa with examples

Africa's diverse and geologically complex terrain has necessitated the strategic deployment of national and regional geophysical mapping programs to enhance mineral exploration and development. These programs have played a transformative role in identifying subsurface mineral resources across the continent, particularly in regions where traditional geological knowledge was limited or fragmented.

In countries such as South Africa, Zambia, and Ghana, geophysical mapping has emerged as a cornerstone of modern mineral exploration strategies. Techniques such as airborne magnetometry, gravity surveys, radiometry, and electromagnetic methods have been widely employed to delineate geological structures, identify mineralized zones, and reduce the uncertainty and costs associated with ground exploration.

A key example is Zambia, where national initiatives have focused on the Central African Copperbelt—one of the world's richest sources of copper. Here, airborne geophysical surveys have successfully targeted subsurface copper mineralization, refining geological models and enhancing drilling efficiency. These efforts have led to significant improvements in exploration success rates, cost optimization, and more strategic land use in the region (Hollis et al., 2018).

Similarly, in Ghana, geophysical mapping has supported both mineral resource assessment and land-use planning, particularly in gold-rich belts such as the Ashanti region. The integration of geophysical data into national exploration strategies has not only enabled the discovery of new deposits but also promoted sustainable mining practices by guiding exploration away from environmentally or socially sensitive areas.

In South Africa, a mature mining jurisdiction, national-scale geophysical datasets have been used extensively to map the Bushveld Complex, Witwatersrand Basin, and other economically significant geological formations. These data continue to support both brownfield and greenfield exploration efforts and are crucial for maintaining the competitiveness of South Africa's mining industry.

Across the continent, regional collaborations such as those under the African Mineral Geoscience Initiative (AMGI) and the Organization of African Geological Surveys (OAGS) are working to standardize methodologies and share geophysical data across borders. These collaborative efforts are helping to build a more coherent continental geoscience framework, driving investment and enabling responsible natural resource development. In summary, Africa's national geophysical mapping programs are playing a vital role in unlocking the continent's mineral wealth. By improving geological understanding and exploration efficiency, these initiatives are contributing to economic development while supporting more sustainable and informed land-use decisions.

11.10.3 National geophysical mapping program of Russia with examples

Russia possesses vast and geologically complex terrains, making geophysical mapping a central component of its mineral exploration strategy. The country's national geophysical programs are primarily overseen by the Federal Agency on Subsoil Use (Rosnedra) and the All-Russian Research Institute of Geology and Mineral Resources (VSEGEI). These institutions coordinate extensive geophysical surveys across Russia's mineral-rich regions, utilizing a combination of airborne magnetometry, radiometry, gravity surveys, and electromagnetic methods to map and assess the subsurface.

A major component of Russia's geophysical mapping strategy is the State Geological Map Program, which integrates geophysical, geological, and geochemical datasets to support national mineral resource management and exploration. Through this program, the government has established comprehensive geophysical coverage, particularly in remote and underexplored regions such as Siberia, the Russian Far East, and the Urals—areas known for their significant reserves of precious metals, base metals, and rare earth elements.

One notable example of success within this framework is the application of airborne geophysical surveys in the Ural Mountains, where high-resolution magnetometry and gravity data contributed to the discovery and delineation of polymetallic ore bodies, including copper, nickel, and platinum group elements. These datasets were crucial in improving the geological models and directing exploration efforts, thereby reducing uncertainty and guiding efficient drilling campaigns (Reeves and Allen, 2003; Kolesov et al., 2017).

In addition to resource targeting, the Russian national mapping initiatives also serve strategic and economic functions. The integration of geophysical data with satellite-based Earth observation platforms enhances the country's capacity to monitor geological hazards, assess groundwater resources, and inform land-use planning in alignment with national development priorities.

Overall, Russia's national geophysical mapping efforts illustrate the strategic application of advanced geophysical methods to facilitate resource discovery, environmental management, and geological understanding across its vast territories. These programs not only support domestic exploration but also contribute to international geological research and global resource knowledge through collaborative projects and data sharing.

11.10.4 National geophysical mapping program of Canada with examples

Canada's National Geophysical Mapping Program has played a crucial role in advancing resource exploration and geological understanding across the vast and diverse landscape of the country. The program is led by Natural Resources Canada (NRCan) through the Geological

Survey of Canada (GSC). The GSC utilizes a variety of geophysical techniques, including airborne surveys (magnetometry, electromagnetic, and radiometric), ground-based surveys, and seismic data acquisition, to enhance the understanding of subsurface conditions, particularly in mineral-rich regions.

One of the notable examples of the program's success is the use of high-resolution airborne geophysical surveys in the Canadian Shield, where extensive exploration for mineral resources like gold, nickel, and copper is conducted. These surveys, which are part of Canada's efforts to map large areas of the country, have been instrumental in identifying new deposits and improving exploration success rates. The data gathered through these efforts helps map geologically complex areas, thus guiding exploration and resource management decisions.

A particularly successful case of the geophysical mapping program is the identification of new mineral deposits in the James Bay and Labrador regions. In these areas, airborne surveys using magnetometry and electromagnetic methods have led to the discovery of previously unrecognized mineralized zones. These surveys have improved the understanding of the regional geology and facilitated more efficient exploration for both base and precious metals. Through such initiatives, Canada's geophysical mapping program contributes significantly to both national and international resource exploration efforts (Hodgson et al., 2016).

11.10.5 National geophysical mapping program of Germany with examples

Germany's national geophysical mapping initiatives are led by the **Federal Institute for Geosciences and Natural Resources (BGR)** in collaboration with regional geological surveys and research institutions. These efforts aim to support natural resource exploration, groundwater assessment, and geological hazard mitigation through high-resolution, multidisciplinary geophysical data acquisition.

One of the prominent examples is the **airborne geophysical survey campaign conducted along the German North Sea coast**. Between 2000 and 2014, BGR (Bundesanstalt fur Geowissenschaften und Rohstoffe) carried out helicopter-borne surveys collecting **electromagnetic, magnetic, and radiometric** data over an area of approximately 5900 km^2. The integration of these methods enabled the identification of subsurface features such as the fresh-saline groundwater interface, buried paleo-valleys, and ancient landscapes. These results have been instrumental for both groundwater management and environmental monitoring (Siemon et al., 2020).

Another significant initiative is the **3D-Deutschland Project**, which aims to develop a high-resolution, lithosphere-scale 3D geological model of Germany. This model incorporates gravity, magnetic, seismic, and geological data to better understand the structure and evolution of the crust and upper mantle, particularly in areas of resource potential and geothermal interest.

Furthermore, Germany's **Geophysical Atlas of the Federal Republic of Germany**, produced by BGR, compiles large-scale geophysical data (gravity, magnetics, and resistivity) and serves as a fundamental reference for national exploration strategies and academic research. These comprehensive and publicly accessible geophysical datasets have proven vital in supporting mineral, geothermal, and groundwater exploration, while also contributing to international geoscientific cooperation.

11.10.6 National geophysical mapping program of Poland with examples

Poland's national geophysical mapping efforts are coordinated by the Polish Geological Institute—National Research Institute (PGI-NRI), which functions as the country's central body for geological and geophysical data collection, analysis, and dissemination. These programs are designed to support mineral exploration, subsurface geological modeling, and environmental monitoring.

A notable initiative is the POLONAISE'97 Project, which focused on deep seismic sounding across Central Europe, with key profiles running through northern and central Poland. This large-scale geophysical program utilized seismic refraction and wide-angle reflection techniques to image the crust and upper mantle. The data obtained revealed complex crustal structures beneath Poland, aiding in the understanding of tectonic settings and enhancing mineral exploration models (Guterch et al., 1999).

Additionally, airborne geophysical surveys conducted by PGI-NRI—using magnetometry and radiometry—have provided critical support in the mapping of sedimentary basins and mineral-bearing formations in regions such as the Lublin Coal Basin and the Lower Silesian Copper Belt. These surveys help in delineating lithological boundaries and detecting geophysical anomalies that guide further geological and drilling investigations.

The PGI-NRI also maintains the Central Geological Database (CBDG), which integrates gravity, magnetic, and seismic datasets. This centralized platform serves as a valuable tool for stakeholders engaged in mineral exploration and geological research, ensuring accessibility to high-quality, standardized geophysical data. These comprehensive mapping programs have significantly improved subsurface understanding in Poland and have been instrumental in optimizing mineral exploration activities, particularly for copper, lignite, and rock salt.

11.10.7 National geophysical mapping program of China with examples

China's National Geophysical Mapping Program has significantly advanced mineral exploration through the application of various geophysical methods. In the Baishiquan-Hongliujing area of Northwest China, an integrated airborne geophysical survey combining electromagnetics and magnetics was conducted using the AeroTEM-IV system and CS3 Cesium Vapor magnetometer. This survey covered approximately 3149 line-kilometers with a 250 m line spacing, leading to the identification of seven anomalies consistent with known ore sites, including a new gold deposit with prospective reserves exceeding 10 tons (Liang et al., 2021).

Additionally, the Xiaorequanzi copper-zinc deposit in Xinjiang was studied using gravity, magnetic, and electromagnetic data. The integration of these datasets revealed that NE-SW trending structures are closely associated with mineralization, providing insights into the subsurface geological framework and guiding further exploration efforts (Shaole et al., 2021).

Furthermore, in the Changting area of Fujian Province, electrical resistivity tomography was employed to investigate the subsurface three-dimensional structure of rare earth ore areas. The results delineated the thickness of the weathered granite layer and the spatial distribution of deep-seated fault structures, which were consistent with drilling data, thereby aiding in the assessment of ion-adsorption-type rare earth ore reserves and the planning of recovery tunnels (Tang et al., 2024). These examples underscore China's commitment to utilizing advanced geophysical techniques in its national mapping programs to enhance the efficiency and sustainability of mineral exploration.

China's national geophysical mapping programs have played a pivotal role in the sustainable exploration and assessment of its vast mineral resources. These initiatives have evolved over decades to encompass both large-scale and high-resolution mapping techniques, designed to better understand subsurface geology and guide mineral resource development. One of the most influential projects is the Regional Geochemistry—National Reconnaissance (RGNR) project, initiated in 1978. This project involved the systematic collection of stream-sediment samples at a density of one sample per square kilometer, later composited into a single sample from four individual collections. Covering over 6 million square kilometers, the RGNR project enabled the analysis of 39 elements and led directly to numerous mineral discoveries, especially in mountainous regions where geological data had previously been sparse.

Building on this foundation, China launched the Deep-penetrating Geochemical Mapping (DEEPMAP) project in 1994, targeting covered terrains such as sedimentary basins. This project operated at a lower density—one sample per 100 square kilometers—but was designed to detect deep-seated mineralization that might not be apparent through conventional surface sampling. By 2007, >800,000 square kilometers had been surveyed under DEEPMAP, contributing valuable insights into concealed ore systems (Wang et al., 2007).

Another significant effort is the 76 Geochemical Element Mapping (76 GEM) project, initiated in 1999. This program expanded the geochemical baseline by analyzing stream-sediment samples for 76 elements across approximately 1 million square kilometers. The high-resolution multi-element data produced by this project have proven instrumental for identifying geochemical anomalies and refining exploration models.

In parallel, China has also emphasized environmental and agricultural sustainability through its Multi-purpose Eco-geochemical Mapping Projects, which began in the Quaternary plains of eastern China in 1999. These projects aim to inform land use planning, pollution monitoring, and agricultural decision-making by providing detailed geochemical data on soil quality and environmental risk.

Collectively, these national geophysical and geochemical mapping programs have significantly improved China's capacity for mineral resource evaluation while promoting sustainable and scientifically informed exploration practices. Furthermore, China's participation in international efforts, such as the Environmental Geochemical Monitoring Network (EGMON), reflects its commitment to global collaboration and leadership in geoscientific innovation.

11.10.8 National geophysical mapping program of Japan with examples

Japan's national geophysical mapping program plays a crucial role in supporting mineral exploration and geological understanding across the country. These efforts are primarily led by the Geological Survey of Japan (GSJ), operating under the National Institute of Advanced Industrial Science and Technology (AIST). Since its inception in 1882, GSJ has published a wide range of geological maps, with scales ranging from 1:50,000 to 1:200,000. These maps form the foundation for geological studies, land use planning, and mineral resource assessments (GSJ, 2021).

Airborne geophysical surveys have been a cornerstone of Japan's exploration strategy, especially under the guidance of the Metal Mining Agency of Japan (MMAJ). Initiated in 1969, MMAJ began with airborne magnetic surveys and by 1975 had expanded to include electromagnetic and radiometric surveys. These airborne methods have been essential in detecting subsurface anomalies and have directly contributed to significant mineral discoveries (Hosoi, 2000). One notable example is the Hishikari gold mine in Kagoshima Prefecture, which stands as one of the richest epithermal gold deposits globally. At Hishikari, a suite of geophysical techniques including gravity surveys, resistivity imaging, induced polarization (IP), and microtremor surveys was utilized to delineate vein systems and guide drilling efforts (Okada and Sumitomo Metal Mining Co., Ltd., 1995; Okada et al., 2018).

In addition to traditional geophysical approaches, Japan has also employed remote sensing and GIS technologies to enhance mineral exploration. For example, in the Kuju volcanic region of Kyushu, researchers used Landsat 7 ETM + imagery and analytical techniques such as color composites, band ratios, and principal component analysis to identify hydrothermal alteration zones, which are indicative of mineralized areas. These methods proved effective in delineating areas of potential mineralization without the need for initial ground-based surveys (Tomiyama et al., 2012).

Recent advances have seen the adoption of Controlled Source Audio-frequency Magnetotellurics (CSAMT) in projects such as the Ohra-Takamine gold prospect. Here, geophysical surveys identified resistive zones corresponding to hydrothermal clay alteration, providing valuable subsurface data that correlated well with surface geological observations (Japan Gold Corp., 2019). These examples highlight how Japan integrates modern geophysical tools with classical mapping techniques to refine mineral exploration targets and improve success rates.

11.11 How international collaborations and data sharing can benefit global mineral exploration efforts

International collaborations and geophysical data sharing play a transformative role in advancing mineral exploration and promoting sustainable resource development worldwide. As mineral demand increases due to global industrialization and the transition to clean energy technologies, cross-border cooperation has emerged as a vital mechanism to overcome exploration challenges and accelerate scientific progress. This section highlights how international partnerships contribute to enhancing the efficiency, reach, and impact of geophysical mapping and mineral exploration efforts across different regions.

11.11.1 Enhanced data access and comprehensive exploration

The integration of geophysical data from multiple countries contributes to the development of global and regional mineral databases. When countries share geophysical and geological datasets—such as airborne magnetic, gravity, and electromagnetic surveys—exploration efforts become more strategic and less redundant. This approach allows for a broader understanding of mineral systems and tectonic settings, especially in transboundary geological belts. Countries with limited survey capabilities can leverage international datasets to identify mineral-rich zones, enabling better-targeted exploration campaigns and reducing costs associated with initial reconnaissance work (Zhang and Yang, 2020).

Global Benefit: Access to shared geophysical data accelerates discovery in underexplored regions, avoids duplication of efforts, and facilitates more informed mineral exploration strategies on a global scale.

11.11.2 Accelerating exploration in remote or challenging environments

Mineral-rich areas located in remote, environmentally sensitive, or geopolitically complex regions often pose logistical and financial barriers to exploration. International collaboration enables joint expeditions and shared infrastructure, allowing multiple countries to collectively explore areas such as the Arctic, deep ocean floors, or high-altitude terrains. These cooperative efforts also help standardize environmental safeguards and promote responsible resource development.

Global Benefit: Pooling resources and expertise makes exploration feasible in challenging environments, distributes costs, and ensures better safety and environmental practices.

11.11.3 Promotion of best practices and standardization

Collaborative programs between national geological surveys and international organizations foster the exchange of geophysical techniques, mapping methodologies, and data interpretation frameworks. Countries with well-established exploration programs often serve as models for others, facilitating capacity building and technological transfer. Harmonization of data formats, classification systems, and survey methodologies ensures interoperability and comparability of geophysical data across regions.

Global Benefit: Adoption of international best practices and standards improves the quality and usability of geophysical data globally, promoting consistency, transparency, and scientific rigor in exploration initiatives.

11.11.4 Risk mitigation through shared expertise and resources

Exploration projects carry inherent risks, including financial investment, environmental impact, and technical uncertainties. International collaborations reduce these risks through shared funding, diversified expertise, and collaborative project design. Joint ventures allow for more comprehensive feasibility studies, improved environmental assessments, and more accurate resource modeling. Countries with complementary capabilities can address challenges collectively, leading to more robust exploration outcomes.

Global Benefit: Shared responsibility and knowledge reduce the financial and operational risks of mineral exploration, increasing the likelihood of successful resource discovery and development.

11.11.5 Global mineral resource security and sustainability

As mineral resources underpin critical infrastructure, technology, and energy systems, their sustainable management has become a shared global concern. Data sharing and international mapping programs support better coordination of exploration efforts and help avoid over-exploitation of particular regions. By promoting equitable access to geological information and encouraging multilateral agreements, countries can align their mineral development policies with global sustainability goals, including responsible sourcing and circular economy principles.

Global Benefit: International collaboration helps ensure that mineral resources are explored and developed responsibly, equitably, and sustainably, contributing to long-term global resource security.

In conclusion, the integration of geophysical data and collaboration among nations enhances the effectiveness, scope, and sustainability of mineral exploration. By participating in international initiatives and adhering to standardized practices, countries contribute to a globally coherent framework for mineral resource development. Such cooperation not only benefits individual nations but also advances the collective scientific understanding of Earth's subsurface, promoting the responsible use of its mineral wealth for current and future generations.

11.12 The role of advanced technologies and international cooperation in shaping the future of mineral investigation

Advanced technologies and international cooperation will be pivotal in shaping the future of global mineral exploration. The combination of cutting-edge technologies and enhanced collaboration among countries and organizations is driving the evolution of geophysical surveying and mineral investigation.

11.12.1 Advanced technologies in geophysical mapping

Unmanned Aerial Vehicles (UAVs) for Geophysical Surveys: UAVs, also known as drones, are increasingly being used to conduct geophysical surveys, especially in difficult-to-reach or hazardous areas. Drones equipped with electromagnetic sensors, LiDAR, or ground-penetrating radar can collect detailed geophysical data from the air, reducing the need for expensive ground-based equipment and improving survey efficiency (Vázquez et al., 2018).

Smart Sensors and Wearable Technology: The use of smart sensors and wearable technology has revolutionized field surveys. Portable geophysical instruments, such as handheld magnetic or electromagnetic devices, allow geologists and geophysicists to quickly analyze geological features and mineralization in real-time. These technologies increase the accuracy of field surveys and provide immediate feedback for further exploration decisions (González et al., 2019).

Blockchain for Data Integrity and Security: As geophysical data sharing becomes increasingly common across borders, the need for secure and reliable data sharing platforms is paramount. Blockchain technology can provide an immutable, transparent, and decentralized system for recording and verifying geophysical data, ensuring data integrity and preventing fraud in exploration activities (Swan, 2015).

11.12.2 International cooperation and data sharing

International cooperation and data sharing are essential to improving the efficiency and sustainability of global mineral exploration. Geophysical exploration is a costly and resource-intensive endeavor, and by collaborating, nations can overcome technical, financial, and logistical challenges. The role of international cooperation in shaping the future of mineral investigation includes:

Global Geophysical Data Repositories: By sharing geophysical data across national and organizational boundaries, countries can create global repositories of geophysical information. These databases facilitate access to valuable geophysical data for countries with limited exploration resources, enabling them to focus their efforts on high-potential areas rather than conducting redundant surveys. International organizations, such as the International Geophysical Year (IGY), play a key role in establishing platforms for data sharing and collaboration (Harrington, 2016).

Cross-Border Exploration Partnerships: Mineral deposits often span multiple countries, and international collaboration is necessary for efficient exploration and management. Joint ventures between geological surveys, research institutions, and private companies from different countries can lead to shared exploration efforts, resource pooling, and a deeper understanding of subsurface geology. For example, the exploration of mineral-rich regions in the Arctic, where vast resources lie beneath ice sheets, requires international cooperation for successful exploration (Hill et al., 2017).

Standardization of Exploration Practices: Global cooperation encourages the adoption of international standards for geophysical data collection, interpretation, and sharing. Standardized practices allow data to be more easily compared and integrated across countries and organizations, promoting consistency and reliability in mineral exploration (Chave, 2017).

Building Capacity in Developing Countries: International cooperation is also vital for strengthening the capacity of developing countries to conduct geophysical surveys and mineral exploration. By sharing knowledge, training resources, and technological expertise, more countries can participate in sustainable exploration efforts, ensuring that resources are explored and managed responsibly (Keller, 2015).

The future of global geophysical mapping is driven by technological advancements and increased international cooperation. Emerging trends in remote sensing, AI, and seismic exploration are enhancing the capabilities of geophysicists to explore mineral resources more efficiently and sustainably. As the global demand for minerals continues to rise, international collaboration and the sharing of geophysical data will become even more crucial for ensuring that mineral exploration meets the challenges of the future while minimizing environmental impact. With these technological innovations and cooperative efforts, the future of mineral investigation will be more precise, cost-effective, and sustainable, ultimately benefiting both the global economy and the environment.

11.13 The role of data standardization and sharing in global geophysical mapping

11.13.1 Importance of standardized geophysical data

The standardization of geophysical data is essential for ensuring the consistency and comparability of data used in mineral exploration across various countries and organizations. Geophysical surveys, which include methods such as magnetometry, seismic, and electrical resistivity, can vary significantly in their techniques, measurement units, and data interpretation models. Without standardization, it becomes difficult to integrate and compare datasets across different regions, reducing the effectiveness of global exploration efforts. Standardized data ensures that data can be interpreted reliably by scientists and industry professionals worldwide, fostering greater collaboration and the ability to identify mineral resources on a global scale. A consistent approach to geophysical data collection and interpretation enhances the quality and transparency of exploration results, which ultimately supports sustainable resource management (Telford et al., 1990).

11.13.2 Data sharing initiatives and platforms

International platforms and initiatives play a critical role in promoting the sharing of geophysical data across countries and organizations. The International Union of Geological Sciences (IUGS), the United Nations Framework Classification for Resources (UNFC), and regional geophysical data repositories contribute to the creation of collaborative frameworks that support the exchange of data. These initiatives help eliminate data silos by encouraging open access to geophysical datasets, fostering transparency, and facilitating the free flow of knowledge. Such platforms also provide guidelines for data quality, ensuring that shared data meets certain standards, and enabling its use for mineral exploration and resource management. The adoption of these initiatives contributes to the creation of global geophysical maps that can guide sustainable mineral exploration while promoting international cooperation in the field (DE Souza and Lima, 2020).

11.13.3 Challenges and solutions for data integration

One of the key challenges in integrating geophysical data from different countries and methodologies is the inconsistency in data formats, measurement techniques, and quality control protocols. Different regions may have distinct approaches to geophysical surveys, which complicates the integration of datasets into a unified global model. Additionally, varying national regulations and data protection laws can hinder the free sharing and access to geophysical data. However, international collaborations and standardized frameworks can address these challenges. For example, by developing common data formats, metadata standards, and methodologies for geophysical surveys, nations can facilitate smoother data integration. Collaborative platforms and joint initiatives, such as the United Nations Framework Classification for Resources (UNFC) or the International Union of Geological Sciences (IUGS), play a pivotal role in fostering international cooperation, which can lead to the harmonization of geophysical data and improved resource management strategies (Barton and Laubach, 2019).

11.14 Policy and governance in geophysical mapping for mineral exploration

11.14.1 National policies and their impact on geophysical mapping

Government policies play a pivotal role in shaping the development and execution of geophysical mapping programs. National governments can influence the scope, focus, and methodologies of geophysical surveys through legislation, funding priorities, and the establishment of regulatory frameworks. These policies help define the legal and operational boundaries for exploration activities, ensuring that they align with national interests, sustainable practices, and environmental protection standards. For example, many countries have established laws that require geophysical surveys to include environmental impact assessments or mandate the use of certain technologies to minimize environmental disturbance during exploration. Furthermore, sustainable resource extraction practices are often embedded in national policies to prevent over-exploitation of mineral resources. In regions with rich mineral deposits, governments may also introduce incentives for using advanced geophysical mapping technologies to ensure efficient and responsible exploration (Pillay and Mota, 2018).

11.14.2 International governance and regulations

International governance frameworks and regulations are crucial for establishing global standards and guiding mineral exploration practices across borders. Numerous international organizations and agreements, such as the United Nations (UN), the World Trade Organization (WTO), and the International Seabed Authority (ISA), have set the stage for mineral exploration and the ethical use of geophysical data. These frameworks are designed to ensure that the extraction of mineral resources is conducted transparently and ethically while respecting environmental and social standards. For instance, the UN's Sustainable Development Goals (SDGs) encourage responsible mining practices that safeguard ecosystems and the rights of indigenous communities. Moreover, international regulations often promote the sharing of geophysical data between nations and regional geophysical organizations, ensuring that these resources are used for the benefit of all, rather than for the exploitation of specific regions or groups. By establishing a global governance structure, these regulations also work to prevent issues like illegal mining, unregulated resource extraction, and environmental degradation (Gusentsova and Kulkova, 2020).

11.14.3 The role of public-private partnerships

Public-private partnerships (PPPs) have become an essential tool in fostering innovation and responsible resource exploration, particularly in the field of geophysical mapping. Governments, private companies, and international organizations often collaborate to share expertise, resources, and funding in the development and implementation of advanced geophysical technologies. These partnerships allow for the pooling of resources and knowledge, enabling more efficient exploration and the development of innovative tools that may not be possible through government or private sector efforts alone. For example, PPPs can facilitate the deployment of cutting-edge geophysical technologies such as remote sensing, satellite imaging, and 3D seismic mapping, which can help reduce the environmental impact of mineral exploration and improve resource estimation accuracy. Additionally, these partnerships often promote transparency and responsible governance by ensuring that mineral exploration is done in accordance with national and international sustainability standards.

The collaboration between government entities, private sector companies, and international organizations can also contribute to the establishment of frameworks that encourage sustainable exploration practices and ensure the ethical use of geophysical data for long-term resource management (Hall and Furlong, 2019).

11.15 Integrating geophysical mapping with environmental and social sustainability

11.15.1 Geophysical mapping for sustainable development goals (SDGs)

Geophysical mapping plays an essential role in supporting the achievement of the United Nations Sustainable Development Goals (SDGs), particularly those related to responsible resource consumption and environmental sustainability. Two key SDGs that are directly relevant to mineral exploration and geophysical mapping are SDG 12 (Responsible Consumption and Production) and SDG 15 (Life on Land). Geophysical surveys, by providing detailed subsurface information, help to ensure that mineral exploration is done with minimal environmental disturbance. For SDG 12, geophysical data support the efficient and responsible extraction of mineral resources, reducing waste and the carbon footprint of exploration activities. For SDG 15, these surveys help identify areas of ecological importance, enabling better planning and reducing the environmental footprint of mining projects. Geophysical mapping also helps identify mineral deposits beneath protected areas or habitats, guiding efforts to balance exploration with the preservation of ecosystems and biodiversity. Integrating these practices into mineral exploration strategies is a step toward sustainable development, promoting resource efficiency and environmental stewardship (Kjellstrom and Yule, 2020).

11.15.2 Environmental impact assessments (EIAs)

Environmental Impact Assessments (EIAs) are critical tools used to evaluate the potential environmental consequences of mining and mineral exploration activities. Geophysical data plays an important role in this process by providing crucial information about subsurface features, such as mineral deposits, groundwater systems, and the geotechnical properties of the land. This information helps environmental consultants assess the potential impact of mining activities on the environment, especially with regard to water resources, soil stability, and biodiversity. Geophysical methods, such as ground-penetrating radar, magnetometry, and seismic surveys, can help identify sensitive environmental areas (e.g., aquifers, wetlands, or forested regions) that need protection during mining activities. By integrating geophysical mapping into the EIA process, mining companies and government regulators can develop strategies to mitigate adverse environmental impacts, such as contamination or habitat destruction, thereby ensuring that mineral exploration and extraction are conducted in an environmentally responsible manner (Veiga and Meech, 2019).

11.15.3 Community engagement and social responsibility

Community engagement and social responsibility are vital components of socially sustainable mining projects. Geophysical mapping helps foster meaningful community engagement by providing local communities with clear, accurate information about potential mining activities in their region. This transparency enables communities to understand the potential social and environmental impacts of mineral exploration, allowing for informed discussions about the project's benefits and risks. Moreover, when communities are involved early in the decision-making process, through consultations based on geophysical data, they can voice concerns and participate in the planning of mitigation strategies. This participatory approach promotes social equity, ensures that the needs of local populations are considered, and helps build trust between mining companies, local governments, and indigenous groups. In addition to contributing to environmental sustainability, such engagement can improve the social license to operate, reduce conflicts, and enhance long-term social outcomes for mining projects, including job creation, infrastructure development, and capacity building (Kemp and Owen, 2018).

11.16 Capacity building and skill development in geophysical mapping

11.16.1 Building capacity in developing nations

International cooperation plays a critical role in strengthening the capacity for geophysical mapping in developing nations. Many of these countries have rich mineral resources but may lack the technical expertise, infrastructure, and financial resources necessary for conducting effective geophysical surveys. Capacity building through international collaboration helps bridge this gap. Programs that focus on **technology transfer**—where developed nations share advanced geophysical technologies and methodologies—are crucial for enhancing exploration efforts in developing countries. Such initiatives may include the transfer of seismic, remote sensing, and geophysical data interpretation tools. In addition to technology, **skill development** is equally important. Training local professionals in the use of these tools, providing knowledge in data collection, analysis, and interpretation, and establishing institutional frameworks for sustainable resource management are key components of capacity-building efforts. Partnerships between governments, international organizations (e.g., UNDP, IAEA), and industry stakeholders can also facilitate these programs, ensuring long-term sustainability by empowering local communities and professionals to independently manage future exploration efforts (Dlamini and Kamutando, 2020).

11.16.2 Educational programs and training

As the demand for geophysical mapping expertise grows, the development of educational programs and training initiatives becomes increasingly important. To address the evolving challenges in resource exploration, it is crucial to equip a new generation of geophysicists, resource managers, and engineers with the necessary technical and analytical skills. Academic institutions play a pivotal role in offering specialized programs in geophysics, geology, and geotechnical engineering, which provide students with the foundational knowledge and practical experience required to engage with modern geophysical methods. Moreover, ongoing **professional development programs** (e.g., workshops, certifications, and online courses) are essential for experienced professionals to stay current with advancements in technology and methodologies. These programs can focus on specific areas of geophysics, such as seismic interpretation, geophysical modeling, and data processing techniques, and can be tailored to different skill levels. Training in interdisciplinary areas like environmental sustainability, resource management, and social responsibility is also increasingly emphasized, ensuring that professionals are prepared to face the multifaceted challenges of modern geophysical mapping (Mugisha and Mungatana, 2021).

11.16.3 Collaborative research and development

Encouraging **joint research and development (R&D)** efforts between countries, universities, and industry stakeholders is vital for solving the complex challenges in geophysical mapping. Geophysical techniques, such as seismic imaging, gravity surveys, and electromagnetic methods, are constantly evolving, requiring collaboration to foster innovation. R&D partnerships can help address the limitations of current geophysical tools, improve data accuracy, and explore new methodologies. For example, collaborative efforts between universities and industry players could lead to the development of more efficient and cost-effective geophysical survey technologies. Furthermore, countries facing similar geological challenges can work together to share insights, best practices, and solutions, enhancing regional cooperation. These partnerships also help pool resources and expertise, which is essential in tackling high-cost or technologically challenging projects. The outcomes of collaborative R&D in geophysical mapping not only advance scientific understanding but also contribute to more efficient and sustainable resource exploration practices globally (Hutchinson and Davis, 2020).

11.17 Conclusions

Geophysical mapping has emerged as a cornerstone in the pursuit of sustainable mineral exploration across the globe. This chapter has provided a holistic overview of the diverse national geophysical mapping programs, highlighting their methodologies, technological advancements, and strategic roles in resource identification and management. From the comprehensive initiatives of countries like India, the USA, and Australia, to the growing efforts in Africa, Asia, and Europe, it is evident that geophysical techniques—ranging from gravity and magnetic surveys to resistivity and seismic imaging—offer unparalleled insights into subsurface geology, aiding in more informed and responsible exploration decisions.

The comparative analysis underscores that while there are variations in policy frameworks, technological capabilities, and resource endowments, common themes such as data integration, environmental sustainability, and the role of public institutions like geological surveys prevail. Particularly, the detailed examples from India's Geological Survey (GSI) reflect how targeted geophysical surveys can contribute significantly to understanding crustal structures, guiding exploration, and reducing environmental footprints.

Global trends point toward a future where international collaboration, data standardization, and advanced technologies such as AI, UAVs, and machine learning will redefine how geophysical mapping supports mineral exploration. Moreover, integrating social and environmental considerations into mapping programs has become critical to aligning with the Sustainable Development Goals (SDGs) and ensuring community trust and participation.

In conclusion, the success of geophysical mapping programs in supporting mineral exploration lies not only in technological prowess but also in strategic governance, institutional collaboration, data transparency, and sustained investment in capacity building. As global demand for critical minerals rises, especially in the context of energy transition and digital economies, the importance of harmonized, innovative, and sustainable geophysical mapping practices will only intensify—demanding continued cooperation among nations, researchers, industries, and communities.

11.18 Key takeaways from global geophysical mapping experiences

Global experiences in geophysical mapping highlight its foundational role in supporting sustainable mineral exploration. Across countries, national geological surveys such as the Geological Survey of India (GSI), the United States Geological Survey (USGS), and Geoscience Australia have demonstrated the critical importance of organized, state-supported mapping initiatives. These programs provide the baseline data necessary for identifying mineral-rich zones, reducing exploration risks, and guiding resource policy decisions (Geoscience Australia, 2020).

A key takeaway is that geophysical methodologies are highly context-specific. Different countries adopt diverse combinations of magnetic, gravity, radiometric, seismic, and resistivity techniques tailored to their unique geological settings and exploration priorities. This diversity underscores the need for flexibility in methodological frameworks while maintaining scientific rigor and consistency. Technological advancements are playing an increasingly transformative role, with innovations such as artificial intelligence, UAV-based surveys, machine learning, and high-resolution sensors significantly enhancing both the efficiency and accuracy of data collection and interpretation (U.S. Geological Survey, 2021).

Another major insight is the value of standardized data and international collaboration. The establishment of shared data platforms and global geoscience partnerships has proven vital in promoting cross-border research and exploration efforts. Public-private partnerships are also emerging as a powerful model for leveraging combined expertise and resources, fostering innovation, and extending the reach of national programs, especially in developing economies.

Sustainability has become a central concern in modern geophysical mapping initiatives. Programs are now more frequently aligned with environmental and social priorities, incorporating Environmental Impact Assessments (EIAs), addressing community concerns, and contributing to broader Sustainable Development Goals (SDGs). These efforts are essential for ensuring that exploration activities are socially responsible and environmentally sound (United Nations Environment Programme [UNEP], 2022).

Capacity building is another critical takeaway from global experiences. Training programs, technical education, and collaborative research initiatives are helping to bridge knowledge gaps, particularly in regions with emerging geoscience infrastructure. Furthermore, case studies from countries like India, Canada, and Australia offer valuable insights into the practical application of geophysical methods, from mapping crustal structures to delineating buried mineralized zones (World Bank, 2023).

Ultimately, the integration of technology, governance, policy, and collaboration will shape the future of geophysical mapping. Success in this domain will depend not only on scientific innovation but also on sustained investment, institutional coordination, and a commitment to environmental stewardship and global knowledge sharing.

11.19 Future directions and recommendations for global collaboration

As the demand for critical minerals and sustainable resource development intensifies, the future of geophysical mapping will increasingly rely on enhanced international collaboration, data harmonization, and shared technological innovation. To meet global exploration challenges and sustainability goals, geoscience institutions, governments, industries, and research organizations must foster deeper partnerships rooted in transparency, mutual benefit, and scientific rigor (International Geophysical Union [IGU], 2021).

One of the most urgent priorities is the establishment of globally accepted standards for geophysical data collection, processing, and interpretation. Standardization not only facilitates data interoperability and cross-border analysis but also enhances the credibility and usability of datasets across disciplines. International geoscience bodies and consortia should take the lead in developing and promoting open-access frameworks that ensure consistency, while also addressing challenges related to data sensitivity and intellectual property rights (UNEP, 2022).

In parallel, the development of centralized, cloud-based geoscience data platforms will be crucial for enabling real-time data sharing and collaborative analysis. These platforms should be designed to support multi-scale, multi-method datasets and allow integration with environmental, geological, and socio-economic data for comprehensive decision-making. Investment in such infrastructure—especially in developing countries—should be prioritized through global funding mechanisms and public-private partnerships (Geoscience Australia, 2020; U.S. Geological Survey, 2021).

Technological transfer and capacity building must also remain at the forefront of international collaboration. Leading nations with advanced geophysical capabilities have a critical role to play in supporting developing countries through training programs, equipment sharing, and collaborative field campaigns. Joint research initiatives and exchange programs can help build local expertise and foster innovation tailored to regional geological contexts.

Furthermore, collaboration should extend beyond technical and scientific realms to include policy alignment and governance. Harmonized regulatory frameworks, agreements on responsible exploration practices, and coordinated environmental safeguards are essential for ensuring that global mineral exploration efforts do not come at the expense of ecological integrity or community well-being (World Bank, 2023).

Lastly, fostering a culture of trust, transparency, and open dialogue among stakeholders is vital for the long-term success of global geophysical mapping initiatives. This includes engaging local communities, respecting indigenous knowledge systems, and prioritizing environmental and social sustainability alongside resource development. In summary, future global collaboration in geophysical mapping will depend on building inclusive, technologically integrated, and ethically grounded networks that support the common goal of sustainable resource discovery and management.

CHAPTER

12

Geophysical applications in mineral exploration—Case studies and insights

12.1 Modern geophysical approaches for concealed mineral exploration

Mineral exploration has evolved significantly with the advancement of geophysical techniques, enabling noninvasive methods to assess subsurface properties and identify potential deposits. Minerals, the fundamental building blocks of rocks, can consist of one or more elements in varying proportions. The Earth's crust is primarily composed of oxygen (O), silicon (Si), aluminum (Al), iron (Fe), calcium (Ca), sodium (Na), potassium (K), and magnesium (Mg). The distribution of economically important minerals across different global regions has strategic implications for resource development (Table 12.1). Some of the major minerals and their key regions of occurrence are as follows:

Geophysical methods are instrumental in mineral exploration by enabling the detection of subsurface anomalies based on physical properties such as gravity, magnetism, electrical conductivity, and seismic wave propagation (Jones and Smith, 2018). The seismic method, which involves transmitting sound waves into the Earth and analyzing their reflections, is particularly effective in identifying geological structures such as faults and folds, especially in hard rock environments. The magnetic method measures variations in the Earth's magnetic field influenced by magnetic minerals like magnetite and is extensively used in the exploration of iron ore, nickel, and copper deposits. The gravity method evaluates subtle changes in the Earth's gravitational field due to density differences and is commonly employed to map large-scale geological structures like sedimentary basins and mountain belts, which may host mineral resources such as coal and oil.

Electrical resistivity and induced polarization (IP) techniques assess the subsurface's resistance or conductivity to electric current, making them valuable in identifying conductive mineral deposits like copper, gold, and sulfide ores. IP surveys provide insights into the presence and depth of mineralized zones. Electromagnetic (EM) methods induce subsurface currents using electromagnetic waves, ideal for detecting conductive minerals, including nickel, copper, and some gold ores. Tools such as gradiometers (for magnetic surveys),

Baseline Geoscience Data Generation for Sustainable Exploration of Natural Resources
https://doi.org/10.1016/B978-0-443-45221-5.00002-3

TABLE 12.1

Important Minerals	Country and Major Center
Aluminum	Australia (Cape York Peninsula, Wipa region), the United States (Arkansas), Jamaica, the Former Soviet Union (Kola Peninsula), Guinea, South Africa (North Natal Province)
Manganese	Nikopol (Ukraine) and Chiatura (Georgia) in the Black Sea region of the Soviet Union, Brazil (Ouro Preto region), Gabon, South Africa (Postmasburg Region, Saras), China (Hunan, Jiangxi, and Kwangtung)
Copper	Chuquicamata (Chile), USA (Arizona province, Montana province), Canada (Sudbury city, Ontario), Australia (Mount Lyell and Mount Isa), Zaire (Katanga), South Africa (Nabadwip, Messina)
Iron ore	Ukraine (Kryvyi Rog), Russia (Magnitogorsk Mountains and Kuznets), Brazil (Itapira Hills in Minas Gerais), the United States (Lake Superior, Mesabi Range), Alabama Province, Australia (Pilbara Region), Sweden (Kiruna and Galbra region), South Africa (Postmasburg region, Transvaal) The Carajas Mine of Brazil is the largest iron ore mine in the world.
Lead	Australia (Broken Hill, Mount Isa (Queensland), Canada (Sudbury), Peru (Cerro-de-Pasco)
Silver	Mexico (Chihuahua, Hindaho), Canada (Ontario, British Columbia, Quebec), the United States (Utah, Montana, Arizona, Colorado), Australia (Mount Isa, Kalgurli, Broken Hill), Bolivia (Potosi), South Africa (Transvaal and Natal Provinces)
Gold	South Africa (Johannesburg, Boxburgh and Orange Free State, Kimberley in Witwatersrand Hill.
Tin	Malaysia (Kinta Valley, Selangor, Yenning island, Perak Valley), Indonesia (Brangadweep, Brittany), China (Yunnan, Hunan province), Myanmar (Shan Plateau, Cayenne Plateau)
Zinc	Australia (Broken Hill and Mount Isa), Canada (British Columbia)
Coal	The United States (Appalachian coal field), the Former Soviet Union (Donetsk Basin [Ukraine], Kuznetsk Basin, and Karaganda), China (Shanxi, Shanxi Szechwan), Australia (New South Wales, Queensland, and Victoria)
Petroleum	Saudi Arabia (Amman, Dhawar, and Qatif [Ras Tanura Refinery]), the United States (Appalachian Territory, Gulf Coast, California Region), Kuwait (Burhan Hills), Iran (Lali, Karamshah, Nafta Safid, Haft Kel Gachsaran), Iran (Kirkuk, Mosul, Basra, Tikrit)
Diamond	South Africa (Kimberley, Johannesburg, Cape Town), Zaire (Katanga Plateau), India (Panna and Golconda Mines)
Atomic Minerals	Canada (Uranium City and Port Radium near Lake Athabasca and Great Bear Lake), the United States (Plateau of Colorado), South Africa (Witwatersrand Hills), Zaire (Katanga Plateau)
Hydro Electricity	The United States (Niagara Falls, St. Anthony Falls, Mount Appalachian, and Alabama), China (Lubok Project, Hurangi River), India (Jog Falls and Tehri Dam), Africa (Akosombo Dam on Volta River (Ghana), Senar Dam (Sudan), Aswan Dam on Nile river (Egypt), Kariba River (Zambezi river), South America (Itaipu Project (Parana River))

seismographs (for seismic analysis), and advanced software for data interpretation enhance the precision of geophysical surveys. The integration of multiple geophysical methods and tools significantly reduces exploration costs and improves the probability of discovering viable mineral resources (Gonzalez et al., 2020).

Recent advances in geophysical data acquisition and interpretation have increased the precision and efficiency of exploration. Remote sensing and satellite-based technologies now offer high-resolution surface mapping, aiding in identifying mineral anomalies in remote regions. Unmanned aerial vehicles (UAVs) equipped with geophysical instruments enable real-time, high-accuracy data collection in inaccessible terrains. 3D geophysical imaging combines multiple datasets—seismic, gravity, and magnetic—into comprehensive subsurface models, allowing more accurate visualization of mineral deposits and geological formations. The use of Artificial Intelligence (AI) and Machine Learning (ML) in processing vast geophysical datasets has improved anomaly detection and pattern recognition, allowing for faster, more informed decision-making. Furthermore, automated acquisition systems and real-time data transmission enhance survey efficiency and reduce operational delays.

The integration of geophysical methods with geological, geochemical, and remote sensing techniques has proven essential in achieving a holistic understanding of subsurface conditions. Geological mapping provides surface context to geophysical anomalies, while geochemical surveys confirm the presence of trace elements related to mineralization. Remote sensing identifies surface expressions of mineralization, such as alterations in vegetation or soil chemistry, which, when cross-referenced with geophysical data, enhance targeting accuracy. Geotechnical studies, including drilling and core sampling, complement geophysical surveys by providing physical property data and supporting resource estimation. These integrated approaches help reduce exploration risks, improve accuracy, and lower costs (Schreiber, 2021).

Environmental considerations are increasingly central to geophysical exploration practices. Surface disturbance from ground surveys can disrupt sensitive habitats, particularly in forests and wetlands. Seismic surveys may generate noise and vibrations that impact wildlife. The use of chemical fluids in electrical methods can lead to contamination if improperly handled. Drilling-based techniques may affect groundwater quality or flow. To mitigate these effects, noninvasive techniques such as UAV surveys and satellite imaging are preferred. Environmental Impact Assessments (EIAs) are essential before exploration begins to identify and manage potential risks. Best Management Practices (BMPs), sustainable waste disposal, and the use of biodegradable chemicals help reduce environmental footprints. Collaboration with local communities and environmental agencies ensures that exploration aligns with conservation goals and promotes responsible resource extraction (Zhou and Li, 2021).

Despite these advances, challenges remain in geophysical mineral exploration. Complex and heterogeneous geological environments complicate data interpretation, and existing geophysical methods have limitations in detecting deep-seated mineralization. Large-scale surveys are still time-consuming and expensive, particularly for smaller firms. Interpretation of geophysical data remains a key challenge due to the indirect nature of measurements and the uncertainty in correlating anomalies to mineral deposits. However, the future holds promise with the continued integration of AI and machine learning, advanced instrumentation, and multitechnique data fusion approaches. Novel technologies like quantum sensors and nanoscale detectors may improve depth penetration and resolution. Sustainability will remain a priority, with environmentally friendly tools and methods becoming central to exploration workflows (Chamberlain and White, 2020).

In summary, the application of geophysics in mineral exploration has transformed the industry through enhanced data acquisition, processing, and interpretation. Case studies and practical insights demonstrate that by integrating geophysical methods with complementary techniques and prioritizing environmental responsibility, exploration activities can become more efficient, accurate, and sustainable. These developments are critical in meeting global demands for mineral resources while preserving ecological integrity for future generations.

Over 5000 years ago, human societies gradually transitioned from the Stone Age to the Metal Age and began to develop and utilize metallic mineral resources (gold, copper, and iron). Particularly, the inhabitants of the Aegean region (one of the seven geographical regions of Turkey) and China discovered and started to use copper and bronze in 3200 BC (Butt and Hough, 2009; Gusentsova and Kulkova, 2020). Large-scale development and the use of various metallic mineral resources ensued. Especially, outcropping ores were primarily developed and mined, and they remained dominant metallic mineral sources (Cheng et al., 2017). However, with continuous economic and social development, particularly since the beginning of the third industrial revolution, there has been an increasing demand for metallic mineral resources. This situation is met with the dilemma that all the outcropping ores have basically been exhausted. Against this background, theoretical prediction methods and prospecting techniques for concealed ore deposits and ore bodies have become a focal but challenging topic of prospecting research. This research is focused primarily on the application of various deep penetrating survey techniques, including fundamental geological, geochemical, and geophysical techniques, to the exploration of concealed ores (Hosseini-Dinani and Aftabi, 2016; Meng et al., 2019; Kang et al., 2020). In particular, geophysical survey techniques have played a tremendous role. Exploration geophysics plays a major role in unlocking mineral reserves. It is well recognized that many easily discovered large mineral deposits with a strong geophysical signature have already been identified. Future discoveries present significant challenges, being located undercover, in remote areas, and with less prominent geophysical signals.

On the basis of the research results and metallogenic theories, numerous concealed metallic ore deposits (Anand and Rajaram, 2006; Wang et al., 2016; Liu et al., 2019) have been discovered, such as the world class pebble porphyry deposit in Alaska (the United States), the Olympic Dam copper-gold-uranium-rare-earth element (Cu–Au–U-REE) deposit and the Ernest Henry copper-gold deposit in Australia, the Nihe and Luohe iron deposits in Anhui (China) (Wu et al., 2011; Shah et al., 2013; Cave et al., 2018), and lead-zinc in Udaipur (Rajasthan), chromite in Odisha, copper in Malanjkhand (Madhya Pradesh), bauxite in East Coast, gold deposits in Hutti and Bharat gold mines in Karnataka (FIMI, 2018). All the above research results concern the detection and discovery of concealed metallic ore deposits of a certain scale in large areas using geophysical methods. By contrast, the application of geophysical survey techniques to the class of concealed ore bodies in the existing metallic ore deposits has been rarely investigated. In fact, as a result of many years of continuous mining, many metal mines are becoming exhausted as their proven resources are depleting, whereas the concealed resources in the deep and surrounding areas of these mines have yet to be effectively detected and evaluated. This phenomenon is most common for iron, the metal with the longest mining history, particularly banded iron formation (BIF) type iron deposits (Fu et al., 2020). BIF-type iron mines, characterized by large scale, as well as easy mining and beneficiation, are the most important iron ore resources in the world. High-grade iron ores in BIF-type iron mines constitute 70% of the world's high-grade iron ore reserves, and BIF-type iron mines produce over 90% of the world's iron ores (Zhang et al., 2012). BIF-type iron mines are found mainly in Russia, Australia, Brazil, Canada, China, Africa, India, and the United States. In China, BIF-type iron mines are mostly in the North China Craton and are concentrated largely in the Anshan-Benxi area, eastern Hebei, the Huoqiu-Wuyang area, Wutai, western Shandong, and Guyang.

The detection and evaluation of concealed mineral resources deep in metallic mines and in the surrounding areas remain technically difficult. In particular, due to the complex topographic and geomorphic conditions on the surface, the detection environments in these areas limit the choices of detection equipment and data collection devices. The modern-day challenges of exploration require developing novel geophysical techniques, which improve exploration success and lead to new discoveries. This state-of-the-art overview of geophysics for minerals focuses on emerging geophysical techniques for mineral exploration, novel modeling and interpretation methods, including joint inversions of multiphysics data, and challenging case studies. It covers a wide range of hidden mineral deposits, including banded iron formations, epithermal gold-silver-copper-iron-molybdenum deposits, iron oxide-copper-gold deposits, and prospecting for groundwater resources. In summary, the chapter on geophysics for mineral exploration provides a snapshot of advanced modeling and inversion methods of geophysical data, and their applications to studying different types of mineral deposits.

12.2 Global mineral exploration: Trends and policy perspectives

Mineral exploration is a critical aspect of securing sustainable access to natural resources, essential for industrial development and economic growth. Over recent years, the global mineral exploration landscape has undergone significant shifts, driven by new technologies, changing demand for resources, and the increasing need for sustainable practices (UNEP, 2020).

12.2.1 Prospects of mineral exploration

The prospects for mineral exploration globally are influenced by several factors, including technological advancements in exploration methods and a growing demand for resources, particularly in emerging markets. Innovations such as remote sensing, geophysical techniques, and AI-powered analysis have made it possible to explore more challenging terrains, such as deep-sea and polar regions, where mineral deposits were once thought to be inaccessible (Huntington, 2019). Moreover, the shift toward electric vehicles and renewable energy technologies has created a surge in demand for specific metals like lithium, cobalt, and rare earth elements, thereby opening new opportunities for exploration in regions previously underexplored (World Bank, 2021).

12.2.2 Promises of sustainable exploration

With increasing awareness about environmental degradation and the social implications of mining, there is a stronger push for sustainable exploration. Promises of sustainable mining practices are being fulfilled through the adoption of cleaner technologies and more efficient use of resources (International Council on Mining and Metals, 2020). The integration of environmental, social, and governance (ESG) criteria into exploration projects is now a fundamental aspect of modern exploration (PwC, 2021). This approach ensures that the exploration and extraction of minerals do not harm ecosystems, respect local communities, and minimize carbon footprints (Global Reporting Initiative, 2020).

12.2.3 Policy reforms in mineral exploration

Policy reforms play a crucial role in shaping the future of mineral exploration. Governments across the world are recognizing the need to implement policies that balance economic interests with environmental protection and social responsibility. Key reforms include simplifying licensing procedures, promoting transparency in governance, and enhancing regulatory frameworks to ensure responsible mining practices (OECD, 2020). In addition, there is growing support for international collaborations and cross-border projects that can address resource shortages in specific regions. Policymakers are also focusing on strengthening legal frameworks around exploration, ensuring that companies adhere to global best practices for sustainability, while providing incentives for new technologies in mineral extraction (World Bank, 2020).

12.2.4 Baseline geoscience data generation for sustainable exploration

One of the most critical components for sustainable exploration is the generation of baseline geoscience data. Accurate, comprehensive, and up-to-date geoscientific data are fundamental for informed decision-making in mineral exploration (Geological Survey of India, 2021). Baseline data provide the foundation for identifying resource-rich areas, assessing the environmental impacts of mining activities, and ensuring that exploration efforts align with sustainability goals. Governments and industry players are increasingly investing in the generation of high-quality geoscience data, often in collaboration with academic institutions and research organizations (United Nations, 2019). These efforts ensure that exploration is conducted with the most accurate information available, minimizing unnecessary environmental damage and optimizing resource extraction. In conclusion, the future of mineral exploration is promising, with technological advances, a commitment to sustainability, and policy reforms guiding the way forward. Through the generation of robust geoscience data, the mining industry can move toward more responsible and efficient practices, ensuring that natural resources are accessed in a way that benefits both the economy and the environment.

12.3 Global spatial patterns of critical mineral resources: A comprehensive analysis

The distribution of mineral resources across the globe is highly uneven, influenced by geological processes, tectonic movements, climate, and time. These resources are critical for industrial development, energy production, and technological advancements. Understanding the spatial pattern of mineral deposits is vital for both exploration and sustainable management. This section discusses the global distribution of some of the most important minerals, highlighting key regions where significant reserves are located and identifying the factors driving their spatial distribution.

12.3.1 Coal

Coal is one of the most widely distributed minerals in the world and plays a key role in energy production. The largest reserves of coal are found in countries like the United States, Russia, China, and India. The geological formation of coal deposits, which typically occur in sedimentary basins, is influenced by ancient vegetation and the climatic conditions that existed during the Carboniferous period. In particular, the mid-latitude regions of the Northern Hemisphere, such as the Appalachian Basin in the United States and the Kuznetsk Basin in Russia, are known for their extensive coal deposits (Cao and Zhang, 2019).

12.3.2 Iron ore

Iron ore, a crucial raw material for steel production, is largely found in countries like Australia, Brazil, China, and India. Australia holds some of the largest and most economically viable deposits of iron ore, particularly in the Pilbara region of Western Australia. Brazil, with its Carajas Mine, also plays a significant role in global iron ore production. Iron ore is typically found in Precambrian rocks, where the presence of iron-rich minerals such as hematite and magnetite is concentrated through natural processes of weathering and erosion (Schreiber, 2021).

12.3.3 Copper

Copper, essential for electrical wiring and other industrial applications, is predominantly found in Chile, the United States, Peru, and China. The Andes mountain range in South America, particularly in Chile, contains some of the world's largest copper reserves, as well as other South American countries like Peru. Copper deposits form primarily in porphyry copper systems, which are associated with tectonic movements and volcanic activity. According to Gonzalez et al. (2020), the mineralization of copper in these regions is closely tied to the geological processes involved in the subduction of oceanic plates beneath continental plates.

12.3.4 Bauxite

Bauxite, the primary ore of aluminum, is distributed across tropical and subtropical regions, with significant deposits in countries such as Australia, Guinea, Brazil, and Jamaica. Australia leads the world in bauxite production, with its reserves concentrated in the northern regions of the country. The formation of bauxite deposits occurs through the weathering of aluminum-rich rocks in hot and humid climates, leading to the leaching of other elements and the accumulation of aluminum hydroxide minerals (Jones and Smith, 2018).

12.3.5 Gold

Gold is widely distributed but often in smaller quantities compared to other minerals. Major gold-producing regions include South Africa, China, Australia, and Russia. South Africa, historically the largest producer, is home to the Witwatersrand Basin, where the world's most extensive gold reserves are located (Chamberlain and White, 2020). Gold is typically found in quartz veins within metamorphic and igneous rocks, often concentrated through geochemical and physical processes over millions of years.

12.3.6 Rare earth elements

Rare earth elements (REEs), critical for modern electronics and green technologies, have a very concentrated distribution. China dominates global production, with significant deposits also found in the United States, Australia, and Russia. The spatial pattern of REEs is closely linked to unique geological processes, such as the formation of carbonatite rocks, which contain high concentrations of these elements. China's dominance is largely due to the massive deposits in the Bayan Obo Mine, which is a primary source of many REEs (Zhou and Li, 2021).

12.3.7 Oil and natural gas

Oil and natural gas reserves are found primarily in sedimentary basins, particularly in regions like the Middle East, Russia, the United States, and Venezuela. The Middle East, with countries like Saudi Arabia, Iraq, and Kuwait, holds the majority of the world's oil reserves, formed through millions of years of organic matter accumulation under specific geological conditions (Khan and Ali, 2020). Similarly, major natural gas reserves are located in Russia, the United States, and Iran, often in conjunction with oil deposits in areas such as the North Sea and the Gulf of Mexico.

The spatial distribution of minerals is influenced by a range of geological factors, from tectonic activity to climatic conditions and the age of the Earth's crust. As the demand for natural resources grows, it is essential to understand the geographical concentration of these resources to guide exploration efforts, optimize extraction methods, and develop policies for sustainable management. Future exploration will increasingly focus on regions with untapped potential, such as the deep sea, polar areas, and underexplored terrestrial zones. By studying these patterns and utilizing advanced exploration technologies, nations can better manage their mineral wealth, mitigate environmental impacts, and ensure equitable distribution for future generations.

12.4 Mineral exploration scenario in India: Prospect, promise, and policy reforms

India has a landmass of 3.3 million square km measuring 3214 km from north to south, 2933 km from east to west, with a multitude of geological constructs and tectonic domains that have evolved over a long geological time scale. Varied potential blocks exist that are rich for mineral exploration, mining, and exploitation of metals for the industries, fuel, energy minerals, and other bulk commodities. India's 7517 km of coastline is also home to rich mineral wealth (Kumar and Pandey, 2022). For centuries, India has mined and produced copper, iron, zinc, gold, silver, diamond, and gemstones, which are well known from our historical architecture. The bulk of the valuable minerals is products of pre-Palaeozoic age and is mainly associated with metamorphic and igneous rocks of peninsular India. The vast alluvial plain tract of north India is devoid of minerals of economic use. The mineral resources provide the country with the necessary base for industrial development. On the basis of chemical and physical properties, minerals may be grouped under two main categories: metallic and nonmetallic, which may further be classified as follows (Figs. 12.1–12.6 and Table 12.2).

A look at the mineral resource status of selected minerals in India indicates abundant and/or adequate resources of iron, chromite, manganese, bauxite, coal, limestone, dolomite, mica, zinc, graphite but deficient to scarce resources of nickel, cobalt, molybdenum, copper, sulfur, potash, apatite, rock phosphate, gold, platinum group elements (PGE)'s, diamond etc., in spite of the vast obvious geological potential (OGP) blocks that exist in India. The OGP blocks have been estimated to be around 100,000 km^2 for gold, 300,000 km^2 for diamond, 160,000 km^2 for base metal, 8000 km^2 for PGEs, and 5000 km^2 for iron ores. By a rough estimate, only 10% of this is explored, and less than 2% is mined. The reserve-to-resource ratio, which, as per mineral exploration practice, should be over 50%, is woefully low for most of the minerals. For instance, it is 17% for bauxite, 14% for copper, 24% for iron, 14% for lead-zinc, 8% for limestone, and 30% for chromite. This calls for a highly

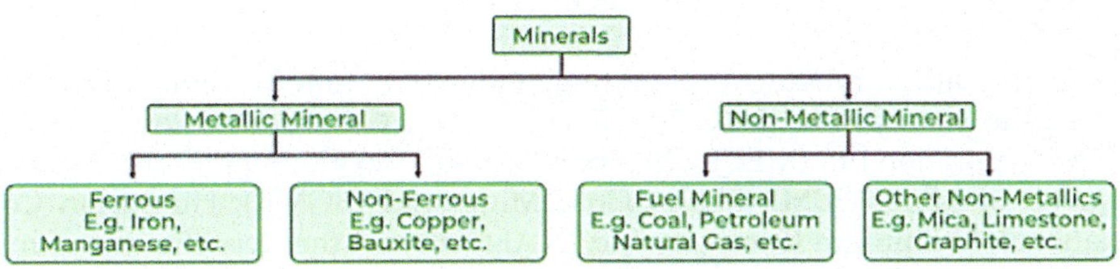

FIG. 12.1 Mineral distribution in India. *Source: Mineral and Energy Resources—Mineral Distribution in India (last updated: January 4, 2024).*

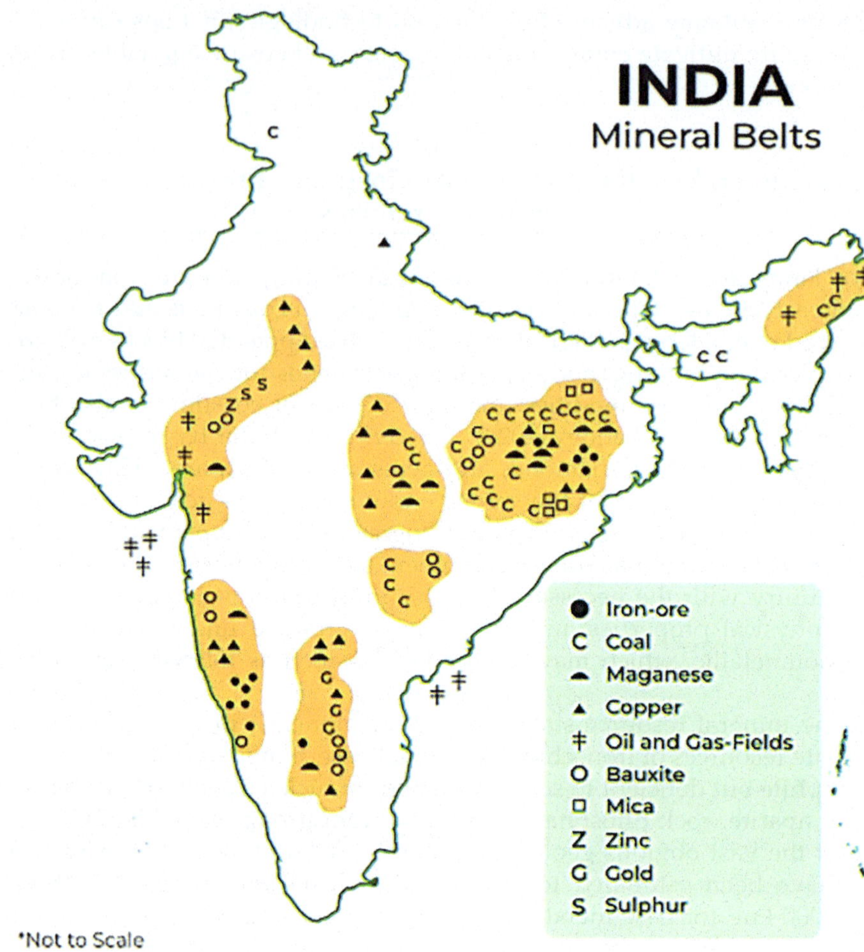

FIG. 12.2 Map of mineral distribution in India. *Source: Mineral and Energy Resources—Mineral Distribution in India (last updated: January 4, 2024).*

focused exploration strategy at a high confidence level (Ministry of Mines 2022, Ministry of Mines Government of India, 2019, document of Ministry of Mines, Government of India; GSI, 2020).

In India, systematic surveying, prospecting, and exploration for minerals are undertaken by the Geological Survey of India (GSI), Oil and Natural Gas Commission (ONGC), Mineral Exploration Corporation Ltd (MECL), National Mineral Development Corporation (NMDC), Indian Bureau of Mines (IBM), Bharat Gold Mines Ltd (BGML), Hindustan Copper Ltd (HCL), National Aluminium Company Ltd (NALCO), and the departments of mining and geology in various states.

India was first to smelt zinc and make high-tech carbon steel. Modern India needs to acquire the same high status in mineral exploration by employing both modern mineral exploration methods and modern technology, exploring the covered mineral wealth and

FIG. 12.3 The distribution of metallic minerals in India. *Source: Mineral and Energy Resources—Mineral Distribution in India (last updated: January 4, 2024).*

suitable policy reforms. As part of the Gondwana land, sitting between the two continents of Australia and Africa, India has unique geological features akin to them and hence a great potential for natural mineral wealth. Western Australia, Canada, and Nevada have such a successful history of mining that India cannot afford to lose sight of it. The Wiluna gold camp in Western Australia has a similar geological environment to that of the Hutti-Maski gold schist belt of Southern India. Odisha, Chattisgarh, and Jharkhand have mineralization

FIG. 12.4 The distribution of nonmetallic minerals in India. *Source: Mineral and Energy Resources—Mineral Distribution in India (last updated: January 4, 2024).*

similar to the iron ore region of Western Australia's Hammerseley. Likewise, bauxite in the Weipa district on the Northern Cape York Peninsula has huge resources, and India has the same potential. China is the world's largest producer of aluminum and steel, though all its production is based on imported raw materials from Australia and Brazil. In a visionary sense, India can also dominate the International steel and aluminum market if we choose to adopt a strategy of undertaking massive exploration to identify and establish all the high-grade iron ores and bauxite resources to sustain annual mining rates of iron and

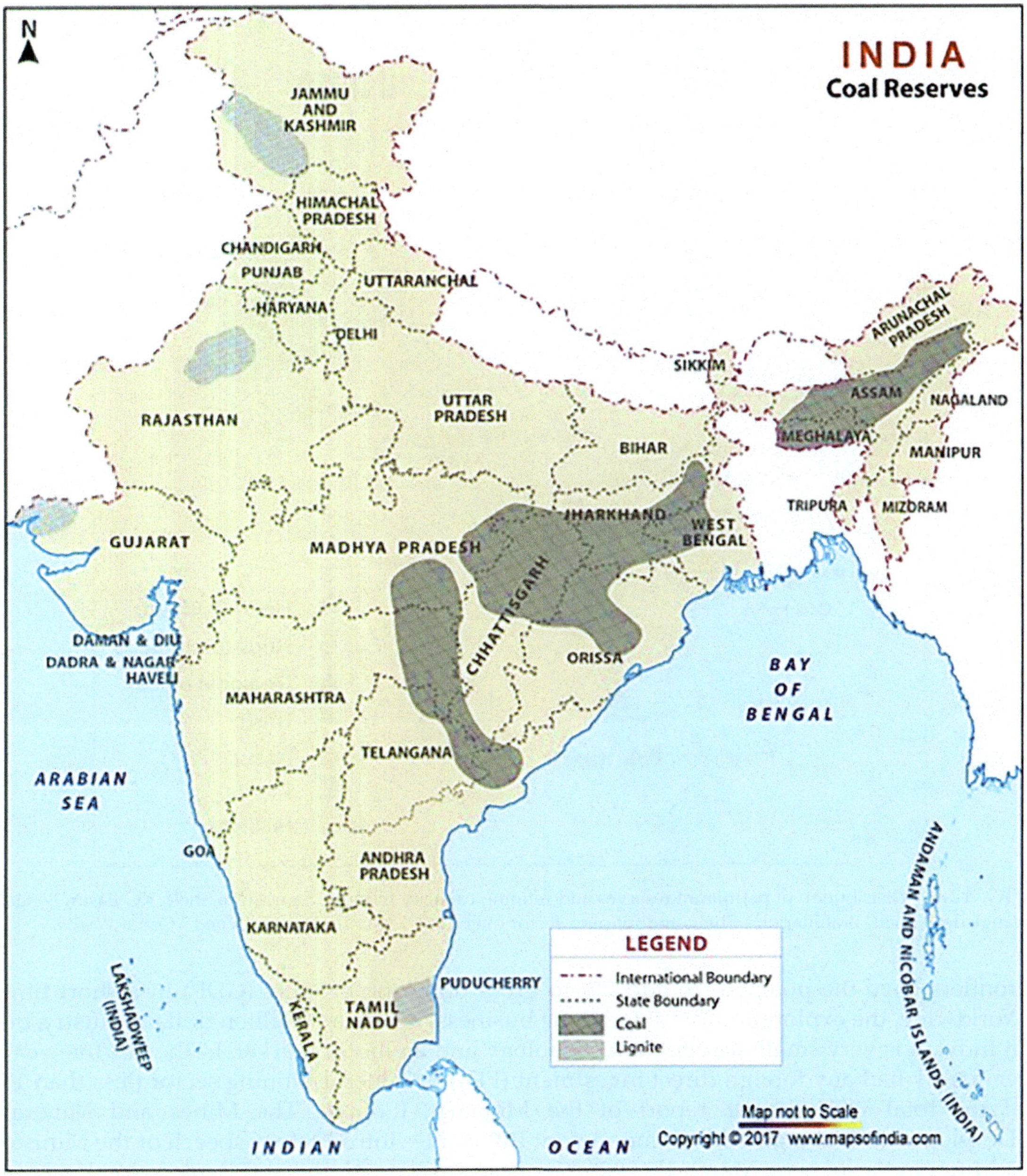

FIG. 12.5 The distribution of coal reserves in India: Rift valley regions like Damodar, Mahanadi, and Godavari. *Source: Indian Mines and Mineral Wealth.*

bauxite. In the last seven decades, India has mostly given emphasis on exploiting bulk commodities only, and very few of the greenfield blocks have been developed, for example, gold schist belts, Dharwar craton, Aravalli, Vindhyans, Singhbhum shear zone, and Himalayas. India has opportunities to add more resources and reserves, as large unexplored

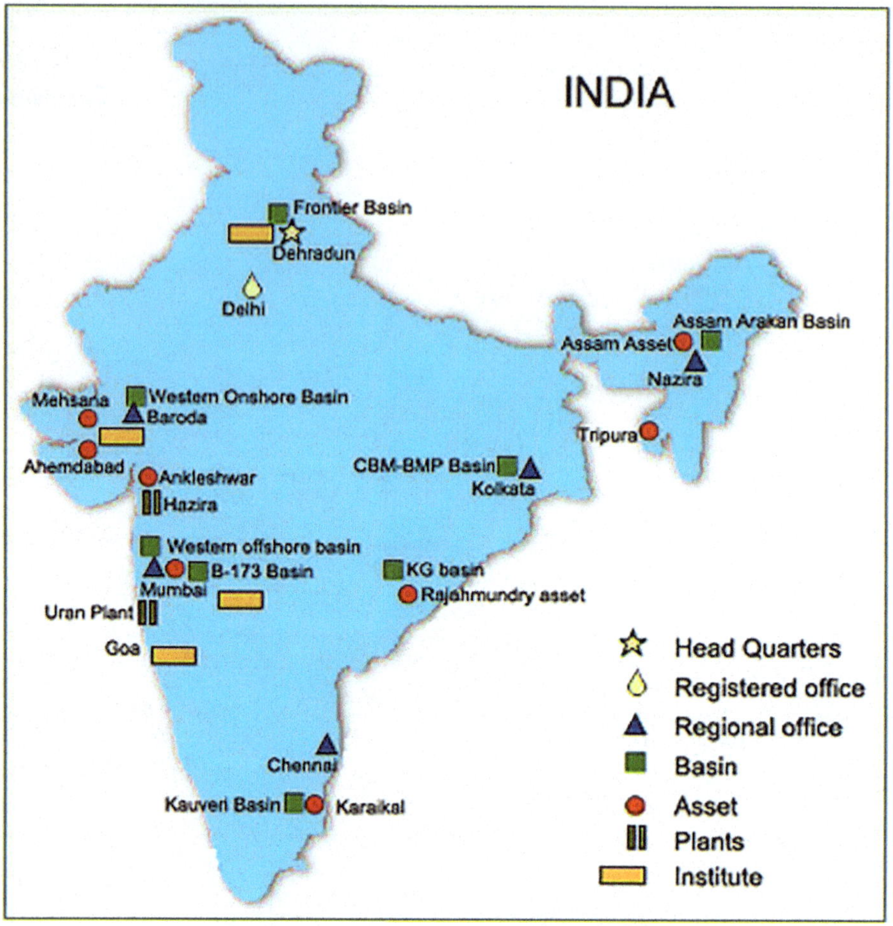

FIG. 12.6 Distribution of petroleum reserves and refining capacity in India: Saurashtra Shelf, KG Basin, Bengal-Bangladesh Shelf, Brahmaputra Shelf, and refining sector overview. *Source: Indian Mines and Mineral Wealth.*

frontiers have the potential to add 2% to gross domestic product (GDP) in a short time. Worldwide, the exploration is a standalone business, nearly a 30 billion dollar industry, but in India, it is very small despite perfect geology and a vibrant market. In the last five years we hardly had any foreign direct investment (FDI) in mineral/mining sector (less than 1% of the total FDI—as per report of the Ministry) [Source: The Mines and Minerals (Development and regulation) Amendment Bill, 2021—Introductory speech of the Minister of Mines in Parliament, March 15, 2021; Geonesis, Indian mining and Exploration Updates, v.7(10) 2020 and v.9(2), 2022. (www.https://geonesis.in); Mineral Exploration Academy-mineral policy discussion. Dr Amit Tripathi, September 2022 (www.https://minexpacad-emy.com).

Minerals are generally concentrated in three broad belts in India. There may be some sporadic occurrences here and there in isolated pockets. These belts include the following:

TABLE 12.2 Classification of minerals.

Minerals			
Metallic mineral		Nonmetallic mineral	
Ferrous (iron ore, manganese, etc.)	Nonferrous (copper, bauxite, etc.)	Fuel minerals (coal, petroleum, natural gas, etc.)	Other nonmetallic minerals (mica, limestone, graphite, etc.)

(1) The north-eastern plateau region: This belt covers Chhotanagpur (Jharkhand), Odisha plateau, West Bengal and parts of Chhattisgarh. Most of India's major iron and steel industries are located in this region. It has a variety of minerals, namely iron ore, coal, manganese, bauxite, and mica; (2) the south-western plateau region: This belt extends over Karnataka, Goa, and contiguous Tamil Nadu uplands and Kerala. This belt is rich in ferrous metals and bauxite. It also contains high-grade iron ore, manganese, and limestone. This belt packs in coal deposits except for Neyveli lignite. This belt does not have as diversified mineral deposits as the north-eastern belt. Kerala has deposits of monazite and thorium, and bauxite clay. Goa has iron ore deposits; (3) the north-western region: This belt extends along the Aravalli in Rajasthan, and part of Gujarat, and minerals are associated with the Dharwar system of rocks. Copper and zinc have been major minerals. Rajasthan is rich in building stones i.e., sandstone, granite, and marble. Gypsum and fuller's earth deposits are also extensive. Dolomite and limestone provide raw materials for the cement industry. Gujarat is known for its petroleum deposits. You may know that Gujarat and Rajasthan both have rich sources of salt; (4) the Himalayan belt region: The Himalayan belt is another mineral belt where copper, lead, zinc, cobalt and tungsten are known to occur. They occur on both the eastern and western parts. Assam Valley has mineral oil deposits. Besides, oil resources are also found in offshore areas near the Mumbai coast (Mumbai High).

India needs to take a leaf out of the strategies of many other countries and frame its own policy based on OGP areas to trigger the mining growth by accelerating the mineral exploration programme. Benefits that flow from the mineral exploration and mining sector have a huge impact on the local populace, industry, especially mineral-dependent industries, employment, and overall prosperity. Recent amendments in the Mines and Minerals (Development and Regulations) Act, 1957, in the year 2021, have sought to bring out the potential of the mineral sector such as to increase the revenue of the State, escalate employment, maintain the continuity of mining operation, increase production, time bound operationalization of mines, increase the speed of exploration and auction of mineral resources and lastly to resolve the long pending issues that have decelerated the growth of this sector. This is an encouraging step and likely to bear fruit. Setting up the district mineral foundation (DMF) will work for the interest and benefit of mining-related operations and overcome the local resistance to the mineral exploration and mining activities. The National Mineral Exploration Trust (NMET) will focus on extending more financial assistance to the government and also the private sectors to augment advanced technology and will surely give a fillip to exploration initiatives [Source: The Mines and Minerals (Development and regulation) Amendment Bill, 2021—Introductory speech of the Minister of Mines in Parliament, March 15, 2021].

12.5 Case studies in geophysical mineral exploration

Case studies in geophysical mineral exploration provide valuable insights into the practical application of various geophysical techniques in real-world scenarios. These studies not only highlight the effectiveness of these methods in detecting mineral deposits but also reveal the challenges and limitations faced during exploration.

One notable example is the exploration of copper deposits in the Andes, where seismic and gravity surveys were combined to map subsurface structures and identify potential ore bodies beneath thick volcanic rocks (Chamberlain and White, 2020). The integration of seismic data with gravity anomalies helped to pinpoint areas with high potential for copper mineralization, leading to the successful discovery of several new deposits in this mineral-rich region.

Another significant case is the exploration for gold in Western Australia, where induced polarization (IP) and magnetic surveys were utilized to map the distribution of disseminated gold ore (Jones and Smith, 2018). The IP method was particularly valuable in detecting low-grade gold deposits, which were previously undetectable with traditional geological methods. Magnetic surveys helped confirm the presence of gold mineralization by identifying the underlying geological structures associated with gold-bearing quartz veins.

In Africa's Democratic Republic of the Congo, a comprehensive electromagnetic (EM) survey was conducted to locate copper-cobalt deposits in the Katanga Basin (Gonzalez et al., 2020). The EM method was successful in detecting conductive zones that were associated with copper and cobalt mineralization. The results of this survey allowed for the identification of drilling targets, which eventually led to the discovery of significant ore bodies in the region.

The North American exploration for nickel in the Sudbury Basin also showcases the successful application of gravity and magnetic surveys (Schreiber, 2021). These surveys helped delineate large-scale geological structures like faults and intrusion zones, where nickel deposits are typically found. The results of these surveys provided geophysicists with a detailed understanding of subsurface geology, leading to the discovery of new nickel-rich areas for mining. These case studies demonstrate how the combination of different geophysical techniques, along with advanced data interpretation, can significantly improve the success rate of mineral exploration. They also highlight the role of geophysical surveys in identifying mineral deposits in challenging terrains, reducing exploration costs, and increasing the efficiency of mineral discovery.

12.6 Case studies in geophysical mineral exploration of India

12.6.1 Auriferous-sulfide mineralization, Gani-Kalva fault zone, Cuddapah basin, India—case study

The Gani-Kalva fault (GKF), located in the western part of the Cuddapah basin, India, is well known for hosting mineral deposits of iron, copper, and gold. In the present study, Vijayakumar et al. (2023) carried out a comprehensive exploration approach, combining semi-detailed gravity and remote sensing surveys followed by detailed geophysical surveys comprising self-potential (SP), electrical resistivity, and induced polarization (IP) to map the potential gold mineralization zones within the GKF. The integration of the linear pattern of gravity anomaly and its derivative with satellite imagery and airborne electromagnetic anomalies suggests that the Gani block between extensional bends and fault splays might be favorable for subsurface mineralization. The inverted resistivity and induced polarization depth sections across these anomalous zones highlighted the presence of the low resistivity and moderate chargeability bodies; typically, at a depth range of 30 to 50 m. Subsequent borehole drilling was conducted at these promising locations, confirming the presence of auriferous sulfide mineralization within quartz veins and veinlets. However, elemental analysis of core samples from this borehole reveals only 2.5 ppm of gold. Although these results may not be very encouraging in the context of current exploration standards, the study clearly emphasized the pivotal role of geophysical techniques in the optimization of the survey area for targeting high mineralization potential zones.

12.6.2 Exploring heavy mineral deposits, Chavara and Manavalakurichi deposits, Southwest coast of India—case study

The coastal tracts of India consist of beach sands showing good concentration of strategic heavy minerals (Rejith et al., 2021). In the present study, the mineralogy of important coastal placer deposits in India, like Chavara deposits in Kerala and Manavalakurichi deposits in Tamil Nadu were studied using Landsat multispectral satellite data for deriving potential targets of these strategic minerals. The ilmenite mineral predominantly exists in these areas, followed by monazite, sillimanite, rutile, zircon, garnet, leucoxene, and kyanite. The distribution of minerals was studied using advanced hyperspectral techniques like minimum noise fraction transformation, pixel purity index, and n-dimensional visualizer, followed by spectral angle mapper (SAM) classification. Two end members of ilmenite mineral and quartz were identified from the Landsat imagery and successfully mapped using SAM algorithms. Detailed investigation of the mineralogy of beach sediments using advanced hyperspectral remote sensing techniques helps to derive potential targets for the eco-friendly exploration of minerals.

12.6.3 Basemetal mineralization in the extension area of Mamandur Polymetals prospect in Villupuram and Thiruvannamalai districts of Tamil Nadu, India—case study

A combination of magnetic, IP, and resistivity methods was attempted in the Mamandur polymetals prospect area of Villupuram and Thiruvannamalai districts of Tamil Nadu, for locating basemetal occurrence (Geological Survey of India, Naskar et al., 2021). The integrated interpretation of the geophysical response of the different geophysical methods revealed more precise information about the subsurface configuration and mineral distribution in the prospect area. From the nature of magnetic anomalies and their analytic signal signatures, like change in gradient, elongation, and behavior of contours, a few linear trends have been inferred in the Mettukuppam and Vadamamandur-Porasapattu mineral block. Strong geophysical anomalies were observed in the western and eastern parts of the study area. The low resistivity zones (<70 Ohm-m) and high chargeability zones (13 mV/V) are situated in the Mettukuppam mineralized zone. However, these low resistivity zones may have been influenced by the presence of clay or weathered soil. In this case, the high chargeability zones will help in confirming the prospective zones, caused by base metal mineralization.

12.6.4 Copper and gold mineralization around Dhani-Basri and Todi Ka Bas area, Dausa and Alwar districts, Rajasthan, India—case study

Geophysical surveys comprising SP, magnetic, IP (time-domain, TD) cum resistivity, gravity, and resistivity sounding have been carried out in Dhani-Basri and Todi Ka Bas area, Dausa and Alwar districts of Rajasthan, for locating basemetal occurrences (Geological Survey of India, Naskar et al., 2019). The IP survey has delineated several anomaly zones, where strike length varies from 400 m to 1500 m. Chargeability values of zone-I, zone-II, and zone-III range from 16 to 22 mV/V. Zone I, characterized by a low SP value ranging from –100 to –200 mV, corroborates well with the known Dhani-Basri deposit. Chargeability values for zones IV, V, VI, and VII range from 15 to 24 mV/V, while a peak chargeability value of 30 mV/V is observed in zone V against a background value of 10 mV/V. These chargeability zones are associated with low SP, low resistivity, and moderate magnetic responses. Further, Bouguer anomaly profiles reflect a region where the borehole intersected a mineralized zone. Bouguer anomaly contour map indicates a fault zone, depending on the trend of the contour, that corroborates well with the regional schistosity. Gravity high closure has been observed mainly for metamorphic content below the alluvium cover. The interpreted layers by Schlumberger resistivity sounding are top soil, hard rock or fractured granite-gneiss, and the Archaean basement. The SP, magnetic, and gravity data have also picked up structural or lithological trends quite well. However, to strengthen the survey results, the resistivity map is compared with chargeability map; those who have intersection area between low resistivity and high chargeability (>24 mV/V) are considered the prospective zones for basemetal mineralization. Further, the mineralization is structurally controlled by a shear zone, and sulfide (chalcopyrite, pyrrhotite, and pyrite) with gold are hosted in the silicified zone.

12.6.5 Basemetal mineralization in Karoi-Rajpura area, Bhilwara district, Rajasthan, India—case study

A combination of SP, IP, resistivity, and magnetic methods was attempted in Karoi-Rajpura area of Bhilwara district of Rajasthan for locating basemetal occurrence (Geological Survey of India, Naskar et al., 2018). The integrated interpretation of the geophysical response of the different geophysical methods revealed more precise information about the subsurface configuration and mineral distribution in the prospect area. Strong geophysical anomalies were observed in the northern, southern, central, and eastern parts of the study area. The low resistivity zones (200 Ohm-m) and high chargeability zones (>24 mV/V) are situated in the southern and eastern parts of the investigated area. However, to strengthen the reliability of the results, the resistivity map is compared with the chargeability map; those who have an intersection area between low resistivity and high chargeability (>24 mV/V) are considered to be the prospective zones of base metal mineralization. However, these low resistivity zones may have been influenced by the presence of clay or weathered soil. In this case, the high chargeability zones will help in confirming the prospective zones caused by base metal mineralization. Recent deep drilling by Hindustan Zinc Limited (HZL) in this area has intersected rich ore lodes, thereby confirming depth-wise continuity of ore zones at some places.

12.7 Case studies in geophysical mineral exploration of South and East Asia

12.7.1 Geological setting, tectonic evolution and spatio-temporal distributions of main mineral resources in South East Asia

Southeast Asia is located at the intersection of the Tethys ore-forming area and the Pacific Rim metallogenic domain, where it has the dual metallogenic characteristics of the Tethys Himalayan metallogenic belt and the Circum Pacific metallogenic belt, and suffers from tectonomagma and tectonometamorphism in different periods of Indochina, Yanshan, and the Alps. Southeast Asia is rich in copper, gold, aluminum, potassium, nickel, chromium, tungsten, tin, lead, zinc, iron, manganese, and antimony (Nevolko et al., 2022). Based on the results of geological surveys and mineral resources research over the past decades, Wang et al. (2023) have systematically summarized the metallogenic geologic settings of Laos, Cambodia, Vietnam, Thailand, Myanmar, Malaysia, Indonesia, and the Philippines in Southeast Asia. A mineral geological database has been established, including regional geological maps, geotectonic units with division maps, magmatic rock distribution maps, and metallogenic regularities of some important metallogenic belts in Southeast Asia. The distribution characteristics of major minerals such as copper, gold, aluminum, potassium, nickel, chromium, tungsten, tin, lead, zinc, iron, manganese, and antimony, and the genetic types of these minerals are summarized. Meanwhile, twenty metallogenic belts are divided, and nine main metallogenic epochs are summarized. Wang et al. (2023) also analyze the influence of the tectonic evolution of the east Tethys, as well as the subduction and collision between the Indian Australian plate, the Pacific plate, the Philippine subplate, and the Eurasian plate on the mineralization in Southeast Asia. This study provides a number of valuable data and geological information for understanding the spatial distribution and temporal evolution of minerals and the prospecting targets in Southeast Asia.

12.7.1.1 Superimposed gold mineralization events in the Tuanshanbei orogenic gold deposit, central Jiangnan Orogen, South China

The Jiangnan orogen, one of the largest gold-producing areas in China, has experienced multiple orogenic events with complex structural overprinting that is marked by multiple stages of magmatism, deformation, metamorphism, and orogenic gold mineralization (Wang et al., 2023; Groves et al., 1998; Goldfarb and Groves, 2015; Wang et al., 2013; Ni et al., 2015; Pirajno and Bagas, 2002; Zhang et al., 2019). Field studies in the orogen indicate the Tuanshanbei gold deposit includes two generations of auriferous quartz-ankerite-pyrite-arsenopyrite veins (Q2 and Q3), with the latter of the two notable for containing more abundant ankerite and base metal sulfides (Wang et al., 2023). Hydrothermal monazite coexists with native gold and gold-bearing metal sulfides in the Q2 and Q3 veins. The Q2 monazite yielded a Tera-Wasserburg (Tera and Wasserburg, 1972) lower intercept age of 415.1 ± 2.1 mA, consistent within error with an ankerite Sm–Nd isochron age of 410 ± 15 mA and a laser ablation-inductively coupled plasma-mass spectrometry hydrothermal zircon 206Pb/238U age of 411.2 ± 4.0 mA. The Q3 monazite yielded a Tera-Wasserburg lower intercept age of 234.3 ± 1.1 mA. These new ages suggest that the Early Devonian gold event was overprinted by hydrothermal activity along the same structural system almost 200 million years later such that the gold resource must be a product of two temporally distinct events. Geologic and structural evidence, coupled with existing published geochemical data, suggests both ore-forming events were related to crustal metamorphism typical of most orogenic gold deposits. Fluids would have been derived from Neoproterozoic metasedimentary basement rocks, most likely from metamorphic devolatilization of the Neoproterozoic Cangxiyan Group greenschist-amphibolite facies metasediments. There is no evidence suggesting any type of magmatic contribution to the ore-forming process at either time. The data are best interpreted to suggest that various parts of the basement were metamorphosed near the greenschist-amphibolite boundary at different times, but during both times, the gold-bearing metamorphic fluids eventually migrated into the same structural conduits.

12.7.1.2 Iron ore concentration area, southern Liaoning province, China

In this study, based on metallogenic theory and the metallogenic geological characteristics of banded iron formation (BIF) type iron ores, equipment for surface geophysical surveys (the high precision ground magnetic survey method, the transient electromagnetic method, and the magnetotelluric method) were employed to detect the concealed iron ore bodies in the transition zone CID-1 between the Hejia and Dumu iron deposits in the Gongchangling iron ore concentration area in the Anshan-Benxi area (Liaoyang, China), a representative area of BIF type iron ores (Fu et al., 2020). The results showed that an optimal combination of these geophysical survey methods accurately determined the anomalous planar spatial locations and anomalous profile morphologies of the concealed iron ore bodies. On this basis, their locations, burial depths, and scales are determined. Two anomalous zones induced by concealed iron ore bodies, YC-1 and YC-2, were discovered in zone CID-1. Two concealed iron-bearing zones—one shallow (0–150 m) and one deep (300–450 m)—were identified in YC-1. A 100 m scale drilling test showed that the cumulative thickness of the shallow iron-bearing zone was over 23.6 m.

12.7.2 An application of integrated geophysical studies in the offshore of Pakistan

Integrated geophysical studies facilitate the delineation of the deep crustal structure of the Earth worldwide. In the present study, gravity, magnetic, bathymetry, and deep seismic reflection data were used to obtain a geologically plausible model from the deep crustal structure and also for the hydrocarbon prospects in the Indus Offshore basin of Pakistan beneath the seismic line PC/9074-86 (Shah et al., 2006). Steep positive gravity gradient toward the southwest of the area, above thick sedimentary strata (greater than 10 Km) in the offshore depression area, is attributed to a prominent rise in the mantle. Intuitively, extending the depth model beyond the present-day shelf break and computing its gravity indicates that the overall gravity effect is one of the typical shelf-edge ("Edge Effect") anomalies. The crustal thickness, which is approximately 24.5 Km thick toward the northeast, is reduced to 6.5 Km on the mantle rise, and it may be the area of transitional crust. Presence of thick sedimentary rocks of varying lithologies, which commonly form source-reservoir-seal trilogies, optimum geothermal gradients, analogies with other basins, oil/gas shows in and around the offshore area, and number of structural traps, including reefs and bright spots in sediments ranging in age from Cretaceous to Early Miocene, show positive evidence of the presence of hydrocarbons. Furthermore, it is speculated that the younger sediments in the offshore area may have attained maturity, due to thinning of the crust as a consequence of mantle rise in the outermost shelf regions, and may produce hydrocarbons.

12.8 Case studies in geophysical mineral exploration of Sweden

12.8.1 Study on joint inversion of gravity, magnetic, and petrophysical data for a gabbro intrusion in Boden, Sweden

Kamm et al. (2015) have studied a gabbro intrusion in northern Sweden, using 3D inversion of airborne magnetic data, ground-based gravity data, and petrophysical measurements on outcrop samples. Gabbro intrusions are of interest because they are potential hosts of Cu–Ni and platinum group element mineralization. Kamm et al. (2015) developed a joint inversion algorithm and applied it to both potential field data sets to obtain spatial distributions of density and magnetic susceptibility. The distributions were coupled through a non-rigidly enforced parameter relationship determined from the petrophysical samples. They managed the problem of balancing the influence of the two data sets by a novel adaptive reweighting scheme that enforced the discrepancy principle for each data set independently. They demonstrated in tests with synthetic data that neither individual nor joint inversions gave reliable estimates for the depth extension of the intrusive body, the near-surface details, or any complex geometrical features. However, the joint inversion improved the image of the interface between the intrusion and the surrounding rocks and revealed that the density and susceptibility models satisfied the observed petrophysical relationship, which, in turn, caused the structures in the models to align. The geometry of the intrusion was an intrinsic result of the inversion, based on the two distinct petrophysical trends for the gabbro and the surrounding rocks. The inferred shape was simple and concise, and was therefore a useful and testable hypothesis about the subsurface geology that agreed with both potential field data sets and the petrophysical information.

12.8.2 Structural evolution of the central Kiruna area for iron oxide, apatite, and epigenetic iron and copper sulfides, northern Norrbotten, Sweden

To guide future exploration, this predominantly field-based study has investigated the structural evolution of the central Kiruna area, the type locality for iron oxide-apatite deposits that stands for a significant amount of the European iron ore production (Andersson et al., 2021). Using a combination of geologic mapping focusing on structures and stratigraphy, petrography with focus on microstructures, X-ray computed tomography imaging of sulfide structure relationships, and structural 2D forward modeling, a structural framework is provided, including spatial-temporal relationships between iron oxide-apatite emplacement, subeconomic Fe and Cu sulfide mineralization, and deformation. These relationships are important to constrain as guidance for exploration in iron oxide-apatite and iron oxide copper-gold prospective terrains and may help to understand the genesis of these deposit types. Results suggest that the iron oxide-apatite deposits were emplaced in an intracontinental back-arc basin, and they formed precrustal shortening under shallow crustal conditions. Subsequent east-west crustal shortening under greenschist facies metamorphism inverted the basin along steep to moderately steep E-dipping structures, often subparallel with bedding and lithological contacts, with reverse, oblique to dip-slip, east-block-up sense of shears. Fe and Cu sulfides associated with Fe oxides are hosted by structures formed during the basin inversion and are spatially related to the iron oxide-apatite deposits, but formed in fundamentally different structural settings and are separated in time. The inverted basin was gently refolded and later affected by hydraulic fracturing, which represents the last recorded deformation-hydrothermal events affecting the crustal architecture of central Kiruna.

12.9 Case studies in geophysical mineral exploration of Australia

12.9.1 Modelled DCIP resistivity and chargeability sections across the Atlantida deposit, Australia

This case study explores the application of Direct Current Induced Polarization (DCIP) methods to map the resistivity and chargeability characteristics of the Atlantida gold deposit (Cu–Au–Mo porphyry deposit). Located in a geological setting conducive to orogenic gold mineralization, the Atlántida deposit provides an ideal environment for utilizing geophysical techniques in mineral exploration. Geophysical data, particularly resistivity and chargeability, are critical in identifying areas of gold-bearing sulfide mineralization and understanding the structural controls influencing the deposit's formation.

Direct Current Induced Polarization (DCIP) is a well-established geophysical method used to measure the resistivity and chargeability of subsurface materials. It works by applying a direct current to the ground and measuring the resulting voltage to identify contrasts in resistivity and chargeability, which provide valuable information about lithology and mineralization. *Resistivity* measures the resistance of subsurface materials to the flow of electric current. Higher resistivity typically correlates with less conductive materials such as quartz and granite, while lower resistivity is indicative of conductive zones like altered rocks and sulfide-rich mineralization (Hawkins et al., 2025). *Chargeability* measures how much charge a rock can store and is sensitive to the presence of metallic sulfides, commonly associated with gold mineralization. Chargeability anomalies typically indicate sulfide-rich zones that are potential targets for gold mineralization (Gibson and Smith, 2022).

At the Atlantida deposit, DCIP surveys were carried out to map the spatial distribution of resistivity and chargeability anomalies. The data from these surveys were processed to generate detailed 2D and 3D sections representing the subsurface properties, which were then integrated with geological and structural models of the area. The resistivity sections helped identify resistive and conductive zones. Resistive zones were interpreted to correlate with unaltered rock formations, while conductive zones were associated with alteration halos and mineralized zones. Key structural features, such as faults and folds, were also identified in the resistivity data, aiding in the interpretation of mineralized zones. Chargeability anomalies were carefully analyzed to locate sulfide-rich mineralization, which is often associated with gold in orogenic deposits. High chargeability values were observed along certain structural features like shear zones and folds, which are typical controls for gold mineralization. These anomalies aligned well with known mineralization trends in the Atlantida deposit, confirming the utility of DCIP chargeability in locating gold-bearing structures (Ammar and Vasquez, 2023).

The DCIP resistivity and chargeability data were integrated with other geophysical datasets, such as airborne magnetics and gravity surveys, to create a more robust exploration model. Airborne magnetic data helped map lithological boundaries and identify magnetically susceptible rock types. Structural features, such as faults and folds, were also identified using magnetic data, refining the interpretation of resistivity and chargeability models (Hawkins et al., 2025). Gravity data were incorporated with the DCIP results to assess density variations related to mineralized zones. This allowed for the identification of deeper structural features that could influence the distribution of gold mineralization, especially in areas covered by younger sediment (Ammar and Vasquez, 2023).

The integration of resistivity, chargeability, and complementary geophysical datasets enabled the identification of high-priority exploration targets at the Atlantida deposit. Based on the resistivity and chargeability models, several high-chargeability anomalies were identified along the limbs of folds and shear zones, which were prioritized for drilling. These zones exhibited both high chargeability (indicating sulfide-rich mineralization) and low resistivity (indicating alteration or mineralization), marking them as strong candidates for further exploration (Gibson and Smith, 2022). The data revealed that gold mineralization was closely associated with specific structural features, such as folds and faults, which acted as conduits for fluid flow and mineral deposition. Identifying these structures was crucial in focusing exploration efforts in areas with high potential for gold mineralization (Fig. 12.7). The modeling of DCIP resistivity and chargeability sections at the Atlantida deposit provided critical insights into the subsurface geology (Neroni and Hunt, 2022).

The success of these methods in identifying gold-bearing sulfide zones emphasizes the importance of integrating geophysical techniques in modern mineral exploration. The approach used in this study has broader implications for the exploration of orogenic gold deposits in similar geological settings. By combining resistivity and chargeability data with structural and lithological models, exploration teams can refine their targeting strategies, improving the efficiency and effectiveness of drilling campaigns.

This case study highlights the significant role of geophysical methods, specifically DCIP resistivity and chargeability, in the exploration of orogenic gold deposits. At the Atlántida deposit, these techniques were instrumental in mapping mineralized zones and understanding the structural controls on gold deposition. The integration of DCIP data with other geophysical surveys enabled more accurate targeting of exploration areas, leading to a more focused and successful exploration program. This method can be applied to other gold deposits in similar geological settings, making it a valuable tool in the exploration geophysicist's toolkit.

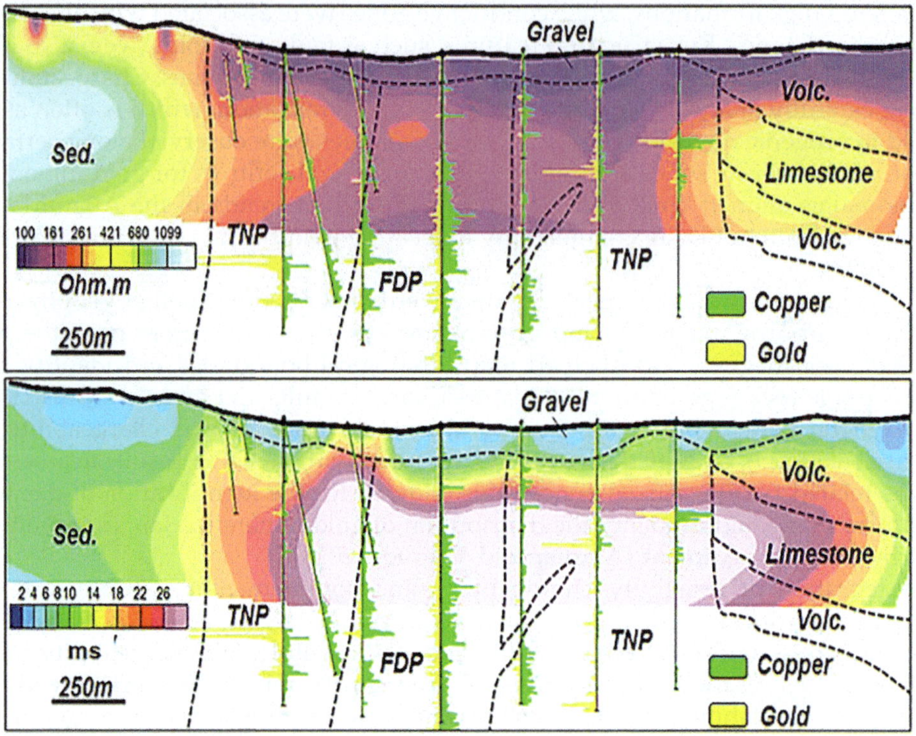

FIG. 12.7 Modeled DCIP resistivity (top) and chargeability (bottom) sections across the Atlántida deposit, Australia. *Source: Modified.*

12.9.2 Summary of K and Th concentrations in the vicinity of the Goonumbla-North Parkes, Australia Cu–Au porphyry deposits

The Goonumbla porphyry copper-gold deposit in New South Wales (NSW), Australia (Fig. 1.10C), is hosted by late Ordovician (439.2 ± 1.2 mA) shoshonitic igneous rocks (Muller et al., 1994) (Fig 12.8).

Potassium is often introduced to host rocks through mineralizing hydrothermal solutions. Several studies have documented the variation of elevated K associated with alterations in porphyry Cu deposits. For instance, Moxham et al., 1965 observed a 1.5% increase in K during alteration in porphyry Cu deposits in the southwest United States, along with an erratic increase in U, but no increase in Th. This could be attributed to the higher mobility of U compared to Th during hydrothermal alteration. Since Th generally remains unaffected by alteration, the K/Th ratio serves as a better indicator of alteration than K alone. Davis and Guilbert, 1973 found similar results. These studies, however, did not account for the effects of weathering. To explore how weathering might influence the gamma-ray signature of porphyry Cu deposits, studies were conducted over two deposits in central New South Wales—one well-exposed (Copper Hill) and one extensively weathered (Goonumbla) (Dickson and Scott, 1997).

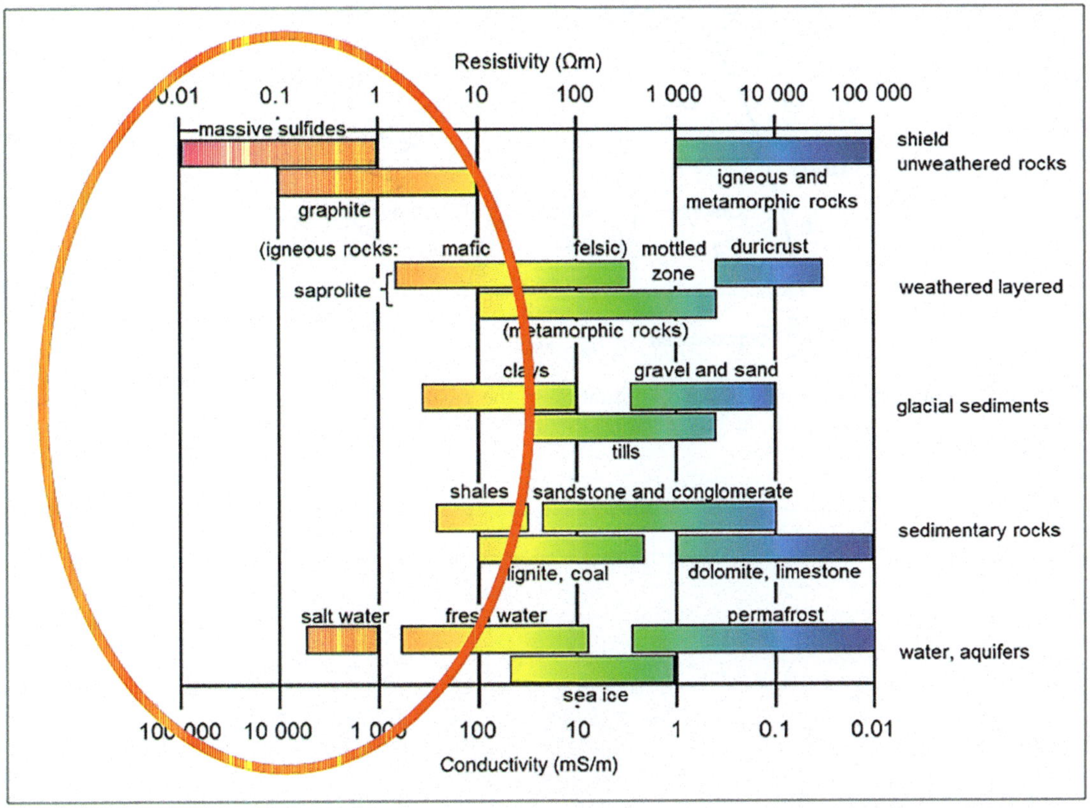

FIG. 12.8 Location of deposits of the Goonumbla porphyry copper-gold deposit in New South Wales (NSW), Australia. *Source: Dickson and Scott, 1997.*

Aerial gamma-ray surveys detect geochemical variations in potassium (K), uranium (U), and thorium (Th) within the top 30 cm of the Earth's surface. This surface layer is influenced by weathering, which typically leads to a loss of K in all rock types and a loss of U and Th in felsic rocks. The extent of this loss is typically around 20% to 30% for all these radioelements. Intermediate and basic rocks show little change in radioelement concentrations during early weathering stages, although pedogenesis can lead to soils with two to three times the U and Th content of the parent rock. However, a wide range of radioelement compositions can occur for a given rock type and its weathered products. Mineralization processes can also impact radioelement concentrations. For instance, K is elevated in altered rocks at the Copper Hill and Goonumbla porphyry Cu deposits in central NSW (Fig. 12.9).

Potassium concentrations in altered dacite were found to be elevated (2.4%–3% K) compared to Ordovician andesites and basalts (~0.6%) and unaltered dacite (~1.6%). Weathering of rocks and pedogenesis lowered the K concentration but did not alter the relative K concentrations at any stage of the weathering profile, as K is predominantly present as muscovite, and the soils remain in situ. Uranium showed a narrow range (1–2 ppm) in both rocks and soils, whereas thorium varied from 1 to 4 ppm in both rocks and soils (Fig. 12.10). It was

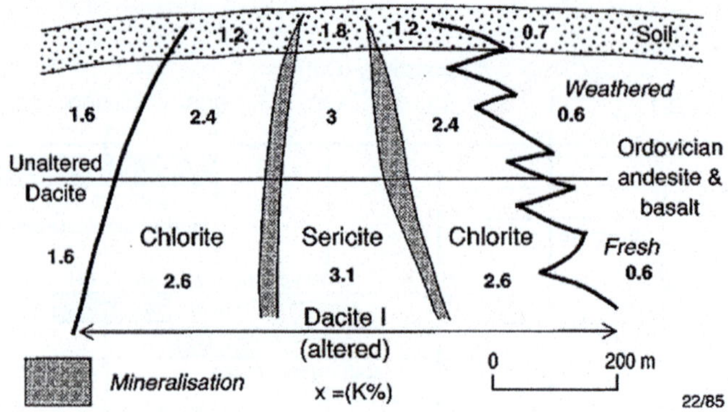

FIG. 12.9 Summary of potassium (K) concentration in the vicinity of the Goonumbla-North Parkes Cu–Au porphyry deposits. Variation of K (in %). *Source: Dickson and Scott, 1997; Dickson and Scott, 1991a.*

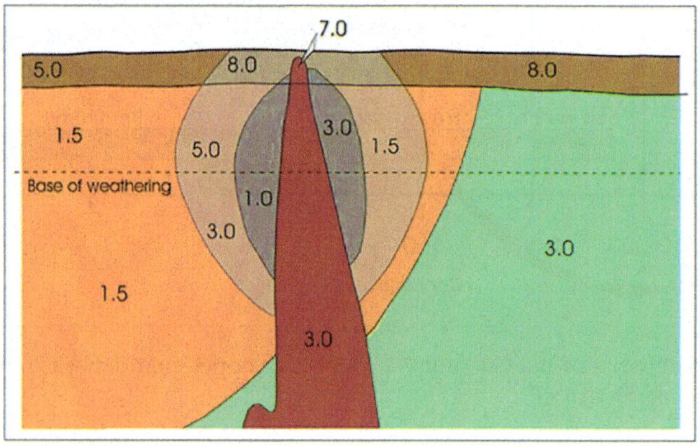

FIG. 12.10 Summary of the thorium (Th) concentration in the vicinity of the Goonumbla-North Parkes Cu–Au porphyry deposits. Variation of eTh (in ppm). *Source: Dentith and Mudge, 2014.*

concluded and confirmed by a subsequent survey that the Copper Hill complex was easily distinguishable by its elevated K content.

Porphyry Cu–Au mineralization at Goonumbla, 25 km north of Parkes in central New South Wales, is associated with a series of prospects within a 30 km-wide Silurian caldera (Heithersay et al., 1990). The mineralization is specifically linked to a number of vertical, finger-like bodies of quartz monzonite that have intruded trachytes and K-rich andesites. Alteration at the site ranges from potassic (with quartz stockworks) to quartz-sericite (phyllic) and propylitic types. As the intrusives are narrow, potassic and phyllic alteration affects both the intrusions and the surrounding rocks. Three nearby deposits, E22, E26, and E27, are currently mined as part of the North Parkes mine. A summary of the potassium (K) and thorium (Th) content of rocks and soils from the E26 deposit, where phyllic alteration largely overlaps the potassic alteration, is shown in Fig. 12.11. Uranium content is not shown as the rocks consistently contain 1–3 ppm U.

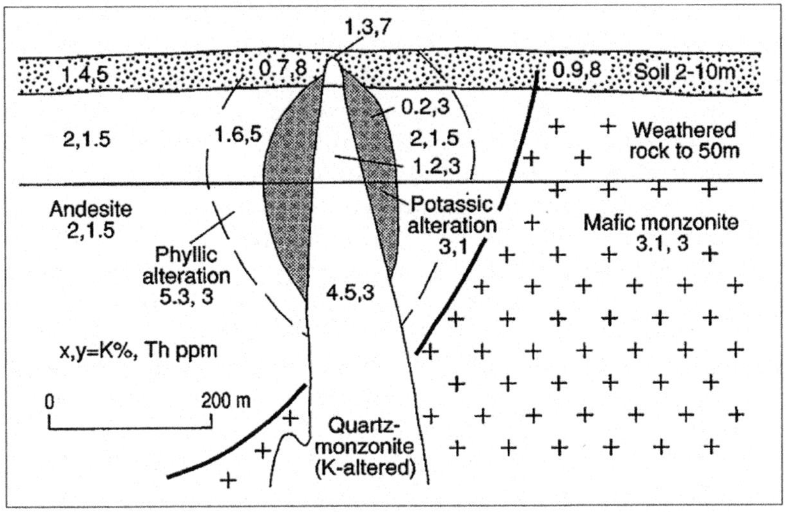

FIG. 12.11 Summary of K and Th concentration in and around North Parkes mine, Goonumbla, New South Wales, Australia. *Source: Dickson and Scott, 1997; Dickson and Scott, 1991a.*

Thorium concentrations show both depletion and enrichment during hydrothermal alteration, as illustrated by the Au prospects at Bimurra, northeast Queensland (see Fig. 12.8). Uranium is more erratically affected by alteration and is generally not a reliable indicator of alteration. Regolith processes can influence these alteration signatures. Highly weathered deposits may lose their K, particularly if the host rock is K-feldspar, as seen at Goonumbla. Transported soils may also obscure or alter rock signatures, often in unexpected ways. For example, the Mt Leyshon gold deposit in northeast Queensland appears potassium-rich in aerial surveys because its signature is not affected by material weathered from late-Silurian dolerites.

Detailed interpretation of aerial gamma-ray surveys for exploration purposes requires the delineation of the major geological units of the survey area, followed by the examination of subtle variations within the most prospective units. This process is aided by other datasets such as magnetics, Landsat TM data, and field validation of identified anomalous areas.

12.10 Case studies in geophysical mineral exploration of the United States

12.10.1 Discovery of the south shoot at Antler volcanogenic massive sulfide (VMS) deposit, Arizona, USA

Antler is a high-grade Cu–Zn VMS deposit located in mountainous terrain in Western Arizona, USA. The deposit has a long mining history dating back to 1916. However, exploration activities slowed in the 1970s, and no work was done for over 40 years. In 2020, New World Resources (NWR) revived the project with a renewed focus on exploration, using advanced geophysical methods to guide their work. Testing on drill core samples showed strong anomalies in density, chargeability, and inductive conductivity in the mineralized areas. Magnetic susceptibility was also high in the mineralized zones and showed a significant amount of remanent magnetization. High-resolution magnetic surveys were carried out using a drone platform. These surveys identified magnetic bodies that matched well with the known mineralized areas. 3D magnetic inversions, combined with drill data, helped define the mineralized zones to depth. Fixed-loop EM was used to detect conductive VMS mineralization directly. However, this method did not meet expectations. 3D IP surveys were then tested and successfully identified the known mineralization to a depth of about 400 m, both in resistivity and chargeability datasets. Drilling in the central zone confirmed the high-grade mineralization found by the IP/resistivity targets. While still looking for the best method to detect targets, Terra Resources then used Controlled Source Audio Magnetotellurics (**CSAMT**) in a Broadside configuration (Neroni and Hunt, 2022). The goal was to map deep conductors over a wide area. The Broadside method emphasizes detecting highly conductive bodies within a conductive host while minimizing less conductive bodies. The CSAMT survey highlighted the known mineralized zones and also revealed a large, conductive body to the south with significant depth—named the "South Shoot" (Fig. 12.12). New World Resources tested this new target with drilling. In March 2021, they announced significant intersections of high-grade copper-zinc-rich massive sulfide mineralization in the South Shoot, confirming it as a major discovery.

12.11 Case studies in geophysical mineral exploration of West Africa

12.11.1 Geophysical expression of the Abra sedimentary replacement Pb–Ag-Cu-Au deposit, WA

The Abra deposit is a high-grade, sedimentary-hosted lead (Pb) deposit located in the Paleoproterozoic Edmund Basin of Western Australia. It was discovered in 1981 but remained undeveloped until underground mining began in late 2021, 40 years after its discovery. The deposit is "blind," meaning it is not exposed at the surface. The top of the deposit starts at 250 m deep. It consists of a stratiform (layered) zone of Pb–Ag–Ba mineralization found in a laminated iron oxide and barite-altered dolomite and siltstone. Below this, there is a "feeder zone" with chlorite-altered, brecciated (broken), and veined carbonaceous siltstone that contains Pb-Ag mineralization in the center. This mineralization changes to Pb–Cu and Cu-Au at greater depths. Abra shows distinct geophysical anomaly responses in several surveys, including magnetic, gravity, Time Domain Electromagnetic (TDEM), and IP (Induced Polarization) surveys. A + 450 nT magnetic anomaly is caused by magnetite in the lower part of the stratiform zone. The dense galena, barite, dolomite, and iron oxide in the apron cause a + 1 mGal gravity anomaly due to the contrast in density with the surrounding lower-density sedimentary rocks.

Airborne, ground, and downhole TDEM surveys detected weak electromagnetic responses caused by galena. Petrophysical testing on core samples confirmed that galena is responsible for these weak conductor responses. PDIP (Dipole-Dipole IP) surveys found a + 20 msec chargeability anomaly on the southern flank of the deposit, which is related to disseminated galena, pyrite, chalcopyrite, and alteration. AMT-MT 2D inversion data showed Abra as a broad, weakly conductive anomaly. The same anomaly was also detected in 2D and 3D

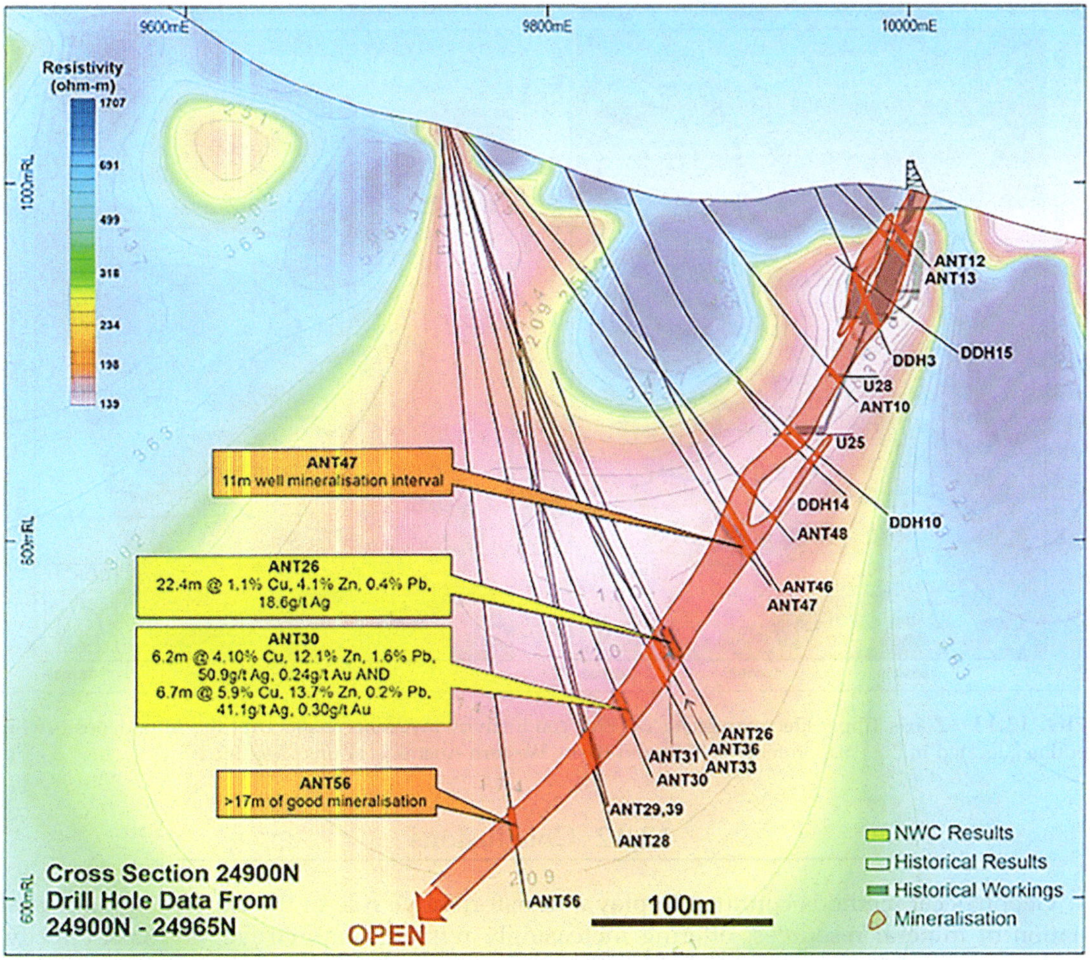

FIG. 12.12 Cross-section of CSAMT from line 24900 N—with drill holes and select significant assay results from the recently discovered "South Shoot" at the Antler VMS deposit, Arizona, USA. Source: Modified.

inversion models of airborne Z-axis Tipper Electromagnetic (ZTEM) data (Fig. 12.13). 2D seismic reflection surveys revealed strong, flat-lying seismic reflectors within the deposit, surrounded by an area with no seismic activity (Neroni and Hunt, 2022). These reflectors are linked to significant density contrasts between high-density mineralization and low-density host rocks. Passive seismic Horizontal-to-Vertical Spectral Ratio (HVSR) surveys detected the top of the deposit as a subtle layer beneath a flat impedance contrast horizon. This horizon is interpreted as weathered siltstone sitting on top of diagenetically cemented siltstone. The deposit remains open at depth, and a new Pb-mineralized zone was discovered just to the north of the known deposit, following the testing of a late-time Versatile Time Domain Electromagnetic (VTEM) anomaly.

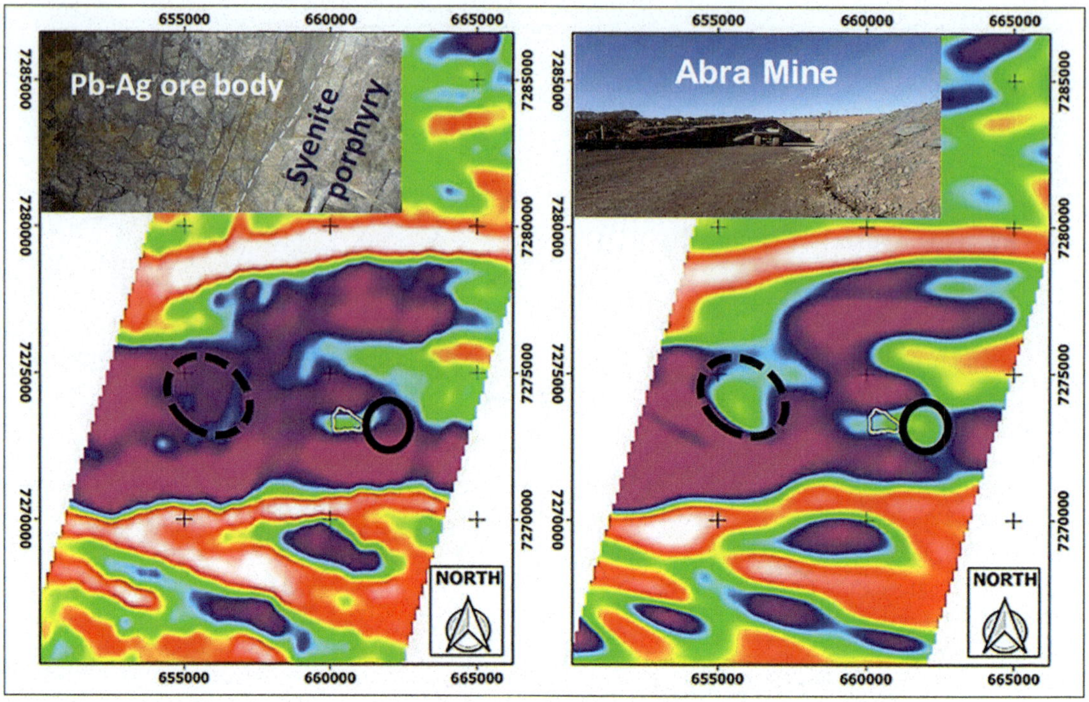

FIG. 12.13 Z-axis Tipper Electromagnetic (ZTEM) conductivity inversion results over the Abra deposit (white outline), located in the Paleoproterozoic Edmund Basin, Western Australia. Source: Modified.

12.12 Conclusions

Geophysical methods continue to play a transformative role in the exploration and evaluation of mineral resources, offering increasingly refined, noninvasive, and cost-effective tools to map subsurface structures, identify prospective zones, and reduce geological uncertainty. The chapter underscores the evolving global and regional exploration scenarios, with particular emphasis on the critical role of policy reforms, data integration, and technological innovation in boosting exploration outcomes.

The wide-ranging case studies presented across India, Southeast Asia, China, Pakistan, Sweden, Australia, the United States, and West Africa illustrate the diversity of geophysical applications in different geological settings. These include gravity, magnetic, electrical, electromagnetic, radiometric, and joint inversion techniques—each tailored to address specific mineralization environments such as auriferous-sulfide zones, base metals, heavy mineral sands, iron ore systems, and polymetallic deposits.

Indian case studies demonstrate the country's vast potential in underexplored terrains like the Cuddapah Basin and Rajasthan's polymetallic provinces, emphasizing the need for sustained geophysical surveys and integrated exploration models. Similarly, studies from Southeast Asia and China highlight how tectonic reconstructions and mineral system modeling can guide deeper exploration strategies. The inclusion of advanced methodologies such as joint inversion, airborne radiometrics, and 3D modeling in Sweden and Australia further reflects the direction in which mineral exploration is heading—toward higher precision and deeper targeting.

Collectively, these insights reinforce that effective geophysical exploration demands a multidisciplinary approach that integrates geological, geochemical, structural, and tectonic data. Additionally, the success of mineral discovery and resource development hinges not only on technical capability but also on favorable policy frameworks, cross-border data sharing, and investment in geoscientific research. Future exploration efforts must therefore continue to embrace data-driven and environmentally sustainable practices to secure critical mineral supplies essential for global energy transitions and technological growth.

CHAPTER

13

Baseline geoscientific data acquisition planning and strategy

13.1 Geological mapping for baseline geoscientific data acquisition

The effective exploration and sustainable management of natural resources depend on acquiring accurate, detailed, and timely baseline geoscientific data. Without this crucial information, it becomes difficult to evaluate the potential for resource extraction, whether it is minerals, water, energy, or other geological materials. This chapter provides a clear framework for planning the acquisition of such data, focusing on the use of various mapping techniques essential for any successful exploration project.

Geological, geophysical, and geochemical mapping are fundamental to modern geoscientific exploration. Each method offers unique insights into the composition, structure, and behavior of the subsurface. Geological mapping focuses on the physical features of rock formations and surface structures. Geophysical methods, such as magnetic, gravity, and seismic surveys, gather data about subsurface features without invasive drilling. Geochemical mapping looks at the chemical properties of rocks, soils, and water to identify elements or compounds that could point to valuable resources.

Integrating geological, geophysical, and geochemical data is key to forming a comprehensive understanding of the area being explored. By combining these different types of data, geoscientists can reduce uncertainty, improve exploration results, and identify potential resource locations. This integration also supports better risk assessment and helps optimize resource extraction, contributing to more sustainable resource management.

Additionally, this chapter will discuss the important precautionary steps needed during data collection to ensure the information is accurate, reliable, and representative. These steps include using high-quality equipment, standardizing data collection methods, calibrating instruments properly, and applying strong analytical techniques. Following these best practices helps reduce the risks of misinterpreting data or inefficient resource allocation.

In conclusion, acquiring baseline geoscientific data is a vital first step in resource exploration. A clear, strategic approach to geological, geophysical, and geochemical mapping, along with careful integration and precautionary measures, ensures that exploration is scientifically sound and supports sustainable resource management. This chapter will guide

Baseline Geoscience Data Generation for Sustainable Exploration of Natural Resources
https://doi.org/10.1016/B978-0-443-45221-5.00016-3

readers through the process, providing the tools and knowledge needed to obtain reliable and comprehensive data for informed decision-making.

13.1.1 Principles and practices of geological mapping

Geological mapping is a fundamental technique in geoscience that involves documenting the distribution, nature, and age of rock formations at Earth's surface. It is the starting point for understanding geological structures, mineral deposits, and the overall geology of a region. The technique provides vital information about the subsurface, which is essential for the exploration of natural resources such as minerals, water, and hydrocarbons.

Geological mapping involves fieldwork, interpretation of rock samples, and the creation of detailed maps that illustrate the spatial distribution of rock types, faults, folds, and other geological features. Accurate geological maps are crucial for resource exploration, land-use planning, and understanding natural hazards such as earthquakes or landslides.

13.1.2 Key components of geological mapping

13.1.2.1 Field mapping

The process begins with geologists conducting surveys in the field. They identify outcrops, take rock samples, and record geological features such as faults, folds, and lithological boundaries. Fieldwork includes measuring strike and dip of rock layers, identifying mineral deposits, and observing structural features such as joints and fractures.

13.1.2.2 Lithological mapping

Lithological mapping involves identifying and mapping different types of rocks, such as igneous, sedimentary, and metamorphic rocks. This is done based on characteristics like color, texture, and mineral content. Rock units are identified and classified based on their physical and chemical properties, which help in understanding their formation and age.

13.1.2.3 Structural mapping

This aspect focuses on the identification and mapping of structural features such as folds, faults, joints, and shear zones. These structures provide information about tectonic activity, stress history, and the potential for resource deposits. Mapping these features helps geologists understand the deformation history of an area and predict subsurface geology.

13.1.2.4 Stratigraphic mapping

Stratigraphic mapping involves studying rock layers (strata) to understand the sequence of geological events in the region. Stratigraphy helps to establish the relative age of rock formations and their depositional history. It also assists in identifying potential locations of natural resources like oil, gas, and coal.

13.1.2.5 Topographic mapping

Involves the creation of maps that show the elevation and contour of the land surface, which helps in identifying geological structures and landforms. This is often combined with geological maps to give a comprehensive picture of the region.

13.1.2.6 *Geochemical analysis*

Often, geological mapping is complemented by geochemical analysis, where rock, soil, and water samples are analyzed for the presence of specific minerals (Specific minerals are naturally occurring inorganic substances with a defined chemical composition and crystal structure). Some common examples include: Quartz (SiO_2), Feldspar ($KAlSi_3O_8$), Mica, Calcite ($CaCO_3$), Magnetite (Fe_3O_4), Pyrite (FeS_2), Halite (NaCl), Gypsum ($CaSO_4 \cdot 2H_2O$), or trace elements (**Trace elements** are elements that are present in very small quantities in a sample, typically <0.1% by weight. Despite their low concentrations, they can have significant chemical, biological, or geological importance. In geology, trace elements in rocks and minerals can provide valuable information about the formation processes, the environment in which they formed, and the presence of certain resources, such as metals and minerals. Common trace elements include gold, silver, copper, arsenic, and rare earth elements) that could indicate the presence of valuable resources.

13.1.3 Geological mapping methods

13.1.3.1 *Traditional mapping*

Traditional mapping involves physically traveling to the field, taking notes, sketching maps, and collecting samples from outcrops. It provides first-hand, high-resolution data from natural outcrops and rock formations. However, it is labor-intensive and limited to accessible areas. It is also time-consuming and may not represent subsurface geology well.

13.1.3.2 *Geophysical techniques*

Geophysical Techniques involve the use of methods such as seismic reflection, magnetic surveys, gravity surveys, and electrical resistivity surveys to infer the subsurface geology. These techniques provide valuable data on subsurface structures without the need for excavation, making them useful for locating resources such as minerals, oil, and gas. However, they can be expensive and may not always be accurate unless validated with ground-truthing from geological mapping.

13.1.3.3 *GIS-based geological mapping*

GIS-Based Geological Mapping involves the use of Geographic Information Systems (GIS) to create, analyze, and present geological data in a spatial format. GIS integrates multiple layers of data, including geological, geophysical, and geochemical information, providing a comprehensive view of the area's geology. One of the main advantages of this method is that it facilitates the integration of various data types, allows for detailed spatial analysis, and makes the data more accessible and shareable. However, it requires technical expertise in GIS software and high-quality input data to ensure accurate results.

Table 13.1 compares different geological mapping methods based on their advantages and disadvantages.

TABLE 13.1 Comparison of geological mapping methods.

Mapping Method	Advantages	Disadvantages
Traditional Mapping	High-resolution, field-based data	Time-consuming, limited to accessible areas
Remote Sensing	Covers large areas quickly, useful for inaccessible regions	Less detail, limited to surface features
Geophysical Techniques	Can infer subsurface geology, locate resources	Expensive, requires calibration and validation
GIS-Based Mapping	Integrates multiple data types, facilitates spatial analysis	Requires specialized software and expertise

13.1.4 Geological mapping applications

13.1.4.1 Mineral exploration

Mineral Exploration involves using geological mapping to identify mineral deposits and delineate ore bodies. By thoroughly understanding the geology of an area, geologists can predict the locations where valuable minerals [Gold (Au), Silver (Ag), Copper (Cu), Diamonds, Platinum (Pt), Rare Earth Elements (REEs), Nickel (Ni), Uranium (U)] are likely to be found. This knowledge helps in targeting specific areas for further exploration and extraction, improving the efficiency and success of mineral resource discovery.

13.1.4.2 Hydrocarbon exploration

Hydrocarbon Exploration involves mapping the distribution of rock formations to help geologists identify potential petroleum and natural gas reservoirs. This process also aids in assessing the risks associated with drilling in certain areas, ensuring that exploration activities are conducted in the most promising and safest locations.

13.1.4.3 Environmental and engineering applications

Environmental and Engineering Applications involve using geological maps to assess soil stability, predict landslides, and plan infrastructure such as roads and tunnels. These maps provide valuable information for ensuring the safety and effectiveness of construction projects by identifying potential geological hazards and helping to design structures that are more resilient to environmental challenges.

13.1.4.4 Natural hazard assessment

Mapping faults, folds, and other structural features is essential for assessing risks associated with earthquakes, landslides, and volcanic activity.

Table 13.2 illustrates the key applications of geological mapping, highlighting its role in resource exploration, hazard assessment, environmental management, and land-use planning.

TABLE 13.2 Key applications of geological mapping.

Application	Description
Mineral Exploration	Locates and maps potential mineral deposits and ore bodies
Hydrocarbon Exploration	Identifies potential petroleum and gas reservoirs
Environmental Impact	Assesses geological hazards such as landslides, erosion, and soil stability
Natural Hazard Mapping	Identifies fault lines, volcanoes, and other natural hazards
Infrastructure Planning	Assists in site selection and stability for construction projects

13.2 Geophysical mapping techniques

This sub-chapter provides a structured overview of geophysical mapping, presenting step-by-step procedures from prefield preparation to advanced data integration. It highlights detailed methods and includes relevant tables as practical references for field and analytical applications. Key aspects covered include:

13.2.1 Prefield preparation

This section outlines the essential steps and resources required for preparing a successful geophysical survey. Prefield preparation ensures efficient field operations, minimizes errors, and optimizes data collection efforts.

13.2.1.1 Procedure for prefield preparation

Review existing data and literature by gathering geological maps, toposheets (Scale: Use 1:50,000 or 1:25,000 scale for detailed mapping and survey planning), and previous geophysical survey reports. Study historical data to understand the geological context and identify any gaps in the information. Compile details on known faults, mineral occurrences, aquifer locations, and potential geohazards to guide the survey planning.

13.2.1.2 Define survey objectives

Establish clear goals based on the specific geological problem, such as resource exploration, hazard assessment, or groundwater mapping. Additionally, identify the target depths and spatial coverage required for the survey to ensure accurate and relevant data collection.

13.2.1.3 Select the survey area

Use toposheets and GIS tools to delineate the area of interest, ensuring precise mapping of boundaries. Additionally, assess the accessibility of the area and address any necessary safety measures or permission requirements before proceeding with the survey.

13.2.1.4 Plan the survey layout

Design survey grids, transect lines, or sampling points to ensure systematic and comprehensive data collection. Mark significant features, such as rivers, roads, and elevation changes, to account for any potential influence on the survey design and execution.

13.2.1.5 *Select and calibrate equipment*

Choose suitable instruments based on the specific geophysical method being used, such as a gravimeter for gravity surveys or a magnetometer for magnetic surveys. Before deployment, perform calibration and functionality tests on the instruments to ensure accurate and reliable data collection during the survey.

13.2.1.6 *Arrange logistics*

Plan transportation, accommodation, and field supplies to support the survey team during the fieldwork. Additionally, obtain any necessary permits for access to restricted or private areas to ensure compliance with legal and regulatory requirements.

13.2.1.7 *Conduct environmental and safety assessments:*

Identify potential environmental impacts that could arise during the survey, such as disruption to ecosystems or wildlife habitats. Develop a comprehensive safety plan that addresses hazards like extreme weather, wildlife encounters, and equipment handling to ensure the well-being of the survey team and minimize environmental disturbance.

The following Table 13.1 summarizes the key components for prefield preparation:

13.2.2 Geophysical techniques in geological mapping

This section explores various geophysical methods used in geological investigations, offering detailed procedures and relevant tabulated data for practical reference. These techniques are essential for characterizing subsurface structures, delineating resource-rich areas, and identifying geohazards. They play a crucial role in geoscientific investigations, enabling the acquisition of spatial and temporal data that supports informed decision-making in natural resource management and hazard mitigation.

13.2.2.1 *Gravity surveys*

Select an appropriate gravimeter and calibrate the instrument to ensure accurate measurements. Establish a base station to correct for any temporal drift during the survey. Measure gravitational acceleration at predetermined grid points across the survey area. After data collection, correct the raw data for factors such as elevation, latitude, and terrain variations. Finally, generate Bouguer anomaly maps to interpret the subsurface density variations and identify geological features. Table 13.2 outlines the key steps involved in the gravity survey procedure, helping to standardize the approach for accurate and consistent data collection.

13.2.2.2 *Magnetic surveys*

In magnetic surveys, the first step is to choose a suitable magnetometer for the survey, such as a proton precession or fluxgate magnetometer, depending on the survey's resolution and depth requirements. A base station check is then conducted to account for diurnal variations in the Earth's magnetic field. The survey is carried out by performing ground or airborne measurements along transects, capturing magnetic field variations across the area. After data collection, the data is corrected for regional trends and noise to ensure the accuracy of the results. Finally, magnetic anomaly maps are created and the anomalies are interpreted

TABLE 13.3 Proposed table for pre-field preparation.

Component	Details
Existing Data Review	Gather geological maps, previous reports, and satellite imagery.
Survey Objectives	Define goals such as mineral exploration, groundwater mapping, or hazard analysis.
Tools for Planning	Use toposheets, GIS software, and historical datasets for survey design.
Survey Area Considerations	Evaluate accessibility, obtain land permissions, and assess terrain challenges.
Instrument Selection	Choose instruments such as gravimeter, magnetometer, GPR, or seismic tools.
Safety Measures	Develop emergency plans, ensure protective gear, and equip field communication tools.

to identify geological structures, ore deposits, or other important features. Table 13.3 summarizes the instruments used in magnetic surveys and their specific applications.

13.2.2.3 *Seismic methods*

In seismic methods, the first step is to plan the seismic survey by determining the appropriate source type, such as explosives or vibroseis, depending on the survey's objectives and geological conditions. Geophones are then placed along a predefined array to record seismic waves generated by the chosen source. Seismic waves are produced and their reflections or refractions are recorded as they travel through different subsurface layers. The seismic data is processed using specialized software to create subsurface models that reveal the geological structures and resource deposits. Finally, seismic sections are interpreted to identify geological features, such as faults, layers, and potential resources. Table 13.4 provides an overview of seismic methods and their specific purposes, helping to guide the application of seismic surveys based on the depth and type of geological features being studied.

13.2.2.4 *Electrical and electromagnetic techniques*

In electrical and electromagnetic techniques, the first step is to deploy electrodes for resistivity surveys or transmitters for electromagnetic surveys, depending on the method being used. Voltage, current, or electromagnetic field responses are then measured at various points in the survey area. The collected data is processed using 2D or 3D inversion software to generate models that represent subsurface resistivity or conductivity. These models are then interpreted to create resistivity or conductivity maps, which help identify geological features such as mineral deposits, groundwater aquifers, and contamination zones. Table 13.5 provides the depth ranges of resistivity and electromagnetic techniques, guiding their application based on the survey objectives and subsurface depth of interest.

13.2.2.5 *Ground-penetrating radar*

Ground-penetrating radar (GPR) surveys begin with selecting an appropriate antenna frequency based on the required survey depth, where lower frequencies are used for deeper subsurface investigations and higher frequencies for shallow features. The survey is

TABLE 13.4 Steps in gravity survey procedure.

Step	Description
Calibration	Performed at a known gravity point to ensure accuracy.
Data Acquisition	Measurements taken at 100 m intervals to ensure spatial coverage.
Data Processing	Terrain and drift corrections are applied to raw data for accurate results.

TABLE 13.5 Instruments and their applications in magnetic surveys.

Instrument	Applications
Proton Precession Magnetometer	Used for general geological mapping to detect magnetic anomalies.
Fluxgate Magnetometer	Suitable for high-resolution surveys to identify detailed magnetic features.

TABLE 13.6 Seismic methods and their purposes.

Seismic Method	Purpose
Reflection	Used for imaging deep subsurface layers to identify geological structures and potential resource zones.
Refraction	Applied for mapping shallow subsurface layers, such as groundwater and near-surface deposits.

conducted by moving the antenna over the surface along transects spaced at regular intervals to ensure comprehensive coverage. The collected radar data is processed to enhance signal reflections and eliminate noise, ensuring clarity and accuracy. Finally, GPR profiles are generated to interpret subsurface features such as buried structures, utility lines, or voids. Table 13.6 outlines the relationship between GPR frequency and its applications, helping to determine the most suitable frequency for specific depth requirements in subsurface investigations (Tables 13.7 and 13.8).

13.2.3 Applications of geophysical techniques

This section provides an overview of the significance of geophysical techniques in geoscientific investigations. It discusses the versatility of these methods in addressing various geological challenges, including resource exploration, environmental studies, and infrastructure development. The section sets the stage by introducing the role of geophysical data in integrating geological, hydrological, and environmental frameworks.

13.2.3.1 Resource exploration

Geophysical methods are extensively utilized in the exploration and evaluation of mineral deposits, hydrocarbon reservoirs, and geothermal energy sources. For mineral exploration, techniques such as magnetic, gravity, and electromagnetic surveys are applied to map ore deposits and understand structural controls. Hydrocarbon exploration relies on seismic

TABLE 13.7 Electrical and electromagnetic techniques and their depth range.

Technique	Depth Range
Resistivity	0–500 m
Electromagnetic (EM)	0–300 m

TABLE 13.8 GPR frequency and applications.

Frequency (MHz)	Application
100	Deep subsurface (>5 m)
500	Shallow subsurface (1–5 m)

reflection and refraction methods to generate high-resolution images of subsurface structures, facilitating oil and gas discoveries. In the context of geothermal energy, electrical resistivity and thermal gradient studies are used to delineate geothermal reservoirs efficiently.

13.2.3.2 Groundwater and aquifer mapping

Geophysical surveys are crucial for identifying and characterizing groundwater resources. Aquifer delineation is achieved through resistivity and electromagnetic methods, which effectively map aquifer boundaries and thickness. For groundwater recharge studies, geophysical techniques are employed to evaluate recharge zones and subsurface porosity. Additionally, water quality investigations utilize induced polarization (IP) surveys to detect contamination and assess salinity levels.

13.2.3.3 Environmental and engineering applications

Geophysical methods are widely used in environmental monitoring, contamination assessment, and infrastructure development. In contamination studies, electrical resistivity and GPR are employed to detect and map subsurface pollution plumes. For infrastructure development, seismic methods and GPR play a key role in foundation studies and utility mapping. Additionally, geophysical imaging is crucial for landslide and sinkhole assessments, as it helps identify weak zones prone to geohazards.

13.2.3.4 Geohazard assessment and mitigation

Understanding and mitigating geohazards necessitate precise geophysical investigations. In earthquake studies, seismic techniques are utilized to monitor fault activity and evaluate seismic risk. For volcanic activity monitoring, gravity and magnetic surveys are employed to detect magma movement beneath the surface. Additionally, slope stability analysis relies on geophysical tools to assess subsurface conditions and identify potential instability in landslide-prone areas.

13.2.3.5 Archaeological and cultural heritage investigations

Noninvasive geophysical methods, such as ground-penetrating radar (GPR) and magnetometry, play a crucial role in archaeological studies. These techniques are highly effective for detecting buried sites, allowing researchers to locate hidden structures and artefacts without excavation. Additionally, high-resolution geophysical mapping supports preservation efforts by aiding in the conservation and management of cultural heritage sites.

13.2.3.6 Case studies and real-world applications

Case studies and real-world applications illustrate the effectiveness of geophysical techniques in addressing diverse challenges. In Chile, the integration of magnetic, gravity, and seismic data has successfully identified copper ore deposits, demonstrating the power of combined methodologies. Groundwater mapping in India, utilizing resistivity and electromagnetic surveys, has proven essential in mitigating water scarcity by accurately identifying aquifers. In Japan, seismic imaging has played a critical role in geohazard assessment, aiding earthquake preparedness and monitoring volcanic activity to ensure public safety.

13.2.3.7 Future directions in geophysical applications

Future directions in geophysical applications are shaped by advancements in technology, sustainability, and data integration. Autonomous drones and AI-driven interpretation are transforming data acquisition and analysis, making surveys more efficient and accurate. The focus on sustainability has led to the development of low-impact, energy-efficient geophysical methods that align with environmental conservation goals. Additionally, integrating geophysical data with machine learning and big data analytics is enhancing predictive capabilities, offering new insights into subsurface structures and processes. These innovations promise to expand the scope and precision of geophysical investigations.

13.2.4 Conclusion

In conclusion, geophysical mapping techniques are indispensable tools for understanding the Earth's subsurface, enabling a wide range of applications from resource exploration to geohazard assessment. The integration of advanced technologies, such as AI and remote sensing, with traditional methods has significantly enhanced the accuracy, efficiency, and scope of these techniques. By providing critical insights into subsurface structures and processes, geophysical mapping supports sustainable exploration, environmental protection, and infrastructure development. As the field continues to evolve, the adoption of innovative, low-impact methods will ensure that geophysical investigations remain both effective and aligned with global sustainability goals.

13.3 Geochemical mapping techniques

Geochemical mapping is a critical aspect of geoscientific exploration, as it provides detailed information about the distribution of chemical elements across large areas. These techniques are applied to assess mineral potential, environmental contamination, and the presence of valuable resources such as metals, oil, and gas. Geochemical data is collected from various sources, such as soils, sediments, water, and rocks, and analyzed to create maps that help guide exploration and resource management strategies.

13.3.1 Geochemical survey types

Geochemical surveys are typically divided into various types based on the medium from which samples are taken. **Stream sediment surveys** involve collecting samples from riverbeds or streams to capture geochemical signals from the underlying geology. **Soil surveys**, on the other hand, involve the collection of surface soil samples, which can reveal elemental compositions indicative of buried mineral deposits. **Rock geochemistry** involves the direct analysis of rock outcrops, often yielding more detailed and localized data on mineral content. Finally, **water sampling** surveys analyze groundwater, lakes, or rivers, providing data on the dissolved chemical elements, which can be indicators of subsurface mineralization.

13.3.2 Sampling techniques

Sampling techniques are fundamental to the success of a geochemical mapping project. **Point sampling** is the most common technique, where individual samples are taken at fixed intervals across the study area, forming a grid to ensure that the data represent the overall geological environment. **Transect sampling** follows a linear path, sampling along a predetermined line to identify geochemical trends across a region. **Composite sampling**, meanwhile, involves combining several smaller samples into one to reduce the influence of local variations and provide a broader regional picture of geochemistry.

13.3.3 Analytical methods

After samples are collected, they are subjected to various analytical methods to determine their chemical composition. Techniques such as **Inductively Coupled Plasma Mass Spectrometry (ICP-MS)** allow for the detection of trace elements at low concentrations. **X-ray Fluorescence (XRF)** is commonly used for on-site, rapid elemental analysis, especially for identifying major elements. **Atomic Absorption Spectroscopy (AAS)** is used primarily for the analysis of metals, while **Neutron Activation Analysis (NAA)** offers sensitivity for detecting even trace amounts of elements in geological samples. These analytical methods are essential for generating accurate geochemical data for mapping.

13.3.4 Data acquisition strategies

Effective data acquisition strategies are necessary to generate reliable and usable geochemical maps. **Grid sampling** is one of the most common strategies, where a uniform grid is used to sample over an area, providing comprehensive and consistent data. This method is effective in regions where mineralization is expected to be spread evenly. In contrast, **targeted sampling** is employed when the geologist has prior knowledge or strong hypotheses about the location of potential deposits. **High-density surveys** may be used in complex geological areas or near known mineralization to ensure that small-scale variations are captured in the data, providing a higher level of resolution.

13.3.5 Data analysis and interpretation

Geochemical data is often analyzed using **statistical and geostatistical techniques**. These methods help to interpret the spatial distribution of elements, highlighting trends, correlations, and anomalies. **Principal component analysis (PCA)** and **cluster analysis** are common methods used to identify patterns in complex datasets, grouping samples with similar geochemical signatures. **Kriging**, a geostatistical interpolation method, is widely used to create continuous maps of geochemical data, filling in gaps between sampled points and providing a detailed representation of element distribution.

13.3.6 Identifying geochemical anomalies

A key objective of geochemical mapping is to identify **geochemical anomalies**, which are areas where the concentration of certain elements deviates significantly from the surrounding region. These anomalies often indicate the presence of mineral deposits or areas of interest for further exploration. For example, elevated levels of copper or gold in stream sediments may suggest a nearby ore body. By analyzing these anomalies, geoscientists can pinpoint target areas for drilling or other detailed investigations.

13.3.7 Integration with remote sensing and geophysical data

Geochemical mapping is often combined with **remote sensing data** such as satellite imagery or aerial photography to help identify surface features that may correlate with subsurface mineralization. Remote sensing can reveal variations in vegetation, soil moisture, and surface temperature, which are linked to underlying geological features. Additionally, **geophysical surveys**, which measure properties such as magnetism, gravity, and electrical resistivity, can complement geochemical data by identifying areas of subsurface alteration or mineralization that are not directly observable on the surface.

13.3.8 Environmental considerations

While conducting geochemical surveys, it is critical to consider the environmental impact of data collection and analysis. Sustainable exploration practices involve minimizing environmental disruption and ensuring that sampling and analytical methods do not introduce harmful contamination to ecosystems. Proper **waste management** procedures must be followed, including the safe disposal of chemical reagents and other materials used during sampling and analysis. Additionally, the results of geochemical surveys are used to assess potential environmental impacts, helping to mitigate risks associated with resource extraction.

13.3.9 Application of geochemical maps in resource exploration

Geochemical maps are invaluable tools in resource exploration. They are used to identify areas that are likely to contain economically viable mineral deposits, allowing exploration teams to focus their efforts on high-potential regions. **Ore deposit exploration** is a primary application, as geochemical anomalies often correlate with the presence of ore bodies. Furthermore, geochemical mapping helps with environmental management by tracking contamination and providing baseline data that can be compared with postexploitation conditions, ensuring that natural resources are extracted in a sustainable and environmentally responsible manner.

13.3.10 Case studies

Case studies provide real-world examples of how geochemical mapping techniques have been applied successfully in resource exploration. These case studies often include detailed discussions of the geochemical methods used, the interpretation of data, and the final outcomes of the exploration efforts. They also highlight the challenges faced during field surveys, including logistical difficulties, environmental considerations, and data interpretation issues. Through these examples, readers can better understand the practical applications of geochemical mapping techniques in diverse geological settings.

13.3.11 Case study 1: Geochemical exploration for gold in the Witwatersrand Basin, South Africa

The Witwatersrand Basin in South Africa is known for its vast gold deposits, making it one of the largest gold-producing regions in the world. A significant portion of the exploration work in this region involved geochemical mapping to identify mineralized zones and gold-bearing formations. Geochemical surveys in the region used stream sediment sampling and soil sampling to analyze the concentration of gold and pathfinder elements such as arsenic, antimony, and copper. These elements are typically associated with gold mineralization. Inductively Coupled Plasma Mass Spectrometry (ICP-MS) was employed to detect trace elements with high precision, and the results were interpreted to highlight anomalies that might indicate the presence of gold ore. The geochemical anomalies detected were consistent with known gold-bearing structures, confirming the accuracy of the sampling methods. The geochemical mapping enabled exploration teams to focus on more promising areas, leading to successful resource delineation (Cawthorn, 2017; Robinson and Chamberlain, 2009).

13.3.12 Case study 2: Geochemical mapping for mineral exploration in the Fraser range, Western Australia

The Fraser Range in Western Australia has become a key exploration area for nickel and gold deposits. In this case, geochemical mapping played an important role in locating mineralized zones beneath the surface, using surface soil and rock samples. The exploration team conducted a soil geochemistry survey in the region, sampling at regular intervals across a grid. The primary focus was on identifying anomalies in nickel (Ni) and cobalt (Co), which are key indicators of nickel sulfide deposits. The team employed X-ray fluorescence (XRF) and ICP-MS for rapid and accurate elemental analysis. The study also used pathfinder elements like chromium and copper to identify areas with potential for nickel mineralization.

Geochemical mapping revealed significant anomalies in areas that aligned with previously known mineralized zones. The team focused on regions with high nickel and cobalt concentrations, which led to the discovery of a large nickel deposit in the Fraser Range (Petrie et al., 2015; Roth and Allen, 2017).

13.3.13 Case study 3: Geochemical mapping for copper in the South American Andes

The Andes mountain range in South America, particularly in countries like Chile and Peru, is rich in copper deposits. Geochemical mapping has been extensively used in these regions to identify potential copper-rich areas for exploration. In this case, stream sediment and soil geochemical surveys were used to detect anomalies in copper concentrations. Inductively Coupled Plasma Optical Emission Spectrometry (ICP-OES) was employed to measure the concentrations of copper and associated elements like molybdenum, arsenic, and silver, which are often found with copper ore bodies. The surveys were conducted in both river valleys and upland areas, and the results were interpreted to delineate geochemical anomalies indicative of hidden copper deposits. The surveys revealed multiple copper anomalies that aligned with known mineralized structures. Geochemical mapping helped to prioritize drilling sites, significantly reducing exploration costs by focusing efforts on the most promising locations. Subsequent drilling led to the discovery of significant copper ore bodies, adding to the region's proven copper reserves (Anderson et al., 2013; Gonzalez and Valdivia, 2016).

13.3.14 Case study 4: Geochemical mapping for environmental monitoring in the Canadian mining regions

Geochemical mapping is also used for environmental monitoring in mining regions to assess the impact of mining activities on surrounding ecosystems. A notable case of this is the Hudson Bay Mining and Smelting Company's geochemical surveys in Northern Manitoba, Canada. In this case, soil sampling and stream sediment sampling were used to monitor the concentrations of contaminants such as mercury, arsenic, and lead. The Atomic Absorption Spectroscopy (AAS) method was applied to determine the levels of toxic metals in water and soil. The survey focused on areas near tailings ponds and historical mining sites to evaluate the environmental impact. The mapping identified elevated concentrations of metals near mining facilities, indicating areas of contamination. The study also highlighted areas where natural processes might be mitigating the effects of contamination. These results were used to develop a remediation plan to restore environmental quality in the affected regions and minimize the impact of future mining operations (Levesque and Brown, 2018;Parker and Hughes, 2020).

13.3.15 Case study 5: Geochemical mapping for uranium exploration in Namibia

Namibia is one of the world's leading producers of uranium, and geochemical mapping has been a key method for exploring uranium deposits in the country's arid desert regions.

In this case, soil sampling and rock geochemistry were used to detect uranium anomalies. The use of Gamma-ray Spectrometry (GRS) allowed for the detection of natural radiation levels in surface samples, which is useful in identifying uranium-bearing mineralization. ICP-MS and XRF were employed for detailed elemental analysis, focusing on uranium and other elements such as vanadium and molybdenum, which are often associated with uranium deposits. Geochemical mapping successfully identified uranium anomalies across a wide area, leading to the discovery of several significant uranium deposits. The mapping helped to define the extent of mineralized zones, facilitating more accurate exploration and mining strategies in this critical resource region (Jansen and Kotze, 2015; Schneider and Van Zyl, 2016).

These case studies showcase the versatility of geochemical mapping techniques across various regions and resources. From gold exploration in South Africa to environmental monitoring in Canada, the ability of geochemical surveys to provide detailed data for exploration and environmental management is essential in sustainable resource extraction. Each of these case studies highlights the importance of integrating different sampling methods, analytical techniques, and data interpretation approaches to achieve accurate and actionable results.

13.4 Integration of geological, geophysical, and geochemical data

13.4.1 Rationale for integrated data acquisition

The integration of geological, geophysical, and geochemical data plays a crucial role in providing a more comprehensive and accurate understanding of geoscientific phenomena. By combining these diverse datasets, researchers and decision-makers can gain a holistic view of the Earth's subsurface and surface conditions, enhancing their ability to make informed decisions regarding resource exploration, environmental monitoring, and geological hazard prediction.

13.4.1.1 Comprehensive understanding of geoscientific phenomena

Geological, geophysical, and geochemical data each provide distinct types of information about the Earth. Geological data typically describe the structure, composition, and history of rock formations, while geophysical data offer insights into the physical properties of the Earth's subsurface, such as density, magnetism, and conductivity. Geochemical data, on the other hand, reveal the chemical composition of rocks, minerals, and fluids in the Earth's interior. When combined, these datasets provide a more detailed and interconnected view of the Earth's processes, making it easier to identify relationships between different layers, materials, and forces that govern geological activity. For example, understanding the link between geological formations and geophysical anomalies can provide insight into the location of mineral deposits or oil reservoirs (Schenk et al., 2014).

13.4.1.2 Enhanced resource exploration

In resource exploration, the integration of geological, geophysical, and geochemical data can significantly improve the identification of potential areas for mineral or hydrocarbon exploration. Geophysical surveys can detect anomalies in the Earth's subsurface that may indicate the presence of valuable resources, while geochemical analysis can confirm the types of minerals or elements present. Geological mapping further helps to constrain the location and depth of these resources. By synthesizing these different data types, exploration teams can focus their efforts more precisely, increasing the likelihood of discovery and minimizing the cost of unsuccessful drilling operations (Wessel et al., 2018).

13.4.1.3 *Improved environmental monitoring*

The integration of these data types also enhances environmental monitoring efforts, particularly when assessing the effects of human activity on the Earth. For instance, combining geochemical data with geophysical surveys can help track contamination in groundwater or soil, and geological data can identify the movement of contaminants over time. Such integrated datasets allow for more accurate environmental assessments and more effective remediation strategies. In addition, the integration of real-time data acquisition systems, such as remote sensing or sensor networks, enables continuous monitoring, helping to predict and mitigate environmental hazards like soil erosion or water pollution (Meyer et al., 2015).

13.4.1.4 *Better geological hazard prediction*

Integrated geoscientific data is also critical in predicting geological hazards, such as earthquakes, landslides, and volcanic eruptions. Geophysical data can provide early warning signals of tectonic activity or ground deformation, while geological data help identify fault lines and historical patterns of seismic activity. Geochemical analysis can monitor changes in gas emissions or magma composition, which may signal volcanic unrest. By integrating these data sources, researchers can create more accurate models of geological hazards and improve risk assessments for vulnerable areas, leading to better preparedness and mitigation strategies (Cox et al., 2012).

13.4.2 Methods of data integration

Data integration is crucial in geoscientific applications, particularly in the exploration of natural resources. The integration of various types of geoscientific data—such as geological, geophysical, and geochemical information—provides a more complete understanding of the subsurface. There are several methodologies that can be employed to effectively integrate these datasets, each contributing to a more informed and comprehensive exploration strategy. Among the most widely used methods are spatial analysis, multi-criteria decision analysis (MCDA), 3D geological modeling, data fusion, and machine learning techniques. Geographic Information Systems (GIS) play a pivotal role in facilitating and enhancing these methods by providing tools for data storage, analysis, and visualization.

Spatial analysis involves the interpretation and manipulation of spatial data, typically using GIS tools (Long, 2016). One of the main techniques in spatial analysis is **overlay analysis**, where different geoscientific data layers—such as geological maps, geophysical data, and geochemical surveys—are overlaid to identify areas of interest for further exploration. This technique helps to highlight areas with favorable conditions for resource potential. Another important spatial technique is **buffer zone analysis**, which helps assess proximity to specific geological features like faults or mineralization zones, critical for understanding resource distribution. **Spatial interpolation**, such as kriging, is also essential in predicting values at unsampled locations, which helps in generating continuous surfaces like mineral concentration maps. GIS plays a central role in spatial analysis by enabling the manipulation of multiple data layers, conducting advanced spatial queries, and visualizing the integration of these datasets in a coherent format (Goodchild, 2007).

Multi-Criteria Decision Analysis (MCDA) is another essential method used to integrate diverse datasets by evaluating multiple criteria and making decisions based on their relative importance. In MCDA, various datasets are assessed according to factors like resource potential, environmental concerns, and accessibility. **Weighting of criteria** is a key step, where each dataset or factor is assigned a weight based on its significance. The data layers are then combined based on these weights to produce a decision model. **Ranking and scoring** techniques allow for the identification of the most suitable exploration targets by assigning scores to different regions based on their overall suitability. Furthermore, **sensitivity analysis** is performed to understand how changes in the weights and input data affect the final decision, ensuring the robustness of the model. GIS plays a fundamental role in this process by providing the platform to integrate different spatial data layers, run analysis models, and visualize decision outcomes as maps that guide further exploration (Malczewski, J., 1999).

3D geological modeling is another powerful tool for integrating geoscientific data. It allows geoscientists to create three-dimensional representations of subsurface features, providing insights into the distribution of mineral resources and structural elements. Geostatistical modeling is commonly used in this context to quantify spatial variability and create accurate 3D models of geological formations. These models help in visualizing and understanding the geometry and volume of mineral deposits, which are essential for resource estimation and planning extraction strategies. Integrating remote sensing and geophysical data into 3D geological models enhances their accuracy, as these data types provide complementary information about subsurface structures. GIS supports 3D geological modeling by providing tools for visualizing and analyzing spatial relationships in three dimensions, and by enabling the combination of various data sources, such as geological surveys, remote sensing data, and geophysical measurements.

Data fusion refers to the process of combining multiple datasets from different sources or sensors to create a unified and more accurate representation of geological conditions. In geoscience, this typically involves combining remote sensing data, geological surveys, and geophysical measurements to improve spatial resolution and capture a broader range of geological features. One technique is **sensor fusion**, where data from various sensors (e.g., satellite imagery, LiDAR, or geophysical surveys) are merged to enhance the accuracy of subsurface models. **Temporal data fusion** is used when datasets are collected over time, such as seismic monitoring or satellite imagery, to track changes in the landscape or subsurface. GIS tools are essential in this process, as they provide a platform for merging different types of data (e.g., raster and vector data) and visualizing the fused datasets, which improves decision-making for exploration (Kang and Ghosh, 2018).

Finally, **machine learning and artificial intelligence (AI)** offer advanced methods for data integration by automating the detection of patterns and relationships within large datasets. These techniques can be applied to identify correlations between different geophysical, geological, and geochemical features that may not be immediately apparent through traditional methods. For example, **pattern recognition algorithms** can help identify areas with high potential for mineral deposits based on historical data, geophysical anomalies, and remote sensing information. **Predictive modeling** uses machine learning algorithms, such as classification and regression techniques, to predict the likelihood of finding resources in unexplored areas based on the integration of various geoscientific datasets. GIS plays a key role in incorporating machine learning models by providing the necessary platform for integrating geospatial data with predictive models, visualizing model outputs such as probability maps, and helping decision-makers interpret these results in the context of exploration planning.

13.4.3 Challenges and opportunities in data integration

13.4.3.1 Challenges in data integration

13.4.3.1.1 Data consistency

One of the primary challenges in data integration is ensuring consistency across different datasets. Data often come from various sources with differing formats, resolutions, and quality levels. Ensuring that data from geological surveys, geophysical data, and remote sensing are comparable and can be merged effectively requires rigorous preprocessing and standardization. Inconsistent units, missing values, and discrepancies in data formats can hinder the integration process, making it difficult to derive meaningful insights from the combined datasets.

13.4.3.1.2 Scaling issues

The scalability of data integration systems is another challenge. Geoscientific datasets are often large, involving vast geographic areas with detailed measurements over time. Integrating such large datasets across different spatial and temporal scales can lead to computational challenges, especially when dealing with high-resolution data. Managing and processing these large datasets requires robust infrastructure and advanced computational tools, which may not always be available or affordable for smaller organizations or projects.

13.4.3.1.3 Integration of heterogeneous datasets

Integrating heterogeneous datasets—data that come from different sources, in different formats, and at different scales—poses significant methodological challenges. For example, combining geological maps, geophysical survey data, and satellite imagery requires aligning datasets with different spatial resolutions and types (e.g., raster vs. vector data). Developing effective methods to harmonize these datasets while preserving their individual characteristics is a complex task, requiring advanced algorithms and specialized software tools.

13.4.3.2 Opportunities in data integration

13.4.3.2.1 Improved decision-making

Integrating multiple geoscientific datasets offers significant opportunities to enhance decision-making. By merging geological, geophysical, and geochemical data, a more comprehensive and accurate picture of the subsurface can be achieved. This integrated approach enables decision-makers to evaluate resource potential more effectively, assess environmental risks, and identify the most promising exploration targets. GIS and other integration tools provide an interactive platform for visualizing combined datasets, making it easier for stakeholders to make informed and data-driven decisions.

13.4.3.2.2 Enhanced predictive accuracy

One of the key benefits of integrated datasets is the potential for improved predictive accuracy. By synthesizing data from different domains, models become more robust and reliable in forecasting the location and characteristics of natural resources. For example, combining geological models with geophysical and remote sensing data can lead to more accurate predictions about the size, shape, and accessibility of mineral deposits. Integrated data enables more precise risk assessments and better resource estimation, which is critical for sustainable exploration and extraction strategies.

13.4.3.2.3 Automation and machine learning opportunities

Data integration opens up opportunities for leveraging machine learning and artificial intelligence (AI) to automate analysis and enhance predictive capabilities. AI algorithms can process large volumes of integrated data to identify patterns and correlations that would be difficult for human analysts to detect. This automation can speed up decision-making processes and improve the accuracy of resource forecasting, particularly when handling complex, multi-dimensional datasets.

13.4.4 Conclusion

13.4.4.1 Value of integrated geoscientific data

Integrated geoscientific data provides a comprehensive and unified view of the subsurface, combining various datasets such as geological, geophysical, and geochemical information. This holistic approach enhances the accuracy and reliability of geological models, enabling more informed exploration and resource management decisions.

13.4.4.2 Emerging techniques for robust geological models

Emerging techniques, such as 3D geological modeling, multi-criteria decision analysis (MCDA), machine learning, and advanced GIS tools, are transforming how geoscientific data is integrated. These methods allow for more dynamic and accurate predictions, improving the understanding of geological features and resource potential.

13.4.4.3 Holistic geological models

By combining diverse datasets with these advanced techniques, geoscientists can build more robust and holistic geological models that account for complex subsurface structures, resource distribution, and environmental factors. This integration leads to better risk assessment, decision-making, and ultimately, more sustainable natural resource exploration.

13.5 Future trends in geoscientific data acquisition and strategy

13.5.1 Advancements in remote sensing and monitoring technologies

In recent years, advancements in remote sensing, satellite technology, drones, and IoT (Internet of Things) sensors have significantly transformed geoscientific data acquisition. These emerging technologies offer enhanced capabilities for gathering real-time data in challenging or hazardous environments, contributing to more efficient, accurate, and timely decision-making in natural resource exploration and monitoring.

13.5.1.1 *Drones and UAVs (unmanned aerial vehicles)*

Drones have become one of the most impactful tools for geoscientific data acquisition. Equipped with high-resolution cameras, LiDAR (Light Detection and Ranging), multispectral sensors, and other geophysical instruments, drones can capture detailed data from remote and difficult-to-reach locations. They are particularly useful for surveying large areas with minimal human intervention, allowing geologists and researchers to gather data in environments that would otherwise be too dangerous or inaccessible, such as active volcanoes, rugged terrains, or contaminated sites. Drones' ability to cover large areas quickly and efficiently while collecting high-resolution imagery and geophysical data has proven invaluable in the exploration of natural resources (Tibuleac and Samulski, 2021). Moreover, UAVs are cost-effective and time-efficient, reducing the need for expensive traditional methods such as manned aircraft or ground-based surveys.

13.5.1.2 *Satellites and remote sensing technologies*

Satellites have been a cornerstone of geoscientific data acquisition for decades, providing valuable information on a global scale. Modern Earth observation satellites equipped with multispectral and hyperspectral sensors offer a broad array of data for mapping terrain, analyzing vegetation health, monitoring soil moisture levels, and detecting mineral deposits. Advances in satellite imaging technologies, such as synthetic aperture radar (SAR) and high-resolution optical sensors, allow for detailed monitoring of surface changes, such as land subsidence or deforestation, in near real-time (Jensen and McDonnell, 2020). The ability to monitor large, inaccessible areas remotely using satellite data enables geoscientists to make informed decisions regarding exploration, conservation, and disaster management. Furthermore, satellite data can be combined with ground-based data, increasing the overall accuracy and comprehensiveness of geoscientific studies.

13.5.1.3 *Internet of Things (IoT) sensors*

The integration of IoT sensors into geoscientific data acquisition systems has significantly improved the ability to collect continuous, real-time data. IoT sensors can be deployed in a variety of environments, from deep-sea exploration to geological fault lines or mining operations. These sensors monitor parameters such as temperature, pressure, vibration, chemical composition, and seismic activity. The data collected by these sensors can be transmitted wirelessly to central databases, enabling real-time monitoring and analysis.

In remote and hazardous areas, IoT sensors help minimize the need for human intervention, reducing risks and enhancing safety. For example, IoT sensors installed in active mining sites can monitor structural integrity and warn of potential collapses, while sensors in oil rigs can detect leaks or irregularities in equipment performance.

13.5.1.4 *The convergence of technologies*

The real power of these emerging technologies lies in their integration. Combining drones, satellite imagery, and IoT sensors provides a more holistic and dynamic approach to geoscientific data acquisition. For instance, satellite data can offer an overview of large-scale environmental changes, while drones can focus on high-resolution, localized data collection. IoT sensors embedded in these systems allow for continuous monitoring, delivering near-real-time updates and facilitating proactive decision-making. This convergence of technologies is enabling the development of "smart" exploration systems that can automatically adapt to changing conditions, process vast amounts of data, and provide actionable insights.

13.5.2 The role of artificial intelligence and machine learning

Artificial Intelligence (AI) and Machine Learning (ML) have become critical tools in geoscientific data analysis, predictive modeling, and automated data processing. By leveraging large datasets, AI and ML algorithms can uncover complex patterns, enabling geoscientists to make more accurate predictions about natural resource deposits, environmental changes, and geological hazards. AI and ML models excel in handling vast amounts of geospatial data, such as satellite imagery and sensor outputs, enabling more efficient data interpretation. For example, AI-based systems can automate the classification of geological features from satellite images, reducing the time required for manual analysis (Jensen and McDonnell, 2020). Furthermore, predictive modeling techniques powered by machine learning help forecast the likelihood of natural events, like earthquakes or landslides, based on historical data and real-time inputs. These models can also assist in resource exploration by identifying areas with high potential for mineral or energy deposits, improving exploration efficiency. The integration of AI and ML into geoscientific workflows not only enhances the accuracy of data analysis but also increases the speed at which results are obtained, offering significant improvements in both research and practical applications, such as risk management and exploration strategies (Gopalakrishnan et al., 2021).

13.5.3 Big data in geosciences

Big data has revolutionized geoscientific research by enabling the collection, storage, and analysis of vast amounts of diverse data from various sources, including satellite imagery, sensor networks, seismic surveys, and geophysical measurements. This massive influx of data has provided geoscientists with unprecedented opportunities to gain insights into Earth processes, resource exploration, and environmental monitoring.

The application of big data in geosciences allows for the integration of various types of data—such as spatial, temporal, and geophysical data—into unified models, improving both accuracy and predictive capabilities. For example, big data analytics can be used to analyze seismic data to identify potential oil and gas reservoirs, track climate change patterns, or study earthquake prediction models. By using sophisticated algorithms and computational tools, researchers can handle complex datasets and extract meaningful patterns that would otherwise remain undetected through traditional methods.

Moreover, big data enables real-time monitoring and decision-making in geoscientific applications. In resource exploration, real-time data from sensors and satellites can be processed and analyzed immediately, allowing for quicker assessments of potential resources and hazards. The integration of big data with AI and machine learning enhances predictive models, making them more reliable and faster at processing large datasets (Gopalakrishnan et al., 2021; Sweeney and Johnson, 2020). The scalability and versatility of big data solutions also allow for more comprehensive analysis of large geographic areas or complex environmental systems, fostering improved strategies for resource management, disaster risk reduction, and environmental sustainability.

13.5.4 Environmental and sustainability considerations

Sustainability has become an essential focus in geoscientific data acquisition, driving practices to minimize environmental impacts, ensure responsible resource extraction, and protect biodiversity. As the need for natural resources grows, geoscientific methodologies are evolving to address these concerns.

13.5.4.1 Minimizing environmental impacts

To reduce the environmental footprint of exploration, advanced geophysical methods like remote sensing, satellite imagery, and noninvasive sensors are increasingly used. These methods allow for detailed data collection without disturbing ecosystems. Drones, in particular, provide an eco-friendly alternative to traditional ground-based surveys, enabling the collection of high-resolution data with minimal disruption to sensitive environments (Miller et al., 2020). By reducing reliance on heavy machinery, these tools help ensure that resource exploration is conducted with minimal ecological impact.

13.5.4.2 Responsible resource extraction

Geoscientific data plays a critical role in enabling responsible resource extraction. Technologies like geophysical surveys and satellite monitoring allow for a better understanding of resource distribution while assessing potential environmental risks before extraction begins (Smith and Johnson, 2021). Predictive modeling techniques, powered by AI and machine learning, are also becoming integral in assessing the long-term environmental consequences of extraction activities. These models help forecast potential damage, such as groundwater contamination or habitat loss, enabling informed decision-making that prioritizes sustainability.

13.5.4.3 Protecting biodiversity

Protecting biodiversity remains a crucial consideration in geoscientific practices. Many natural resources are located in ecologically sensitive areas, where exploration can pose risks to wildlife. Noninvasive technologies, such as IoT sensors and satellite monitoring, enable real-time tracking of environmental health, providing critical information to prevent damage. These technologies allow for early intervention to mitigate adverse effects on biodiversity (Gopalakrishnan et al., 2021). By incorporating biodiversity assessments into geoscientific studies, potential risks to ecosystems can be minimized during resource extraction.

13.6 Conclusion: Summary and future directions

13.6.1 Key takeaways

This chapter highlights the essential role of geological, geophysical, and geochemical data in shaping geoscientific planning and strategy. These datasets are fundamental in understanding natural resource distribution, environmental changes, and potential geological hazards. The key techniques and methods discussed include advanced remote sensing, geophysical surveys, satellite monitoring, and predictive modeling using AI. These tools enable more precise and sustainable resource extraction, real-time environmental monitoring, and risk assessment. Additionally, the integration of these data types supports informed decision-making in resource management and conservation.

13.6.2 Challenges and areas for improvement

Geoscientific data acquisition faces several challenges, primarily related to data quality, interoperability, and accessibility. The vast and complex nature of geospatial data can lead to inconsistencies and difficulties in integrating information from different sources. Data accessibility remains a critical issue, particularly in developing regions where infrastructure for data collection and storage is limited. These challenges can be mitigated through technological innovations, such as the development of standardized data formats, advanced data-cleaning algorithms, and improved data-sharing platforms. Collaboration between researchers, government bodies, and industry leaders will also play a pivotal role in addressing these barriers.

13.6.3 Future trends and opportunities

Emerging trends in geoscience, such as artificial intelligence (AI), machine learning (ML), and real-time monitoring, are set to transform the field of data acquisition and analysis. AI and ML are revolutionizing the processing of large-scale geoscientific datasets, enabling faster, more accurate predictions and identifying patterns that were previously undetectable. Real-time monitoring systems, powered by IoT sensors and satellite technology, offer continuous data collection, allowing for timely interventions in resource management, environmental protection, and disaster response. These advancements will significantly enhance geoscientific research and planning, making it more efficient, accurate, and sustainable.

13.6.4 Final thoughts

In conclusion, continued investment in geoscientific data acquisition strategies, technology, and collaboration is critical to addressing global challenges such as climate change, resource depletion, and environmental degradation. Data-driven decision-making is essential for sustainable management of the Earth's resources, ensuring that future generations can benefit from a well-managed and resilient planet. By embracing innovation and fostering collaboration, we can overcome current challenges and unlock new opportunities for the geosciences.

CHAPTER

14

Practical exercises in geophysical and remote sensing data analysis

14.1 Practical approaches to geoscience data analysis

This chapter will serve as a bridge between theoretical knowledge and practical application. It will focus on providing computer-based exercises, programming code examples (such as Python, R, or MATLAB), and datasets that will allow students to apply various geophysical and remote sensing techniques discussed in the previous chapters. These exercises will help students become proficient in using tools and methods for analyzing geophysical data, interpreting results, and solving real-world exploration problems.

14.1.1 Building essential data analysis skills for geoscience students

In the rapidly advancing field of geoscience, particularly within mineral exploration and geophysical survey techniques, students must develop a deep understanding of not only theoretical principles but also practical application. The ability to analyze, interpret, and visualize geophysical data through computational methods is essential for modern geoscientists. While traditional geoscience education has focused heavily on theoretical knowledge and fieldwork, the increasing complexity of datasets and the growing importance of sustainable resource exploration demand that students gain practical experience with data analysis tools. Practical skills in data analysis empower students to:

(i) **Apply Theoretical Knowledge**: Bridging the gap between theory and practice, students will better understand how the geophysical methods learned in earlier chapters (e.g., seismic, geoelectrical, remote sensing) are implemented in real-world scenarios.
(ii) **Enhance Problem-Solving Abilities**: Working with actual data enables students to face real-world challenges and apply critical thinking to interpret complex geophysical results.
(iii) **Prepare for Industry Needs**: Geoscientists are increasingly expected to handle large datasets, model data, and use computational techniques in various industries such as mineral exploration, environmental monitoring, and oil and gas exploration.

Baseline Geoscience Data Generation for Sustainable Exploration of Natural Resources
https://doi.org/10.1016/B978-0-443-45221-5.00018-7

Practical skills are no longer supplementary but fundamental to geoscience education. As the field grows more data-centric, learning to manipulate and interpret data using programming languages and analytical tools will set students apart in the professional world.

14.1.2 Introduction to the tools and programming languages used for data analysis in geophysical exploration

Several programming languages and software tools are widely used in the field of geophysical data analysis. Each tool or language offers unique advantages depending on the type of analysis and the complexity of the dataset. Below is an overview of the most commonly used tools in the field:

14.1.2.1 Python

Why Python? Python has become one of the most widely used languages in geoscience data analysis due to its simplicity, versatility, and robust libraries. With packages like **NumPy**, **Pandas**, **Matplotlib**, and **SciPy**, Python is well-suited for data manipulation, visualization, and statistical analysis. Python's versatility makes it a popular choice for tasks ranging from simple data processing to complex machine learning applications.

Applications in Geoscience: In geophysics, Python is used for processing seismic, gravity, magnetic, and electromagnetic data. For example, Python can be used to filter and analyze noise in seismic data or to model gravity anomalies in mineral exploration.

Example: "Geophysical Data Analysis with Python: Principles and Applications" by Smith and O'Neill (2017) discusses Python's role in data analysis in geophysics, with a focus on seismic and gravity data processing.

14.1.2.2 MATLAB

Why MATLAB? MATLAB is particularly popular for numerical computing and data visualization, providing an extensive environment for matrix-based operations. It is often favored for its built-in functions tailored to geophysical data analysis, including spectral analysis, data inversion, and filtering techniques.

Applications in Geoscience: MATLAB is commonly used in seismic processing, electromagnetic modeling, and geophysical data inversion. MATLAB's extensive toolbox support (e.g., Signal Processing Toolbox) allows geoscientists to perform high-level analysis with less custom coding.

Example: MATLAB is widely used for seismic data processing, including the application of Fourier transforms to analyze seismic waves in mineral exploration (MathWorks, 2021).

14.1.2.3 R

Why R? R is a powerful programming language for statistical computing and data visualization. It is often used in geospatial data analysis, statistical modeling, and graphical presentations. R's rich set of libraries, like **ggplot2** for data visualization, **sp** for spatial data analysis, and **geoR** for geostatistical analysis, makes it highly suited for geophysical exploration.

Applications in Geoscience: In geophysics, R is utilized to process remote sensing data, analyze mineral distributions, and perform geostatistical analysis on geospatial data.

Example: "Applied Spatial Data Analysis with R" by Bivand et al. (2013) highlights the use of R for spatial data analysis, which is essential in many geophysical survey applications, especially in mineral exploration.

14.1.2.4 GIS software

Why GIS? Geographic Information Systems (GIS) software is essential for visualizing and analyzing spatial data. Popular GIS platforms, such as **ArcGIS** and **QGIS** (QGIS Documentation, 2021**)**, allow geoscientists to overlay geophysical data with geographic features, creating valuable visual representations of mineral deposits, geological structures, and other resources.

Applications in Geoscience: GIS is widely used in mapping mineral resources, analyzing geospatial survey data, and creating detailed resource maps to aid in exploration.

14.2 Geophysical data analysis using Python

14.2.1 Overview

Python has become one of the most popular programming languages for geophysical data analysis due to its simplicity, flexibility, and wide range of scientific libraries. With packages such as **NumPy**, **Pandas**, **Matplotlib**, and **SciPy**, Python provides powerful tools for processing, analyzing, and visualizing geophysical data. In this section, we will focus on using Python to process geophysical survey data, such as gravity and magnetic data, and demonstrate how to apply basic data analysis techniques like filtering, normalization, and anomaly detection.

14.2.2 Code example: Basic Python code for processing geophysical survey data

Here's an example of Python code for processing geophysical data (e.g., gravity or magnetic data). In this example, we will use a dataset that represents gravity anomaly values along a profile.

14.2.2.1 Python code

Python

```
# Importing necessary libraries
import NumPy as np
import pandas as pd
import matplotlib.pyplot as plt
from SciPy.Signal import butter, filtfilt
# Sample gravity_data (anomalies in microGal) and corresponding distance (m)
distance = np.Linspace (0, 100, 100) # Distance along the profile in meters
gravity data = np.random.normal (0, 0.5, 100) + np.sin (distance 10) # Simulated
gravity data with noise
# Create a DataFrame
data = pd.DataFrame ({'Distance': distance, 'Gravity_Anomaly': gravity_data})
# Plot the raw gravity data
```

```
plt.figure (figsize = (10, 6))
plt.plot (data['Distance'], data['Gravity_Anomaly'], label = 'Raw Gravity Data',
color = 'b')
plt.title ("Raw Gravity Anomaly Data")
plt.xlabel ("Distance (m)")
plt.ylabel ("Gravity Anomaly (microGal)")
plt.legend ()
plt.show ()
# Function to apply a Butterworth filter for noise reduction
def butter_lowpass_filter (data, cutoff, fs, order = 4):
Nyquist = 0.5 * fs
normal_cutoff = cutoff/Nyquist
b, a = butter (order, normal_cutoff, btype = 'low', analog = False)
y = filtfilt (b, a, data)
return y
# Apply a low-pass filter to remove high-frequency noise
filtered_data = butter_lowpass_filter (data ['Gravity_Anomaly'], cutoff = 0.1, fs
= 1.0)
# Plot the filtered gravity data
plt.figure (figsize = (10, 6))
plt.plot (data ['Distance'], filtered_data, label = 'Filtered Gravity Data', color
= 'g')
plt.title ("Filtered Gravity Anomaly Data")
plt.xlabel ("Distance (m)")
plt.ylabel ("Gravity Anomaly (microGal)")
plt.legend ()
plt.show ()
# Normalizing the data to a 0-1 range
normalized_data = (filtered_data - np . min (filtered_data))/(np . max (filtered_
data) - np.min (filtered_data))
# Plot the normalized data
plt.figure (figsize = (10, 6))
plt.plot (data ['Distance'], normalized_data, label = 'Normalized Gravity Data',
color = 'r')
plt.title ("Normalized Gravity Anomaly Data")
plt.xlabel ("Distance (m)")
plt.ylabel ("Normalized Gravity Anomaly")
plt.legend ()
plt.show ()
# Anomaly detection using a simple threshold (finding values greater than a threshold)
threshold = 0.7 # Threshold for anomaly detection
anomalies = data [normalized_data > threshold]
# Plot anomalies on the data
plt.figure (figsize = (10, 6))
plt.plot (data ['Distance'], normalized_data, label =''Normalized Gravity Data',
color = 'r')
plt.scatter (anomalies ['Distance'], normalized_data [normalized_data > threshold],
color = 'orange', label = 'Anomalies')
plt.title ("Anomaly Detection in Gravity Data")
```

```
plt.xlabel ("Distance (m)")
plt.ylabel ("Normalized Gravity Anomaly")
plt.legend ()
plt.show ()
print ("Anomalies Detected at: ", anomalies ['Distance']. values)
```

14.2.2.2 *Explanation of code*

14.2.2.2.1 Simulated gravity data

We generate a sample dataset representing gravity anomalies along a profile using NumPy. The dataset includes both random noise and a simple sinusoidal trend to simulate real-world gravity anomalies.

14.2.2.2.2 Plotting raw data

We plot the raw gravity anomaly data along the profile using matplotlib to visualize the data before any processing.

14.2.2.2.3 Low-pass filtering

We apply a low-pass Butterworth filter to remove high-frequency noise from the data using the scipy.signal.butter and scipy.signal.filtfilt functions. This is a common technique to smooth noisy geophysical data.

14.2.2.2.4 Normalization

We normalize the filtered gravity data so that it falls within the range [0, 1]. This technique is useful when comparing datasets with different ranges or scales.

14.2.2.2.5 Anomaly detection

We perform anomaly detection by identifying values that exceed a specified threshold. In this case, anomalies are defined as values greater than 0.7 after normalization.

14.2.2.2.6 Plotting anomalies

The anomalies are highlighted on the normalized gravity data plot in orange to visually represent the locations of potential geophysical anomalies (such as mineral deposits or faults).

14.2.3 Exercise: applying Python for geophysical data analysis

14.2.3.1 *Objective*

In this exercise, students will use Python to process and analyze a sample dataset of gravity anomalies, applying filtering, normalization, and anomaly detection techniques. They will also visualize the results to better understand how these methods can be used to identify geophysical features.

14.2.3.2 *Step-by-step instructions*

14.2.3.2.1 Download the data

Provide students with a dataset of gravity anomalies. You can either create a sample dataset or provide real geophysical data (such as gravity data from a specific survey region).

14.2.3.2.2 Plot the raw data

Have students plot the raw gravity data and visually inspect it for noise and potential anomalies. Ask them to describe the data characteristics (e.g., noise, trends).

14.2.3.2.3 Apply a low-pass filter

Instruct students to apply a low-pass Butterworth filter (or another filtering technique of their choice) to the data. Ask them to compare the filtered data with the raw data and describe how the filtering process has reduced noise.

14.2.3.2.4 Normalize the data

Have students normalize the filtered data and plot the results. Encourage them to explain why normalization is important in geophysical data analysis.

14.2.3.2.5 Perform anomaly detection

Ask students to apply an anomaly detection method (such as using a threshold value) to the normalized data. They should highlight the locations of anomalies and discuss what these anomalies could represent in a real-world geophysical survey.

14.2.3.2.6 Interpret the results

Finally, ask students to interpret the results. What might the identified anomalies indicate? How could these anomalies be relevant for further exploration or resource assessment?

14.2.3.3 Deliverables

Python code for processing the data. Plots of raw, filtered, and normalized data, with anomalies highlighted. A brief report summarizing their findings, including the locations of anomalies and any observations they made during the analysis.

This section introduces students to the fundamentals of geophysical data analysis using Python. They learn how to filter, normalize, and detect anomalies in gravity survey data, all essential tasks for processing real-world geophysical datasets. The exercises provide a hands-on opportunity for students to gain practical experience with Python, preparing them for more advanced geophysical data analysis in future chapters.

14.3 Remote sensing data processing with R

14.3.1 Overview

Remote sensing involves collecting data about the Earth's surface from a distance, usually via satellites or aerial platforms. Remote sensing data is commonly used for a variety of geospatial analyses, such as land cover classification, vegetation analysis, and mineral mapping. R, with its powerful libraries for geospatial analysis such as **raster**, **sp**, **rgdal**, and **ggplot2**, is an excellent tool for processing remote sensing data. In this section, we will explore how to use R for basic remote sensing data processing tasks, including vegetation index calculations and land cover classification (Jensen, 2015).

14.3.2 Code example: R code for processing satellite imagery or aerial data

In this example, we will process a sample remote sensing image (e.g., a satellite image) using R to calculate the **Normalized Difference Vegetation Index (NDVI)**, a commonly used index for vegetation analysis, and visualize the results.

14.3.2.1 R code

```
R
# Loading necessary libraries
Library (raster) # For working with raster data
Library (ggplot2) # For plotting
Library (RColorBrewer) # For color schemes
# Load satellite image (assume we have a multispectral image with bands: Red and
NIR)
# Example: You can use your own raster data file or download publicly available
satellite imagery
# For demonstration, we will create a simple example with artificial data
red_band <- raster (matrix (runif (100, min = 0, max = 255), nrow = 10)) # Simulated
Red band
nir_band < - raster (matrix (runif (100, min = 0, max = 255), nrow = 10)) # Simulated
NIR band
# Plot the red and NIR bands
par (mfrow = c (1,2))
plot (red_band, main = "Red Band")
plot (nir_band, main = "NIR Band")
# Calculate NDVI: (NIR - Red)/(NIR + Red)
NDVI < - (nir_band - red_band)/(nir_band + red_band)
# Plot NDVI
plot (NDVI, main = "NDVI (Normalized Difference Vegetation Index)", col = terrain
. colors (100))
# Color NDVI into a better range for visualization
NDVI_colored < - reclassify (NDVI, c (-Inf, -0.1, 1, 0.1, Inf)) # Reclassifying NDVI
to better show different ranges
# Plot the classified NDVI image with color scheme
plot (NDVI_colored, main = "Classified NDVI", col = brewer .pal (9, "YlGn"))
# Save the NDVI image as a new raster file (optional)
writeRaster (NDVI, "NDVI_image.tif", format = "GTiff")
```

14.3.2.2 Explanation of code

14.3.2.2.1 Loading libraries

We start by loading necessary libraries: **raster** for working with raster data, **ggplot2** for plotting, and **RColorBrewer** for advanced color schemes.

14.3.2.2.2 Loading the data

In this example, we simulate two raster bands (Red and NIR) using random data. These bands represent a multispectral image where each band corresponds to different wavelengths of light. In practice, you would load actual satellite data (e.g., from a Landsat or Sentinel-2 image) using the raster () function.

14.3.2.2.3 Plotting the bands

We plot both the Red and NIR bands to visualize their pixel values.

14.3.2.2.4 Calculating NDVI

The NDVI is calculated using the formula:

$$NDVI = \frac{(NIR - RED)}{(NIR + RED)} \tag{14.1}$$

This index is a measure of vegetation health, with higher values indicating more vegetation.

14.3.2.2.5 Visualizing NDVI

We visualize the NDVI using a color scheme that represents different vegetation states. This helps in understanding vegetation distribution across the study area.

14.3.2.2.6 Reclassifying NDVI

We reclassify the NDVI to group the values into different categories for easier interpretation, such as distinguishing between healthy vegetation and barren areas.

14.3.2.2.7 Saving the NDVI image

Finally, we save the NDVI image as a new raster file for further analysis or use.

14.3.3 Exercise: Students using R to analyze remote sensing data

14.3.3.1 Objective

In this exercise, students will process and analyze remote sensing data using R. They will calculate vegetation indices (like NDVI), classify land cover, and analyze different aspects of the environment from satellite imagery.

14.3.3.2 Step-by-step instructions

14.3.3.2.1 Download the remote sensing data

Provide students with access to a sample remote sensing dataset, such as **Landsat 8** imagery, which includes multiple bands (Red, NIR, Green, Blue, etc.). Students can download public satellite data from sources like USGS Earth Explorer or use free datasets from NASA Earth Science.

14.3.3.2.2 Load the data into R

Students should load the Red and NIR bands of the satellite imagery into R using the raster () function. They will then visualize these bands using the plot () function.

```
R
red_band <- raster (“path_to_red_band.tif”)
nir_band <- raster(“path_to_nir_band.tif”)
```

14.3.3.2.3 Calculate the NDVI

Students will compute the NDVI for the dataset using the same formula described in the code example. This will help them understand how vegetation can be assessed using satellite data.

```
R
NDVI <- (nir_band - red_band)/(nir_band + red_band)
```

14.3.3.2.4 Classify land cover

Students will reclassify the NDVI into categories such as: **Healthy Vegetation**: NDVI > 0.5, **Sparse Vegetation**: 0.2 < NDVI < 0.5, and **Bare Soil/Water**: NDVI < 0.2.

```
R
land_cover <- reclassify (NDVI, c (-Inf, 0.2, 1, 0.2, 0.5, 2, 0.5, Inf, 3))
```

14.3.3.2.5 Visualize the results

After classification, students should plot the NDVI and land cover categories to visually interpret the data. They can use color schemes to highlight different land cover types.

```
R
plot (land_cover, main = "Classified Land Cover", col = c ("brown", "green", "yellow"))
```

14.3.3.2.6 Map mineral deposits

If students are working with a mineral exploration dataset, they will attempt to map mineral deposits by combining different bands of the satellite data (such as using NIR and SWIR bands) and performing classification based on spectral signatures.

14.3.3.2.7 Interpret the results

Once the classification is complete, students should interpret the results. They can assess the amount of healthy vegetation, barren land, or water in the area based on the NDVI or land cover classification.

14.3.3.3 *Deliverables*

R code for data processing and analysis; Plots of NDVI and land cover classifications; A brief report summarizing the findings, such as areas with high vegetation health, the location of barren land, or areas for potential mineral exploration.

This section introduces students to the fundamentals of remote sensing data processing using R. By calculating the NDVI and performing land cover classification, students gain hands-on experience with remote sensing data, which is widely used for environmental monitoring, land use planning, and resource exploration. The exercise allows students to apply these concepts to real-world data, preparing them for more advanced geospatial analysis in geoscience and remote sensing applications.

This chapter will equip students with essential skills to analyze remote sensing data using R and prepare them for more complex geospatial and environmental studies. The exercises will provide students with practical knowledge that can be directly applied to real-world projects in remote sensing.

14.4 Geospatial data analysis with GIS software

14.4.1 Introduction

Geospatial data analysis plays a pivotal role in the exploration of natural resources, especially in the context of geology, mineral exploration, and environmental monitoring. Geographic Information Systems (GIS) are powerful tools used for spatial data management and analysis. This section will guide you through the essentials of using GIS software to analyze geospatial data for sustainable resource exploration.

14.4.2 Section 1: Tutorial on using GIS software for geospatial data analysis

This tutorial is designed to provide students with a hands-on understanding of how to use GIS software for analyzing geospatial data in the context of resource exploration. The tutorial will cover the basic operations of GIS software, the process of importing geospatial data, and essential analysis techniques.

14.4.2.1 Introduction to GIS software

GIS software provides a platform to map, analyze, and interpret spatial data. The two most commonly used GIS tools are **ArcGIS** (by ESRI) and **QGIS** (an open-source platform). Both tools support a variety of functions for geospatial data management, analysis, and visualization.

ArcGIS Overview: ArcGIS is a comprehensive GIS software suite with powerful spatial analysis and visualization capabilities. It supports both vector and raster data types and offers various extensions tailored to specific fields like geology, hydrology, and environmental science.

QGIS Overview: QGIS is a free, open-source GIS tool that allows users to perform similar tasks as ArcGIS. Its user-friendly interface makes it an ideal choice for students and professionals on a budget. Additionally, it supports a wide variety of plugins for different analytical tasks.

14.4.2.2 Importing and organizing geospatial data

The first step in using GIS software is importing data. Geospatial data can be found in various formats, such as shapefiles (.shp), GeoTIFFs, or CSV files containing geographic coordinates.

Steps to import data:

1. **Shapefiles**: A shapefile is a widely used format in GIS that stores vector data (points, lines, polygons). To import a shapefile:
   - (i) Go to the "Add Data" option in ArcGIS or QGIS.
   - (ii) Browse to the location of your shapefile and load the data.
2. **Raster Data**: Raster data, such as satellite imagery, elevation models, and geophysical data, are stored in GeoTIFFs or other raster formats.
   - (i) Use the "Add Raster" function to load raster data into the GIS software.

3. **CSV Data**: Datasets like geological sample points, mineral locations, or other survey data can be stored in CSV files with latitude and longitude coordinates.

   (i) Import CSV data by selecting "Add Layer" and choosing "Add Delimited Text Layer."

   Data Organization:

   (i) After importing the data, organize it into different layers based on categories (e.g., geological features in one layer, mineral deposits in another).
   (ii) Ensure all layers are in the same coordinate system to avoid misalignment.

14.4.2.3 *Data visualization*

Visualizing geospatial data is key to understanding patterns and relationships. GIS software provides multiple visualization tools to display your data:

1. **Symbology**: Assign different colors, symbols, and labels to each data layer. For instance, mineral deposits can be shown in varying colors based on concentration levels.
2. **Map Design**: Customize the layout by adjusting the map scale, adding titles, legends, north arrows, and scale bars.
3. **D Visualization**: For terrain and elevation data, GIS software like ArcGIS offers 3D visualization to view the topography of the landscape.

14.4.2.4 *Conducting spatial analysis*

GIS provides several tools to analyze spatial data, which is crucial for resource exploration. Key analysis tools include:

1. **Buffering**: Create buffer zones around specific features, such as mineral deposits, to understand their proximity to other important areas like roads or water bodies.
2. **Overlay Analysis**: Combine multiple layers to identify intersections or areas of interest. For example, overlay a geological map with a mineral distribution map to identify regions with potential for resource extraction.
3. **Spatial Interpolation**: Predict values at locations where data is not available by using techniques such as kriging or inverse distance weighting.

14.4.2.5 *Creating reports and exporting results*

After completing the analysis, you can generate reports and export your results. This includes:

(i) **Creating Printable Maps**: Customize map elements like legends, labels, and titles to create clear, publishable maps.
(ii) **Exporting Data**: Export your results as GIS-compatible formats (e.g., shapefiles, GeoTIFFs) for further analysis or sharing.

14.4.3 Section 2: Exercise on creating geospatial maps using real-world datasets

This exercise allows students to apply the concepts learned in the tutorial by creating their own geospatial maps using real-world datasets. They will focus on creating thematic maps that represent the distribution of mineral deposits and geological features.

14.4.3.1 *Dataset overview*

Students will be provided with several real-world geospatial datasets, which may include:

(i) **Geological Data**: Maps showing rock formations, faults, and sedimentary layers.
(ii) **Mineral Distribution Data**: Locations of known mineral deposits, often represented as point data.
(iii) **Satellite Imagery**: Raster data that shows land use, vegetation, or topographic features.
(iv) **Elevation Data**: Digital Elevation Models (DEMs) for terrain analysis.

14.4.3.2 *Importing data into GIS*

Using the techniques learned in the tutorial, students will import and organize the datasets into GIS software. They will:

(i) Import the geological, mineral, and elevation data.
(ii) Ensure the data layers are correctly aligned in the coordinate system.
(iii) Organize the data into separate layers for easier analysis.

14.4.3.3 *Data visualization and thematic mapping*

Next, students will create maps that display various geological and mineral features:

1. **Geological Map**: Use symbology to color different geological units based on their composition or age.
2. **Mineral Distribution Map**: Visualize the distribution of mineral deposits using symbols or color gradients based on mineral concentration.
3. **Combined Map**: Overlay the geological and mineral maps to identify regions where certain geological features correlate with the presence of mineral deposits.

14.4.3.4 *Conducting spatial analysis*

Students will apply spatial analysis techniques to gain insights into resource exploration:

1. **Buffer Zones**: Create buffer zones around mineral deposits to identify areas with high potential for exploration.
2. **Overlay Analysis**: Use overlay analysis to determine areas where geological features (e.g., fault lines or sediment layers) align with mineral deposits.
3. **Terrain Analysis**: Use DEM data to assess terrain features such as slope and aspect, which are important in resource exploration.

14.4.3.5 *Creating a map layout*

Students will learn how to design and export high-quality maps for presentation:

1. **Adding Elements**: Include a title, legend, scale bar, and north arrow.
2. **Customization**: Adjust the layout for clarity and readability.
3. **Exporting**: Export the final map as an image or PDF for use in reports.

14.4.3.6 Final report

Finally, students will compile their findings into a report that includes:

(i) The objectives and methodology of their analysis.
(ii) A discussion of the results and how they can inform resource exploration decisions.
(iii) High-quality maps and visualizations of the analyzed data.

By completing this chapter, students will have gained foundational skills in using GIS software for geospatial data analysis. These skills are vital for anyone involved in the exploration of natural resources, as they allow for data-driven decisions that promote sustainable and efficient resource management. The integration of GIS in resource exploration is a powerful approach to understanding complex spatial relationships in the environment, and these tools will help shape the future of natural resource exploration and management.

This expanded and detailed version provides both theoretical explanations and practical applications, which should help students grasp the fundamental GIS concepts and techniques necessary for resource exploration.

14.5 Electromagnetic and seismic data interpretation with MATLAB

14.5.1 Introduction

In natural resource exploration, **electromagnetic (EM)** and **seismic surveys** are fundamental techniques for understanding subsurface structures and identifying potential resources. These surveys provide critical data, including resistivity measurements and seismic wave travel times, that can be analyzed to interpret geological features (Telford et al., 2012). In this section, we focus on practical applications of MATLAB for processing and interpreting **electromagnetic** and **seismic data**, enabling users to identify resource-rich zones and subsurface anomalies.

Through **hands-on exercises** and **code examples**, students will develop the skills necessary to process raw geophysical data, interpret anomalies, and apply these techniques to real-world exploration scenarios.

14.5.2 MATLAB for electromagnetic data interpretation

Electromagnetic surveys measure the resistivity of the subsurface, which is crucial for detecting materials like minerals, water, and other geological features. In this section, we demonstrate how to process **electromagnetic data** using MATLAB, focusing on visualizing and fitting models to the data.

14.5.2.1 Key concepts in electromagnetic data

Apparent Resistivity: This refers to the measured resistance to electrical current flow in the ground, which varies based on the material's composition (e.g., minerals, groundwater).

Geophysical Anomalies: Sudden changes in resistivity often indicate subsurface features like mineral deposits, water bodies, or fault lines.

14.5.2.2 MATLAB code example: Processing electromagnetic data

The following MATLAB code processes and visualizes electromagnetic resistivity data measured at different depths.

```
MATLAB
% Example Electromagnetic Data: Resistivity at different depths
depth = [0, 10, 20, 30, 40, 50]; % Depth in meters
resistivity = [100, 150, 200, 250, 300, 350]; % Resistivity in Ohm-meters
% Plot the resistivity data
figure;
plot (depth, resistivity, '-o', 'LineWidth', 2);
title ('Electromagnetic Survey: Resistivity vs Depth');
xlabel ('Depth (m)');
ylabel ('Resistivity (Ohm-meters)');
grid on;
% Fit a polynomial model to the resistivity data
p = polyfit (depth, resistivity, 1);
fitted_resistivity = polyval (p, depth);
% Plot the fitted model
hold on;
plot (depth, fitted_resistivity, '- - r', 'LineWidth', 2);
legend ('Measured Resistivity', 'Fitted Model');
```

Explanation:

(i) The code plots the resistivity values at various depths and uses **polyfit** to apply a linear model to the data.
(ii) The **fitted model** helps identify any resistivity trends, which could be used to highlight resource-rich zones or geological structures.

14.5.3 MATLAB for seismic data interpretation

Seismic surveys help explore subsurface geological formations by analyzing seismic wave travel times. By measuring the time it takes for seismic waves to travel through different layers of rock, we can interpret subsurface features and determine the material properties of various layers.

14.5.3.1 Key concepts in seismic data

Seismic Velocity: The velocity at which seismic waves travel through different materials, helping to infer the properties of subsurface layers.

Travel Time: The time taken for seismic waves to travel from the source to the receiver. Changes in travel time indicate changes in the material composition or geological structure.

14.5.3.2 MATLAB code example: Processing seismic data

The following MATLAB code processes seismic travel time data to calculate seismic velocity and visualize the relationship between travel time and distance.

```
MATLAB
% Example Seismic Data: Travel time at different distances
distance = [0, 100, 200, 300, 400, 500]; % Distance in meters
```

```
travel_time = [0.5, 0.9, 1.3, 1.7, 2.1, 2.5]; % Travel time in seconds
% Plot the travel time data
figure;
plot (distance, travel_time, '-o', 'LineWidth', 2);
title ('Seismic Survey: Travel Time vs Distance');
xlabel ('Distance (m)');
ylabel ('Travel Time (s)');
grid on;
% Estimate the seismic velocity (average velocity for simplicity)
velocity = mean (distance./ travel_time); % Seismic velocity = Distance/Time
% Display the estimated seismic velocity
Fprintf ('Estimated Seismic Velocity: %.2f m/s\n', velocity);
```

Explanation:

(i) The code uses distance and travel time data to estimate seismic velocity using the formula

$$\text{Velocity} = \frac{\text{Distance}}{\text{Time}} \tag{14.2}$$

(ii) Plotting the relationship between travel time and distance helps visualize how seismic waves travel through different materials, and the seismic velocity is calculated and displayed.

14.5.4 Interactive exercise: Identifying geophysical anomalies

This exercise gives students the opportunity to apply the concepts learned in Sections 14.5.2 and 14.5.3 to real-world data, helping them identify subsurface anomalies that could indicate valuable resources.

14.5.4.1 Exercise overview

In this hands-on exercise, students are provided with **real seismic and electromagnetic data**. The goal is to process the data using MATLAB, identify key anomalies, and interpret the geological features of the subsurface.

14.5.4.2 Exercise tasks

Task 1: Electromagnetic Data Analysis

1. **Import the provided resistivity data** into MATLAB.
2. **Plot the resistivity** vs. **depth** using the example code provided above.
3. **Analyze the resistivity data:** Look for regions where resistivity changes sharply, which might indicate mineral deposits or groundwater.
4. **Fit a model to the data:** Use a polynomial fit to better visualize trends and anomalies.

Task 2: Seismic Data Analysis

1. **Import the provided seismic data** (distance and travel time).
2. **Plot the travel time data** and calculate the seismic velocity using the formula provided.
3. **Identify any anomalies** in the travel time that may indicate geological features such as faults, water reservoirs, or mineral deposits.

14.5.4.3 *Deliverables*

(i) **Plots and Visualizations**: Students should submit the plots of the processed electromagnetic and seismic data.
(ii) **Interpretation Report**: A brief report that includes:

 (a) The analysis of resistivity and travel time data.
 (b) Interpretation of anomalies, including potential resource locations (e.g., mineral-rich zones, water bodies).
 (c) A discussion of how these findings can inform further exploration or excavation.

This section has demonstrated the practical applications of **MATLAB** in processing and interpreting **electromagnetic** and **seismic data**. By working through the provided code examples and exercises, students have gained hands-on experience in:

(i) Analyzing resistivity data from electromagnetic surveys.
(ii) Interpreting travel time data from seismic surveys.
(iii) Identifying geological features and anomalies in subsurface structures.

The ability to process and analyze these types of geophysical data is crucial for natural resource exploration and geological studies. With the tools and techniques presented here, students are well-equipped to tackle real-world challenges in geophysical data analysis.

14.6 Using drones for survey data analysis

14.6.1 Introduction

The integration of drones into survey data collection has revolutionized the way geospatial data is gathered for resource exploration. Drones are capable of collecting high-resolution imagery, LiDAR data, thermal images, and other sensor data from difficult-to-reach or hazardous areas. The advantage of drones lies in their ability to capture data quickly and cost-effectively compared to traditional methods.

In resource exploration, drones are used to map geological features, identify mineral deposits, assess vegetation health, and monitor environmental changes. This section introduces the basics of drone-based data collection and analysis, focusing on how open-source software can be utilized to process and analyze this data for resource exploration applications.

14.6.2 Drone-based data collection

Drones equipped with various sensors, such as cameras, LiDAR, and thermal sensors, are widely used in survey data collection for applications such as environmental monitoring, geological mapping, and vegetation analysis. The ability to capture both high-resolution aerial imagery and 3D data allows for in-depth analysis of large areas.

Types of Data Collected by Drones:

1. **Aerial Imagery**: High-resolution photographs or videos taken from a drone, useful for mapping geological features, vegetation, and more.

2. **LiDAR (Light Detection and Ranging)**: A sensor that uses laser pulses to generate detailed 3D maps of terrain and vegetation. It is especially useful in dense forests or areas with thick vegetation.
3. **Thermal Imaging**: Captures infrared radiation to create temperature maps, which can be useful for monitoring vegetation health, water bodies, and geological features such as hot springs or mineral deposits.
4. **Multispectral Imagery**: Provides information across multiple wavelengths of light, useful for vegetation health analysis and identifying mineral signatures.

Drone Platforms:

(i) **Quadcopters**: Small and agile drones commonly used for aerial imagery and video.
(ii) **Fixed-wing Drones**: More suitable for large-area surveys, offering longer flight times and better coverage.

14.6.3 Open-source software for drone data processing

After drone data is collected, it needs to be processed and analyzed. Several open-source software platforms provide powerful tools to handle drone-captured data. These tools allow for tasks such as data stitching, filtering, and analysis.

Popular Open-source Software for Drone Data:

(i) **QGIS**: A popular geographic information system (GIS) that supports the processing of drone imagery and geospatial data. It can be used to perform spatial analysis, create maps, and visualize results.
(ii) **OpenDroneMap**: A tool that converts drone imagery into 3D models, point clouds, and orthophotos. It's useful for generating detailed terrain models and performing photogrammetric analysis.
(iii) **MicMac**: A photogrammetric software suite used for processing drone imagery to create 3D reconstructions, including dense point clouds and textured models.
(iv) **Python (Geospatial Libraries)**: Python libraries such as **Geopandas**, **Rasterio**, and **OpenCV** can be used for processing and analyzing geospatial data, including drone imagery. These libraries allow users to create custom scripts for specialized analysis.

14.6.4 Tutorial: Drone data collection and analysis

This section provides a **step-by-step tutorial** on how to collect and analyze drone-based data using open-source software tools. Students will learn the process of importing drone imagery, processing the data, and performing analysis to explore resources.

Steps for Drone Data Collection:

1. **Plan the Survey**:

   (i) Select the area to be surveyed based on resource exploration needs.
   (ii) Choose the appropriate drone and sensors (e.g., RGB camera, multispectral sensor, or LiDAR).

(iii) Plan the flight path and ensure it covers the entire survey area.

2. **Collect Drone Data**:

   (i) Deploy the drone and ensure stable flight conditions.
   (ii) Capture imagery, LiDAR, or thermal data based on the survey goals.

3. **Download Data**:
   (i) After completing the flight, download the collected imagery or data from the drone.

Steps for Processing Drone Data:

1. **Import Data into QGIS or OpenDroneMap**:

   (i) Import the raw imagery or LiDAR data into QGIS or OpenDroneMap.
   (ii) For QGIS, load the imagery as a raster layer and the LiDAR data as a point cloud or vector layer.

2. **Stitching and Alignment**:

   (i) Use photogrammetry tools to align and stitch drone images together to create a continuous image or orthophoto.

3. **Creating 3D Models (if applicable)**:

   (i) For LiDAR or photogrammetric data, use OpenDroneMap or QGIS to generate 3D models of the terrain or structures.

4. **Perform Spatial Analysis**:

   (i) Analyze the data by calculating various metrics, such as elevation profiles, vegetation indices (e.g., NDVI for vegetation health), or mineral signatures.

5. **Visualization and Interpretation**:

   (i) Visualize the results using QGIS to identify key geological features, mineral deposits, or environmental concerns.
   (ii) Overlay the drone imagery with existing maps to enhance analysis and interpretation.

14.6.5 Exercise: Analyzing drone-captured imagery

In this hands-on exercise, students will process and analyze drone-captured imagery. The exercise focuses on identifying potential resource-rich areas by analyzing geospatial patterns in the imagery, such as vegetation health or mineral signatures.

Exercise Steps:

1. **Download the Imagery**:

   (i) Students will be provided with a sample set of drone-captured images that cover an area of interest.

2. **Import Imagery into QGIS**:
   (i) Students will import the drone imagery into QGIS for processing and analysis.
3. **Stitching the Images**:
   (i) If the images are not already stitched, students will use QGIS tools or OpenDroneMap to stitch them together into a single orthophoto or mosaic.
4. **Analyze Vegetation or Geological Features**:
   (i) For vegetation analysis: Students will calculate the NDVI (Normalized Difference Vegetation Index) to assess vegetation health and identify potential areas for resource exploration.
   (ii) For geological analysis: Students will look for surface features like rock outcrops, soil color variations, or fault lines that may indicate the presence of resources.
5. **Interpret the Results**:
   (i) Based on their analysis, students will interpret the imagery to identify areas that could be of interest for resource exploration.

Deliverables:

(i) **Processed Imagery**: Submit the processed drone imagery with overlaid analysis results (e.g., NDVI maps, geological features).
(ii) **Analysis Report**: A report summarizing the analysis steps, results, and potential resource areas identified from the imagery.

This section demonstrated the powerful capabilities of drones in modern survey data collection for resource exploration. By utilizing drones equipped with various sensors, such as cameras, LiDAR, and thermal imagers, it is possible to gather high-resolution data from otherwise inaccessible areas. Open-source software tools, such as QGIS and OpenDroneMap, provide the necessary platforms for processing, analyzing, and interpreting drone-captured data.

By completing the exercises and tutorials, students have gained hands-on experience in processing drone imagery and analyzing it to identify potential resources. This skill set is invaluable for modern geospatial analysis and resource exploration, providing a cost-effective and efficient approach to survey work.

14.7 Hands-on case study: Integrating geophysical, remote sensing, and geospatial data

14.7.1 Introduction

In resource exploration, integrating multiple types of data—such as geophysical data, remote sensing imagery, and geospatial data—is crucial for making accurate assessments and identifying potential mineral deposits or other natural resources (Gitas and Koutsou, 2019). This hands-on case study guides students through a comprehensive project where they integrate different data types to solve a real-world exploration problem. The case study simulates the kind of data analysis tasks professionals face in the resource exploration industry. By using tools like **Python**, **GIS (Geographic Information Systems)**, and **MATLAB**, students will learn how to combine various data sources to uncover insights about subsurface features, mineral deposits, and other exploration targets.

14.7.2 Case study overview: Identifying mineral deposits

This case study focuses on using integrated data to identify potential **mineral deposits** in a given region. Students will work with the following data sources:

(i) **Geophysical Data**: Gravity or magnetic survey data used to infer subsurface geology and identify potential mineralization zones.
(ii) **Remote Sensing Data**: Satellite or drone imagery (e.g., multispectral imagery) used to detect surface features indicative of mineralization, such as vegetation anomalies or soil composition changes.
(iii) **Geospatial Data**: Geological maps, topographic data, and location coordinates that provide essential context and spatial reference for the exploration project.

The goal is to combine these data sets and analyze them to identify locations that may be promising for further exploration or resource extraction.

14.7.3 Data integration process

The integration of multiple data types into a unified model requires careful alignment, pre-processing, and analysis. This section outlines the general steps students will follow during the case study:

Step 1: Data Collection and Pre-processing

(i) **Geophysical Data (Magnetic/Gravity Surveys)**: Students will be provided with raw geophysical data (e.g., magnetic anomalies or gravity measurements). This data will be processed and transformed into maps showing anomalies that may suggest the presence of mineral deposits.
(ii) **Remote Sensing Data (Satellite/Drones)**: Satellite or drone-captured imagery (e.g., multispectral, thermal) will be used to identify surface features. Pre-processing steps will include radiometric calibration, geometric correction, and mosaicking (if necessary) to create a continuous image.
(iii) **Geospatial Data (GIS)**: Geological maps, topographic data, and other relevant geospatial layers (e.g., fault lines, mineral deposit locations) will be integrated into the GIS for contextual understanding and to provide spatial reference for the analysis.

Step 2: Data Alignment and Transformation

(i) **Geospatial Registration**: Align geophysical and remote sensing data with geospatial layers using geographic coordinates. This ensures all datasets are in the same spatial reference system.

(ii) **Data Transformation**: If needed, students will use tools such as **QGIS, MATLAB**, or **Python** to transform the datasets into comparable formats (e.g., converting raw geophysical data into grids, converting imagery into raster layers).

Step 3: Data Analysis and Interpretation

(i) **Geophysical Data Analysis**: Using tools like **MATLAB** or **Python**, students will analyze geophysical data to identify anomalies. They might use techniques such as **gridding, filtering**, and **contouring** to visualize potential mineral deposits or subsurface geological structures.
(ii) **Remote Sensing Data Analysis**: Students will use **Python** or **GIS** software to process remote sensing data and generate indices such as **NDVI** (Normalized Difference Vegetation Index) to analyze vegetation health and detect surface features that may indicate mineralization or environmental stress.
(iii) **Geospatial Data Analysis**: Students will overlay all the processed geophysical and remote sensing data on GIS maps, helping to interpret spatial relationships between different data layers (e.g., mineral deposits in relation to geological features such as faults or lithological boundaries).

14.7.4 Tutorial: Case study workflow

The following is a step-by-step workflow that will guide students through the process of integrating data to solve the exploration problem.

Step 1: Importing and Preparing Data

(i) Import the geophysical, remote sensing, and geospatial datasets into **QGIS** or **MATLAB**.
(ii) Ensure all data is aligned geographically by converting to the same coordinate system and format.
(iii) Clean the datasets by removing noise or irrelevant information (e.g., outliers in geophysical data, cloud cover in remote sensing data).

Step 2: Visualizing the Data

(i) In **QGIS**, display geophysical data as contour maps or heatmaps to highlight areas of interest based on anomaly values.
(ii) In **Python**, use libraries like **Matplotlib** and **Folium** for visualization of geospatial layers and remote sensing data (e.g., plotting satellite imagery with NDVI overlays).
(iii) Use **MATLAB** to generate 3D visualizations if geophysical data (e.g., gravity or magnetic anomalies) includes depth information.

Step 3: Performing Data Analysis

(i) **Geophysical Data**: Use **MATLAB** to perform anomaly detection, including filtering and gridding, to visualize magnetic or gravity anomalies.

(ii) **Remote Sensing Data**: Process multispectral satellite images in **Python** using libraries like **NumPy**, **OpenCV**, and **Matplotlib** to calculate vegetation indices or surface reflectance.
(iii) **Geospatial Data**: Perform spatial analysis in **QGIS** or **Python** to overlay geological maps with geophysical and remote sensing data, enabling better interpretation of geological features.

Step 4: Data Interpretation

(i) Combine all datasets to identify promising mineral deposit areas. Use the integrated data to identify potential target zones based on geological features, surface vegetation patterns, and geophysical anomalies.
(ii) Generate a final report that outlines the methodology, data processing steps, and results, including maps and visualizations of the identified resource-rich areas.

14.7.5 Exercise: Integrating data for exploration

In this exercise, students will work with real-world exploration data to complete the case study. They will integrate geophysical, remote sensing, and geospatial data to solve the problem of identifying potential mineral deposits.

Exercise Steps:

1. **Data Acquisition**:
   (i) Download the provided geophysical, remote sensing, and geospatial datasets.
2. **Data Pre-processing**:
   (i) Import the datasets into the chosen software tools (e.g., QGIS, Python, MATLAB).
   (ii) Pre-process the data (aligning datasets, transforming formats, cleaning noise).
3. **Data Analysis**:
   (i) Perform anomaly detection in geophysical data using **MATLAB** or **Python**.
   (ii) Analyze remote sensing data to detect vegetation patterns or mineral signatures using **Python** libraries (e.g., calculating NDVI).
   (iii) Overlay geospatial data in **QGIS** to combine all layers and interpret geological features.
4. **Interpretation and Reporting**:
   (i) Identify mineral-rich areas based on the integrated data.
   (ii) Submit a report that includes visualizations, analysis steps, and final interpretation of the results.

Deliverables:

(i) **Integrated Maps and Visualizations**: Submit maps displaying the geophysical anomalies, remote sensing analysis, and geospatial layers.
(ii) **Final Report**: A comprehensive report detailing the analysis process, findings, and potential exploration targets.

This case study demonstrates the value of integrating multiple data types for resource exploration. By combining geophysical, remote sensing, and geospatial data, students gain a holistic understanding of how these different data sources complement each other in real-world exploration scenarios. Using software like **Python**, **MATLAB**, and **GIS**, students learn how to process, analyze, and interpret these data types to make informed decisions about resource-rich areas.

14.8 Student resources and further reading

14.8.1 Links to data repositories

For students to practice their skills in geophysical, remote sensing, and geospatial data analysis, here are some valuable public datasets and repositories. These resources provide real-world data that students can use to explore the practical applications discussed throughout the book.

1. **USGS Earth Explorer**
   (i) **Link**: https://earthexplorer.usgs.gov/
   (ii) **Description**: A comprehensive resource for accessing satellite imagery, aerial photography, and geophysical data. It includes remote sensing data such as Landsat, MODIS, and SRTM data, which can be used for geospatial analysis, vegetation mapping, and more.
2. **NASA Earth Observing System Data and Information System (EOSDIS)**
   (i) **Link**: https://earthdata.nasa.gov/
   (ii) **Description**: Provides access to a wide range of Earth science data from NASA's Earth Observing System satellites. It includes datasets related to land surface, atmosphere, oceans, and more, suitable for remote sensing analysis.
3. **European Space Agency (ESA) Copernicus Open Access Hub**
   (i) **Link**: https://scihub.copernicus.eu/
   (ii) **Description**: Offers free access to Sentinel satellite data, which includes multispectral imagery, radar data, and more. Sentinel data is highly relevant for environmental monitoring, land use change analysis, and resource exploration.
4. **Geophysical Data Center—NOAA**
   (i) **Link**: https://www.ngdc.noaa.gov/
   (ii) **Description**: The National Geophysical Data Center (NGDC) provides various datasets related to Earth's physical properties, including gravity, magnetic, and seismic data. These can be used for geophysical analysis and exploration studies.
5. **Open Topography**
   (i) **Link**: https://opentopography.org/
   (ii) **Description**: A platform for accessing high-resolution topographic data, including LiDAR data, which can be used for creating detailed terrain models and performing geospatial analysis.

6. **Google Earth Engine**
   (i) **Link**: https://earthengine.google.com/
   (ii) **Description**: A cloud-based platform for planetary-scale environmental data analysis. Students can access a variety of remote sensing datasets, including Landsat, MODIS, and Sentinel data, for performing geospatial analysis using the Google Earth Engine API.
7. **USGS National Map**
   (i) **Link**: https://www.usgs.gov/centers/national-geospatial-program/national-map
   (ii) **Description**: A mapping platform providing a range of geospatial data, including topography, elevation, hydrography, and land use, which can be valuable for resource exploration and geological analysis.
8. **The Mineral Resources Data System (MRDS)**
   (i) **Link**: https://mrdata.usgs.gov/mrds/
   (ii) **Description**: A database from the USGS containing information about known mineral resources worldwide. The MRDS can be used to correlate geophysical and remote sensing data with existing resource locations.

14.8.2 Additional resources

In addition to datasets, here are several recommended online tutorials, courses, and software tools that will help students improve their practical knowledge and skills in geophysical, remote sensing, and geospatial data analysis.

1. Online Tutorials and Courses
Coursera—Geographic Information Systems (GIS) Specialization

(i) **Link**: https://www.coursera.org/specializations/gis
(ii) **Description**: A comprehensive series of courses offered by UC Davis on Coursera, covering the fundamentals of GIS, spatial analysis, and geospatial data processing using QGIS and ArcGIS.

edX—Remote Sensing and GIS for Natural Resources Management

(i) **Link**: https://www.edx.org/course/remote-sensing-and-gis-for-natural-resources-management.
(ii) **Description**: This online course, offered by the University of Twente, teaches how remote sensing and GIS can be applied in resource management, including practical exercises in data analysis.

Udemy—Introduction to Remote Sensing and GIS

(i) **Link**: https://www.udemy.com/course/remote-sensing-and-gis/
(ii) **Description**: A beginner-friendly course that introduces remote sensing techniques and GIS tools, focusing on data collection, analysis, and interpretation.

Future Learn—Introduction to Remote Sensing

(i) **Link**: https://www.futurelearn.com/courses/remote-sensing
(ii) **Description**: This course introduces remote sensing concepts and technologies, including practical applications in environmental monitoring and resource exploration.

2. Software Tools for Data Analysis
(A) QGIS

(i) **Link**: https://qgis.org/en/site/
(ii) **Description**: An open-source Geographic Information System (GIS) used for spatial data visualization, analysis, and mapping. QGIS is essential for working with geospatial data, including processing remote sensing and geophysical datasets.

(B) MATLAB

(i) **Link**: https://www.mathworks.com/products/matlab.html
(ii) **Description**: A powerful platform for numerical computing, widely used for processing and analyzing geophysical data. MATLAB provides tools for signal processing, image analysis, and visualization.

(C) Python (Geospatial Libraries)

(i) **Link**: https://www.python.org/
(ii) **Description**: Python is a versatile programming language with powerful libraries for geospatial analysis, including **Geopandas**, **Matplotlib**, **Rasterio**, and **NumPy**. These libraries are widely used for remote sensing and geophysical data processing.

(D) OpenDroneMap

(i) **Link**: https://www.opendronemap.org/
(ii) **Description**: Open-source software for processing drone imagery into 3D models, point clouds, and orthophotos. Useful for analyzing drone-captured data in resource exploration and environmental monitoring.

(E) Google Earth Engine

(i) **Link**: https://earthengine.google.com/
(ii) **Description**: A cloud-based platform that allows users to access and analyze vast amounts of satellite imagery and geospatial data, ideal for environmental monitoring and resource exploration tasks.

(F) GDAL (Geospatial Data Abstraction Library)

(i) **Link**: https://gdal.org/
(ii) **Description**: An open-source library for reading, writing, and manipulating geospatial raster and vector data. GDAL is useful for handling large datasets and performing spatial operations.

These resources will enable students to further develop their skills in geophysical, remote sensing, and geospatial data analysis. By utilizing public datasets, online tutorials, and open-source software tools, students can enhance their understanding of real-world applications in resource exploration, environmental monitoring, and geospatial analysis. Continued practice and exploration of these resources will help students prepare for careers in these dynamic fields.

CHAPTER

15

Emerging technologies and ethics in geoscience exploration

15.1 AI/ML-driven geophysical data analysis for sustainable exploration

The exploration and extraction of natural resources have long been at the core of economic development, with mineral, water, and energy resources powering the global economy. However, as these resources become increasingly scarce and environmental pressures mount, there is a growing need to adopt more efficient, sustainable, and ethical practices in the exploration and extraction industries. Over the past few decades, technological advancements have paved the way for more sophisticated approaches in geoscience. Among these advancements, Artificial Intelligence (AI) and Machine Learning (ML) are standing out as transformative tools. This chapter explores the role of AI/ML in geophysical data generation and interpretation, the application of geostatistics in resource extraction decisions, and ethical considerations for sustainable resource exploration. The goal is to demonstrate how these technologies can not only improve efficiency but also ensure that the exploration of natural resources aligns with ethical and environmental considerations.

In geosciences, the process of gathering, processing, and interpreting data is crucial to successful resource exploration. Traditionally, geophysicists rely on seismic surveys, magnetic measurements, and other remote sensing techniques to map subsurface resources. However, the volume and complexity of the data have grown exponentially, making it difficult to extract meaningful insights using conventional methods. Here, AI and ML techniques offer significant advantages by automating and enhancing data generation and interpretation processes.

This title emphasizes the integration of modern technologies such as AI and Machine Learning (ML) with geophysical data, highlighting their potential to enhance the mineral exploration process. The inclusion of AI and ML brings a forward-looking dimension to exploration, underscoring their role in improving data analysis, predictive modeling, and decision-making, ultimately contributing to more efficient and sustainable exploration practices.

AI and ML algorithms significantly accelerate the analysis of large and complex geophysical datasets. These technologies can identify hidden patterns and anomalies that may be

Baseline Geoscience Data Generation for Sustainable Exploration of Natural Resources
https://doi.org/10.1016/B978-0-443-45221-5.00005-9

missed through traditional methods, allowing for more accurate predictions of mineral deposits. By leveraging AI and ML, mineral exploration becomes more effective, as these technologies help make quicker and more informed decisions regarding where to focus exploration efforts.

Furthermore, AI-driven models are key in assessing the likelihood of discovering mineral resources based on historical geophysical and environmental data. This predictive capability ensures that exploration is targeted in areas with the highest potential, thereby reducing the unnecessary disturbance of environmentally sensitive regions. The integration of AI and ML with geophysical mapping allows for smarter decision-making that minimizes exploration costs and environmental impacts.

In terms of resource extraction, machine learning can optimize drilling processes by analyzing vast amounts of geophysical data, improving efficiency while reducing waste. This not only enhances the overall effectiveness of resource extraction but also ensures that operations are conducted in a manner that minimizes environmental disruption. By utilizing these advanced technologies, the exploration process becomes more resource-efficient and aligned with sustainable practices.

The combination of AI and ML technologies also facilitates the fusion of multiple types of data, such as geophysical, geological, and environmental information, into cohesive models. This integrated approach enhances the accuracy and utility of baseline geoscience data, providing a clearer understanding of subsurface conditions and helping to refine exploration strategies. This data fusion supports a more holistic approach to exploration, leading to more reliable and informed decision-making.

Additionally, AI plays a critical role in environmental monitoring during exploration activities. Continuous monitoring enabled by AI technologies allows for the early detection of potential environmental risks, such as changes in local ecosystems or geological stability. Machine learning algorithms can assess these risks more accurately, allowing for the development of targeted mitigation strategies. This real-time risk assessment ensures that exploration activities remain responsible and sustainable, further promoting the long-term preservation of natural resources. By incorporating AI and ML into geophysical mapping and baseline geoscience data analysis, the mineral exploration process is not only made more efficient but also aligned with sustainable practices that minimize ecological impact while maximizing resource potential.

15.1.1 The role of AI/ML in geophysical data interpretation for mineral exploration

AI/ML can substantially improve the efficiency of geophysical data generation. Data collection through methods like seismic surveys, magnetic field measurements, and satellite imagery generates massive amounts of raw data. Traditional approaches to managing and processing these datasets can be slow and prone to human error. AI can automate many of these tasks, leading to quicker, more accurate results.

Mineral exploration is inherently complex and expensive, requiring precise and accurate interpretation of vast amounts of geophysical data. Traditional methods of data analysis can be time-consuming and prone to human error, especially when dealing with large datasets generated from seismic surveys, geochemical tests, or remote sensing technology. The advent of Artificial Intelligence (AI) and Machine Learning (ML) has brought significant improvements to this field, enhancing the speed and accuracy of geophysical data interpretation.

AI/ML algorithms have been successfully implemented in various aspects of geophysical data interpretation, including seismic data analysis, mineral modeling, and prediction of mineral deposits. These technologies can process large, high-dimensional datasets far more quickly than traditional methods, allowing geoscientists to uncover patterns that may not be immediately obvious. For instance, AI-powered algorithms can automatically identify anomalies in seismic or geochemical datasets, reducing the time spent manually sifting through raw data and enabling more timely decision-making in the exploration process.

One prominent application of AI/ML in mineral exploration is the use of neural networks to interpret seismic data. Seismic waves, which are used to map subsurface geological formations, can be difficult to analyze due to the complexity of the data. AI models, particularly deep learning techniques, can be trained to recognize specific geological features within seismic signals, enabling a more accurate interpretation of rock formations, faults, and potential mineral deposits.

Another application is the use of ML models in geochemical data analysis. Geochemical surveys, which involve analyzing soil, rock, or water samples for the presence of trace metals and minerals, produce large volumes of data. AI/ML techniques can automate the analysis of this data, helping to identify correlations between the presence of certain elements and the likelihood of valuable mineral deposits. For example, AI can assist in mapping mineralization zones by identifying patterns in geochemical anomalies, guiding exploration teams to more promising areas for further investigation.

In addition to improving data interpretation, AI/ML also aids in predictive modeling. Machine learning algorithms can learn from historical exploration data and geophysical surveys to make predictions about where undiscovered mineral resources are most likely to be found. These predictive models can consider a wide range of factors, such as geological features, geochemical data, and even environmental variables, to provide a more holistic view of potential exploration targets.

Furthermore, AI and ML help in enhancing the efficiency of resource discovery by optimizing the exploration process. By applying AI-driven approaches to data integration, exploration teams can combine multiple sources of geophysical, geochemical, and remote sensing data into a single, unified model. This integrated model provides a clearer understanding of the subsurface, improving the accuracy of resource estimation and reducing the risks associated with exploration.

Despite the potential of AI/ML, there are important ethical considerations to keep in mind when applying these technologies in resource exploration. Issues such as data privacy, environmental impact, and the ethical implications of AI decision-making must be addressed to ensure responsible and sustainable exploration practices. Additionally, while AI/ML can significantly enhance data interpretation, human expertise remains crucial in guiding the application of these technologies, ensuring that the results are used appropriately in the context of real-world exploration challenges.

In summary, AI and ML have become invaluable tools in the field of mineral exploration, allowing geoscientists to interpret geophysical data more accurately and efficiently. As these technologies continue to evolve, they promise to revolutionize the exploration process, leading to more effective and sustainable resource discovery.

15.2 Recent case studies in AI/ML for geophysical mapping

Several recent case studies highlight the use of AI and ML in enhancing global geophysical mapping and its integration with baseline geoscience data.

15.2.1 Case study 1: AI-assisted geophysical data processing in Canada

A team of geophysicists in Canada applied machine learning algorithms to process data from airborne electromagnetic surveys used in mineral exploration. The AI models were trained on large datasets, enabling them to identify mineral signatures in the geophysical data with a higher degree of precision than traditional methods. This led to the identification of unexplored mineral zones, reducing the need for extensive ground surveys. The project highlighted the value of AI in generating accurate baseline geophysical data while optimizing resource exploration (Smith and Harper, 2023).

15.2.2 Case study 2: Deep learning for subsurface mapping in Australia

Researchers in Australia utilized deep learning techniques to interpret gravity and magnetic data for subsurface mapping in a mineral-rich region. By applying neural networks to geophysical datasets, they were able to create detailed 3D models of the subsurface, offering new insights into the distribution of mineral resources. The approach not only enhanced the accuracy of baseline geophysical data but also provided a sustainable method for mineral exploration by reducing the environmental impact of traditional exploration techniques (Brown and Martin, 2022).

15.2.3 Case study 3: AI in exploration of rare earth elements (REE) in Africa

In Africa, a mining company integrated AI into its exploration process for rare earth elements (REE). By combining geophysical data with geological datasets and applying machine learning models, the company was able to predict areas with high concentrations of REE more effectively. The AI-driven models helped minimize drilling efforts and avoided environmentally sensitive areas, thus aligning the exploration with sustainable practices (Williams and Chirwa, 2021).

15.2.4 Case study 4: GIS-based mineral prospectivity mapping using machine learning in the Zhuonuo ore district, Tibet

The Zhuonuo Ore Concentration Area (ZOCA), located in the southern subterranean of Tibet, is considered one of the most prospective regions for copper-gold (Cu–Au) and molybdenum (Mo) mineral deposits (Cheng et al., 2023). Traditional methods of prospectivity, such as stream sedimentary geochemistry, often result in large and inaccurate predictions, particularly in the Gangdese Orogenic Belt, where high altitude and intense weathering obscure accurate data. Therefore, there is an urgent need for more precise and efficient methods to map prospective areas for Cu resources. In this case study, a **GIS-based machine learning**

approach is combined with a **mineral system model** to create more accurate and geologically meaningful mineral prospectivity maps for the region.

The key steps in the methodology include establishing a comprehensive Mineral System Model for Porphyry Copper Deposits (PCDs). This model is based on understanding the geological processes that govern mineralization, such as the metal source, transport pathways, and deposition zones. It forms the foundation for the spatial mapping and prediction of potential mineral deposits in the study area. Next, Spatial Proxy Transformation is applied to represent key ore-forming processes. The spatial proxies are derived from various geological components, such as proximity to intrusive rocks (acting as the source of metals), NE orientation faults (pathways for metal transport), hydrothermal alteration zones (fluids involved in mineralization), and a metallogenic strength diagram (depicting the deposition of metals). These proxies form the input for the predictive models. Fig. 15.1 shows the spatial proxies used for mineral system mapping: (a) Proximity to Eocene to Miocene intrusive rocks (source) (b) Metallogenic strength diagram for Cu–Mo–W–Bi–Au–Ag–Pb–Zn (deposition) (c) Proximity to NE orientation faults (pathway) (d) Fe-oxide and propylitization alteration zones (hydrothermal fluids).

In the Machine Learning Approaches phase, three different techniques are employed to capture complex mineralization patterns: Radial Basis Function Link Networks (RBFLN), a supervised learning technique; Random Forest (RF), another supervised learning method; and Fuzzy Clustering (FC), an unsupervised learning approach. These methods are used to analyze and classify the spatial data based on the known deposits and geological features. To evaluate the predictive accuracy of each model, several performance metrics are used, including the Receiver Operating Characteristic (ROC) curve, Predictive Area (P–A) plotting, and Normalized Density (Nd). These metrics help in assessing the reliability and effectiveness of each machine learning model in identifying prospective areas. Fig. 15.2 illustrates the ROC curve, which shows the performance of each model: RBFLN model: AUC = 0.99; RF model: AUC = 0.96; FC model: AUC = 0.94. This methodology ensures a rigorous and systematic approach to identifying potential mineral deposits using advanced machine learning techniques combined with geological knowledge.

The machine learning models provided reliable predictions of prospective areas, with the RBFLN model demonstrating the highest accuracy. The RBFLN model predicted that 12% of the study area contained 88% of the known deposits, indicating a highly effective prediction. The RF model and FC model also showed strong performance, predicting 14% and 21% of the area, respectively, which corresponded to 86% and 79% of known deposits. The predictive results for each model were as follows: RBFLN Model: AUC = 0.99, Predicted area = 12% of the total area with 88% of known deposits; RF Model: AUC = 0.96, Predicted area = 14% of the total area with 86% of known deposits; FC Model: AUC = 0.94, Predicted area = 21% of the total area with 79% of known deposits; The Normalized Density (Nd) values further supported the strength of these models: RBFLN Model: Nd = 7.33; RF Model: Nd = 6.14; FC Model: Nd = 3.76. These results demonstrate that the **RBFLN model** is the most robust and reliable approach for predicting mineralization areas, providing strong evidence for its application in future exploration. Fig. 15.3 shows predicted prospective areas from machine learning models: (A) RBFLN predicted area, (B) FC predicted area, (C) RF predicted area.

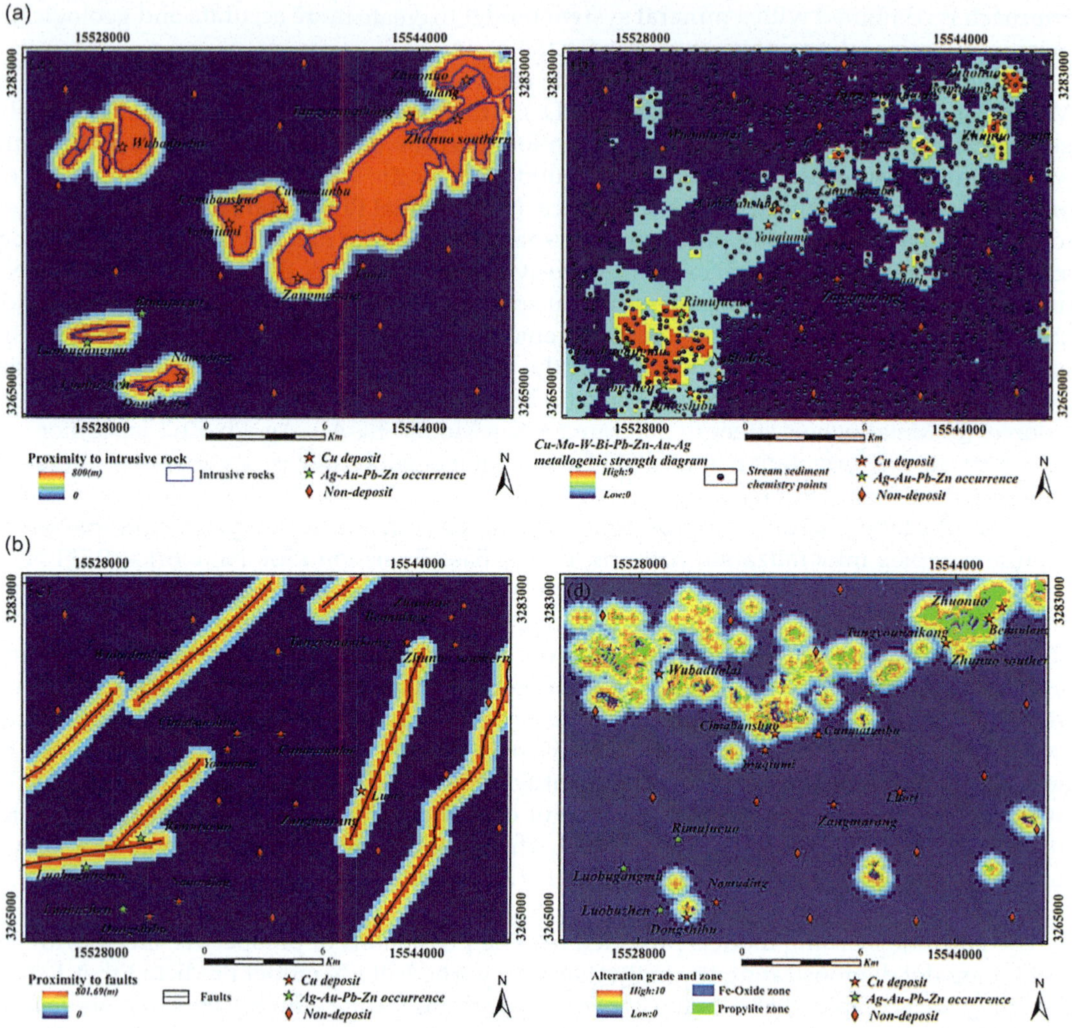

FIG. 15.1 Spatial proxies (evidence layers) used for mineral prospectivity mapping (MPM): (A) Proximity to Eocene to Miocene rocks (source); (B) Metallogenic strength diagram for Cu–Mo–W–Bi–Ag–Au–Pb–Zn (deposition); (C) Proximity to NE-oriented faults (pathways); (D) Fe-oxide and propylitization alteration zones (hydrothermal fluids). *Source: Cheng et al. (2023)—Open Access.*

This study demonstrates the effectiveness of using GIS-based machine learning techniques combined with a mineral system approach for mineral prospectivity mapping in the **Zhuonuo Ore Concentration Area**. The findings indicate that the **RBFLN model** is the most accurate for predicting prospective mineralized zones, providing valuable targets for further exploration. **High-potential areas** for **Porphyry Copper Deposits (PCDs)** were identified in the northern **Cimabanshuo**, northwest **Wubaduolai**, and southwestern and western **Zhuonuo** regions. Additionally, **epithermal Ag–Au–Pb–Zn** deposits were predicted along the NE

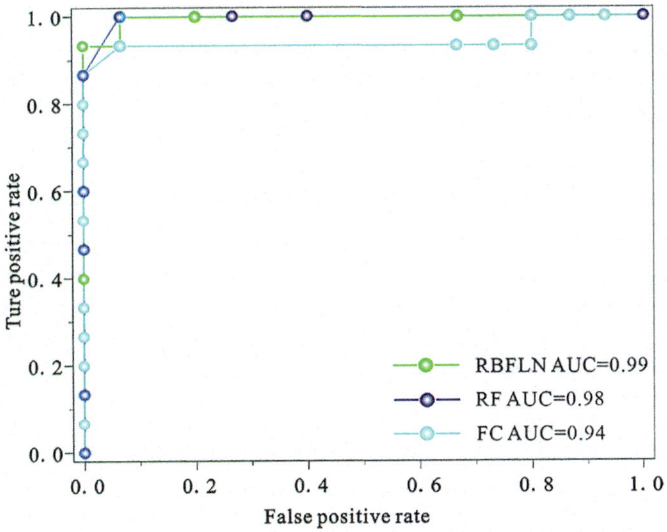

FIG. 15.2 ROC curve comparison of the RBFLN, RF, and FC models. *Source: Cheng et al. (2023)—Open Access.*

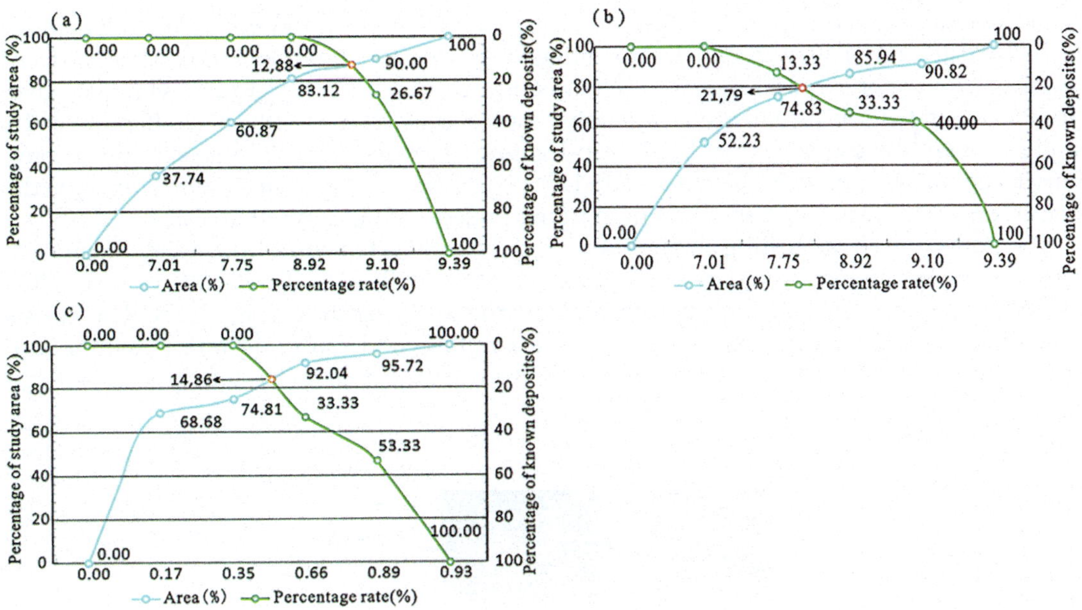

FIG. 15.3 (A) Predictive Area (P–A) plotting of the Radial Basis Function Link Networks (RBFLN) model; (B) Predictive Area (P–A) plotting of the Fuzzy Clustering (FC) model; (C) Predictive Area (P–A) plotting of the Random Forest (RF) model. *Source: Cheng et al. (2023)—Open Access.*

orientation faults and their extensions in areas like **Dongshibu** and **Luobuzhen**. The approach utilized in this study is applicable not only to this specific mineral system but also to other districts and mineral types, offering a robust tool for efficient and sustainable exploration in geologically complex regions.

15.2.5 Case study: Seismic data enhancement with deep learning

In a case study conducted by Schlumberger, deep learning algorithms were employed to enhance seismic data interpretation in offshore oil exploration. The project demonstrated how convolutional neural networks (CNNs) were trained to automatically detect subsurface structures like fault lines, gas reservoirs, and hydrocarbon deposits. The AI model reduced the need for manual interpretation by over 50% while maintaining high accuracy (Schlumberger, 2020).

15.2.6 Seismic event detection using deep learning for DAS data: A case study from the Otway Stage 3 CO_2 injection project

Distributed Acoustic Sensing (DAS) is increasingly used for efficient seismic data acquisition in various applications like vertical seismic profiling (VSP), reservoir monitoring, and microseismicity detection. Utilizing low-cost fiber optic cables, DAS systems provide a flexible solution for challenging environments. Unlike traditional seismic sensors, DAS systems are immune to electromagnetic interference and do not require recalibration. However, the large volume of data generated by DAS systems presents significant computational challenges, requiring innovative methods for data handling and analysis.

In mineral resource exploration, active seismic methods have not been widely adopted due to high costs and shallower exploration targets. As demand for minerals increases, deeper resource exploration requires more efficient seismic methods. Passive seismic techniques, using drilling-generated seismic energy as a passive source, provide a cost-effective alternative. This study explores the use of a supervised deep-learning approach (Fig. 15.4), specifically a convolutional neural network (CNN), to detect and locate passive seismic sources, with a focus on data from the Otway Stage 3 CO_2 injection monitoring project (Al-Hemyari et al., 2024).

The dataset used for testing the methodology was collected from the Otway Stage 3 CO_2 injection project, which involves DAS-equipped wells (CRC-3, CRC-4, CRC-6, and CRC-7) recording passive seismic data before and during CO_2 injection (Fig. 15.5). The passive seismic records were acquired using DAS at a 1-millisecond sampling rate. A synthetic 1D elastic properties model based on a well log from CRC-3 was extended to a 2D setup to

FIG. 15.4 A high-level diagram illustrating the application of supervised deep learning for passive seismic event detection in the Otway Stage 3 CO_2 injection monitoring project, Australia. Source: Al-Hemyari et al. (2024)—Open Access.

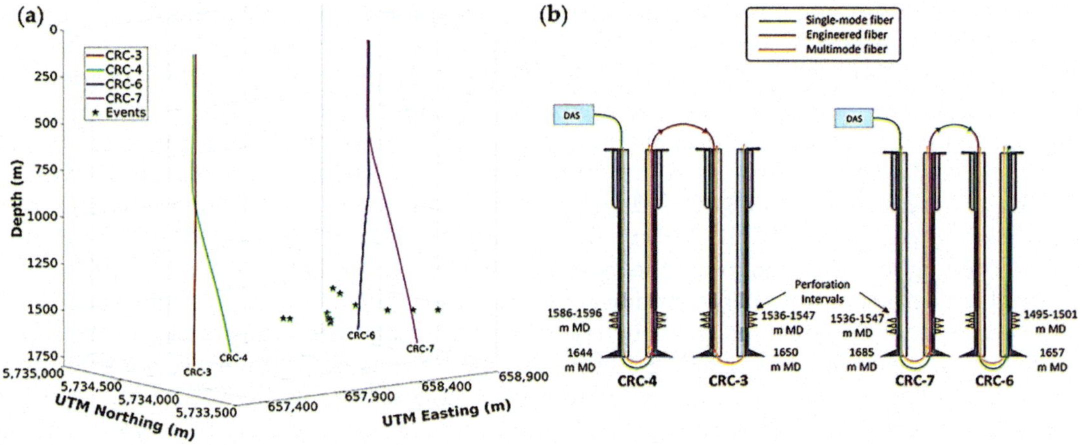

FIG. 15.5 (A) Diagram of the four wells (CRC-3, CRC-4, CRC-6, CRC-7) used to record induced events at the Otway Stage 3 CO_2 injection monitoring site, Australia, with green stars indicating event locations. (B) Configuration schematics of the fiber-optic cable installations at the site. *Source: Al-Hemyari et al. (2024)—Open Access.*

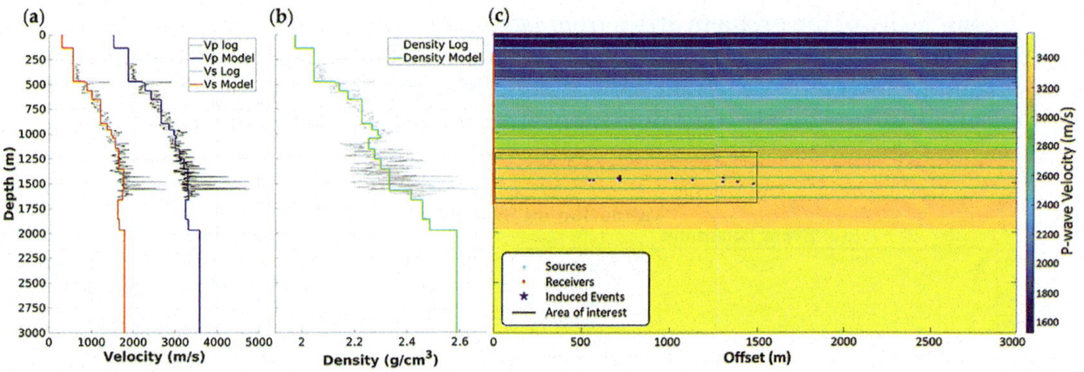

FIG. 15.6 Upscaled 1D model of (A) P-wave and S-wave velocities and (B) density from the vertical CRC-3 well. (C) An extended 2D model over 3000 m of offset with an overlay of source and receiver locations used for modeling, focusing on an area of interest around the induced event locations at the Otway Stage 3 CO_2 injection monitoring site, Australia. *Source: Al-Hemyari et al. (2024)—Open Access.*

simulate passive seismic data from several event locations around the reservoir for neural network training (Fig. 15.6).

The study uses the ResNet50 architecture, a 50-layer CNN effective in image recognition tasks. The neural network processes 512 × 512-pixel grayscale images designed to handle DAS data. With over 24 million parameters, ResNet50 learns patterns in seismic data to predict event locations, depths, and subsurface properties. The data preparation workflow for both training and application datasets involves several steps. The synthetic data is time-shifted and standardized, while noise data is collected from non-injection periods and pre-processed to ensure consistency. Semi-synthetic patches are created by combining synthetic data with noise and labeled with relevant information, including event location, depth, and subsurface properties.

For both training and application datasets, the common pre-processing steps include data loading, geometry matching, amplitude clipping, standardization, bandpass filtering, resizing, and amplitude fitting. The final training dataset consists of 12,364 semi-synthetic and 14,736 noise patches, all converted into grayscale images. The model was trained over 300 epochs, using a batch size of 8, the Adam optimizer, and a decaying learning rate. The training took approximately 39.5 h, with validation loss stabilizing after 150 epochs, indicating model convergence. After training, the model was tested on synthetic data, where predictions showed a clear separation between noise and event locations. However, slight biases were observed in depth predictions, particularly for larger offsets. The performance of the model could be improved by incorporating a more diverse training dataset with various depths.

When applied to the CRC-3 field data, the model successfully predicted 17 induced events out of 450 patches. Most patches were identified as noise, and predictions were completed in 29.8 s. Some noise patches were unexpectedly close to the area of interest (AOI), and four induced events had predicted depths reasonably close to manual detections. However, 10 events were classified as noise. Fig. 15.7A shows the predicted locations of synthetic and noise patches, while Fig. 15.7B demonstrates the reconstruction of subsurface properties using the predicted events.

The complexity of the problem arises from factors like varying receiver geometries, source mechanisms, and elastic properties, all of which increase the size of the training dataset and model complexity. While using 2D models simplifies predictions to depth and offset, azimuthal anisotropy and real-world deviations may require extending the model to 3D. Time

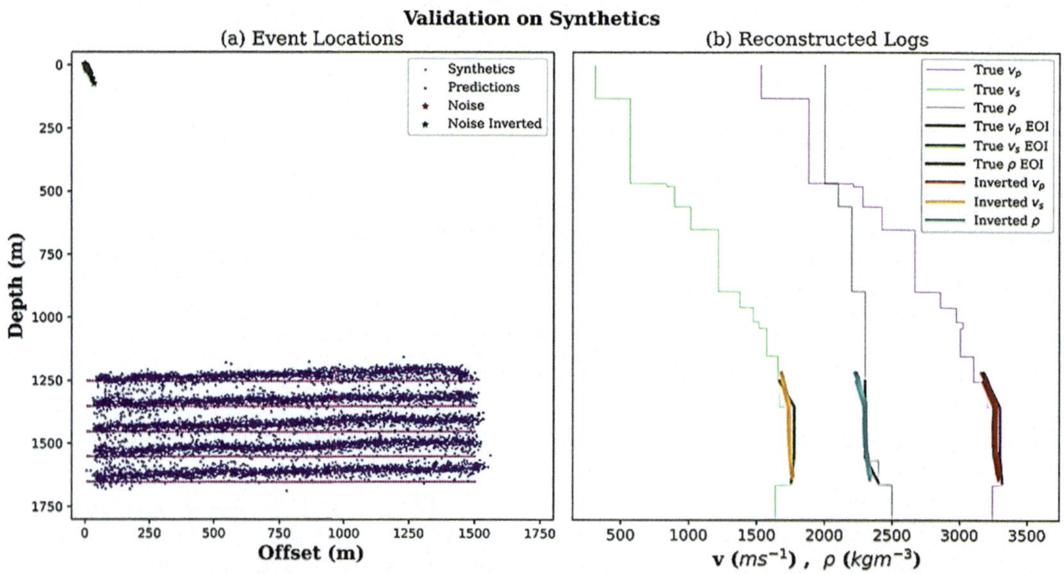

FIG. 15.7 (A) Location predictions of synthetic events compared to the ground truth locations used for training in the Otway Stage 3 CO_2 injection monitoring project, Australia. (B) Reconstructed subsurface properties overlaid on the true models from the same region. Source: Al-Hemyari et al. (2024)—Open Access.

shifts in the training data significantly affected model accuracy, particularly for offset predictions. By including time shifts in the training dataset, prediction errors decreased, showing the importance of this factor for improving model performance. Additionally, the model struggled with low signal-to-noise ratios (SNR), particularly below a threshold of 9%. The ResNet50 architecture performed poorly at low SNRs, suggesting the need for alternative models or denoising techniques to improve event detection.

Finally, the model was applied to data from nearby wells (CRC-4 and CRC-6). The predictions for CRC-4 and CRC-6 showed good separation between events and noise, with CRC-6 performing better due to higher SNR. The model performed less effectively for CRC-4 due to lower SNR and deviations in well geometry. In conclusion, the study demonstrates that synthetic seismic data can be used to train a deep learning model for passive seismic event detection. The ResNet50 neural network, with proper data pre-processing and augmentation techniques, is capable of effectively detecting and locating seismic events in DAS data. While promising, the model could benefit from improvements in handling time shifts, low SNR data, and well-specific geometries, offering a valuable approach for mineral exploration and monitoring.

Example code (Python—seismic data pre-processing)

Below is a simplified example in Python showing how one might pre-process seismic data using deep learning for feature extraction.

```
Python
import numpy as np
import tensorflow as tf
from tensorflow.keras.models import Sequential
from tensorflow.keras.layers import Conv2D, MaxPooling2D, Flatten, Dense
from sklearn.model_selection import train_test_split
# Example seismic data pre-processing
def pre-process_data (seismic_data):
seismic_data = seismic_data/np.max (seismic_data) # Normalize
return seismic_data
# Create a simple CNN model for seismic feature extraction
def create_cnn_model ():
model = Sequential ([Conv2D (32, (3, 3), activation = 'relu', input_shape = (256,
256, 1)), MaxPooling2D ((2, 2)), Flatten (), Dense (64, activation = 'relu'), Dense
(1, activation = 'sigmoid)])
model.compile (optimizer = 'adam', loss = 'binary_crossentropy', metrics =
['accuracy'])
return model
# Example seismic data (dummy data)
seismic_data = np.random.rand (100, 256, 256, 1) # 100 samples, 256 × 256 image size
labels = np.random.randint (2, size = (100, 1)) # Binary classification (e.g., pres-
ence or absence of resource)
# Pre-process the data
seismic_data = pre-process_data (seismic_data)
# Train-test split
X_train, X_test, y_train, y_test = train_test_split (seismic_data, labels, test_
size = 0.2, random_state = 42)
```

```
# Create and train the model
model = create_cnn_model ()
model.fit (X_train, y_train, epochs=10, batch_size = 16, validation_data = (X_test,
y_test))
```

15.2.7 Case study: Predicting mineral deposits with machine learning

A landmark project conducted by BHP Billiton applied machine learning techniques to predict the location of mineral deposits based on geological, geophysical, and geochemical data. The team used Random Forest (RF) and Support Vector Machine (SVM) algorithms to classify exploration regions as "high probability" or "low probability" for containing mineral resources. The model improved the efficiency of the exploration process by 30%, reducing both time and cost (BHP Billiton, 2019).

15.2.8 Integrating knowledge-driven and data-driven approaches for gold exploration: A case study of the Hatu region, Xinjiang, China

Gold exploration is essential for economic growth and national defense security. Traditional exploration methods can be divided into knowledge-driven and data-driven approaches. Knowledge-driven methods rely on expert interpretation, while data-driven methods use geostatistics and machine learning algorithms to identify patterns in geological, geophysical, and geochemical data. This case study focuses on combining both approaches, applying the Fuzzy Comprehensive Evaluation Knowledge-Driven Data-Driven (FCEKDD) model to enhance the prediction of gold mineralization in the Hatu region, Xinjiang, China, known for its complex geological settings.

Located in the Central Asian Orogenic Belt (CAOB) in the western Junggar Basin, Xinjiang, the Hatu region is rich in volcanic-sedimentary formations, such as the Lower Carboniferous Xibeikulas, Baogutu, and Tailigula formations. The Hatu Gold Mineralization Belt, influenced by tectonic movements and faulting, is one of the most significant zones in the region. Several gold deposits exist here, with the Hatu deposit being the largest.

For a comprehensive evaluation of gold mineralization potential, various datasets were collected. Geological maps (1:50,000) provided vital geological characteristics, stratigraphy, and structural features. Geophysical data, including gravity data from 136 samples, and geochemical data covering 39 elements, including Ag, As, Au, and Cu, were collected and corrected for outliers. Remote sensing data from Landsat 8, with imagery from June 3, 2023, offered crucial information on land cover and alteration zones relevant to mineral exploration.

The FCEKDD model integrates geological knowledge and data-driven methods. Key methodologies, including Principal Component Analysis (PCA) for alteration extraction, geostatistical analysis (Kriging) for geochemical anomalies, and structural analysis using gravity data, were applied. Gravity data revealed fluctuations in the gravity field in the X direction, indicating structural anomalies. The fault systems, especially northeast-trending faults like the Hatu and Anqi faults, were identified as major controls for gold mineralization (Fig. 15.8). Faults were categorized as first-order (major), second-order (smaller), and third-order (minor) based on their size and influence, highlighting faulting as a significant geological factor for gold deposit distribution.

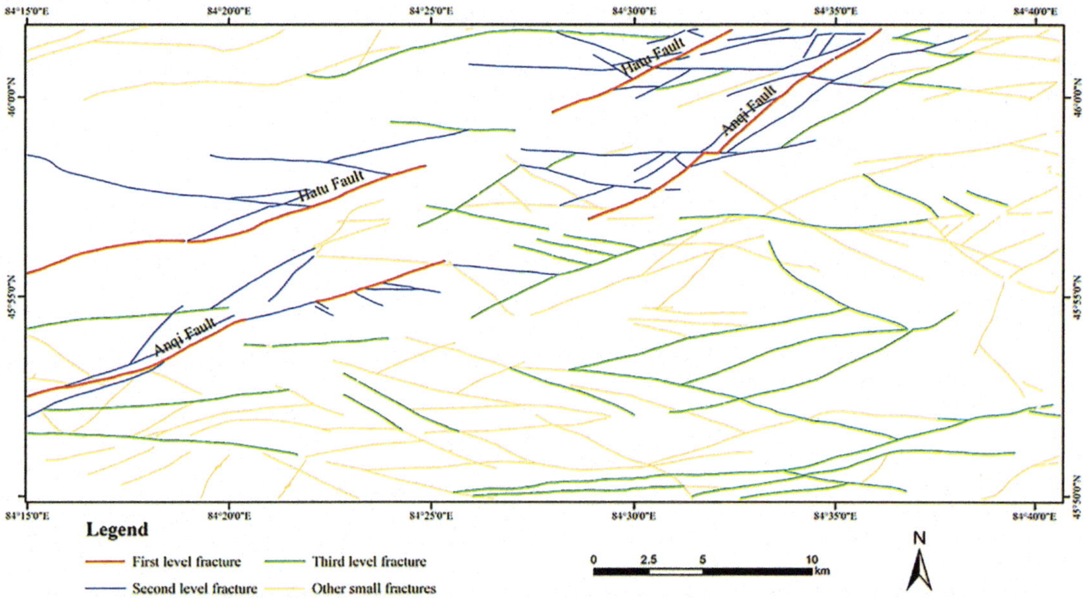

FIG. 15.8 Structural map and tectonic fracture classification of the Hatu region, Xinjiang, China, obtained by geophysical and geological methods. *Source: Feng et al. (2024)—Open Access.*

Geochemical analysis identified key elements related to gold mineralization, such as Au, Sb, Ag, As, and Cu. Statistical analysis confirmed a normal distribution for these elements, and Kriging interpolation created anomaly maps. C–A fractal analysis helped define lower limits for anomalies, and factor analysis revealed five main clusters of elements. The Au–Sb, Au–Ag, and W–As clusters strongly correlated with known gold deposits, indicating their importance in gold mineralization.

Alteration zones, crucial for identifying gold mineralization, were extracted from remote sensing data. PCA enhanced the spectral signatures of iron-bearing and hydroxyl-bearing minerals, which were found to correlate with the major fault systems, particularly the Anqi and Hatu faults. Iron-bearing alterations (Iron Staining Index and SWIR) and hydroxyl-bearing alterations (Re-SWIR ratio) provided valuable indicators of gold mineralization (Fig. 15.9).

The FCEKDD model integrated all data, including geological, geophysical, geochemical, and remote sensing information. The area was divided into 21,507 grid units, and mineralization anomalies were quantified. The output of the model produced a mineralization favourability map, categorizing the region into three prospectivity levels: high potential (first-level), moderate potential (second-level), and low potential (third-level). The model successfully predicted the known gold deposits, with most falling within the second-level zone. Of 56 known mineralization belts, 32 were in the high-potential zone, 14 in the moderate zone, and 8 in the low zone. Unpredicted sites had weaker structural control and less favorable geochemical conditions, dominated by copper and cobalt anomalies rather than gold (Fig. 15.10).

This study shows that integrating geological, geophysical, geochemical, and remote sensing data through the FCEKDD model enhances gold mineralization prediction, especially in complex

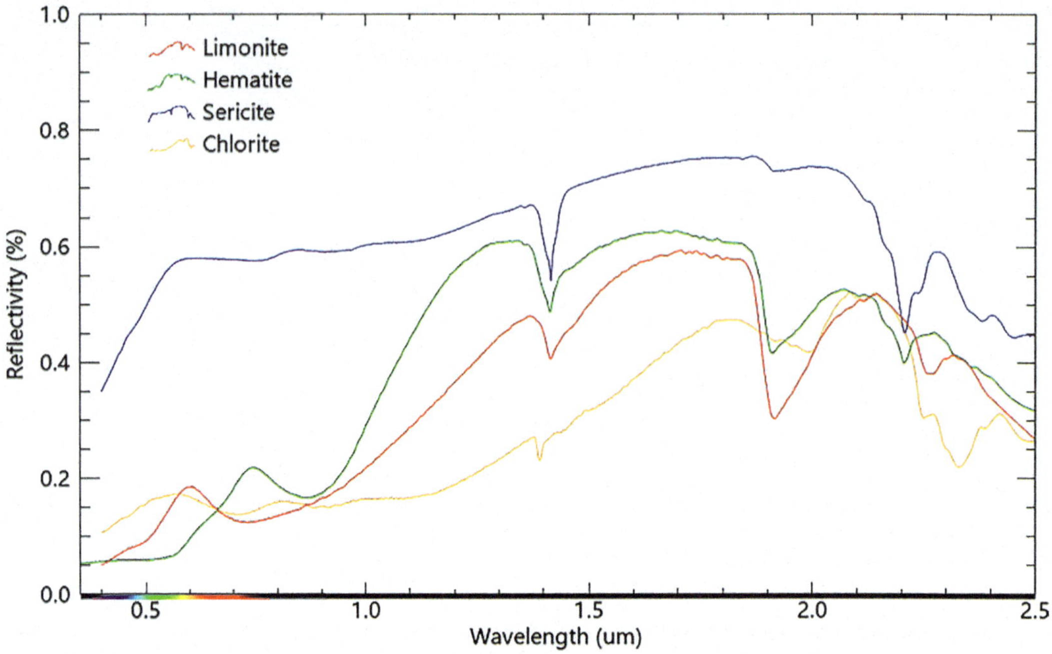

FIG. 15.9 Spectral curves of iron-stained and hydroxylated minerals in the Hatu region, Xinjiang, China, extracted using Principal Component Analysis (PCA). *Source: Feng et al. (2024)—Open Access.*

geological regions. The model effectively identified high-priority areas for gold exploration, particularly those influenced by major faulting and favourable geochemical and alteration characteristics. The findings offer a robust framework for targeting potential gold resources in the Hatu region and other similar areas. Future research could further improve the FCEKDD model by adding additional data sources, such as detailed remote sensing imagery or seismic data, to enhance its predictive accuracy. Expanding the model's application to other regions with different geological characteristics could help assess its versatility for diverse mineral exploration scenarios.

15.2.9 Case study: Machine learning for uncertainty in prospectivity mapping of gold deposits from Juruena, Brazil

Mineral prospectivity mapping (MPM) is a crucial tool used to identify areas with high potential for discovering new mineral deposits. It is a multi-criteria process that can be driven by geological knowledge, geoscientific data, or a hybrid approach. Traditional MPM relies heavily on Geographic Information Systems (GIS) for analyzing evidential layers. However, manual data processing in GIS is limited by the human brain's capacity to handle multiple variables, which can reduce reproducibility and introduce dependency on expert assumptions.

To address these challenges, machine learning (ML) methods have recently been applied to MPM, enabling the automatic identification of relationships in complex datasets. Among

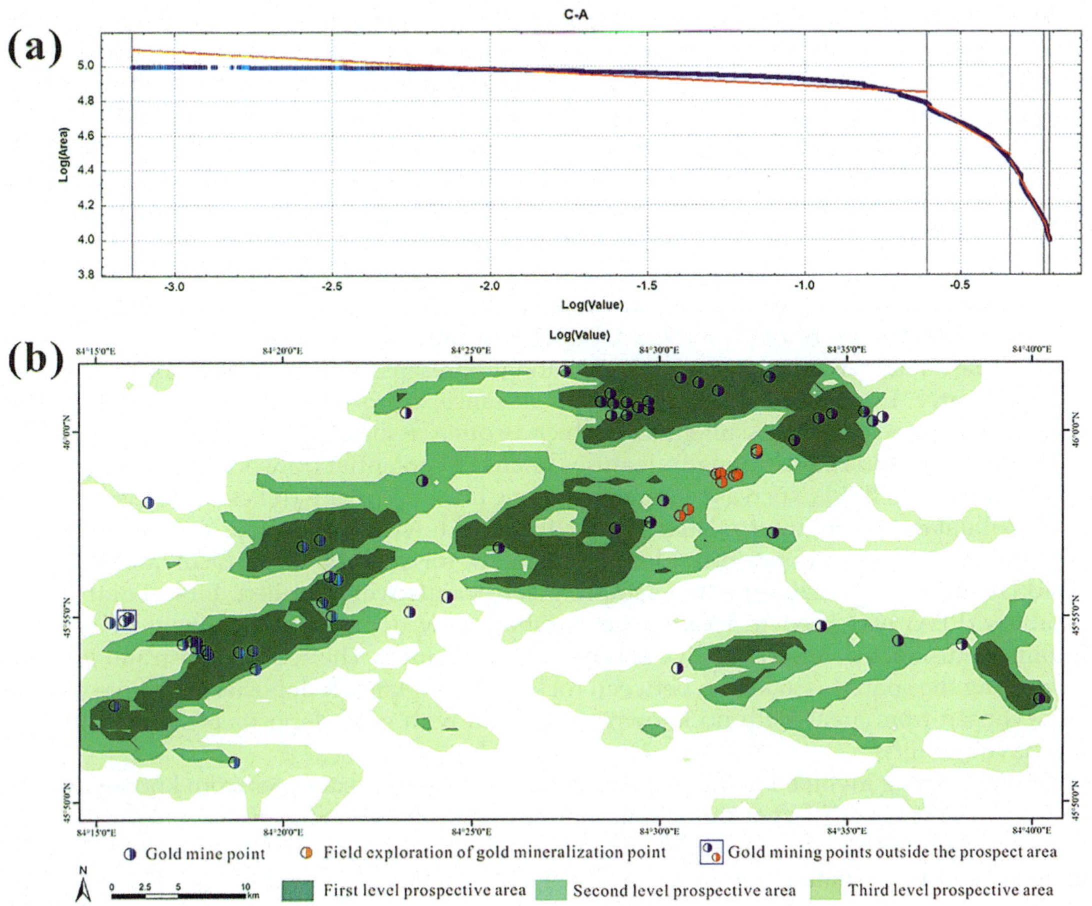

FIG. 15.10 Double-Logarithmic plot of the Cumulative-Area (C–A) method (A) and grading map of mineralization prospects based on the Cumulative-Area (C–A) method (B) in the Hatu region, Xinjiang, China. *Source: Feng et al. (2024)—Open Access.*

these, supervised learning methods have shown great potential by classifying locations as prospective or non-prospective based on known deposit locations and random negative locations. However, defining true "negative" sites is problematic, as geological processes are often non-random, and deposits are rare events.

Previous studies have explored several methods for generating negative training datasets, such as utilizing the spatial distribution of deposits or sampling from barren lithology areas. Yet, uncertainties persist, especially regarding the effectiveness of negative sampling in mineral exploration. In this study, developed an ML-based framework to quantify uncertainty in gold prospectivity mapping. The framework was applied to the Juruena Mineral Province (JMP), located in northern Mato Grosso, Brazil, by generating hundreds of potential maps using three common ML methods—Random Forest (RF), Support Vector Machine (SVM), and K-Nearest Neighbors (KNN). For each method, created 400 negative training datasets constrained by spatial and geological criteria.

The Juruena Mineral Province, rich in gold, is situated in the southern Amazonian Craton, a tectonically stable region that spans over 4.5 million km^2. The province is divided into several geochronological units, with significant gold mineralization associated with shear zones and veins in granites, faults, and porphyries. Gold mineralization in the JMP is linked to magmatic-hydrothermal systems at varying crustal depths, with primary deposits occurring in veins and stockworks within shallow-emplaced granites and porphyry-hosted epithermal veins. The region's geological features, such as the Peixoto de Azevedo Domain and the Colíder Group, played a key role in selecting relevant features for the ML models used in this study.

Silva dos Santos et al. (2022) used freely available airborne and geological data from the Geological Survey of Brazil (GeoSGB-CPRM) and Shuttle Radar Terrain Mission (SRTM) data. These datasets were instrumental in creating representations of key ore-forming factors, such as source, transport, and alteration. For instance, airborne gamma-ray data helped to highlight areas with potassic alteration, which is often associated with gold mineralization in the JMP (Fig. 15.11). Additionally, features related to potential transport paths and sources for hydrothermal fluids, such as buffered magnetic lineaments and dykes, were incorporated. The final datasets were interpolated to a uniform grid, with gamma-ray and elevation data being oversampled and undersampled, respectively, for analysis using QGIS and Python.

To create negative datasets, two sampling constraints were applied. First, Point Pattern Analysis (PPA) was used to identify areas where new deposits are unlikely by setting a minimum distance between positive and negative examples. The second constraint involved measuring the spatial association between rock units and known deposits. Negative samples were chosen from areas with no known association with gold deposits based on lithologic unit favourability.

Model training involved using 32 positive and 32 negative examples from known deposits and occurrences, with hyperparameters optimized using a random search algorithm. The three ML models (RF, SVM, and KNC) were evaluated with 400 realizations each. All models were trained using data from the GeoSGB-CPRM database, with a total of 11 deposits and 48 occurrences used for training and testing. Hyperparameters were tuned through 5-fold cross-validation to identify optimal settings.

In total, 200 potential maps were generated for each negative dataset criterion. The results showed that the maps generated using the lithologic unit favorability criterion predicted higher probabilities of gold deposits in the test area compared to the first criterion (based on PPA). Specifically, areas with very high probability (>0.9) increased from 8.2%–21.4% to 16.6%–26.4% when the lithologic criterion was applied. Additionally, predictions using the lithologic criterion showed more stable results, with reduced standard deviation in the test area.

The model performance was further evaluated using the Area Under the Curve (AUC) scores, which were higher for the models trained with the second negative dataset criterion. For instance, the Random Forest model's AUC score improved from 0.77 to 0.86, and K-Nearest Neighbors achieved an AUC of 0.86, demonstrating improved reliability and predictive performance (Fig. 15.12). The AUC scores standard deviations also decreased with the lithologic unit favorability criterion, indicating that geological knowledge enhances model stability.

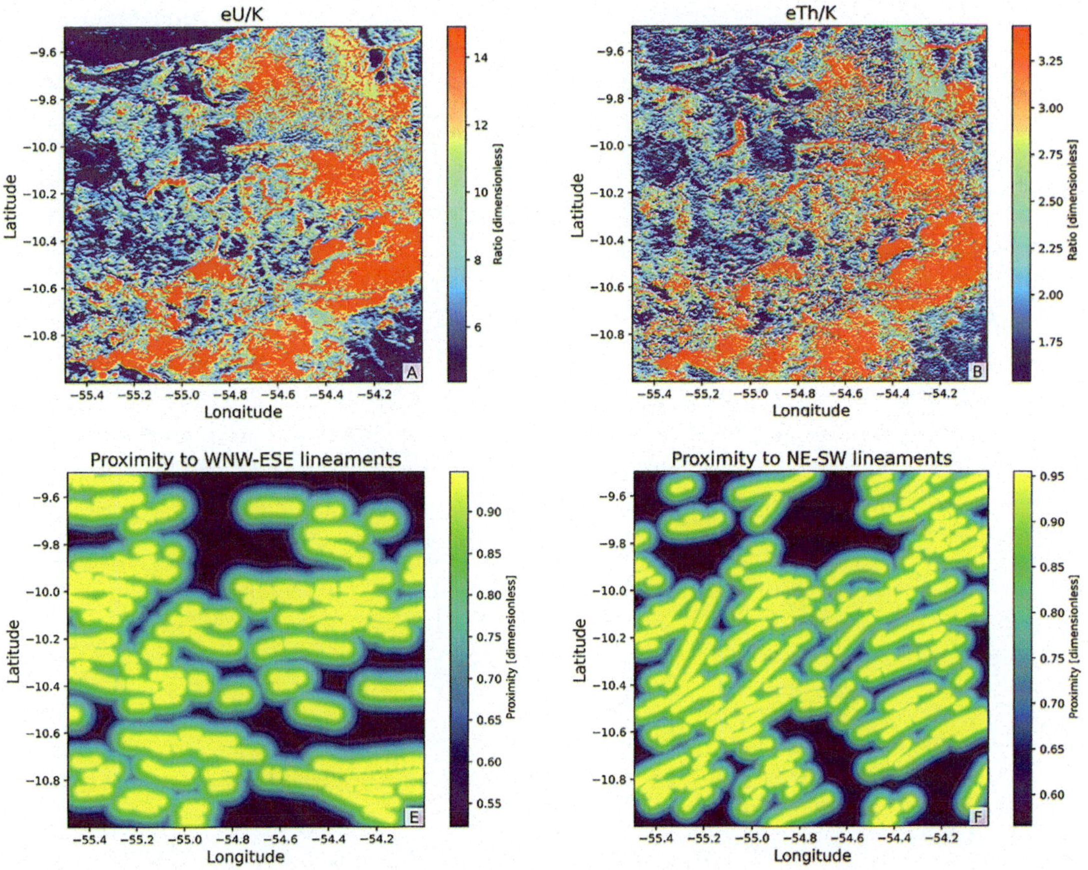

FIG. 15.11 Features used to map prospective areas for magmatic-hydrothermal gold deposits in the Juruena Mineral Province (JMP). (A) Ratio between equivalent uranium and potassium; (B) Ratio between equivalent thorium and potassium; (C) Elevation from Shuttle Radar Topography Mission (SRTM) data; (D) Proximity to mapped dykes; (E) Proximity to WNW-ESE magnetic lineaments; (F) Proximity to NE-SW magnetic lineaments. These features were derived using airborne and geological data, including gamma-ray data and SRTM elevation data, to identify key ore-forming factors in the region. *Source: Silva dos Santos (2022)—Open Access.*

To interpret the feature importance, Silva dos Santos et al. (2022) used SHAP (SHapley Additive exPlanations) values for the SVM and KNC models. The results indicated that the most important features in predicting gold deposits were elevation and proximity to dykes, regardless of the negative sampling method (Fig. 15.13). This finding reinforces the relevance of these geological factors in gold mineralization, offering valuable insights for future exploration.

Overall, the study demonstrates that incorporating geological knowledge into the negative sampling process improves the accuracy and stability of machine learning models in mineral prospectivity mapping. By generating multiple maps and analyzing the uncertainty associated

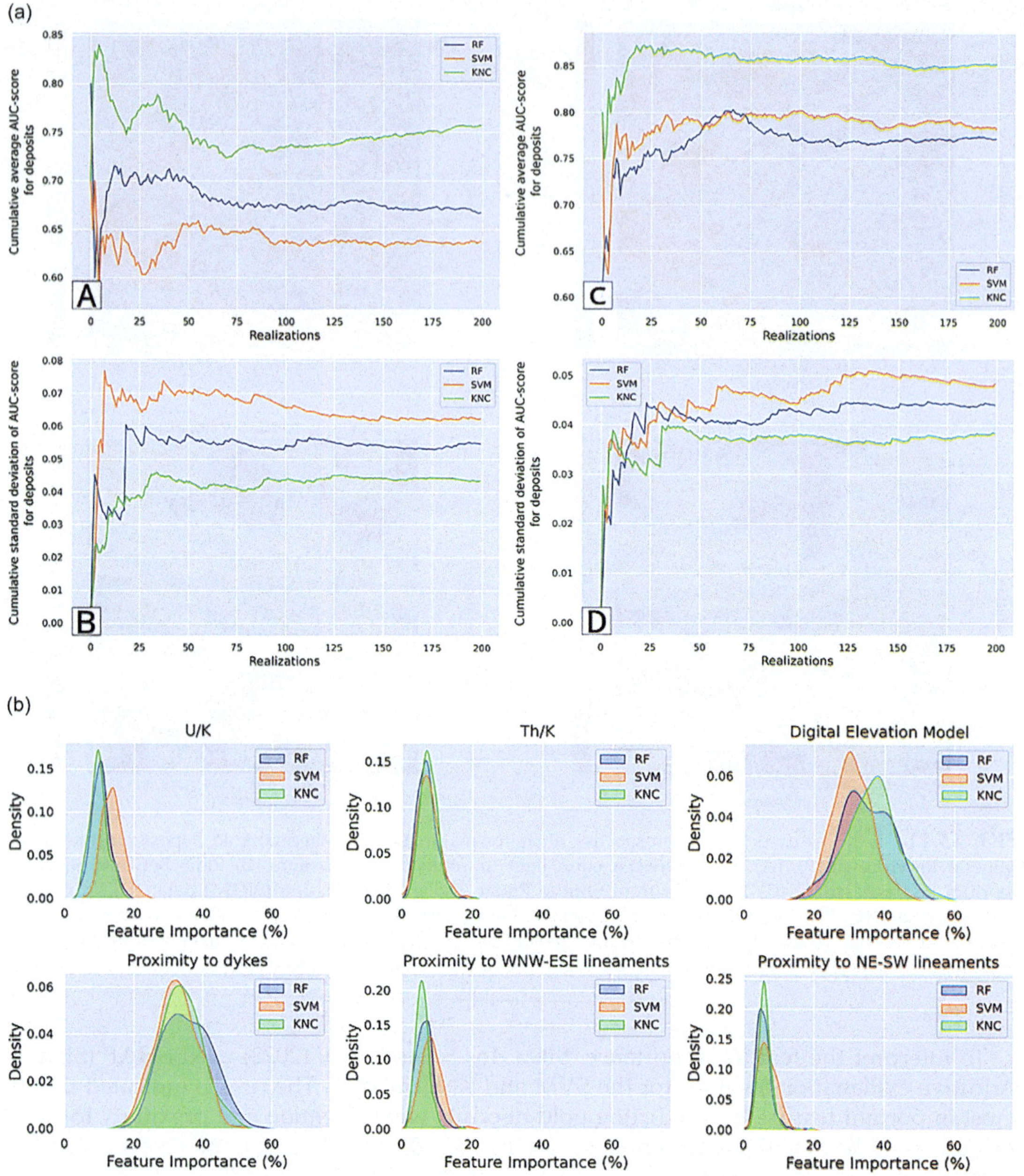

FIG. 15.12 Area under the curve (AUC) scores for deposits in the test area, along with the realizations, for the Juruena Mineral Province (JMP). The left column shows cumulative statistics for AUC using the first criterion for generating negative datasets: (A) Cumulative average AUC; (B) cumulative standard deviation of AUC. The right column shows cumulative statistics for AUC using the second criterion: (C) Cumulative average AUC; (D) cumulative standard deviation of AUC. *Source: Silva dos Santos (2022)—Open Access.*

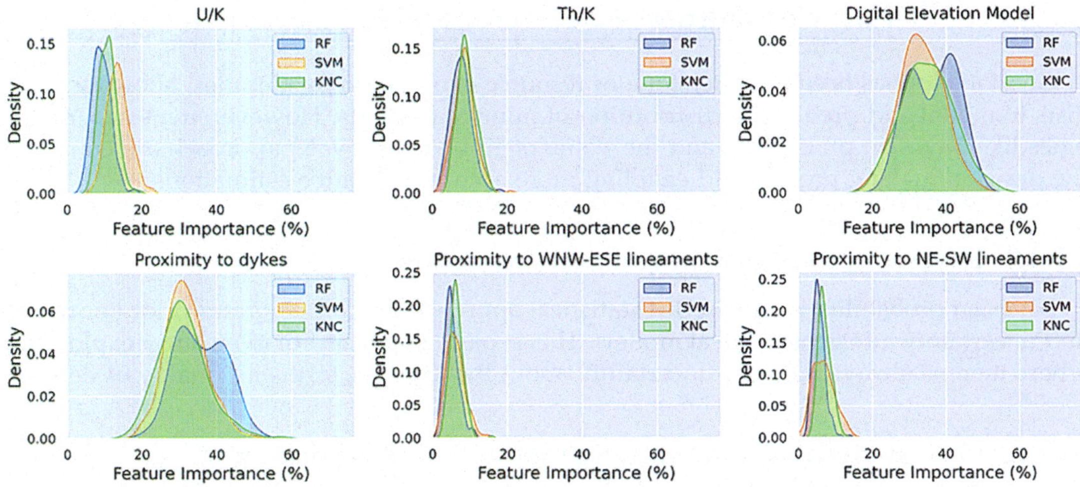

FIG. 15.13 Kernel density estimation of feature importance for the Juruena Mineral Province (JMP). (A) Estimated density of feature importance for all realizations using the first criterion (Point Pattern Analysis) to create negative datasets. (B) Estimated density of feature importance for all realizations using the second criterion (Lithologic Unit Favorability) to create negative datasets. Feature importance was calculated based on SHAP (SHapley Additive exPlanations) values derived from the trained machine learning models. *Source: Silva dos Santos (2022)—Open Access.*

with each model, it is possible to better understand the underlying geological processes and provide more reliable predictions for gold exploration in the Juruena Mineral Province.

Example code (Python—ML model for mineral deposit prediction)

```
Python
from sklearn.ensemble import RandomForestClassifier
from sklearn.model_selection import train_test_split
from sklearn.metrics import accuracy_score
import numpy as np
# Example geological data (features: geophysical, geochemical, geological factors)
features = np.random.rand (1000, 10) # 1000 samples, 10 features
labels = np.random.randint (2, size = 1000) # 0 for non-resource, 1 for resource
# Train-test split
X_train, X_test, y_train, y_test = train_test_split (features, labels, test_size =
0.2, random_state = 42)
# Initialize and train the model
rf_model = RandomForestClassifier (n_estimators = 100, random_state = 42)
rf_model.fit (X_train, y_train)
# Predictions and evaluation
y_pred = rf_model.predict (X_test)
accuracy = accuracy_score (y_test, y_pred)
print (f" Model accuracy: {accuracy * 100:.2f} %")
```

15.3 Geostatistics and resource extraction decisions

Geostatistics has been a cornerstone of resource exploration for decades, allowing geologists to model and predict the distribution of mineral deposits. However, newer methodologies like Bayesian probability and the Value-of-Knowledge (VoK) approach are enhancing the decision-making process and enabling more accurate resource estimations.

15.3.1 Bayesian probability in resource extraction

Bayesian probability is a powerful statistical approach that incorporates prior knowledge and uncertainty into geostatistical models. This is particularly useful in resource exploration, where there is always inherent uncertainty about the location, size, and quality of deposits.

15.3.2 Case study: Bayesian modeling for resource estimation

A case study by Rio Tinto employed Bayesian methods to model the uncertainty in copper ore deposit sizes. The model integrated prior knowledge from historical drilling data and geological surveys to update predictions about the copper content in specific areas of a mine site. The Bayesian model provided a more refined resource estimate, allowing the company to optimize its extraction plans and reduce waste (Rio Tinto, 2018).

15.3.3 Case study: Bayesian approach to the zinc extraction curve of soil with sewage sludge in the treatment plant of Franca, São Paulo State, Brazil

Sewage sludge disposal is a challenge for sanitation companies. Using it in agriculture can supply essential nutrients, such as zinc (Zn), which are crucial for plant growth and crop yield (Carvalho et al., 2015; Silva and Andrade Pinto, 2010). Zinc plays a key role in plant health and is vital during peak plant demand to avoid yield loss (Crusciol et al., 2016). However, excessive zinc can be toxic to plants (Yagi et al., 2006), and in Brazil, soils often lack sufficient zinc for optimal crop yields (Menezes et al., 2010). Several extraction solutions are used to measure soil zinc (Bortolon and Gianello, 2010; Rodrigues et al., 2012), and sequential extractions track element variations over time (Oliveira et al., 2003; Souza et al., 2010). The non-linear first-order kinetic model is commonly used to study soil processes like zinc release (Oliveira et al., 2013; Zeviani et al., 2012) and has been applied to zinc extraction from sewage sludge-treated soil (Souza et al., 2010). However, this method requires large sample sizes for accuracy (Martins Filho et al., 2008). The Bayesian approach, suitable for small samples, allows incorporating prior data and provides probability distributions for model parameters, based on Bayes Theorem:

$$P(\theta|\,\text{data}) = \frac{P(\text{data}\,|\,\theta)\cdot P(\theta)}{P(\text{data})} \tag{15.1}$$

where P(θ|data) is the posterior probability of the parameter θ given the data, P(data|θ) is the likelihood of observing the data given θ, P(θ) is the prior probability of θ, and P(data) is the marginal likelihood. This study uses the Bayesian method to apply the nonlinear first-order kinetic model for zinc extraction from sewage sludge-treated soil.

The data analyzed were from Souza et al. (2010), which involved 15 sequential zinc extractions from sewage sludge-treated soil using seven extraction solutions. The experiment was conducted in a laboratory with a randomized design and three replicates. The soil (Kandiudox) from Jaguariuna, São Paulo, had 25.5 g/kg organic matter, 450 g/kg clay, pH 5.5, and a Zn concentration of 71.74 mg/kg. Sewage sludge was applied annually for five years at doses totaling 150.53 t/ha (tons per hectare).

Seven extraction solutions were used: four organic acids (MOAS, SOAS, lactic acid, acetic acid) and three routine methods (Mehlich I, NH_4OAc, DTPA). Zinc accumulation was modeled using the nonlinear first-order kinetic model:

$$C_i = C_l\left(1 - e^{-kt_i}\right) + \epsilon \tag{15.2}$$

where C_i is the zinc extracted (mg/kg) at extraction t_i, C_l is the equilibrium zinc amount (mg/kg), k is the extraction rate, t_i is the number of extractions, and ϵ is the error with normal distribution. The likelihood for Bayesian inference was:

$$P(C_l \mid \text{data}) \sim N\left(\mu, \sigma^2\right) \tag{15.3}$$

The prior distributions were chosen based on previous studies.

C_l followed a normal distribution:

$$C_l \sim N\left(\mu, \sigma^2\right) \tag{15.4}$$

Precision σ followed a gamma distribution:

$$P(\sigma) \sim \text{Gamma}(\alpha, \beta) \tag{15.5}$$

The extraction rate k followed a beta distribution:

$$P(k) \sim \text{Beta}(\alpha_k, \beta_k) \tag{15.6}$$

Bayesian inference was applied using the joint posterior distribution obtained by multiplying the likelihood by the prior distributions. The Gibbs sampler and Metropolis-Hastings algorithms were used for computation in R.

Inferences for each parameter were made using marginal distributions derived from the joint posterior distribution. The posterior full conditional distributions for C_l and σ followed normal and gamma distributions, respectively. For the extraction rate parameter k, the Metropolis-Hastings algorithm was used.

The chain convergence analysis showed no evidence against convergence, with $p > 0.05$ and dependence factor (DF) always < 5. Table 15.1 presents the posterior means and credible intervals for each parameter. All parameters were significant, as no credible intervals included zero, validating the Bayesian approach. The estimates were consistent with those from Souza et al. (2010) using classical methods.

The extracted Zn (C_l) followed the order: Mehlich I = lactic acid > MOAS = SOAS > acetic acid > DTPA > NH_4OAc, consistent with findings by Araujo and Nascimento (2005) and Menezes et al. (2010). The extraction rate k was highest for DTPA and Mehlich I, aligning with Souza et al. (2010). A left-asymmetric distribution for k was observed for DTPA and

TABLE 15.1 Posterior mean for the parameters and highest posterior density (HPD).

Extraction solution	Parameter	Posterior mean HPD 95%		
			LL	UL
Mehlich I	C_1	36.2520	35.4314	36.9987
	k	0.9382	0.8580	0.9982
	τ	0.5096	0.1787	0.8923
	C_1	10.9212	9.2158	12.6657
NH_4OAc (ammonium acetate)	k	0.1254	0.0877	0.1636
	τ	3.2442	1.1876	5.4576
	C_1	21.7316	20.9246	22.5406
DTPA (diethylenetriaminepentaacetic acid)	k	0.8932	0.7642	0.9948
	τ	0.5259	0.1895	0.9114
	C_1	31.1650	30.1250	32.1784
MOAS (modified organic acid solution)	k	0.4687	0.3888	0.5544
	τ	0.5888	0.2269	1.0072
	C_1	31.7904	30.8301	32.7242
SOAS (simplified organic acid solution)	k	0.4075	0.3539	0.4658
	τ	0.7362	0.2765	1.2463
	C_1	27.1902	26.4479	27.9740
Acetic acid	k	0.3564	0.3142	0.3988
	τ	1.4602	0.5686	2.5114
	C_1	35.0520	33.9295	36.3727
Lactic acid	k	0.4610	0.3831	0.5515
	τ	0.3952	0.1577	0.6841

LL represents the lower limit, and UL represents the upper limit.

Mehlich I, suggesting an asymmetric prior distribution for these in future studies. Fig. 15.14 shows the distribution of the parameter k for different extraction solutions.

The Bayesian method effectively modeled zinc extraction from soil with sewage sludge using seven extraction solutions. It enabled the use of prior distributions and

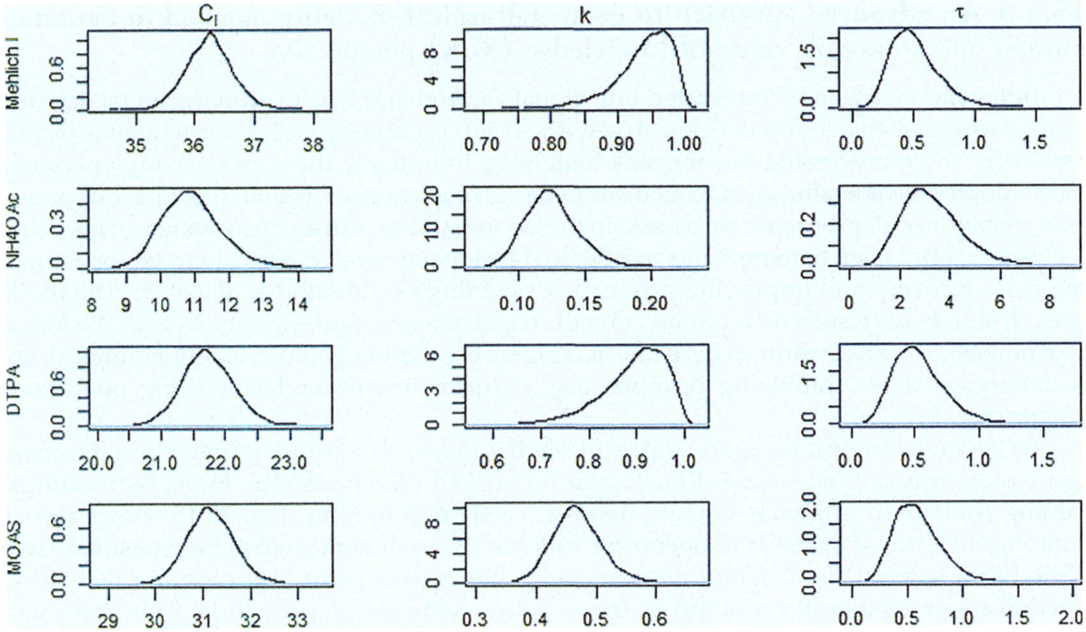

FIG. 15.14 Marginal distribution probability density functions for the model parameters (C_l, accumulated zinc; k, extraction rate; and precision) using the Bayesian method. *Source: Marcelino et al., 2020—Open Access.*

provided probability distributions for model parameters. The posterior distributions for C_l and σ followed normal and gamma distributions, respectively. For the extraction rate parameter k, future studies may consider an asymmetric prior distribution for DTPA and Mehlich I.

15.3.4 Case study: Value-of-knowledge (VoK) in resource exploration

The VoK approach evaluates the value of acquiring additional data in the decision-making process. It is used to assess the potential improvement in decisions based on new information, allowing exploration teams to prioritize data acquisition efforts.

15.3.5 Case study: VoK in mining projects

In a project led by Anglo American, VoK was applied to identify the most cost-effective areas for drilling in a copper exploration project. By considering the value of new geological data at various drilling locations, the company was able to prioritize drilling sites that offered the highest return on investment in terms of resource discovery, thus improving exploration efficiency by 25% (Anglo American, 2021).

15.3.6 An advanced approach to geometallurgical modeling applied in bauxite mining operations: A value-of-knowledge (VoK) perspective

Mining has increasingly integrated into global value chains due to growing metal demand, but it faces challenges in balancing this with social, economic, and environmental factors, especially when processing lower-grade materials. To mitigate these risks, many operations have adopted geometallurgical, which combines geological exploration, resource estimation, and metallurgical processes to assess material variability during processing. This study focuses on a Brazilian bauxite mine, aiming to develop a geometallurgical model for estimating mass recovery and improving production planning (Rodovalho et al., 2025). The model uses drill hole data such as alumina content, reactive silica, and mass recovery. The Value-of-Knowledge (VoK) improves decision-making by combining geological, mineralogical, and metallurgical data, optimizing resource use, reducing waste, and enhancing operational efficiency.

The geometallurgical program aimed to predict mass recovery variations during mineral processing, focusing on coarse (RECG) and fine (RECF) fractions. The mine, with multiple mining fronts and a tabular bauxite deposit, used strip mining. Fig. 15.15 shows the 2D contour of the three plateaus in operation, and Fig. 15.16 illustrates drill hole positions with varied grid spacing for different plateaus. A specific geometallurgical program was developed to assess geological variability in the feed, involving key steps such as geometallurgical tests, defining domains through exploratory data analysis, generating multivariate models,

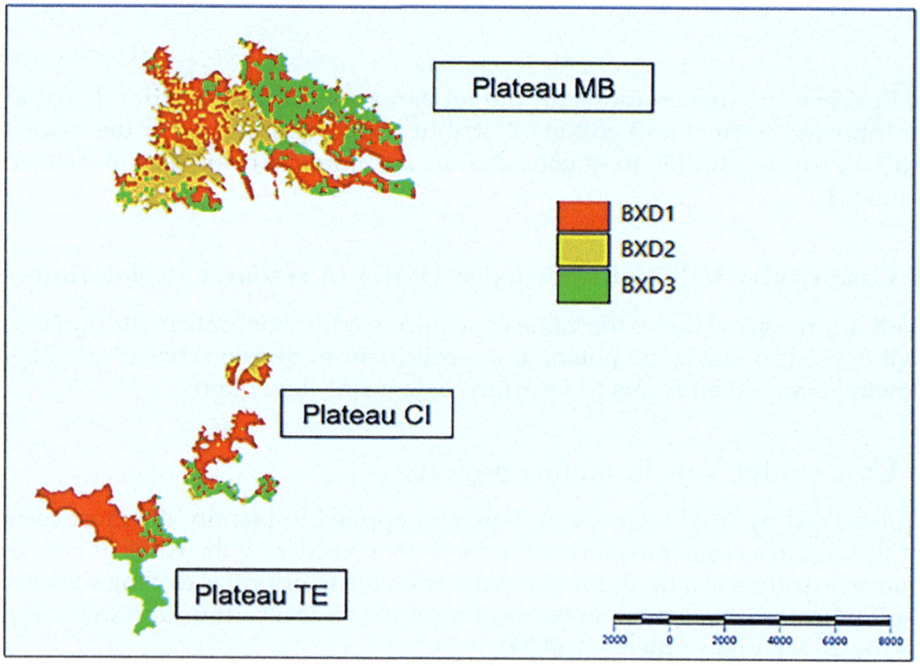

FIG. 15.15 Bauxite mining areas with geochemical classification using 2D contouring (TE Plateau). *Source: Rodovalho et al. (2025)—Open Access.*

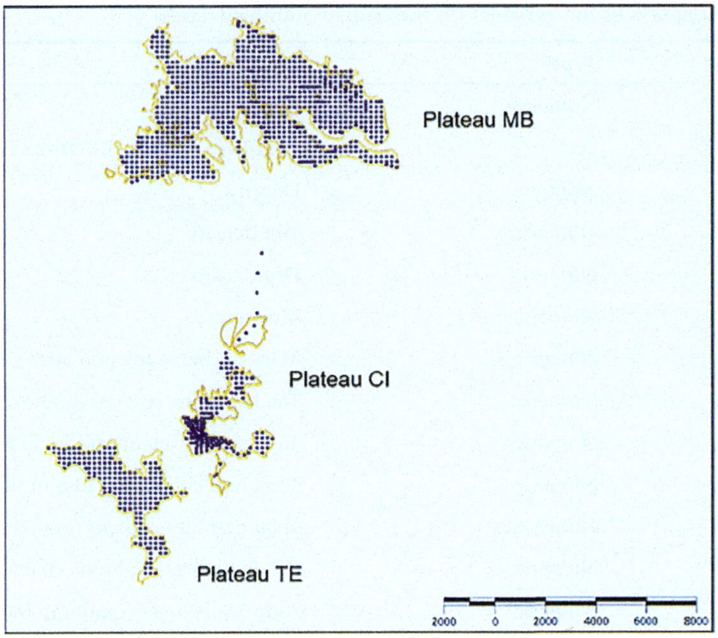

FIG. 15.16 Spatial distribution of drill holes across the bauxite mining site (TE Plateau). *Source: Rodovalho et al. (2025)—Open Access.*

and estimating coarse recovery with linear regression. Finally, the model reconciled actual and estimated recovery values in the mine plan.

For each plateau in Fig. 15.15, a drilling hole database with variables like grades, retained material in fine and coarse fractions, densities, moisture content, and geological/size classifications was available. Table 15.2 lists the variables evaluated using multivariate linear regression. The goal was to define geometallurgical domains based on the behavior of RECG and RECF variables. The first step involved analyzing histograms for RECG and RECF within the geological sections: BXD1, BXD2, and BXD3. Fig. 15.17 shows these histograms, revealing significant deviations in the BXD2 population compared to BXD1 and BXD3. The coarse fraction shows higher material retention, but BXD2 has a lower average compared to the combined population.

This trend was consistent across the CI and MB plateaus. Boxplot analysis further confirmed that BXD1 and BXD3 showed similar sieving behavior, indicating higher mass recovery when fed from these regions. Fig. 15.18 shows the boxplot segregating data into two groups: GMD1 (BXD1 + BXD3) and GMD2 (BXD2). The analysis suggested two distinct geometallurgical domains. A normality test showed P-values for GMD1 and GMD2 below 0.005, indicating non-normal distributions. A Kruskal-Wallis test, with a P-value below 5%, confirmed significant variability in mass recovery between the two domains.

To generate solids corresponding to the GMD1 and GMD2 populations, 3D modeling was necessary. Implicit modeling using the Radial Basis Function (RBF) in Micromine 2024 was applied, incorporating geological context such as mapping, geological exploration, and

TABLE 15.2 Description of the variables for the bauxite mining dataset.

Variables	Type	Description
LITO	Numeric	Lithology code
LITOREV	Text	Lithology domain (BXD1, BXD2, and BXD3)
DENSITY	Numeric	Density
DENSU	Numeric	Wet density
DENSS	Numeric	Dry density
UMID	Numeric	Moisture
RECG	Numeric	Mass of coarse fraction after screening
ALTG	Numeric	Total alumina content (grade) in the coarse fraction
SITG	Numeric	Total silica content (grade) in the coarse fraction
FETG	Numeric	Total iron content (grade) in the coarse fraction
TITG	Numeric	Total titanium content (grade) in the coarse fraction
PPCG	Numeric	Calcination grade in the coarse fraction
ALAPG	Numeric	Alumina content (grade) in the coarse fraction
SIRG	Numeric	Reactive silica content (grade) in the coarse fraction
RECF	Numeric	Mass of fine fraction after screening
ALTF	Numeric	Total alumina content (grade) in the fine fraction
SITF	Numeric	Total silica content (grade) in the fine fraction
FETF	Numeric	Total iron content (grade) in the fine fraction
TITF	Numeric	Total titanium content (grade) in the fine fraction
PPCF	Numeric	Calcination grade in the fine fraction
ALAPF	Numeric	Alumina content (grade) in the fine fraction
SIRF	Numeric	Reactive silica content (grade) in the fine fraction

secondary variables. This method ensured the positions of contacts and solids were well-aligned with real-world data. Fig. 15.19 shows the 2D contouring of each plateau, formed by depositional planes based on the region's stratigraphy. Contact analysis was conducted to validate the implicit modeling results and test the contacts between geometallurgical domains. This method evaluates the number of samples within each solid, their distance to the contact surface, and content variation with distance. The analysis revealed hard contacts, where recovery variations were abrupt at domain boundaries, confirming the model's validity and its ability to accurately segregate populations.

The variable selection process began with a correlation matrix to evaluate relationships between all physical and chemical variables across each domain, fraction, and plateau. Fig. 15.20 shows the correlation matrix for the TE plateau, GMD1 in the coarse fraction, highlighting the correlation between variables and RECG. Variables with the highest

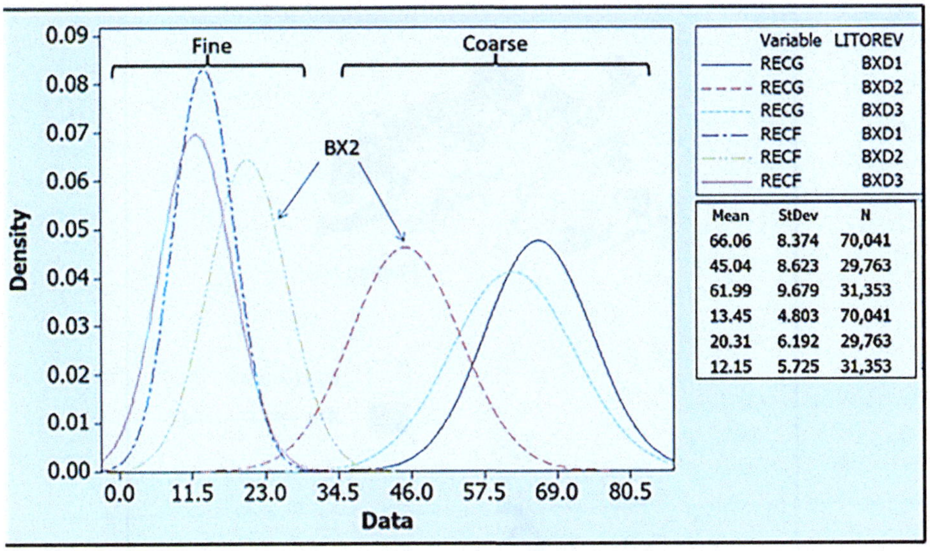

FIG. 15.17 Histograms for fine and coarse fractions from bauxite mining (TE Plateau). *Source: Rodovalho et al. (2025)—Open Access.*

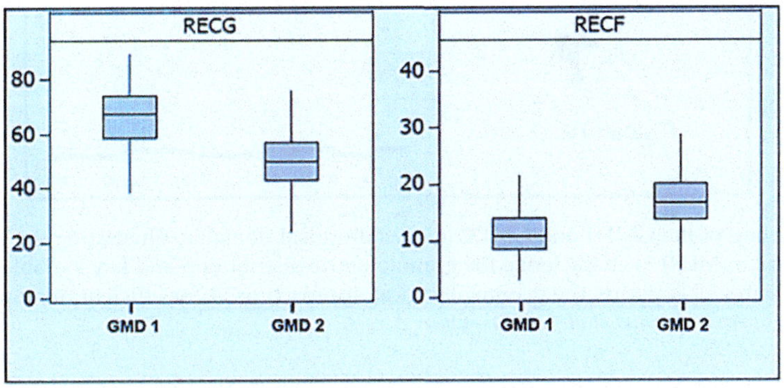

FIG. 15.18 Boxplot of GMD 1 and GMD 2 populations (TE plateau), showing granulometric distribution (GMD) variations. The method used involves multivariate linear regression (MLR) for estimating key variables in bauxite mining. *Source: Rodovalho et al. (2025)—Open Access.*

correlations were selected for stepwise regression analysis. Minitab 16 software was used for statistical analysis to select variables based on significance levels. The focus was on mass recovery in the coarse fraction as the response variable, with all other variables as predictors. Stepwise regression (both forward and backward) was applied, excluding variables with weak correlations until a satisfactory equation was achieved. The fine fraction was excluded from further analyses due to limited data in the coarse fraction. Table 15.3 presents the results of the stepwise regression for the coarse fraction of GMD1 on the TE plateau.

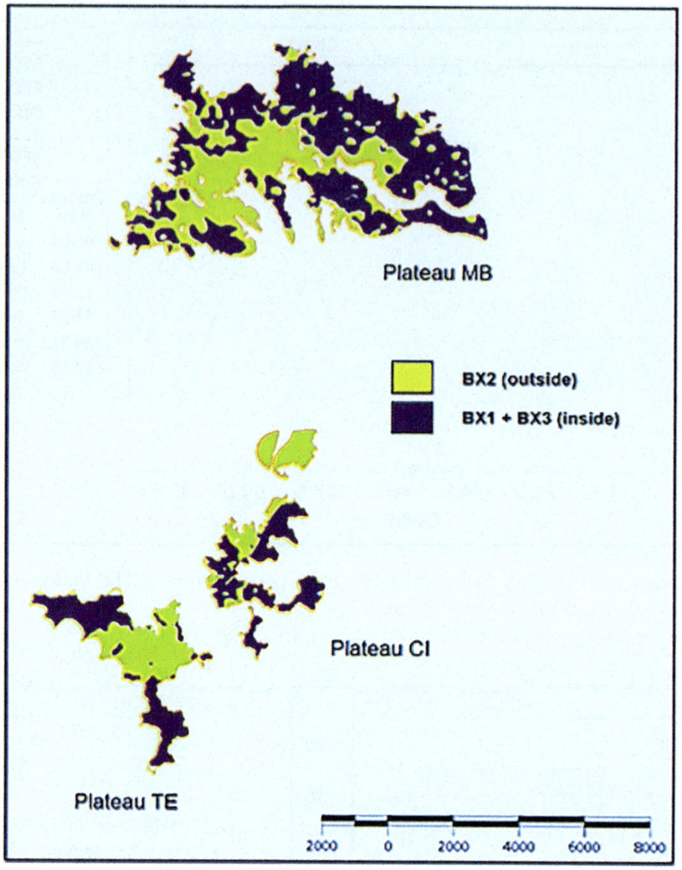

FIG. 15.19 Modeling of the GMD1 and GMD2 geometallurgical domains, illustrating the application of multivariate linear regression (MLR) to characterize the granulometric distribution and key variables for bauxite mining in the domains. This model supports the geometallurgical approach used for efficient mine planning and performance prediction. *Source: Rodovalho et al. (2025)—Open Access.*

Based on the results from Table 15.3, parameters that best explained mass recovery in the coarse fraction (TE Plateau/GMD1) were identified, as shown in **Eq.** (15.7).

$$\begin{aligned}\text{RECG} = 84.9 - 2.85\times[\text{UMID}] + 8.09\times[\text{TITG}] + 0.441\times[\text{FETG}] \\ - 0.0475\times[\text{ALAPG}] - 0.970\times[\text{SIRG}]\end{aligned} \tag{15.7}$$

A residual analysis was conducted to check for autocorrelation using the Durbin-Watson statistic and Cook's distance analysis. The Durbin-Watson statistic value was 1.25, indicating no autocorrelation, and Cook's distance analysis did not reveal any influential points.

For further validation, a graphical residual analysis confirmed that the assumptions of normal distribution, constant variance, and independence were met. Fig. 15.21 a shows a normal probability plot with p-values >0.15, indicating no significant deviation from

	RECG	UMID	TITG	FETG	ALAPG	SIRG	PPCG
UMID	-0,778 0,000						
TITG	0,067 0,001	-0,328 0,000					
FETG	-0,680 0,000	0,584 0,000	0,088 0,000				
ALAPG	0,121 0,000	0,170 0,000	-0,946 0,000	-0,335 0,000			
SIRG	0,425 0,000	-0,326 0,000	0,171 0,000	-0,305 0,000	-0,117 0,000		
PPCG	0,092 0,000	-0,383 0,000	0,819 0,000	-0,056 0,000	-0,850 0,000	0,174 0,000	
SITG	-0,570 0,000	0,526 0,000	0,161 0,000	0,869 0,000	-0,379 0,000	-0,098 0,000	0,028 0,000

Cell contents: Pearson Correlation
P-Value

FIG. 15.20 Correlation matrix for RECG, GMD1, and the TE plateau, showcasing the relationships between the recovery, granulometric distribution, and the TE plateau's geometallurgical characteristics. This analysis, based on multivariate linear regression (MLR), aids in understanding how key variables influence bauxite mining performance. *Source: Rodovalho et al. (2025)—Open Access.*

TABLE 15.3 Stepwise regression for bauxite mining plateau TE/GMD 1/RECG.

Response	**Mass recovery**				
Step	1	2	3	4	5
Constant	89.1	91.2	1508.9	−1384	84.9
Variables	Coefficients				
UMID	−1.9	−2.7	−2.3	−2.82	−2.85
TITG		7.1	7.91	8.15	8.09
FETG			0.45	0.47	0.44
ALPG				−0.04	−0.05
SIRG					−0.97
R^2_{ADJ}	75.1	76.5	82.3	83.1	85.2

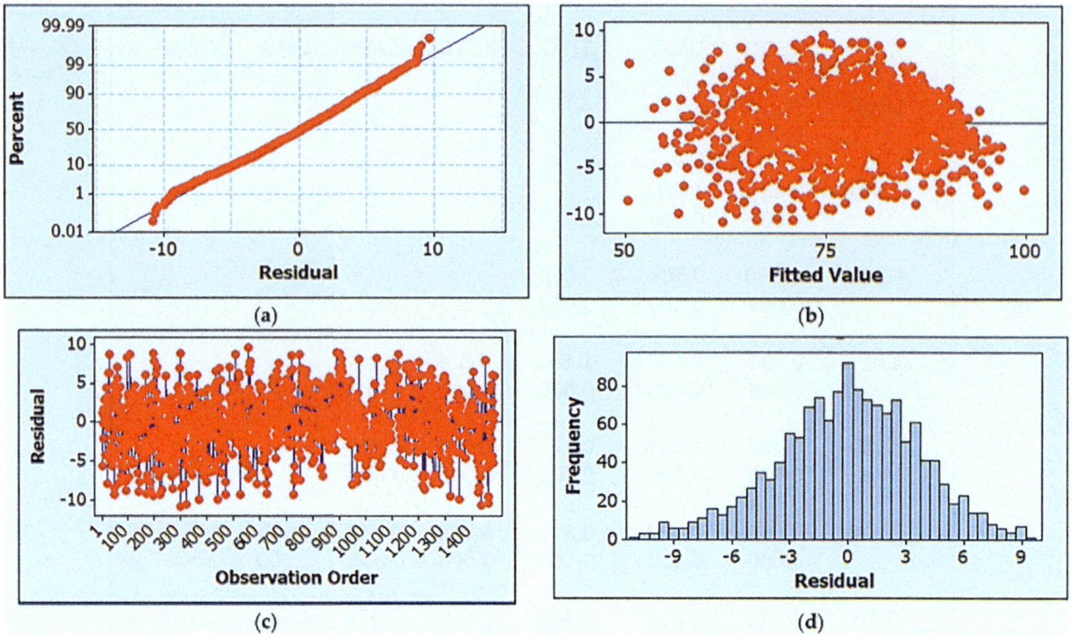

FIG. 15.21 Residual analysis for RECG/GMD1/TE plateau: (A) Residual normality check chart for RECG/GMD1/TE plateau; (B) Residual evaluation graph vs. adjusted RECG/GMD1/TE plateau values; (C) Graph of residuals (red) vs. observations (blue line) for RECG/GMD1/TE plateau; (D) Histogram analysis of RECG/GMD1/TE plateau residuals. This analysis evaluates the accuracy of the geometallurgical model for the TE plateau, using residuals to assess model fit and identify any patterns or biases in the data. Source: Rodovalho et al. (2025)—Open Access.

normality. Fig. 15.21B presents residuals vs. fitted values, showing constant variance with no outliers. Fig. 15.21C shows residuals vs. the order of observation, confirming that errors are independent. Finally, Fig. 15.21D presents a histogram supporting the normal distribution of residuals. These results confirm that **Eq.** (15.7) accurately explains mass recovery in the coarse fraction (TE Plateau/GMD1).

The procedures from the previous section were applied to all plateaus in Fig. 15.15, resulting in twelve multivariate linear regression (MLR) models—one for each fraction, domain, and plateau. These models were incorporated into the block model to estimate RECG and REC values for the geometallurgical model. Fig. 15.22 summarizes the process, starting with MLR generation, followed by the estimation of RECF and RECG values. From phase 3, mine operations implemented the geometallurgical model for mine planning and compared the estimated results with those from a mineral processing unit, a procedure called reconciliation.

Fig. 15.23 shows the mass recovery estimates for the MB, TE, and CI plateaus, with cold colors indicating lower recoveries for the coarse fraction and warm colors indicating higher recoveries. The blended feed from these areas was used by the mineral processing unit. The geometallurgical model was used to develop a short-term mine plan. While estimates were available for all fractions, only the coarse fraction was validated due to operational

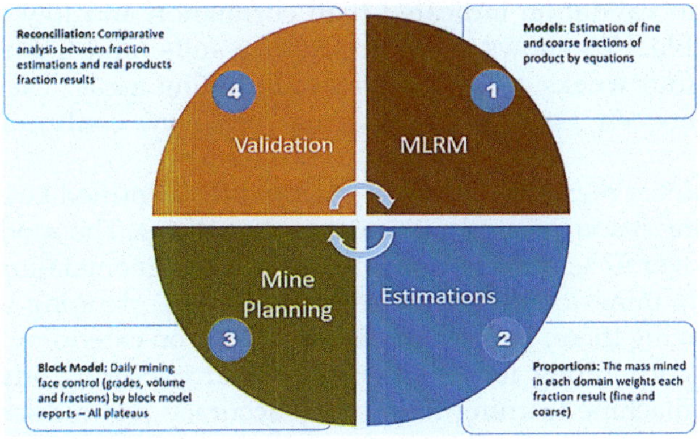

FIG. 15.22 Methodology for the practical application of the geometallurgical model, outlining the steps involved from the generation of multivariate linear regression models (MLR) to the estimation of RECG and RECF values. This approach supports mine planning and operational decision-making, ensuring accurate predictions and efficient bauxite processing in real-time mining scenarios. Source: Rodovalho et al. (2025)—Open Access.

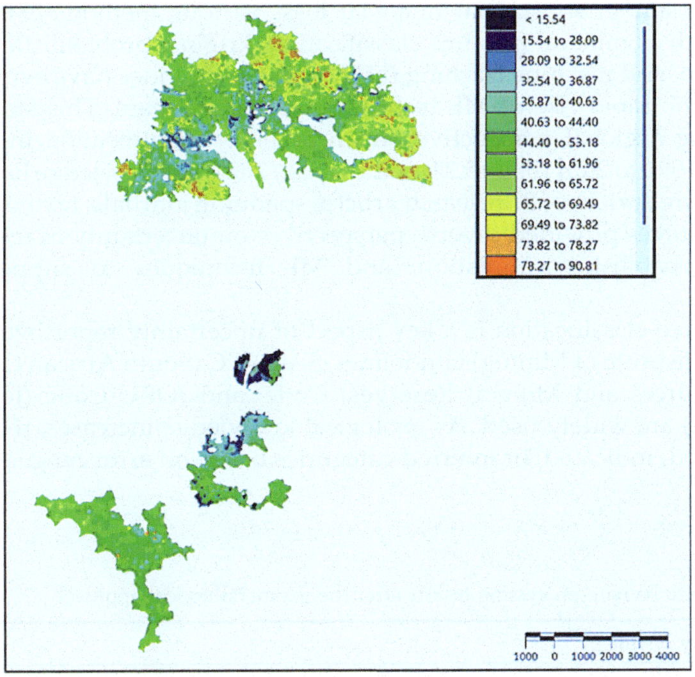

FIG. 15.23 Estimated geometallurgical models (2D contouring), presenting mass recovery estimates for the MB, TE, and CI plateaus. The contour map uses color gradients, where cold colors represent the lowest mass recoveries for the coarse fraction, and warm colors indicate the highest values, aiding in mine planning and mineral processing optimization. *Source: Rodovalho et al. (2025)—Open Access.*

constraints. Statistical validation indicated high correlation and low error, confirming the model's validity. Table 15.3 shows the correlation results, and Table 15.4 presents mass recovery data from four weeks of testing across all mining areas. The model showed only a 2.4% deviation between estimated and actual recovery, confirming its accuracy and reliability.

The multivariate linear regression method successfully identified key variables for bauxite processing, optimizing resources and minimizing deviations. The model's accuracy for the coarse fraction was over 97%, making it a valuable tool for geometallurgical programs. This method, applicable to mines of all sizes, can improve mine planning by identifying critical variables and prioritizing them. Further studies will focus on extending validation to include longer-term data, incorporating fine fraction data, and exploring advanced methods like random forests and machine learning to improve accuracy and reduce test times. The ultimate goal is to enhance the predictability of metallurgical responses, boosting productivity and sustainability.

15.3.7 Case study: Uncertainty quantification in mineral resource estimation

Mineral resource estimation is crucial for sustainable mining operations, involving the assessment of geological features. It is often conducted using drillhole data or geophysical tools, but uncertainties can lead to significant financial risks. A 15% discrepancy in grade estimation can result in substantial financial loss. Various methods exist to assess these uncertainties, such as mineral resource classification, kriging, probabilistic methods, geostatistical simulation, and machine learning (ML). These techniques have evolved from kriging in the 1950s to the rapid rise of ML techniques in recent years. This study reviews these methods, offering a global perspective on uncertainty quantification in mineral resource estimation from 1991 to 2024 (Fig. 15.24) (Lindi et al., 2024). A knowledge base was developed through a literature review of 87 research articles spanning journals, textbooks, and technical reports. These sources provided diverse perspectives on uncertainty in resource estimation, focusing on probabilistic, simulation, and ML techniques to support effective risk management.

Mineral resource classification is a key aspect of uncertainty reporting. Systems like the CIM (Canadian Institute of Mining) Guidelines, SAMREC (South African Code for Reporting of Mineral Resources and Mineral Reserves) Code, and JORC Code (Joint Ore Reserves Committee, 2012) are widely used. As geological knowledge increases, resources are classified into measured, indicated, or inferred categories based on error ranges like the 15% rule,

TABLE 15.4 Coarse bauxite processing results after the geometallurgical approach.

Coarse mass recovery results	
Real	66.2%
Estimation	67.8%
Variation	2.4%

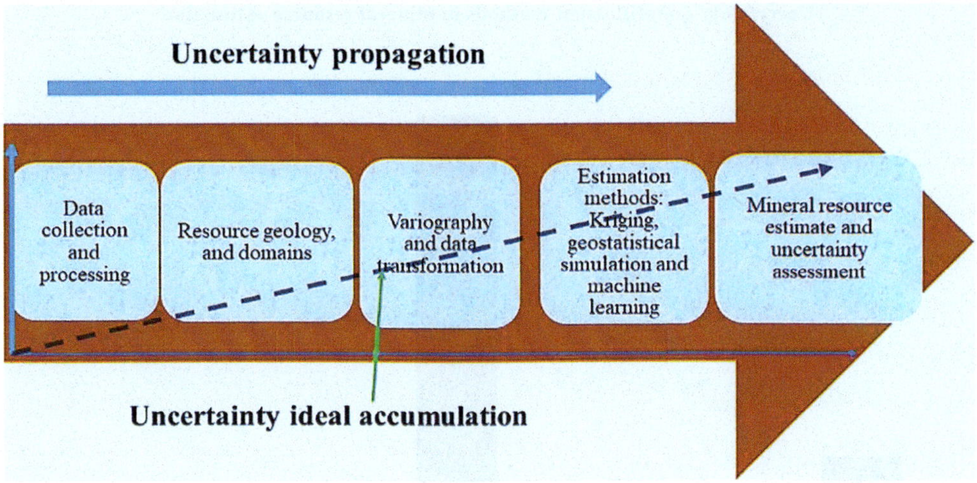

FIG. 15.24 Stepwise process of uncertainty propagation and accumulation in mineral resource estimation, including the Volume of Knowledge (VoK) considerations (not drawn to scale). *Source: Lindi et al. (2024)—Open Access.*

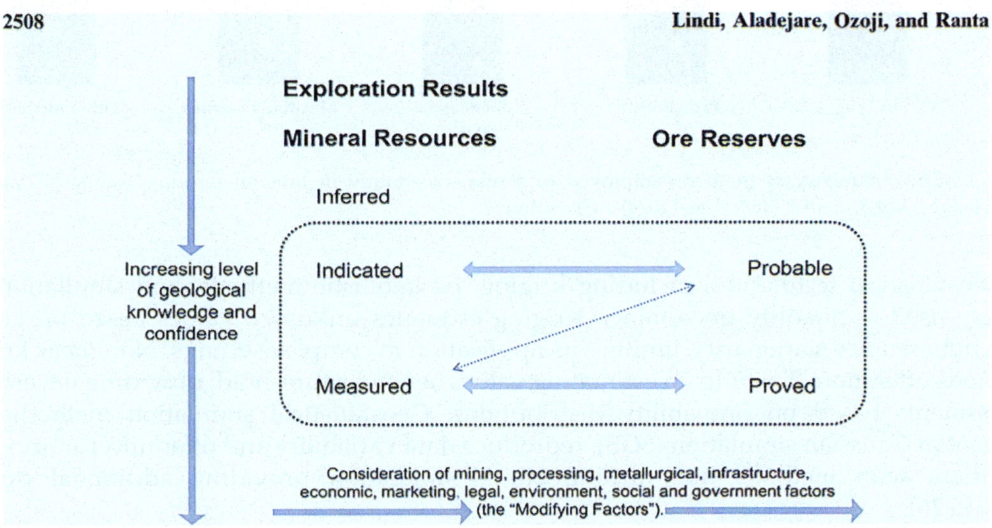

FIG. 15.25 Simplistic relationship between exploration results, mineral resources, and reserves, considering Volume of Knowledge (VoK). *Source: Lindi et al. (2024)—Open Access.*

used to categorize resources at a 90% confidence interval (Fig. 15.25). Machine learning methods like random forests and support vector machines have been applied in classification tasks, offering techniques for assessing uncertainty in the classification process. However, uncertainty reporting often excludes factors like geological variability and grade spatial uncertainty.

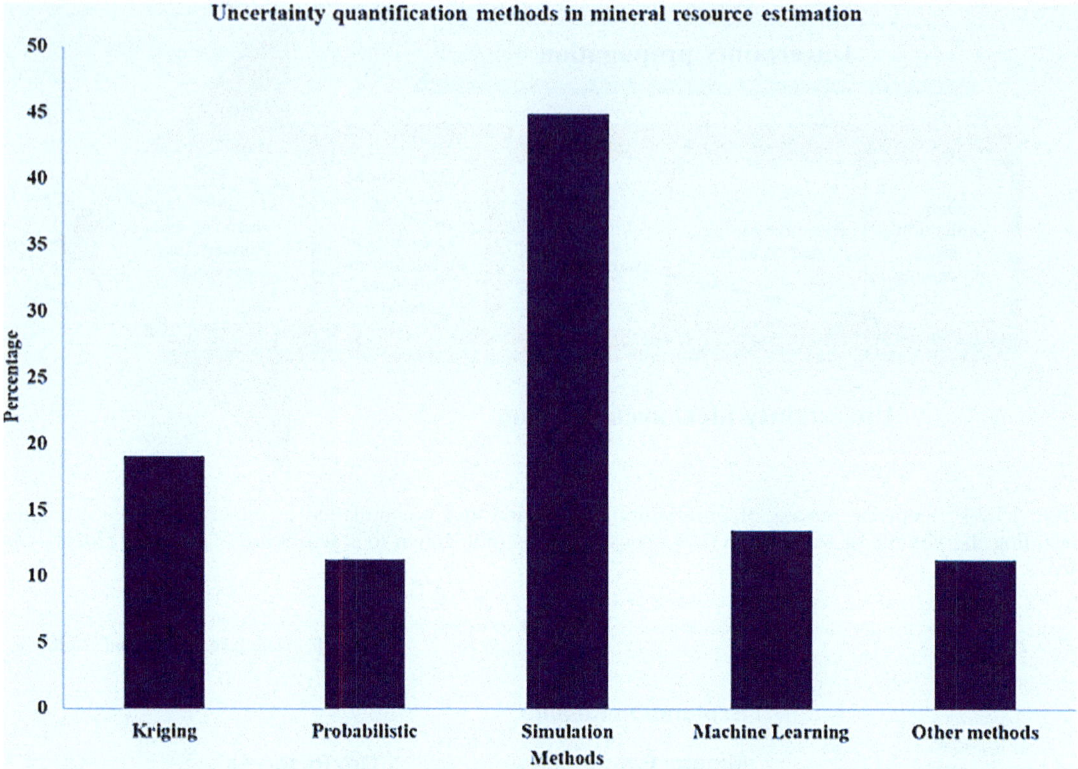

FIG. 15.26 Frequency of methods employed to assess uncertainty in mineral resource estimates (Value of Knowledge—VoK). *Source: Lindi et al. (2024)—Open Access.*

Geostatistical techniques, including kriging, probabilistic methods, and simulation, are widely used to quantify uncertainty. Kriging estimates unknown values based on known data but assumes stationarity, limiting its application in complex settings. Nonlinear kriging methods offer more flexibility in estimating values beyond a threshold, providing uncertainty assessments based on probability distributions. Geostatistical simulation methods, like sequential Gaussian simulation (SGS), reproduce data variability and offer uncertainty quantification, with methods like turning bands simulation providing additional options (Fig. 15.26).

Machine learning techniques, such as artificial neural networks (ANN), random forests (RF), and support vector machines (SVM), have been adopted due to their flexibility and ability to handle various types of data without strict assumptions. These methods reduce human bias and are alternatives to traditional geostatistical methods. However, their application in uncertainty quantification is still developing. ML models quantify errors between predicted and observed data, expressing uncertainty through confidence intervals, though they are susceptible to noise and inference errors, especially if the data used for training is erroneous.

The advantages and limitations of each method depend on deposit type, data availability, and technical expertise. Kriging is simple to implement but can underestimate data variability, while probabilistic methods like Bayesian approximation are useful when data is limited. Geostatistical simulation methods are useful for complex deposits but require specialized skills. Machine learning methods, while flexible, are still evolving in terms of uncertainty quantification.

Uncertainty arises at both sampled and unsampled locations, influenced by geological logging, geochemical and geophysical data, and drillhole positioning. Sparse data collection increases uncertainty, while geological logging introduces subjectivity and variability. Errors in drillhole positioning impact resource volume and grade estimates. Advanced methods like the distance function can assess uncertainty in drillhole positioning and mineral vein intersection (Table 15.4).

Geological models, using geophysical, geochemical, and remote sensing data, introduce uncertainty due to sparse data, poor geological continuity, and expert interpretation. Methods like Sequential Indicator Simulation (SIS) and Multi-Point Simulation (MPS) assess geological uncertainty. Continuous variables, like grade and tonnage, contribute to uncertainty, with nugget effects and testing equipment accuracy influencing estimates. Kriging, MPS, and SGS methods are commonly used to assess uncertainty in continuous variables.

Interrelationships between continuous variables, such as grade, tonnage, and density, introduce additional uncertainty. Multi-element deposits are particularly affected. Techniques like Multi-Gaussian simulation and MPS model these interrelationships, quantifying associated uncertainties. Categorical variables, such as rock types and lithology, add complexity to resource estimation. Geostatistical simulations define geological domains and simulate continuous variables within each domain, considering both continuous and categorical variables.

Mathematical models used in mineral resource estimation inherently carry uncertainty due to assumptions in their formulation. These models depend on the quality of geological data, and uncertainty propagates through multiple sub-models. Methods like Bayesian inference, conditional simulation, and machine learning models help address these uncertainties, but each method has its limitations. The estimation of recoverable mineral resources involves determining mineable orebodies with confidence intervals. Uncertainty in grade and tonnage poses challenges, and simulation techniques like Conditional Simulation (CS) and Multiple-Point Simulation (MPS) assess the impact of uncertainties on mining planning. These simulations help model grade-tonnage curves, considering mining selectivity and new data collected during mining operations (Table 15.5).

Uncertainty quantification is crucial in mineral resource estimation, helping reduce its impact and improving resource estimate accuracy. This study compiles various methods used to assess uncertainty, guiding the selection of appropriate techniques based on data, deposit type, and user experience. As estimation progresses, uncertainty accumulates and propagates, making managing uncertainty at each step essential. The study also highlights the need for developing new methods to improve the reliability of resource estimates (Table 15.6).

TABLE 15.5 Techniques for uncertainty assessment at sampled locations, including methods for addressing uncertainty due to geological modeling.

Type of uncertainty	Cause of uncertainty	Methods used for assessment	References
Uncertainty due to geological logging	Core loss which affects sample grade and mineralization structures	Quality Assurance (QA)/Quality Control (QC)	Dominy and Petersen (2005)
	Sampling spacing may mismatch mineralization structures	QA/QC	Sketchley (1998)
	Interpretation uncertainty of classification of drillhole intervals into different domains and subjective logging results	Bayesian method	McManus et al. (2020, McManus et al. (2021a)
Uncertainty in geochemical assays and other variables	Reduced sample size to match laboratory protocols	QA/QC	Sketchley (1998)
	Uncertainties introduced by input data quality in complex 3D geological models	Probability distribution	Wellmann et al. (2010)
Uncertainty due to positioning errors	Uncertainty due to data spacing during collection	SGS	Wilde (2010)
	Uncertainties introduced by lithology and fault boundary points and drillhole orientation measurements	Probability distribution	Wellmann et al. (2010)
	Drillhole distances and vein intersection	Distance function	Rossi and Deutsch, (2014)

TABLE 15.6 Techniques for uncertainty assessment of recoverable mineral resources.

Cause of uncertainty method	Uncertainty quantification	References
Mineral resource uncertainty due to acquisition of additional data	MCS (Monte Carlo simulation)	Manchuk et al. (2015)
Uncertainty due to varying grades across the rock-type boundaries	Multi-Gaussian and plurigaussian modeling	Maleki et al. (2022
Varying tonnage in different classes of ore material	Direct sampling multiple-point simulation	Talebi et al. (2019)
Uncertainty in overall mineral resources at a given cutoff grade	CS	Emery et al. (2006)
Grade inconsistency	TBS (turning bands simulation), SGS, and SIS	Abzalov (2016)
Mineral resource model selection and grade uncertainty assessment due to lack of spatial information	MPS with SGS	Hong and Oh (2021)

Cause of uncertainty method	Uncertainty quantification	References
Uncertainty about grade model due to transitional domain boundaries	SGS	Harrison et al. (2009)
Uncertainty in grade–tonnage curves	CS using minimum/maximum auto correlation factors	Tajvidi et al. (2015)
Recoverable mineral resource estimation at certain cutoff grades	Direct block-support sequential co-simulation	Hosseini et al. (2017)
Grade and tonnage uncertainty due to spatial complexity of an orebody	MPS and SGS	Paithankar and Chatterjee (2018)
Grade–tonnage curve uncertainty	MLR (Multiple Linear Regression) and a multilayer neural network	Mery and Marcotte (2022b)

15.4 Ethical considerations in sustainable resource exploration

While technological advancements have greatly improved the efficiency of resource exploration, they also raise important ethical considerations. It is essential that the exploration and extraction of natural resources be conducted in a manner that is not only economically viable but also socially responsible and environmentally sustainable.

15.4.1 Minimizing environmental impact

One of the central ethical concerns in resource exploration is minimizing its impact on the environment. AI/ML technologies can play a key role in achieving this goal by enabling more efficient exploration processes that reduce the number of exploration sites and limit the environmental footprint of extraction activities.

15.4.1.1 Minimizing environmental impact through AI and machine learning: Case studies in environmental monitoring

A case study from TotalEnergies used AI-powered drones to monitor the environmental impacts of oil exploration activities in the Amazon rainforest. AI was used to analyze imagery data and detect signs of deforestation, land degradation, and water contamination in real-time, helping the company to mitigate environmental harm and ensure compliance with sustainability protocols (TotalEnergies, 2022).

Zhu et al. (2001) introduced the Soil-Land Inference Model (SoLIM). This model uses geographic information systems (GIS), expert knowledge, and fuzzy logic to improve soil mapping. By combining AI with traditional environmental sciences, the study showed how AI can create more accurate and useful models. It emphasized that AI can improve soil and land management, especially in creating better soil maps for sustainable agriculture and land use planning.

Khatami et al. (2016) conducted a detailed meta-analysis on land-cover classification methods, exploring how adding texture data can improve classification accuracy. The study found that support vector machines work better than other algorithms when texture data is included. This advancement is especially important for remote sensing, where land-cover classification is crucial for tracking environmental changes over time.

In a more recent study, Ren et al. (2024) proposed the dual-attention multiscale network (DAMFANet) for remote sensing image change detection. The model combines feature fusion and semantic information integration, greatly improving change detection performance. This approach addresses the challenge of handling large, diverse data in environmental monitoring, providing a scalable and efficient solution for remote sensing image processing. By combining attention mechanisms with multiscale feature learning, the model opens new possibilities for AI applications in environmental science, particularly for deforestation monitoring and land-use change detection.

Cao et al. (2025) used a Bayesian optimization algorithm to predict pollutant levels in urban environments. Their study showed that combining ensemble models with real-time sensor data improves air quality predictions, especially in densely populated urban areas. The Bayesian optimization approach fine-tuned model parameters, improving both predictive performance and model interpretability.

Palani et al. (2008) applied artificial neural networks (ANNs) to forecast water quality in Singapore's coastal waters. The model achieved high predictive accuracy, with Nash–Sutcliffe efficiency coefficients ranging from 0.8 to 0.9. This demonstrates the effectiveness of AI in predicting key water quality parameters like nutrient concentrations and dissolved oxygen levels, which are critical for managing coastal ecosystems and addressing environmental and economic impacts.

Reichstein et al. (2019) applied deep learning techniques to simulate climate dynamics and predict long-term environmental changes. The study highlighted the potential of recurrent neural networks (RNNs) and convolutional neural networks (CNNs) to model time series data and capture complex spatio-temporal relationships in climate systems. By integrating high-resolution satellite data with ground-based measurements, this study offers a scalable approach to climate modeling that can be applied in various regions and environmental contexts.

Phillips et al. (2006) developed the Maxent algorithm for species distribution modeling, which has become a widely used tool in biodiversity research. Their study showed that Maxent outperforms other algorithms, like the Genetic Algorithm for Rule-set Production (GARP), in predicting species ranges based on environmental variables. This research has significant implications for conservation planning, helping predict how species distributions will shift due to environmental changes like habitat loss and climate change.

Chapi et al. (2017) used the Bagging-LMT model to predict flood susceptibility in Iran. Their model was more accurate than traditional methods in identifying high-risk areas, providing a valuable tool for disaster preparedness and risk management. The study demonstrates how AI/ML techniques can enhance disaster prediction models, offering decision-makers actionable insights for resource allocation and emergency planning.

The integration of AI/ML with emerging technologies, like the Internet of Things (IoT), unmanned aerial vehicles (UAVs), and big data platforms, has opened new opportunities for environmental monitoring and disaster management (Alotaibi and Nassif, 2024). These applications combine AI/ML with advanced data collection and processing technologies, enabling real-time and comprehensive environmental assessments. Asha et al. (2022) developed the IoT-based Environmental Toxicology for Air Pollution Monitoring with Artificial

Intelligence Techniques (ETAPM-AIT) model, which uses IoT sensors and AI algorithms to monitor air quality and detect toxic pollutants. Their system provides real-time alerts to industrial facilities and government agencies, enabling faster responses to air pollution events. This integration of IoT with AI represents a significant advancement in real-time environmental monitoring, offering scalable solutions for cities and industries addressing air pollution impacts.

Fig. 15.27A shows the global distribution of research publications on AI and ML in environmental monitoring. China leads with 6170 publications, followed by the United States

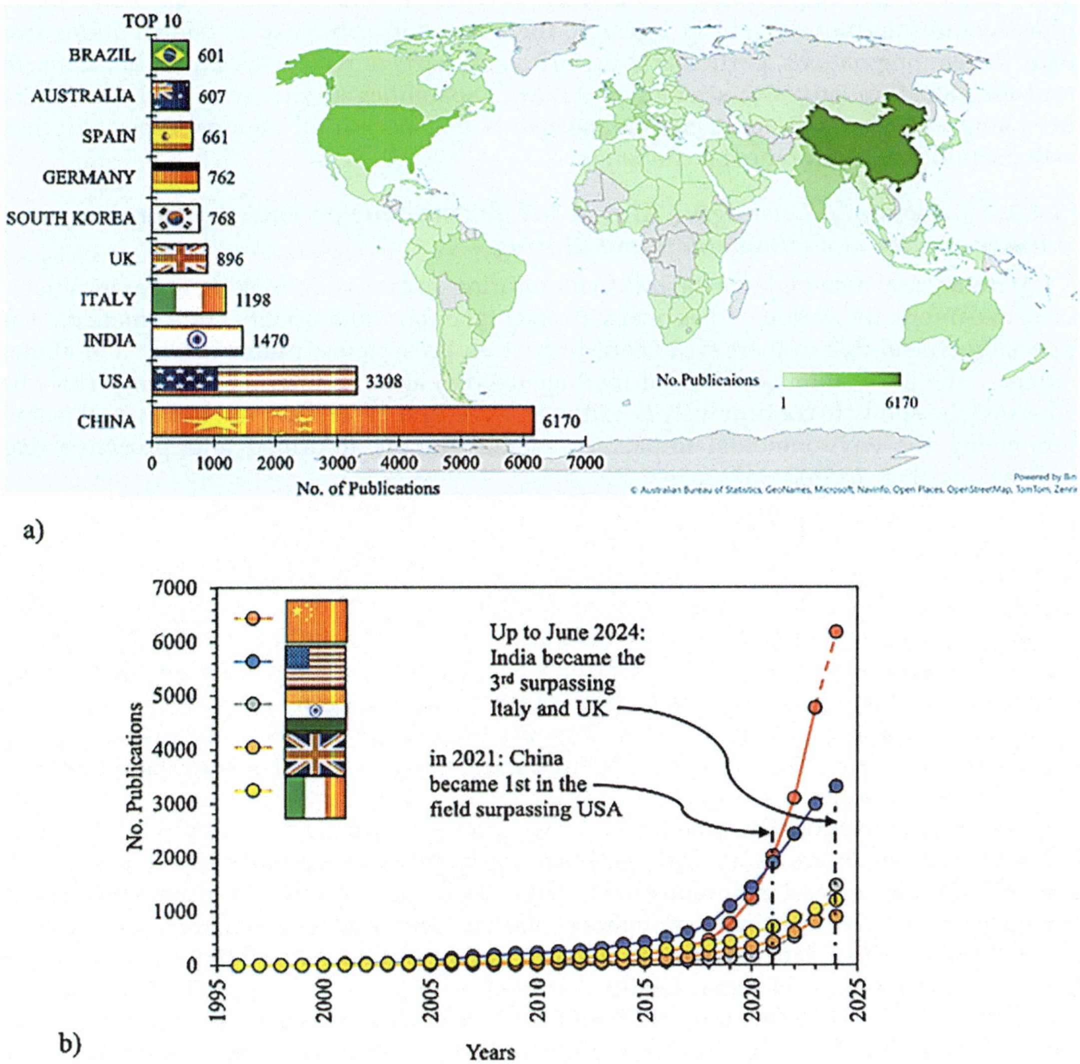

FIG. 15.27 Research contributions in AI and ML for environmental monitoring and data analysis: (A) Geographical distribution, (B) Temporal evolution of the top five countries. *Source: Alotaibi and Nassif (2024)—Open Access.*

with 3308. Other notable contributors include India (1,470), the United Kingdom (836), and Italy (1,198). South Korea, Germany, Spain, Australia, and Brazil also make significant contributions, with publication numbers ranging from 601 to 768. These numbers do not account for co-authorship, so if a paper involves both US and China authors, each country is credited.

Fig. 15.27B displays the contributions of the top five countries (China, the United States, India, the United Kingdom, and Italy) over time. It shows a sharp rise in publications from 2015 onward, with China surpassing the United States in 2021 to become the top contributor. By August 2024, India moved up to third place, overtaking Italy and the United Kingdom. While countries like China and the United States dominate due to their strong technological infrastructures and investment in research, there is a noticeable gap in contributions from many developing nations, particularly in Africa and parts of Asia. This gap underscores the need for global initiatives to strengthen AI/ML capabilities in environmental monitoring, promoting inclusive and equitable advancements that benefit all regions, especially those facing significant environmental challenges.

15.4.1.2 Case study: Leveraging AI and IoT for minimizing environmental impact in water quality and climate data monitoring

Environmental data plays a vital role in monitoring and decision-making in environmental sciences. Among the various types of environmental data, water quality and climate data are especially crucial due to their significant impact on ecosystems, public health, and climate patterns. The integration of Artificial Intelligence (AI) and the Internet of Things (IoT) has enhanced the ability to continuously monitor, analyze, and predict key environmental factors, minimizing the environmental impact by enabling more informed and proactive decision-making (Fig. 15.28).

Water quality monitoring is essential for maintaining healthy aquatic ecosystems and ensuring safe water for human use. Several parameters are critical to assessing water quality, providing a comprehensive understanding of its condition. pH measures water acidity or alkalinity. Significant deviations from neutral pH can indicate contamination, affecting aquatic life and water safety. Turbidity indicates water clarity, and high turbidity can result from soil erosion, algal blooms, or industrial waste, harming aquatic organisms and complicating water treatment. Dissolved Oxygen (DO) measures oxygen levels, critical for aquatic life. Low DO can lead to hypoxic conditions, disturbing ecosystems, and causing fish kills. Other factors, such as heavy metals, nutrients, and pathogens, are also important to monitor for ensuring safe water.

AI improves water quality monitoring by enhancing predictive capabilities and automating detection processes. AI algorithms, particularly predictive models, use historical and real-time data to forecast pollution events, such as industrial spills or agricultural runoff, and enable proactive measures to mitigate contamination. AI also automates anomaly detection in water quality data, triggering alerts to authorities and facilitating swift actions to reduce environmental and public health risks (Fig. 15.29).

IoT devices, including pH, turbidity, and DO sensors, play a critical role in real-time water quality monitoring. These sensors are deployed in key areas to collect data on various water quality parameters. For example, in regions like the Great Lakes, IoT sensors continuously monitor water quality, while AI analyzes the data to predict pollution trends and

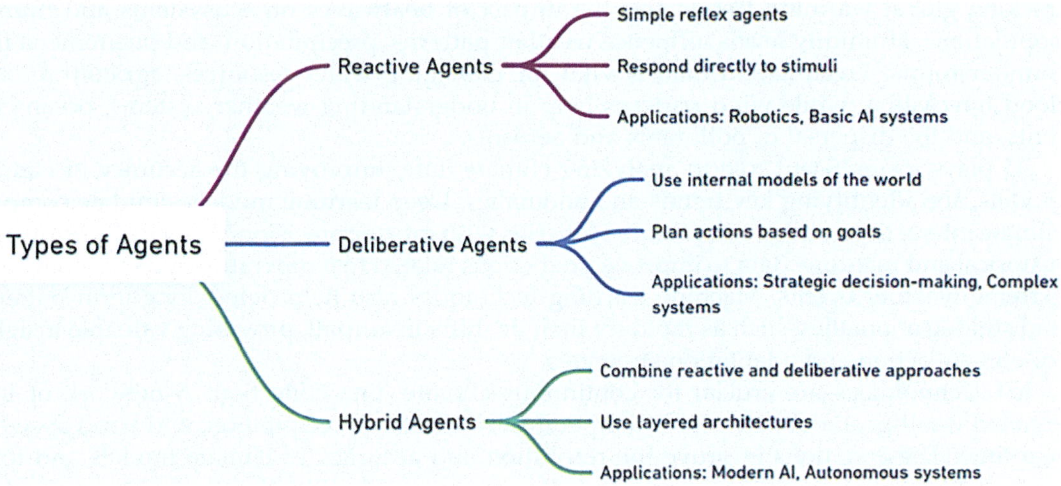

FIG. 15.28 Illustrates the classification of AI agents in environmental sciences, emphasizing their role in enhancing the monitoring and management of water quality and climate data. *Source: Miller et al. (2025)—Electronics—Open Access.*

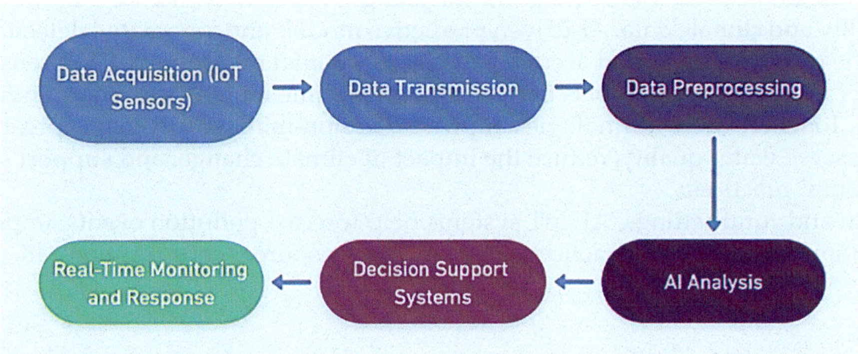

FIG. 15.29 Depicts the integrated AI-IoT system workflow, showing how AI and IoT technologies collaborate to collect, analyze, and process environmental data in real-time, improving decision-making and proactive management in water quality and climate monitoring. *Source: Miller et al. (2025)—Electronics—Open Access.*

improve water management practices. Additionally, in agricultural settings, smart irrigation systems use IoT sensors to monitor water quality, with AI optimizing irrigation practices to minimize runoff and improve crop yields.

Climate data encompasses various indicators that reflect the state and dynamics of the global climate system. Monitoring and analyzing climate data is crucial for understanding climate change, predicting future trends, and developing effective mitigation and adaptation strategies. Key climate indicators include temperature (surface and atmospheric), humidity, precipitation (rainfall and snowfall), and wind patterns. Temperature data is essential for tracking global warming trends and the impact of heatwaves on ecosystems and human populations. Humidity levels influence weather patterns, precipitation, and phenomena like thunderstorms. Precipitation data is vital for managing water resources, agriculture, and flood forecasting, while wind patterns help in understanding weather systems, ocean currents, and the dispersal of pollutants and aerosols.

AI plays a significant role in analyzing climate data, improving the accuracy of climate models, and identifying key trends and anomalies. Deep learning models simulate complex climate interactions and predict future scenarios with greater precision. AI algorithms process historical and real-time data to improve predictions related to temperature, precipitation, and extreme weather events. Machine learning techniques also help detect long-term patterns and sudden anomalies, such as rapid ice melt or shifts in rainfall, providing valuable insights for climate action and adaptation strategies.

IoT technologies are crucial for continuous climate data collection. A network of IoT-enabled weather stations measures temperature, humidity, precipitation, and wind speed in real-time. These stations improve the resolution and accuracy of climate models and forecasts. Additionally, satellite imaging and drones equipped with climate monitoring instruments collect data from remote regions, measuring atmospheric composition, sea surface temperatures, and ice coverage. These technologies, integrated with AI-driven analytics, enable efficient data collection and provide actionable insights for climate scientists and policymakers (Fig. 15.30).

The combination of AI and IoT significantly improves environmental monitoring in both water quality and climate data. AI-driven predictive models and automated detection systems enhance the responsiveness and accuracy of environmental management. IoT sensors ensure continuous and detailed data collection, enabling real-time monitoring of key environmental indicators. Together, these technologies improve decision-making and enable proactive measures to preserve water quality, reduce the impact of climate change, and support sustainable environmental practices.

In urban and rural settings, AI–IoT systems help forecast pollution events, improve water safety, optimize irrigation practices, and reduce resource consumption. In remote or

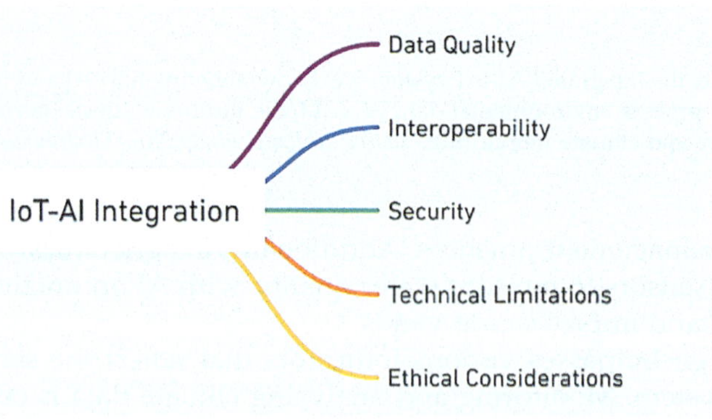

FIG. 15.30 Provides an overview of the challenges in AI-IoT integration, highlighting the complexities involved in combining AI and IoT technologies for effective environmental monitoring, including issues related to data management, system compatibility, and scalability. *Source: Miller et al. (2025)—Electronics—Open Access.*

disaster-prone regions, AI models on low-power IoT devices can be dynamically updated through 5G networks, enhancing the efficiency of environmental monitoring. Moreover, AI and IoT support biodiversity conservation by providing real-time ecological data, allowing for better-informed conservation strategies.

The integration of AI and IoT in environmental monitoring for water quality and climate data has proven to be highly effective in minimizing environmental impacts. These technologies enable continuous monitoring, enhance prediction accuracy, and support proactive environmental management, all of which are essential for sustaining ecosystems, protecting public health, and addressing climate change.

15.4.2 Ethical sourcing and community impact

Another ethical consideration is the potential impact on indigenous communities and local populations. Resource exploration and extraction activities can often affect the land, livelihoods, and cultural heritage of local people. AI and ML models must be developed with careful consideration of these social factors. It is essential to incorporate community consultations and participatory decision-making processes to ensure that local and indigenous voices are heard. Furthermore, respecting land rights and engaging in fair compensation schemes is crucial for maintaining social responsibility.

15.4.2.1 Example: Social impact assessment using AI

Newmont Mining utilized AI to assess the social impact of its exploration activities on indigenous communities in Canada. AI models were used to simulate potential outcomes of various exploration methods and identify the least disruptive approaches for both the environment and the communities involved. The data-driven approach led to more respectful and informed negotiations between Newmont and local communities (Newmont Mining, 2021).

15.4.2.2 Example: Ethical sourcing and community impact through AI-driven supply chain transparency—case study: AI in fair trade coffee sourcing

In the coffee industry, ethical sourcing is a key concern, particularly regarding the fair treatment of farmers and sustainable environmental practices. A notable case study involves the use of AI to improve transparency and traceability in the coffee supply chain.

A global coffee company partnered with AI and blockchain technology to create a system that tracks the coffee's journey from the farm to the consumer. This system uses AI algorithms to analyze data collected from sensors and blockchain records, ensuring the coffee beans are ethically sourced, free from child labour, and produced under fair trade conditions. By providing real-time information on the origins of each batch of coffee, the AI-driven system promotes transparency, allowing consumers to make more informed choices (Jain and Shah, 2020).

The social impact of this initiative is significant. Farmers are paid fairly for their crops, and the transparency created by AI ensures that consumers know their purchases support sustainable and ethical practices. Additionally, the use of AI helps in monitoring environmental factors, such as water use and pesticide application, which encourages more sustainable farming methods and reduces environmental harm. This case exemplifies how AI can be harnessed to ensure ethical sourcing and positively impact communities by fostering fair labor practices and sustainable agricultural practices (Smith and Harwell, 2021; Thompson and Miller, 2019).

15.4.3 Resource management and depletion

While AI/ML can optimize the identification of mineral deposits and improve resource extraction processes, there is a risk of overexploitation of resources. Ethical exploration requires ensuring that resources are not extracted beyond sustainable limits, leading to long-term environmental and societal harm. Technologies should be employed with foresight to ensure responsible management of resources, preventing depletion and ensuring that natural resources are available for future generations.

15.4.3.1 *Example: AI-driven resource management in water conservation*

In the context of resource management, particularly regarding water resources, the management of water use and the prevention of depletion has become a critical concern globally. A notable example is the use of AI-driven solutions to optimize water usage in agriculture, where water scarcity is a growing issue due to climate change and unsustainable practices.

15.4.3.2 *Case study: AI in precision irrigation for water conservation*

In California, a region that has faced severe water shortages in recent years, a farming cooperative implemented an AI-powered precision irrigation system to address water depletion challenges. The system integrates IoT sensors, weather data, and AI models to monitor soil moisture levels, predict water needs for crops, and adjust irrigation schedules in real time (Zhang et al., 2020).

The system uses machine learning algorithms to analyze data on weather conditions, soil types, crop types, and historical water usage to develop highly accurate irrigation plans. This enables farmers to apply water only when and where it is needed, reducing water waste and ensuring that crops receive optimal moisture levels for growth (Sharma and Singh, 2021).

Through AI-driven management, the cooperative has seen a significant reduction in water usage—by up to 30%—while maintaining crop yields (Jones et al., 2021). Moreover, the system helps prevent over-irrigation, which can lead to soil degradation, and under-irrigation, which can stress crops. This AI-enabled approach also ensures that farmers can use water more efficiently, contributing to the long-term sustainability of the region's water resources (Sharma et al., 2020).

This case demonstrates how AI can be effectively used in resource management, particularly in addressing the depletion of vital resources like water. By enabling precise, data-driven decisions, AI can help ensure the responsible use of resources and prevent their unsustainable depletion.

15.4.3.3 *Case study: AI-driven water management and precision irrigation for agriculture in India*

In India, water scarcity remains a critical challenge, especially in regions where agriculture is the primary livelihood. Many states, including Maharashtra, Rajasthan, and Punjab, face significant water depletion issues, mainly due to inefficient irrigation practices. To address these challenges, AI-powered water management systems are being increasingly adopted to optimize irrigation and conserve water resources in agriculture.

A prominent example is the implementation of AI and IoT-based precision irrigation systems in Maharashtra, a state frequently affected by droughts. This system integrates IoT sensors, weather data, and satellite imagery to monitor real-time soil moisture levels and predict water needs for crops. The system uses machine learning algorithms to process this data, providing farmers with precise irrigation schedules tailored to each field's requirements (Srinivas et al., 2021).

The AI system also incorporates local weather forecasts and climatic conditions to dynamically adjust irrigation schedules, optimizing water usage during dry periods. By utilizing historical weather data, soil moisture levels, and crop-specific water needs, the system has enabled farmers to reduce water consumption by 20%–30% while maintaining high crop yields (Kumar and Rathi, 2020).

The use of AI and IoT has significantly improved water management in the region. Farmers can now make more informed decisions, ensuring that water is not over-extracted and that irrigation is more efficient. This technology not only supports water conservation but also helps maintain food security by ensuring adequate water supply for crops, even in water-scarce areas. Through this approach, AI and IoT are playing a crucial role in enhancing the sustainability of agriculture and promoting the responsible use of water resources in India (Srinivas et al., 2021).

15.4.4 Transparency and accountability

With AI/ML systems influencing decision-making in resource exploration, it is vital to maintain transparency and accountability. These systems must be explainable, and the decision-making process should be open to scrutiny. Stakeholders, including local communities, environmental organizations, and government bodies, should be able to understand how AI models influence exploration decisions. Transparency in how data is collected, analyzed, and interpreted is critical for building trust and ensuring that the exploration process is carried out with integrity.

15.4.5 Fair distribution of benefits

Ethical resource exploration should also address the fair distribution of benefits from resource extraction. AI/ML technologies can optimize extraction processes, increasing profitability for companies, but it is important to ensure that the economic benefits are equitably distributed among all stakeholders. Governments, local communities, and workers involved in the exploration process should receive fair compensation and have access to the opportunities created by the exploration activities.

15.4.6 Human and labour rights

As technology plays a larger role in resource exploration, the impact on workers and labour conditions cannot be ignored. The implementation of AI and automation in exploration processes could displace workers or alter labour markets. Ethical exploration requires ensuring that workers' rights are upheld, providing them with training and opportunities to transition into new roles created by the evolving technological landscape. Moreover, companies must adhere to labour laws and maintain ethical working conditions throughout the supply chain.

15.4.7 Long-term sustainability

Lastly, the ethical consideration of long-term sustainability must guide resource exploration efforts. While AI/ML technologies improve short-term exploration success, they should be used with a mindset of sustainability, aiming to preserve the planet's resources, ecosystems, and biodiversity for future generations. Ethical exploration practices involve integrating sustainability into every aspect of the exploration process, from resource discovery and extraction to post-extraction land rehabilitation and ecosystem restoration.

In conclusion, the ethical considerations surrounding sustainable resource exploration are multifaceted and complex. While AI/ML technologies offer unprecedented efficiency and accuracy in mineral exploration, their application must be guided by principles of environmental stewardship, social responsibility, and fairness. By addressing these ethical concerns, we can ensure that resource exploration is carried out in a way that benefits both current and future generations while protecting the natural world.

15.5 Vision for the future: The role of AI/ML in the geoscience industry

While it is still early to have a dedicated chapter on AI/ML, it is undeniable that these technologies will play an increasingly prominent role in shaping the future of resource exploration. Over time, the integration of AI/ML will likely result in the full automation of exploration processes, where algorithms make real-time decisions about where and how to extract resources.

15.5.1 Fully automated exploration and decision-making

One of the most transformative aspects of AI/ML in the future of geoscience is the possibility of fully automated exploration. Currently, human expertise is essential in making decisions about exploration strategies, interpreting complex geophysical data, and choosing drilling locations. However, as AI/ML algorithms become more advanced, they will increasingly take over these tasks, using vast datasets from geological surveys, satellite imagery, seismic readings, and other sources to autonomously identify potential exploration targets. In real-time, AI could assess and adjust exploration strategies based on updated data, optimizing resource extraction and minimizing environmental disruption.

For example, AI could use predictive analytics to assess geological conditions and recommend the best drilling locations, significantly reducing the need for trial and error. Machine learning models, continuously trained on new data, would adapt to changing conditions, offering dynamic, data-driven exploration plans. This level of automation could dramatically reduce human error, increase the speed of exploration, and improve overall accuracy in identifying valuable resource deposits.

15.5.2 Integration of AI with advanced remote sensing technologies

In the future, AI and ML will increasingly be integrated with cutting-edge remote sensing technologies, such as drones, satellites, and unmanned aerial vehicles (UAVs). These technologies, combined with AI/ML, will allow for real-time data collection and analysis of large and difficult-to-reach areas. Drones equipped with AI-powered sensors could conduct high-resolution geological surveys, capturing detailed imagery and geophysical data, which is then processed by machine learning models to identify mineral deposits or environmental risks.

Satellite data, for instance, will be analyzed in conjunction with AI algorithms to monitor land degradation, detect changes in vegetation or soil composition, and assess the impact of resource exploration on surrounding ecosystems. The combination of AI and remote sensing will enable faster and more accurate environmental monitoring, allowing for the implementation of more sustainable and environmentally responsible exploration practices.

15.5.3 Enhanced predictive modeling for resource estimation

As AI and ML techniques improve, they will play a significant role in predictive modeling for resource estimation. Traditional methods of estimating resource quantities often rely on a combination of geological assumptions and sampling data, which can sometimes lead to inaccuracies or inefficiencies. However, AI/ML can help create more precise and dynamic models by analyzing a broader range of data, including geological surveys, mineral assays, remote sensing, and environmental factors.

Machine learning algorithms will be able to build complex models that predict not only the quantity of resources but also the quality and viability of extraction. This will allow for more informed decision-making, helping exploration teams identify the most promising areas for development while avoiding costly, unsuccessful drilling operations. AI's ability to process and analyze large amounts of diverse data will significantly enhance the accuracy of resource assessments, reducing exploration risks and costs.

15.5.4 AI-driven sustainability in resource management

A major focus in the future of AI/ML in the geoscience industry will be improving sustainability in resource exploration and extraction. The push for environmentally responsible practices has never been more urgent, and AI/ML technologies will be critical in driving sustainable resource management. By using real-time data analysis, AI will be able to monitor and control resource extraction processes to ensure that they do not exceed sustainable limits.

Machine learning models could optimize extraction methods to reduce waste, energy consumption, and environmental degradation. For instance, AI could recommend the most efficient mining techniques, assess the potential for recycling by-products, and predict the environmental impact of resource extraction. Additionally, AI-powered systems could assist in land reclamation efforts after resource extraction, guiding restoration projects by identifying the most effective techniques for rebuilding ecosystems.

15.5.5 AI in risk mitigation and safety

The future of AI/ML in the geoscience industry will also include advancements in safety and risk mitigation. Resource exploration and extraction often involve significant risks, such as geological hazards (e.g., earthquakes, landslides), worker safety concerns, and environmental disasters. AI/ML systems will be key in predicting and mitigating these risks by analyzing vast datasets from geological surveys, environmental sensors, and weather systems.

For example, AI could predict seismic events or identify early signs of subsidence or other geological hazards, allowing companies to take preventive actions before accidents occur. AI systems could also monitor worker conditions and ensure safety protocols are followed, reducing the risk of human injury during exploration activities. Furthermore, AI could assist in real-time monitoring of environmental conditions, providing early warnings of issues such as water contamination or air pollution caused by exploration activities.

15.5.6 Collaborative AI for global resource management

In the future, AI will play a key role in collaborative and global resource management. As resource extraction becomes more integrated with global supply chains, AI-driven platforms will allow for cross-border collaboration between countries, research institutions, and private companies. AI-powered platforms can help manage resource allocation, optimize extraction rates, and share best practices for sustainability and safety across regions. By facilitating global cooperation, AI/ML technologies will promote the equitable distribution of resources, reduce conflicts over resource access, and foster transparency in the supply chain. Additionally, AI-driven models could be used to simulate and predict future resource needs, helping to plan for the long-term sustainability of natural resources.

15.5.7 Ethical considerations in AI/ML-driven exploration

As AI/ML technologies continue to shape the future of geoscience, it is essential to address ethical considerations surrounding their application. Issues such as data privacy, equity in resource access, and the displacement of workers due to automation must be carefully considered. As AI becomes more autonomous in decision-making, human oversight will remain critical to ensure that exploration activities are conducted ethically, responsibly, and in a way that benefits all stakeholders. Future AI/ML systems must be designed with transparency, fairness, and accountability in mind. This will include ensuring that algorithms are explainable, that the data used to train these models is diverse and unbiased, and that the rights of affected communities are respected throughout the exploration process.

15.6 Conclusion

The integration of emerging technologies such as Artificial Intelligence (AI), Machine Learning (ML), and Geostatistics has profoundly transformed the landscape of geoscience, particularly in the realm of sustainable resource exploration. This chapter has explored the multifaceted roles of AI/ML in geophysical data generation, interpretation, and integration, demonstrating how these technologies enhance the efficiency, accuracy, and predictability of mineral exploration processes. From seismic data analysis and geochemical modeling to predictive resource estimation, AI/ML has proven to be a powerful catalyst in uncovering hidden geological patterns and optimizing exploration strategies.

Recent case studies from diverse global contexts—ranging from AI-assisted geophysical data processing in Canada to deep learning applications for rare earth element exploration in Africa—highlight the practical successes and transformative potential of AI/ML. These examples underscore the ability of AI/ML to not only accelerate data processing but also to improve decision-making processes, reduce exploration costs, and mitigate risks.

Geostatistics, particularly Bayesian probability modeling, has emerged as a complementary tool to AI/ML, enhancing resource estimation, uncertainty quantification, and decision-making in resource extraction. Case studies illustrate how advanced statistical methods can improve the reliability of resource models, optimize extraction strategies, and support sustainable resource management practices.

However, the advancement of these technologies comes with ethical responsibilities. Sustainable resource exploration demands careful consideration of environmental impacts, social equity, resource management, and transparency. The chapter emphasized the importance of ethical sourcing, community impact assessments, and the responsible use of AI/ML to ensure that technological progress aligns with environmental sustainability, social justice, and long-term resource stewardship.

Looking forward, the future of geoscience is poised for revolutionary changes driven by AI/ML. Fully automated exploration, integration with advanced remote sensing technologies, enhanced predictive modeling, and AI-driven sustainability initiatives are set to redefine the industry. Nevertheless, as we embrace these innovations, it is imperative to maintain a balanced approach that prioritizes ethical considerations, human rights, and ecological conservation.

In conclusion, the convergence of AI/ML, geostatistics, and ethical frameworks offers a promising pathway toward more effective, sustainable, and responsible resource exploration. As these technologies evolve, they hold the potential to not only transform the geoscience industry but also to contribute to the broader goals of global sustainability and resource security.

CHAPTER

16

Equation dictionary for geoscientific exploration

16.1 Introduction

This chapter serves as a consolidated reference of equations and mathematical expressions essential for understanding, applying, and analyzing geoscientific data in mineral and natural resource exploration. Derived from disciplines such as geophysics, geostatistics, remote sensing, and data integration, these equations underpin the methodologies explained across the book. For ease of reference, the equations are grouped into thematic sections, with definitions, units, and brief contextual explanations to aid practical application by geoscientists, students, and exploration professionals of different geophysical methods worldwide (Mishra and Naskar, 2024, 2025 and 2026).

16.2 Gravity method equations

16.2.1 Bouguer anomaly

$$\Delta g_{\mathrm{B}} = \Delta g_0 - \delta g_{\mathrm{E}} - \delta g_{\mathrm{L}} + \delta g_{\mathrm{B}}$$

Where: Δg_{B} = Bouguer anomaly (mGal), Δg_0 = Observed gravity, δg_{E} = Earth tide correction, δg_{L} = Latitude correction, and δg_{B} = Bouguer correction.

The Bouguer correction accounts for the gravitational attraction of the rock mass between the station and sea level. Since this mass increases the observed gravity, the correction removes that extra gravitational effect to estimate what the gravity would be. So, it's added in the equation to counterbalance (remove) the mass effect.

Baseline Geoscience Data Generation for Sustainable Exploration of Natural Resources
https://doi.org/10.1016/B978-0-443-45221-5.00014-X

TABLE 16.1 Summary of Bouguer correction formulas showing added and subtracted correction terms, where the Bouguer anomaly (Δg_B) is expressed in terms of observed gravity, elevation, latitude, free-air, and Bouguer slab corrections.

Form	Bouguer correction
$\Delta g_B = \Delta g_0 - \delta g_E - \delta g_L + \delta g_B$	Added
$\Delta g_B = \Delta g_{FA} - \delta g_B$	Subtracted

Sometimes you'll see:

$$\Delta g_B = \Delta g_{FA} - \delta g_B$$

Where: Δg_{FA} = Free-air anomaly. In this case, δg_B is subtracted because the Bouguer anomaly is defined as free-air anomaly minus Bouguer correction.

16.2.1.1 *Summary: Table 16.1*

Both are correct—it depends on how the equation is structured. Just stay consistent with the definitions of the terms you're using.

16.2.2 Free-air correction

$$\delta g_{FA} = 0.3086 \times h$$

Where: h = Elevation above sea level (m).

16.2.3 Bouguer slab correction

$$\delta g_B = 2\pi G \rho h$$

Where: G = Gravitational constant, ρ = Density of rock (kg/m^3), and h = Elevation (m).

16.2.4 Gravity method example

Problem:

A gravity reading at a station shows an observed gravity (Δg_0) of 980.132 mGal. The Earth tide correction is 0.004 mGal, the latitude correction is 0.012 mGal, and the Bouguer correction is 0.085 mGal. Calculate the Bouguer anomaly.

Solution:

$$\Delta g_B = \Delta g_0 - \delta g_E - \delta g_L + \delta g_B$$

$$\Delta g_B = 980.132 - 0.004 - 0.012 + 0.085 = 980.201 \text{mGal}$$

16.3 Magnetic method equations

16.3.1 Magnetic anomaly

$$\Delta T = T_{\text{obs}} - T_{\text{ref}}$$

Where: T_{obs} = Observed magnetic intensity, and T_{ref} = Reference magnetic field.

16.3.2 Magnetic susceptibility

$$M = \chi H$$

Where: M: Magnetization, χ: Magnetic susceptibility, and H: Magnetic field strength.

16.4 Electrical and electromagnetic methods

16.4.1 Apparent resistivity (Schlumberger method)

$$\rho_{\text{a}} = \frac{\pi (\text{AB})^2 - (\text{MN})^2}{4\,\text{MN}} \cdot R$$

Where: AB: Current electrode separation, MN: Potential electrode separation, R: Measured resistance.

16.4.2 Skin depth

$$\delta = \sqrt{\frac{2\rho}{\omega\mu}}, \; \omega = 2\pi f$$

Where: ρ: Resistivity (Ωm), ω: Angular frequency, and μ: Magnetic permeability.

16.4.3 Self-potential (SP) equation

$$\text{SP} = \frac{k \cdot \Delta T}{\Delta x}$$

Where: SP: Potential difference, k: Electrokinetic coupling coefficient, ΔT: Temperature gradient, and Δx: Distance.

16.5 Seismic method equations

16.5.1 Travel time (two-layer model)

$$t = \frac{x}{v_1} + \frac{2h \cdot \cos i}{v_2}$$

Where: t: Travel time, x: Offset, v_1, v_2: Velocities of upper and lower layers, h: Depth to interface, and i: Angle of incidence.

16.5.2 Normal moveout (NMO)

$$\Delta t = \frac{x^2}{2v^2 t_0}$$

Where: x: Receiver offset, v: Velocity, and t_0: Zero-offset time.

16.5.3 Reflection coefficient

$$R = \frac{Z_2 - Z_1}{Z_2 + Z_1}, Z = \rho v$$

Where: Z: Acoustic impedance, ρ: Density, and v: Velocity.

16.6 Remote sensing and spectral analysis

16.6.1 Normalized difference vegetation index (NDVI)

$$\text{NDVI} = \frac{\text{NIR} - \text{Red}}{\text{NIR} + \text{Red}}$$

Where: NIR: Near-infrared reflectance, and Red: Red reflectance.

16.6.2 Band ratio (e.g., iron oxide mapping)

$$\text{BR} = \frac{\text{Band}_3}{\text{Band}_1}$$

Where: Band_3: Spectral band sensitive to iron oxides, and Band_1: Reference band.

16.6.3 Remote sensing example—NDVI

Problem:

Using a satellite image, the NIR reflectance is 0.60 and Red reflectance is 0.20. Calculate NDVI.

Solution:

$$\text{NDVI} = \frac{\text{NIR} - \text{Red}}{\text{NIR} + \text{Red}} = \frac{0.60 - 0.20}{0.60 + 0.20} = \frac{0.40}{0.80} = 0.50$$

Interpretation: Indicates healthy vegetation.

16.7 Geostatistics and resource estimation

16.7.1 Kriging estimator

$$Z^*(x_0) = \sum_{i=1}^{n} \lambda_i Z(x_i)$$

Where: $Z^*(x_0)$: Estimated value at location x_0, λ_i: Weight, and $Z(x_i)$: Observed value.

16.7.2 Variogram

$$\gamma(h) = \frac{1}{2N(h)} \sum_{i=1}^{N(h)} [Z(x_i) - Z(x_i + h)]^2$$

Where: $\gamma(h)$: Variogram for lag h, and $N(h)$: Number of pairs at lag h.

16.7.3 Kriging example

Problem:

Estimate the value at x_0 using Kriging:

Measurements:

$$Z(x_1) = 5.2, \quad \lambda_1 = 0.3$$
$$Z(x_2) = 6.0, \quad \lambda_2 = 0.5$$
$$Z(x_3) = 5.8, \quad \lambda_3 = 0.2$$

Solution:

$$\mathbf{Z}^*(\mathbf{x}_0) = \sum \lambda_{\mathbf{i}} \mathbf{Z}(\mathbf{x}_{\mathbf{i}}) = 0.3(5.2) + 0.5(6.0) + 0.2(5.8) = 1.56 + 3.0 + 1.16 = 5.72$$

16.8 AI and machine learning in geosciences

16.8.1 Gradient descent

$$\theta := \theta - \alpha \cdot \nabla J(\theta)$$

Where: θ: Model parameters, α: Learning rate, and $\nabla J(\theta)$: Gradient of the loss function. The symbol: = means "is updated to" or "is assigned as."

16.8.2 Decision tree splitting—Gini index

$$\text{Gini} = 1 - \sum_{i=1}^{n} p_i^2$$

Where: p_i: Proportion of class i.

16.8.3 Confusion matrix—Accuracy

$$\text{Accuracy} = \frac{\text{TP} + \text{TN}}{\text{TP} + \text{TN} + \text{FP} + \text{FN}}$$

Where: TP: True positives, TN: True negatives, FP: False positives, and FN: False negatives.

16.8.4 Comparative overview of common AI/ML algorithms in geosciences

This section provides a quick-reference comparison of widely used AI and machine learning algorithms in geoscientific applications. It outlines their key features, advantages, limitations, and suitability for different data types and exploration tasks (Table 16.2).

TABLE 16.2 Comparative overview of commonly used artificial intelligence and machine learning (AI/ML) algorithms in geosciences, highlighting their key features, advantages, limitations, and suitability for different geoscientific data types and exploration applications.

Algorithm	Key features	Pros	Cons	Best suited for
Decision tree	Rule-based model that splits data into decision paths	Easy to interpret; handles non-linear data well	Prone to overfitting; may be unstable	Lithological classification, mineral prediction, landslide zoning
Random forest	Ensemble of decision trees using bootstrapped datasets	High accuracy; reduces overfitting; works with large feature sets	Less interpretable; computationally heavier than a single tree	Mineral prospectivity, hydrocarbon play classification, risk mapping
Support vector machine (SVM)	Finds optimal hyperplane to separate classes	Effective in high-dimensional spaces; good for small-to-medium datasets	Less effective with noisy or large datasets; needs careful tuning	Rock type discrimination, remote sensing spectral data classification
K-Nearest neighbors (KNN)	Classifies based on nearest neighbors in feature space	Simple; no training phase; intuitive	Sensitive to noisy data and scaling; slower with large datasets	Groundwater quality classification, anomaly detection
Artificial neural network (ANN)	Mimics brain-like learning with layers of nodes	Good for complex, non-linear relationships; flexible	Requires large datasets; "black box" interpretation	Seismic facies classification, mineral grade prediction
Gradient boosting (e.g., XGBoost)	Ensemble learning via sequential tree optimization	High accuracy; handles missing data; robust	Long training time; complex hyperparameters	Ore body prediction, geohazard modeling

16.8.5 Notes for global application

(i) Data Requirements: Algorithms like ANN and XGBoost typically require larger and cleaner datasets. In regions with limited data availability, Decision Trees or SVM may be more appropriate.
(ii) Computational Resources: Random Forests and Gradient Boosting perform well but need more computing power—Consider available infrastructure.
(iii) Interpretability: Simpler models like Decision Trees and KNN are more transparent, which can be important in regulatory or stakeholder communication in resource-sensitive regions.

16.9 Airborne and marine survey equations

16.9.1 Altitude correction (airborne surveys)

$$A_{\text{corr}} = A_{\text{meas}} \cdot e^{-kh}$$

Where: A_{corr}: Corrected anomaly, A_{meas}: Measured value, k: Decay constant, and h: Flight height.

16.9.2 Marine CSEM response

$$E_x(r,t) = \frac{I}{2\pi\sigma r^2} \cdot e^{-\sqrt{\mu\sigma\omega r}}$$

Where: E_x: Electric field, I: Source current, σ: Conductivity, r: Distance, μ: Permeability, and ω: Angular frequency.

16.10 Integration and modeling

16.10.1 Data normalization

$$x' = \frac{x - \mu}{\sigma}$$

Where: x': Normalized value, μ: Mean, and σ: Standard deviation.

16.10.2 Weighted overlay analysis (GIs modeling)

$$S = \sum_{i=1}^{\sigma} w_i \cdot x_i$$

Where: S: Suitability index, w_i: Weight of layer i, and x_i: Normalized score of layer i.

TABLE 16.3 Units, physical constants, and commonly used parameters in geophysical and geoscientific applications, including their symbols, descriptions, units, and typical values.

Symbol	Description	Unit	Typical value
G	Gravitational constant	Nm^2/kg^2	6.674×10^{-11}
ρ	Density	kg/m^3	2500–2700
μ_0	Magnetic permeability of free space	H/m	$4\pi \times 10^{-7}$
ω	Angular frequency	Rad/s	$2\pi f$
f	Frequency	Hz	–

16.11 Units, constants, and quick reference table (Table 16.3)

16.12 Software references/tools

Suggested Open Tools:

- QGIS—For GIS and raster data overlay
- Oasis Montaj or Geosoft—Gravity and magnetic data processing

Python libraries:
- geopandas, rasterio, scikit-learn, pykrige
- SAGA GIS—Geostatistical mapping
- Google Earth Engine—Remote sensing spectral index computation

16.13 Data integration workflow example

Flowchart: Integrating Remote Sensing + Geophysics + ML

1. Data Acquisition (Satellite + Ground geophysics)
2. Preprocessing (Noise removal, corrections)
3. Feature Extraction (NDVI, anomalies, elevation, conductivity)

   Data Normalization $\left(\text{Use } x' = \frac{x - \mu}{\sigma}\right)$
4. Weighted Overlay or ML Model Input
5. Prediction or Target Mapping

16.14 Visual aids and diagrams

To complement the equation-based approaches outlined in this chapter, this section presents a collection of simplified diagrams that illustrate key geoscientific exploration concepts. Visual aids are essential for clarifying abstract mathematical relationships, data acquisition methods, and interpretation techniques across disciplines such as geophysics, geochemistry, remote sensing, and AI/ML applications. These schematics and plots are designed to enhance conceptual understanding, support field application, and serve as quick-reference tools for practitioners and students alike (Table 16.4).

Fig. 16.1A displays the gravity and Bouguer anomaly profile across the study area. Elevation correction has been applied to account for topographic effects. The slab model

TABLE 16.4 Summary of visual aids and schematic diagrams used in this chapter to illustrate key geoscientific exploration concepts across gravity, magnetic, electrical, seismic, remote sensing, geostatistics, and AI/ML applications.

Section	Diagram type	Description
Gravity	Bouguer anomaly profile	Showing elevation correction, slab model
Magnetic	Field line maps	Show anomaly locations
Electrical	Electrode configurations	Schlumberger vs. Wenner array
Seismic	Ray path	Incident and reflected waves in 2-layer model
Remote Sensing	Spectral response curves	NDVI and band ratio visualization
Geostatistics	Variogram plot	Nugget, sill, range visual
AI/ML	Decision tree	Splitting criterion visual (Gini index)

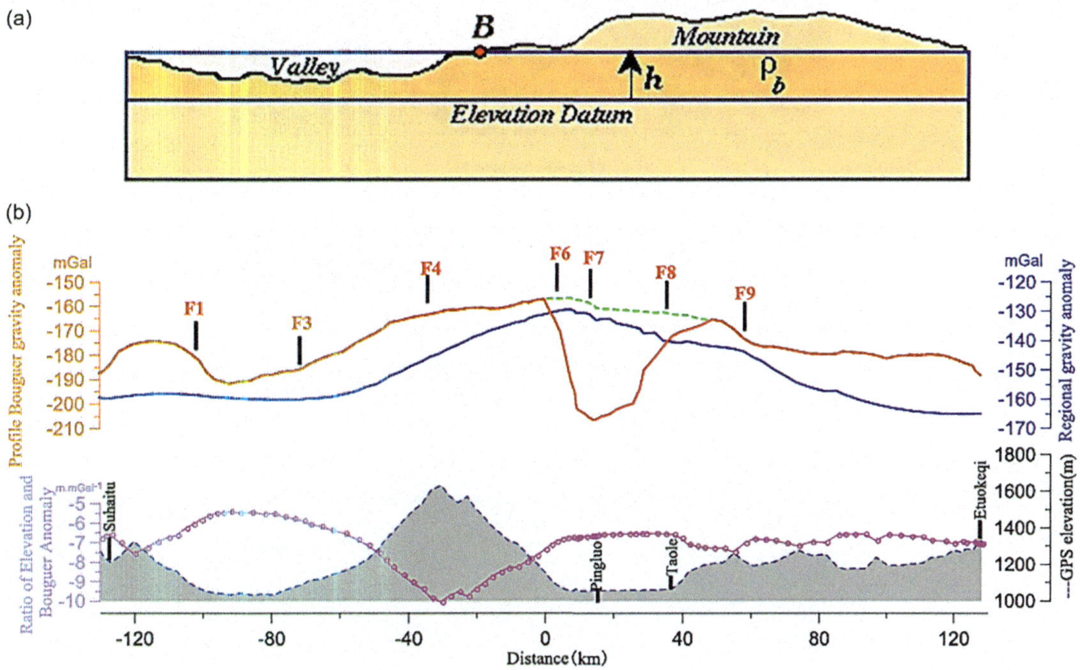

FIG. 16.1 (A) Gravity and Bouguer anomaly profile with elevation correction and slab model. (B) Bouguer gravity anomaly, GPS elevation, and elevation-to-anomaly ratio along the profile. *Panel A: From Mishra and Naskar (2025). Panel B: From Wu et al. (2016)—Open Access.*

represents the density contrast between surface features and subsurface structures, aiding in geological interpretation. Fig. 16.1B illustrates the Bouguer gravity anomaly along the profile, accompanied by the corresponding GPS-derived elevation data. The plot also includes the ratio between elevation and Bouguer anomaly, providing insights into the relationship between surface topography and subsurface density variations. This comparison aids in identifying geological structures and variations in crustal composition along the profile.

Fig. 16.2A presents the total magnetic field anomalies derived from the survey. Yellow points indicate the orientation of survey profiles, while points P1 and P2 mark the locations of the PWW and the dry well, respectively. Two distinct anomaly patterns are visible: a central zone with densely spaced contour lines indicating strong magnetic anomalies, surrounded by a broader area with smoother magnetic variations. These patterns help in identifying contrasting subsurface features and geological structures.

Fig. 16.2B illustrates the magnetic field distribution across the Kluane Region, Canada. Magnetic field line patterns reveal the presence and extent of subsurface anomalies, which may correspond

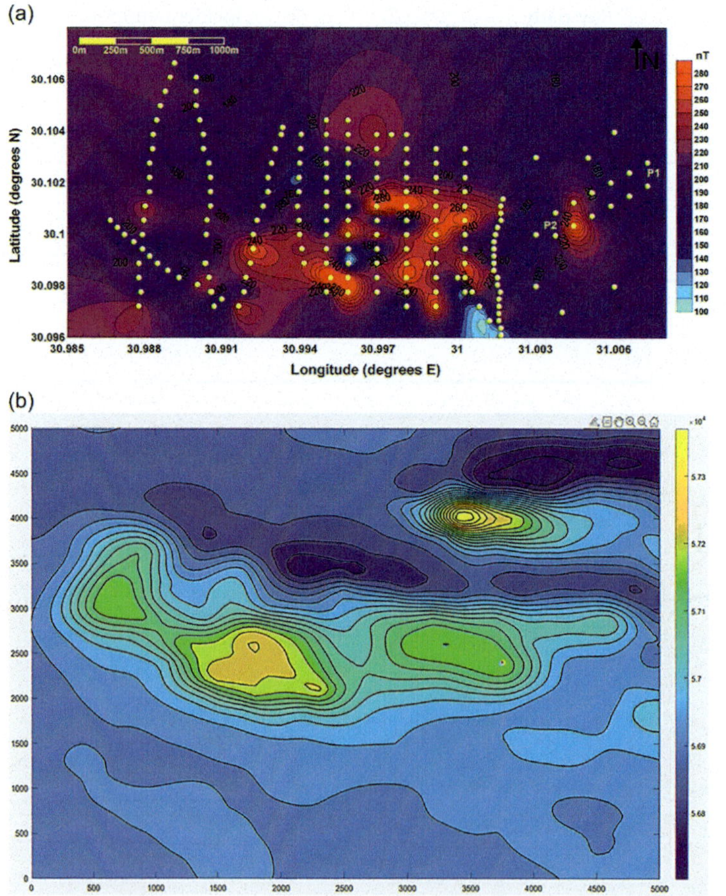

FIG. 16.2 (A) Total magnetic field anomaly map with survey profile trends. (B) Magnetic field map of the Kluane region, Canada. *Panel A: From Arafa-Hamed, 2014, in press). Panel B: From Li et al. (2024)—Open Access.*

to variations in rock type, structure, or mineral content. The identified anomalies assist in understanding the geological framework and potential mineralization zones in the region.

Electrode configurations such as Wenner, Schlumberger, and Dipole-Dipole are widely utilized in geoelectrical resistivity surveys for subsurface investigations in mineral exploration (Fig. 16.3A). A and B represent the current electrodes, while M and N are the potential electrodes. These configurations facilitate the measurement of apparent resistivity by injecting electrical current into the ground and recording the resulting potential differences. Each array offers distinct advantages in terms of depth penetration, spatial resolution, and sensitivity to lateral and vertical resistivity variations. Such features make them effective for mapping ore bodies, detecting mineralized zones, and interpreting the structural framework of the subsurface.

In the context of baseline geoscience data generation for resource exploration, **Continuous Vertical Electrical Sounding (CVES)** has proven to be an effective technique for detailed subsurface characterization. This method employs a multichannel resistivity system (van Overmeeren and Ritsema, 1988; Dahlin, 1996), which automatically increases the electrode

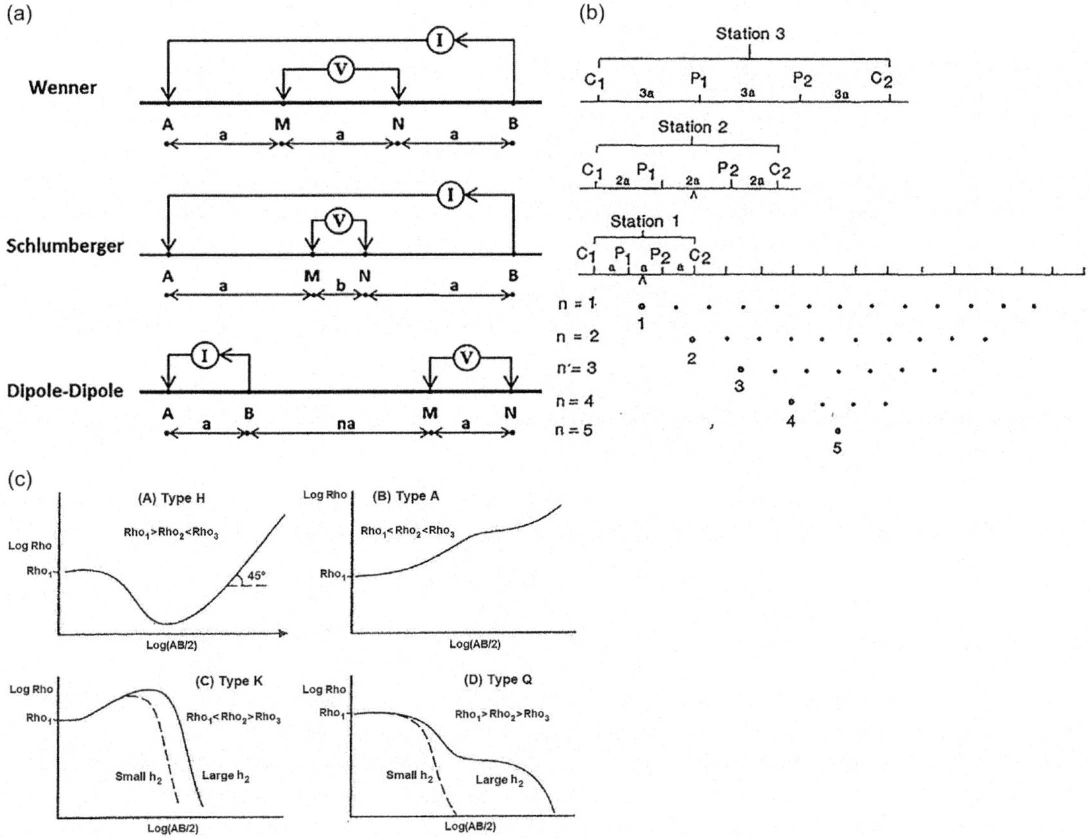

FIG. 16.3 (A) Electrode configurations for mineral exploration: Wenner, Schlumberger, and Dipole-Dipole arrays. (B) Example of a measurement sequence used to construct a resistivity pseudosection for subsurface mineral investigation. (C) Typical apparent resistivity curve types for three-layer models. *Panel A: From Naskar et al. (2002). Panel B: From Naskar et al., 2008. Panel C: From Reynolds (1997).*

spacing during data acquisition. By progressively expanding the current and potential electrode separations, CVES enhances the depth of investigation while maintaining high spatial resolution. This approach allows for the efficient construction of resistivity pseudosections (Fig. 16.3B), which are critical for identifying and delineating mineralized zones and associated geological structures.

A preliminary interpretation of the apparent resistivity curve is usually based on the shape of the curve. For a system with three layers, each defined by its resistivity (ρ) and thickness (h), four main types of curves can be observed:

H-type curve: The second layer has lower resistivity than the layers above and below it ($\rho_1 > \rho_2 < \rho_3$) (Fig. 16.3C).

A-type curve: Resistivity increases with depth ($\rho_1 < \rho_2 < \rho_3$).

K-type curve: The second layer has higher resistivity than the first and third layers ($\rho_1 \rho_2 \rho_3$).

Q-type curve: Resistivity decreases with depth ($\rho_1 > \rho_2 > \rho_3$).

These curve types help in identifying different subsurface layering patterns and are commonly used in geophysical surveys for resource exploration.

16.14.1 Seismic: ray path and wave behavior in a two-layer model

In seismic investigations, understanding how seismic waves travel through different layers of the Earth is essential for interpreting subsurface structures. In a basic two-layer model, each layer has its own seismic wave velocity—usually increasing with depth. When a seismic wave (generated by a source such as a hammer or explosive) travels downward and reaches the boundary between two layers, several things happen (Fig. 16.4A):

(i) The wave that approaches the boundary is called the incident wave.
(ii) A portion of the wave energy reflects back toward the surface—This is the reflected wave.
(iii) The remaining energy passes into the second layer, changing direction due to the difference in velocity. This bending of the wave is known as refraction, and the wave that continues is the refracted wave.

This change in direction follows a rule called Snell's law, which explains how wave speed differences between layers affect the angle of refraction. According to Snell's law:

(i) Angle of incidence = angle of reflection (P to P or S to S reflection),

$$\frac{\sin i_1}{V_1} = \frac{\sin i_2}{V_2}$$

Where: i_1 is the angle of the incident wave, i_2 is the angle of the refracted wave, V_1 and V_2 are the wave velocities in the first and second layers, respectively.

If the wave hits the boundary at a certain angle—called the critical angle—the refracted wave travels along the boundary between the two layers instead of going deeper. This wave then produces head waves that return to the surface and are recorded by seismic receivers (geophones) (Fig. 16.4B). These head waves help identify the depth and velocity of subsurface layers. This basic model of wave behavior is fundamental to seismic refraction and reflection methods, which are widely used in geophysical surveys to locate natural resources such as minerals, groundwater, and hydrocarbons.

When a compressional (P) wave travels straight down and hits a boundary between two different layers of the Earth, part of the wave bounces back (reflects), and part of it may go through (transmits). The **reflection coefficient** tells us how much of the wave energy is reflected back. For **vertical incidence**, this reflection depends mainly on the **density** and **P-wave velocity** of the two layers. The formula becomes simpler in this case and depends on the acoustic impedance of each layer, which is just:

$$\text{Impedance}(Z) = \text{density}(\rho) \times \text{velocity}(V)$$

So, the reflection coefficient (R) is (Fig. 16.4C):

$$R = \frac{Z_2 - Z_1}{Z_2 + Z_1} = \frac{V_2\rho_2 - V_1\rho_1}{V_2\rho_2 + V_1\rho_1}$$

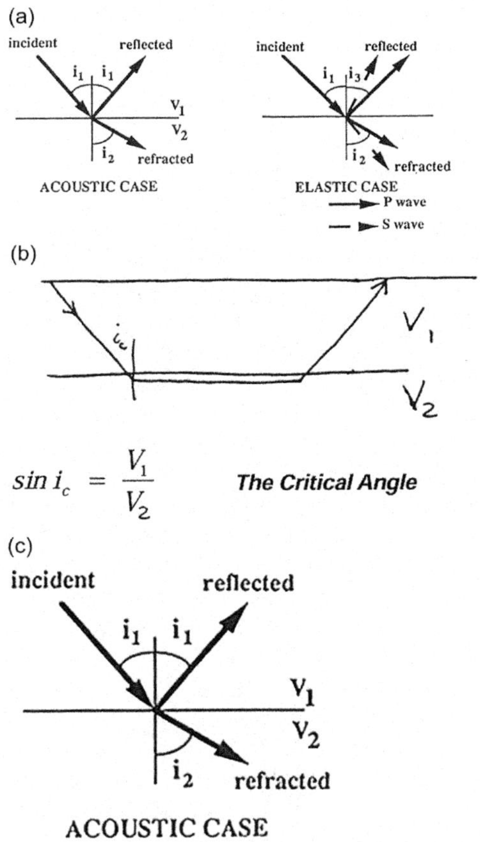

FIG. 16.4 (A) Seismic wave travel times and ray paths in a two-layer model: Reflections, refractions, and wave behavior illustrated (any appropriate velocity and subscripts may be used; for example, for P incident and P reflected or refracted, P incident and S reflected or refracted, etc.). (B) Ray path behavior at critical angles: Generation of head waves in a two-layer seismic velocity model. (C) Illustration of the simple reflection coefficient at vertical incidence in a two-layer seismic model. *Panels A–C: Wikipedia—Open Access.*

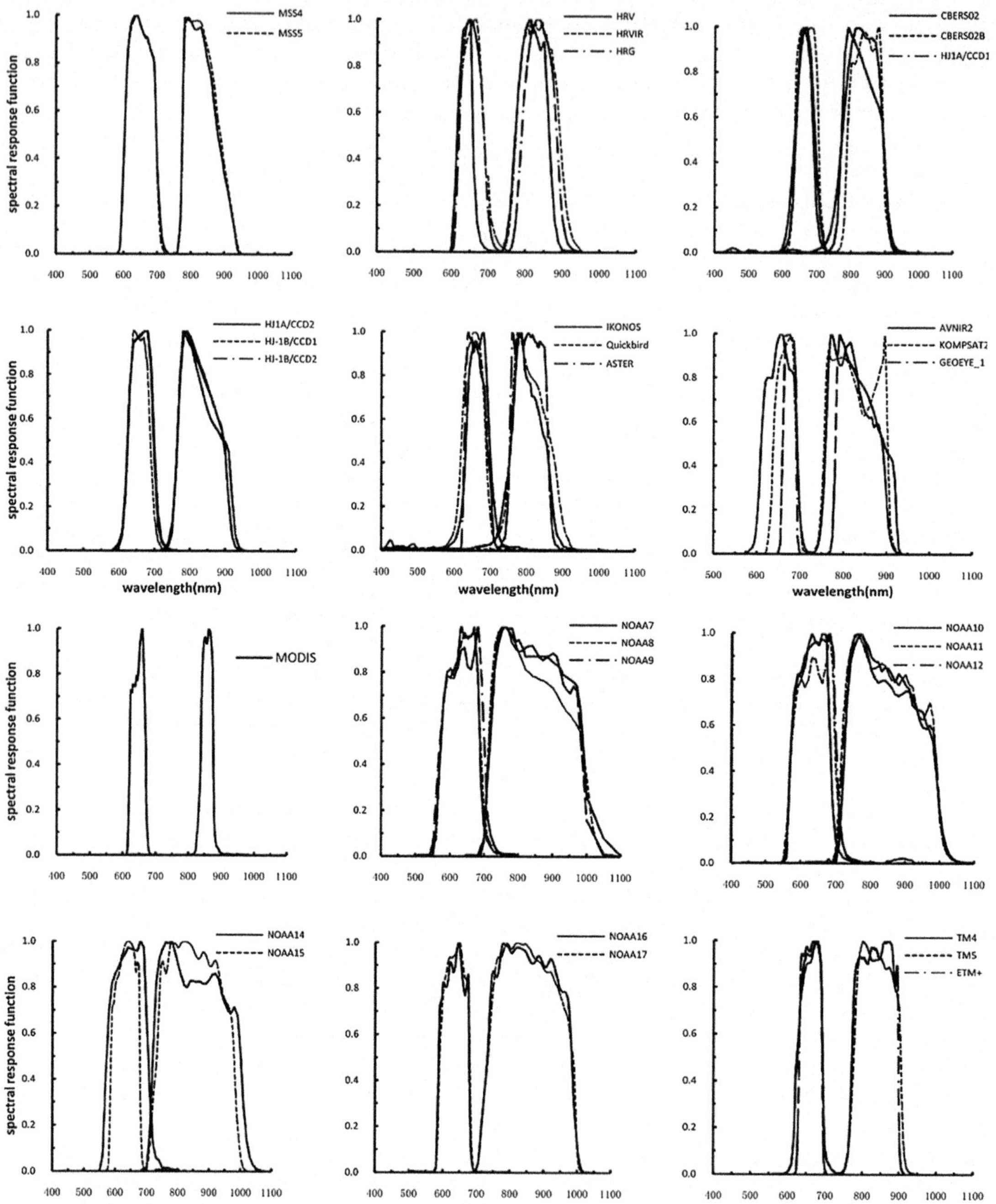

FIG.16.5A Simulated relative spectral response functions of red and near-infrared (NIR) channels in sensor systems *(Source: Huang et al. (2013)—Open Access.)*

Where: $\mathbf{Z}_1$ = impedance of the top layer, $\mathbf{Z}_2$ =impedance of the bottom layer, V_1 and ρ_1 are the velocity and density, respectively, above the interface, and V_2 and ρ_2 are the velocity and density below the interface. Note that the reflection coefficient is proportional to the acoustic impedance contrast. Because very large change in velocity and density between adjacent layers is rare, the reflection coefficient is usually very small (~0.02–0.10).

Similarly, the simple transmission coefficient is:

$$\mathbf{T} = \frac{2\mathbf{V}_1\rho_1}{\mathbf{V}_2\rho_2 + \mathbf{V}_1\rho_1}$$

If the density is unknown, one can compute reflection and transmission coefficients using the velocity only.

Remote Sensing is the science of gathering information about the Earth's surface using sensors on satellites or aircraft. It captures data across various wavelengths of light to help analyze natural features like vegetation, water, and soil. Spectral Response Curves describe how different materials reflect or absorb light at different wavelengths (Fig. 16.5A). Each surface has a unique spectral signature that helps in identification. While there's no single formula here, the key concept is the reflectance (ρ) at wavelength λ, often shown as (Fig. 16.5B):

$$\rho(\lambda) = \frac{\text{reflected light at } \lambda}{\text{incident light at } \lambda}$$

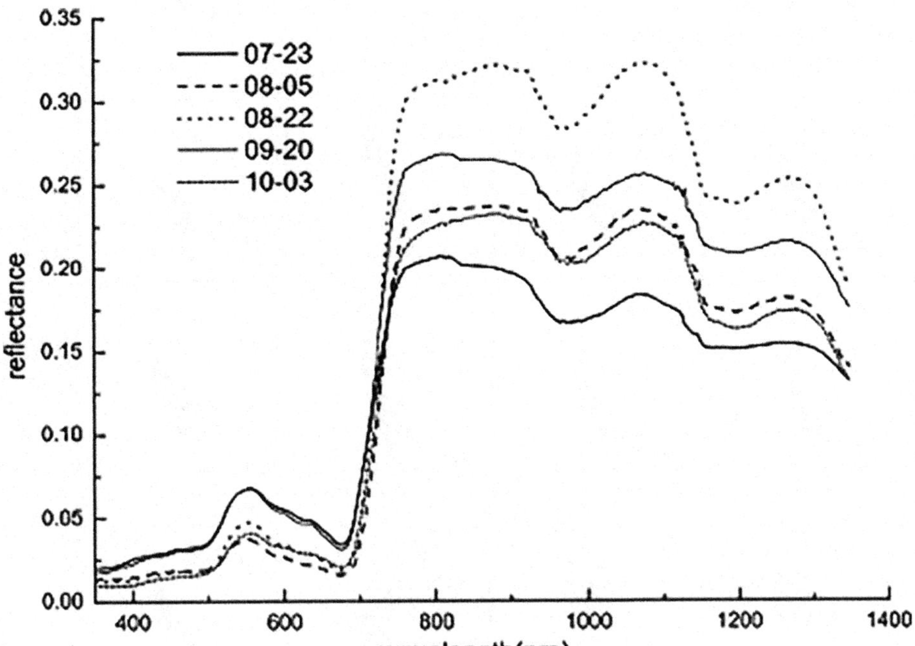

FIG.16.5B Average rice canopy reflectance at N1 nitrogen fertilization level across growth stages in 2002 *(Source: Huang et al. (2013)—Open Access.)*

NDVI (normalized difference vegetation index) quantifies vegetation health by comparing reflectance in the near-infrared (NIR) and red (RED) bands (Fig. 16.5C):

$$\text{NDVI} = \frac{\rho_{\text{NIR}} - \rho_{\text{RED}}}{\rho_{\text{NIR}} + \rho_{\text{Red}}}$$

Where: ρ_{NIR} and ρ_{RED} are the reflectance values in the near-infrared and red bands, respectively. NDVI values range from −1 to +1, with higher values indicating healthier vegetation.

Band Ratio Visualization uses the ratio of reflectance values from two different bands to highlight specific features:

$$\text{Band ratio} = \frac{\rho_{\text{B and A}}}{\rho_{\text{B and B}}}$$

This technique helps emphasize contrasts such as soil moisture or vegetation differences in remote sensing images.

Description: The variogram is a fundamental tool in geostatistics used to quantify spatial continuity or variability of a geoscientific variable (e.g., ore grade, soil contamination, groundwater level). It expresses how data similarity decreases with increasing distance between sample points. The standard variogram plot displays semi-variance ($\gamma(h)$) on the vertical axis against lag distance (spatial separation between sample pairs) (h) on the horizontal axis. Key components visualized in the diagram include:

Geostatistics: Variogram plot Nugget, sill, range

Diagram Type: Variogram plot
Visual Elements: Nugget, sill, range

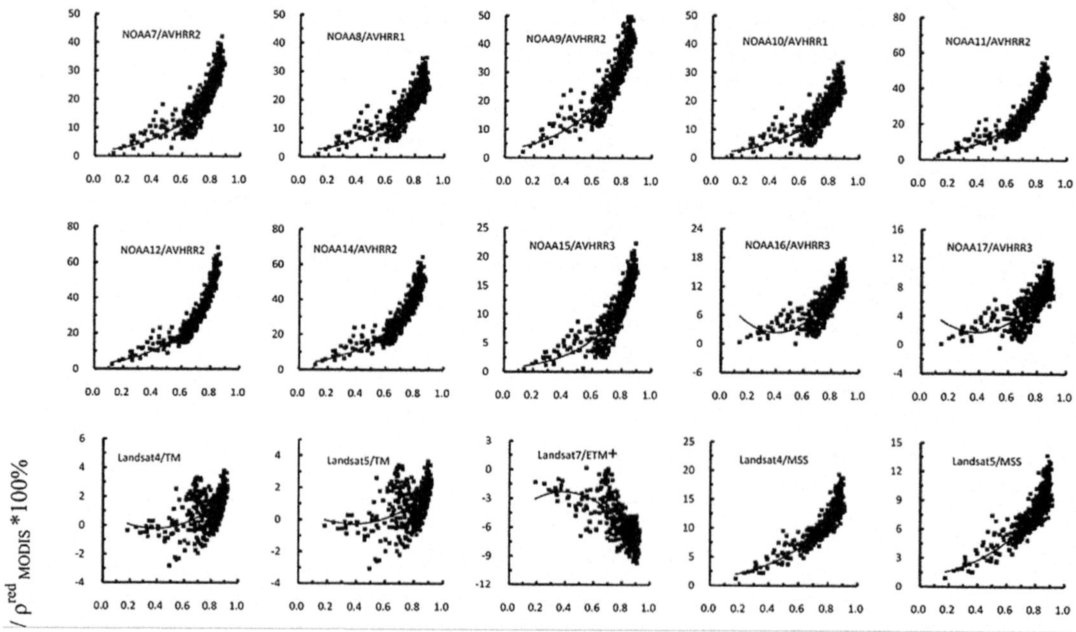

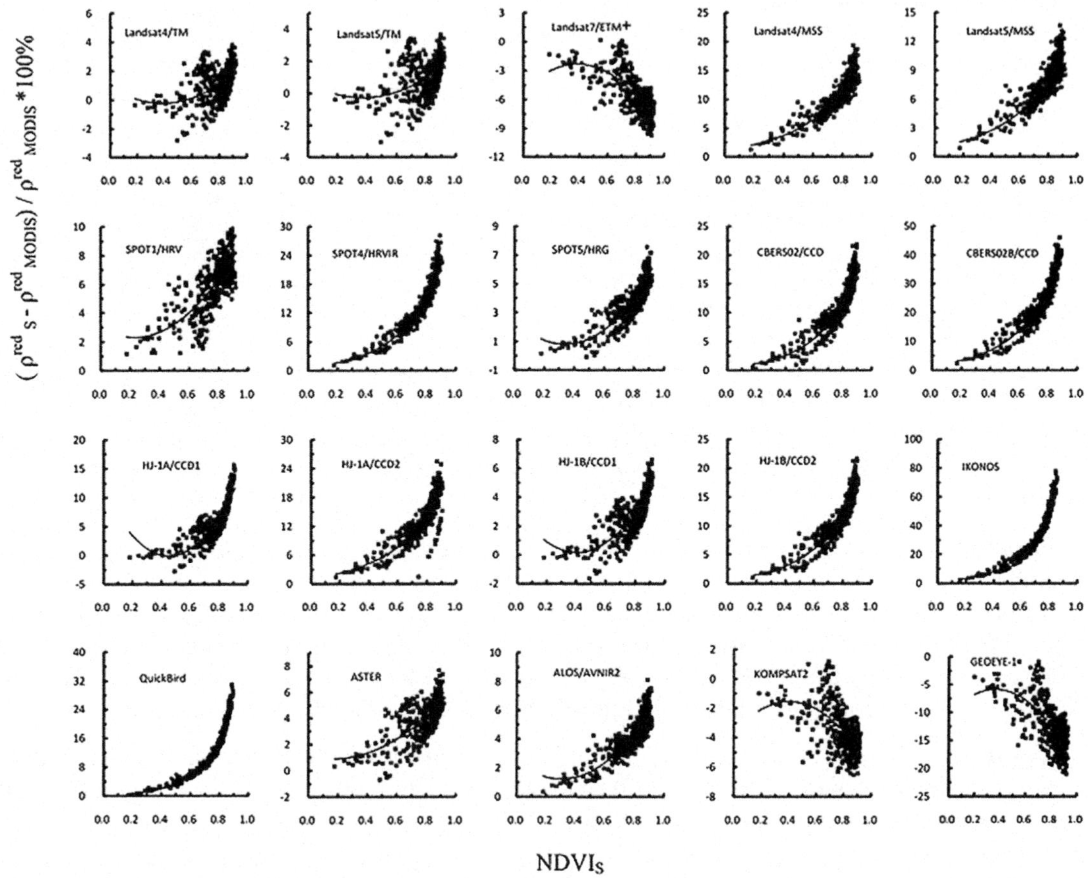

FIG.16.5C Relative red reflectance differences vs. NDVI for various sensors compared to Terra MODIS *(Source: Huang et al. (2013)—Open Access.)*

(i) Nugget: The y-intercept, representing measurement error or micro-scale variability.
(ii) Sill: The value at which the variogram levels off, indicating the total variance of the dataset.
(iii) Range: The distance at which the variogram reaches the sill, beyond which sample values are no longer spatially correlated.

This visual aid helps in selecting appropriate models (e.g., spherical, exponential) for spatial interpolation techniques like kriging, and in understanding the scale and structure of spatial dependence in exploration datasets.

Fig. 16.6 presents the omnidirectional empirical variograms for the log-transformed SWL (Static Water Level) and TDS (Total Dissolved Solids) data, computed using a 500 m lag distance. The plots illustrate key variogram parameters: nugget, sill, and range, which define the spatial structure of variability. Near the origin, the nugget effect captures microscale variability and measurement error, while the range indicates the distance over which spatial correlation persists. The convergence toward the sill suggests a clear spatial dependence structure.

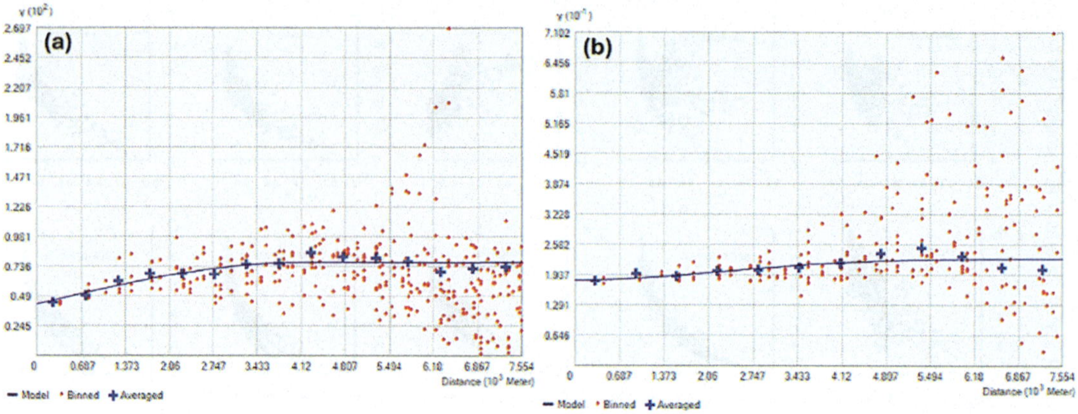

FIG. 16.6 Omnidirectional empirical variogram models for the log-transformed. (A) Static water level (SWL) and (B) Total dissolved solids (TDS) datasets. *From Mostafa et al. (2025)—Open Access.*

Visual inspection of the variograms reveals no significant anisotropy, justifying the use of an isotropic spherical model. This model was selected as the best fit based on both visual trends and quantitative comparison against exponential and Gaussian models. Cross-validation metrics (RMSE [Root Mean Square Error], MAE [Mean Absolute Error], mean standardized error) confirmed that the spherical model yielded the most accurate and unbiased predictions, supporting its use for spatial interpolation of both SWL and TDS variables.

RMSE: A measure of the average magnitude of the prediction errors. It gives higher weight to large errors due to squaring. Lower RMSE values indicate better model accuracy. It is sensitive to outliers.

$$\mathrm{RMSE} = \sqrt{\frac{1}{n}\sum_{i=1}^{n}(y_i - \hat{y}_i)^2}$$

Where: n = Total number of observations (or validation points), y_i = Actual (observed) value at location iii (e.g., measured SWL or TDS), $\hat{y}_i$ = Predicted value at location iii (from kriging or interpolation), $y_i - \hat{y}_i$ = Prediction error (residual) at each point, $(y_i - \hat{y}_i)^2$ = Squared error to penalize large deviations more, $\frac{1}{n}\Sigma$ = Average of squared errors, and $\sqrt{\ldots}$ = Square root to bring back to original units (e.g., meters, mg/L).

MAE: The average of the absolute differences between predicted and observed values. Also measures accuracy, but treats all errors equally (less sensitive to outliers than RMSE).

$$\mathbf{MAE} = \frac{1}{n}\sum_{i=1}^{n}|y_i - \hat{y}_i|$$

Where: n = Number of observations, y_i = Observed (true) value at point i, $\hat{y}_i$ = Predicted value at point i, and $\frac{1}{n}\Sigma$ = Average of absolute errors.

AI/ML:	*Decision tree*	*Splitting criterion visual (Gini index)*

A decision tree diagram using the **Gini index** illustrates how data is split at each node based on feature values that maximize class purity. The visual shows decision points (nodes) where the dataset is divided to reduce impurity, with Gini values labeled to indicate the degree of class separation. Lower Gini index values at splits reflect more homogeneous subsets, guiding optimal feature selection during geoscientific classification tasks, such as lithology or mineral potential mapping.

Fig. 16.7A presents a simplified decision tree used in AI/ML applications for **mineral classification**. The tree begins with a **root node**, progresses through **internal decision nodes**, and ends at **leaf nodes**, each representing a classified mineral type. **Subtrees** illustrate decision paths formed by applying the **Gini index** as the splitting criterion. This visual aid supports understanding of how machine learning models make hierarchical decisions in geological datasets.

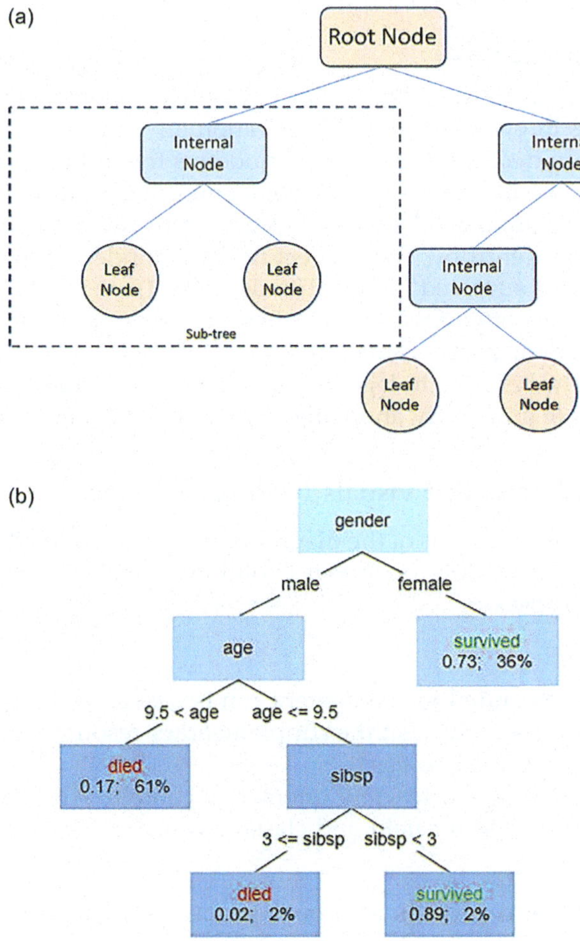

FIG. 16.7 (A) Schematic of a simple decision tree for mineral classification using Gini Index. (B) Decision tree for passenger survival on the Titanic based on gender, age, and family connections. *Panel A and B: Wikipedia—Open Access.*

Fig. 16.7B (decision tree) visualizes the survival probabilities of Titanic passengers using features such as gender, age, and number of siblings/spouses aboard ("sibsp"). Each leaf node displays the probability of survival alongside the percentage of passengers that fall into that category. The tree splits the dataset into subgroups based on these characteristics, highlighting how different factors influenced survival chances. This figure is adapted from Stephen Milborrow's work and demonstrates how decision trees can interpret complex survival patterns in historical datasets.

16.14.2 How to interpret visual aids and diagrams

Visual representations such as decision trees, geophysical models, and AI workflows are essential for translating complex equations and data patterns into intuitive formats. Understanding how to interpret these diagrams enhances comprehension and practical application in the field or during data analysis.

16.14.2.1 Key interpretation tips

(i) Color Codes: Most diagrams use color to distinguish elements (e.g., root nodes in yellow, internal nodes in green, or high vs. low anomaly zones).
(ii) Node Types in Trees: In decision trees, the root node is the initial split, internal nodes are decision points, and leaf nodes represent outcomes or classifications.
(iii) Splitting Criteria: For AI models like decision trees, splits are made based on criteria such as the Gini index or entropy. These affect how the model separates data classes (e.g., "survived" vs. "not survived" in the Titanic dataset).
(iv) Flow Direction: Diagrams often follow a top-down or left-right structure indicating the flow of computation or decision-making.
(v) Legend and Labels: Always refer to legends, axis labels, and units—especially in geophysical diagrams (e.g., gravity anomalies in mGal or seismic velocity in m/s).

16.14.3 Interactive and extended visuals (optional)

Where available, interactive versions of the diagrams are provided to allow zooming, layer toggling, and animation of data flow or model evolution. These are especially useful for educational or training purposes.

16.14.3.1 Scan to explore

Interactive diagrams or expanded visual content can be accessed via QR codes provided next to selected figures. Alternatively, visit the supplementary resources page at: www.baselinegeodata.org/visuals (*example placeholder*).

Note: Ensure stable internet access to view interactive content. All supplemental visuals are mobile-friendly and optimized for field and classroom use.

16.15 Common pitfalls and best practices (Table 16.5)

This section highlights frequent mistakes and practical guidance when applying geophysical equations and machine learning methods in geoscientific studies. Awareness of these issues can enhance data quality, interpretation accuracy, and model reliability in exploration projects.

16.15.1 Geophysical methods

TABLE 16.5 Common pitfalls and recommended best practices in applying geophysical methods, highlighting frequent errors, their implications, and practical guidance for improving data quality and interpretation accuracy.

Common pitfalls	Best practices
Using incorrect units or inconsistent coordinate systems	Always standardize units (e.g., SI) and verify coordinate references (e.g., WGS84)
Oversimplifying subsurface models	Use layered or heterogeneous models where justified by geology
Ignoring environmental noise (e.g., urban EM interference)	Pre-survey planning and noise filtering are critical
Inadequate station spacing or data density	Optimize survey design for resolution and target depth
Misinterpreting anomalies without geological context	Integrate geophysical results with field mapping, drilling, and lithology

16.16 Case study snapshots (mini examples)

This section presents concise case study summaries demonstrating the practical integration of geophysical methods and AI/ML techniques. These mini-examples are drawn from published research and illustrate how combining data types improves exploration accuracy and decision-making.

16.16.1 Gravity + machine learning for mineral prospectivity

Location: Central Iran (Yousefi and Carranza, 2015d)

Objective: Identify prospective zones for iron and copper mineralization.

Methods: Bouguer gravity data were processed to map density contrasts, followed by the application of a decision tree classifier to integrate geological, geochemical, and geophysical indicators.

Outcome: The hybrid approach improved target prediction accuracy compared to single-method analysis.

16.16.2 Seismic + geostatistics for reservoir characterization

Location: Krishna-Godavari Basin, India (Sen and Stoffa, 2013)

Objective: Estimate porosity distribution in a hydrocarbon-bearing sandstone reservoir.

Methods: 3D seismic attributes were extracted and co-kriged with well-log porosity data using geostatistical methods (e.g., sequential Gaussian simulation).

Outcome: The integration improved the spatial prediction of porosity in unsampled areas.

16.16.3 Remote sensing + SVM for lithological mapping

Location: Eastern Desert, Egypt (Gad and Kusky, 2006)

Objective: Map lithological units in a Precambrian shield region.

Methods: Landsat-8 and ASTER imagery were processed using band ratios and principal component analysis. An SVM classifier was trained on field-verified samples to generate the final map.

Outcome: The classification accuracy exceeded 85%, outperforming traditional unsupervised methods.

16.16.3.1 Summary tip

When integrating multiple methods:

(i) Ensure data alignment (e.g., spatial resolution, projection).
(ii) Select complementary techniques to reduce uncertainty.
(iii) Validate with ground-truth data or known targets.

16.17 Glossary of key terms

16.17.1 Geophysical and mathematical symbols: Table 16.6

TABLE 16.6 Geophysical and mathematical symbols used in exploration studies, including definitions and typical units.

Symbol/Term	Definition
g	Gravitational acceleration (typically 9.81 m/s^2)
Δg	Gravity anomaly (mGal); difference between observed and theoretical gravity
B	Magnetic field intensity (nT or T)
P (rho)	Density of a material (kg/m^3)
V_p, V_s	Primary (P-wave) and Secondary (S-wave) seismic velocities (m/s)
σ (sigma)	Electrical conductivity (S/m); inverse of resistivity
R	Electrical resistivity (Ωm)
λ (lambda)	Wavelength (m); used in remote sensing and seismic analysis
Z	Impedance ($Z = \rho V$) or depth, depending on context
φ (phi)	Porosity (fraction or %)

16.17.2 Geostatistics and spatial analysis: Table 16.7

TABLE 16.7 Key terms and concepts in geostatistics and spatial analysis, with definitions relevant to geoscientific data interpretation.

Term	Definition
Variogram	A function describing the degree of spatial dependence between sample points
Kriging	A geostatistical interpolation method that minimizes estimation error
Co-kriging	An extension of kriging that uses secondary variables to improve primary variable estimation
Nugget effect	The variogram value at very small lag distances, indicating measurement error or micro-scale variation
Sill	The plateau value of the variogram indicating total variance
Range	The distance over which spatial correlation exists between samples
Semivariance (γ)	Half the variance of the difference between field values at two locations

16.17.3 AI/ML and data science in geosciences: Table 16.8

TABLE 16.8 Common AI/ML and data science terms in geosciences, including definitions and their relevance to modeling and analysis.

Term	Definition
Gini index	A measure of node impurity used in decision tree algorithms (lower is better)
Entropy	A metric of information disorder used in decision tree splits (ID3 algorithm)
Feature	An input variable or attribute used in machine learning models
Label/target	The output variable the model is trying to predict (e.g., mineral presence, lithology class)
Overfitting	A model that performs well on training data but poorly on new, unseen data
Normalization	Scaling input data to a common range (e.g., 0–1) to improve algorithm performance
Support vector machine (SVM)	A supervised learning model that finds the best boundary between classes
Random forest	An ensemble of decision trees used for classification or regression
SHAP values	A method to explain the contribution of each feature to the model's output
Confusion matrix	A performance metric showing true/false positives and negatives in classification models

16.17.4 Remote sensing and spectral terms: Table 16.9

TABLE 16.9 Remote sensing and spectral terms used in geoscientific applications, with definitions and typical usage in data interpretation.

Term	Definition
Spectral signature	A material's unique reflectance pattern across different wavelengths
NDVI (normalized difference vegetation index)	A ratio of near-infrared to red reflectance used to detect vegetation health
SWIR	Shortwave Infrared; used to identify mineralogy in multispectral imagery
Band ratio	Ratio of values in two spectral bands to enhance specific features (e.g., iron oxides)
Principal component analysis (PCA)	A statistical method to reduce dimensionality and highlight variance in image data

16.18 Cross-disciplinary applications

Geoscientific equations form the bedrock of subsurface exploration and interpretation. While their primary use is in mineral, hydrocarbon, and geothermal resource exploration, these equations are equally vital in various non-resource-focused fields. The adaptability of these methods makes them instrumental in civil engineering, archaeology, and environmental monitoring. This section outlines how key geoscientific equations and methods transcend disciplines, highlighting real-world applications in each area.

16.18.1 Civil engineering applications

Geophysical equations are extensively used in subsurface investigations prior to construction projects (e.g., tunnels, bridges, dams).

16.18.1.1 Gravity and seismic methods

(i) Bouguer anomaly and density contrast equations help in mapping subsurface voids or sinkholes.
(ii) Seismic refraction and reflection equations determine bedrock depth, soil layering, and rippability.

Key Equation Example: Seismic travel time:

$$t = \frac{x}{v}$$

Where: t = travel time, x = offset distance, and v = wave velocity in the medium. These are used in geotechnical site assessments to ensure safe construction practices.

16.18.1.2 Electrical resistivity

Equations for apparent resistivity (ρ_a) help detect water tables, fault zones, or clay layers.

$$\rho_a = k\frac{\Delta V}{I}$$

Where: k = geometric factor, ΔV = potential difference, and I = injected current. Used in foundation design, slope stability, and ground improvement projects.

16.18.2 Archaeological applications

Geoscientific methods allow non-invasive exploration of archaeological sites, preserving heritage while investigating subsurface features.

16.18.2.1 Magnetic methods

Equations calculating magnetic anomalies (e.g., total magnetic intensity) are used to detect fired clay structures, walls, or metal artifacts.

16.18.2.2 Ground-penetrating radar (GPR)

Time-domain reflectometry equations help in identifying buried foundations, tombs, or pits.

$$d = \frac{vt}{2}$$

Where: d = depth, v = velocity of EM wave, and t = two-way travel time. GPR is favored in urban archaeology due to high resolution and shallow penetration depth.

16.18.3 Environmental monitoring

Many environmental studies require accurate subsurface characterization to assess contamination, water content, or land degradation.

16.18.3.1 Electromagnetic methods

Conductivity and permittivity equations help in detecting saline intrusion, contaminant plumes, or soil moisture.

16.18.3.2 Remote sensing and spectral analysis

Equations in spectral indices (e.g., NDVI, NDBI) derived from satellite data are used to monitor:

(a) Vegetation health
(b) Urban expansion
(c) Erosion zones

$$\mathrm{NDVI} = \frac{\mathrm{NIR} - \mathrm{Red}}{\mathrm{NIR} + \mathrm{Red}}$$

TABLE 16.10 Cross-disciplinary applications of geoscientific methods, highlighting their use in civil engineering, archaeology, and environmental monitoring.

Method	Civil engineering	Archaeology	Environmental monitoring
Gravity	Foundation stability, sinkhole detection	Large buried structures	Subsurface void detection
Seismic	Bedrock mapping, site classification	Paleoseismology, buried roads	Subsurface stratigraphy
Electrical Resistivity	Groundwater, clay mapping	Pit/grave detection	Soil contamination, landfill mapping
Electromagnetic	Soil salinity, seepage paths	Burned areas, ditches	Moisture, pollution mapping
Magnetic	Iron content, rebar location	Kilns, walls, fired structures	Mine tailings, buried drums
Remote Sensing	Land-use classification	Vegetation masking of ruins	Deforestation, erosion, land-use change
Geostatistics	Site zoning, risk maps	Artefact density mapping	Spatial pollution modeling
AI/ML	Subsurface classification	Automated feature recognition	Automated hazard detection

16.18.3.3 *Geostatistics*

Kriging and variogram models are used for spatial interpolation of environmental parameters such as pollutant concentration, groundwater levels, or soil quality.

$$Z^*(x_0)=\sum_{i=1}^{n}\lambda_i Z(x_i)$$

Where: $Z^*(x_0)$ = estimated value at location i_0, λ_i = weights, and $Z(x_i)$ = known values.

16.18.4 Summary table of cross-disciplinary applications (Table 16.10)

The versatility of geoscientific equations makes them foundational tools across numerous disciplines. Their integration with AI/ML and modern remote sensing technologies further amplifies their potential to address urban challenges, heritage conservation, and environmental sustainability. Interdisciplinary collaboration will continue to enhance the relevance of these equations in the coming decades.

16.19 Conclusions

This chapter has provided a comprehensive compilation of equations, tools, and conceptual frameworks that underpin modern geoscientific exploration. From foundational methods

References

Abdella, Y., Alfredsen, K., 2010. A GIS Toolset for Automated Processing and Analysis of Radar Precipitation. Comput. Geosci., 36(4), 422–429.

Abdelouhed, F., Algouti, A., Algouti, A., Algouti, A., Mohammed, I., Mourabit, Z., 2021. Contribution of GIS and remote sensing in geological mapping, lineament extractions and hydrothermal alteration minerals mapping using ASTER satellite images: case study of central Jebilets – morocco. Disaster Advances, 14(1), 15–25.

Abdelrahman, E.M., Ammar, A.A., Sharafeldin, S.M., Hassanein, H.I., 1997a. Shape and depth solutions from numerical horizontal self-potential gradients. J. Appl. Geophys., 37, 31–43.

Abdelrahman, E.M., El-Araby, T.M., Ammar, A.A., Hassanein, H.I., 1997b. A least-squares approach to shape determination from residual self-potential anomalies. Pure Appl. Geophys., 150, 121–128.

Abdelrahman, E.M., Ammar, A.A., Hassanein, H.I., Hafez, M.A., 1998a. Derivative analysis of SP anomalies. Geophysics, 63, 890–897.

Abdelrahman, E.M., Hassaneen, A.G., Hafez, M.A., 1998b. Interpretation of self-potential anomalies over two-dimensional plates by gradient analysis. Pure Appl. Geophys., 152, 773–780.

Abdelrahman, E.M., El-Araby, T.M., Essa, K.S., 2003. A least-squares minimisation approach to depth, index parameter, and amplitude coefficient determination from magnetic anomalies due to thin dykes. Explor. Geophys., 34, 241–248.

Abdelrahman, E.M., Abo-Ezz, E.R., Essa, K.S., El-Araby, T.M., Soliman, K.S., 2006. A least-squares variance analysis method for shape and depth estimation from gravity data. J. Geophys. Eng., 3, 143–153.

Abdelrahman, E.M., Essa, K.S., 2013. A new approach to semi-infinite thin slab depth determination from second moving average residual gravity anomalies. Explor. Geophys., 44, 185–191.

Abdelrahman, E.M., Sharafeldin, M.S., 1997. A least-squares approach to depth determination from self-potential anomalies caused by horizontal cylinders and spheres. Geophysics, 62, 44–48.

Abedi, M., Norouzi, G.H., Bahroudi, A., 2012. Support vector machine for multi-classification of mineral prospectivity areas. Comput. Geosci., 46, 272–283.

Abedi, M., Torabi, S.A., Norouzi, G.H., 2013. Application of fuzzy-AHP method to integrate geophysical data in a prospect scale, a case study: seridune copper deposit. Boll. Geofis. Teor. Appl., 54, 145–164.

Abedi, M., Norouzi, G.H., 2012. Integration of various geophysical data with geological and geochemical data to determine additional drilling for copper exploration. J. Appl. Geophys., 83, 35–45.

Abrams, M.J., Ashley, R.P., Rowan, L.C., Goetz, A.F.H., Kahle, A.B., 1977. Mapping of hydrothermal alteration in the Cuprite mining district, Nevada, using aircraft scanner images for the spectral region 0.46-2.36 μm. Geology, 5, 713–718.

Abrams, M.J., Rothery, D.A., Pontual, A., 1988. Mapping in the Oman ophiolite using enhanced Landsat Thematic Mapper images. Tectonophysics, 151, 387–401.

Abrams, M.J., Hook, S.J., 1995. Simulated Aster data for geologic studies. IEEE Transactions of Geosciences and Remote Sensing, 33, 692–699.

Abzalov, M., 2016. Applied Mining Geology. Springer, vol. 12.

Acevedo, W., Masuoka, P., 1997. Time-Series Animation Techniques for Visualizing Urban Growth. Comput. Geosci., 23(4), 423–435.

Adam, E., Perron, G., Grant, A., Matthews, L., Milkereit, B., 2003. 3D seismic imaging for VMS deposit exploration, Matagami, Quebec. In: Hardrock Seismic Exploration. Society of Exploration Geophysicists, 229–246.

Adams, J.A.S., Osmond, Y.K., Rogers, J.J.W., 1959. The geochemistry of uranium and thorium Physics and Chemistry of the Earth, 3, pp. 298–343

Adams, J.A.S., Gasparini, P., 1970. Gamma-ray Spectrometry of Rocks. Amsterdam, Elsevier, 295p.

Adams, J.A.S., Lowder, W.M., 1964. The Natural Radiation Environment. Chicago, Il., University of Chicago Press, 1069p.

Adams, J.B., Smith, M.O., Johnson, P.E., 1986. Spectral mixture modeling: a new analysis of rock and soil types at the Viking lander site. J. Geophys. Res., 91, 8098–8112.

Adiat, K.A.N., Nawawi, M.N.M., Abdullah, K., 2012. Assessing the accuracy of GIS-based elementary multi criteria decision analysis as a spatial prediction tool – a case of predicting potential zones of sustainable groundwater resources. J. Hydrol., 440-441, 75–89.

Adimurthy, V., 2015. Concept design and planning of India's first interplanetary mission. Curr. Sci., 109(6), 1050–1054.

Agterberg, F., 2011. A modified weights-of-evidence method for regional mineral resource estimation. Nat. Resour. Res., 20, 95–101.

Agterberg, F.P., 2011. Geospatial Tools for Resource Management. Springer.

Agterberg, F.P., 2011. Quantitative Methods in Mineral Exploration. Springer.

Agterberg, F.P., Cheng, Q., 2002. Conditional Independence test for weights-of-evidence modeling. Nat. Resour. Res., 11(4), 249–255.

Ahmad Tajuddin, H.J., Ibrahim, Noorliza, C., 2004. Detecting subsurface voids using the microgravity method – a case study from Kuala Lipis, Pahang. Geol. Soc. Malaysia Bull., 48, 31–35.

Ahmadi, O., 2015. Application of the seismic reflection method in mineral exploration and crustal imaging: contributions to hard rock seismic imaging. Digital comprehensive summaries of Uppsala dissertations from the Faculty of Science and Technology 1269. Acta Universitatis Upsaliensis Uppsala, 76p. ISSN 1651-6214 978-91-554-9290-8 urn: nbn:se:uu: diva-259396.

Ahmed El-Rabbany, A.E., 2002. Introduction to GPS: The Global Positioning System. Artech House, 176p.

Ahmed, M., et al., 2021. Constructing a geophysical test site for a coastal community's research and education activities. Lead. Edge 40(3), 1938–3789.

Ahnert, A., Borowski, C., 2000. Environmental risk assessment of anthropogenic activity in the deep sea. Journal of Aquatic Ecosystem Stress and Recovery, 7(4): 299–315, doi: 10.1023/A:1009963912171. S2CID 82100930

Aiken, C.L.V., Hastings, D., 1973. Numerical modelling of the induced polarisation. In 11th International Symposium of Computer Applications in the Mining Industry, College of Mines, University of Arizona.

Airo, M., 2015. Geophysical signatures of mineral deposit types- synopsis. Geophysical Survey of Finland. 58 (special paper), 5–70.

Aitcin, P.C., Houle, J., 1986. Design and construction of concrete sea structures. Can. J. Civ. Eng., 13(4):498–499.

Aitken, A.R.A., Occhipinti, S.A., Lindsay, M.D., Trench, A., 2018. A role for data richness mapping in exploration decision making. Ore Geol. Rev., 99, 398–410.

Aitken, F., Foulc, J.N., 2019. From Deep Sea to Laboratory 1: The First Explorations of the Deep Sea by HMS Challenger (1872-1876). John Wiley & Sons

Aizebeokhai, A.P., 2010. 2D and 3D geoelectrical resistivity imaging: theory and field design. Sci. Res. Essays, 5(23), 3592–3605, 4 December, 2010 Available online at http://www.academicjournals.org/SRE ISSN 1992-2248 ©2010 Academic Journals.

Akhavi, M.S., Webster, T.L., Raymond, D.A., 2001. RADARSAT-1 imagery and GIS modeling for mineral exploration in Nova Scotia, Canada. Geocarto Int., 16, 55–61.

Aki, K., Christoffersson, A., Husebye, E.S., 1977. Determination of the three—dimensional seismic structure of the lithosphere. J. Geophys. Res 82, 277–296.

Alarifi, S.S., Abdelkareen, M., Abdalla, F., Abdelsadek, I.S., Galilan, H., Al-Saleh, A.M., Alotaibi, M., 2022. Fusion of multispectral remote sensing data through GIS based overlay method for revealing potential areas of hydrothermal minerals resources. Minerals, 12.

Aldouri, R., Keller, G.R., Gates, A.Q., Rasillo, J., Salayandia, L., Kreinovich, V., Seeley, J.M., Taylor, P., Holloway, S., 2004. GEON: geophysical Data Add the 3rd Dimension in Geospatial Studies. Proceedings of the ESRI International User Conference, paper 1898.

Al-Emran, M., Javed, M., Ahmed, F., 2018. Role of GIS in managing geological exploration data. J. Geol. Syst., 25(3), 151–165.

Aleotti, J., Micconi, G., Caselli, S., Benassi, G., Zambelli, N., Bettelli, M., Calestani, D., Zappettini, A., 2019. Haptic teleoperation of UAV equipped with gamma-ray spectrometer for detection and identification of radioactive materials in industrial plants. In: Tolio, T., Copani, G., Terkaj, W. (Eds.), Factories of the Future: The Italian Flagship Initiative. Springer International Publishing, Cham, 197–214.

Al-Hemyari, E., Collet, O., Tertyshnikov, K., Pevzner, R., 2024. Supervised deep learning for detecting and locating passive seismic events recorded with DAS: a case study. Sensors, 24, 6978.

Ali, A., Pour, A., 2014. Lithological mapping and hydrothermal alteration using Landsat 8 data: a case study in ariab mining district, red sea hills, Sudan. Int. J. Basic Appl. Sci., 3(3).

Ali, I., Greifeneder, F., Stamenkovic, J., Neumann, M., Notarnicola, C., 2015. Review of machine learning approaches for biomass and soil moisture retrievals from remote sensing data. Remote Sens., 7, 16398–16421. 10.3390/rs71215841.

Al-Khayyat, M., Nasr, T., 2010. GPS and GIS integration in petroleum exploration: case studies from the Middle East. Journal of Petroleum Geology, 33(2), 155–168.

Allen, R., Ripa, M., Jansson, N., 2008. Palaeoproterozoic volcanic- and limestone hosted Zn-Pb-Ag-(Cu-Au)

massive sulphide deposits and Fe oxide deposits in Bergslagen, Sweden. IGCP Project 502.

Allis, R.G., 1990. Geophysical anomalies over epithermal systems. J. Geochem. Explor., 36, 339–374.

Almakki Mohamed, M.T., Al-Naimi, L.S., Mgbeojedo, T.I., Agoha, C.C., 2021. Geological mapping and mineral prospectivity using remote sensing and GIS in parts of Hamissana, Northeast Sudan. J. Petrol. Explor. Prod., 11, 1123–1138.

Almasi, A., Yousefi, M., Carranza, E.J.M., 2017. Prospectivity analysis of orogenic gold deposits in Saqez-Sardasht Goldfield, Zagros Orogen, Iran. Ore Geol. Rev., 91, 1066–1080.

Alonso, E., et al., 2012. Evaluating rare earth element availability: a case with revolutionary demand from clean technologies. Environ. Sci. Technol. Lett., 46, 3406–3414.

Alotaibi, E., Nassif, N., 2024. Artificial intelligence in environmental monitoring: in-depth analysis. Discov. Artif. Intell., 4, 84.

Al-Sayid, K.Z., 2019. Geological and Geochemical Exploration Techniques. Egyptian Geological Survey, Department of Exploration, Egyptian Mineral Resources Authority.

Alvarado Molina, M., 2018. Design and development of a methodology to monitor PM10 dust particles produced by industrial activities using UAV's. Master Thesis, the University of Queensland, Brisbane, Australia.

American, A., 2021. Optimizing Exploration Costs with the Value of Knowledge Approach. Resource Evaluation Journal, 22(2), 112–125.

American Bureau of Shipping, 2020. Guide for subsea mining. https://ww2.eagle.org/content/dam/eagle/rules/guides/current/offshore/318_guideforsubseamining-guide-oct20.pdf.

Ammar, M., Vasquez, A., 2023. Application of DCIP and resistivity models in orogenic gold exploration. Journal of Geophysical Research: Solid Earth, 128(9), 2234–2251.

An, Z.G., Di, Q.Y., Wu, F.Q., Wang, G.J., Wang, R., 2012. Geophysical exploration over a long deep shield tunnel, for the west route diversion project from the Yangtze River to the Yellow River. Bulletin of Engineering Geology and the Environment, 71, 195–200.

An, Z.G., Di, Q.Y., 2010. Application of the CSAMT method for exploring deep coal mines in Fujian province, southeastern China. Journal of Environmental and Engineering Geophysics, 15, 243–249.

Anand, S.P., Rajaram, M., 2006. Aeromagnetic data analysis for the identification of concealed uranium deposits: a case history from Singhbhum uranium province, India. Earth Planets Space, 58, 1099–1103.

Anbazhagan, S., Arivazhagan, S., 2010. Reflectance spectra of analog anorthosites: implications for lunar highland mapping. Planetary and Space Science, no. 58, 752–760.

Andersen, C., Theissen-Krah, S., Hannington, M., Rupke, L., Petersen, S., 2017. Faulting and off-axis submarine massive sulfide accumulation at slow spreading mid-ocean ridges: a numerical modeling perspectivce. Geochem. Geophys. Geosyst., 18(6), 2305–2320.

Anderson, C., Dyke, A., Noble, M., Raetz, M., 2006. The BHEI Falcon gravity survey and Goldfinger BHT Project - exploration geophysics, geochemistry and geology integrated in 3D: in Korsch, R., Barnes, R., eds., Broken Hill Exploration Initiative: Abstracts for the September 2006 Conference, Geoscience Australia Record 2006/21, 1–3.

Anderson, D.A., Smith, R.J., Martinez, P., 2013. Geochemical surveys for copper exploration in the Andes: a case study from Peru., Geochemistry: Exploration, Environment, Analysis, 13(2), 145–160.

Anderson, L.A., 1984. Self-potential investigations in the Puhimau thermal area, Kilauea Volcano, Hawaii. 54th Annual Internat. Mtg. Soc. Expl. Geophys., Expanded Abstracts, Soc. Expl. Geophys. Tulsa, Session: EM.3.5.

Anderson, R., 2010. Geologic mapping and characterization of Gale Crater and implications for its potential as a Mars Science Laboratory landing site. Mars, J5, 76–128.

Anderson, S.E., McNeill, J.D., 2018. Advances in Drone-Based Geophysics for Mineral Exploration. Geophys. J. Int., 212(3), 1281–1295.

Andersson, J.B.H., Bauer, T.E., Martinsson, O., 2021. Structural evolution of the Central Kiruna area, Northern Norrbotten, Sweden: implications on the geologic setting generating iron oxide-apatite and epigenetic iron and copper sulfides. Econ. Geol., 116(8), 1981–2009.

Andrada de Palomera, P., Gonzalez-Alvarez, L., Sadeghi, J.A., 2014. GIS applications in exploration targeting. Earth Sci. Rev., 137, 27–44.

Andrada de Palomera, P., Frank, J.A., Emmanuel, J.M., 2014. Prospectivity for epithermal gold-silver deposits in the Deseado Massif, Argentina. Ore Geol. Rev, 69p, doi: 10.1016/j.oregeorev.2014.12.007.

Andreeva, E.D., Matsueda, H., Okrugin, V.M., Takahashi, R., Ono, S., 2013. Au-Ag-Te mineralization of the low-sulfidation epithermal Aginskoe deposit, Central Kamchatka, Russia. Resour. Geol., 63, 337–349.

Anjirwala, H., Bhatia, M., 2016. Shale gas scenario in India and comparison with USA. International Journal of Science and Research (IJSR) 2319-7064, 5(8).

Anjos, R.M., Veiga, R., Carvalho, C., Macario, K.D., Gomes, P.R.S., 2007. Geological provenance of Quaternary deposits from the southeastern Brazilian coast. Nucl. Phys., A 787, 642–647.

Arafa-Hamed, T., 2014. Comprehensive clues provided by popular free remote sensing imagery to interpretation of geophysical studies explained by example cases of magnetic surveys. 2090-9977ª2014 Production and hosting by Elsevier B.V. on behalf of National Research Institute of Astronomy and Geophysics. NRIAG Journal of Astronomy and Geophysics, p.10.

Araujo, J.C.T., Nascimento, C.W.A., 2005. Zinc fractionation and availability by different extractants in sewage sludge-incubated soils. Rev. Bras. Cienc. Solo, 29(6), 977–985.

Araya, A., Kanazawa, T., Shinohara, M., Yamada, T., Fujimoto, H., Iizasa, K., Ishihara, T., 2012. Gravity gradiometer implemented in AUV for detection of seafloor massive sulfides 2012 Oceans. VA, USA: IEEE, 1–4.

Archie, G.E., 1942. The electrical resistivity log as an aid in determining some reservoir characteristics. Petroleum Transactions of AIME, 146, 54–62.

Ardron, J.A., Rayfuse, R., Gjerde, K., Warner, R., 2014. The sustainable use and conservation of biodiversity in ABNJ: what can be achieved using existing international agreements"? Mar. Policy, 49, 98–108. doi:10.1016/j.marpol.2014.02.011. ISSN 0308-597X.

Arnous, M.O., ElMowafy, A.A., Azzaz, S.A., Omar, A.E., Hafeez, W.M.A., 2021. Exploration radioactive mineralization using mappable data integration approach: example from Wadi Dahab area, Southeastern Sinai, Egypt. Arab. J. Geosci., 14, 599.

Art, G., Kaiser, S., Smith, T., 1994. Center for Earth Observation Users Guide. In: 8th Revision August 2004 by Larry Bonneau, Yale University.

Arulbalaji, P., Padmalal, D., Sreelash, K., 2019. GIS and AHP Techniques Based Delineation of Groundwater Potential Zones: a case study from Southern Western Ghats, India. Scientific Reports, 9:2082 10.1038/s41598-019-38567-x.

Asha, P., Natrayan, L., Geetha, B.T., Beulah, J.R., Sumathy, R., Varalakshmi, G., Neelakandan, S., 2022. IoT enabled environmental toxicology for air pollution monitoring using AI techniques. Environ. Res.

Ashutosh, S., Roy, P.S., 2021. Three Decades of Nationwide Forest Cover Mapping Using Indian Remote Sensing Satellite Data: a Success Story of Monitoring Forests for Conservation in India. J. Indian Soc. Remote Sens., 49, 61–70.

Atchuta Rao, D., Ram Babu, H.V., Silvakumar Sinha, G.D.J., 1982. A Fourier transform method for the interpretation of self-potential anomalies due to the two-dimensional inclined sheets of finite depth extent. Pure Appl. Geophys., 120, 365–374.

Atchuta Rao, D., Ram Babu, H.V., 1983. Quantitative interpretation of self-potential anomalies due to two-dimensional sheet-like bodies. Geophysics, 48, 1659–1664.

Audisio, C., Nigrelli, G., Lollino, G., 2009. A GIS Tool for Historical Instability Processes Data Entry: an Approach to Hazard Management in two Italian Alpine River Basins. Comput. Geosci., 35(8), 1735–1747.

Auken, E., Christiansen, A.V., Jacobsen, B.H., et al. 2005. Piecewise 1D laterally constrained inversion of resistivity data. Geophys. Prospect. 53(4), 497–506.

Auken, E., Christiansen, A.V., Kirkegaard, C., et al. 2015. An overview of a highly versatile forward and stable inverse algorithm for airborne, ground-based and borehole electromagnetic and electric data. Expl Geoph. 46(3), 223–235.

Babu, S.R., Nair, S.K., 2021. Coastal zone management using geospatial technologies in Kerala. Environ. Monit. Assess., 193(5), 1–16.

Bagari, U.M., Ugarkar, A.G., Chandan Kumar, B., Puranik, S.C., 2013. Application of Remote Sensing and GIS in the Hydrogeological Study of Lower Bennihalla Basin, Dharwar and Gadag Districts of Karnataka, India. International Journal of Earth Sciences and Engineering, 6(2), 1809–1814.

Bailin, Y., Xingli, W., 1991. Spectral reflectance features of rocks and ores and their Applications. Chin. J. Geochem., 10(2), 188–195.

Balaram, V., Anjaiah, K.V., Reddy, M.R.P., 1995. Comparative study on the trace and rare earth element analysis of an Indian polymetallic nodule reference sample by inductively coupled plasma atomic emission spectrometry and inductively coupled plasma mass spectrometry. Analyst, 120:1401.

Balaram, V., 1999. Determination of precious metal concentrations in polymetallic nodule reference sample from the Indian Ocean by ICPMS. Mar Georesour Geotec., 17:17–26.

Balaram, V., Satyanarayanan, M., Murthy, P.K., Mohapatra, C., Prasad, K.L., 2013. Chemical characterization of cobalt crust from afanasy-nikitin seamount in the eastern Indian Ocean by inductively coupled plasma time-of-flight mass spectrometry. Mapan, 28:63–77.

Balaram, V., 2022. Rare earth element deposits: sources, and exploration strategies. J. Geol. Soc. India, 98, 1210–1216.

Balaram, V., 2023. Deep sea mineral deposits as a future source of critical metals, and environmental issues – a brief review. Minerals and Mineral Materials, 2:5, DOI: 10.20517/mmm.2022.12.

Balaram, V., 2023. Potential future alternative resources for rare earth elements: opportunities and challenges. Minerals, 13:425.

Balaram, V., Banakar, V.K., Subramanyam, K.S.V., et al., 2023. Yttrium and rare earth element contents in seamount cobalt crusts in the Indian Ocean. Available from: https://www.jstor.org/stable/24089153 [Last accessed on 28 Apr 2023].

Balaram, V., Mathur, R., Banakar, V.K., et al., 2023. Determination of the platinum group elements and gold in manganese nodule reference samples by nickel sulfide fire assay and Te-coprecipitation with ICP-MS. Available from: https://pubs.er.usgs.gov/publication/70029015 [Last accessed on 28 Apr 2023].

Balaram, V., Sawant, S.S., 2022. Indicator minerals, pathfinder elements, and portable analytical instruments in mineral exploration studies. Minerals, 12, 394.

Baldo, M., Buldrini, F., Chiarucci, A., Rocchini, D., Zannini, P., Ayushi, K., Ayyappan, N., 2023. Remote sensing analysis on primary productivity and forest cover dynamics: a Western Ghats India case study. Ecol. Inform., 73, 101922.

Ballhorn, U., 2012. Airborne and spaceborne LiDAR data as a measurement tool for peatland topography, peat fire burn depth, and forest above ground biomass in Central Kalimantan, Indonesia. Munchen, Ph.D. Thesis.

Ballu, V.S., Hildebrand, J.A., Webb, S.C., 1998. Seafoor gravity evidence for hydrothermal alteration of the sediments in middle valley. Juan De Fuca Ridge Mar. Geol., 150:99–111.

Banakar, V.K., Hein, J.R., 2000. Growth response of a deep water ferromanganese crust to evolution of the Neogene Indian Ocean. Mar. Geol., 162:529–540.

Banerjee, B., 1971. Quantitative interpretation of self-potential anomalies of some specific geometric bodies. Pure Appl. Geophys., 90, 138–152.

Banos, A., 1966. Dipole Radiation in the Presence of a Conducting Half-space. Pergamon Press, New York.

Baral, H., 2004. Applications of GIS in community-based forest management in Australia (and Nepal). Masters Research thesis, School of Forest and Ecosystem Science, University of Melbourne.

Barmpounakis, E.N., Golias, J.C., 2016. Unmanned aerial aircraft systems for transportation engineering: current practice and future challenges. Int. J. Transp. Sci. Technol., 5, 111–122.

Barrow, B., Nelson, H.H., 1998. Collection and analysis of multi-sensor ordnance signatures with MTADS. J. Environ. Eng. Geophys., 3, 71–79.

Barton, M.D., Laubach, S.E., 2019. Integration of geophysical data for sustainable mineral exploration. Minerals, 9(8), 473.

Bath, A.R., 1989. Deep sea mining technology: recent developments and future projects. Offshore Technol. Conf. 10.4043/5998-MS.

Bayat, B., 2007. Indirect exploration for ore deposits in weathered terrains with airborne gravity gradiometry. PhD. thesis, Curtin University.

Becker, J.J., et al., 2009. Global bathymetry and elevation data at 30 arc seconds resolution: SRTM30_PLUS. Marine Geodesy, 32, 355–371.

Bedini, E., 2011. Mineral mapping in the Kap Simpson complex, central East Greenland, using HyMap and ASTER remote sensing data. Adv. Space Res., 47, 60–73.

Bedini, E., 2019. Application of WorldView-3 imagery and ASTER TIR data to map alteration minerals associated with the Rodalquilar gold deposits, southeast Spain. Adv. Space Res., 63, 3346–3357.

Begemann, H.K.S.P., 1953. Improved method of determining resistance to adhesion by sounding through a loose sleeve placed behind the cone. In: The 3rd International conference on soil mechanics and foundation engineering. https://www.issmge.org/publications/publication.

Behera, M.D., 1999. Remote sensing and environment. Employment News (Government of India) XXIV(13), 1–2.

Behera, M.D., Tripathi, P., Mishra, B., Kumar, S., Chitale, V., Behera, S.K., 2015. Above-ground biomass and carbon estimates of Shorea robusta and Tectona grandis forests using QuadPOL ALOS PALSAR data. Adv. Space Res.

Behnia, P., 2007. Application of radial basis functional link networks to exploration for Proterozoic mineral deposits in Central Iran. Nat. Resour. Res., 16, 147–155.

Bellefleur, G., Schetsellar, E., White, D., Miah, K., Dueck, P., 2015. 3D seismic imaging of the Lalor volcanogenic massive sulphide deposit, Manitoba, Canada: 3D seismic imaging of Lalor deposit. Geophys. Prospect., 63(4), DOI:10.1111/1365-2478.12236.

Bellefleur, G., Malinowski, M., Urosevic, M., 2019. Editorial for special issue seismic methods in mineral exploration. Minerals, 9, 630; doi:10.3390/min9100630.

Beltenev, V., Ivanov, V., Rozhdestvenskaya, I., Cherkashov, G., Stepanova, T., Shilov, V., Pertsev, A., Davydov, M., Egorov, I., Melekestseva, I., Narkevsky, E., Ignatov, V., 2007. A new hydrothermal field at 13°30′N on the Mid-Atlantic Ridge. Inter Ridge News, 16, 9–10.

Beltenev, V., Ivanov, V., Rozhdestvenskaya, I., Cherkashov, G., Stepanova, T., Shilov, V., Pertsev, A., 2008. New data about hydrothermal fields on the Mid-Atlantic Ridge between 11–14°N: 32nd Cruise of R/V Professor Logatchev. Geochem. Geophys. Geosyst., 10, 1029.

Beltenev, V., Ivanov, V., Rozhdestvenskaya, I., Cherkashov, G., Stepanova, T., Shilov, V., Davydov, M., Laiba, A., Kaylio, V., Narkevsky, E., Pertsev, A., Dobretzova, I., Gustaytis, A., Popova, Y.E., Amplieva, Y.E., Evrard, C., Moskalev, L., Gebruk, A., 2009. New data about hydrothermal fields on the Mid-Atlantic Ridge between 11°-14°N: 32nd cruise of R/V Professor Logatchev. InterRidge News, 18, 13–17.

Benoit, M., Van der Meer, S.R.P., Le Moigne, C., 2013. Nuclear magnetic resonance in geological exploration: a review. Geophys. J. Int., 195(1), 124–136.

Benson, A.K., Payne, K.L., Stubben, M.A., 1997. Mapping groundwater contamination using DC resistivity and VLF geophysical methods - A case study. Geophysics, 62(1), 80–86.

Beraton, K.K., Peer, B., Dunbar, N.W., Blom, R., 1997. A remote sensing approach to alteration mapping: AVIRIS data and extension related potassium metasomatism, Socorro, New Mexico. Int. J. Remote Sens., 18, 3595–3609.

Berdichevsky, M.N., Vanyan, L.L., Fainberg, E.B., 1969. Magnetovariational sounding using spatial field derivatives. Geomag. Aer., 9, 369–371.

Best, M.E., 2015. Mineral resources. Treatise on Geophysics (second edition), 11, 525–556.

Bhattacharya, B.B., Shalivahan, J.A., Bera, A., 2007. Three-dimensional probability tomography of self-potential anomalies of graphite and sulphide mineralization in Orissa and Rajasthan, India. Near-Surface Geophys., 5, 223–230.

Bhattacharya, B.B., Roy, N., 1981. A note on the use of a nomogram for self-potential anomalies. Geophys. Prospect., 29, 102–107.

Billiton, B.H.P., 2019. Using Machine Learning in Mineral Exploration. Geology and Technology, 45(1), 33–48.

Biswas, A., 2015. Interpretation of residual gravity anomaly caused by a simple shaped body using very fast simulated annealing global optimization. Geoscience Frontiers, 6, 875–893.

Bivand, R., Pebesma, E., Gomez-Rubio, V., 2013. Applied Spatial Data Analysis with R. Springer.

Blakely, R.J., 1995. Potential Theory in Gravity and Magnetic Applications. Cambridge University Press, Cambridge, doi: 10.1017/CBO9780511549816.

Blom, R.G., Daily, M., 1982. Radar image processing for rock type discrimination. IEEE Transactions of Geoscience and Remote Sensing, 20, 343–351.

Boddice, D., Metje, N., Tuckwell, G., 2022. Microgravity surveying before, during and after distant large earthquakes. J. Appl. Geophys., 197, 104542.

Boggs, D.B., Maddever, R.A.M., Lee, J.B., Turner, R.J., Downey, M.A., Dransfield, M.H., 2007. First test survey results from the Falcon helicopter-borne airborne gravity gradiometer system. Preview, 126, 26–28.

Bohlen, T., Muller, C.F., Milkereit, B., 2003. Elastic wave scattering from massive sulphide orebodies: on the role of composition and shape. In: Eaton, D., Milkereit, B., Salisbury, M. Hardrock Seismic Exploration, (SEG Developments in Geophysics Series), 10, 70–89.

Boleneus, D.E., Raines, G.L., Causey, J.D., Bookstrom, A.A., Frost, T.P., Hyndman, P.C., 2001. Assessment method for epithermal gold deposits in northeast Washington State using weights-of-evidence GIS modeling. US Geological Survey, Open-File Report 01-501, 52.

Bonham-Carter, G.F., Agterberg, F.P., Wright, D.F., 1989. Weights of evidence modeling: a new approach to mapping mineral potential. In: Agterberg, F.P., Bonham-Carter, G.F. (eds.), Statistical Applications in the Earth Sciences, Geological Survey of Canada, Paper 89-9, 171–183.

Bonham-Carter, G.F., 1994. Geographical Information Systems for Mineral Exploration. GIS Research.

Bonham-Carter, G.F., 1994. Geographic Information Systems for Geoscientists: Modeling with GIS, 398p, Pergamon, UK.

Booysen, R., Jackisch, R., Lorenz, S., Zimmermann, R., Kirsch, M., Paul, A.M.N., Gloaguen, R., 2020. Detection of REEs with lightweight UAV-based hyperspectral imaging. Scientific Reports, 10:17450, 10.1038/s41598-020-74422-0, nature research.

Booysen, R., Lorenz, S., Jakish, R., Globaguen, R., Madriz, Y., 2021. How drones contribute to mineral exploration. Int. Geosci. Remote Sens. Symp. 10.1109/IGRASS47720.2021.9554041.

Bortolon, L., Gianello, C., 2010. Copper and zinc extracted by multielement solution for Southern Brazilian soils. Ciencia Rural, 40(3), 670–673.

Bosch, M., Mukerji, T., Gonzalez, E.F., 2010. Seismic inversion for reservoir properties combining statistical rock physics and geostatistics: a review. Geophysics, 75(5), 75A165-75A176.

Boschen, R.E., Rowden, A.A., Clark, M.R., Gardner, J.P.A., 2013. Mining of deep sea seafloor massive sulphides: a review of the deposits, their benthic communities, impacts from mining, regulatory frameworks and management strategies. Ocean Coast Manage, 84, 54–67.

Boucher, A., 2014. Geological Surveys and Sustainability in Natural Resource Exploration. Earth Science Journal, 32(4), 89–101.

Bouwmeester, D., Peuchen, J., Van, D., Sarata, B., Willemse, C., Van Baars, S., Peelen, R., 2009. Offshore technology conference – prediction of breakout forces for deep water seafloor objects. In Proceedings of the Offshore Technology Conference Offshore Technology Conference, Houston, TX, USA, 4 May 2009.

Brekovskikh, L.M., 1980. Waves in Layered Media, 2nd edn. Academic Press, New York.

Brewer, A.K. 1937. Radioactivity of potassium and geologic time. Science, 86(2226), 198–199.

Brewitt-Taylor, C.R., 1975. Self-potential prospecting in the deep oceans. Geology, 3(9), 541–542.

Briaud, J.L., 2013. In situ tests. Wiley. 10.1002/9781118686195.ch7.

Brodic, B., et al., 2015. Multicomponent broadband digital-based seismic land streamer for near-surface applications. J. Appl. Geophys., 123, 227–241.

Brown, L., Martin, D., 2022. Use of deep learning for subsurface mapping in mineral exploration. Geophys. Prospect., 70(5), 915–930.

Brownell, G.L., Maletskos, C.J., 1950. Half-life of Fe55 and Co6°. Phys. Rev., 80(6), 1102.

Brownell, G.M., 1950. Radiation surveys with a scintillation counter. Econ. Geology, 45, 167–174.

Bruton, A.M., 2000. Improving the accuracy and resolution of SINS/DGPS airborne gravimetry. PhD thesis, University of Calgary.

Bu, W., Shi, X., Peng, J., Qi, L., 2003. Geochemical characteristics of seamount ferromanganese nodules from mid-Pacific Ocean. Chin Sci Bull., 48:98–105.

Buckingham, W.F., Sober, S.E., 1983. Mineralogical characterization of rock surfaces by hydrothermal alteration and weathering: application to remote sensing. Econ. Geol., 78, 664–674.

Budkewitsch, P., Newton, G., Hynes, A.J., 1994. Characterization and Extraction of Linear Features from Digital Images. Canadian Journal of Remote Sensing, 20, 268–279.

Burns, R.E., Erickson, B.H., Lavelle, J.W., Ozturgut, E., 1978. Observation and Measurements during the Monitoring of Deep Ocean Manganese Nodule Mining Tests in the North Pacific, University of California Libraries, 140p.

Buselli, G., O'Neill, B., 1977. SIROTEM: a new portable instrument for multichannel transient electromagnetic measurements. Bulletin of the Australian Society of Exploration Geophysics, 8(3), 82–87.

Butler, D.K., 1991. Tutorial-Engineering and environmental applications of microgravimetry. Proceedings of the Symposium on the Application of Geophysics to Engineering and Environmental Problems. The Soc. Engg. Miner. Explor. Geophys, 139–246.

Butler, D.K., Llopis, J.L., 1990. Assessment of anomalous seepage conditions. In: Ward, S.H. (Ed.), Geotechnical and Environmental Geophysics, 2: Environmental and Groundwater. Society of Exploration Geophysicists, Tulsa, OK.

Butt, C.R.M., Hough, R.M., 2009. Why gold is valuable? Elements, 5, 277–280.

Cadling, L., Odenstad, S., 1950. Vane borer. An apparatus for determining the shear strength of clay soils directly in the ground (https://www.sgi.se/en/research-learning/publikationer/sgi-proceedings/vane-borer).

Cagniard, L., 1953. Basic theory of the magneto-telluric method of geophysical prospecting. Geophysics, 18(3), 605–635

Cai, G., Lum, K.Y., Chen, B.M., Lee, T.H., 2010. A brief overview on miniature fixed-wing unmanned aerial vehicles. In proceedings of the IEEE ICCA, Xiamen, China, 9-11 June 2010; IEEE: Piscataway, NJ, USA, 285–290.

Cai, H.Z., Xiong, B., Zhdanov, M., 2015. Three-dimensional marine controlled-source electromagnetic modelling in anisotropic medium using finite element method (in Chinese). Chin. J. Geophys., 58, 2839–2850.

Cairns, G.W., Evans, R.L., Edwards, R.N., 1996. A time domain electromagnetic survey of the TAG hydrothermal mound. Geophys. Res. Lett., 23, 3455–3458.

Campbell, F.M., Ghisetti, F., Kaiser, A.E., Green, A.G., Horstmeyer, H., Gorman, A.R., 2010. Structure and evolution of the seismically active Ostler Fault Zone (New Zealand) based on interpretations of multiple high resolution seismic reflection profiles. Tectonophysics, 495, 195–212. doi: 10.1016/j.tecto.2010.09.

Candansayar, M.E., Tezkan, B., 2006. A comparison of different radiomagnetotelluric data inversion methods for buried waste sites. J. Appl. Geophys., 58, 218–231.

Cao, H., Han, L., Liu, M., Li, L., 2025. Spatial differentiation of carbon emissions from energy consumption based on machine learning algorithm: a case study during 2015-2020 in Shaanxi, China. Journal of Environmental Sciences, 149, 358–373.

Cao, X., Zhang, L., 2019. Coal Resources and Distribution Patterns in Global Sedimentary Basins. J. Energy Resour., 47(3), 129–146.

Carranza, E.J.M., 2004. Weights of evidence modeling of mineral potential: a case study using small number of prospects, Abra, Philippines. Nat. Resour. Res., 13(3), 173–187.

Carranza, E.J.M., 2004. GIS-based Mineral Prospectivity Mapping. Springer.

Carranza, E.J.M., Woldai, T., Chikambwe, E.M., 2005. Application of data-driven evidential belief functions to prospectivity mapping for aquamarine-bearing pegmatites, Lundazi District, Zambia. Nat. Resour. Res., 14, 47–63.

Carranza, E.J.M., 2008. Geochemical anomaly and mineral prospectivity mapping in GIS. In: Handbook of Exploration and Environmental Geochemistry, vol. 11, Elsevier, Amsterdam.

Carranza, E.J.M., 2009a. Controls on mineral deposit occurrence inferred from analysis of their spatial pattern and spatial association with geological features. Ore Geol. Rev., 35, 383–400.

Carranza, E.J.M., 2009b. Improved wildcat modeling of mineral prospectivity. Resour. Geol., 60, 129–149.

Carranza, E.J.M., 2017. Natural resources research publications on geochemical anomaly and mineral potential mapping, and introduction to the special issue of papers in these fields. Nat. Resour. Res., 26(4), 379–410.

Carranza, E.J.M., Hale, M., 2001a. Logistic regression for geologically constrained mapping of gold potential, Baguio district, Philippines. Explor. Min. Geol., 10, 165–175.

Carranza, E.J.M., Hale, M., 2001b. Geologically-constrained fuzzy mapping of gold mineralization potential, Baguio district, Philippines. Nat. Resour. Res., 10, 125–136.

Carranza, E.J.M., Hale, M., 2002. Spatial association of mineral occurrences and curvilinear geological features. Math. Geol., 34, 203–221.

Carranza, E.J.M., Hale, M., 2003. Evidential belief functions for data-driven geologically constrained mapping of gold potential, Baguio district, Philippines. Ore Geol. Rev, 22, 117–132.

Carranza, E.J.M., Laborte, A.G., 2015a. Random forest predictive modeling of mineral prospectivity with small number of prospects and data with missing values in Abra (Philippines). Comput. Geosci., 74, 60–70.

Carranza, E.J.M., Laborte, A.G., 2015b. Data-driven predictive mapping of gold prospectivity, Baguio district, Philippines: application of Random Forests algorithm. Ore Geol. Rev, 71, 777–787.

Carranza, E.J.M., Laborte, A.G., 2016. Data-driven predictive modeling of mineral prospectivity using random forests: a case study in Catanduanes Island (Philippines). Nat. Resour. Res., 25, 35–50.

Carranza, E.J.M., Sadeghi, M., 2010. Predictive mapping of prospectivity and quantitative estimation of undiscovered VMS deposits in Skellefte district (Sweden). Ore Geol. Rev, 38, 219–241.

Carroll, K.A., 2011. Gravity gradiometry for lunar surface exploration ASTRO2010. In: 15th CASI Canadian Astronautics Conf. (Toronto, 4-6 May 2010).

Carvalho, C.S., Ribeirinho, V.S., Andrade, C.A., Grutzmacher, P., Pires, A.M.M., 2015. Chemical composition of sewage sludge organic matter. Revista Brasileira de Ciencias Agrarias, 10(3), 413–419.

Carvalho, F.P., 2017. Mining industry and sustainable development: time for change. Food Energy Secur., 6, 61–77.

Cave, B.W., Lilly, R., Glorie, S., Gillespie, J., 2018. Geology, apatite geochronology, and geochemistry of the Ernest Heny Inter-lens: implications for a re-examined deposit model. Minerals, 8, 405.

Cawthorn, R.G., 2017. The Witwatersrand goldfields, South Africa: geochemical mapping and exploration. J. Geochem. Explor., 177, 50–60.

Census of India, 2001. Primary census abstract for Uttaranchal, Bihar and Jharkhand, vol 4. Office of the Registrar General, India, Government of India, New Delhi.

Cetin, H., Warner, T.A., Levandowski, D.W., 1993. Data classification, visualization, and enhancement using n-dimensional probability density functions (nPDF): AVIRIS, TIMS, TM and geophysical applications. Photogramm. Eng. Remote Sens., 59, 1755–1764.

CGLS, 2022. ESA Copernicus Global Land Service. VITO Products Hub. https://land.copernicus.vgt.vito.be[2022-07-04T00:00:00+03:00].

Chamberlain, A., White, D., 2020. Gold mining and its distribution: an overview of major global reserves. International Journal of Mining Economics, 39(3), 154–171.

Chandra, P.C., 2016. Groundwater Geophysics in Hard Rock. CRC Press, London.

Chang, K.T., 2010. Introduction to Geographic Information Systems. 5th ed. McGraw-Hill higher education: New York City, NY, USA, 448p.

Chapi, K., Singh, V.P., Shirzadi, A., Shahabi, H., Bui, D.T., Pham, B.T., Khosravi, K., 2017. A novel hybrid artificial intelligence approach for flood susceptibility assessment. Environ Model Softw. 95, 229–245.

Charette, M.A., Smith, W.H.F., 2010. The volume of earth's ocean. Oceanography, 23, 112–114.

Charlu, T.K., Kalluraya, V.K.K., 2023. Ferromanganese encrustation from Lakshadweep area, Arabia Sea. Last accessed on 28 April 2023. Available from https://indianjournalofmarketing.com/index.php/jgsi/article/view/68307.

Chatterjee, R.S., Prabakaran, B., Jha, V.K., 2003. Fusion of surface relief data with high spectral and spatial resolution satellite remote sensor data for deciphering geological information in a mature topographic terrain. Int. J. Remote Sens., 24, 4761–4775.

Chattoraj, S.L., Prasad, G., Sharma, R.U., Ray, P.K.C., Van der Meer, F.D., Guha, A., Pour, A.B., 2020. Integration of remote sensing, gravity and geochemical data for exploration of Cu mineralization in Alwar basin, Rajasthan, India. International Journal of Applied Earth Observation and Geoinformation, 91, 102162.

Chauhan, P., Kaur, P., Ajai, K.K., 2012. Lunar geosciences using Chandrayaan-1: indian perspective. Proc. Indian Natl. Sci. Acad., 78(3), 239–247.

Chave, A., 2017. Standardization in Geophysical Exploration: a Global Perspective. International Journal of Exploration Geophysics, 45(2), 135–148.

Chen, C., 2004. Searching for intellectual turning points: progressive knowledge domain visualization. Proc. Natl. Acad. Sci., USA 101.

Chen, C., 2006. Cite Space II: detecting and visualizing emerging trends and transient patterns in scientific literature. J. Am. Soc. Inf. Sci. Technol., 57(3):359–377. https://doi.org/10.1002/asi.20317.

Chen, D.F., Su, Z., Cathles, L.M., 2006. Types of Gas Hydrates in Marine Environments and Their Thermodynamic Characteristics. Terr. Atmos. Ocean. Sci., 17(4), 723–737.

Chen, F., Wang, Z., Wu, X., 2021. Predictive models in mineral exploration using machine learning techniques. J. Appl. Geophys., 171, 78–90.

Chen, G.B., Wang, H.N., Yao, J.J., Han, Z.Y., 2009. Three-dimensional numerical modeling of marine controlled source electromagnetic responses in a layered anisotropic seabed using integral equation method (in Chinese). Acta Phys. Sin., 58, 3848–3857.

Chen, J., Oldenburg, D.W., 2003. 3-D inversion of magnetic induced polarization data. 3DEMIII Workshop, February 2003, Adelaide.

Chen, Q., Shi, Y.H., Pan, Y., 2007. Downhole technology for submarine CPT and its device development. Ocean Eng., 25, 73–76.

Chen, S., 2006. Application of GIS to the 2nd Round of Iron Ore Exploration in the East Hubei Province. Contrib. Geol. Miner. Resour. Res., 21(3), 214–219.

Chen, T., van Galder, C., Dransfield, M., 2017. Full spectrum Falcon: measuring broadband airborne gravity. In: SEG Technical Program Expanded Abstracts 2017, Society of Exploration Geo Physicists, 1823–1828.

Chen, W.Y., 2015. Mineral potential mapping with a restricted Boltzmann machine. Ore Geol. Rev, 71, 749–760.

Chen, W.Y., Xue, G.Q., Muhammad, Y.K., Gelius, L.J., Zhou, N., Li, H., Zhong, H., 2015. Application of short-offset TEM (SOTEM) technique in mapping water-enriched zones of coal stratum, an example from East China. Pure Appl. Geophys., 172, 1643–1651.

Chen, W.Y., Xue, G.Q., Cui, J.W., Zhong, H.S., 2016. Study on the response and optimal observation area for SOTEM (in Chinese). Chin. J. Geophys., 59, 739–748.

Chen, W.Y., Khan, M.Y., Xue, G.Q., 2017. Response of surface-to-borehole SOTEM method on 2D earth. J. Geophys. Eng., 14, 987–997.

Chen, W.Y., Han, X., Xue, G.Q., 2019. Analysis on the full-component response and detectability of electric source surface-to-borehole TEM method (in Chinese). Chin. J. Geophys., 62, 1969–1980.

Chen, X., 2015. The Application and Research of Magnetic Exploration in Exploration of Iron Mining Area in Wenchuan. Chengdu University of Technology, Chengdu (in Chinese).

Chen, Y., Wu, W., 2017. Mapping mineral prospectivity using an extreme learning machine regression. Ore Geol. Rev., 80, 200–213.

Chen, Z., Mou, L., Meng, X., 2016. The horizontal boundary and top depth estimates of buried source using gravity data and their applications. J. Appl. Geophys., 124, 62–72.

Cheng, H., Zhenga, Y., Wu, S., Lin, Y., Gao, F., Lin, D., Wei, J., Wang, S., Shu, D., Wei, S., Chen, L., GIS-based mineral prospectivity mapping using machine learning methods: a case study from Zhuonuo ore district, Tibet. Ore Geol. Rev, 161(2023), 105627.

Cheng, H.J., Chen, C., Zhan, X.Z., Chang, J.Y., Ding, Y.L., Kuwanixibieke, M., Yang, R.H., Jia, N.E., Fu, H.Z., 2017. New progress in the prediction theory and prospecting method for concealed deposits. Geol. Explor., 53, 456–463, (In Chinese with English abstract).

Cheraghi, S., Malehmir, A., Bellefleur, G., 2011. Crustal-scale reflection seismic investigations in the Bathurst Mining Camp, New Brunswick, Canada. Tectonophysics, 506, 55–72. doi: 10.1016/j.tecto.2011.04.011.

Cherkashov, G., Poroshina, I., Stepanova, T., Ivanov, V., Beltenev, V.E., Lazareva, L., Rozhdestvenskaya, I., Samdvarov, M., Shilov, V., Glasby, G.P., Fouquet, Y., Kuznetsov, V., 2010. Seafloor massive sulfides from the northern equatorial mid-Atlantic ridge: new discoveries and perspectives. Mar. Georesour. Geotech., 28(3), 222–239.

Cherkashov, G.A., Ivanov, V.N., Beltenev, V.I., Lazareva, L.I., Rozhdestvenskaya, I.I., Samovarov, M.L., Poroshina, I.M., Sergeev, M.B., Stepanova, T.V., Dobretsova, I.G., Kuznetsov, V.Y., 2013. Massive sulfide ores of the northern equatorial Mid-Atlantic Ridge. Oceanology, 53(5), 607–619.

Chernicoff, C.J., Richards, J.P., Zappettini, E.O., 2002. Crustal lineament control on magmatism and mineralization in northwestern Argentina: geological, geophysical, and remote sensing evidence. Ore Geol. Rev, 21, 127–155.

Chetty, T.R.K., 1996. Proterozoic shear zones in Southern Granulite Terrain, India. The Archaean and Proterozoic Terranes in Southern India within East Gondwana, Gond. Res. Group Mem.3:77–90.

Chin, A., Hari, K., 2023 (last accessed on 28 April 2023). Predicting the impacts of mining of deep sea polymetallic nodules in the Pacific Ocean: a review of scientific literature. Available from https://miningwatch.ca/sites/default/files/nodule mining in the pacific ocean.pdf.

China Association of Oceanic Engineering, 2021. Guidelines for environment protection and preservation of deep sea mining activities, T/CAOE 41-2021a (in Chinese). http://caoe.org.cn/upload/202112/2021121014431611269540.pdf.

Choi, Y., Yoon, S.Y., Park, H.D., 2009. Tunnelling Analyst: a 3D GIS Extension for Rock Mass Classification and Fault Zone Analysis in Tunnelling. Comput. Geosci., 35(6), 1322–1333.

Choi, Y., Baek, J., Park, S., 2020. Review of GIS-Based Applications for Mining: planning, Operation, and Environmental Management. Applied Sciences, 10, 2266.

Chowdary, V.M., Ramakrishnan, D., Srivastava, Y., Chandran, V., Jeyaram, A., 2009. Integrated Water Resource Development Plan for Sustainable Management of Mayurakshi Watershed, India using Remote Sensing and GIS. Water Resource Management, 23, 1581–1602.

Christensen, A., Rajagopalan, S., 2000. The magnetic vector and gradient tensor in mineral and oil exploration. Preview, 84, 77p.

Christensen, A.N., Mahanta, A.M., Boggs, D.B., Dransfield, M.H., 2001. Falcon gravity gradiometer survey results over the Cannington Ag-Pb-Zn deposit: Preview, 93, 85.

Christensen, N.I., 1990. Seismic properties of rocks, In: Encyclopedia of Earth Science. Geophysics: Boston, MA, Springer, doi: 10.1007/0-387-30752-4_133.

Christensen, P.R., Bandfield, J.L., Hamilton, V.E., Howard, D.A., Lane, M.D., Piatek, J.L., Ruff, S.W., Stefanov, W.L., 2000. A thermal emission spectral library of rock forming minerals. J. Geophys. Res., 105, 9735–9739.

Christie, A.B., Brathwaite, R.L., 2003. Epithermal gold in New Zealand: mineral deposit models. Institute of Geological and Nuclear Sciences science report, 2003/11.

Chu, Y., Cai, G.J., Liu, S.Y., 2014. Study on liquefaction estimation method based on state parameters of CPTU in-situ measurement. China Earthq. Eng. J., 36, 905–910.

Chung, C.F., Fabbri, A.G., 1993. The Representation of Geoscience Information for data integration. Non-Renewable Resour., 2, 122–139.

Chung, J.S., 2003. Deep ocean technology: learning curve. In: Proceedings of 5th Ocean mining symposium, Tsukuba, Japan. https://www.researchgate.net/publication/290362656.

Cichostepski, K., Dec, J., Kwietniak, A., 2019. Simultaneous inversion of shallow seismic data for imaging of sulphurized carbonates. Minerals, 9, 203.

Clark, A.J., 1996. Seeing beneath the Soil: Prospecting Methods in Archaeology. B.T, Batsford Ltd.

Clark, R.N., King, T.V.V., Klejwa, M., Swayze, G., Vergo, N., 1990. High Spectral Resolution Reflectance Spectroscopy of Minerals. J. Geophys. Res., 95, 12653–12680.

Clark, R.N., Swayze, G.A., Gallagher, A., King, T.V.V., Calvin, W.M., 1993. The US Geological Survey Digital Spectral Library, Version 1:0.2 to 3.0 mm. US Geological Survey Open File Report 93-592.

Clark, R.N., 1999. Spectroscopy of Rocks and Minerals, and Principles of Spectroscopy. In: Rencz, A.N. (ed) Manual of Remote Sensing, 3, Remote Sensing for the Earth Sciences. John Wiley and Sons, New York, 3–58.

Clark, R.N., Swayze, G.A., Livo, K.E., Kokaly, R.F., Sutley, S.J., Dalton, J.B., McDougal, R.R., Gent, C.A., 2003. Imaging spectroscopy: earth and planetary remote sensing with the USGS Tetracorder and expert systems. J. Geophys. Res., 108(E12), 5131, doi: 10.1029/2002JE001847.

Coelho, C.V.S., Moreira, C.A., Rosolen, V., Bueno, G.T., Salles, J., Furlan, L.M., Govone, J.S., 2020. Analyzing the spatial occurrence of high-alumina clays (Brazil) using electrical resistivity tomography (ERT). Pure Appl. Geophys. (doi: 10.1007/s0002020-02444-w).

Coggon, J.A., 1971. Electromagnetic and electrical modelling by the finite element method. Geophysics, 36, 137–155.

Coggon, J.A., 1973. A comparison of IP electrode arrays. Geophysics, 38, 737–761.

Cole, K.S., Cole, R.H., 1941. Dispersion and adsorption in dielectrics. Alternating Current Field. J. Chem. Phys., 9, 341.

Collett, T.S., Riedel, M., Cochran, J.R., Boswell, R., Pushpendra Kumar, A.V., Sathe, 2008. Indian continental margin gas hydrate prospects: results of the Indian National Gas Hydrate Program (NGHP) Expedition 01; 6th International Conference on Gas Hydrates (ICGH).

Collett, T.S., Riedel, M., Cochran, J., Boswell, R., Presley, J., Kumar, P., Sathe, A., Sethi, A., Lall, M., Sibal, V., 2008. The NGHP Expedition 01 Scientists National Gas Hydrates Program Expedition 01 Initial Reports.

Colman-Sadd, S.P., Ash, J.S., Nolan, L.W., 1997. GeoLegend: a Database System for Managing Geological Map Units in a Geographic Information System. Comput. Geosci., 23(7), 715–724.

Constable, S., 2006a. Marine electromagnetic methods-a new tool for offshore exploration. Lead. Edge, 25(4), 438–444. 10.1190/1.2193225.

Constable, S., 2010. Ten years of marine CSEM for hydrocarbon exploration. Geophysics, 75(5), 7567–7581.

Constable, S., Kannberg, P., Callaway, K., Mejia, R.D., 2012. Mapping shallow geological structure with towed marine CSEM receivers. SEG Technical Program Expanded Abstracts, 1–5. 10.1190/segam2012-0839.1.

Constable, S., 2013. Review paper: instrumentation for marine magnetotelluric and controlled-source electromagnetic sounding. Geophys. Prospect., 61(s1), 505–532.

Constable, S., Kowalczyk, P., Bloomer, S., 2018. Measuring marine self-potential using an autonomous underwater vehicle. Geophys. J. Int., 215, 49–60.

Constable, S.C., Parker, R.L., Constable, C.G., 1987. Occam's inversion: a practical algorithm for generating smooth models from electromagnetic sounding data. Geophysics, 52(3), 289–300.

Constable, S.C., Kannberg, P.K., Weitemeyer, K., 2016. Vulcan: a deep-towed CSEM receiver. Geochemistry, Geophysics, Geosystems, 17, 1042–1064.

Coolbaugh, M.F., Raines, G.L., Zehner, R.E., 2007. Assessment of exploration bias in data-driven predictive models and the estimation of undiscovered resources. Nat. Resour. Res., 16, 199–207.

Corry, C.E., DeMoully, G.T., Gerety, M.T., 1983. Field procedure manual for self-potential surveys. Revised and Published by Zonge Engineering & Research Organization Tucson, Arizona, 55p.

Corry, C.E., 1985. Spontaneous polarization associated with porphyry sulfide mineralization. Geophysics, 50, 1020–1034.

Corwin, R.F., 1973. Offshore application of self-potential prospecting [D]. Berkeley University of California.

Corwin, R.F., 1976. Offshore use of the self-potential method. Geophys. Prospect., 24(1), 79–90.

Corwin, R.F., 1984. The self-potential method and its engineering applications; an overview. 54th Annual Internat. Mtg. Soc. Expl. Geophys. Expanded Abstracts, Soc. Expl. Geophys. Tulsa, Session: SP.1.

Corwin, R.F., Hoover, D.B., 1979. The self-potential method in geothermal exploration. Geophysics, 44, 226–245.

Corwin, R.F., Morrison, H.F., 1977. Self-potential variations preceding earthquakes in central California. Geophys. Res. Lett., 4, 171–174.

Costa, E.D.D., Starkey, J., 2001. Photolin: a program to identify and analyze linear structures in aerial photographs, satellite images and maps. Comput. Geosci., 27, 527–534.

Cox, C.S., Constable, S.C., Chave, A.D., Webb, S.C., 1986. Controlled-source electromagnetic sounding of the oceanic lithosphere. Nature, 320, 52–54. 10.1038/320052a0.

Cox, D.R., Snell, E.J., 1989. Analysis of Binary Data, 2nd ed. Chapman and Hall, London.

Cox, S.J., Baird, J.D., Brown, M.T., 2012. The role of integrated geophysical and geochemical data in volcanic hazard forecasting. Geophys. J. Int., 191(1), 250–267.

Crain, E.R., 2006. Crain's Petrophysical Pocket Pal. Ontario: ER Ross.

Criss, R.E., Champion, D.E., 1984. Magnetic properties of granitic rocks from the southern half of the Idaho batholith: influences of hydrothermal alteration and implications for aeromagnetic interpretation. J. Geophys. Res., 89(B8), 7061–7076.

Crowley, J.K., Brickey, D.W., Rowan, L.C., 1989. Airborne imaging spectrometer data of the Ruby Mountains, Montana: mineral discrimination using relative absorption band-depth images. Remote Sens. Environ., 29, 121–134.

Crusciol, C.A.C., Carmeis Filho, A.C.A., Fernandes, A.M., Alvarez, R.C.F., 2016. Micronutrient and silicon uptake and removal by upland rice cultivars with different plant architecture. Rev. Bras. Cienc. Solo, 40(1), 1–16.

Cui, Y., Liu, J., Ren, X., Shi, X., 2009. Geochemistry of rare earth elements in cobalt rich crusts from the Mid-Pacific M seamount. J Rare Earths., 27:169–176.

Cuperus, R., Flokstra, F.F., Droogendijk, H., Wiegerink, R.J., Flokstra, J., 2009. Gravity gradiometer system for Earth exploration MicroNano Conf. (Delft, The Netherlands, 2009).

Curray, J.R., 2005. Tectonics and history of the Andaman Sea region. J. Asian Earth Sci., 25, 187–232.

Curtis, J.B., 2002. Fractured shale-gas systems. AAPG Bull., 86(11), 1921–1938.

Cuyvers, L., Berry, W., Gjerde, K., Thiele, T., Wilhem, C., 2018. Deep sea bed mining, a rising environmental challenge. IUCN, Gland. 10.2305/IUCN.CH.2018.16.en.

Dabas, M., 2009. Theory and practice of the new fast electrical imaging system ARP. In: Campana, S., Piro, S. (eds.), Seeing the Unseen. Geophysics and Landscape Archaeology. London: CRC Press, pp. 105–126.

Dahlin, T., 2001. The development of DC resistivity imaging techniques. Comput. Geosci., 27, 1019–1029.

Dahlin, T., Loke, M.H., 1997. Quasi-3D resistivity imaging-mapping of three-dimensional structures using two-dimensional DC resistivity techniques. Proceedings of the 3rd Meeting of the Environ. Eng. Geophy. Soc., 143–146.

Daniels, D.J., 1996. Surface penetrating radar. Electronics & Communication Engineering Journal, 8(4), 165–182.

Darnley, A.G., 1972. Airborne gamma-ray survey techniques. In: Bowie, S.H., et al., eds., Uranium Prospecting Handboo, London, The Institution of Mining and Metallurgy p. 174–211

Darnley, A.G., 1993. Some notes on the importance of airborne gamma-ray spectrometry in International Geochemical Mapping. J. Geochem. Explor., 49(1-2), 201–212.

Darnley, A.G., Fleet, M., 1968. Evaluation of airborne gamma-ray spectrometry in the Bancroft and Elliott Lake areas of Ontario, Canada. Proc. 5th Symposium on remote sensing of environment, University of Michigan, Ann Arbor, 833p.

Darracott, B.W., McCann, D.M., 1986. Planning engineering geophysical surveys. In: Hawkins, A.B. (Ed.), Site Investigation Practice: Assessing BS5930 Geological Society Engineering Geology, Special Publications No. 2. (pp. 85–90)

Daryono, L.R., Rahmadi, A., Suyanto, I., Suryanto, W., 2019. Design and implementation of seismic data acquisition system using PC/104. SN Applied Sciences, 1(9), 1182.

Das, S., Behera, S.C., Kar, A., Narendra, P., Guha, S., 1997. Hydrogeomorphological mapping in groundwater exploration using remotely sensed data – a case study in Keonjhar district in Orissa. J. Indian Soc. Remote Sens., 25, 247–259.

Dayal, U., Allen, J.H., 1975. The effect of penetration rate on the strength of remoulded clay and sand samples. Can. Geotech. J. 12(12):336–348.

de la Torre, Daniel Marc, G., Gao, J., Macinnis-Ng, C., 2021. Remote sensing-based estimation of rice yields using various models: a critical review. Geospat. Inf. Sci., 24p.

de la Vega-marquez Fernandez, T., Prol-Ledesma, R.M., Orozco, G., 2001. Hydrothermal alteration and main structures mapping using TM images in La Primavera geothermal field (Mexico). Geofisica Internacional, 40(3), 147–162.

De Meijer, R. 1998. Heavy minerals: from Edelstein to Einstein. J. Geochem. Explor. 81–103.

de Quadros, T.P., Koppe, J., Strieder, A., Costa, J.C.L., 2006. Mineral-potential mapping: a Comparison of weights-of-evidence and fuzzy methods. Nat. Resour. Res., 15, 49–65.

De Santo, E.M., Asgeirsdottir, A., Barros-Platiau, A., Biermann, F., Dryzek, J., Goncalves, L.R., Kim, R.E., Mendenhall, E., Mitchell, R., Nyman, E., Scobie, M., Sun, K., Tiller, R., Webster, D.G., Young, O., 2019. "Protecting biodiversity in areas beyond national jurisdiction: an earth system governance perspective". Earth Syst. Gov. 2100029. doi:10.1016/j.esg.2019.100029. ISSN 2589-8116. S2CID 199907536.

De Souza, D., Lima, E.P., 2020. Geophysical data sharing and sustainability in resource exploration: a global approach. Geophys. J. Int., 221(3), 1582–1596.

Deekshatulu, B.L., Roy, P.S., 2000. Human resources development in space technology applications. Space Forum 5, 239–257.

DeGroot-Hedlin, C., Constable, S., 1990. Occam's inversion to generate smooth, two-dimensional models form magnetotelluric data. Geophysics, 55, 1613–1624.

Dehgahnnejad, M., Malehmir, A., Svensson, M., Linden, M., Moller, H., 2017. High resolution refection seismic imaging for the planning of a double-train-track tunnel in the city of Varberg, southwest Sweden. Near-Surface Geophys., 15(3), 226–240.

Deng, H., Yan, P., Liu, H.L., 2004. Seismic characteristics of gas hydrate in the Nansha waters. Marine Geology and Quaternary Geology, 24(4), 89–94 (in Chinese with English Abstract).

Deng, M., 2009. A conditional dependence adjusted weights of evidence model. Nat. Resour. Res., 18, 249–258.

Deng, Y., Chen, Y., Wang, P., Essa, K.S., Xub, T., Liang, X., et al., 2016. Magmatic underplating beneath the Emeishan large igneous province (South China) revealed by the COMGRA-ELIP experiment. Tectonophysics, 672-673, 16–23.

Dentith, P.M., Mudge, S.T., 2014. Geophysics for the Mineral Exploration Geoscientist, 1 edition. Cambridge University Press, Cambridge, United Kingdom.

DeWitte, L., 1948. A new method of interpretation of self-potential data. Geophysics, 13, 600–608.

Dey, S., Ghosh, P., 2008. GRDM-A Digital Field Mapping Tool for Management and Analysis of Field Geological Data. Computers & Geosciences, 34(5), 464–478.

DGH (Directorate General of Hydrocarbons), 2012. Hydrocarbon exploration and production activities, India. Directorate General of Hydrocarbons, Under Ministry of Petroleum & Natural Gas, Govt. of India, 1–162.

DGH (Directorate General of Hydrocarbons), 2018. Policy framework for exploration and exploitation of unconventional hydrocarbons under existing production sharing contracts (PSCs), Coal Bed Methane (CBM) contracts and nomination fields. Directorate General of Hydrocarbons, Under Ministry of Petroleum & Natural Gas, Government of India, 1–4.

Di, Q., Xue, G., Wang, Z., An, Z., Fu, C., Wang, R., Lei, D., Olalekan, F., 2018. Development of the emerging electromagnetic methods for deep earth exploration. International Symposium on Deep Earth Exploration and Practices Beijing, China October 24-26, 2018, p. 5.

Di, Q., Xue, G., Yin, C., Li, X., 2020. New methods of controlled-source electromagnetic detection in China. Sci. China Earth Sci., 63(9), 1268–1277.

Di, Q.Y., Wang, M.Y., Shi, K.F., Zhang, G.L., 2002. An applied study on prevention of water bursting disaster in mines with the high resolution V6 system. Chinese Journal of Geophysics, 45, 787–792.

Di, Q.Y., Unsworth, M., Wang, M.Y., 2004. 2.5D CSAMT modeling with the finite element method over 2D complex earth media. Chinese Journal of Geophysics, 47, 825–829.

Di, Q.Y., Xue, G.Q., Lei, D., Wang, Z.X., Li, H., Olalekan, F., 2020. Demonstration of the newly developed MTEM systems for gold detection in China. Geol. J., 55, 1763–1770.

Dickson, B.L., Scott, K.M., 1991a.Interpretation of aerial gamma-ray surveys. CSIRO Division of Exploration Geosciences, Restricted Report 301R.

Dickson, B.L., Scott, K.M., 1997. Interpretation of aerial gamma-ray surveys – adding the geochemical factors. AGSO Journal of Australian Geology and Geophysics, 17(2), 187–200.

DiFrancesco, D., Meyer, T., Christensen, A., FitzGerald, D., 2009. Gravity gradiometry – today and tomorrow. 11th SAGA Biennial Technical Meeting and Exhibition Swaziland, 16-18 September 2009, 80–83.

Dinesh, A.C., Nisha, N.V., Varghese, S., et al., 2020. Extensive occurrence of Fe-Mn crusts and nodules on seamounts in the southern Andaman Sea, India. Curr. Sci.; 119:704–708.

Ding, L., Chen, X., Gao, Y., 2009. Seafoor massive sulfides: the frontier of deep ocean miming. Ocean Technol., 28(1):126–132.

Diorio, P., Mahanta, A., Rose, M., Lockhart, G., 2003. Examples of the application of airborne gravity gradiometry to natural resource exploration. Geophysical Research Abstracts, 5.

DJI, 2019. The world leader in camera drones/quadcopters for aerial photography. Available online: https://www.dji.com/ (accessed on 31 August 2019).

Dlamini, N.L., Kamutando, A.M., 2020. Technology transfer and capacity building in geophysical mapping in Africa. Geosciences, 10(5), 175.

DMA, 1987. World Geodetic System 1984. DMA Technical Report TR 8350.2.

Domenikiotis, C., Spiliotopoulos, F., Tsiros, E., Dalezios, N.R., 2004. Early cotton yield assessment by the use of the NOAA/AVHRR derived drought Vegetation Condition Index in Greece. Int. J. Remote Sens., 25, 2807–2819.

Dominy, S.C., Petersen, J.S., 2005. Sampling coarse gold bearing mineralisation – developing effective protocols and a case study from the Nalunaq deposit, Southern Greenland. In Second world conference on sampling and blending, Brisbane, QLD.

Dong, B., Zhou, B., 2004. Application and Trend of GIS Technique Used in Oil and Gas Exploration. Earthquake Research in Plateau, 16(3), 28–33.

Dossing, A., Kolster, M.E., Rasmussen, T.M., Petersen, J.T., da Silva, E.L.S., 2023. UAV-towed scalar magnetic gradiometry: a case study in relation to iron oxide copper-gold mineralization, Nautanen (Arctic Sweden). Lead. Edge, 42(2), 103–111.

Dransfield, M., Airborne gravity gradiometry in the search for mineral deposits. In: Proceedings of Exploration 07: Fifth Decennial International Conference on Mineral Exploration edited by Milkereit, B., 2007, 341–354.

Dransfield, M., Milkereit, B., 2007. Airborne gravity gradiometry in the search for mineral deposits. In: Milkereit, B. (ed) Proceedings of Exploration 07: Fifth Decennial International Conference on Mineral Exploration, 341–354.

Dransfield, M.H., Christensen, A.N., Liu, G., Rose, M., 2001b. Airborne gravity gradiometry in mineral exploration. In: Presented at the Gravity Gradiometry Workshop, SEG Conference, San Antonio.

Dransfield, M.H., Walker, G., 2005. The Falcon airborne gravity gradiometer system: performance and discovery. In: Presented to the KEGS Symposium. New Developments in Airborne Geophysics.

Drummond, B.J., Goleby, B.R., Own, A.J., Yeates, A.N., Swager, C., Zhang, Y., Jackson, J.K., 2000. Seismic reflection imaging of mineral systems: three case histories. Geophysics, 65, 1852–1861.

Du, X.G., Guan, L.Q., Zhuo, W.X., 2016. Analysis of the current situation of deep sea mining development and demand for deep sea mining vessels in China. Strait Sci 12, 62–67.

Ducart, D.F., Crosta, A.P., Filho, C.R.S., Coniglio, J., 2006. Alteration mineralogy at the Cerro La Mina epithermal prospect, Patagonia, Argentina: field mapping, shortwave- infrared spectroscopy, and ASTER images. Econ. Geol., 101, 981–996.

Dunnington, L., Nakagawa, M., 2017. Fast and safe gas detection from underground coal fire by drone fly over. Environ. Pollut., 229, 139–145.

Durrance, E.M., 1986. Gamma-ray methods in exploration geophysics. In: Introduction to Applied Geophysics, 123–145, Elsevier.

Dutta, D., Kundu, A., Patel, N.R., Saha, S.K., Siddiqui, A.R., 2015. Assessment of agricultural drought in Rajasthan (India) using remote sensing derived Vegetation Condition Index (VCI) and Standardized

Precipitation Index (SPI). The Egyptian Journal of Remote Sensing and Space Science, 18(1), 53–63.

Dutta, T., et al., 2016. Global demand for rare earth resources and strategies for green mining. Environ. Res., 150, 182–190.

Duuring, P., Hagemann, S., 2013a. Genesis of superimposed hypogene and supergene Fe orebodies in BIF at the Madoonga deposit, Yilgarn Craton, Western Australia. Miner. Deposita, 48, 371–395. doi: 10.1007/s00126-012-042.

Duval, J.S., 1980, Radioactivity method. Geophysics, 45(11), 1690–1694.

Eadic, T., 1981. Detection of hydrocarbon accumulations by surface electrical methods: a feasibility study. Research in Applied Geophysics, 15, University of Toronto.

Eaton, D., Guest, S., Milkereit, B., Bleeker, W., Crick, D., Schmitt, D., Salisbury, M.H., 1996. Seismic imaging of massive sulfide deposits; Part III, borehole seismic imaging of near-vertical structures. Econ. Geol., 91, 835–840. doi: 10.2113/gsecongeo.91.5.835.

Eaton, D.W., Milkereit, B., Salisbury, M.H., 2003. Hard rock seismic exploration. Geophysical Development Series, 10, Soc. Explor. Geophys.

Edwards, R.N., 1997. On the resource evaluation of marine gas hydrate deposits using seafloor transient electric dipole-dipole methods. Geophysics, 62(1), 63–74. 10.1190/1.1444146.

Edwards, R.N., 2005. Marine controlled-source electromagnetic: principles, methodologies, future commercial applications. Surv. Geophys., 26(6), 675–700. http://dx.doi.org/10.1007/s10712-005-1830-3.

Egmond, F.M., Van, V., Van Der, S., Knotters, M., Koomans, R.L., Limburg, J., 2018. Mapping soil texture with a gamma-ray spectrometer: comparison between UAV and proximal measurements and traditional sampling. WOt-technical report 137, 58p.

Ehsan, S.A., Carbonell, R., Ayarza, P., Marti, D., Perez-Estaun, A., Martinez-Poyatos, D.J., Simancas, J.F., Azor, A., Mansilla, L., 2014. Crustal deformation styles along the reprocessed deep seismic reflection transect of the Central Iberian Zone (Iberian Peninsula). Tectonophysics, 621, 159–174. doi: 10.1016/j.tecto.2014.02.014.

EIA (Energy Information Administration), 2011. An Initial Assessment of 14 Regions outside the United States.

EIA (U.S. Energy Information Administration), 2016. Annual energy outlook 2016 with projections to 2040.

EIA (U.S. Energy Information Administration), 2017. Annual energy outlook 2017: with projections to 2050. U.S. Energy Information Administration, 1–127.

Eidesmo, T., Ellingsrud, S., MacGregor, L., Constable, S., Sinha, M., Johansen, S., Kong, F., Westerdahl, H., 2002. Seabed logging (SBL), a new method for remote and direct identification of hydrocarbon filled layers in deep water areas. First Break, 20, 144–152.

Eiken, O., Degutsch, M., Riste, P., Rod, K., 1989. Snow streamer: an efficient tool in seismic acquisition. First Break, 7(1223), doi:10.3997/1365–2397.1989021.

El-Araby, H.M., 2004. A new method for complete quantitative interpretation of self-potential anomalies. J. Appl. Geophys., 55, 211–224.

Elijah, O., Rahman, T.A., Orikumhi, I., Leow, C.Y., Hindia, M.N., 2018. An overview of internet of things (IoT) and data analytics in agriculture: benefits and challenges. IEEE Internet Things J., 5, 3758–3773.

Elyasi, G.R., Bahroudi, A., Abedi, M., 2019. Risk-based analysis in mineral potential mapping: application of quantifier-guided ordered weighted averaging method. Nat. Resour. Res., 28, 931–951.

Emery, X., Ortiz, J.M., Rodrıguez, J.J., 2006. Quantifying uncertainty in mineral resources by use of classification schemes and conditional simulations. Mathematical Geology, 38, 445–464.

Emond, A.M., Zhdanov, M.S., Petersen, E.U., 2006. Electromagnetic modeling based on the rock physics description of the true complexity of rocks: applications to porphyry copper deposits: 76th Annual International Meeting, SEG, Expanded Abstracts, 1313–1317a.

Emond, A.M., 2007, Electromagnetic modeling of porphyry systems from the grain-size to the deposit-scale using the generalized effective medium Theory of Induced Polarization: m.S. Thesis, University of Utah.

Emsbo, P., McLaughlin, P.I., Breit, G.N., et al., 2015. Rare earth elements in sedimentary phosphate deposits: solution to the global REE crisis? Gondwana Res., 27:776–785.

Eppelbaum, L.V., 2019. Advanced analysis of self-potential data in ore deposits of the South Caucasus. ANAS Transactions, Earth Sciences, 2, 21–35.

Ercan, O.A., Seren, A., Elmas, A., 2014. Gold and Silver prospection using magnetic, radiometry and microgravity methods in the Kısladag Province of Western Turkey. Resour. Geol., 64(1), 25–34.

Esdale, D., Pridmore, D., Coggon, J., Muir, P., Williams, P., Fritz, F., 1987. Olympic Dam deposit? Geophysical case history. Explor. Geophys., 18(2), 47–49.

Esdale, D., Pridmore, D.F., Coggon, J., Muir, P., Williams, P., Fritz, F., 2003. The Olympic Dam copper-uranium-gold-silver-rare earth element deposit, South Australia: a geophysical case history. In: Dentith, M. (ed) Geophysical Signatures of South Australian Mineral Deposits. The University of Western Australia, ASEG and PIRSA, 147–168.

Esmaili, S.R., Telford, F., Tyler, J.R., 2016. Spectral gamma ray logging and its application to mineral exploration. Explor. Geophys., 47(3), 144–157.

Essa, K.S., 2012. A fast interpretation method for inverse modeling of residual gravity anomalies caused by simple geometry. Journal of Geological Research, 2012, 327037.

Essa, K.S., 2014. New fast least-squares algorithm for estimating the best-fitting parameters of some geometric-structures to measured gravity anomalies. Journal of Advanced Research, 5, 57–65.

Essa, K.S., Mehanee, S.A., Soliman, K.S., Diab, Z.E., 2020. Gravity profile interpretation using the R-parameter imaging technique with application to ore exploration. Ore Geol. Rev., 126:103695.

Essa, K.S., Munschy, M., 2019. Introductory Chapter: mineral Exploration from the Point of View of Geophysicists. Minerals, DOI: 10.5772/intechopen.84830.

European Commission, 2023. Protecting the ocean, time for action: ocean and fisheries. High ambition coalition on biodiversity beyond national jurisdiction. eing the ocean, time for action

Evans, D.L., Farr, T.G., Ford, J.P., Thompson, T.W., Werner, C.L., 1986. Multipolarization radar images for geologic mapping and vegetation discrimination. IEEE Transactions of Geosciences and Remote Sensing, GE-24, 246–257.

Evans, R.D., Mugele, R.A., 1936. Increased gamma ray sensitivity of tube counters and the measurement of the thorium content of ordinary materials. Review of Scientific Instruments, 7(12), 441–449.

Everett, J.E., Hyndman, R.D., 1967. Geomagnetic variations and electrical conductivity structure in south-western Australia. Phys. Earth Planet. Inter., 1, 24–34.

Fan, N., Nian, T.K., Guo, X.S., Jiao, H.B., Lu, S., 2020a. Piecewise strength model for three types of ultra-soft fine-grained soils. Soils Found 60(4):778–790.

Fan, N., Jiang, J., Dong, Y., Guo, L., Song, L., 2022. Approach for evaluating instantaneous impact forces during submarine slide pipeline interaction considering the inertial action. Ocean Eng., 245:110466.

Farewell, T.S., Farewell, D.M., 2010. Knowledge-based Soil Attribute Mapping in GIS: corrections and Extensions to the Expector Method. Transactions in GIS, 14(2), 183–192.

Farfour, M., Gaci, S., El-Ghali, M., Mostafa, M., 2021. A review about recent seismic techniques in shale-gas exploration. Methods and Applications in Petroleum and Mineral Exploration and Engineering Geology, 65–80.

Farhadi, H., Najafzadeh, M., 2021. Flood Risk Mapping by Remote Sensing Data and Random Forest Technique. Water, 13:3115. https://doi.org/10.3390/w13213115.

Farrand, W.H., Harsanyi, J.C., 1997. Mapping the distribution of mine tailings in the Coeur d'Alene River valley, Idaho, through the use of a Constrained Energy Minimization Technique. Remote Sens. Environ., 59, 64–76.

Feebrey, C.A., Hishida, H., Yoshioka, K., Nakayama, K., 1998. Geophysical expression of low-sulfidation epithermal Au-Ag deposits and exploration implications; examples from the Hokusatsu region of SW Kyushu, Japan. Resour. Geol., 48, 75–86.

Feng, Y., Wang, Y., Wang, C., Tian, J., Wen, S., Zhou, Y., Cheng, Y., A Fuzzy gold mineral prediction model integrating with Knowledge-Driven and Data-Driven: a case study of the Hatu region in Xinjiang, China. Minerals 2024, 14, 1209.

Fernandez-Hernandez, J., Gonzalez-Aguilera, D., Rodriguez-Gonzalvez, P., Mancera-Taboada, J., 2015. Image-based modelling from unmanned aerial vehicle (UAV) photogrammetry: an effective, low-cost tool for archaeological applications. Archaeometry, 57, 128–145.

Ferreir, G., White, K., Griffiths, G., Bryant, R., Stefouli, M., 2002. The mapping of hydrothermal alteration zones on the island of Lesvos, Greece using an integrated remote sensing dataset. Int. J. Remote Sens., 23, 341–356.

Feseker, T., Dahlmann, A., Foucher, J.P., Harmegnies, F., 2009. In-situ sediment temperature measurements and geochemical pore water data suggest highly dynamic fluid flow at Isis mud volcano, eastern Mediterranean Sea. Mar. Geol., 261, 128–137.

FIMI (Federation of Indian Mineral Industries), 2018. Exploration for deep-seated/concealed minerals. 50 years of dedicated service to the mining industry, 9p.

Finn, R.L., Wright, D., 2016. Privacy, data protection and ethics for civil drone practice: a survey of industry, regulators and civil society organisations. Comput. Law Secur. Rev., 32, 577–586.

Fisher, R., O'Leary, R.A., Low-Choy, S., Mengersen, K., Knowlton, N., Brainard, R.E., Caley, M.J., 2015. Species richness on coral reefs and the pursuit of convergent global estimates. Curr Biol., 25(4):500–505. https://doi.org/10.1016/j.cub.2014.12.022.

Fitterman, D.V., Corwin, R.F., 1982. Inversion of self-potential data from the Cerro-Prieto geothermal field Mexico. Geophysics, 47, 938–945.

Fitts, C.R., 2024. Field Exploration and Wells. Groundwater Science (Book, Third Edition), 661p.

FitzGerald, D., Christensen, A., Paterson, R., 2011. Direct Detection of Buried Orebodies Using Gradiometry. 35th APCOM Symposium/WOLLONGONG, NSW, 24-30 September 2011.

Flovenz, O.G., Hersir, G.P., Saemundsson, K., Armannsson, H., Fridriksson, T., 2012. Geothermal energy exploration techniques. In: Sayigh, A., (ed.), Comprehensive Renewable Energy, 7. Elsevier, Oxford, 51–95.

Foell, E.J., Thiel, H., Schriever, G., 1990. DISCOL: a long term, large scale, disturbance-recolonization experiment in the abyssal eastern tropical South Pacific Ocean. Offshore Technology Conference. OnePetro, 108.

Fofonoff, N., 1985. Physical properties of seawater: a new salinity scale and equation of state for seawater. J. Geophys. Res., 90, 3332–3342.

Foley, C.P., Leslie, K.E., Binks, R., Lewis, C., Murray, W., Sloggett, G.J., Lam, S., Sankrithyan, B., Savvides, N., Katzaros, A., Muller, K.-H., Mitchell, E.E., Pollock, J., Lee, J., Dart, D.L., Barrow, R.R., Asten, M., Maddever, A., Panjkovic, G., Downey, M., Hoffman, C., Turner, R., 1999. Field trials using HTS SQUID magnetometers for ground-based and airborne geophysical applications, IEEE Trans. Appl. Supercond.

Foley, C.P., Leslie, K.E., 1998. Potential use of high Tc SQUIDs for airborne electromagnetics. Explor. Geophys. 29, 30–34.

Fontbote, L., Kouzmanov, K., Chiaradia, M., Pokrovski, G.S., 2017. Sulphide minerals in hydrothermal deposits [J]. Elements, 13, 97–103.

Ford, A., Miller, J.M., Mol, A.G., 2016. A comparative analysis of weights of evidence, evidential belief functions, and fuzzy logic for mineral potential mapping using incomplete data at the scale of investigation. Nat. Resour. Res., 25, 19–33.

Ford, A., Hart, C.J.R., 2013. Mineral potential mapping in frontier regions: a Mongolian case study. Ore Geol. Rev, 51, 15–26.

Ford, A., McCuaig, T.C., 2010. The effect of map scale on geological complexity for computer-aided exploration targeting. Ore Geol. Rev., 38, 156–167.

Ford, R.L., McCuaig, T.C., Li, H., 2016. Prospectivity Mapping in the Mineral Exploration Process. Elsevier.

Forsberg, R., Olesen, A.V., Einarsson, I., 2015. Airborne gravimetry for geoid determination with Lacoste Romberg and Chekan gravimeters. Gyrosc. Navig., 6(4):265–270.

Foti, S., 2017. Geophysical tests for geotechnical site characterization. ISSMGE IT Administrator: international Society for Soil Mechanics and Geotechnical Engineering: https://www.issmge.org/education/recorded-webinars.

Fouda, Y.E.-S., 2004. Urban planning advanced analytic techniques and the human role: a challenge or a complement. Alex. Eng. J., 43(6), 791–801.

Foulger, G.R., Peirce, C., 2007. Geophysical Methods in Geology. http://www.earthsci.unimelb.edu.au/ES304. Teaching Handbook, University of Durham, UK, 78p.

Fox, R.W., 1830. On the electromagnetic properties of metalliferous veins in the mines of Cornwall. Philos. Trans. R. Soc. Lond., 120, 399–414.

Frahm, C.P., 1972. Inversion of the magnetic field gradient equation for a magnetic dipole field. NCSL Informal Report, 135–172.

Franklin, J.M., Gibson, H.L., Jonasson, I.R., Galley, A.G., 2005. Volcanogenic massive sulphide deposits. Econ. Geol., 100th Anniversary Volume: 523–560.

Fraser, A., Huggins, P., Reens, J., Cleverly, P., 1997. A satellite remote sensing technique for geological horizon mapping. Int. J. Remote Sens., 18, 1607–1615.

Fraser, S.J., 1991. Discrimination and identification of ferric oxides using satellite Thematic Mapper data: a Newman case study. Int. J. Remote Sens., 12, 635–641.

Frasheri, A., Lubonja, L., Alikaj, P., 1995. On the application of geophysics in the exploration for copper and chrome ores in Albania. Geophys. Prospect. 43, 743–757. (doi: 10.1111/j.13652478.1995.tb00278.x).

Freden, S.C., Gordon Jr, F. Manual of Remote Sensing (American Society of Photogrammetry, Falls Church, Va., 1983), chapter 12.

Freire, G., Cota, R., 2017. Capture of images in inaccessible areas in an underground mine using an unmanned aerial vehicle. In proceedings of the 1st International Conference on Underground Mining Technology, Sudbury, ON, Canada, 11-13 October 2017.

Frid, M., Frid, V., A case study of the integration of ground-based and drone-based ground-penetrating radar (GPR) for an archaeological survey in Hulata (Israel): advancements, challenges, and applications. Appl. Sci. 2024, 14, 4280.

Friedel, S., Jacobs, F., 1998. Large scale resistivity imaging at Merapi volcano. Annales Geophysicae, 16(1), C180.

Friend, C.R.L., Janardhan, A.S., 1984. Mineral chemistry of coexisting phases from shonkinitic rocks, Salem, Tamil Nadu, India. Mineral. Mag., 48, 181–193.

Frischknecht, F.C., 1967. Fields about an oscillating magnetic dipole over a two-layer earth, and application to ground and airborne electromagnetic surveys. Q. Colorado School Mines, 62(1), 1–326.

Fu, C., Qingyun Di, Q., An, Z., 2013. Case History: application of the CSAMT method to groundwater exploration in a metropolitan environment. Geophysics, 78(5), B201–B209.

Fu, C.M., Di, Q.Y., Wang, M.Y., 2009. Three-dimensional numerical simulation of marine controlled source electromagnetic method (in Chinese). Oil Geophys. Prospect., 44, 358–363.

Fu, J., Jia, S., Wang, E., 2020. Combined magnetic, transient electromagnetic, and magnetotelluric methods to detect a BIF-type concealed iron ore body: a case study in Gongchangling iron ore concentration area, Southern Liaoning Province, China. Minerals, 10, 1044; doi: 10.3390/min10121044.

Fu, L., 2011. Induced polarization effect in time domain: theory, modelling, and applications. Department of Geology and Geophysics, the University of Utah. A thesis submitted to the faculty of the University of Utah in partial fulfilment of the requirements for the degree of Master of Science in Geophysics.

Fu, L.K., Meng, D.H., Song, Y.C., 1993. Some new advances in the application of the induced polarization method to water prospecting. Geophys. Geochem. Explor., 17, 435–440.

Fujimoto, H., Koizumi, K., Watanabe, M., Oshida, A., Furuta, T., Takamura, N., Ura, T., 2000. Underwater gravimeter on board the R-One Robot. In: The 2000a international symposium on underwater technology. IEEE Catalog No.00EX418

Fujinaga, K., et al., 2016. Geochemistry of REY rich mud in the Japanese Exclusive Economic Zone around Minamitorishima Island. Geochem. J., 50, 575–590.

Fujinaga, K., Nakamura, K., Ohta, J., et al., 2022. Umber as a lithified REY rich mud in Japanese accretionary complexes and its implications for the osmium isotopic composition of middle cretaceous seawater. Ore Geol. Rev., 142:104683.

Furness, P., 1992. Modelling spontaneous mineralization potentials with a new integral equation. J. Appl. Geophys., 29, 143–155.

Gabr, S., Ghulam, A., Kusky, T., 2010. Detecting areas of high potential gold mineralization using ASTER data. Ore Geol. Rev., 38, 59–69.

Gad, S., Kusky, T., 2006. ASTER spectral ratio techniques for lithological mapping in the Arabian-Nubian Shield. J. Afr. Earth Sci., 44(3), 196–202.

Gaffey, S.J., 1985. Reflectance spectroscopy in the visible and near-infrared (0.35-2.55 microns): applications in carbonate petrology. Geology, 13, 270–273.

Gaffey, S.J., 1986. Spectral reflectance of carbonate minerals in the visible and near infrared (0.35-2.55 microns): calcite, aragonite, and dolomite. Am. Mineral., 71, 151–162.

Gajos, M., Sierka, E., 2012. GIS Technology in Environmental Protection: research Directions Based on Literature Review. Pol. J. Environ. Stud., 21, 241–248.

Galley, A.G., Hannington, M.D., Jonasson, I.R., 2007. Volcanogenic massive sulphide deposits, in mineral deposits of Canada: a synthesis of major deposit-types, District metallogeny, the evolution of geological provinces, and exploration methods, ed. Goodfellow, W.D., Geological Accociation of Canada, doi:10.2113/gsecongeo.102.6.1182.

Gandhi, S.M., Sarkar, B.C., 2016. Geophysical exploration. Essent. Miner. Explor. Eval., 97–123.

Gao, X., 2002. Integration of GPS with remote sensing and GIS: reality and prospect. Photogrammetric Engineering & Remote Sensing, 68(1), 23–28.

Gao, Y., Zhang, Z., Xiong, Y., Zuo, R., 2016. Mapping mineral prospectivity for Cu polymetallic mineralization in southwest Fujian Province, China. Ore Geol. Rev., 75, 16–28.

Garnett, R.H.T., 2000. Marine placer gold, with particular reference to Nome, Alaska. In: Cronan, D.S. (Ed.), Hand Book of Marine Mineral Deposits. CRC Press, Boca Raton, FL, 67–101.

Gausepohl, F., Hennke, A., Schoening, T., Koser, K., Greinert, J., 2020. Scars in the abyss: reconstructing sequence, location and temporal change of the 78 plough tracks of the 1989 DISCOL deep sea disturbance experiment in the Peru Basin. Biogeosciences, 17:1463–1493.

Gauthier, M., Charles, C., 1979. French technological development in nodule mining. In: Offshore technology conference. OnePetro. 10.4043/3369-MS.

Gauthier, S., Drapeau, P., 2006. Remote sensing and GIS applications for forest resource management in Canada. Canadian Journal of Forest Research, 36(6), 1362–1375.

Gehrmann, R.A.S., North, L.J., Graber, S., Szitkar, F., Petersen, S., Minshull, T.A., Murton, B.J., 2019. Marine mineral exploration with controlled source electromagnetics at the TAG hydrothermal field, 26°N mid-Atlantic ridge. Geophysical Research Letters, 46(11), 5808–5816.

Gehrmann, R.A.S., Haroon, A., Morton, M., Djanni, A.T., Minshull, T.A., 2020. Seafloor massive sulphide exploration using deep-towed controlled source electromagnetics: navigational uncertainties. Geophys. J. Int., 220, 1215–1227.

Gendzwill, D.J., 1969. Underground applications of seismic measurements in a Saskatchewan potash mine. Geophysics, 34, 906–915.

Gendzwill, D.J., Stead, D., 1992. Rock mass characterization around Saskatchewan potash mine openings using geophysical techniques: a review. Can. Geotech. J., 29, 666–674.

Geological Survey of India (GSI), 2006. Geology and Mineral Resources of the States of India. Miscellaneous Publication, Part VI - Tamil Nadu and Pondicherry, 1–59.

Geological Survey of India (GSI), 2011. Govt. of India Geology and Mineral Resources of Rajasthan (3rd Edition). Miscellaneous Publication, GSI: Kolkata, 30(12), 130p.

Geological Survey of India (GSI), 2017. Geological Mapping and Mineral Exploration in the Eastern Himalayas Using GIS and Remote Sensing. GSI Technical Report.

Geological Survey of India (GSI), 2018. Mineral Exploration in the Aravalli Range Using GPS, GIS, and Remote Sensing. GSI Research Report.

Geological Survey of India (GSI), 2019. Groundwater Resource Mapping Using GIS and Remote Sensing in Rajasthan. GSI Annual Report.

Geological Survey of India (GSI). (2020). Annual Report on Geophysical and Geological Mapping. GSI Publications.

Geological Survey of India (GSI), 2020. Coastal Erosion Monitoring in India Using Remote Sensing, GIS, and GPS. GSI Technical Report.

Geological Survey of India (GSI), Annual Reports and Data (2020).

Geological Survey of India (GSI), 2020. Institutional Report. Proc. Indian Natn. Sci. Acad., 86(1), 651–674. DOI: 10.16943/ptinsa/2020/49827.

Geological Survey of India (GSI), 2021. Geoscientific Data for Sustainable Mineral Exploration in India. GSI Publication.

Geological Survey of India (GSI), 2021. Seismic Hazard Mapping in Northeast India Using GIS and Remote Sensing. GSI Q. J.

Geological Survey of India (GSI) (n.d.) Annual Report (2008-2009). Mission Mode Restructuring and Geoscientific Surveys.

Geoscience Australia, 2020. Geophysical surveys for mineral exploration: a national initiative. Geoscience Australia. Retrieved from https://www.ga.gov.au.

German, C.R., Yoerger, D.R., Jakuba, M., Shank, T.M., Langmuir, C.H., Nakamura, K., 2008. Hydrothermal exploration with the autonomous benthic explorer [J]. Deep Sea Research I, 55, 203–219.

Ghomsi, F.E.K., Pham, L.T., Stefen, R., Ribeiro-Filho, N., Tenzer, R., 2022. Delineating structural features of North Cameroon using the EIGEN6C4 high resolution global gravitational model. Geol. J., 57(10), 4285–4299.

Ghosh, R., Sain, K., Ojha, M., 2010. Estimating the amount of gas-hydrate using effective medium theory: a case study in the Blake Ridge. Mar. Geophys. Res., 31, 29–37.

Giao, P.H., Chung, S.G., Kim, D.Y., Tanaka, H., 2003. Electric imaging and laboratory electric resistivity testing for geotechnical investigation of Pusan clays. Int. J. Appl. Geophys. 52(4), 157–175.

Giao, P.H., Vichalai, C., 2006. Construction of a geophysical testing site on soft clays. In: Conference Paper. Proc. of the 8th SEGJ Intl' Symposium – Imaging and Interpretation, Kyoto, Japan, 277–282.

Gibson, R., Smith, C., 2022. The role of induced polarization in mapping mineralization in sedimentary rock sequences. Geophys. Prospect., 70(4), 1187–1205.

Gilbert, L.A., McDuf, R.E., Johnson, H.P., 2007. Porosity of the upper edifice of axial seamount. Geology, 35:49–52.

Gill, T., Chew, C., 2018. Simplifying complex systems in mineral exploration. J. Min. Geospat. Sci., 34(2), 98–112.

Gitas, I., Koutsou, S., 2019. Remote Sensing and GIS Applications in Agricultural Meteorology. Springer.

Gjerde, K.M., Clark, N.A., Harden-Davies, H.R., 2019. Building a platform for the future: the relationships of the expected new agreement for marine biodiversity in areas beyond national jurisdiction and the UN convention on the law of the sea. Ocean Yearbook Online, 33(1), 1–44. doi:10.1163/9789004395633_002. 0191-8575. S2CID 211421146.

Gladwell, D.R., Lett, R.E., Lawrence, P., 1983. Application of reflectance spectroscopy to mineral exploration. Econ. Geol., 78, 699–710.

Global Reporting Initiative, 2020. Sustainability Reporting for the Mining Sector. GRI.

Glover, P.W.J., 2015. Geophysical Properties of the Near Surface Earth: electrical Properties. In Treatise on Geophysics, 89–137.

Godson, G., Faizan, M., Sanjeevi, S., 2021. Forest Fire Risk Assessment Using GIS Science – a Case Study of South India. In: Book: Geographic Information Science for Land Resource Management (283–300). DOI:10.1002/9781119786375.ch15.

Goetz, A.F.H., Rowan, L.C., Kingston, M.J., 1982. Mineral identification from orbit: initial results from the Shuttle Multispectral Infrared Radiometer. Science, 218, 1020–1031.

Goldfarb, R.J., Groves, D.I., 2015. Orogenic gold: common or evolving fluid and metal sources through time: Lithos., 233, 2–26.

Goldstein, M.A., 1971. Magnetotelluric experiments employing an artificial dipole source. Ph.D. Thesis, University of Toronto, Toronto, Ontario, Canada.

Goldstein, M.A., Strangway, D.W., 1975. Audio-frequency magnetotelluric with a grounded dipole source. Geophysics, 40, 669–683.

Goleby, B.R., Blewett, R.S., Fomin, T., Fishwick, S., Reading, A.M., Henson, P.A., Kennett, B.L.N., Champion, D.C., Jones, L., Drummond, B.J., Nicoll, M., 2006. An integrated multi-scale 3D seismic model of the Archaean Yilgarn Craton, Australia. Tectonophysics, 420, 75–90. doi: 10.1016/j.tecto.2006.01.028.

Goleby, B.R., et al. (2014). Geophysical Methods for Exploration in the Pilbara Region: successes in Mineral Exploration. Australian Journal of Earth Sciences, 61(4), 514–528.

Gollner, S., Kaiser, S., Menzel, L., et al., 2017. Resilience of benthic deep-sea fauna to mining activities. Mar. Environ. Res., 129:76–101.

Gong, P., Tang, X.B., Huang, X., Wang, P., Wen, L.S., Zhu, X.X., Zhou, C., 2019. Locating lost radioactive sources using a UAV radiation monitoring system. Appl. Radiat. Isot. 150, 1–13.

Gonzalez, A., Ortiz, M., Silva, J., 2019. Real-time geophysical surveying with smart sensors and wearable devices. J. Geophys. Technol., 33(3), 267–278.

Gonzalez, F., Garcia, J., Perez, D., 2020. Copper deposits and tectonic settings: the role of subduction zones. J. Geol., 58(5), 1012–1025.

Gonzalez, F.J., Somoza, L., Hein, J.R., et al., 2016. Phosphorites, CO rich Mn nodules and Fe-Mn crusts from Galicia Bank, NE Atlantic: reflections of Cenozoic tectonics and paleoceanography: mineralizations from Galicia Bank. Geochem. Geophys. Geosyst., 17:346–374.

Gonzalez, F.J., Medialdea, T., Schiellerup, H., et al., 2023. MINDeSEA: exploring seabed mineral deposits in European seas, metallogeny and geological potential for strategic and critical raw materials. Spec. Publ., 526: SP526–2022.

Gonzalez, F.J., Somoza, L., Maldonado, A., et al., 2023. High technology elements in Co rich ferromanganese crusts from the Scotia Sea. Available from https://core.ac.uk/downloadpdf/19714312. Last accessed on 28 April 2023.

Gonzalez, M.L., Valdivia, A.D., 2016. Copper exploration in the Andes: geochemical mapping and resource assessment. Journal of South American Earth Sciences, 70, 50–62.

Gonzalez-Alvarez, I., Porwal, A., Beresford, S.W., McCuaig, T.C., Maier, W.D., 2010. Hydrothermal Ni prospectivity analysis of Tasmania, Australia. Ore Geol. Rev, 38, 168–183.

Gonzalez-Alvarez, S., Carranza, F., Laborte, J.F., 2010. Prospectivity mapping for mineral exploration using GIS. Geomatics Research, 18(3), 255–272.

Goodchild, M.F., 2007. GIS and Spatial Data Integration. Wiley.

Gopalakrishnan, S., Sharma, V., Smith, A., 2021. Machine Learning in Geoscience: Applications and Trends. Wiley.

Goto, T., Kasaya, T., Machiyama, H., Takagi, R., Matsumoto, R., Okuda, Y., Satoh, M., Watanabe, T., Seama, N., Mikada, H., Sanada, Y., Kinoshita, M., 2007. A marine deep-towed DC resistivity survey in a methane hydrate area, Japan Sea. Explor. Geophys., 39, 52–59.

Gozlan, E., Cull, J., 2001. Microgravity for mineral exploration. ASEG 15th Geophysical Conference and Exhibition, August 2001, Brisbane, ASEG Extended Abstracts, 1, 1–6, DOI: 10.1071/ASEG2001ab046 and Published online on 2019, Taylor & Francis.

Granian, H., Tabatabaei, S.H., Asadu, H.H., Carranza, E.J.M., 2015. Multivariate regression analysis of lithogeochemical data to model subsurface mineralisation. A case study from the Sari Gunay epithermal gold deposit NW Iran. J. Geochem. Explor. 148, 249–258.

Grasty, R.L., Cox, J.R., 1997. A car-borne gamma ray spectrometer system for natural radioactivity mapping and environmental monitoring. In: RESUME 95, Rapid Environmental Surveying Using Mobile Equipment, Report, Nordic Nuclear Safety Research Secretariat, 71–90.

Grauch, V.J.S., Jachens, R.C., Blakely, R.J., 1995. Evidence for a basement feature related to the Cortez disseminated gold trend and implications for regional exploration in Nevada. Econ. Geol., 90, 203–207.

Graziano, F., Anita, M., 2005. Robert mallet and the 'Great Neapolitan Earthquake' of 1857. Notes Rec. R. Soc. 10.1098/rsnr.2004.76.

Grebe, H., 1997. General Mathematical Model for String Connections between Mobile Deep Sea Devices and Their Mother Stations.. Siegen Universitat-GH Siegen https://www.amazon.deenHenning-Grebe/dp/382652778X.

Green, J., Mine rescue robots requirements: outcomes from an industry workshop. In Proceedings of the 2013 6th Robotics and Mechatronics Conference (RobMech), Durban, South Africa, 30-31 October 2013; IEEE Computer Society: Washington, DC, USA, 2013; 111–116.

Greenhalgh, S.A., Mason, I.M., 1997, Seismic imaging with application to mine lay out and development, In: Gubins, A., Ed., Geophysics and Geochemistry at the Millennium.

Griffiths, D.H., Turnbull, J., Olayinka, A.I., 1990. Two-dimensional resistivity mapping with a complex controlled array. First Break, 8, 121–129.

Griffiths, D.H., Turnbull, J., 1985. A multi-electrode array for resistivity surveying. First Break, 3, 16–20.

Grissemann, C., Ludwig, R., 1986. Recherche sur la fracturation profonde en zone de socle cristllin a partir de forage a gros debit et de lineaments landsat a l aide de methodes geophysiques avancees. Cooperation techniques projet no. 82.2060.0, Bundesanstalt für Geowissenschaften und Rohstoffe, Hannover.

Groten, E., 2004. Fundamental parameters and current (2004) best estimates of the parameters of common relevance to astronomy, geodesy, and geodynamics. J. Geodesy, 77, 724–797.

Grove, C.I., Hook, S.J., Payloe II., E.D. 1992. Laboratory Reflectance Spectra of 160 Minerals, 0.4 to 2.5 Micrometers. 300p, NASA-JPL Publication 92-2, Psadena.

Groves, D.I., Goldfarb, R.J., Gebre-Mariam, M., Hagemann, S.G., Robert, F., 1998. Orogenic gold deposits – a proposed classification in the context of their crustal distribution and relationship to other gold deposit types. Ore Geol. Rev, 13, 7–27.

Groves, D.I., Goldfarb, R.J., Robert, F., Hart, C.J.R., 2003. Gold deposits in metamorphic belts: overview of current understanding, outstanding problems, future research, and exploration significance. Econ. Geol., 98, 1–29.

GSJ, 2021. Geological Mapping on Land – to make them more familiar to our life. Geological Survey of Japan, AIST. Retrieved from https://www.gsj.jp/en/about/overview/geological-map.html.

Gu, Z., Guo, X., Nian, T., Fu, C., Zhao, W., 2022. An improved evaluation method for the undrained shear strength of uniform soft clay in the non-full flow state based on ball penetration simulations. Appl Ocean Res 128:103365.

Gunderson, B.M., Newman, G.A., Hohmann, G.W., 1986. Three-dimensional transient electromagnetic responses for a grounded source. Geophysics, 51, 2117–2130.

Gunn, P.J., Maidment, D., Milligan, P.R., 1997a. Interpreting aeromagnetic data in areas of limited outcrop. J. Aust. Geol. Geophys., 17, 175–186.

Gunn, P.J., Milligan, P., Mackey, T., Liu, S., Murray, A., Maidment, D., Haren, R., 1997b. Geological mapping using the national airborne and gravity datasets: an example focussing on Broken Hill. J. Aust. Geol. Geophys., 17, 127–136.

Guo, S.Z., Liu, R., 2015. Application of cone penetration test in offshore engineering. Chin. J. Geotech. Eng., 37, 207–211.

Guo, X., Fan, N., Liu, Y., Liu, X., Wang, Z., Xie, X., Jia, Y., Deep seabed mining: frontiers in engineering geology and environment. International Journal of Coal Science & Technology (2023) 10:23 10.1007/s40789-023-00580-x.

Guo, X.S., Nian, T.K., Gu, Z.D., 2021a. A fluid mechanics approach to evaluating marine soft clay strength by a ball full flow penetrometer. Appl Ocean Res 116:10286.

Guo, X.S., Nian, T.K., Gu, Z.D., Li, D.Y., Fan, N., Zheng, D.F., 2021b. Evaluation methodology of laminar-turbulent flow state for fluidized material with special reference to submarine landslide. J. Waterw. Port Coast Ocean Eng. 147(1):04020048.

Guo, X.S., Nian, T.K., Wang, D., Gu, Z.D., 2022a. Evaluation of undrained shear strength of surficial marine clays using ball penetration based CFD modelling. Acta Geotech. 17(5):1627–1643.

Guo, X.S., Nian, T.K., Fu, C.W., Zheng, D.F., 2023a. Numerical investigation of the landslide cover thickness effect on the drag forces acting on submarine pipelines. J. Waterw. Port Coast Ocean Eng. 149(2):04022032.

Guo, Z., Dong, H., Liu, J., 2016. Comparison of marine controlled-source electromagnetic data acquisition systems by a reservoir sensitivity index: analyzing the effect of water depths. Acta Oceanol. Sin., 35, 113–119.

Guo, Z., Xue, G., Liu, J., Wu, X., Electromagnetic methods for mineral exploration in China: a review. Ore Geol. Rev, 118 (2020), 103357.

Gupta, A., Mehta, S., 2021. Remote sensing and GIS applications in agriculture: a case study in Punjab. Indian J. Agric. Sci., 91(7), 986–992.

Gupta, R.P., 2005. Remote Sensing Geology, 2nd edn. Springer Verlag, Berlin, 627p.

Gupta, R.P., 2018. Remote Sensing Geology. Springer Berlin, Heidelberg, Technology & Engineering, 428p.

Guptasarma, D., 1983. Effect of surface polarization on resistivity modeling. Geophysics, 48, 98–106.

Gusentsova, T.M., Kulkova, M.A., 2020. The subsistence strategy and paleoenvironment on the Stone Age site Podolye 1 in the southern Ladoga Lake region (Eastern Baltic). Quat. Int., 541, 41–51.

Gustavson, T., Sanderson, S.E., 2020. International regulations and governance for ethical Mineral exploration. International Journal of Environmental Policy and Governance, 22(4), 318–330.

Guterch, A., Grad, M., Janik, T., Keller, G.R., Thybo, H., 1999. POLONAISE'97 – an international seismic experiment in Poland. Tectonophysics, 314(1-3), 101–121.

H., L., Li, Z., Wang, Z., 2005. Application Research of GIS in Petroleum Exploration Data Management. Application Research of Computers.

Hack, H.R.G.K., 1982. Seismic methods in engineering geology. M.Sc. thesis, University of Utrecht;

University of Technology Delft, Section Engineering Geology, Delft, The Netherlands.

Hagemann, S.G., Jansen, H., Meinert, L., 2016. Ore-forming Processes in Mineral Systems. Springer, vol. 1.

Hagemann, S.G., Angerer, T., Duuring, P., Rosiere, C.A., Figueiredo, E., Silva, R.C., Lobato, L., Hebsler, A.S., Walde, D.H.G., 2016. BIF-hosted iron mineral system: a review. Ore Geol. Rev., 76, 317–359.

Hagemann, S.G., St. John, C.W.A., Brown, R.H.S., 2016a. Ore-forming Processes in Mineral Systems. Springer.

Hagemann, S.G., Lisitsin, V.A., Huston, D.L., 2016a. Mineral system analysis: quo vadis. Ore Geol. Rev., 76, 504–522.

Haile, T., Atsbaha, S., 2014. Electrical resistivity tomography, VES and magnetic surveys for dam site characterization, Wukro, Northern Ethiopia. J. Afr. Earth Sci. 97, 67–77. (doi: 10.1016/j.jafrearsci.2014.03.023).

Halbach, P., Kriete, C., Prause, B., Puteanus, D., 1989. Mechanisms to explain the platinum concentration in ferromanganese seamount crusts. Chem. Geol., 76:95–106.

Halbach, P., Marbler, H., 2009. Marine ferromanganese crusts: contents, distribution and enrichment of strategic minor and trace elements. Bundesanstalt Fur Geowissenschaften Und Rohstoffe, 1–73.

Halbach, P.E., Jahn, A., Cherkashov, G., 2017. Marine Co rich ferromanganese crust deposits: description and formation, occurrences and distribution, estimated world-wide resources. In: Sharma, R., editor. Deep Sea Mining. Cham: Springer International Publishing, 65–141.

Haldar, S., (editor) 2018. Mineral Exploration: Principles and Applications. 1st ed. Netherlands: Elsevier, 378p.

Hall, R.B., 1988. Hardrock versus softrock geology. In: General Geology. Springer US, 331–331.

Hall, R.H., Furlong, K.L., 2019. Public-Private partnerships in mineral exploration: a path to responsible resource development. Minerals Economics and Policy, 28(5), 60–73.

Halliday, Resnick, 2015. Fundamental of Physics 10th Edition, 1.

Hammer, S., 1939. Terrain corrections for gravimeter stations. Geophysics, 4(3), 184–194.

Han, B., Hu, X., Schultz, A., Zuo, B.X., Li, J.H., 2015. Three-dimensional forward modeling of the marine controlled-source electromagnetic field with complex source geometries (in Chinese). Chin. J. Geophys., 58, 1059–1071.

Han, S.L., Zhang, S.G., Liu, J.X., Hu, H.J., Zhang, W.S., 2013. Integrated interpretation of dual frequency induced polarization measurement based on wavelet analysis and metal factor methods. Transactions of Nonferrous Metals Society of China, 23(5), 1465–1471.

Hanna, W.F., 1969. Negative aeromagnetic anomalies over mineralized areas of the Boulder batholith, Montana. U.S. Geological Survey Professional Paper 650-D, 159–167.

Hannington, M., de Ronde, C., Petersen, S., 2005. Seafloor tectonics and submarine hydrothermal systems Economic Geology 100th Anniversary Volume. USA: Society of Economic Geologists, 111–141.

Hannington, M.D., Jonasson, I.R., Herzig, P.M., Petersen, S., Physical and chemical processes of seafloor mineralization at mid-ocean ridges in seafloor hydrothermal systems: physical, chemical, biological and geological interactions (Humphris, S.E., Zierenberg, R.A., Mullineaux, L.S., Tomson, R.E.) AGU Monograph 91, 115–157 (American Geophysical Union, 1995).

Hannington, M.D., Jamieson, J., Monecke, T., Petersen, S., Beaulieu, S., 2011. The abundance of seafloor massive sulfide deposits. Geology, 39(12), 1155–1158.

Hanzawa, H., Tanaka, H., 1992. Normalized undrained strength of clay in the normally consolidated state and in the field. Soils Found 32(1):132–148.

Hardin, P.J., Jensen, R.R., 2011. Small scale unmanned aerial vehicles in environmental remote sensing: challenges and opportunities. GISci. Remote Sens., 48, 99–111.

Hari, K., Alex, T., Mujib, M., Kenneth, C., 2023. India Moon Landing: in latest Moon Race, India lands first in Southern Polar Region. The New York Times. ISSN 0362-4331. Archived from the original on 26 August 2023. Retrieved 26 August 2023.

Harini, P., Sahadevan, D.K., Das, I.C., Manikyamba, C., Durgaprasad, M., Nandan, M.J., 2018. Regional Groundwater Assessment of Krishna River Basin Using Integrated GIS Approach. J. Indian Soc. Remote Sens., 46, 1365–1377.

Harrington, D., 2016. Geophysical Data Sharing for Sustainable Exploration: challenges and Opportunities. Geophys. J. Int., 174(4), 1234–1245.

Harris, D., Pan, G., 1999. Mineral Favourability Mapping: a Comparison of Artificial Neural Networks, Logistic Regression, and Discriminant Analysis. Nat. Resour. Res., 8, 93–109.

Harris, D.P., 1990. Mineral Exploration Decisions: A Guide to Economic Analysis and Modelling. (1st edition), John Wiley, 436p.

Harris, D.P., Zurcher, L., Stanley, M., Marlow, J., Pan, G., 2003. A comparative analysis of favourability mappings by weights of evidence, probabilistic neural networks, discriminant analysis, and logistic regression. Nat. Resour. Res., 12, 241–255.

Harris, J.R., 1991. Mapping of regional structures of eastern Nova Scotia using remotely sensed images: implications for regional tectonics and gold exploration. Canadian Journal of Remote Sensing, 17, 122–135.

Harris, J.R., Grunsky, E., Behnia, P., Corrigan, D., 2015. Data and knowledge-driven mineral prospectivity maps for Canada's North. Ore Geol. Rev, 71, 788–803.

Harris, J.R., Grunsky, E.C., 2015. Predictive lithological mapping of Canada's north using Random Forest classification applied to geophysical and geochemical data. Comput. Geosci., 80, 9–25.

Harris, R., Pan, J., 1999. GPS-based Surveying Techniques and Applications. Geomatics Press.

Harrison, C.B., Urosevic, M., 2012. Seismic processing, inversion and AVO for gold exploration – case study from Western Australia. Geophysics, 77, WC235–WC243.

Harrison, S., Leuangthong, O., Crawford, B., Oshust, P., 2009. Uncertainty-based grade modelling of kimberlite: a case study of the Jay kimberlite pipe, EKATI Diamond Mine, Canada. Lithos, 112, 73–82.

Harvey, T., 2022. Minerals geophysics. Preview, 2022:216, 35–35.

Hassanalian, M., Khaki, H., Khosravi, M., 2015. A new method for design of fixed wing micro air vehicle. Proc. Inst. Mech. Eng. Part G J. Aerosp. Eng., 229, 837–850.

Hassanalian, M., Rice, D., Abdelkefi, A., 2018. Evolution of space drones for planetary exploration: a review. Prog. Aerosp. Sci., 97.

Hassanalian, M., Rice, D., Abdelkefi, A., 2018. Aerodynamic performance analysis of fixed wing space drones in different Solar System bodies. In proceedings of the 2018 AIAA Aerospace Sciences Meeting, Kissimmee, FL, USA, 8-12 January 2018; American Institute of Aeronautics and Astronautics: Reston, VA, USA.

Hassanalian, M., Rice, D., Johnstone, S., Abdelkefi, A., 2018. Performance analysis of fixed wing space drones in different Solar System bodies. Acta Astronaut., 152.

Hassanalian, M., Abdelkefi, A., 2017. Classifications, applications, and design challenges of drones: a review. Prog. Aerosp. Sci., 91.

Hassanalian, M., Abdelkefi, A., 2017. Design, manufacturing, and flight testing of a fixed wing micro air vehicle with Zimmerman platform. Meccanica, 52.

Hatch, D., Kuna, S., Fecher, J., 2006a. Evaluation of an airship platform for airborne gravity gradiometry. In: Proceedings of the Australian Earth Sciences Convention 2006, Melbourne Australia.

Hatch, D., Murphy, C., Mumaw, G., Brewster, J., 2006b. Performance of the Air-FTG® system aboard an airship platform. Preview, 127, 17–22.

Hatherly, P.J., Urosevic, M., Lambourne, A., Evans, B.J., 1994. A simple approach to calculating refraction statics corrections. Geophysics, 59(1), 156–160.

Haugen, G.U., Schoenberg, M.A., 2000. The echo of a fault or fracture. Geophysics, 65, 176–189. doi: 10.1190/1.1444708.

Hawkins, A., Dohey, T., Crawford, A., Maidment, D., 2025. Geophysical response and exploration methods for callie-style targets in Tanami, NT, Australia. Newmont Corporation Technical Report.

He, J.S., 2010. Wide Field Electromagnetic Method and Pseudo Random Signal Method (In Chinese) Higher Education Press, Beijing.

He, L.F., Feng, M.H., He, Z.X., Wang, X.B., 2006. Application of EM methods for the investigation of Qiyueshan tunnel, China. Journal of Environmental and Engineering Geophysics, 11, 151–156.

He, X.F., Santosh, M., Zhang, Z.M., Tsunogae, T., Chetty, T.R.K., Ram Mohan, M., Anbazhagan, S., 2015. Shonkinites from Salem, southern India: implications for Cryogenian alkaline magmatism in rift-related setting. J. Asian Earth Sci., 113, 812–825.

Hearst, R.B., 1998. Reflections on kimberlite: a seismic adventure: SEG expanded abstracts.

Heath, P., Heinson, G., Greenhalgh, S., 2003. Some comments on potential field tensor data. Explor. Geophys., 34, 57–62.

Hein, J., Koschinsky, A., Mikesell, M., Mizell, K., Glenn, C., Wood, R., 2016. Marine phosphorites as potential resources for heavy rare earth elements and yttrium. Minerals, 6:88.

Hein, J.R., Mizell, K., Koschinsky, A., Conrad, T.T., 2013. Deep ocean mineral deposits as a source of critical metals for high and green technology applications: comparison with land-based resources. Ore Geol. Rev., 51:1–14.

Hein, J.R., Koschinsk, A., Bau, M., Manheim, F.T., Kang, J., Roberts, L., 2017. Cobalt rich ferromanganese crusts in the Pacific. Handbook of Marine Mineral Deposits. Routledge, 239–279.

Hein, J.R., Koschinsky, A., Kuhn, T., 2020. Deep ocean polymetallic nodules as a resource for critical materials. Nat. Rev. Earth Environ., 1, 158–169, doi: 10.1038/s43017-020-0027-0.

Hein, J.R., Koschinsky, A., 2014. Deep ocean ferromanganese crusts and nodules.

Heincke, B., Jackisch, R., Saartenoja, A., Salmirinne, H., Rapp, S., Zimmermann, R., Pirttijarvi, M., Sorensen, E.V., Gloaguen, R., Ek, L., Bergstrom, J., Karinen, A., Salehi1, S., Madriz, Y., Middleton, M., 2019. Developing multi-sensor drones for geological mapping and mineral exploration: setup and first results from the MULEDRO project. Geological Survey of Denmark and Greenland Bulletin 43, e2019430302.

Heinonen, S., Malinowski, M., Hlousek, F., Gislason, G., Buske, S., Koivisto, E., Wojdyla, M., 2019. Cost-effective seismic exploration: 2D reflection imaging at the Kylylahti sulphide deposit, Finland. Minerals, 9, 263.

Heinson, G., White, A., Constable, S., Key, K., 1999. Marine self potential exploration. Explor. Geophys., 30, 1–4.

Heinson, G., White, A., Robinson, D., Fathianpour, N., 2005. Marine self-potential gradient exploration of the continental margin. Geophysics, 70(5), G109–G118.

Henley, R.W., Adams, D.P.M., 1992. Strike-slip reactivation as a control on epithermal: vein-style gold mineralization. Geology, 20, 443–446.

Hennage, D.H., Nopola, J.R., Haugen, B.D., 2019. Fully Autonomous Drone for Underground Use; American Rock Mechanics Association: Alexandria, VA, USA.

Hensen, C., Nuzzo, M., Hornibrook, E., Pinheiro, L.M., Bock, B., Magalhaes, V.H., Bruckmann, W., 2007. Sources of mud volcano fluids in the Gulf of Cadiz – indications for hydrothermal imprint. Geochim. Cosmochim. Acta, 71, 1232–1248.

Heron, A.M., 1953. Geology of Central Rajasthan. Memoirs Geol. Surv. India, 79, 339p.

Hesthammer, J., Stefatos, A., Boulaenko, M., Fanavoll, S., Danielsen, J., 2010. CSEM performance in light of well results. Lead. Edge, 29(1), 34–41. 10.1190/1.3284051.

Hill, D., Roberts, D., Li, Y., 2017. International Collaboration in Arctic Mineral Exploration: case Studies and Best Practices. Journal of Arctic Geology, 44(1), 101–115.

Hill, J.A., 2000. Testing the resolution of very shallow geophysics: data from the EIGG test site, Leicester UK. Conference. 6th EAGE/EEGS, Meeting of Environmental and Engineering Geophysics, September 3-7, 2000, Bochum – Germany.

Hinze, W.J., 1990. The role of gravity and magnetic methods in engineering and environmental studies Geotechnical and Environmental Geophysics: volume I, Review and Tutorial, 4, Society of Exploration Geophysicists, 75–126.

Hinze, W.J., von Frese, R.R.B., Saad, A.H., 2013. Gravity and Magnetic Exploration – Principles, Practices, and Applications. 1st ed. Cambridge University Press, 525p.

Hoagland, P., Beaulieu, S., Tivey, M.A., et al., 2010. Deep sea mining of seafloor massive sulfides. Mar. Policy, 34:728–732.

Hobbs, B., Li, G., Clarke, C., Linfoot, J., 2005. Inversion of multi-transient electromagnetic data. 68th Conference & Technical Exhibition, EAGE, Extended Abstracts, A015.

Hobbs, R.W., Drummond, B.J., Goleby, B.R., 2006. The effects of three-dimensional structure on two-dimensional images of crustal seismic sections and on the interpretation of shear zone morphology. Geophys. J. Int., 164, 490–500.

Hodgson, J.C., Baylis, S.M., Mott, R., Herrod, A., Clarke, R.H., 2016. Precision wildlife monitoring using unmanned aerial vehicles. Sci. Rep., 6, 22574.

Hodgson, L., Paterson, S., Baird, C. (2016). Geophysical mapping of the Canadian Shield: advancing mineral exploration. Canadian Journal of Earth Sciences, 53(2), 121–132.

Hoffmann, R., McAllister, I., 2018. Use of Unmanned Aerial Systems Reduces HES Risks. Society of Petroleum Engineers, Calgary, AB, Canada.

Hoffman-Wellenhof, B., et al., 1997. GPS Theory and Practice. 4th Rev. Edition, Springer Wien, New York.

Hofstadter, R., 1949. The detection of gamma rays with thallium activated sodium iodide crystals. Phys. Rev. 75, 796–810.

Hogg, S., 2004. Practicalities, pitfalls and new developments in airborne magnetic gradiometry. First Break, 22(7).

Holbrook, W.S., Hoskins, H., Wood, W.T., Stephen, R.A., Lizarralde, D., 1996. Methane hydrate and free gas on the Blake Ridge from vertical seismic profiling. Science, 273(5283), 1840–1843.

Holbrook, W.S., Gorman, A.R., Hornbach, M.J., Hackwith, K.L., Nealon, J.W., Lizarralde, D., et al (2002) Seismic detection of marine methane hydrate. Lead. Edge 21:686–689.

Holbrook, W.S., Gorman, A.R., Hornbach, M.J., Hackwith, K.L., Nealon, J.W., Lizarralde, D., et al (2002) Seismic detection of marine methane hydrate. Lead. Edge 21:686–689.

Hollis, J.M., et al. (2018). Airborne Geophysical Surveys in Africa: a Tool for Mineral Resource Exploration in Challenging Environments. Geophys. Prospect., 66(2), 511–526.

Hollister, C.D., Ewing, J.I., et al., 1972. Initial reports of the Deep Sea Drilling Project, v. XI. Washington, D.C., U.S. Govt. Printing Office, 1078p.

Holz, S., Jegen, M., Petersen, S., Hannington, M., 2015. How to find buried and inactive seafloor massive sulfides using transient electromagnetics – a case study from Palinuro Seamount in the Tyrrhenian Sea, Underwater Mining Institute Conference, Tampa Bay, FL.

Hong, J., Oh, S., 2021. Model selection for mineral resource assessment considering geological and grade uncertainties: application of multiple-point geostatistics and a cluster analysis to an iron deposit. Nat. Resour. Res., 30(3), 2047–2065.

Hong, S., Kim, H.W., Yeu, T., Choi, J.S., Lee, T.H., Lee, J.K., 2019. Technologies for safe and sustainable mining of deep seabed minerals. In: Environmental Issues of Deep Sea Mining: Impacts, Consequences and Policy Perspectives. Springer, Cham, 95–14. 10.1007/978-3-030-12696-4_5.

Honkamo, M., 1988. Pre-Quaternary rocks of the Haukipudas and Kiiminki map-sheet areas. Geol. Surv. Finland, Geological map of Finland, 1:100000, Explanation to the maps of pre-Quartenary rock, (in Finnish), 46.

Honsho, C., Yamazaki, T., Ura, T., Okino, K., Morozumi, H., Ueda, S., 2016. Magnetic anomalies associated with abundant production of pyrrhotite in a sulphide deposit in the Okinawa Trough, Japan. Geochem. Geophys. Geosyst., 17, 4413–4424.

Hood, P., 1981. Aeromagnetic gradiometry: a superior geological mapping tool for mineral exploration programs. In: Weinstock, H., Overton, W.C.) SQUID Applications to Geophysics, Proceedings of the Workshop held 2-4 June 1980, Los Alamos, Soc. Expl. Geophys., Tulsa, Oklahoma, 72–76.

Hordt, A., Bairlein, K., Spagnoli, G., Jegen, M., Hannington, M., Petersen, S., Laurila, T.E., 2016. Induced polarization of seafloor massive sulfides. International Workshop on Induced Polarization. Aarhus, Denmark : University of Aarhus. 10.1016/j.cageo.2015.09.015.

Horo, D., Pal, S.K., Singh, S., Srivastava, S., 2020. Combined self-potential, electrical resistivity tomography and induced polarization for mapping of gold prospective zones over a part of Babaikundi-Birgaon Axis, North Singhbhum Mobile Belt, India. Explor. Geophys., 51(5), 507–522.

Horo, D., Pal, S.K., Singh, S., 2021. Mapping of gold mineralization in Ichadih, North Singhbhum Mobile Belt, India using electrical resistivity tomography and self-potential methods. Min. Metall. Explor., 38, 397–411.

Hosoi, Y., 2000. Application of airborne geophysics to mineral exploration - Present and Future. Shigen-to-Sozai, 116(10), 805–813.

Hosseini, S.A., Asghari, O., Emery, X., 2017. Direct block support simulation of grades in multi-element deposits: application to recoverable mineral resource estimation at Sungun porphyry copper-molybdenum deposit. Journal of the Southern African Institute of Mining and Metallurgy, 117(6), 577–585.

Hosseini-Dinani, H., Aftabi, A., 2016. Vertical lithogeochemical halos and zoning vectors at Goushfil Zn-Pb deposit, Irankuh district, southwestern Isfahan, Iran: implications for concealed ore exploration and genetic models. Ore Geol. Rev., 72, 1004–1021.

Hou, D., O'Connor, D., Nathanail, P., Tian, L., Ma, Y., 2017. Integrated GIS and multivariate statistical analysis for regional scale assessment of heavy metal soil contamination: a critical review. Environ. Pollut., 231, 1188–1200.

Howell, L.G., Frosch, A., 1939. Gamma ray and well-logging. Geophysics, 4(2), 106–114.

Howland-Rose, A.W., Linford, G., Pitcher, D.H., Seigel, H.O., 1980. Some recent magnetic induced-polarization developments – part II: survey results. Geophysics, 45(1):55–74.

Hronsky, J.M.A., 2004. The science of exploration targeting. Muhling, J., Goldfarb, R., Vielreicher, N., Bierlein, F., Stumpfl, E., Groves, D., Kenworthy, S. (Eds.), SEG2004 – Predictive Mineral Discovery Under Cover. SEG2004 Conference, Perth, Centre for Global Metallogeny, The University of Western Australia, Publication, 33, 129–133.

Hronsky, J.M.A., 2011. Self-organized critical systems and ore formation: the key to spatial targeting? Soc. Econ. Geol. Newsl., 84, 14–16.

Hronsky, J.M.A., Groves, D.I., 2008. Science of targeting: definition, strategies, targeting and performance measurement. Aust. J. Earth Sci., 55, 3–12.

Hronsky, J.M.A., Kreuzer, O.P., 2019. Applying spatial prospectivity mapping to exploration targeting: fundamental practical issues and suggested solutions for the future. Ore Geol. Rev., 107, 647–653.

Hsu, S.K., Chiang, C.W., Evans, R.L., Chen, C.S., Chiu, S.D., Ma, Y.F., Wang, Y., 2014. Marine controlled source electromagnetic method used for the gas hydrate investigation in the offshore area of SW Taiwan. J. Asian Earth Sci., 92:224–232.

Hu, B., Xu, Y.Y., Wan, B., Wu, X.C., Yi, G.H., 2018. Hydrothermally altered mineral mapping using synthetic application of Sentinel-2A MSI, ASTER and Hyperion data in the Duolong area, Tibetan Plateau, China. Ore Geol. Rev., 101, 384–397.

Huang, H., Fraser, D.C., 1999. Airborne resistivity data levelling. Geophysics, 64, 378–385.

Huang, W., Huang, J., Wang, X., Wang, F., Shi, J., 2013. Comparability of red/near-infrared reflectance and NDVI based on the spectral response function between MODIS and 30 other satellite sensors using rice canopy spectra. Sensors, 13, 16023–16050.

Huizar-Alvarez, R., et al., 2003. Geologic structure and groundwater flow in the Pachuca-Zumpango sub-basin Central Mexico. Environ. Geol. 43(4), 385–399.

Hunt, G.R., 1979. Near-infrared (1.3-2.4 μm) spectra of alteration minerals-potential for use in remote sensing. Geophysics, 44, 1974–1986.

Hunt, G.R., Ashley, R.P., 1979. Spectra of alteRed Rocks in the visible and near infrared. Econ. Geol., 74, 1613–1629

Hunt, G.R., Salisbury, J.W., 1970a. Visible and near-infrared spectra of minerals and rocks: i silicate minerals. Mod. Geol., 1, 283–300.

Hunt, G.R., Salisbury, J.W., 1970b. Visible and near - infrared spectra of minerals and rocks: III oxides and hydroxides. Mod. Geol., 2, 195–205.

Hunt, G.R., Salisbury, J.W., 1971. Visible and near-infrared spectra of minerals and rocks: II carbonates. Mod. Geol., 2, 23–30.

Hunt, G.R., Salisbury, J.W., 1976. Visible and near-infrared spectra of minerals and rocks: XII. Metamorphic rocks. Mod. Geol., 5, 219–228.

Huntington, T., et al. (2019). Geophysical Exploration for Mineral Deposits in Polar Regions. J. Geophys. Res., 126(5), 1450–1468.

Hurich, C.A., 1996. Kinematic evolution of the lower plate during intracontinental subduction: an example from the Scandinavian Caledonides. Tectonics, 15, 1248–1263. doi: 10.1029/96TC00828.

Huston, D.L., Stevens, B., Southgate, P.N., Muhling, P., Wyborn, L., 2006. Australian Zn-Pb-Ag ore-forming systems: a review and analysis. Econ. Geol., 101, 1117–1158.

Hutchinson, I., Davis, R.A., 2020. Collaborative research in geophysical mapping: advancing solutions for sustainable mineral exploration. Journal of Geophysical Research: Solid Earth, 125(11), e2020JB019461.

Hutchinson, P., Barta, L.S., 2002. VLF surveying to delineate long wall mine-induced fractures. Lead. Edge, 21, 491–493.

Hyakudome, T., Yoshida, H., Ishibashi, S., Ochi, H., Sawa, T.K., Nakano, Y., Watanabe, Y., Nakatani, T., Sugesawa, M., Ohta, Y., Watanabe, K., Oomoka, S., Nanbu, Y., Komuku, M.M., 2012. Development of AUV for scientific observation Oceans. VA, USA: IEEE, 1–4.

IAEA (International Atomic Energy Agency), 1979. Gamma Ray Surveys in Uranium Exploration. Technical Reports Series No. 186, IAEA, Vienna.

IAEA (International Atomic Energy Agency), 1989. Construction and Use of Calibration Facilities for Radiometric Field Equipment. Technical Reports Series No. 309, IAEA, Vienna.

IAEA (International Atomic Energy Agency), 1991. Airborne Gamma Ray Spectrometer Surveying. Technical Reports Series, No. 323, IAEA, Vienna.

IAG (International Association of Geodesy), 1980. The geodesist's handbook. Bull. Geod., 54(3).

ICRU, 1994. Gamma ray Spectrometry in the Environment. ICRU Report 53, International Commission on Radiation Units and Measurements, Bethesda, USA.

Iijima, K., et al., 2016. Discovery of extremely REY rich mud in the western North Pacific Ocean. Geochem. J., 50, 557–573.

Iizasa, K., Fiske, R.S., Ishizuka, O., Yuasa, M., Hashimoto, J., Ishibashi, J., Naka, J., Horii, Y., Fujiwara, Y., Imai, A., Koyama, S., 1999. A kuroko-type polymetallic sulfide deposit in a submarine silicic caldera. Science, 283, 975–977.

Inazaki, T., 2004. High resolution seismic refection surveying at paved areas using an S-wave type land streamer. Explor. Geophys., 35, doi: 10.1071/EG04001.

International Council on Mining and Metals, 2020. Mining and sustainable development: the role of technology and innovation. ICMM.

International Geophysical Union (IGU), 2021. Global standards in geophysical mapping for sustainable exploration. International Geophysical Union. Retrieved from https://www.igu.org.

International Seabed Authority, 2003. Establishment of a geological model of the polymetallic nodule resources in the Clarion-Clipperton Fracture Zone of the Equatorial North Pacific Ocean. In Proceedings of the International Seabed Authority Workshop, Nadi, Fiji, 13-20 May 2003; International Seabed Authority: Kingston, Jamaica,.

Irvine, R.J., Smith, M.J., 1990. Geophysical exploration for epithermal gold deposits. J. Geochem. Explor., 36, 375–412.

Ishikawa, H., Nogami, S., Endo, M., Sunwall, D., Feijth, J., Hope, J., Mouri, T., Nakamoto, G., 2020. Targeting geothermal exploration using airborne multi-geophysics, Kutcharo caldera, Teshikaga area. Nat. Resour. Res. 10.1007/s11053-020-09704-5.

Ishizu, K., Goto, T., Ohta, Y., Kasaya, T., Iwamoto, H., Vachiratienchai, C., Siripunvaraporn, W., Tsuji, T., Kumagai, H., Koike, K., 2019. Internal structure of a seafloor massive sulphide deposit by electrical resistivity tomography, Okinawa trough. Geophysical Research Letters, 46, 11025–11034.

Isles, D., Moody, I., 2004. Examples of Falcon data from diamond exploration projects in Northern Australia: in R.J.L. Lane, ed. Airborne Gravity 2004 - Abstracts from the ASEG-PESA Airborne Gravity 2004 Workshop, Geoscience Australia Record 2004/18, 121–124.

Isles, D.J., Harman, P.G., Gunneen, J.P., 1989. The contributions of high resolution aeromagnetic to Archean gold exploration in the region, Western Australia. Econ. Geol. Monogr., 6, 389–397.

IUCN, 2020. Western Ghats – 2020 Conservation Outlook Assessment. IUCN World Heritage Outlook.

Ivanovich, M., Harmon, R.S., 1982. Uranium Series Disequilibrium: Applications to Environmental Problems, Clarendon Press, Oxford.

Jackisch, R., 2020. Drone-based surveys of mineral deposits. Nature Reviews Earth & Environment. 10.1038/s43017-020-0042-1.

Jackisch, R., 2023. Drone-Based Magnetic Surveys for Mineral Exploration and Mapping. Helmholtz Institute Freiberg for Resource Technology, Freiberg, Saxony, Germany. Photo courtesy of Richard Gloaguen, Helmholtz Institute Freiberg for Resource Technology, Germany.

Jackish, R., Heinke, B.H., Zimmeman, R., Sorensen, E.V., Porltijarvi, M., Kirsh, M., Salmivine, H., Lodi, S., Kuronen, U., Gloaguen, R., 2021. Drone based magnetic and multispectral surveys to develop a 3D model for mineral exploration at Quill sat, Disko Island Greenland. Solid Earth. 10.5194/se-2021-33.

Jackson, C.A.L., Jackson, M.P.A., Hudec, M.R., Rodriguez, C.R., 2015. Enigmatic structures within salt walls of the Santos Basin-Part 1: geometry and kinematics from 3D seismic reflection and well data. J. Struct. Geol., 75, 135–162. doi: 10.1016/j.jsg.2015.01.010.

Jain, A., Shah, R., 2020. Artificial intelligence in sustainable supply chains: enhancing transparency and ethical sourcing. Journal of Supply Chain Management, 56(2), 78–94.

James, C.K., Maurice, A.T., Dana, R.Y., 2013. Dynamics and navigation of autonomous underwater vehicles for submarine gravity surveying. Geophysics, 78(3):55–68.

Jansen, P.M., Kotze, T.E., 2015. Geochemical mapping for uranium exploration in Namibia. Mineral Exploration and Mining Journal, 28(4), 189–203.

Japan Gold Corp., 2019. Retrieved from https://www.newsfilecorp.com/release/54261. Japan gold announces positive results from geophysical surveys at the Ohra-Takamine Gold Project.

Jaques, A.L., Wellman, P., Whitaker, A., Wyborn, D., 1997. High-resolution geophysics in modern geological mapping. J. Aust. Geol. Geophys., 17, 159–186.

Jarvie, D.M., Hill, R.J., Ruble, T.E., Pollastro, R.M., 2007. Unconventional shale gas systems: the Mississippian Barnett shale of north-central Texas as one model for thermogenic shale gas assessment. AAPG Bull., 91, 475–499.

Jarvis, K., Terblanche, O., Brown, S., 2022. Building 3D rock models from seismic at the Tropicana gold mine. In: Modern Applications of Geophysics – Mineral Case Studies: Reflections on the MAG21 symposium presented by the WA Branch of the ASEG. Taylor and Francis Group Published by Neroni and Hunt, 2022. Preview, 2022, 216, 19–25, DOI: 10.1080/14432471.2022.2034311.

Jenkin, G.R.T., Lusty, P.A.J., McDonald, I., Smith, M.P., Boyce, A.J., Wilkinson, J.J., 2014. Ore deposits in an evolving earth: an introduction. Geological Society, 393(1), 1–8.

Jensen, J.R., 2015. Introductory Digital Image Processing: A Remote Sensing Perspective (4th ed.). Pearson.

Jensen, J.R., McDonnell, J.R., 2020. Remote Sensing of the Environment: An Earth Resource Perspective. Pearson Education.

Ji, S., Long, C., 2006. Seismic reflection response of folded structures and implications for the interpretation of deep seismic reflection profiles. J. Struct. Geol., 28, 1380–1387. doi: 10.1016/j.jsg.2006.05.003.

Jiang, B., Liu, C., 2009. Street-Based Topological Representations and Analyses for Predicting Traffic Flow in GIS. International Journal of Geographical Information Systems, 23(9), 1119–1137.

Jiang, Y.Y., 2011. The Study of Cone Penetration Testing Methods and the Application in Offshore Engineering; Tianjin University: Tianjin, China.

Jocobson, R.S., Dorman, L.M., Purdy, G.M., Schultz, A., Solomon, S.C., 1991. Ocean bottom seismometer facilities available. EOS Trans. Am. Geophys. Union, 72(46):506–515.

John, D.A., Garside, L.J., Wallace, A.R., 1999. Magmatic and tectonic setting of late Cenozoic epithermal gold-silver deposits in northern Nevada, with an emphasis on the Pah Rah and Virginia Ranges and the northern Nevada rift. Geol. Soc. Nev. Spec. Publ., 29, 64–158.

John, P.R., ed. 2011. Minerals; p. 1. In: the Series Geology: Landforms, Minerals, and Rocks. Rosen Publishing Group. 978-1-61530-489-9.

Joint Ore Reserves Committee, 2012. The JORC code 2012 edition effective 20 December 2012 and mandatory from 1 December 2013 Australasian code for reporting of exploration results, mineral resources and ore reserves prepared by the Joint Ore Reserves Committee of The Australasian Institute of Mining and Metallurgy, Australian Institute of Geoscientists and Minerals Council of Australia (JORC).

Joly, A., Porwal, A.K., McCuiag, T.C., 2012. Exploration targeting for orogenic gold deposits in the Granites-Tanami Orogen: mineral system analysis, targeting model and prospectivity analysis. Ore Geol. Rev., 48, 349–383.

Joly, A., Porwal, A., McCuaig, T.C., Chudasama, B., Dentith, M.C., Aitken, A.R.A., 2015. Mineral systems approach applied to GIS-based 2D-prospectivity modelling of geological regions: insights from Western Australia. Ore Geol. Rev., 71, 673–702.

Jones, A., 2020. India back in action with launch of Earth Observation Satellite, nine rideshare small sats. SpaceNews, retrieved 2021-01-06.

Jones, E.J.W., 1999. Marine Geophysics. John Wiley & Sons Ltd., Chichester, 466p.

Jones, H., Patel, P., Johnson, R., 2021. Precision agriculture and the role of artificial intelligence in sustainable water use. Journal of Agricultural Technology, 29(6), 1505–1519.

Jones, R., Smith, P., 2018. Bauxite Formation and Distribution: a Tropical Weathering Process. Geol. Soc. Aust. Bull., 132(2), 211–230.

Juhlin, C., Wahlgren, C.H., Stephens, M.B., 2000. Seismic imaging in the frontal part of the Sveconorwegian orogen, south-western Sweden. Precambrian Res., 102, 135–154. doi: 10.1016/S0301-9268(00)00064-4.

Juhlin, C., Dehghannejad, M., Lund, B., Malehmir, A., Pratt, G., 2010. Reflection seismic imaging of the end-glacial Parvie Fault system, northern Sweden. J. Appl. Geophys., 70, 307–316.

Juhlin, C., Lund, B., 2011. Reflection seismic studies over the end-glacial Burtrask fault, Skelleftea, Sweden. Solid Earth, 2, 9–16. doi: 10.5194/se-2-9-2011.

Juhlin, C., Palm, H., 2003. Experiences from shallow reflection seismics over granitic rocks in Sweden, D.W., E., Milkereit, B., Salisbury, M.H. (eds.) Hardrock Seismic Exploration. SEG Developments in Geophysics Series.

Juhojuntti, N., Juhlin, C., Dyrelius, D., 2001. Crustal reflectivity underneath the Central Scandinavian Caledonides. Tectonophysics, 334, 191–210.

Junno, N., Koivisto, E., Kukkonen, I., Malehmir, A., Montonen, M., 2019. Predicting missing seismic velocity values using self-organizing maps to aid the interpretation of seismic reflection data from the Kevitsa Ni-Cu-PGE deposit in Northern Finland. Minerals, 9, 529.

Kahle, A.B., Morrison, A.D., Tsu, H., Yamaguchi, Y., 1996. Geologic remote sensing in the thermal infrared. Int. Arch. Photogramm. Remote Sens., 31, 327–330.

Kamguia, J., Manguelle-Dicoum, E., Tabod, C.T., Tadjou, J.M., 2005. Geological models deduced from gravity data in the Garoua basin, Cameroon. J. Geophys. Eng., 2(2), 147–152.

Kaminski, V.F., Kuzmin, P., Legault, J.M., 2010. AirMt – passive airborne EM system. Presented at 3rd CMOS-CGU Congress, Ottawa.

Kamm, J., Lundin, I.A., Bastani, M., Sadeghi, M., Pedersen, L.B., 2015. Joint inversion of gravity, magnetic, and petrophysical data — a case study from a gabbro intrusion in Boden, Sweden. Geophysics, 80(5), B131–B152.

Kamto, P.G., Oksum, E., Yap, L., Kandel, L.H., Kamguia, J., 2023. High precision structural mapping using advanced gravity processing methods: a case study from the North region of Cameroon.

Kang, H., Chen, Y.L., Li, D.P., Zhao, J.X., Cui, F.R., Xu, Y.L., 2020. Deep-penetrating geochemistry for concealed sandstone-type uranium deposits: a case study of Hadatu uranium deposit in the Erenhot Basin, North China, J. Geochem. Explor., 211, 106464.

Kang, H., Ghosh, M., 2018. Geospatial Data Fusion for Resource Exploration. Springer.

Kaplan, E.D., Hegarty, C., 2005. Understanding GPS Principle and Applications. Artech House Publishers, Norwood, 706p.

Karimi, H.A., Houston, B.J., 1996. Evaluating Strategies for Integrating Environmental Models with GIS: current Trends and Future Needs. Computers Environment and Urban Systems, 20(6), 413–425.

Karjalainen, M., Ahokas, E., Hyyppa, J., Heinonen, T., Jaakkola, A., Matikainen, L., Lehtomaki, M., Kukko, A., 2016. Remote sensing methods for power line corridor surveys. ISPRS J. Photogramm. Remote Sens., 119, 10–31.

Kasaya, T., Iwamoto, H., Kawada, Y., Hyakudome, T., 2020. Marine DC resistivity and self-potential survey in the hydrothermal deposit areas using multiple AUVs and ASV. Terr. Atmos. Ocean. Sci., 31(5), 579–588.

Kasaya, T., Iwamoto, H., Kawada, Y., 2021. Deep sea DC resistivity and self-potential monitoring system for environmental evaluation with hydrothermal deposit mining. Frontiers in Earth Science, 9, 608381.

Kato, Y., Fujinaga, K., Nakamura, K., et al., 2011. Deep sea mud in the Pacific Ocean as a potential resource for rare earth elements. Nat. Geosci.; 4:535–539.

Katz, M., 1982. Lineament analysis of Landsat imagery applied to mineral exploration. In: Mineral Exploration Techniques in Tropical Forest Areas (Laming, D.I.C., Gibbs, A.K. Eds.). Hiden Wealth, 157–166.

Kaufinann, R.D., 1996. Microgravity survey of the Oak Ridge K-25 Site, Oak Ridge, Tennessee, Energy Systems Environmental Restoration Program. U.S. Department of Energy, 3–34.

Kaufman, A.A., 1989. A paradox in geoelectromagnetism, and its resolution, demonstrating the equivalence of frequency and transient domain methods. Geoexploration, 25(4), 287–317.

Kaufman, A.A., Keller, G.V., 1983. Frequency and Transient Soundings. Elsevier, Amsterdam.

Kaufman, A.A., Keller, G.V., 1983. Frequency and transient sounding. Methods in Geochemistry and Geophysics, 16. Elsevier Scientific Publishing Co., Amsterdam, 685p.

Kaufman, A.A., Morozova, G.M., 1970. Theoretical basis for the transient sounding method in the near-zone: nauka.

Kaur, H., Srinivas, S., Rao, N.B., Bhushan, K., Golani, P.R., Balakrishna, R., 2018. Analysis of gravity and magnetic data of Peddavoora area in deciphering the structural controls of uranium mineralization in the northeastern part of Cuddapah Basin, Nalgonda. J. Geophys., 39(1), 2018–2021.

Kawada, Y., Kasaya, T., 2017. Marine self-potential survey for exploring seafloor hydrothermal ore deposits. Scientific Reports, 7, 1–12.

Kawada, Y., Kasaya, T., 2018. Self-potential mapping using an autonomous underwater vehicle for the Sunrise deposit, Izu-Ogasawara arc, southern Japan. Earth, Planets and Space, 70, 1–15.

Ke, L., Yu, K.S.H., Wong, Y.S., Tam, N.F.Y., 2005. Spatial and vertical distribution of polycyclic aromatic hydrocarbons in mangrove sediments. Sci Total Environ., 340(1-3):177–187. 10.1016/j.scitotenv.2004.08.015.

Keane, J.F., Carr, S.S., 2013. A brief history of early unmanned aircraft. Johns Hopkins APL Tech. Dig., 32, 558–571.

Kearey, P., Brooks, M., Ian, H., 2002. An Introduction to Geophysical Exploration. Third edition, Blackwell Publishing.

Keller, G., 2015. Capacity building in geophysical exploration: addressing the needs of developing countries. J. Glob. Geophys., 59(2), 202–215.

Keller, G.V., Pritchard, J.I., Jacobson, J.J., Harthill, N., 1984. Mega-source time domain electromagnetic sounding methods. Geophysics, 49, 993–1009.

Keller, G.V., Harris, L.M.S., Brown, D.O., 2017. Borehole nuclear magnetic resonance: new techniques for fluid detection in geological formations. Journal of Geophysical Research: Solid Earth, 122(10), 8593–8610.

Keller, G.V., Frischknecht, F.C., 1966. Electrical Methods in Geophysical Prospecting: Oxford, Pergamon Press, 517p.

Kemna, A., 2000. Tomographic Inversion of Complex Resistivity. Ruhr Universitat Bochum, 169.

Kemp, D., Owen, J.R., 2018. Community engagement in mining projects: using geophysical data for social responsibility. Resources Policy, 59, 57–66.

Khalil, M.H., Ahmed, K.S., Elnahry, A.H., Hasan, A.N., 2014. Integrated geophysical, remote sensing and GIS studies for groundwater assessment, Abu Zenima Area, West Sinai, Egypt. International Journal of Geosciences, 5, 882–907.

Khan, K.A., 2000. Integrated Geo Systems – a Computational Environment for Integrated Management, Analysis and Presentation of Petroleum Industry Data. In: Coburn, T.C., Yarus, J.M., Eds., Geographic Information Systems in Petroleum Exploration and Development, AAPG Book on Computers in Geology, no. 4, Chapter 16, 215–226.

Khan, K.A., Akhter, G., Ahmad, Z., 2011. GIS as Project Manager in Geophysical Applications Software. Journal of Geographic Information System, 3, 166–172.

Khan, M., Ali, S., 2020. Oil and Gas Reserves in the Middle East: a Global Geopolitical Overview. Energy and Geopolitics, 28(2), 35–49.

Kharol, S.K., Badarinath, K.V.S., Roy, P.S. 2008. Studies on emissions from forest fires using multi-satellite datasets over north east region of India. In: The International Archives of the Photogrammetry, Remote Sensing and Spatial Information Sciences, 37, 473–478 Part B8 Beijing.

Khatami, R., Mountrakis, G., Stehman, S.V., A meta-analysis of remote sensing research on supervised pixel-based land-cover image classification processes: general guidelines for practitioners and future research. Remote Sens. Environ. 2016, 177, 89–100.

Kiberu, J., 2002. Induced polarization and Resistivity measurements on a suite of near surface soil samples and their empirical relationship to selected measured engineering parameters. International Institute for Geoinformation Science and Earth Observation Enschede, The Netherlands.

Kienzle, S., 2004. The effect of DEM Raster Resolution on First Order, Second Order, and Compound Terrain Derivatives. Transactions in GIS, 8(1), 83–111.

Killeen, P.G., 1997a. Borehole geophysics: exploring the third dimension. In "Proceedings of Exploration 97: Fourth Decennial Conference on Mineral Exploration" edited by Gubins, A.G., 31–42.

Killeen, P.G., 1997b. Nuclear techniques for ore grade estimation. In "Proceedings of Exploration 97: Fourth Decennial Conference on Mineral Exploration" edited by Gubins, A.G., 677–684.

Kilty, K.T., 1983. Werner deconvolution of profile potential field data. Geophysics, 48, 234–237.

Kim, M.G., Hyeong, K., Yoo, C.M., 2022. Distribution of rare earth elements and yttrium in sediments from the Clarion-Clipperton fracture zone, Northeastern Pacific Ocean. Geochem. Geophys. Geosyst., 23.

Kindervater, K.H., 2016. The emergence of lethal surveillance: watching and killing in the history of drone technology. Secur. Dialogue, 47, 223–238.

Kirsch, R., 2006. Groundwater Geophysics: A Tool for Hydrogeology. Springer-Verlag Berlin Heidelberg Printed in Germany, 500p.

Kjellstrom, T., Yule, B., 2020. Geophysical mapping and the sustainable development goals: aligning resource extraction with environmental stewardship. Environ. Geol., 74(6), 1183–1195.

Knoll, C., Kerschner, H., Heller, A., Rastner, P., 2009. A GIS-Based Reconstruction of Little Ice Age Glacier Maximum Extents for South Tyrol, Italy. Transactions in GIS, 13(5-6), 449–463.

Knox, A., O'Neill, J.F., Smith, R.N., 2016. Innovations in gamma-ray spectrometry for borehole geophysics. Geophys. Prospect., 64(2), 437–449.

Knox-Robinson, C.M., 2000. Vectorial fuzzy logic: a novel technique for enhanced mineral prospectivity mapping, with reference to the orogenic gold mineralisation potential of the Kalgoorlie Terrane, Western Australia. Aust. J. Earth Sci., 47, 929–941.

Knox-Robinson, C.M., Wyborn, L.A.I., 1997. Towards a holistic exploration strategy: using geographic information systems as a tool to enhance exploration. Aust. J. Earth Sci., 44, 453–463.

Koefoed, O., 1979. Geosounding Principles 1: Resistivity Sounding Measurements. Amsterdam: Elsevier Science.

Koike, K., Nagano, S., Ohmi, M., 2001. Lineament analysis of satellite images using a Segment Tracing Algorithm (STA). Comput. Geosci., 21, 1091–1104.

Kolesov, V.V., Pavlenkin, A.D., Goryachev, N.A., Ivanov, A.N., 2017. Using the digital models of gravity anomalies for zoning of the Earth's gravity field. Izv. Phys. Solid Earth, 53(6), 935–945.

Kolodny, Y., 1969. Are marine phosphorites forming today? Nature, 224:1017–1019.

Koomans, R.L., van der Wal, P., Kamsma, T.P.J., 2014. A geophysical technique for the verification and quantification of zinc slag in roads. In: Near Surface 2009 – 15th EAGE European Meeting of Environmental and Engineering Geophysics.

Kopf, A., 2002. Significance of mud volcanism. Rev. Geophys., 40, 1–52.

Koschinsky, A., Hein, J.R., 2003. Uptake of elements from seawater by ferromanganese crusts: solid-phase associations and seawater speciation. Mar. Geol., 198:331–351.

Koschmann, A.H., Bergendahl, M.H., 1968. Principal gold producing areas of the United States. Geological Survey Professional Paper 610.

Kowalczyk, P., 2008. Geophysical prelude to first exploitation of submarine massive sulphides. First Break, 26, 99–106.

Krauss, C., 2009. New way to tap gas may expand global supplies. The New York Times. Retrieved 6 August 2013.

Krawczyk, C., Polom, U., Trabs, S., Dahm, T., 2012. Sinkholes in the city of Hamburg – new urban shear wave refection seismic system enables high resolution imaging of subrosion structures. J. Appl. Geophys., 78, 133–143, doi:10.1016/j.jappgeo.2011.02.003.

Kreuzer, O.P., Yousefi, M., Nykanen, V., 2020. Introduction to the special issue on spatial modelling and analysis of ore forming processes in mineral exploration targeting. Ore Geol. Rev., 119, 103391p.

Krishnamurthy, J., 1997. The evaluation of digitally enhanced Indian Remote Sensing Satellite (IRS) data for lithological and structural mapping. Int. J. Remote Sens., 18, 3409–3437.

Kruse, F.A., Kierein-young, K.S., Boardman, J.W., 1990. Mineral mapping at Cuprite, Nevada with a 63-channel imaging spectrometer. Photogramm. Eng. Remote Sens., 56, 83–92.

Kruse, F.A., Lefkof, A.B., Boardman, J.W., Heidebrecht, K.B., Shapiro, A.T., Barloon, J.P., Goetz, A.F.H. 1993. The spectral image processing system (SIPS)-interactive visualization and analysis of imaging spectrometer data. Remote Sens. Environ., 44, 145–163.

Kruse, F.A., Baugh, M.W., Perry, S.L., 2015. Validation of Digital Globe Worldview-3 earth imaging satellite shortwave-infrared bands for mineral mapping. J. Appl. Remote Sens., 9, 1–18.

Kuhn, T., Bau, M., Blum, N., Halbach, P., 1998. Origin of negative Ce anomalies in mixed hydrothermal-hydrogenetic Fe-Mn crusts from the Central Indian Ridge. Earth Planet. Sci. Lett., 163:207–220.

Kuhn, T., Wegorzewski, A., Ruhlemann, C., Vink, A., 2017. Correction to: composition, formation, and occurrence of polymetallic nodules. In: Sharma, R., editor. Deep Sea Mining. Cham: Springer International Publishing, E1–E1.

Kumar, A., Singh, P., 2022. Application of GPS, GIS, and remote sensing in mineral exploration: a case study from Jharkhand. J. Geospat. Sci., 15(2), 125–139.

Kumar, D., Singh, R.B., Kaur, R., 2019. SDG 9: case Study – infrastructure Assessment for Sustainable Development. Spatial Information Technology for Sustainable Development Goals, 157–167.

Kumar, J., Sain, K., Arun, K.P., 2019. Seismic attributes for characterizing gas hydrates: a study from the Mahanadi offshore, India. Marine Geophysical Research, 40, 73–86.

Kumar, P., Dash, S.K., Thakur, R.R., Jonna, S., Tripathi, S., 2014. Space based information support for decentralized planning (SIS-DP) – a case study of Balangir district, Odisha, India. The International Archives of the Photogrammetry, Remote Sensing and Spatial Information Sciences, XL-8, 2014 ISPRS Technical Commission VIII Symposium, 9-12 December 2014, Hyderabad, India.

Kumar, P., Pandey, P., 2022. Mineral Exploration Scenario in India: prospect, Promise and Policy Reforms. J. Geol. Soc. India, 98, 1178–1181.

Kumar, S., 2021. Insights of shale gas exploration and development in India: review. International Journal of Scientific Research, 10(5), 14–16, 2277-8179, doi: 10.36106/ijsr.

Kumar, S., Rathi, M., 2020. AI-driven water conservation in Indian agriculture: a case study of Maharashtra. J. Water Resour. Rural Dev., 8(2), 45–56.

Kumar, V.A., Mondal, N.C., Ahmed, S., 2020. Identification of Groundwater Potential Zones Using RS, GIS and AHP Techniques: a Case Study in a Part of Deccan Volcanic Province (DVP), Maharashtra, India. Journal of the Indian Society of Remote Sensing (Springer), Research Article, 10.1007/s12524-019-01086-3.

Kuras, O., Beamish, D., Meldrum, P.I., Ogilvy, R.D., 2006. Fundamentals of the capacitive resistivity technique. Geophysics, 71, G135–G152.

Kuras, O., Meldrum, P.I., Beamish, D., Ogilvy, R.D., Lala, D., 2007. Capacitive resistivity imaging with towed arrays. Journal of Environmental and Engineering Geophysics, 12, 267–279.

Kuras, O., Pritchard, J.D., Meldrum, P.I., Chambers, J.E., Wilkinson, P.B., Ogilvy, R.D., Wealthall, G.P., 2009. Monitoring hydraulic processes with automated time-lapse electrical resistivity tomography (ALERT). C. R. Geosci., 341, 868–888.

Kurbatov, L.M., 1937. On the radioactivity of bottom sediments. American Journal of Science, s5-33(194), 147–153.

Kutina, J., 1969. Hydrothermal ore deposits in the western United States: a new concept of structural control of distribution. Science, 165, 1113–1119.

Kuzmin, P.V., Borel, G., Morrison, E.B., Dodds, J., 2010. Geophysical prospecting using rotationally invariant parameters of natural electromagnetic fields: WIPO International Patent Application No. PCT/CA2009/001865.

Kwang, C., Osei, E.M., Duker, A.A., 2014. Application of remote sensing and geographic information system for gold potential mapping in birim north district of eastern region of Ghana. Intl J Rem Sens Appli 4 (1).

Labson, V.F., Becker, A., Morrison, H.F., Conti, U., 1985. Geophysical exploration with audio-frequency natural magnetic fields. Geophysics, 50, 656–664.

Lahren, M.M., Schweickert, R.A., Taranik, J.V., 1988. Analysis of the northern Sierra accreted terrain, California, with airborne thermal infrared multispectral scanner data. Geology, 16, 525–528.

Landsberg, H., Klepper, M.R., 1939. Measurements of radioactivity for stratigraphic studies. American Geophysical Union, 20(3), 277–280.

Lane, R., 2006. Investigations stemming from the 2003 Broken Hill airborne gravity gradiometer survey. In: Korsch, R., Barnes, R. eds., Broken Hill Exploration Initiative: Abstracts for the September 2006 Conference., Geoscience Australia Record 2006/21, 116–128.

Lang, F., Singh, J.S.P., Pereira, M.T.R., 2018. Optimizing Data Selection in Mineral Exploration Using GIS. Exploration and Mining Geology, 27(3), 239–252.

Langenberg, C.W., 1983. Polyphase deformation in the Canadian shield of northeastern Alberta; Alberta Geological Survey, Bulletin 45, 33p.

Langenberg, C.W., Nielsen, P.A., 1982. Polyphase metamorphism in the Canadian shield of northeastern Alberta. Alberta Geological Survey, Bulletin 42, 80p.

Larkin, S.P., Levander, A., Okaya, D., Goff, J.A., 1996. A deterministic and stochastic velocity model for the Salton Trough/Basin and Range transition zone and constraints on magmatism during rifting. J. Geophys. Res. Solid Earth, 101, 27883–27897. doi: 10.1029/96JB02535.

Latef, M.Z.B.A., Kahar, R.B., 2008. Oil and Gas Exploration in Ethiopia Using GIS. GIS development, Kaula Lumpur, Malaysia.

Latwal, A., Saxena, S., Dubey, S.K., Choudhary, K., 2019. Evaluation of pre-harvest production forecasting of mustard crop in major producing States of India, under FASAL project. The International Archives of the Photogrammetry Remote Sensing and Spatial Information Sciences XLII-3/W6:115–122. DOI:10.5194/isprs-archives-XLII-3-W6-115-2019.

Leal Filho, W., Abubakar, I., Nunes, C., et al. Deep seabed mining: a note on some potentials and risks to the sustainable mineral extraction from the oceans. JMSE 2021; 9521.

Leber, S., Feng, T., Zhao, L., 2017. Remote sensing for mineral exploration: satellite geophysics and Its applications. Earth Science Frontiers, 24(3), 89–102.

Leclair, A.D., Lucas, S.B., Broome, H.J., Viljoen, D.W., Weber, W., 1997. Regional mapping of Precambrian basement beneath Phanerozoic cover in southeastern Trans-Hudson Orogen, Manitoba and Saskatchewan. Can. J. Earth Sci., 34, 618–634.

Lee, J.B., 2001. FALCON gravity gradiometer technology. In: Extended Abstracts from ASEG 15th Geophysical Conference and Exhibition, Brisbane, Australia.

Lee, J.B., Liu, G., Rose, M., Dransfield, M.H., Mahanta, A., Christensen, A.N., Stone, P.M., 2001. High resolution gravity surveys from a fixed wing aircraft. In: Geoscience and Remote Sensing Symposium, 2001. IGARSS '01. IEEE 2001 International, 3, 1327–1331.

Lee, S., Choi, Y., 2016. Reviews of unmanned aerial vehicle (drone) technology trends and its applications in the mining industry. Geosyst. Eng., 19, 197–204.

Legault, J.M., 2012. Ten years of passive airborne AFMAG EM development for mineral exploration. 82nd Annual International Meeting, SEG, Expanded Abstracts, 4p.

Legault, J.M., Wilson, G.A., Gribenko, A.V., Zhdanov, M.S., Zhao, S., Fisk, K., 2012. An overview of the ZTEM and AirMt airborne electromagnetic systems: a case study from the Nebo–Babel Ni–Cu–PGE deposit, West Musgrave, Western Australia. Feature Paper, 10.1071/PVv2012n158p26.

Leica Geosystems, A.G., 1999. GPS Basics 1.0.0en, Heerbrugg, Switzerland, 64p.

Leitloff, J., Rosenbaum, D., Kurz, F., Meynberg, O., Reinartz, P., 2014. An operational system for estimating road traffic information from aerial images. Remote Sens., 6, 11315–11341.

Lenoble, J.P., 1981. Polymetallic nodules resources and reserves in the north Pacific from the data collected by Afernod. Ocean Manag. 7(1-4): 9–24.

Levesque, M.C., Brown, C.F., 2018. Geochemical mapping for environmental Monitoring in Northern Manitoba. Environmental Geochemistry and Health, 40(5), 1341–1356.

Lewis, P., 1997. A Review of GIS Techniques for Handling Geoscience Data within Australian Geological Surveys. In Geophysics and geochemistry at the millenium. Proceedings of Exploration 97: Fourth Decennial International Conference on Mineral Exploration, Toronto, Canada, 14-18 September 1997; Gubins, A.G., Ed.; Prospectors and Developers Associaton: Toronto, Canada, 81–86.

Li, H., Xue, G.Q., Di, Q., 2018. Numerical analysis of land-based inline-source configuration for the controlled-source electromagnetic method. J. Environ. Eng. Geophys., 23, 47–59.

Li, H., Di, Q.Y., Xue, G.Q., 2019. A comparative study of inline and broadside time domain CSEM methods for mapping resistive targets on land. Geophysics, 84, 1–44.

Li, H., Li, X., Yuan, F., Jowitt, S.M., Zhang, M., Zhou, J., Zhou, T., Li, X., Ge, C., Wu, B., 2020. Convolutional neural network and transfer learning based mineral prospectivity modeling for geochemical exploration of Au mineralization within the Guandian–Zhangbaling area, Anhui Province, China. Appl. Geochem., 122, 104747.

Li, H., Zhang, Q., Pan, M., Chen, D., Yu, Z., Xu, Y., Ding, Z., Liu, X., Wan, K., Dai, W., 2024. Enhancing precision in magnetic map interpolation for regions with sparse data. Appl. Sci. 14, 756.

Li, N., Song, X., Li, C., Xiao, K., Li, S., Chen, H., 2019. 3D geological modeling for mineral system approach to GIS-based prospectivity analysis: case study of an MVT Pb-Zn deposit. Nat. Resour. Res., 28(3), 995–1019.

Li, Q., Wang, W., Li, C., 2008. Application of MapGIS in Geophysical Data Mapping. J. Earth Sci. Environ., 30(2), 200–203.

Li, S., Chen, J., Xiang, J., 2019. Applications of deep convolutional neural networks in prospecting prediction based on two-dimensional geological big data. Neural Computing and Applications.

Li, S., Gu, C., Yang, J., Zhang, Y., Diao, S., Ji, Y., 2022. A review of marine controlled-source electromagnetic data pre-processing technology. AIP Advances, 12, 090701.

Li, T., Cai, G., Wang, C., Liang, K., Ma, S., Luo, W., 2021. Quantifying clay mineral sources in marine sediments by using end-member mixing analysis. Geo-Marine Letters, 41, 6.

Li, Y., Wu, X.P., Lin, P.R., Han, S.X., Liu, W.Q., 2017. Three-dimensional modeling of marine controlled-source electromagnetism using the vector finite element method for arbitrary anisotropic media (in Chinese). Chin. J. Geophys., 60, 1955–1978.

Li, Y., Liu, C., 2019. Applications of multirotor drone technologies in construction management. Int. J. Constr. Manag., 19, 401–412.

Liang, S., Sun, S., Lu, H., 2021. Application of airborne electromagnetics and magnetics for mineral exploration in the Baishiquan-Hongliujing area, Northwest China. Remote Sens., 13(5), 903.

Liao, J., Chen, J., Sun, X., et al., 2022. Quantifying the controlling mineral phases of rare earth elements in deep sea pelagic sediments. Chem. Geol., 595:120792.

Liao, S.L., Tao, C.H., Li, H.M., Barriga, F.J.A.S., Liang, J., Yang, W.F., Yu, J.Y., Zhu, C.W., 2018. Bulk geochemistry, sulphur isotope characteristics of the Yuhuang-1 hydrothermal field on the ultraslow-spreading Southwest Indian Ridge. Ore Geol. Rev, 96, 13–27.

Lichoro, C.M., 2016. Gravity and magnetic methods. Presented at SDG Short Course I on Exploration and Development of Geothermal Resources, organized by UNU-GTP, GDC and KenGen, at Lake Bogoria and Lake Naivasha, Kenya, Nov. 10-31, 7p.

Lie, J.E., Andersson, M., 1998. The deep seismic image of the crustal structure of the Tornquist zone beneath the Skagerrak Sea, northWestern Europe. Tectonophysics, 287, 139–155. doi: 10.1016/S0040-1951(98)80065-6.

Lillesand, T.M., Kiefer, R.W., Chipman, J.W., 2004. Remote Sensing and Image Interpretation, Fifth edition. Wiley, New York.

Lillesand, T.M., Kiefer, R.W., Chipman, J.W., 2008. Remote Sensing and Image Interpretation. John Wiley & Sons Inc., Hoboken, NJ, USA.

Lillesand, T.M., Kiefer, R.W., Chipman, J.W., 2015. Remote Sensing and Image Interpretation. Seventh edition. Wiley, New York, 736p.

Lillesand, T.M., Kiefer, R.W., 1979. Remote Sensing and Image Interpretation. Wiley, New York, 612p.

Lillesand, T.M., Kiefer, R.W., 2000. Remote Sensing and Image Interpretation. John Wiley & Sons, New York.

Linde, N., Revil, A., 2007. Inverting self-potential data for redox potentials of contaminant plumes. Geophys. Res. Lett., 34, L14302.

Lindi, O.T., Aladejare, A.E., Ozoji, T.M., Ranta, J.-P., 2024. Uncertainty Quantification in Mineral Resource Estimation. Nat. Resour. Res., 33(6), 2503–2526.

Linfoot, J., Mattsson, J., Price, D., et al., 2011. Case study of a towed streamer EM survey over the Troll field, North Sea. In: 2011 SEG Annual Meeting, Society of Exploration Geophysicists, 594–598.

Linzen, S., Schultze, V., Chwala, A., Schuler, T., Schulz, M., Stolz, R., Meyer, H.G., 2009. Quantum detection meets archaeology – magnetic prospection with SQUIDs, highly sensitive and fast, In: New Technologies for Archaeology, 71–85 Reindel, M., Wagner, G.A., Springer.

Lipton, I., Nimmo, M., Parianos, J., 2016. TOML Clarion Clipperton Zone Project, Pacific Ocean; AMC Consultants Pty Ltd.: brisbane, Australia, 2016.

Lisitsin, V.A., Gonzalez-Alvarez, I., Porwal, A., 2013. Regional prospectivity analysis for hydrothermal-remobilised nickel mineral systems in western Victoria, Australia. Ore Geol. Rev., 52, 100–112.

Liu, B., Chen, J., Pinheiro, L.M., Yang, L., Liu, S., Guan, Y., Song, H., Wu, N., Xu, H., Yang, R., 2021. An insight into shallow gas hydrates in the Dongsha area, South China Sea. Acta Oceanol. Sin., 40, 136–146.

Liu, C., Guo, J., Tian, Y., et al., 2020. Development and field tests of a deep sea laser-induced breakdown spectroscopy (LIBS) system for solid sample analysis in sea water. Sensors, 207341. DOI PubMed PMC.

Liu, C.C., Sousa Jr, M.A., Gopinath, T.R., 2000. Regional structural analysis by remote sensing for mineral exploration, Paraiba state, Northeast Brazil. Geocarto Int., 15, 69–75.

Liu, H., Yao, Y., Deng, H., Geological and geophysical conditions for potential natural gas hydrate resources in Southern South China Sea waters. Journal of Earth Science, 22(6), 718–725, 2011 1674-487X Printed in China DOI: 10.1007/s12583-011-0222-5.

Liu, H.F., Liu, J.X., Guo, R.W., Tong, X.Z., Gong, L., Peng, Y.H., 2014. Development of multi-channel observation and inversion for IP electrical sounding method. Transactions of Nonferrous Metals Society of China, 24(3), 816–823.

Liu, J., Zhang, J.Z., Jiang, L., Lin, Q., Wan, L., 2019. Polynomial-based density inversion of gravity anomalies for concealed iron-deposit exploration in North China. Geophysics, 84, 325–334.

Liu, L., Yin, C.C., Liu, Y.H., Qiu, C.K., Huang, X., Zheng, B., 2018. Spectral element method for 3D frequency-domain marine controlled-source electromagnetic forward modeling (in Chinese). Chin. J. Geophys., 61, 756–766.

Liu, T., Cui, F., Zhang, M.X., 2015. Research progress of deep seabed pore pressure observation techniques in situ. J. Hydraul. Eng., 46(S1):111–116 (in Chinese).

Liu, W., Zhang, L., 2023. Emerging trends in GNSS and GPS technologies for resource exploration and environmental monitoring. J. Geospat. Technol., 34(2), 124–136.

Liu, Y., Li, J.B., Wang, Y.J., Tao, C.H., Zeng, Z.G., Han, X.Q., Li, C.S., Zhou, Y.D., Liang, J., 2021b. Development and prospect of exploration and research on seafloor polymetallic sulfides of China. Chin J Nonferrous Met 31(10):2624–2637.

Liu, Y., Yonghang, L., 2022. Mineral Prospectivity Mapping of Tungsten Polymetallic Deposits Using Machine Learning Algorithm and Comparison of Their Performance in the Gannan Region. China University of Geosciences, China.

Liu, Z., Zhao, Y., Colin, C., Stattegger, K., Wiesner, M.G., Huh, C.A., Zhang, Y., Li, X., Sompongchaiyakul, P., You, C.F., Huang, C.Y., Liu, J.T., Siringan, F.P., Le, K.P., Sathiamurthy, E., Hantoro, W.S., Liu, J., Tuo, S., Zhao, S., Zhou, S., He, Z., Wang, Y., Bunsomboonsakul, S., Li, Y., 2016. Sourceto sink transport processes of fluvial sediments in the South China Sea. Earth-Science Reviews, 153, 238–273.

Locke, C.A., De Ronde, C.E.J., 1987. Delineation of gold-bearing hydrothermally alteRed Rocks using gravity data – a New Zealand example. Geoexploration, 24, 471–481.

Logn, O., Bolviken, B., 1974. Self-potentials at the Joma pyrite deposit, Norway. Geoexploration, 12, 11–28.

Loke, M.H., 1994. The inversion of two-dimensional resistivity data. Unpubl. PhD thesis, Un. Of Birmingham.

Loke, M.H., 1997, 1999, 2000 (copyright). Electrical imaging surveys for environmental and engineering studies. A practical guide to 2D and 3D surveys.

Loke, M.H., Chambers, J.E., Rucker, D.F., Kuras, O., Wilkinson, P.B., 2013. Recent developments in the direct current geoelectrical imaging method. J. Appl. Geophys., 95, 135–156.

Loke, M.H., Lane Jr, J.W., 2005. Inversion of data from electrical resistivity imaging surveys in water-covered areas. Explor. Geophys., 35, 266–271.

Long, P., 2016. Geospatial Data Integration for Natural Resources Exploration. Elsevier.

Longhi, I., Sgavetti, M., Chiari, R., Mazzoli, C., 2001. Spectral analysis and classification of metamorphic rocks from laboratory reflectance spectra in the 0.4-2.5 μm interval: a tool for hyperspectral data integration. Int. J. Remote Sens., 22, 3763–3782.

Longley, P.A., Goodchild, M.F., Maguire, D.J., Rhind, D.W., 2005. Geographic Information Systems and Science. John Wiley & Sons Inc., Hoboken, NJ, USA.

Louvari, E.K., Kiratzi, A.A., 1997. RAKE: a Windows Program to Plot Earthquake Focal Mechanisms and the Orientation of Principal Stresses. Comput. Geosci., 23(8), 851–857.

Low, H.E., Randolph, M.F., Rutherford, C., Bernie, B.B., Brooks, J.M., 2008. Characterization of near seabed surface sediment. In: proceedings of offshore technology conference, offshore technology, Houston, paper OTC 19149- 10.4043/19149-MS.

Lowrie, W., 2007. Fundamentals of Geophysics. Second Edition, Published in the United States of America by Cambridge University Press, New York.

Lu, F.C., Qu, Y.D., Liao, M.H., 2004. The development status and applications of in-situ cone penetration test technology. Ocean Technol., 23, 32–36.

Lukasheh, A.F., Droste, R.L., Warith, M.A., 2001. Review of Expert System (ES), Geographic Information System (GIS), Decision Support System (DSS), and their applications in landfill design and management. Waste Manag. Res., 19, 177–185.

Lundberg, E., Juhlin, C., Nasuti, A., 2012. High resolution reflection seismic profiling over the Tjellefonna fault in the More-Trondelag Fault Complex, Norway. Solid Earth, 3, 175–188. doi: 10.5194/se-3-175-2012.

Lunne, T., Robertsen, P.K., Powell, J., 2009. Cone penetration testing in geotechnical practice. Soil Mech Found Eng 46:237.

Lunne, T., 2010. The CPT in offshore soil investigations – a historic perspective. Proc. CPT, 10, 71–113.

Luo, M., Li, Y.G., 2015. Effects of the electric anisotropy on marine controlled-source electromagnetic responses (in Chinese). Chin. J. Geophys., 58, 2851–2861.

Luo, X., Dimitrakopoulos, R., 2003. Data-driven fuzzy analysis in quantitative mineral resource assessment. Comput. Geosci., 29, 3–13.

Luo, Y.C., 2009. Improve the Method Theory and the Applied Research of Seismic Survey Resolution – Deconvolution. Chengdu University of Technology, Chengdu (in Chinese). https://cdmd.cnki.com.cn/Article/CDMD-10616-2009221416.htm.

Lusty, P.A.J., Scheib, C., Gunn, A.G., Walker, A.S.D., 2012. Reconnaissance-scale prospectivity analysis for gold mineralisation in the Southern Uplands-DownLongford Terrane, Northern Ireland. Nat. Resour. Res., 21, 359–382.

Lusty, P.A.J., Hein, J.R., Josso, P., 2018. Ferromanganese crusts: earth's store house for critical metals. Elements, 14, 313–318.

Lyatsky, H.V., 1996. Continental-crust Structures on the Continental Margin of Western North America. Springer-Verlag, Heidelberg 352p.

Lyatsky, H.V., Pana, D.I., Grobe, M., 2005. Basement structure in central and southern Alberta: insights from gravity and magnetic maps. Alberta Geological Survey, Special Publication 72, 76p.

Lyatsky, H.V., 2010. Magnetic and gravity methods in mineral exploration: the value of well-rounded geophysical skills. Recorder – Canadian Society of Exploration Geophysicist, 35(8), 47p.

Lyklema, J., 1995. Fundamentals of Interface and Colloid Science: Solid Liquid Interfaces, Volume II. Academic Press, New York, NY.

Lykousis, V., Sakellariou, D., Locat, J., 2007. Submarine Mass Movements and Their Consequences: 3rd International Symposium. Kluwer Academic Publishers, Amsterdam. https://link.springer.com/content/pdf/bfm:978-1-4020-6512-5/1.pdf.

MacFarlane, J.W., Payton, O.D., Keatley, A.C., Scott, G.P.T., Pullin, H., Crane, R.A., Smilion, M., Popescu, I., Curlea, V., Scott, T.B., 2014. Lightweight aerial vehicles for monitoring, assessment and mapping of radiation anomalies. J. Environ. Radioact.

Maciąg, L., Zawadzki, D., Kozub-Budzyn, G.A., Piestrzynski, A., Kotlinski, R.A., Wrobel, R.J., 2019. Mineralogy of cobalt rich ferromanganese crusts from the Perth Abyssal Plain (E Indian Ocean). Minerals, 9(2): 84, doi: 10.3390/min9020084, ISSN 2075-163X.

Macnae, J.C., 1979. Kimberlites and exploration geophysics. Geophysics, 44, 1395–1416.

Macnae, J.C., Lamontagne, Y., West, G.F., 1984. Noise processing techniques for time domain EM systems. Geophysics, 49, 934–948.

Mahanta, A.M., Boggs, D.B., Dransfield, M.H., Whittal, M., Liu, G., Scott, R.L., 2001. King George: measured and modelled AGG response over an IOCG terrain. Preview, 93, 79–80.

Mahanta, A.M., 2003. Application of airborne gravity gradiometer technology in coal exploration. In: ASEG CD 16th Geophysics Conference and Exhibition Adelaide Convection Centre 16-19th February 2003.

Malczewski, J., 1999. GIS and Multi-Criteria Decision Analysis. Wiley.

Malehmir, A., Urosevic, M., Bellefleur, G., Juhlin, C., Milkereit, B., 2012. Seismic methods in mineral exploration and mine planning – introduction.

Geophysics, 77(5), WC1–WC2. 10.1190/2012-0724-SPSEIN.1.

Malehmir, A., Durrheim, R., Bellefleur, G., Urosevic, M., Juhlin, C., White, D.J., Milkereit, B., Campbell, G., 2012. Seismic methods in mineral exploration and mine planning: a general overview of past and present case histories and a look into the future. Geophysics, 77, WC173–WC190.

Malehmir, A., et al., 2015. Planning of urban underground infrastructure using a broadband seismic land streamer – tomography results and uncertainty quantifications from a case study in southwest of Sweden. Geophysics, 80, B177–B192.

Malehmir, A., Dynesius, L., Paulusson, K., Paulusson, A., Lohansson, H., Bastani, M., Wedmark, M., Marsden, P., 2017. The Potential of Rotary-wing UAV-based Magnetic Surveys for Mineral Exploration: A Case Study from Central Sweden. The Leading Edge.

Malehmir, A., Maries, G., Backstrom, E., Schon, M., Marsden, P., 2017. Developing cost effective seismic mineral exploration methods using a land streamer and a drop hammer. Scientific Reports, 7, 10325, doi: 10.1038/s41598-017-10451-6.

Malehmir, A., et al., 2017a. Land streamer seismics and physical property measurements in the Siilinjarvi open-pit apatite (phosphate) mine, central Finland. Geophysics, 82, B29–B48.

Maleki, M., Mery, N., Soltani-Mohammadi, S., Khorram, F., Emery, X., 2022. Geological control for in-situ and recoverable resources assessment: a case study on Sarcheshmeh porphyry copper deposit Iran. Ore Geol. Rev, 66, 105133.

Malinowski, M., White, D.J., Mwenifumbo, C.J., Salisbury, M., Bellefleur, G., Schmitt, D., Dietiker, B., Schetselaar, E., Duxbury, A., 2008. Seismic exploration for VMS deposits within the Paleoproterozoic Flin Flon Belt, Trans-Hudson Orogen, Canada. 70th EAGE Conference & Exhibition 216p.

Malinowski, M., Schetselaar, E., White, D.J., 2012. 3D seismic imaging of volcanogenic massive sulfide deposits in the Flin Flon mining camp, Canada: part 2 – forward modeling. Geophysics, 77, WC81–WC93. doi: 10.1190/geo2011-0474.1.

Malinowski, M., White, D., 2011. Converted wave seismic imaging in the Flin Flon mining camp, Canada. J. Appl. Geophys., 75, 719–730.

Manchuk, J.G., Stiefenhofer, J., Thurston, M., Deutsch, C.V., 2015. Framework for resource uncertainty prediction and data valuation: an application to diamond deposits. Canadian Institute of Mining, Metallurgy and Petroleum, 66, 178–190.

Mansoor, N., Slater, L., 2007. Aquatic electrical resistivity imaging of shallow-water wetlands. Geophysics, 72, F211–F221.

Marcelino, S.E., Ribeiro, F.T.D., Campos, F.A., Augusto, M.J., Jesus, F.T., 2020. Bayesian approach to the zinc extraction curve of soil with sewage sludge. Acta Sci. Technol., vol. 42, 1806-2563 1807-8664.

Maries, G., Malehmi, A., Backstrom, E., Downhole physical properties measurements supporting iron-oxide deep exploration and mining in Blotberget, Sweden. Near-surface Geoscience, Geophysics on Mineral Exploration and Mining, Barcelona, Spain, 2016, doi:10.3997/2214-4609.201602092.

Mariita, N.O., 2009. The gravity method. Presented at Short Course IV on Exploration for Geothermal Resources, organized by UNU-GTP, KenGen and GDC, at Lake Naivasha, Kenya, November 1-22.

Marino, E., Gonzalez, F., Somoza, L., et al., 2017. Strategic and rare elements in Cretaceous-Cenozoic cobalt-rich ferromanganese crusts from seamounts in the Canary Island Seamount Province (northeastern tropical Atlantic). Ore Geol. Rev., 87:41–61.

Marino, E., Gonzalez, F.J., Kuhn, T., et al., 2019. Hydrogenetic, diagenetic and hydrothermal processes forming ferromanganese crusts in the canary island seamounts and their influence in the metal recovery rate with hydrometallurgical methods. Minerals, 9:439.

Marjoribanks, R., 2010. Mineral Exploration: Principles and Applications. Springer.

Marjoribanks, R., 2010. Geological Methods in Mineral Exploration and Mining, 2nd ed. Springer-Verlag, Berlin Heidelberg.

Markcsc, 2011. Comparison to gravity. Comparison of gz and Gzz anomalies for a point source at 1 km depth. Created 27th June 2011.

Markiewicz, R.D., Davenport, G.C., Randall, J.A., 1984. The use of self potential surveys in geotechnical investigations. 54th Annual Internat. Mtg. Soc. Expl. Geophys. Expanded Abstracts, Soc. Expl. Geophysics, Tulsa, Session: SP.6.

Markl, R.G., Bryan, G.M., Ewing, J.I., 1970. Structure of the Blake-Bahama Outer Ridge. J. Geophys. Res., 75, 4539–4555.

Markwitz, V., Maier, W.D., Gonzalez-Alvarez, I., McCuaig, T.C., Porwal, A., 2010. Magmatic nickel sulfide mineralization in Zimbabwe: review of deposits and development of exploration criteria for prospectivity analysis. Ore Geol. Rev., 38, 139–155.

Mars, J.C., 2018. Mineral and lithologic mapping capability of World View 3 data at Mountain Pass, California, using true and false color composite images, band ratios, and logical operator algorithms. Econ. Geol., 113, 1587–1601.

Mars, J.C., Rowan, L.C., 2006. Regional mapping of phyllic-and argillic-alteRed Rocks in the Zagros magmatic arc, Iran, using Advanced Spaceborne

Thermal Emission and Reflectance Radiometer (ASTER) data and logical operator algorithms. Geosphere, 2, 161–186.

Marshall, D.J., Madden, T.R., 1959. Induced polarisation, a study of its causes. Geophysics, 24, 790–816.

Martin, L., 2013. Using gravity to detect underground threats. DARPA selects Lockheed Martina to develop sensor system that can locate and identify underground targets by spotting gravity-based effects from an airborne platform.

Martin, P.G., Payton, O.D., Fardoulis, J.S., Richards, D.A., Scott, T.B., 2015. The use of unmanned aerial systems for the mapping of legacy uranium mines. J. Environ. Radioact.

Martin, P.G., Payton, O.D., Fardoulis, J.S., Richards, D.A., Yamashiki, Y., Scott, T.B., 2016. Low altitude unmanned aerial vehicle for characterising remediation effectiveness following the FDNPP accident. J. Environ. Radioact. 151.

Martinez, C., Li, Y., Krahenbuhi, R., Braga, M.A., 2013. 3D inversion of airborne gravity gradiometry data in mineral exploration: a case study in the Quadrilatero Ferrifero, Brazil Geophysics, 78(1), B1–B11.

Martinez, P.D., Carbonell, R., Palomeras, I., Simancas, J.F., Ayarza, P., Martí, D., Azor, A., Jabaloy, A., Gonzalez Cuadra, P., Tejero, R., Martin Parra, L.M., Matas, J., Gonzalez Lodeiro, F., Perez-Estaun, A., Garcia Lobon, J.L., Mansilla, L., 2012. Imaging the crustal structure of the Central Iberian Zone (Variscan Belt): the ALCUDIA deep seismic reflection transect. Tectonics, 31. doi: 10.1029/2011TC002995.

Martins, A.C., Elis, V., De Tomi, G., Bettencourt, J., Marin, T., 2016. Resistivity and induced polarization to support morphological modeling in limestone mining. Geofísica Internacional 55, 227–238. (doi:10.19155/geofint.2016.055.4.1).

Martins, A.C., Terenci, E.R., Tomi, G.D., Tichauer, R.M., 2016. Impact of geophysics in small-scale mining. J. Remote Sensing & GIS. 5, 172. (doi:10.4172/2469-4134.C1.003).

Martins Filho, S., Silva, F.F., Carneiro, A.P.S., Muniz, J.A., 2008. Bayesian approach in the growth curves of two cultivars of common bean. Ciencia Rural, 38(6), 1516–1521.

Masek, J.G., Wulder, M.A., Markham, B., McCorkel, J., Crawford, C.J., Storey, J., Jenstrom, D.T., 2020. Landsat 9: empowering open science and applications through continuity. Remote Sens. Environ., 248:111968, doi: 10.1016/j.rse.2020.111968.

Masouka, P.M., Harris, J., Lowman Jr, P.D., Blodget, H.W., 1988. Digital processing of orbital radar data to enhance geologic structure: examples from the Canadian shield. Photogramm. Eng. Remote Sens., 54, 621–632.

Massari, S., Ruberti, M., 2013. Rare earth elements as critical raw materials: focus on international markets and future strategies. Resour. Policy 38, 36–43.

Massons, J., Domingo, D., Grau, J., 1996. Automatic Classification of VIS-IR METEOSAT Images. Comput. Geosci., 22(10), 1137–1146.

Masuda, Y., Cruickshank, M.J., Mero, J.L., 1971. Continuous bucket line dredging at 12000 feet. In: Offshore technology conference, Houston, Texas, 94–99.

Mataragio, J.P., Braga, M.A., Dayan, H., Jadir, P.N., 2006. Exploring for iron oxide deposits in Brazil using airborne full tensor gradiometry (AIR-FTG®), presented at PDAC 2006, http://www.pdac.ca/pdac/conv/2006/pdf/papers-ts-mataragio.pdf.

MathWorks, 2021. MATLAB for Geophysics: Data Processing and Interpretation. MathWorks.

Matt, B., 2011. How Do Geophysical Data and Geographic Information Systems (GIS) Relate to Each Other. Sensors & Systems.

Maurya, S.P., Singh, N.P., 2018. Application of LP and ML sparse spike inversion with probabilistic neural network to classify reservoir facies distribution – a case study from the Blackfoot field, Canada. J. Appl. Geophys., 159, 511–521.

Mavko, G., Mukerji, T., 1998. Bounds on low frequency seismic velocities in partially saturated rocks. Geophysics, 63(3), 918–924.

Mayer, D., 2018. Cost-effective geophysical surveys using remote sensing technologies. Journal of Exploration Geophysics, 60(5), 45–58.

Mayne, P.W., Coop, M.R., Springman, S.M., Huang, A., Zornberg, J.G., 2009. Geomaterial behavior and testing. In: Presented at the 17th international conference on soil mechanics and geotechnical engineering, Alexandria, Egypt. International Society for Soil Mechanics and Geotechnical Engineering, London 4, 2777–2872.

Mazac, O., Kelly, W.E., Landa, I., 1985. A hydrogeophysical model for relations between electrical and hydraulic properties of aquifers. J. Hydrol., 79(1-10).

Mazzarini, F., Pareschi, M.T., Sbrana, A., Favalli, M., Fulignati, P., 2001. Surficial hydrothermal alteration mapping at Vulcano Island using MIVIS data. Int. J. Remote Sens., 22, 2045–2070.

McCafferty, A.E., Phillips, J.D., Driscoll, R.L., 2016. Magnetic and gravity gradiometry framework for Mesoproterozoic iron oxide-apatite and iron oxide-copper gold deposits, southeast Missouri. Econ. Geol., 111(8), 1859–1882.

McCarthy, D.D., Petit, G., (Eds) 2004. IERS Conventions (2003), Technical Note 32, Frankfurt am Main: Verlag des Bundesamts fur Kartographie und Geodasie.

McCracken, K.G., Oristaglio, M.L., Hohmann, G.W., 1986, A comparison of electromagnetic exploration systems. Geophysics, 51, 810–818.

McCuaig, T.C., Beresford, S., Hronsky, J., 2010. Translating the mineral systems approach into an effective exploration targeting system. Ore Geol. Rev., 38, 128–138.

McCuaig, T.C., Hronsky, J.M.A., 2000. The current status and future of the interface between the exploration industry and economic geology research. Rev. Econ. Geol., 13, 553–559.

McCuaig, T.C., Hronsky, J.M.A., 2014. The mineral system concept: the key to exploration targeting. Soc. Econ. Geol. Spec. Publ., 18, 153–176.

McKay, A., Mattsson, J., Du, Z.J., 2015. Towed streamer EM – reliable recovery of subsurface resistivity. First Break, 33(4), 75–86.

McKay, G., Harris, J.R., 2016. Comparison of the data-driven Random Forests model and a knowledge-driven method for mineral prospectivity mapping: a case study for gold deposits around the Huritz Group and Nueltin Suite, Nunavut, Canada. Nat. Resour. Res., 25, 125–143.

McKinney, R.A., Pappas, M., 2017. Coastal resource management and wetland restoration using GPS, GIS, and remote sensing: a case study of the Florida Everglades. Ecological Applications, 27(8), 2334–2346.

McLeod, T., Samson, C., Labrie, M., Shehata, K., Mah, J., Lai, P., Wang, L., Elder, J.H., 2013. Using video acquired from an unmanned aerial vehicle (UAV) to measure fracture orientation in an open-PIT mine. Geomatica, 67, 173–180.

McMahon, R., Mulligan, P., 2015. Integrating remote sensing and geospatial technologies in mineral exploration: a case study in Western Australia. J. Geophys. Res., 120(3), 125–139.

McManus, S., Rahman, A., Horta, A., Coombes, J., 2020. Applied Bayesian modeling for assessment of interpretation uncertainty in spatial domains. Statistics for Data Science and Policy Analysis, 66, 3–13.

McManus, S., Rahman, A., Coombes, J., Horta, A., 2021a. Comparison of interpretation uncertainty in spatial domains using portable X-ray fluorescence and ICP data. Appl. Comput. Geosci., 12, 100067.

McNamara, J., 2008. GPS for Dummies, 2nd Ed., Willey Publishing Inc., River Street Hoboken, 408p.

McNamara, S.P., 2008. GPS for Surveyors. Springer Science & Business Media.

McNeill, J.D., 1980. Electromagnetic terrain conductivity measurements at low induction numbers. Geonics Technical Note TN-6.

McNeill, J.D., 1990. Use of Electromagnetic Methods for Groundwater Studies. In: Ward, S.H. (ed) Geotechnical and Environmental Geophysics, 1, Soc Expl Geophysics, Tulsa, 191–218.

McNeill, J.D., Labson, V.F., 1991. Geological Mapping using VLF Radiofields. In: Nabighian, M.C. (ed), Geotechnical and Environmental Geophysics, 1, Review and Tutorial. Tulsa: Society of Exploration Geophysicists, 191–218.

McNoleg, O., 1996. The Integration of GIS, Remote Sensing, Expert Systems and Adaptive Co-Kriging for Environmental Habitat Modelling of Highland Haggis Using Object-Oriented, Fuzzy Logic and Neural-Network Techniques. Comput. Geosci., 22(5), 585–588.

McWilliam, N., Teeuw, R., Whiteside, M., Zukowskyj, P., 2005. GIS, GPS and Remote Sensing Field Techniques Manual 2005, Section A, chapter 1, 1–8.

Mehanee, S.A., Essa, K.S., 2015. 2.5D regularized inversion for the interpretation of residual gravity data by a dipping thin sheet: numerical examples and case studies with an insight on sensitivity and non-uniqueness. Earth, Planets and Space., 67, 130.

Meiser, P., 1962. A method of quantitative interpretation of self-potential measurements. Geophys. Prospect., 10, 203–218.

Meju, M.A., 2002. Geoelectromagnetic exploration for natural resources: models, case studies and challenges. Surv. Geophys. 23, 133–206.

Memmi, J.M., Pride, D.E., 1997. The application of Diamond Exploration Geoscientific Information System (DEGIS) technology for integrated diamond exploration in the north-central United States of America. Int. J. Remote Sens., 18, 1439–1464.

Menard, H.W., Shipek, C.J., 1958. Surface concentrations of manganese nodules. Nature, 182, 1156–1158.

Mendenhall, P., Cooper, R., Yuen, D., 2020. AI-driven geophysical data integration in exploration. Geophys. J. Int., 211(6), 1075–1088.

Mendhe, V.A., Kumar, S., Kamble, A.D., Mishra, S., Varma, A.K., Bannerjee, M., Mishra, V.K., Sharma, S., Buragohain, J., Tiwari, B., 2018b. Organo-mineralogical insights of shale gas reservoir of Ib-River Mand-Raigarh Basin, India. Journal of Natural Gas Science and Engineering, 59, 136–155.

Mendonca, C.A., 2008. Forward and inverse self-potential modeling in mineral exploration. Geophysics, 73(1), F33–F43.

Menezes, A.A., Dias, L.E., Neves, J.C.L., Silva, J.V.O., 2010. Zinc availability to maize in different soils with and without liming determined by mehlich-1, mehlich-3 and DTPA. Rev. Bras. Cienc. Solo, 34(2), 417–424.

Meng, G.X., Lv, Q.T., Yan, J.Y., Deng, Z., Qi, G., Xue, R.H., 2019. The research and application of explorational technology of "Penetrating" to geology and mineral investigation in overburden area. Acta Geosci. Sin., 40, 637–650, (In Chinese with English abstract).

Menke, 1984. Geophysical Data Analysis: Discrete Inverse Theory. Elsevier, MATLAB Edition, 3rd Edition. Paperback ISBN: 9780128100486. eBook ISBN: 9780123977847 330p.

Menon, S., 2008. India's rise to moon. Nature, 455, 874–875.

Merchant, J., Narumalani, S., 2009. Technology integration for environmental monitoring. Taylor & Francis, 1, 45–67.

Merchant, J.W., Sunil, N., 2009. Integrating remote sensing and geographic information systems – university of Nebraska – lincoln.

Merkel, W.H., Kaushika, R.M., Gorman, E., 2008. NRCS GeoHydro – a GIS Interface for Hydrologic Modeling. Comput. Geosci., 34(8), 918–930.

Mero, J., 1965. The Mineral Resources of the Sea. Elsevier Oceanography Series Elsevier Amsterdam, Holland.

Mery, N., Marcotte, D., 2022b. Quantifying mineral resources and their uncertainty using two existing machine learning methods. Math. Geosci., 54(2), 363–387.

Meyer, M.S., Anderson, R.E., Becker, A.L., 2015. Geochemical and geophysical data integration for environmental monitoring: a case study on groundwater contamination. Environ. Monit. Assess., 187(3), 23–36.

Michaud, S.C., Covel, C.A., 1998. The Application of Very Low Frequency Electromagnetics and Downhole Geophysical Techniques to an Island Hydrogeologic Study: rhode Island Formation, Narragansett Basin, 333p.

Mickus, K., Johnson, E., 2001. Mapping sedimentary and volcanic units within and surrounding Petrified Forest National Park, Arizona, using Landsat-5 and SPOT panchromatic data. Int. J. Remote Sens., 22, 1919–1935.

Mihalasky, M.J., Bonham-Carter, G.F., 2001. Lithodiversity and its spatial association with metallic mineral sites, Great Basin of Nevada. Nat. Resour. Res., 10, 209–226.

Milkereit, B., Berrer, E.K., King, A.R., Watts, A.H., Roberts, B., Adam, E., Eaton, D.W., Wu, J., Salisbury, M.H., 2000. Development of 3-D seismic exploration technology for deep nickel-copper deposits – a case history from the Sudbury basin, Canada. Geophysics, 65, 1890–1899.

Miller, C., Routh, P., Donaldson, P., Oldenburg, D., 2006. Imaging a geothermal system using controlled-source electromagnetic. Geothermal Resources Council Transactions Bulletin, 30, 447–451.

Miller, C., Barrash, W., Clement, W., Routh, P., 2008. Controlled-source electromagnetic survey at New Hyde Park, New York. Technical Report for U.S. Environmental Protection Agency Grant X-96004601-0.

Miller, K.A., Thompson, K.F., Johnston, P., Santillo, D., 2018. An overview of seabed mining including the current state of development, environmental impacts, and knowledge gaps. Front. Mar. Sci., 4:418.

Miller, M., Yong, F., Hu, L., Sergey, K., Vladimir, A., Evgeniy, T., 2023. Prediction of subway vibration values on the ground level using Machine Learning. Geotech. Geol. Eng. 41, 3753–3766.

Miller, R., Pullan, S., Waldner, J., Haeni, F., 1986. Field comparison of shallow seismic sources. Geophysics, 51(11), 2067–2092.

Miller, R.K., Rajendran, S., Thomas, S., 2020. Sustainable Resource Management and Geoscience Technology. Elsevier.

Miller, T., Durlik, I., Kostecka, E., Kozlovska, P., Lobodzinska, A., Sokołowska, S., Nowy, A., Integrating artificial intelligence agents with the internet of things for enhanced environmental monitoring: applications in water quality and climate data. Electronics 2025, 14, 696.

Mining, 2019. Available online: https://www.flyability.com/mining (accessed on 26 March 2019).

Mining, N., 2021. Social Impact Assessment with AI in Indigenous Communities. Journal of Sustainable Mining, 30(1), 25–36.

Ministry of Mines, Government of India, 2021. Status of National Mineral Exploration Trust.

Ministry of Mines, Government of India, 2022. Geoscientific survey and mineral exploration initiatives

Ministry of Mines, Government of India, 2022. Mission-I & Mission-II: Mineral Exploration Initiatives.

Mishra, O.P., 2025. Seismic tomography: a versatile tool for multi-scale subsurface imaging. J. Geol. Soc 101 (6), 770–775. doi:10.17491/jgsi/2025/174158.

Mishra, O.P., Naskar, D.C., 2024. Geophysical Methods: Fundamentals, Applications, and Case Studies. first ed., CRC Press, pp. 593. doi:10.1201/9781032661902.

Mishra, O.P., Naskar, D.C., 2025. A Field Manual of Magnetotelluric (MT) Surveys with Case Studies for Earth Scientists and Engineers. first ed., CRC Press, pp. 376. doi:10.1201/9781003638377.

Mishra, O.P., Naskar, D.C., 2025. Geophysical Methods: fundamentals, Applications, and Case Studies (Book: first Edition). Taylor & Francis, Published by CRC Press, 2385 NW Executive Center Drive, Suite 320, Boca Raton FL 33431, First Edition, ISBN: 9781032599298<isbn> (hbk) ISBN: 9781032661896 (pbk) ISBN: 9781032661902 (ebk) DOI: 10.1201/9781032661902, pp. 593.

Mishra, O.P., Naskar, D.C., 2026. Geothermal Exploration and Development: Geophysical Techniques for Sustainable Energy Security. first ed., CRC Press, pp. 708. doi:10.1201/9781003672425.

Mishra, O.P., Singh, A.P., Singh, O.P., 2023. Seismicity and décollement geometry beneath the Sikkim - Darjeeling Himalaya using Local earthquake tomography: implications for seismotectonics and Seismogenesis. Phys. Earth Plant. Inters 336. 10.1016/j.pepi.2023.106999

Mishra, O.P., Zhao, D., 2003. Crack density, saturation rate and porosity at the 2001 Bhuj, India, earthquake hypocenter: a fluid-driven earthquake? Earth Planet. Sci. Lett 212, 393–405.

Mishra, O.P., Zhao, D., 2004. Seismic evidence for dehydration embrittlement of the subducting Pacific slab. Geophys. Res. Lett 31, L09610. doi:10.1029/2004GL019489.

Mishra, R., Sahu, N., 2020. Remote sensing and GIS in forest resource management: a case study in Central India. Indian J. For., 43(1), 72–81.

Mishra, S., Mendhe, V.A., Kamble, A.D., Varma, A.K., Singh, B.D., Pandey, A.K., 2016. Prospects of shale gas exploitation in Lower Gondwana of Raniganj Coalfield (West Bengal), India. Palaeobotanist, 65, 31–46.

Mishra, S., Shrivastava, P., Dhurvey, P., 2017. Change Detection Techniques in Remote Sensing: a Review. In: International Journal of Wireless and Mobile Communication for Industrial Systems, 4(1), 1–8.

Misra, R.S., Mathur, S.K., Singh, B., 2000. Final Report on Creation of Theme-Based Database of Degree Sheet No. 53F Covering parts of Uttaranchal, Uttar Pradesh, Punjab, Haryana and Himachal Pradesh. Unpublished G.S.I. Report for F.S. 1998-2000.

Mobley, C.D., 1994. Light and Water: Radiative Transfer in Natural Waters. Academic Press, San Diego.

Mochizuki, S., Kataoka, J., Tagawa, L., Iwamoto, Y., Okochi, H., Katsumi, N., Kinno, S., Arimoto, M., Maruhashi, T., Fujieda, K., Kurihara, T., Ohsuka, S., 2017. First demonstration of aerial gamma-ray imaging using drone for prompt radiation survey in Fukushima.

Mohammed Aslam, M.A., Balasubramanian, A., Kondoh, A., Rokhmatuloh, Mustafa, A.J., 2003. Hydrogeomorphological Mapping Using Remote Sensing Techniques for Water Resource Management around Palaeochannels. Conference Paper 2003, 0-7803-7929-2/03/$17.00 (C) 2003 IEEE.

Mohr, P.J., Taylor, B.N., 2005. CODATA recommended values of the fundamental physical constants: 2002. Rev. Mod. Phys., 77, 1–107.

Moitra, A.K., Bhaduri, A., Mukhopadhayay, G.C., Das, A.K., 2022. Subchapter 1.3 – exploration of shale gas in India – prospects and challenges. Innovative Exploration Methods for Minerals, Oil, Gas, and Groundwater for Sustainable Development, 29–42.

Moldoveanu, N., Combee, L., Egan, M., Abriel, W., Hampson, G., 2012. Over/under towed-streamer acquisition: a method to extend seismic bandwidth to both higher and lower frequencies. Seg Technical Program Expanded Abstracts 25(1):35–41.

Moller, I., 2001. Ohm Mapper field tests at sandy and clay till sites in Denmark. In Proceedings of the EEGS-ES 7th Annual Meeting. Environmental and Engineering Geophysical Society-European Section, 100–101.

Molnar, D.K., Julien, P.Y., 1998. Estimation of Upload Erosion Using GIS. Comput. Geosci., 24(2), 183–192.

Monsouri, E., Fie, F., Rad, A.J., Arran, M., 2018. Remote Sensing data processing with multivariate regression analysis method for iron mineral resource potential mapping: a case study in the Sarvian area Central Iran. Solid Earth 9, 373–384.

Montanaro, et al., 2022. Landsat 9 thermal-infrared sensor 2 (TIRS-2) stray light mitigation and assessment. IEEE Transactions on Geoscience and Remote Sensing, 60, 1–8, 5002408, doi: 10.1109/TGRS.2022.3177312.

Monteleone, V., Marin-Moreno, H., Bayrakci, G., Best, A., Shaon, F., Hossain, M.M., Al Karim, A., Alam, M.K., 2022. Seismic characterization and modelling of the gas hydrate system in the northern Bay of Bengal, offshore Bangladesh. Marine and Petroleum Geology, 141, 105690.

Moon, C., Lee, J., Park, H., 2006. Advanced surveying techniques: GPS and RTK systems. In: Advanced Geomatics and Surveying Techniques. Elsevier.

Moon, C., et al. (2006). Geochemical and Geophysical Methods in Mineral Exploration. Elsevier.

Moon, C.J., Whateley, M.K.G., Evans, A.M., 2006. Introduction to Mineral Exploration, 2nd ed. Blackwell Publishing.

Moore, D.H., Vander Berg, A.H.M., William, C.E., Magart, A.P.M., 1998. Paleozoic geology and resources of Victoria. J. Aust. Geol. Geophys., 13, 107–122.

Moore, G.K., 2010. What is a picture worth? A history of remote sensing/Quelle est la valeur d'une image? Un tour d'horizon de teledetection. Hydrol. Sci. Bull., 24, 477–485.

Moore, R., 2023. What is mineral exploration: stages, methods, laws & regulations? www.pheasantenergy.com/mineral-exploration/. Last updated on July 24, 2023.

Mora, P., Tarantola, A., 1989. Seismic Imaging. Springer, New York.

Moreira, C.A., Borssatto, K., Ilha, L.M., Dos Santos, S.F., Rosa, F.T.G., 2016. Geophysical modeling in gold deposit through DC resistivity and induced polarization methods. REM 69, 293–299. (doi:10.1590/0370-44672016690001).

Moreira, C.A., Paes, R.A.S., Ilha, L.M., Bittencourt, J.C., 2019. Reassessment of copper mineral occurrence through electrical tomography and pseudo 3D modeling in Camaquã Sedimentary Basin, Southern Brazil. Pure App. Geophys. 176, 737–750. (doi:10.1007/s00024018-2019-2).

Morelli, C., et al. (1971). The International Gravity Standardization Net 1971. IUGG IAG Special Publication, 4, 194p.

Moritz, H., 1980. Geodetic reference system 1980. Bull. Geod., 54, 395–40.

Morrell, A.E., 2004. Geophysical characteristics of low sulphidation epithermal gold-silver deposits in the Waihi Waitekauri region, New Zealand. Unpublished MSc thesis, The University of Auckland.

Mostafa, A.E., Ali, M.A.M., Ali, F.A., Rabeiy, R., Saleem, H.A., Shebl, A., Ali, M.A.H., 2025. Groundwater potential mapping in semi-arid areas using integrated remote sensing, GIS, and geostatistics techniques. Water, 17, 1909.

Mouzong, M.P., Kamguia, J., Nguiya, S., Shandini, Y., Manguelle-Dicoum, E., 2014. Of Garoua sedimentary basin, Benue Trough, North Cameroon, using gravity data. J. Biol., 4(1), E25–E33.

Mugisha, J.M., Mungatana, E., 2021. Educational initiatives in geophysical mapping: training the next generation of resource managers. J. Appl. Geophys., 187, 104–114.

Mukhopadhyay, R., Chosh, A.K., Iyer, S.D., 2018. The Indian Ocean Nodule Field – Geology and Resource Potential, 2nd ed.; Elsevier: Amsterdam, The Netherlands 978-0-12-805474-1.

Muller, D., Groves, D.I., Heithersay, P.S., 1994. The shoshonite porphyry Cu-Au association in the Goonumbla District, N.S.W., Australia. Mineralogy and Petrology, 51, 299–321.

Munoz-Royo, C., Ouillon, R., El Mousadik, S., Alford, M.H., Peacock, T., An in situ study of abyssal turbidity-current sediment plumes generated by a deep seabed polymetallic nodule mining preprototype collector vehicle. Sci. Adv. 2022; 8:eabn1219.

Munschy, M., Boulanger, D., Ulrich, P., Bouiflane, M., 2007. Magnetic mapping for the detection and characterization of UXO: use of multi-sensor fluxgate 3-axis magnetometers and methods of interpretation. J. Appl. Geophys., 61, 168–173.

Murphy, C., Brewster, J., Robinson, J., 2007. Evaluating Air-FTG® survey data: bringing value to the full picture. Preview, 126, 24–28.

Murray, J., Renard, A.F., 1891. Deep Sea Deposits (Based on the Specimens Collected during the Voyage of HMS Challenger in the Years 1872 to 1876); Eyre and Spottiswoode: London, UK.

Mursu, J., 1991. VLF and VLF-R measurements using a moving VLF transmitter. MSc thesis (in Finnish), University of Oulu.

Mussett, A.E., Khan, M.A., 2000. Well Logging and Other Subsurface Geophysics. In: Looking into the Earth: An Introduction to Geological Geophysics. New York, Cambridge University Press, 285–305.

Mustard, J.F., Pieters, C.M., 1987. Abundance and Distribution of Ultramafic Microbreccia in Moses Rock Dike: quantitative Application of Mapping Spectroscopy. J. Geophys. Res., 92(B10), 376–390.

Muttiah, R.S., Engel, B.A., Jones, D.D., 1996. Waste Disposal Site Selection Using GIS-Based Simulated Annealing. Comput. Geosci., 22(9), 1014–1017.

Nabighian, et al., 2005. The historical development of the magnetic method in exploration. Geophysics, 70, 33ND–61ND.

Nabighian, M.N., 1979. Quasi-static transient response of a conducting half-space – an appropriate representation. Geophysics, 44, 1700–1705.

Nabighian, M.N., ed., 1988. Electromagnetic Methods in Applied Geophysics: Theory, vol. 1: SE

Nabighian, M.N., et al., 2005. Historical development of the gravity method. Explor. Geophys., 70(6), 63ND–89ND.

Nabighian, M.N., Macnae, J.C., 1991. Time domain electromagnetic prospecting methods: Electromagnetic Methods in Applied Geophysics – Vol. 2 Applications. Investigations in Geophysics, No. 3. 0-931830-46-X, 427–520.

Nageswara, R.P.P., Ranganath, B.K., Chandrashekhar, M.G. 1991. Remote sensing applications in sericulture, Indian Silk, 30, 7–15.

Naghizadeh, M., Snyder, D., Cheraghi, S., Foster, S., Cilensek, S., Floreani, E., Mackie, J., 2019. Acquisition and processing of wider bandwidth seismic data in crystalline crust: progress with the metal earth project. Minerals, 9, 145.

Naskar, D.C., Choudhury, K., Roy, K.K., Singh, O.P., Majumdar, R.K., 2002. Geophysical approachs for delineation of shallow crustal structure along Borgaon- Sanwer transect, Madhya Pradesh. Geological Society of India, 60, 173–181.

Naskar, D.C., 2005. Geophysical studies of crust and upper mantle structures in parts of Madhya Pradesh and Maharashtra, Central India. Ph.D. Thesis, Jadavpur University, p.150.

Naskar, D.C., Das, L.K., Roy, K.K., Singh, O.P., 2008. Structural studies of crust and upper mantle in parts of Madhya Pradesh, Central India. Geological

Survey of India, Pan-African Event: India & Antarctica, Spl. Publ. No. 91, pp. 211–228.

Naskar, D.C., Singh, O.P., Das, L.K., Syiem, S.M., 2011. Integrated geophysical survey for delineation of active fault in the foothill of Bhutan Himalaya, northeast India. Indian Journal of Geoscience (Geological Survey of India Publication), 65(3), 211–222.

Naskar, D.C., Das, L.K., Roy, K.K., Majumdar, R.K., Choudhury, K., 2012. A re-examination of the crustal structure across Narmada-Son-Lineament (NSL), Madhya Pradesh, Central India using magnetotelluric and gravity modelling, vol. XXXIII, no. 3, pp. 69–77.

Naskar, D.C., Kulkarni, A.V., Lakshmana, M., 2018. Geophysical investigation for basemetal mineralization in Karoi-Rajpura area, Bhilwara district, Rajasthan. J. Ind. Geophys. Union, 22(4), 379–388.

Naskar, D.C., Chakraborty, P.K., Thota, R., Anurag Tripathi, A., 2020. Crustal structure from gravity and magnetic Data in parts of Nalgonda and Khammam Districts, Telangana (India). Journal of Indian Geophysical Union, 24 (2), 22–32.

Naskar, D.C., Nagaraja, P.N., Suru, P., Babu, R., 2021. Geophysical studies in parts of Kolar, Dharmapuri, Chittoor and Thiruvannamalai districts, Karnataka and Tamilnadu. Journal of Indian Geophysical Union, 25(4), 25–40.

Naskar, D.C., Kumar, V., Muktibodh, P., 2021. Geophysical investigation for base metal mineralization in extension area of Mamandur polymetals prospect in Villupuram and Thiruvannamalai districts of Tamil Nadu. J. Ind. Geophys. Union vol. 25, no. 2, pp. 30–44.

Naskar, D.C., Nagaraja, P.N., Ramesh, M., Suru, P., 2022. Gravity and magnetic studies in parts of Odisha and Chhattisgarh: its implication in regional geology. Bulletin of Geophysics and Oceanography (BGO) (Bollettino Di Geophysica Teorica Ed Applicata – BGTA), 63(2), 289–310.

Naskar, D.C., Chandrasekhar, S.V.N., Sharma, A., Purushottam, M., 2022. Regional gravity and magnetic surveys in parts of Belgaum, Bijapur, Dharwar and Raichur districts, Karnataka (India), vol. 26, no. 1, pp. 23–40.

Naskar, D.C., Regar, R.L., Gupta, D., Das, P.C., Chopra, M., 2022. Integrated geophysical survey for delineating aquifer zones in Sagar area, South 24 Parganas, West Bengal. Innovative Exploration Methods for Minerals, Oil, Gas, and Groundwater for Sustainable Development (Moitra, A.K., Bhattacharya, J., Kayal, J.R., Mukerji, B., Das, A.K.), pp. 225–240.

Naskar, D.C., Chaudhuri, A., Muktibodh, P., Maurya, V.C., 2023. Structural fabric and crustal architecture by geophysical surveys in Dharwar, north Canara and Shimoga districts of Karnataka (India). Journal of Indian Geophysical Union, 27(5), 1–20.

Naskar, D.C., Yadav, P.K., Ray, J., Maurya, D., Insight into structural fabric and crustal architecture across Eastern Dharwar Craton, Southern India from combined analysis of gravity and magnetic data. J. Ind. Geophys. Union, 28(5) (2024), 350–370.

Naskar, D.C., Kulkarni, A.V., 2019. Geophysical investigation for copper and gold mineralization around Dhani-Basri and Todi ka Bas area, Dausa and Alwar districts, Rajasthan. J. Ind. Geophys. Union, 23(2), 120–133.

Naskar, D.C., Lahiri, A.K., 2014. Electrical imaging of shallow and deep crustal structure using magnetotelluric method in the eastern part of Birbhum coalfield, West Bengal, India, vol. XXXV, no. 4, pp. 147–157.

Naskar, D.C., Rai, M.K., 2018. Delineation of geometry and spatial distribution of basic bodies using magnetic and resistivity methods in the Sohagpur coalfield area, Madhya Pradesh, Journal of Research, Granthaalayah vol. 6, no. 8, pp. 146–158. ISSN-2350-0530(O), ISSN-2394-3629(P) DOI: 10.5281/zenodo.1403846.

Naskar, D.C., Saha, D.K., 2015. Geophysical investigations for delineation of Gondwana sediments below Deccan trap beyond the western limit of Wardha valley coalfields, Yeotmal and Wardha districts, Maharashtra-a comprehensive analysis of case studies. J. Ind. Geophys. Union, 19(4), 433–446.

Nath, B., Balaram, V., Sudhakar, M., Pluger, W., 1992. Rare earth element geochemistry of ferromanganese deposits from the Indian Ocean. Mar Chem., 38:185–208.

National Oceanic and Atmospheric Administration Seabed Management (n.d.) Available online: http://www.gc.noaa.gov/gcil_seabed_management.html (accessed on 4 April 2016).

Naudet, V., Revil, A., Bottero, J.Y., Begassat, P., 2003. Relationship between self-potential (SP) signals and redox conditions in contaminated groundwater. Geophys. Res. Lett., 30, 2091.

Nayak, P.N., 1990. Deep crustal configuration of Central India. Geological Survey of India, Spl Publ 28, 67–98.

Nedeljkovic, V., Prndock, N., 1996. Finding spectral anomalies in hyperspectral imagery using nonparametric probability density estimation techniques. Proceedings of the Eleventh Thematic conference and Workshop on Applied Geologic Remote Sensing, Las Vegas, Nevada, 1, 197–204.

Nelson, A., Holland, D., Yogi, O., Heidron, D., 2004. Integrated use of seismic, ground and airborne gravity/gravity gradiometer and ground geological mapping methods in the Eastern Papuan basin, PNG. Preview, 111, 84.

Nelson, H.H., McDonald, J.R., 2001. Multisensor towed array detection system for UXO detection. IEEE Trans. Geosci. Remote Sens., 39, 1139–1145.

Nelson, P.H., 2001. Introduction to borehole geophysics. SEG Geophysical Monograph Series, 1, 27–42.

Neroni, R., Hunt, D., 2022. Modern Applications of Geophysics – mineral Case Studies: reflections on the MAG21 symposium presented by the WA Branch of the ASEG, Preview, 2022:216, 19–25, DOI: 10.1080/14432471.2022.2034311.

Nettleton, L.L., 1939. Determination of density for reduction of gravitimeter observations. Geophysics, 4, 8176–8183.

Nettleton, L.L., 1971. Elementary gravity and magnetics for geologists and seismologists. Society of Exploration Geophysicists, 121p.

Neuendorf, K.K.E., Mehl Jr., J.P., Jackson, J.A., eds. 2011. Glossary of Geology. American Geological Institute, 799p.

Neville, R.A., Levesque, J., Staenz, K., Nadeau, C., Hauff, P., Borstad, G.A., 2003. Spectral unmixing of hyperspectral imagery for mineral exploration: comparison of results from SFSI and AVIRIS. Canadian Journal of Remote Sensing, 29, 99–110.

Nevolko, P.A., Svetlitskaya, T.V., Nguyen, T.H., Pham, T.D., Fominykh, P.A., Tran, T.H., Tran, T.A., Shelepaev, R.A., 2022. Genesis of the Thien Ke tungsten deposit, Northeast Vietnam: evidence from mineral composition, fluid inclusions, S-O isotope systematics and U-Pb zircon ages. Ore Geol. Rev., 143:104791.

Ni, P., Wang, G.G., Chen, H., Xu, Y.F., Guan, S.J., Pan, J.Y., Li, L., 2015. An early Paleozoic orogenic gold belt along the Jiangshao fault, South China: evidence from fluid inclusions and Rb-Sr dating of quartz in the Huangshan and Pingshui deposits. J. Asian Earth Sci., 103, 87–102.

Nian, T.K., Guo, X.S., Fan, N., Jiao, H.B., Li, D.Y., 2018. Impact forces of submarine landslides on suspended pipelines considering the low temperature environment. Appl Ocean Res., 81:116–125.

Nian, T.K., Song, X.L., Zhao, W., Jiao, H.B., Guo, X.S., 2022. Submarine slope failure due to overpressure fluid associated with gas hydrate dissociation. Environ. Geotech., 9(2):108–123.

Nichols, E.A., Morrison, H.F., Lee, S., 1994. Controlled-source magnetotellurics for groundwater. 64th Annual International Meeting, SEG, Expanded Abstracts, 553–554.

Noorollahi, Y., Itoi, R., Fujii, H., Tanaka, T., 2007. GIS Model for Geothermal Resource Exploration in Akita and Iwate Prefectures, Northern Japan. Computers & Geosciences, 33(8), 1008–1021.

Nuis, 2006. National Urban Information System (NUIS) Scheme Memorandum of Understanding (MOU). Town & Country Planning Organisation Government of India Ministry of Urban Development.

Nuis, 2015. National urban information system (NUIS) scheme. Status report as on August 2015. Town & Country Planning Organisation Ministry of Urban Development Government of India, 18p along with 8 annexure.

Nykanen, V., 2008. Radial basis functional link nets used as a prospectivity mapping tool for orogenic gold deposits within the Central Lapland Greenstone Belt, Northern Fennoscandian Shield. Nat. Resour. Res., 17, 29–48.

Nykanen, V., Groves, D.I., Ojala, V.J., Eilu, P., Gardoll, S.J., 2008a. Reconnaissance scale conceptual fuzzy-logic prospectivity modeling for iron oxide copper – gold deposits in the northern Fennoscandian Shield, Finland. Aust. J. Earth Sci., 55, 25–38.

Nykanen, V., Groves, D.I., Ojala, V.J., Gardoll, S.J., 2008b. Combined conceptual/ empirical prospectivity mapping for orogenic gold in the Northern Fennoscandian Shield, Finland. Aust. J. Earth Sci., 55, 39–59.

Nykanen, V., Ojala, V.J., 2007. Spatial analysis techniques as successful mineral potential mapping tools for orogenic gold deposits in the northern Fennoscandian Shield, Finland. Nat. Resour. Res., 16, 85–92.

Nyquist, J.E., Corry, C.E., 2002. Self-potential: the ugly duckling of environmental geophysics. Lead. Edge 21(5), 446–451.

O'Brien, J., Rodriguez, A., Sixta, D., Davies, M.A., Houghton, P., 2005. Resolving the K-2 salt structure in the Gulf of Mexico: an integrated approach using prestack depth imaging and full tensor gravity gradiometry. Lead. Edge, 24, 404–409.

O'Dowd, C., Wood, G., Brisbin, D., Powell, B., 2006. Enhancing uranium exploration through seismic methods and potential field modeling at the McArthur River mine site, Saskatchewan, Canada. 56th Annual International Meeting, SEG, Expanded Abstracts. 1253–1257

O'Leary, D.W., Friedman, J.D., Pohn, H.A., 1976. Lineament, linear and lineation: some proposed new standards for old terms. Geological Society of America Bulletin, 87, 1463–1469.

O'Reilly, M., Smith, G., Johnson, L., 2023. Advances in Airborne Electromagnetic Surveys for Mineral Exploration: enhancements and Applications. J. Geophys. Res., 129(4), 1837–1852.

Obhodas, J., Sudac, D., Meric, I., et al., 2018. In-situ measurements of rare earth elements in deep sea sediments using nuclear methods. Sci. Rep. 8:4925.

OECD, 2020. Policy Frameworks for Responsible Mineral Resource Development. Organisation for Economic Co-operation and Development.

Ogilvy, R.D., Meldrum, P.I., Kuras, O., Wilkinson, P.B., Chambers, J.E., Sen, M., Pulido-Bosch, A., Gisbert, J., Jorreto, S., Frances, I., Tsourlos, P., 2009. Automated monitoring of coastal aquifers with electrical resistivity tomography. Near Surface Geophysics, 7, 367–375.

Ohta, J., Yasukawa, K., Nakamura, K., et al., 2021. Geological features and resource potential of deep sea mud highly enriched in rare earth elements in the Central Pacific Basin and the Penrhyn Basin. Ore Geol. Rev., 139:104440.

Ok, A.O., Akar, O., Gungor, O., 2012. Evaluation of random forest method for agricultural crop classification. Eur. J. Remote Sens., 45, 421–432.

Okada, K., Minami, Y., Ono, M., 2018. Microtremor survey for exploration targeting epithermal vein systems at the Hishikari Gold Mine, Kagoshima, Japan. EarthDoc. 10.3997/2214-4609.201802740.

Okada, K., 2022. Breakthrough technologies for mineral exploration. Miner. Econ., 35, 429–454 10.1007/s13563-022-00317-3.

Okada, K., Iwashita, A., 1992. Hyper-multispectral image analysis based on waveform characteristics of spectral curve. Advanced Space Research, 12, 433–442.

Zurmotai, N.H., 2016. GIS, remote sensing and GPS: their activity, integration and field work. Conference: The collection of materials of the International scientific-practical conference on 28-29 April 2016 Almaty, Kazakh Leading Academy of Architecture and Civil Engineering, Almaty, Kazakhstan.

Okada, K., Sumitomo, M., etal Mining Co., Ltd., 1995. Geophysical exploration for epithermal gold deposits: case studies from the Hishikari Gold Mine, Kagoshima, Japan. Explor. Geophys., 26(2-3), 78–83.

Olayinka, A.I., Yaramanci, U., 1999. Choice of the best model in 2-D geoelectrical imaging: case study from a waste dump site. Eu. J. Environ. Eng. Geophy., 3, 221–244.

Oldenborger, G.A., Knoll, M.D., Routh, P.S., LaBrecque, D.J., 2007. Time-lapse ERT monitoring of an injection/withdrawal experiment in a shallow unconfined aquifer. Geophysics, 72, F177–F187.

Oldenburg, D.W., Li, Y., 1999. Estimating depth of investigation in dc resistivity and IP surveys. Geophy., 64, 403–416.

Oliveira, C., Amaral Sobrinho, N.M.B., Mazur, N., 2003. Solubility of heavy metals in soils treated with enriched sewage sludge. Rev. Bras. Cienc. Solo, 27(1), 171–181.

Oliveira, W.J., Silva, C.A., Muniz, J.A., Savian, T.V., 2013. Nitrogen mineralization in latosols fertilized with organic residues. Rev. Bras. Cienc. Solo, 37(3), 715–725.

Omeiza, A.J., Lawal, K.M., Dewu, B., Raimi, J., 2023. Development of geophysical test sites and its impacts on the research and education activities. Bull Eng Geol Environ 82:32. 10.1007/s10064-023-03076-9.

Omer, S., Deshmukh, S., Tomar, P.S., Mishra, R.K., Harilal, 2020. Source facies characterization for shale gas exploration using seismic inversion & AVO analysis: a case study from Krishna-Godavari basin, India 13th Biennial International Conference and Exhibition, Kochi.

Onojasun, O.E., 2016. A case for large scale mineral exploration in hard rock environments using seismic reflection method. IOSR Journal of Applied Geology and Geophysics (IOSR-JAGG) e-ISSN: 2321-0990, p-ISSN: 2321-0982. 4, 1 Ver. I (Jan-Feb 2016), PP 93–98 www.iosrjournals.org.

Osinowo, O.O., Application of Landsat enhanced thematic mapper plus (ETM+) and radar imageries for geological boundary delineation and lineament mapping in a transition zone around Ijebu-Ode, southwestern Nigeria. Africa Geosci. Rev., 20(3 & 4) (2013), 83–96.

Osinowo, O.O., Gomy, A., Isseini, M., 2021. Mapping hydrothermal alteration mineral deposits from Landsat 8 satellite data in Pala, Mayo Kebbi Region, south-western Chad. Scientific African, 11, e00687.

Othman, A.A., Gloaguen, R., 2017. Integration of spectral, spatial and morphometric data into lithological mapping: a comparison of different Machine Learning Algorithms in the Kurdistan Region, NE Iraq. J. Asian Earth Sci., 146, 90–102.

Ouchi, K., 2013. Recent trend and advance of synthetic aperture radar with selected topics. Remote Sens. 10.3390/rs5020716.

overview, F.T., 2023. "Japan deep sea impact experiment=Jet". Available from: https://onepetro.org/ISOPEOMS/proceedingsabstract/OMS95/All-OMS95/25538.

Ozturgut, E., Lavelle, J.W., Steffin, O., Swift, S.A., 1978. Environmental investigations during manganese nodule mining tests in the north equatorial pacific in November 1978; 1980, 1–48..

P.T.C., H., Clowes, R.M., Ramachandran, K., 2004. Seismic reflection imaging of thin, kimberlite dykes and sills: exploration and deposit characterization of the Snap Lake dyke, Canada. Lithos, 76, 259–367.

Padhy, P.K., Kumar, A., Chandra, R.Y., Das, S.K., JHA, S.K., Advani, D.R., 2016. Shale oil exploration from

Palaeocene-Early Eocene sequence in Cambay rift basin, India. Proc. Indian Natn. Sci. Acad., 82(3) Special Issue, 945–963.

Paillet, F.L., Cheng, C.H., 1991. Acoustic Waves in Boreholes: Boca Raton, Florida, CRC Press, 264p.

Paithankar, A., Chatterjee, S., 2018. Grade and tonnage uncertainty analysis of an African copper deposit using multiple-point geostatistics and sequential Gaussian simulation. Nat. Resour. Res., 27, 419–436.

Pal, M., Rasmussen, T., Porwal, A., 2020. Optimized lithological mapping from multispectral and hyperspectral remote sensing images using fused multi-classifiers. Remote Sens., 12, 177.

Palacky, G., 2015. Electromagnetic and resistivity techniques in mineral exploration: a review. Geophysics, 80(1), 39–50.

Palani, S., Liong, S.Y., Tkalich, P., 2008. An ANN application for water quality forecasting. Mar Pollut Bull. 56(9), 1586–1597.

Palmer, S.D., 1997. Vbscript 2 & Activex Programming: Master the Art of Creating Interactive Web Pages. 11 Edition, Coriolis Group Books, 8p.

Pan, G.C., Harris, D.P., 2000. Information Synthesis for Mineral Exploration. Oxford University Press Inc., New York.

Pan, J.P., Zhang, D.W., Yue, L.Q., Wang, Y., Hu, W., 2006. Status of global offshore oil and gas exploration and development and its trends. China Min 11:1–4 (in Chinese).

Panigrahy, S., 2017. Mapping of Wetlands using Satellite Remote Sensing Data: indian Experience. Wetland Science, 423–448.

Panissod, C., Michel, D., Hesse, A., Joivet, A., Tabbagh, J., Tabbagh, A., 1998. Recent developments in shallow depth electrical and electrostatic prospecting using mobile arrays. Geophysics, 63, 1542–1550.

Paramasivam, C.R., 2016. Remote sensing and spectral studies of magnesite mining area, Salem, India. Ph.D Thesis, Periyar University, Periyar Palkalai Nagar, Salem, Tamil Nadu, India, 193p.

Parasnis, D.S., 1962. Principles of Applied Geophysics, 84.

Parasnis, D.S., 1997. Principles of Applied Geophysics. Publisher Springer Netherlands, 429p, ISBN 0412640805 9780412640803.

Pare, P., Legault, J.M., 2010. Ground IP-resistivity and airborne SPECTREM and helicopter ZTEM survey results over Pebble copper-moly-gold porphyry deposit, Alaska. 80th Annual International Meeting, SEG, Expanded Abstracts, 1734-1738.

Parianos, J., Lipton, I., Nimmo, M., 2021. Aspects of Estimation and Reporting of Mineral Resources of Seabed Polymetallic Nodules: a Contemporaneous Case Study. Minerals, 11, 200. https://doi.org/10.3390/min11020200.

Park, S., Choi, Y., 2020. Application of unmanned aerial vehicles in mining from exploration to reclamation: a review. Minerals, 663. https://doi.org/10.3390/min10080663.

Parker, R.T., Hughes, S.E., 2020. Environmental monitoring using geochemical mapping in mining areas: a case study from Canada. Journal of Environmental Management, 244, 228–239.

Parkinson, W., 1959. Directions of rapid geomagnetic variations. Geophys. JR Astr. Soc., 2, 1–14.

Paropkari, A.L., Ray, D., Balaram, V., et al., 2010. Formation of hydrothermal deposits at Kings Triple Junction, northern Lau back-arc basin, SW Pacific: the geochemical perspectives. J. Asian Earth Sci., 38:121–130.

Parveen, M.T., Ilahi, R.A., 2022. Assessment of land-use change and its impact on the environment using GIS techniques: a case of Kolkata Municipal Corporation, West Bengal, India. GeoJournal, 87, 551–566.

Patel, P., Kumar, S., 2019. Geospatial technologies for water resource management in Rajasthan. Water Resources Management Journal, 33(8), 2345–2358.

Paul, M.K., 1965. Direct interpretation of self-potential anomalies caused by inclined sheets of infinite horizontal extensions. Geophysics, 30, 418–423.

Paul, M.K., Datta, S., Banerjee, B., 1965. Interpretation of self-potential anomalies due to localized causative bodies. Pure Appl. Geophys., 61, 95–100.

Pecher, I.A., Holbrook, W.S., 2000. Seismic methods for detecting and quantifying marine methane hydrate/free gas reservoirs. In: Max, M.D. (eds) Natural Gas Hydrate. Coastal Systems and Continental Margins, 5. Springer, Dordrecht.

Pedroso, E.C., Rivard, B., Crosta, A.P., Souza Filho, C.R., Miranda, F.P., 2001. Reconnaissance geologic mapping in the Tapajos mineral province, Brazilian Amazon, using spaceborne SAR imagery and airborne geophysics. Canadian Journal of Remote Sensing, 27, 669–678.

Pellerin, L., 2002. Applications of electrical and electromagnetic methods for environmental and geotechnical investigations. Surveys in Geophysics, 23, 101–132.

Pellerin, L., Wannamaker, P.E., 2005. Multi-dimensional electromagnetic modeling and inversion with application to near-surface earth investigations. Computers and Electronics in Agriculture, 46, 71–102.

Pelton, W.H., Ward, S.H., Hallof, P.G., Nelson, P.H., 1978. Mineral discrimination and removal of inductive coupling with multifrequency IP. Geophysics, 43, 588–609.

Pendock, N., 1990. Gold Exploration in Greenstone Belts by Use of Landsat TM. Institution of Mining & Metallurgy, Remote Sensing: An Operational Technology for the Mining and Petroleum Industries, Springer Netherlands, Dordrecht, 243–247.

Perron, G., Calver, A.J., 1998. Shallow, high resolution seismic imaging at the Ansil mining camp in the Abitibi greenstone belt. Geophysics, 63, 379–391.

Petrie, S.S., Dixon, T.M., Hatch, C.E., 2015. Geochemical mapping for nickel and cobalt exploration in the Fraser Range. Australian Journal of Earth Sciences, 62(6), 675–688.

Petterson, M.G., Kim, H.J., Gill, J.C., 2021. Gill, J.C., Smith, M. (eds.): conserve and sustainably use the oceans, seas, and marine resources. Geosciences and the Sustainable Development Goals, Sustainable Development Goals Series, Cham: Springer International Publishing, 339–367, doi: 10.1007/978-3-030-38815-7_14, ISBN 978-3-030-38814-0, S2CID 234955801, archived from the original on 17 February 2023.

Phaneuf, D., Cote, I., Dumas, P., 1999. Evaluation of the contamination of marine algae (seaweed) from the St. Lawrence River and likely to be consumed by humans. Environ. Res. Sect., 80:175–182. 10.1006/enrs.1998.3915.

Phani, R.C., Joseph, C., Bommu, N.R., Prabhat, S., 2025. Application of UAV-borne magnetic survey in diamond exploration: a case study over kimberlite-carbonatite intrusion from Khaderpet, Eastern Dharwar Craton, South India. Journal of the Geological Society of India, 101(2), 198–207.

Phillips, S.B., Aneja, V.P., Kang, D., Arya, S.P., 2006. Modelling and analysis of the atmospheric nitrogen deposition in North Carolina. Int J Glob Environ Issues, 6(2-3), 231–252.

Piikki, K., Wetterlind, J., Soderstrom, M., Stenberg, B., 2015. Three-dimensional digital soil mapping of agricultural fields by integration of multiple proximal sensor data obtained from different sensing methods. Precis. Agric. 16, 29–45.

Pillay, S.S., Mota, R., 2018. National mineral policy and sustainable development in geophysical exploration. Resources Policy, 59, 287–295.

Piper, D.Z., 1974. Rare earth elements in ferromanganese nodules and other marine phases. Geochim. Cosmochim. Acta, 38:1007–1022.

Pirajno, F., 1992. Hydrothermal Mineral Deposits, Principles and Fundamental Concepts for the Exploration Geologists. Springer, Berlin.

Pirajno, F., Bagas, L., 2002. Gold and silver metallogeny of the South China fold belt: a consequence of multiple mineralizing events? Ore Geol. Rev, 20, 109–126.

Pirttijarvi, M., 2014. AEMINV-Laterally Constrained 1-D Inversion of EM Profile Data, Version 1.3 User's Guide. University of Oulu, 31.

Pirttijarvi, M., Saartenoja, A., Korkeakangas, P., 2021. Louhi EM Test Data (Version 1.0) [Data set]. Zenodo.

Pirttijarvi, M., Saartenoja, A., Korkeakangas, P., 2022. Drone-based electromagnetic survey system for geophysical applications. Open Research Europe, 2:3 Last updated: 13 Nov 2024, 27p.

Pirttijarvi, M., Oy, R., 2016. Radai's UAV based radiometric measurements at Rautuvaara Mine in Kolari. Radai Survey Report.

Podwysocki, M.H., Segal, D.B., Abrams, M.J., 1983. Use of Multispectral scanner images for assessment of hydrothermal alteration in the Marysvale, Utah, Mining area. Econ. Geol., 78, 675–687.

Pomykala, R., Tora, B., 2017. Circular economy in mineral processing. Published by EDP Sciences, E3S Web Conferences, 18, 01024.

Ponzoni, F.J., Albuquerque, B.F.C., 2008. Pre-launch absolute calibration of CCD/CBERS-2B sensor. Sensors, 8, 6557–6565.

Porras, D., Carrasco, J., Carrasco, P., Alfageme, S., Gonzalez Aguilera, D., Gui Jarro, R.L., 2021. Drone magnetometry in mining research: an application in the study of Triassic Cu-Co-Ni mineralisation in the estancias mountain range, Almeria (Spain). Drones 5, 151. https://doi.org/10.3390/drones5040151.

Portoghese, I., Uricchio, V., Vurro, M., 2005. A GIS Tool for hydrogeological water balance evaluation on a regional scale in semi-arid environments. Computers Geosciences, 31(1), 15–27.

Porwal, A., Carranza, E.J.M., Hale, M., 2003a. Extended weights-of-evidence modelling for predictive mapping of base metal deposit potential in Aravalli province, western India. Explor. Min. Geol., 10, 155–163.

Porwal, A., Carranza, E.J.M., Hale, M., 2003c. Knowledge-driven and data-driven fuzzy models for predictive mineral potential mapping. Nat. Resour. Res., 12, 1–25.

Porwal, A., Carranza, E.J.M., Hale, M., 2004. A hybrid neuro-fuzzy model for mineral potential mapping. Math. Geol., 36, 803–826.

Porwal, A., Carranza, E.J.M., Hale, M., 2006b. A hybrid fuzzy weights-of-evidence model for mineral potential mapping. Nat. Resour. Res., 15, 1–14.

Porwal, A., Deb Das, R., Chaudhary, B., Gonzalez-Alvarez, I., Kreuzer, O., 2015. Fuzzy inference systems for prospectivity modeling of mineral systems and a case-study for prospectivity mapping of surficial Uranium in Yeelirrie Area, Western Australia. Ore Geol. Rev., 71, 839–852.

Porwal, A.K., Carranza, E.J.M., Hale, M., 2003. Knowledge-driven and data-driven fuzzy models for predictive mineral potential mapping. Natural Resource Research, 12, 1–25.

Porwal, A.K., Carranza, E.J.M., Hale, M., 2006. Bayesian network classifiers for mineral potential mapping. Computers & Geosciences, 32, 1–16.

Porwal, A.K., Gonzalez-Alvarez, I., Markwitz, V., McCuaig, T.C., Mamuse, A., 2010. Weights-of evidence and logistic regression modeling of magmatic nickel sulfide prospectivity in the Yilgarn Craton, Western Australia. Ore Geol. Rev, 38, 184–196.

Porwal, A.K., Kreuzer, O.P., 2010. Introduction to the special issue: mineral prospectivity analysis and quantitative resource estimation. Ore Geol. Rev., 38, 121–127.

Pradham, P., Younan, N.H., King, R.L., 2008. Concepts of Image Fusion in Remote Sensing Applications. Image Fusion, Academic Press, 393–428.

Prado, E.M.G., de Souza Filho, C.R., Carranza, E.J.M., Motta, J.G., 2020. Modeling of Cu-Au prospectivity in the Carajas mineral province (Brazil) through machine learning: dealing with imbalanced training data. Ore Geol. Rev., 124, 103611.

Prasanth Kumar, P., Sairam, M., Shankar, T., Praharaj, S., Maitra, S., 2021. Application of Remote Sensing in Agriculture. Indian J. Nat. Sci., 12(69), 37475–37479.

Prensky, S.E., 1999. Advances in borehole imaging technology and applications, in Lovell, M.A., Williamson, G., Harvey, P.K., eds., Borehole Imaging: Applications and Case Histories: London, England, Geological Society of London Special Publications, 159, 1–43, doi:10.1144/GSL.SP.1999.159.01.01.

Priest, J.A., Best, A.I., Clayton, C.R.I., 2005. A laboratory investigation into the seismic velocities of methane gas hydrate-bearing sand. J. Geophys. Res., 110, B04102, doi:10.1029/2004JB003259.

Pringle, R.W., Roulston, K.L., Brownell, G.M., 1950, Ultrasensitive portable gamma-ray spectrometer. Nature, 165 (4196), 527p.

Prost, G., 1980. Alteration mapping with airborne multispectral scanners. Econ. Geol., 75, 894–906.

Pugin, A.J.M., Pullan, S.E., Hunter, J.A., 2009. Multicomponent high resolution seismic reflection profiling. Lead. Edge, 28(10), 1248–1261.

Pugin, A.J.M., Pullan, S.E., Hunter, J.A., Oldenborger, G.A., 2009. Hydrogeological prospecting using P-and S-wave land streamer seismic reflection methods. Near-Surface Geophys., 7(5-6), 315–327.

PwC, 2021. Mining for the Future: Integrating ESG into Exploration Projects. PricewaterhouseCoopers LLP.

QGIS Documentation, 2021. QGIS User Guide (Version 3.16). QGIS.org.

Qiu, Z., Tao, C., Ma, W., et al., 2022. Material source of sediments from West Clarion-Clipperton Zone (Pacific): evidence from rare earth element geochemistry and clay minerals compositions. JMSE, 10:1052.

Quamar, M.M., Al-Ramadan, B., Khan, K., Shafiullah, M., El Ferik, S., 2023. Advancements and Applications of Drone-Integrated Geographic Information System Technology – a Review. Remote Sens. 15, 5039.

Quiring, S.M., Ganseh, S., 2010. Evaluating the utility of the Vegetation Condition Index (VCI) for monitoring meteorological drought in Texas. Agric. For. Meteorol., 150, 330–339.

Radiological health handbook, 1970. United States, Bureau of Radiological Health. Publisher: rockville, Md.: u.S. Dept. of Health, Education, and Welfare, Public Health Service, Food and Drug Administration, Bureau of Radiological Health; Washington: for sale by the Supt. of Docs., U.S. G.P.O., 458p.

Radogna, P.V., Olivier, R., Logean, P., 2004. A Subway Project In Lausanne, Switzerland, As An Urban Microgravimetry Test Site: acquisition, Correction Of Buildings Influence And Modeling. European Association of Geoscientists & Engineers, Source: Conference Proceedings, 17th EEGS Symposium on the Application of Geophysics to Engineering and Environmental Problems, Feb 2004, cp 186 00088 DOI 10.3997/2214-4609-pdb186INF01.

Raghavan, V., Masumoto, S., Koike, K., Nagano, M., 1995. Automatic lineament extraction from digital images using a segment tracing and rotation transformation approach. Comput. Geosci., 21, 555–559.

Rahimi, H., Abedi, M., Yousefi, M., Bahroudi, A., Elyasi, G.R., 2021. Supervised mineral exploration targeting and the challenges with the selection of deposit and non-deposit sites thereof. Appl. Geochem., 104940.

Rahimi, M., Rad, A.R.J., Almasi, A.R., Samani, B., Rad, F.T., 2016. Gold metallogenesis and detecting areas of high potential gold mineralization using large Spatio-temporal datasets, Northwest Iran, Western sector of Tethyan metallogenic belt. Proceedings of the Conference Sept, Turkey, Society of Economic Geologists, Inc., 25–28.

Raines, G.L., 1999. Evaluation of weights of evidence to predict epithermal-gold deposits in the Great Basin of the Western United States. Nat. Resour. Res., 8(4), 257–276.

Rajani, P.R., Manoj, R., Joshi, R., Jishnu, B., Nagasundaram, M., 2022. Base metals and Lithium rich ferromanganese oxide deposits from the South Andaman Sea, Northeastern Indian Ocean: mode of occurrence and genesis. J. Asian Earth Sci., 234:105272.

Rajendran, S., A case study of the minerals on the seabed of the West Coast of India. In: Ardus, D.A., Champ, M.A. (eds) Ocean Resources (Book). Springer,

Dordrecht. 10.1007/978-94-009-2131-3_13, © 1990 Kluwer Academic Publishers, 1, 159–167.

Rajesh, H.M., 2004. Application of remote sensing and GIS in mineral resource mapping – an overview. Journal of Mineralogical and Petrological Sciences, 99, 83–103.

Rajesh, J., Pande, C.B., Kadam, S.A., Gorantiwar, S.D., Shinde, M.G., 2021. Exploration of groundwater potential zones using analytical hierarchical process (AHP) approach in the Godavari river basin of Maharashtra in India. Applied Water Science, 11:182.

Raju, P.L.N., Sarma, K.K., Barman, D., Handique, B.K., 2016. Operational remote sensing services in north eastern region of India for natural resources management, early warning for disaster risk reduction and dissemination of information and services. The International Archives of the Photogrammetry, Remote Sensing and Spatial Information Sciences, Volume XLI-B4, XXIII ISPRS Congress, 12–19 July 2016, Prague, Czech Republic.

Ramachandra, M., Raghu Babu, K., Pradeep Kumar, B., Ganesh, P., Ranganath, P., 2020. Remote sensing and GIS applications in mineral detection and mapping using band rationing technique. Journal of Shanghai Jiaotong University, 16(7), 424–439, ISSN: 1007-1172.

Ramana Murthy, M.V., Usha, T., Kankara, R.S., 2022. Three Decades of Indian Remote Sensing in Coastal Research. J. Indian Soc. Remote Sens., 50, 599–612.

Rambat, S., 2011. A Los-cost remote sensing system for agricultural applications. Ph.D. dissertation, Aston University, Birmingham, UK.

Ranganathan, P.C., Siddan, A., 2021. Application of spectral signature to analyze quality of magnesite ore mineral deposits and alteRed Rocks of Salem, India. Arabian Journal of Geosciences (Springer), 14:651. http://doi.org/10.1007/s12517-021-06963-1.

Rani, N. (GSI), 2022. Application of Remote Sensing Techniques in Geological Mapping and Mineral Exploration – a Case Study from Dharwar Craton. Journal of Geological Society of India, 98, 144–145.

Ranjan, A., Panigrahi, B., Sahu, H.B., Misra, P., 2018. SkyHelp: UAV Assisted Emergency Communication in Deep Open Pit Mines. In Proceedings of the 1st International Workshop on Internet of People, Assistive Robots and ThingS-IoPARTS'18, Munich, Germany, 10 June 2018; 31–36.

Rao, A., Sarma, V.S., 1981. A modified pseudo depth section as a tool in resistivity and IP prospecting. Geophysical Research Bulletin, 187–208 (NGRI).

Rao, M., Jaiswal, R.K., Singh, P., Raghawaswamy, Pathan, S.K., Gurumukhi, K.T., 2005. A national urban information system – EO and GIS concept. 56th IAC, Fukuoka: session B.4. Data Environ. Appl. Glob. Change Stud., 21p.

Rast, M., Hook, S.J., Elvidge, C.D., Alley, R.E., 1991. An evaluation of techniques for the extraction of mineral absorption features from high spectral resolution remote sensing data. Photogramm. Eng. Remote Sens., 57, 1303–1309.

Rawal, D., Chhabra, A., Pandya, M., Vyas, A., 2020. Land use and land cover mapping – a case study of Ahmedabad district. The International Archives of the Photogrammetry, Remote Sensing and Spatial Information Sciences XLIII-B3-2020, XXIV ISPRS Congress (2020 edition), 189–193.

Ray, S.S., Saxena, S., Choudhary, K., Tahlani, P., Vadrevu, K.P., 2022. Agricultural Drought Assessment Using Remote Sensing, with Special Emphasis on India. Remote Sens. Agric. Land Cover/Land Use Changes S. SE Asian Countries, 309–332.

Rebillard, P., Evans, D., 1983. Analysis of coregistered Landsat, Seasat and SIR-A images of varied terrain types. Geophysical Research Letters, 10, 277–280.

Reeve, J., 1990. The discovery and evaluation of the Olympic Dam deposit. In: Blaikie, A., Woodcock, J., Glasson, K. Geological Aspects of the Discovery of Important Minerals in Australia, Australasian Institute of Mining and Metallurgy, 125–133.

Reeves, C.C., Allen, M.M., 2003. Geophysical Mapping and Resource Identification in the Rocky Mountains: USGS National Geophysical Surveys. U.S. Geological Survey Professional Paper 1702.

Reichstein, M., Camps-Valls, G., Stevens, B., Jung, M., Denzler, J., Carvalhais, N., Prabhat, N., 2019. Deep learning and process understanding for data driven Earth system science. Nature, 566(7743), 195–204.

Rein, B., Kaufmann, H., 2003. Exploration for gold using panchromatic stereoscopic intelligence satellite photographs and Landsat TM data in the Hebei area, China. Int. J. Remote Sens., 24, 2427–2438.

Rejith, R.G., Sundararajan, M., Venkatesan, S., Mohammed-Aslam, M.A., 2021. Remote sensing for exploring heavy mineral deposits: a case study of Chavara and Manavalakurichi deposits, southwest coast of India. Remote Sensing of Ocean and Coastal Environments (Earth Observation), 177–188.

Rejith, R.G., Sundararajan, M., Gnanappazham, L., Seenipandi, K., Nainarpandian, C., 2022. Exploring beach placer minerals in the east coast of Tamil Nadu, India, using EO-1 Hyperion data. Journal of Applied Remote Sensing, 16(1), 012017.

Rekhibi, S., Wadi, M., Said, A., 2015. Remote sensing & GIS techniques for gold exploration. Proceedings of the International Conference on Advances in Science, Engineering, Technology and Natural Resources (ICASETNR-15) Aug, Kota Kinabalu (Malaysia),

International Institute of Chemical, Biological & Environmental Engineering, 27–28.

Ren, H., Xia, M., Weng, L., Hu, K., Lin, H., 2024. Dual-attention-guided multiscale feature aggregation network for remote sensing image change detection. IEEE J Sel Top Appl Earth Obs Remote Sensing, 17, 4899–4916.

Ren, J., Liu, Y., Wang, F., et al., 2021. Mechanism and influencing factors of REY enrichment in deep sea sediments. Minerals, 11:196.

Resmini, R.G., Kappus, M.E., Aldrich, W.S., Harsanyi, J.C., Anderson, M., 1997. Mineral mapping with HYperspectral Digital Imagery Collection Experiment (HYDICE) sensor data at Cuprite, Int. J. Remote Sens., 18, 1553–1570.

Revil, A., 1999. Ionic diffusivity, electrical conductivity, membrane and thermoelectric potentials in colloids and granular porous media: a unified model. J. Coll. Int. Sci., 212, 503–522.

Revil, A., Ehouarne, L., Thyreault, E., 2001. Tomography of self-potential anomalies of electrochemical nature [J]. Geophysical Research Letters, 28(23), 4363–4366.

Revil, A., Trolard, F., Bourrie, G., Castermant, J., Jardani, A., Mendonça, C.A., 2009. Ionic contribution to the self-potential signals associated with a redox font. J. Cont. Hydrol., 109, 27–39.

Revil, A., Karaoulis, M., Johnson, T., Kemna, A., 2012. Review: some low frequency electrical methods for subsurface characterization and monitoring in hydrogeology [J]. Hydrogeol. J., 20(4), 617–658.

Revil, A., Jardani, A., 2013. The Self-potential Method: Theory and Applications in Environmental Geosciences. Cambridge University Press, Cambridge.

Reynolds, J.M., 1997. An Introduction to Applied and Environmental Geophysics. Chichester, England: John Wiley and Sons Ltd.

Reynolds, J.M. 2011. An Introduction to Applied and Environmental Geophysics. John Wiley & Sons Ltd, Chichester, 2nd ed., 712p.

Richa, Maurya, S.P., Singh, K.H., Singh, R., Kumar, R., Kushwaha, P.K., 2022. Application of maximum likelihood and model-based seismic inversion techniques: a case study from KG basin, India. Journal of Petroleum Exploration and Production Technology, 12, 1403–1421.

Ridley, W.I., 2010. Weathering processes in volcanogenic massive sulphide occurrence model (eds Shanks III, W.C.P., Turston, R.) Scientific Investigations Report 2010-5070-C 195–201 (U.S. Geological Survey

Ridwan, M.A., Radzi, M., Asyikin, N., Ahmad, W., Munirah, W.S.H., Mustafa, I.S., Din, N., Jalil, Y., Isa, A.M., Othman, N., Zaki, W., Diyana, W.M., 2018. Applications of Landsat-8 Data: a survey. Int. J. Eng. Technol. (UAE), 7, 436–441. 10.14419/ijet.v714.35.22858.

Riedel, M., Cosma, C., Enescu, N., Koivisto, E., Komminaho, K., Vaittinen, K., Malinowski, M., 2018. Underground vertical seismic profiling with conventional and fiber-optic systems for exploration in the Kylylahti polymetallic mine, Eastern Finland. Minerals, 8, 538.

Ritterbush, S.W., 2023. Marine resources and the potential for conflict in the South China Sea. Available from: http://hdl.handle.net/10427/76180 [Last accessed on 28 Apr 2023].

Robb, L.J., 2005. Introduction to Ore-forming Processes. Blackwell Pub, Malden, MA.

Robert, F., Poulsen, K.H., Cassidy, K.F., Hodgson, C.J., 2005. Gold metallogeny of the superior and Yilgarn Cratons. In: Economic Geology 100th Anniversary volume, 1001–1034.

Roberts, B., Zaleski, E., Perron, G., Adam, E., Petrie, L., Salisbury, M., 2003a. Seismic exploration of the Manitouwadge Greenstone belt, Ontario: a case history, Eaton, D.W., Milkereit, B., Salisbury, M.H. (eds.) Hardrock Seismic Exploration. SEG Developments in Geophysics Series, 110–126.

Robinson, E.S., Coruh, C., 1988. Basic Exploration Geophysics, New York: John Wiley and Sons.

Robinson, G.L., Chamberlain, C.J., 2009. Geochemical surveys and gold exploration: a case study from the Witwatersrand basin. Exploration Geochemistry Journal, 18(3), 200–212.

Rodovalho, E., Lima, J., Campos, P., Casagrande, P., Mazzinghy, D., 2025. An advanced approach for geometallurgical modeling applied to bauxite mines. Mining, 5, 11.

Rodrigues, F.A.V., Barros, N.F., Neves, J.C.L., Alvarez, V.H., Novais, R.F., 2012. Zinc availability for eucalypt seedlings in cerrado soils. Rev. Bras. Cienc. Solo, 36(4), 1249–1257.

Rodriguez-Galiano, V., Sanchez-Castillo, M., Chica-Olmo, M., Chica-Rivas, M., 2015. Machine learning predictive models for mineral prospectivity: an evaluation of neural networks, random forest, regression trees and support vector machines. Ore Geol. Rev, 71, 804–818.

Rogers, J.J.W., Adams, J.A.S., 1969. Uranium. In: Wedepohl, K.H. (Ed.), Hand Book of Geochemistry, II/5, Springer, Berlin, 92E–92M.

Rogers, J.J.W., Adams, J.A.S., 1969b. Thorium. In: Wedepohl, K.H. (Ed.), Handbook of Geochemistry, II/5, Springer, Berlin, 90E–90D.

Rogerson, D., 1996. Inside COM. Microsoft Press, Redmons, WA, 159–210.

Rohrbach, A., Rohrbach, M., Hu, R., Darrell, T., Schiele, B., 2016. Learning social etiquette: human trajectory understanding in crowded scenes. In Computer Vision – ECCV; Springer: Cham, Switzerland 9905

Roonwal, G.S., 2017. Remote Sensing in Mineral Exploration. Chapter in the part of the Springer Geology Book Series (SPRINGERGEOL) – Mineral Exploration: Practical Application, 119–153.

Rosado-Fuentes, A., et al., 2021. Geophysical characterization, design and construction of the Teoloyucan geophysical test site for archaeological and engineering applications, Central Mexico. J. Appl. Geophys. 194, 104459.

Rose, M., Zeng, Y., Dransfield, M.H., 2006. Applying Falcon gravity gradiometry to hydrocarbon exploration in the Gippsland basin, Victoria. Explor. Geophys., 37, 180–190.

Rossi, B.B., Staub, H.H., 1949. Ionization Chambers and Counters. McGraw-Hill Book Comp., New York.

Rossi, M.E., Deutsch, C.V., 2014. Mineral resource estimation. In: Mineral Resource Estimation.

Roth, J.L., Allen, S.M., 2017. Soil geochemistry and exploration for nickel in the Fraser Range. Miner. Explor. Geochem., 14(4), 244–257.

Rothery, D., 1987. Improved discrimination of rock units using Landsat Thematic Mapper imagery of the Oman ophiolite. Journal of the Geological Society of London, 144, 587–597.

Rouse, M., Earls, A., Sharon Shea, I.W., 2019. What is Drone (Unmanned Aerial Vehicle, UAV). Available online https://internetofthingsagenda.techtarget.com/definition/drone.

Roy, A., 1966. The method of continuation in mining geophysical interpretation. Geoexploration, 4, 65–83.

Roy, A., Chowdhury, D.K., 1959. Interpretation of self-potential data for tabular bodies. Journal Science and Engineering, 3, 35–54.

Roy, B., Kasemi, N., 2021. Monitoring urban growth dynamics using remote sensing and GIS techniques of Raiganj Urban Agglomeration, India. Egypt. J. Remote Sens. Space Sci., 24, 221–230.

Roy, I.G., 2015. On computing first and second order derivative spectra. J. Comput. Phys., 295, 307–321.

Roy, I.G., 2018. A simple technique of robust estimation of derivative of noisy data. J. Ind. Geophys. Univ., 22, 371–378.

Roy, I.G., 2022. Self-potential: a low-cost geophysical method in investigating groundwater and contaminant plume. Elsevier 3, 193–210, Eds. Moitra, A.K., Bhattacharya, J., Kayal, J.R., Mukerji, B., Das, A.K. "Innovative Exploration Methods for Minerals, Oil, Gas, and Groundwater for Sustainable Development".

Roy, P.S., Behera, M.D., Srivastav, S.K., 2017. Satellite Remote Sensing: sensors, Applications and Techniques. In: Proc. Natl. Acad. Sci. India Sect. A Phys. Sci., 87(4), 465–472.

Roy, P.S., Ravan, S.A., 1996. Biomass estimation using satellite remote sensing data – an investigation on possible approaches for natural forest. J. Biosci., 21(4), 535–561.

Roy, R., Cassard, D., Cobbold, P.R., Rossello, E.A., Billa, M., Bailly, L., Lips, A.L.W., 2006. Predictive mapping for copper-gold magmatic-hydrothermal systems in NW Argentina: use of a regional-scale GIS, application of an expert-guided data-driven approach, and comparison with results from a continental-scale GIS. Ore Geol. Rev., 29, 260–286.

Roy, S.V.S., Mohan, N.L., 1984. Spectral interpretation of self-potential anomalies of some simple geometric bodies. PAGEOPH, 78, 66–77.

Rudd, J., Chubak, G., Larnier, H., Stolz, R., Schiffler, M., Zakosarenko, V., Schneider, M., Schulz, M., Meyer, M., 2022. Commercial operation of a SQUID-based airborne magnetic gradiometer. Lead. Edge, 41(7), 486–492.

Ruiz-Armenta, J.R., Prol-Ledesma, R.M., 1998. Techniques for enhancing the spectral response of hydrothermal alteration minerals in Thematic Mapper image of central Mexico. Int. J. Remote Sens., 19, 1981–2000.

Rupprecht, S.M., Pieters, J.E., 2018. Re-opening of old gold mines for small scale mining in South Africa- the process of creating a small scale mine in a historically mined out South African gold field. University of Johannesburg: johannesburg, South Africa, 2018. Available online: https://core.ac.uk/download/pdf/161412655.pdf.

Ruter, H., Strack, K.M., 1991. Bedrock exploration system using transient electromagnetic measurements. Patent PCT/DE91/00238.

Rutter, H., Esdale, D., 1985. The geophysics of the Olympic Dam discovery. Explor. Geophys., 16(3), 273–276.

Rymer, H., 1996. Microgravity monitoring. Springer, In: monitoring and mitigation of volcano hazards (Book) by Roberto Scarpa & Robert I. Tilling, 169-197, 842p.

Rymer, H., Brown, G.C., 1989. Gravity changes as a precursor to volcanic eruptions at Poasvolcano, CostaRica. Nature, 342, 902–905.

S., D., Elizabeth, M., 2018. Implementation challenges of area based management tools (ABMTs) for biodiversity beyond national jurisdiction (BBNJ). Mar. Policy, 9734–43. doi:10.1016/j.marpol.2018.08.034. ISSN 0308-597X. S2CID 239830990.

Sa, R., Sun, X., He, G., et al., 2018. Enrichment of rare earth elements in siliceous sediments under slow deposition: a case study of the central North Pacific. Ore Geol. Rev., 94:12–23.

Saadia, Y., Nedjai, R., Haddouche, O., Boutaleb, A., Chemam, M., 2019. Contribution of GIS in Mineral Exploration through Mineralization Distribution Characterization: application to The Setifien/Hodna Massifs and The Eastern Saharan Atlas (NE Algeria). Aspects Min Miner Sci., 3(2), 373–381, AMMS.000556.2019. DOI: 10.31031/AMMS.2019.03.000556.

Sabine, C., Realmuto, V.J., Taranik, J.V., 1994. Quantitative estimation of granitoid composition from Thermal Infrared Multispectral Scanner (TIMS) data, Desolation Wilderness, northern Sierra Nevada, California. J. Geophys. Res., 99, 4261–4271.

Sabins, F.F., 1999. Remote sensing for mineral exploration. Ore Geol. Rev., 14, 157–183.

Safdary, R., Soffianian, A., Pourmanafi, S., 2022. Application of Landscape Metrics and Object-Oriented Remote Sensing to Detect the Spatial Arrangement of Agricultural Land. Quaestiones Geographicae, 41(1).

Safipour, R., Holz, S., Halbach, J., Jegen, M., Petersen, S., Swidinsky, A., 2017. A self-potential investigation of submarine massive sulfides: palinuro Seamount, Tyrrhenian Sea Geophysics, 82(6), A51–A56.

Sahoo, P., Pandit, M., 2020. Application of remote sensing and GIS for bauxite and limestone exploration in Odisha. Indian Journal of Applied Geology, 39(4), 455–464.

Saibi, H., 2018. Microgravity and Its Applications in Geosciences. Chapter3. In: Gravity - Geoscience Applications, Industrial Technology and Quantum Aspect. InTech. 10.5772/intechopen.71223.

Said, K.O., Onifade, M., Akinseye, P., Kolapo, P., Abdulsalam, J., 2023. A review of geospatial technology-based applications in mineral exploration. Geojournal, 88, 2889–2911.

Sain, K., Rai, M., Sen, M.K., 2014. A review on shale gas prospect in Indian sedimentary basins. J. Ind. Geophys. Union, 18(2), 183–194.

Sain, K., Gupta, H.K., 2012. Gas hydrates in India: potential and development. Gondwana Res., 22, 645–657.

Sakellariadou, F., Gonzalez, F.J., Hein, J.R., Rincon-tomas, B., Arvanitidis, N., Kuhn, T., 2022. Seabed mining and blue growth: exploring the potential of marine mineral deposits as a sustainable source of rare earth elements (MaREEs) (IUPAC Technical Report). Pure Appl. Chem., 94:329–351.

Salehi, T., Tangestani, M.H., 2018. Large-scale mapping of iron oxide and hydroxide minerals of Zefreh porphyry copper deposit, using Worldview-3 VNIR data in the north-eastern Isfahan, Iran. Int. J. Appl. Earth Obs. Geoinf., 73, 156–169.

Salek, O., Matolín, M., Gryc, L., 2018. Mapping of radiation anomalies using UAV mini- airborne gamma-ray spectrometry. J. Environ. Radioact. 182, 101–107.

Salisbury, J., Walter, L., Vergo, N., D'Aria, D., 1991. Infrared (2.1-2.5 μm) Spectra of Minerals. Johns Hpokins University Press, Baltimore, 267p.

Salisbury, J.W., Hunt, G.R., 1974. Remote sensing of rock types in the visible and near-infrared. Proceedings of the Ninth International Symposium on the Remote Sensing of the Environment, III, 1953–1958.

Salisbury, M.H., Milkereit, B., Ascough, G.L., Adair, R., Schmitt, D., Matthews, L., 1997. Physical properties and seismic imaging of massive sulphides, In: Canada, P.A.D.A.O., eds., Geophysics and Geochemistry at the Millennium, 383–390, GEO F/X Division of AG Infromation Systems Ltd.

Salisbury, M.H., Milkereit, B., Ascourgh, G., Adair, R., Matthews, L., Schmitt, D.R., Mwenifumbo, J., Eaton, D.W., Wu, J., 2000. Physical properties and seismic imaging of massive sulfides. Geophysics, 65, 1882–1889.

Sanada, Y., Torii, T., 2015. Aerial radiation monitoring around the Fukushima Dai-ichi nuclear power plant using an unmanned helicopter. J. Environ. Radioact. 139, 294–299.

Sanchez, M., Rojas, F., 2014. Monitoring environmental impact from mining operations: integrating GPS, GIS, and remote sensing in the Andes. Environ. Monit. Assess., 186(7), 4395–4409.

Satish Kumar, J., Sanjeevi, S., Govindan, S., 2011. Hyperspectral Radiometry to Characterise Dunite Alteration and Magnesite Deposits of Salem, South India. Journal of Indian Society Remote Sensing, Springer, 39(4), 497–505.

Sato, M., Mooney, H.M., 1960. The electrochemical mechanism of sulphide self-potentials. Geophysics, 25, 226–249.

Sattarova, V., Astakhov, A., Aksentov, K., et al., 2023. Geochemistry of the Laptev and East Siberian seas sediments with emphasis on rare-earth elements: application for sediment sources and paleoceanography. Cont Shelf Res., 254:104907.

Satyanarayana Murty, B.V., Haricharen, P., 1985. Nomogram for the complete interpretation of spontaneous potential profiles over sheet-like and cylindrical two-dimensional sources. Geophysics, 50, 1127–1135.

Satyavani, N., Sain, K., Lall, M., Kumar, B.J.P., 2008. Seismic attribute study for gas hydrates in the Andaman Offshore India. Marine Geophysical Research, 29, 167–175.

Sawatzky, D.L., Raines, G.L., Bonham-Carter, G.F., Looney, C.G., 2009. Spatial Data Modeller (SDM): arcMAP 9.3 geoprocessing tools for spatial data modelling using weights of evidence, logistic regression, fuzzy logic and neural networks, 15341.

Sawatzky, W., Jones, M., Davis, S., 2009. Advanced correction techniques in GPS surveying. J. Geospat. Sci., 14(3), 145–159.

Schenk, C.J., Menzie, W.D., Ziegler, T.L., 2014. Geophysical and geochemical data integration for mineral resource exploration. Journal of Geophysical Research: Solid Earth, 119(4), 3292–3308.

Schetselaar, E., Bellefleur, G., Hunt, P., 2019. Elucidating the effects of hydrothermal alteration on seismic reflectivity in the footwall of the Lalor volcanogenic massive sulphide deposit, Snow Lake, Manitoba, Canada. Minerals, 9, 384.

Schlumberger, 2020). Deep Learning for Seismic Data Interpretation. Journal of Geophysics, 88(3), 152–160.

Schmidt, G., 1959. Results of underground-seismic reflection investigations in the Siderite District of the Siegerland. Geophys. Prospect., 7, 287–290.

Schmidt, P.W., Clark, D.A., Leslie, K., Bick, M., Tilbrook, D., Foley, C., 2004. GETMAG – a SQUID magnetic tensor gradiometer for mineral and oil exploration. Explor. Geophys., 35, 297–305.

Schmidt, P.W., Clark, D.A., 2000. Advantages of measuring the magnetic gradient tensor. Preview, 85, 26–30.

Schmidt, S., Hensley, A., Liu, Z., 2019. Advancements in Electromagnetic Surveys for Deep Subsurface Exploration. Explor. Geophys., 50(4), 18–28.

Schneider, D.P., Van Zyl, C.J., 2016. Uranium exploration using geochemical and radiometric surveys in Namibia. J. Afr. Earth Sci., 118, 91–103.

Schreiber, M., 2021. Geological distribution and economic impact of Iron Ore in the World. Mineral Explor. Rep., 34(4), 78–92.

Schriever, W., 1952. Reflection seismograph prospecting – how it started. Geophysics 17(4):936–942.

Scott, W.J., 2014. Geophysics for Mineral Exploration. A Manual for Prospectors Prepared for Matty Mitchell Room Newfoundland and Labrador Department of Natural Resources.

Seed, H.B., De Alba, P., 1986. Use of SPT and CPT tests for evaluating the liquefaction resistance of sands. Use of in-situ tests in geotechnical engineering. ASCE, 281–30.

Seguin, M.K., Gupta, O.P., Nagrajan, V., Jain, A.K., Sen, N., Mahapatra, S., Mishra, O.P., 1994. Identification and delineation of abandoned, unknown and unsurveyed galleries in Raniganj coalfield by geophysical methods, Exploration Technology for modern coal mines: coal Exploration, Seminar Vol., VI-49-VI-70.

Seigel, H.O., Nabighian, M., Parasnis, D.S., Vozof, K., 2007. The early history of the induced polarization method. Lead. Edge, 3, 312–321.

Seigel, H.O., Howland-Rose, A.W., 1990. Magnetic Induced Polarization method. In: Induced Polarization: Applications and Case Histories, Fink, J.B., Sternberg, B.K., McAlister, E.O., Wieduwilt, W.G. (ed.), 23–56, Society of Exploration Geophysicists, Tulsa, Oklahoma.

Seigel, H.O., Howland-Rose, A.W., 1983. The magnetic induced polarisation method. Fitch, A.A. (ed.), Developments in Geophysical Exploration Methods –4 © Applied Science Publishers Ltd.

Seigel, H.O., 1974. The magnetic induced polarization method. Geophysics, 39, 321–339.

Seigel, H.O., 1959. Mathematical formulation of type curves for induced polarization. Geophysics, 24(3), 546–565.

Seiler, R.A., Kogan, F., Sullivan, J., 1998. AVHRR-based vegetation and temperature condition indices for drought detection in Argentina. Adv. Space Res., 21, 481–484.

Seiler, R.A., Kogan, F., Wei, G., 2000. Monitoring weather impact and crop yield from NOAA AVHRR data in Argentina. Adv. Space Res., 26, 1177–1185.

Sekhar, G., Goswami, J., Naveen, J., Priyanka, E., 2022. Crop Acreage and Yield Estimation using Remote Sensing and GIS and Various Related Schemes. Indian J. Nat. Sci., 13(72), 42666–42673.

Sen, M.K., Stoffa, P.L., 2013. Global Optimization Methods in Geophysical Inversion. Cambridge University Press. [ISBN: 9781107018772].

Serson, P.H., 1973. Instrumentation for induction studies on land. Phys. Earth Planet. Inter., 7, 313–322.

Seshunarayana, T., Viswaja, D., Srinivas, K.N.S.S.S., Srinivasa Rao, P., Trupti, S., Pavan Kishore, P., 2008. Application of high-resolution seismic survey in CBM exploration – a case study, Sohagpur west block, Madhya Pradesh. The 7th International Conference and Exposition on Petroleum Geophysics, Hyderabad.

Sha, Z.B., Zhang, M., Zhang, G.X., Liang, J.Q., Su, P.B., 2015. Using 4C OBS to reveal the distribution and velocity attributes of gas hydrates at the northern continental slope of South China Sea. Appl. Geophys., 12(4):555–563.

Shah, Z.U.H., Naveed, A., Ali, M., 2006. An application of integrated geophysical studies in Pakistan offshore. J. Virtual Explor., Electronic Edition, ISSN 1441-8142, volume 23, paper 3 In: (Eds.) Shahina Tariq, Zia Ul Hasan Shah, M Zafar, M Shah, and S Qureshi, Geology and Geophysics of Pakistan, 2006, 9p.

Shah, A., Bedrosian, P.A., Anderson, E.D., Kelley, K.D., Lang, J., 2013. Integrated geophysical imaging of a concealed mineral deposit: a case study of the world-class Pebble porphyry deposit in southwestern Alaska. Geophysics, 78, 317–328.

Shahmoradi, J., Talebi, E., Roghanchi, P., Hassanalian, M., 2020. A comprehensive review of applications of drone technology in the mining industry. Drones, 4, 34; doi:10.3390/drones4030034.

Shahriari, H., Honarmand, M., Mirzaei, S., Saffari, A., 2022. Application of drone-based data for direct exploration activities and estimating resources in emperador marble quarry, Kerman province Iran. J. Min. Environ., 13 (1), 253–267.

Shahsavani, H., Smith, R.S., 2024. Aeromagnetic Gradiometry with UAV, a Case Study on Small Iron Ore Deposit. Drone Systems and Applications (Canadian Science Publishing), 12, 1–9.

Shaole, S., et al. (2021). Subsurface structures of the Xiaorequanzi deposit, NW China: new insights from gravity, magnetic and electromagnetic data. Geophys. Prospect. https://doi.org/10.1111/1365-2478.13049.

Sharma, R., Singh, G., 2021. Optimizing water usage with AI-driven irrigation systems: a case study. Water Resour. Manag., 35(4), 1059–1072.

Sharma, R., Kumar, S., Singh, A., 2020. Water conservation through AI in agriculture: solutions for a sustainable future. Journal of Environmental Management, 270, 110887.

Sharma, K.N.M., Singhroy, V.H., Madore, L., Levesque, J., Hebert, C., Hinse, M., 1999. Use of radar images in the identification of major regional structures in the Grenville Province, Western Quebec. Canadian Journal of Remote Sensing, 25, 278–290.

Sharma, S.K., 2018. Role of Remote Sensing and GIS in Integrated Water Resources Management (IWRM). Chapter in Book: Ground Water Development – Issues and Sustainable Solutions, 211-227. Editor: S.P. Sinha Ray.

Sharma, R., Smith, S., 2019. Deep sea mining and the environment: an introduction. In: Sharma, R., editor. Environmental Issues of Deep Sea Mining. Cham: Springer International Publishing, 3–22.

Sheard, S.N., Ritchie, T.J., Christopherson, K.R., Brand, E., Mining, environmental, petroleum, and engineering industry applications of electromagnetic techniques in geophysics. Survey in Geophysics, 26(2005), 653–669.

Sheriff, R.E., 1991. "Geophysics". Encyclopedic Dictionary of Exploration Geophysics (3rd ed). Society of Exploration, 978-1-56080-018-7.

Sheriff, R.E., Geldart, L.P., 1995. Exploration Seismology. Cambridge University Press, Cambridge.

Shi, W., Morgan, F.D., 1996. Non-uniqueness in self-potential inversion. 66th Ann. Internat. Mtg. Soc. Expl. Geophys. Ext. Abstr., 950–953.

Shinohara, M., Yamada, T., Ishihara, T., Araya, A., Kanazawa, T., Fujimoto, H., Uehira, K., Tsukioka, S., Omika, S., Iizasa, K., 2015. Development of an underwater gravity measurement system using autonomous underwater vehicle for exploration of seafloor deposits. Oceans.

Shirmard, H., Farahbakhsh, E., Muller, D., Chandra, R., 2021. A review of machine learning in processing remote sensing data for mineral exploration. arXiv, arXiv:2103.07678.

Showstack, R., 2019. Momentum grows for mapping the seafloor. Eos, 100.

Shrestha, A.B., Oli, P.P., Shrestha, S., 2022. Application of UAV drone technology in mining industries of Nepal. Volunteering for the future – geospatial excellence for a better coding 1-10.

Shrivastava, P.K., Roy, S.K., Bhai, H.Y. 2017. Antarctic glacier monitoring. Proc. Indian Natn. Sci. Acad., 83, No. 2, Thematic Issue, pp. 255–267.

Shuyuan, M., 2014. GIS and Geophysics. Environ. Sci. Manag. NRS 509.

Sidorov, V.A., Tickshaev, V.V., 1969. Electrical Prospecting with Transient Field in near Zone: Saratov University Press.

Siemon, B., Steuer, A., Wiederhold, H., 2020. Helicopter-borne geophysical mapping along the German North Sea coast: an overview. Remote Sens., 12(10), 1629.

Sikakwe, G.U., 2023. Mineral exploration employing drones, contemporary geological satellite remote sensing and geographical information system (GIS) procedures: a review. Remote Sensing Applications: Society and Environment, 31, 100988, 21p.

Silcox, W.H., Bodine, J.A., Burns, G.E., Reeds, C.B., Wilson, D.L., Sauve, E.R., 1987. Offshore operations. In: Petroleum Engineering Handbook, Chapter 18, 2nd edn. SPE, Richardson. https://www.osti.gov/biblio/5929149.

Silva Dos Santos, V., Gloaguen, E., Hector Abud Louro, V., Blouin, M., Machine Learning methods for quantifying uncertainty in prospectivity mapping of magmatic-hydrothermal gold deposits: a case study from Juruena mineral province, Northern Mato Grosso, Brazil. Minerals 2022, 12, 941.

Silva, B.V.N., Andrade Pinto, L.V., 2010. Potential use of the sewage sludge as organic manuring in covering forestation native arboreal species planted in area degraded by intensive pasturing. Revista Agrogeoambiental, 2(1), 50–56.

Simonetti, E., Simonetti, D., Preatoni, D., 2014. Phenology-based land covers classification using Landsat-8 time series. 10.2788/15561.

Singer, A., 1984. The paleoclimatic interpretation of clay minerals in sediments – a review. Earth Sci. Rev., 21, 251–293.

Singer, D., Kouda, R., 1999. A comparison of the weights-of-evidence method and probabilistic neural networks. Nat. Resour. Res., 8, 287–298.

Singh, A.P., Mishra, O.P., 2015. Seismological evidence for Monsoon induced micro to moderate earthquake sequence. Tectonophysics 661, 38–48.

Singh, A.P., Mishra, O.P., Rastogi, B.K., 2011. 3-D Seismic structure of the Kachchh, Gujarat and its implications for the earthquake hazard mitigation. Nat. Hazards 57 (1), 83–105. doi:10.1007/s11069-010-9707-2.

Singh, T., Jhariya, D.C., Sahn, M., Dewngan, P., Dhekne, P.Y., 2022. Classifying minerals using deep learning algorithms. IOP Conf. Ser. Earth Environ. Sci. 1032.

Singh, R., Gupta, M., 2022. The role of drones in modern resource exploration: a GPS-integrated approach. Remote Sensing Applications: Society and Environment, 26, 100451.

Singh, B., Malinowski, M., Hlousek, F., Koivisto, E., Heinonen, S., Hellwig, O., Buske, S., Charmarczuk, M., Juurela, S., 2019. Sparse 3D seismic imaging in the Kylylahti mine area, Eastern Finland: comparison of time vs depth approach. Minerals, 9, 305.

Sinha, M., Patel, P., Unsworth, M., Owen, T., MacCormack, M., 1990. An active source electromagnetic sounding system for marine use. Mar. Geophys. Res., 12(1-2), 59–68.

Sinharay, R.K., 2019. Applications and future of ground penetrating radar (GPR) as a near surface geophysical technique: a review. Journal of Engineering Geology – A Bi-annual Journal of ISEG, XLIV(1 & 2), 12–17.

Sinha Roy, S., Mathai, T., 1979. Professional Paper No. 5, Centre for Earth Science Studies.

Sionneau, T., Bout-Roumazeilles, V., Biscaye, P.E., Vliet-Lanoe, B.V., Bory, A., 2008. Clay mineral distributions in and around the Mississippi River watershed and Northern Gulf of Mexico: sources and transport patterns. Quaternary Science Reviews, 27, 1740–1751.

Siva Sankar, E., Simhadri Rao, B., Abdul Hakeem, K., 2021. Snowmelt Runoff Forecasting in Himalayan Basins Using Remote Sensing Inputs. In: Part of the Geography of the Physical Environment Book Series (GEOPHY) (Chapter – Water, Cryosphere and Climate Change in the Himalayas, 19–37).

Sjodahl, P., Dahlin, T., Johansson, S., Loke, M.H., 2008. Resistivity monitoring for leakage and internal erosion detection at Hallby embankment dam. J. Appl. Geophys., 65, 155–164.

Sketchley, D.A. 1998. Gold deposits: establishing sampling protocols and monitoring quality control. Exploration and Mining Geology, 7(1-2), 129–138.

Skidmore, A.K., 1989. A Comparison of Techniques for Calculating Gradient and Aspect from a Gridded Digital Elevation Model. International Journal of Geographical Information Systems, 3(4), 323–344.

Slichter, L.B., 1955. Geophysics applied to prospecting for ores. Economic Geology, 50th Anniversary Volume.

Smit, D., Saleh, S., Voon, J., Costello, M., Moser, J., 2006. Recent controlled-source EM results show positive impact on exploration at shell. SEG/New Orleans 2006 Annual Meeting, 25, 3536–3541. 10.1190/1.2370274.

Smith, J., Harper, M., 2023. Application of machine learning in airborne geophysical surveys. J. Geophys. Res., 128(6), 1085–1101.

Smith, T.L., Harwell, R.L., 2021. Blockchain and AI integration in fair trade: a case study of the coffee industry. Journal of Business Ethics, 163(1), 113–130.

Smith, P., Johnson, M., 2021. Advanced Geophysics and Sustainability. Springer.

Smith, M.D., O'Neill, T.R., 2017. Geophysical Data Analysis with Python: Principles and Applications. Springer.

Smith, R., Shore, M., Rainsford, D., 2012. How to make better use of physical properties in mineral exploration: the exploration site measurement. Lead. Edge, 31(3), 249–368.

Smith, R., 2014. Electromagnetic induction methods in mining geophysics from 2008 to 2012. Surv. Geophys., 35(1), 123–156.

Snyder, D.B., Cary, P., Salisbury, M., 2009. 2D-3C high resolution seismic data from the Abitibi Greenstone Belt, Canada. Tectonophysics, 472, 226–237.

Soderstrom, M., Eriksson, J., 2013. Gamma-ray spectrometry and geological maps as tools for cadmium risk assessment in arable soils. Geoderma, 192, 323–334.

Soderstrom, M., Sohlenius, G., Rodhe, L., Piikki, K., 2016. Adaptation of regional digital soil mapping for precision agriculture. Precis. Agric. 17, 588–607.

Sommer, M., Holz, S., Moorkamp, M., Swidinsky, A., Heincke, B., Scholl, C., Jegen, M., 2013. GPU parallelization of a three-dimensional marine CSEM code. Comput. Geosci., 58, 91–99.

Son, Y.S., Kang, M.K., Yoon, Y.J., 2014. Lithological and mineralogical survey of the Oyu Tolgoi region, South-eastern Gobi, Mongolia using ASTER reflectance and emissivity data. Int. J. Appl. Earth Obs. Geoinf., 26, 205–216.

Son, Y.S., Kim, K.E., Yoon, W.J., Cho, S.J., 2019. Regional mineral mapping of island arc terranes in south-eastern Mongolia using multispectral remote sensing data. Ore Geol. Rev., 113, 103106.

Son, Y.S., You, B.W., Bang, E.S., Cho, S.J., Kim, K.E., Baik, H., Nam, H.T., 2021. Mapping Alteration Mineralogy in Eastern Tsogttsetsii, Mongolia, Based on the WorldView-3 and Field Shortwave-Infrared Spectroscopy Analyses. Remote Sens., 13, 914.

Song, H.B., Matsubayashi, O., Kuramoto, S., 2003. Full waveform inversion of gas hydrate related bottom simulating reflectors. Chi. J. Geophys., 46, 44–52.

Song, H.B., Wu, S.G., Jiang, W.W., 2007. The characteristics of BSRs and their derived heat flow on the profile 973 in the northeastern South China Sea. Chinese Journal of Geophysics, 50(5), 1508–1517.

Sonkamble, S., Chandra, S., GPR for earth and environmental applications: case studies from India. J. Appl. Geophys., 193(2021), 104422.

Sonkamble, S., Satishkumar, V., Amerender, B., Sethurama, S., 2013. Combined application of GPR, MERI and VES exploring groundwater potential zones in dry land environs: a study from Hyderabad, Southern India. Arab. J. Geosci. 7 (8), 3109–3117.

Sorensen, K., 1996. Pulled Array Continuous Profiling. First Break, 14, 85–90.

Souza, E.M., Muniz, J.A., Marchi, G., Guilherme, L.R.G., 2010. Non-linear modeling of zinc extracted from a sewage sludge-treated. Acta Sci. Technol., 32(2), 193–199.

Spatz, D.M., 1997. Remote sensing characteristics of the sediment and volcanic hosted precious metal system: imagery selection for exploration and developments. Int. J. Remote Sens., 18, 1413–1438.

SPC, 2023. Deep sea minerals: manganese nodules, a physical, biological, environmental, and technical review. Last accessed on 28 April 2023. https://static1.squarespace.com/static5cca30fab2cf793ec6d94096/t/5d9e42fd9ba4e7058d7e68a2/1570653017813/DSMvol1B_screen.pdf.

Spencer, C., Thurlow, G., Wright, J., White, D., Carroll, P., Milkereit, B., Reeds, L., 1993. A vibroseis reflection seismic survey at the Buchans mine in central Newfound-land. Geophysics, 58, 154–166.

Spies, B.R., 1989. Depth of investigation in electromagnetic sounding methods. Geophysics, 54, 872–888.

Sprenke, K.F., Wavra, C.S., Godfrey, J.D., 1986. Geophysical expression of the Canadian shield in northeastern Alberta; Alberta Geological Survey, Bulletin 52, 54p.

Srinivas, M., Verma, P., Sharma, A., 2021. Artificial intelligence and IoT in precision irrigation systems for sustainable water management. Indian J. Agric. Sci., 91(3), 330–341.

Srinivasarao, B., Gupta, A.K., Shukla, A.K., Markandeyulu, A., Rajendran, R., Choudhury, D.K., 2022. Application of microgravity survey in the exploration of rare metal bearing pegmatites – a case study from Marlagalla, Mandya District, Karnataka. J. Geol. Soc. India, 98, 1126–1130 10.1007/s12594-022-2133-8.

Srivastava, G.S., Malaviya, A.K., Misra, R.S., Bajpai, I.P., Srivastava, R., Kumar, B., Singh, B., Srivastava, V., Verma, R., Srivastava, M., Shukla, M.K., 1998. Report on the Setting up of Geoscientific Data Centre in Northern Region under GSI-BRGM Collaboration. Unpublished G.S.I. Report for F.S. 1996-1998.

Srivastava, G.S., Malaviya, A.K., Misra, R.S., Bajpai, I.P., Srivastava, R., Singh, B., 2000. Multicriteria analysis for Mineral Targeting in a part of Amtiar Gad South block, Tons Valley, Uttar Pradesh, using GIS Techniques. Proceedings of the Seminar ISRS-2000, IIT, Kanpur.

Stanton, T.E., Casagrande, A., Housel, W.S., Tschebotaref, G.P., 1939. Penetration devices bearing. In: Proceedings of the eighteenth annual meeting of the highway research board Washington, D.C. November 28-December 2, 1938 Part II: Soil mechanics and soil stabilization, 18, 421–422.

Stavrev, P.Y., 1997. Euler deconvolution using differential similarity transformations of gravity or magnetic anomalies. Geophys. Prospect., 45, 207–246.

Steeples, D.W., Miller, R.D., 1990. Seismic reflection methods applied to engineering, environmental, and groundwater problems. In: Ward, S.H. (Ed.), Geotechnical and Environmental Geophysics (vol. 1, pp.1–30), Society of Exploration Geophysicists.

Stern, R.J., Beyth, M., Bodechtel, J., Wetzel, H.-U., 2002. Potential of the International Space Station for imaging Earth: lessons from MOMS-2P aboard Mir. Geology, 30, 851–854.

Stewart, D.P., Randolph, M.F., 1991. A new site investigation tool for the centrifuge. A Balkema, Rotterdam. https://www.semanticscholar.org/paper/A-new-site-investigation-tool-for-the-centrifuge-Stewart-Randolph/9d13e147fd01e1dd0549794ba9428b753eefb729.

Stoll, J., Bigalke, J., Grabner, E.W., 1995. Electrochemical modelling of self-potential anomalies. Surv. Geophys., 16, 107–120.

Stolz, R., Schiffler, M., Becken, M., Thiede, A., Schneider, M., Chubak, G., Marsden, P., Bergshjorth, A.B., Schaefer, M., Terblanche, O., 2022. SQUIDs for magnetic and electromagnetic methods in mineral exploration. Miner. Econ., 35, 467–494 10.1007/s13563-022-00333-3.

Strack, K.M., Hanstein, T., LeBrocq, K., Moss, D.C., Vozoff, K., Wolfgram, P.A., 1989. Case histories of LOTEM surveys in hydrocarbon prospective areas. First Break, 7(12), 467–477.

Strack, K.M., Luschen, E., Kotz, A.W., 1990. Long-offset transient electromagnetic (LOTEM) depth soundings applied to crustal studies in the Black Forest and Swabian Alb, Federal Republic of Germany. Geophysics, 55(7), 834–842.

Strack, K.M., 1992. Exploration with Deep Transient Electromagnetics. Elsevier science Publishing Co.

Strack, K.M., 1992. Chapter2: theoretical background: exploration with deep transient electromagnetics. Methods in Geochemistry and Geophysics, 30, 0–444–89541–8,21–46.

Strack, K.M., 1999. Exploration with Deep Transient Electromagnetics. Elsevier Scientific Publ. Co., Inc.

Strangway, D.W., Koziar, A., 1979. Audio-frequency magnetotelluric sounding: a case history at the Cavendish geophysical test range. Geophysics, 44(8), 1429–1446.

Stummer, P., Maurer, H.R., 2001. Real-time experimental design applied to high-resolution direct-current resistivity surveys. Expanded Abstracts, In The International Symposium on Optical Science and Technology, SPIE 4491, 143–150.

Su, Z., Chen, D.F., 2007. Calculation of methane hydrate solubility in marine environment and its constraints on gas hydrate occurrence. Chinese J. Geophys., (in Chinese) 50(5), 1332–1340.

Su, Z., Cao, Y.C., Yang, R., et al., 2012. Analytical research on evolution of methane hydrate deposits at Shenhu Area, northern South China Sea. Chinese J. Geophys. (In Chinese), 55(5), 1764–1774.

Su, Z., Cao, Y., Wu, N., et al., 2012. Numerical investigation on methane hydrate accumulation in Shenhu Area, northern continental slope of South China Sea. Marine Petroleum and Geology, 38(1), 158–165.

Su, Y., 2020. Deep sea exploitation under equipment perspective. Wuhan University (in Chinese). https://doi.org/10.27379/d.cnki.gwhdu.2020.000706.

Subrahmanyam, C., Chand, S., 2006. Evolution of the passive continental margins of India – a geophysical appraisal. Gondwana Res., 167–178.

Sujith, P.P., Ramanan, D., Gonsalves, M.J.B.D., Lokabharathi, P.A., 2017. Microbial activity promotes the enrichment of cobalt over nickel on hydrogenetic ferromanganese crusts. Mar Georesources Geotechnol., 35:1158–1167.

Sun, H., et al., 2017. Characterization of shallow karst using electrical resistivity imaging in a limestone mining area. Environ. Earth Sci. 76, 767. (doi:10.1007/s12665-017-7112-9).

Sun, T., Chen, F., Zhong, L., Liu, W., Wang, Y., 2019. GIS-based mineral prospectivity mapping using machine learning methods: a case study from Tongling ore district, eastern China. Ore Geol. Rev., 109, 26–49.

Sun, T., Li, H., Wu, K., Chen, F., Zhu, Z., Hu, Z., 2020. Data-driven predictive modelling of mineral prospectivity using machine learning and deep learning methods: a case study from southern Jiangxi province, China. Minerals, 10, 102.

Sundararajan, N., 2012. A general perspective on geophysical methods in mineral exploration. In: On a Sustainable Future of the Earth's Natural Resources, (Part of the Springer Earth System Sciences, Book Series), 53–83, (Springer Earth).

Supper, R., Romer, A., Jochum, B., Bieber, G., Jaritz, W., 2008. A complex geo-scientific strategy for landslide hazard mitigation – from airborne mapping to ground monitoring. Advances in Geosciences, 14, 195–200.

Surya Prakash, L., Ray, D., Paropkari, A.L., et al., 2012. Distribution of REEs and yttrium among major geochemical phases of marine Fe-Mnoxides: comparative study between hydrogenous and hydrothermal deposits. Chem. Geol., 312-313:127–137

Sussman, G.J., Stallman, R.M., 1975. Heuristic techniques in computer aided circuit analysis. IEEE Trans. Circuits Syst. 22(11):857–865.

Swan, M., 2015. Blockchain: Blueprint for a New Economy. O'Reilly Media.

Sweeney, K., Johnson, R., 2020. Big Data Applications in Geosciences. Springer.

Szitkar, F., Petersen, S., Caratori Tontini, F., Cocchi, L., 2015. High-resolution magnetics reveal the deep structure of a volcanic-arc-related basalt-hosted hydrothermal site (Palinuro, Tyrrhenian Sea). Geochem. Geophys. Geosyst., 16, 1950–1961.

Tajvidi, E., Monjezi, M., Asghari, O., Emery, X., Foroughi, S., 2015. Application of joint conditional simulation to uncertainty quantification and resource classification. Arab. J. Geosci., 8, 455–463.

Takahashi, T., 2004. ISRM suggested methods for land geophysics in rock engineering. Int. J. Rock Mech. Min. Sci. Geomech. Abstr., 41, 885–914.

Takaya, Y., Yasukawa, K., Kawasaki, T., et al., 2018. The tremendous potential of deep sea mud as a source of rare earth elements. Sci. Rep., 8:5763. DOI PubMed PMC.

Talebi, H., Mueller, U., Tolosana-Delgado, R., 2019. Joint simulation of compositional and categorical data via direct sampling technique–application to improve mineral resource confidence. Computers & Geosciences, 122, 87–102.

Tang, J.T., He, J.S., 2005. Theory and Application of CSAMT Method. Central South University Press.

Tang, Y., et al. (2024). Exploration of the subsurface three-dimensional structure in rare earth ore areas using electrical resistivity tomography: a case study in Changting, Fujian, China. Minerals, 14(11), 1164.

Tang, A., 2014. A guide to penetration testing. Netw. Secur., 8–11.

Tangestani, M.H., Moore, F., 2002. Porphyry copper alteration mapping at the Meiduk area, Iran. Int. J. Remote Sens., 23, 4815–4825.

Tang, X.B., Meng, J., Wang, P., Cao, Y., Huang, X., Wen, L.S., Chen, D., 2016. Efficiency calibration and minimum detectable activity concentration of a real-time UAV airborne sensor system with two gamma spectrometers. Appl. Radiat. Isot. 110, 100–108.

Tao, C.H., Xiong, W., Xi, Z.Z., Deng, X.M., Xu, Y.X., 2013. TEM investigations of South Atlantic Ridge 13.2°S hydrothermal area. Acta Oceanol. Sin., 32, 68–74.

Tarantola, A., 1984. Inversion of seismic reflection data in the acoustic approximation. Geophysics, 49(8), 1259–1266.

Ted, M., 2014. Worldwide survey of deepwater drilling rigs [EB/OL]. (2014b-07-01). Offshore Magazine, Houston. https://www.ofshore-mag.com.

Telford, W.M., Geldart, L.P., Sheriff, R.E., 2012. Applied Geophysics (2nd ed.). Cambridge University Press.

Telford, W.M., Geldart, L.P., Sheriff, R.E., 1990. Applied Geophysics, 2nd edn . Cambridge: Cambridge University Press.

Telford, W.M., Geldart, L.P., Sheriff, R.E., Keys, D.A., 1976. Applied Geophysics. Cambridge, Cambridge University Press.

Tera, F., Wasserburg, G.J., 1972. U-Th-Pb systematics in three Apollo 14 basalts and the problem of initial Pb in lunar rocks. Earth Planet. Sci. Lett., 14, 281–304.

Tessnow-von, W.I., Vadrot, A.B.M., 2020. The voice of science on marine biodiversity negotiations: a systematic literature review. Frontiers in Marine Science, 7, doi: 10.3389/fmars.2020.614282. 2296-7745.

Thiel, H., Schriever, G., 1990. Deep-sea mining, environmental impact and the DISCOL project. Ambio, 19(5):245–250. https://www.researchgate.net/publication/279903214_Deep-sea_mining_environmental_impact_and_the_DISCOL_project.

Thompson, R., Miller, K., 2019. Technology-Driven social impact: using AI to monitor and ensure fair trade practices. Int. J. Bus. Technol., 8(3), 45–58.

Thompson, S., Fueten, F., Bockus, D., 2001. Mineral identification using artificial neural networks and the rotating polarizer stage. Comput. Geosci., 27, 1081–1089.

Tibuleac, I., Samulski, M., 2021. Drone Technology and Applications for Environmental and Geoscientific Monitoring. Springer.

Tiller, R., Mendenhall, E., De Santo, E., Nyman, E., 2023. Shake it off: negotiations suspended, but hope simmering, after a lack of consensus at the fifth intergovernmental conference on biodiversity beyond national jurisdiction. Mar. Policy, 148:105457. Doi: 10.1016/j.marpol.2022.105547, ISSN 0308-597X. S2CID 255023969.

Timoshenko, A., Yashin, G., Serpiva, V., Hamadov, R., Fedotov, D., Kartashova, M., Golikov, P., 2025. Seismic data acquisition utilizing a group of UAVs. Drones, 9, 156.

Tinto, R., 2018. Bayesian Methods in Resource Estimation. Min. Econ. Rev., 39(4), 290–305.

Tiwari, D., Nair, M.R., Rao, P.S., 2014. Crustal deformation studies in India using GNSS. National Geophysical Research Institute.

Tiwari, V.M., Veenadhari, B., Gahalaut, V.K., Mukherjee, S., Dimri, V.P., 2014. Gravity, GPS and Geomagnetic Data in India. Proc. Indian Natn. Sci. Acad., 80(3), 705–712.

Todd, D.K., 1980. Groundwater Hydrology (2nd ed.). Hoboken: Wiley.

Tokam, A.P.K., Tabod, C.T., Nyblade, A.A., Julia, J., Wiens, D.A., Pasyanos, M.E., 2010. Structure of the crust beneath Cameroon, West Africa, from the joint inversion of Rayleigh wave group velocities and receiver functions. Geophys. J. Int., 183(2), 1061–1076.

Tomiyama, T., Mia, B., Fujimitsu, Y., Nishijima, J., 2012. Mapping hydrothermal altered mineral deposits using Landsat 7 ETM+ image in and around Kuju volcano, Kyushu, Japan. J. Earth Syst. Sci., 121, 1049–1057.

Torge, W., 1989. Gravimetry. Walter de Gruyter, Berlin.

Toro, N., Robles, P., Jeldres, R.I., 2020. Seabed mineral resources, an alternative for the future of renewable energy: a critical review. Ore Geol. Rev., 126, 103699, doi:10.1016/j.oregeorev.2020.103699.

Toro, N., Jeldres, R.I., Ordenes, J.A., Robles, P., Navarra, A., 2020. Manganese nodules in Chile, an alternative for the production of Co and Mn in the future – a review. Minerals, 10:674.

Torres-Verdín, C., Bostick, F.X., 1992, Principles of spatial surface electric field filtering in magnetotellurics: electromagnetic array profiling (EMAP). Geophysics, 57, 603–622.

TotalEnergies, 2022. AI for Environmental Monitoring in Amazon Exploration. Environ. Stud. Q., 56(1), 75–82.

Toth, M., Sander, E., Voss, B., 2020. Satellite geophysics for large-scale geological mapping. Geophysical Surveys, 79(4), 66–80.

Townsend, T.E., 1987. Discrimination of iron alteration minerals in visible and near-infrared reflectance data. J. Geophys. Res., 92, 1441–1454.

Troiano, A., Giuseppe, M.G.D., Petrillo, Z., Patella, D., 2009. Imaging 2D structures by the CSAMT method: application to the Pantano di S. Gregorio Magno faulted basin (Southern Italy). J. Geophys. Eng., 6, 120–130.

Tsoukalas, L.H., Uhrig, R.E., 1997. Fuzzy and Neural Approaches in Engineering. John Wiley & Sons, New York, NY, 606p.

Tuckwell, G., et al., The use of microgravity to detect small distributed voids and low density ground. Q. J. Eng. Geol. Hydrogeol., 41(2008), 371–380.

Tyler, A.N., Davidson, D.A., Grieve, I.C., 2001. In situ radiometric mapping of soil erosion and field-moist bulk density on cultuvated fields, 88–96.

United Nations, 2023. Intergovernmental conference of marine biodiversity of areas beyond national jurisdiction. General Assembly Resolution 72/249.

United Nations, 2023. Statement attributable to the Spokesperson for the Secretary General – on Int'l Legally Binding Instrument under the UN Convention on the Law of the Sea. Stephane Dujarric, Spokesman for the Secretary General.

United Nations, 2023. Draft agreement under the United Nations Convention on the Law of the Sea on the conservation and sustainable use of marine biological diversity of areas beyond national jurisdiction. General Assembly – a/CONF.232/2023/CRP.2/Rev.1.

United Nations (UN), 2021. The 17 Goals, Department of Economic and Social Affairs, Sustainable Development. https://sdgs.un.org/goals (Accessed 25 December, 2021).

United Nations, 2019. Geoscience and Mineral Resources: the Role of Data in Sustainable Development. United Nations Department of Economic and Social Affairs.

United Nations Environment Programme (UNEP), 2022. Guidelines on sustainable resource management in geophysical exploration. United Nations Environment Programme. Retrieved from https://www.unep.org.

United Nations Environmental Programme, 2020. Mineral Resources and Environmental Sustainability: a Global Perspective. United Nations Environment Programme.

United Nations Environment Programme, 2018. Global Resource Assessment: sustainable Exploration of Minerals. UNEP Report.

Uniyal, A., Gond, A., Tiwari, T., Pandit, S., Gupta, C., Saikhom, V., Dinkar, G., 2022. UAV/Drone-based identification of geomorphic and structural anomalies as probable indicators of economic mineral deposits in Girar area of Lalitpur District, U.P., (India). J. Indian Soc. Remote Sens., 50(5), 937–947 10.1007/s12524-021-01457-9.

Unsworth, M., Lu, X.Y., Watts, M.D., 2000. CSAMT exploration at Sellafield: characterization of a potential radioactive waste disposal site. Geophysics, 65, 1070–1079.

Urosevic, M., Evans, B.J., Hatherly, P.J., 1992. The improvement in seismic resolution by map and trace attributes analysis. Explor. Geophys., 23, 387–392.

Urosevic, M., Stoltz, E., Massey, S., 2005. Seismic exploration for gold in a hard rock environment Yilgarn Craton, Western Australia. Presented at EAGE 67th meeting.

Urosevic, M., Bona, A., Ziramov, S., Pevzner, R., Kepic, A., Egorov, A., Kinkela, J., Pridmore, D., Dwyer, J., 2017. Seismic for mineral resources – a mainstream method of the future. In proceedings of the exploration 17 seismic methods & exploration workshop, Toronto, ON, Canada, 21-25 October 2017.

United States Department of the Interior and United States Geological Survey (USGS), 2018. Mineral commodity summaries (US Government Printing Office).

United States Department of Energy (USDOE), Critical materials strategy. DOE/PI-0009 (http://energy.gov/sites/prod/files/DOE_CMS2011_FINAL_Full.pdf) (2011).

United States Geological Survey (USGS), 2023. Global marine mineral resources. Available from: https://www.usgs.gov/globalmarinemineralresources. Last accessed on 28 April 2023.

United States Geological Survey (USGS), 2021. National geophysical data and resource mapping. United States Geological Survey. Retrieved from https://www.usgs.gov.

Vajk, R., 1949. Baron Roland Eotvos. Geophysics, 14(1):6–9.

Vallee, M.A., Smith, R.S., Keating, P., 2011. Metalliferous mining geophysics: state of the art after a decade in the new millennium. Geophysics, 76, W31–W50.

van Kann, F., 2004. Requirements and general principles of airborne gravity gradiometers for mineral exploration. In: Lane, R.J.L. (ed) Abstracts from the ASEG-PESA, Airborne Gravity 2004 Work Shop, 1–6.

van Leeuwen, E.H., 2000. BHP develops airborne gravity gradiometer for mineral exploration. Lead. Edge, 19(12), 1296–1297.

van der Graaf, E.R., Koomans, R.L., Limburg, J., de Vries, K., 2007. In situ radiometric mapping as a proxy of sediment contamination: assessment of the underlying geochemical and geophysical principles. Appl. Radiat. Isot.

van der Meer, F., Vazquez Torres, M., Van Dijk, P.M., 1997. Spectral characterization of ophiolite lithologies in the Troodos Ophiolite complex of Cyprus and its potential in prospecting for massive sulphide deposits. Int. J. Remote Sens., 18, 1245–1257.

van der Meer, F., Bakker, W., 1997. Cross Correlogram Spectral Matching (CCSM): application to surface mineralogical mapping using AVIRIS data from Cuprite, Nevada. Remote Sens. Environ., 61, 371–382.

van der Meer, F., 1998. Imaging spectrometry for geological remote sensing. Geologie En Mijnbouw, 77, 137–151.

van der Meer, F., 2000. Geophysical inversion of imaging spectrometer data for geologic modeling. Int. J. Remote Sens., 21, 387–393.

van der Meer, F., De Jong, S., 2003. Spectral mapping methods: many problems, some solutions. In: 3rd EARSeL Workshop on Imaging Spectroscopy, Herrsching, 146–162.

van der Meer, F.D., van der Werff, H.M.A., van Ruitenbeek, F.J.A., Hecker, C.A., Bakker, W.H., Noomen, M.F., van der Meijde, M., Carranza, E.J.M., 2012. Multi- and hyper-spectral geologic remote sensing: a review. Int. J. Appl. Earth Obs. Geoinf., 14, 112–128.

van der Meijde, M., Pailb, R., Binghamc, R., Floberghagen, R., 2015. GOCE data, models, and applications: a review. Int. J. Appl. Earth Obs. Geoinf., 35, 4–15.

Vander Velpen, B.P.A., 1988. A computer processing package for DC resistivity interpretation for a IBM compatibles. It Vander Velpen, B.P.A., 1988. A computer processing package for DC resistivity interpretation for a IBM compatibles. ITC Jour. 4, the Netherlands. C Jour. 4, The Netherlands.

Varma, A.K., Hazra, B., Samad, S.K., Panda, S., Mendhe, V.A., 2014. Methane sorption dynamics and hydrocarbon generation of shale samples from West Bokaro and Raniganj basins India. Journal of Natural Gas Science and Engineering, 21, 1138–1147.

Veiga, M.M., Meech, J.A., 2019. Environmental impact assessments and the role of geophysical data in sustainable mining. Environmental Impact Assessment Review, 74, 132–141.

Vengala, J., Dharek, M.S., Sunagar, P., Sreekeshava, K.S., Pramod, K., Baliga, D., Haripriya, P.S., Thejaswi, P., 2021. Flood Damages Assessment Using Remote Sensing and GIS: a Case Study of 2018 Kodagu Floods. In: Chintalacheruvu, M.R., Patra, K.C., Jhajharia, D., Kumari, S. Chapter in Conference Proceedings: Advanced Modelling and Innovations in Water Resources Engineering (AMIWRE), Part of the Book Series: Lecture Notes in Civil Engineering (LNCE), 463–475, 176).

Vergouw, B., Nagel, H., Bondt, G., Custers, B., 2016. The future of drone use. Asser Press: The Hague, the Netherlands, 27, 21–46.

Verma, R., Kumar, S., 2021. Integration of GPS, GIS, and remote sensing for coal exploration and mining in Chhattisgarh. Journal of Mineral Exploration and Mining, 42(6), 759–766.

Verma, R.K., Mallik, K., 1979. Detectability of intermediate conductive and resistive layers by time domain electromagnetic sounding. Geophysics, 44, 1862–1878.

Verma, A.K., Singh, T.N., 2013. Prediction of water quality from simple field parameters. Environmental Earth Sciences, 69(3), 821–829.

Veryaskin, A.V., 2018. Gravity, magnetic and electromagnetic gradiometry: strategic technologies in the 21st century. Morgan & Claypool Publishers, IOP Concise Physics, doi:10.1088/978-1-6817-4700-2ch1, 1–56.

Vichabian, Y., Morgan, F.D., 2002. Self-potentials in cave detection. Lead. Edge 21, 866–871.

Vickers, R.S., Lyon, R.J.P., 1967. Infrared sensing from spacecraft-a geological interpretation. Proceedings, Thermophysics Special Conference, American Institute of Aeronautics and Astronautics, 67–284.

Viger, R.J., 2008. The GIS Weasel: an Interface for the Development of Geographic Information Used in Environmental Simulation Modeling. Computers & Geosciences, 34(8), 891–901.

Vijayakumar, D., Mohan Babu, M., Srinivasa Rao, G., Rohit, G., Madhusudhan, D.G., 2023. Optimization of Survey Area for Auriferous-Sulphide Mineralization Using Integrated Geophysical Study: a Case Study from Gani-Kalva Fault Zone, Cuddapah Basin, India. Pure Appl. Geophys., 10.1007/s00024-023-03354-3.

Vincent, R.K., 1997. In: Fundamentals of Geological and Environmental Remote Sensing, Prentice Hall, Upper Saddle River, NJ, 80–121.

Virieux, J., Operto, S., 2009. An overview of full-waveform inversion in exploration geophysics. Geophysics, 74(6), WCC1–WCC26.

von Herzen, R.P., Kirklin, J., Becker, K., 1996. Geoelectrical measurements at the TAG hydrothermal mound. Geophys. Res. Lett., 23, 3451–3454.

von Hippel, A.R., 1954a. Dielectrics and Waves. MIT Press, Cambridge MA, USA.

Vozoff, K., 1972. The magnetotelluric method in the exploration of sedimentary basins. Geophysics, 37(1), 98–141.

Vozoff, K., 1991. The magnetotelluric method. In: Nabighian, M.N. (ed.), Electromagnetic Methods. Applied Geophysics, 2, 641–711.

Vrba, J., 1996. SQUID gradiometers in real environments: in Weinstock, H. (ed.), SQUID Sensors:

Fundamentals and Applications, Kluwer Academic, 117p.

Wadhawan, S.K., 2016. Marine geology and coastal surveys for seabed mapping and mineral resource evaluation in India – an overview. Proc. Indian Natn. Sci. Acad., 82(3), 675–683. DOI: 10.16943/ptinsa/2016/48476.

Wadhawan, S.K., Hariprasad, H., Dinesh, A.C., Karthikeyan, M., Verghese, S., Drishya, G., 2013. Strategizing mineral exploration for offshore Obvious Geological Potential (OGP) areas within exclusive economic zone (EEZ) of India. Indian J. Geosci., 67, 191–198.

Wadhawan, A., Kothari, D., Verma, S. (2013). Remote Sensing Techniques for Mineral Exploration in India. Indian Journal of Geophysics, 45(2), 231–247.

Wait, J.R., 1956. Transient fields of a vertical dipole over a homogenous curved ground. Can. J. Phys., 34(1), 27–35.

Wait, J.R., 1959a. Overvoltage research and geophysical application. In: International Series of Monographs on Earth Sciences, 4, 22–83. Pergomon press, Oxford, England.

Wall, F., 2014. Rare earth elements. Critical Metals Handbook, 312–339.

Walter, C., Braun, A., Fotopoulos, G., 2019.Spectral analysis of magnetometer swing in high-resolution UAV-borne aeromagnetic surveys. In 2019 IEEE Systems and Technologies for Remote Sensing Applications through Unmanned Aerial Systems (STRATUS) 1–4.

Wan, Y.Q., Fan, Y.H., Jin, M.S., 2021. Application of hyperspectral remote sensing for supplementary investigation of polymetallic deposits in Huaniushan ore region, northwestern China. Scientific Reports (Nature Research), 11,440.

Wang, D., Lin, F., Shi, M., Wang, H., Yang, X., 2023. Geological setting, tectonic evolution and spatio-temporal distributions of main mineral resources in South East Asia: a comprehensive review. Solid Earth Sciences, 8, 34–48.

Wang, C., Shao, Y.J., Goldfard, R., Tan, S.M., Sun, J., Zhou, C., Zheng, H., Liu, Q.Q., Xiong, Y.Q., 2023. Superimposed gold mineralization events in the Tuanshanbei orogenic gold deposit, Central Jiangnan orogen, South China. Econ. Geol., XXX(XX), X–X. ISSN 0361-0128; doi:10.5382/econgeo.5034; 25p.

Wang, Z., Zuo, R., Liu, H., 2021. Lithological Mapping Based on Fully Convolutional Network and Multi-Source Geological Data. Remote Sens., 13(23), 4860.

Wang, Z., Zuo, R., Jing, L., 2021. Fusion of Geochemical and Remote Sensing Data for Lithological Mapping Using Random Forest Metric Learning. Math. Geosci., 53, 1125–1145.

Wang, F.Y., Mao, X.C., Deng, H., Zhang, B.Y., 2020. Manganese potential mapping in western Guangxi-southeastern Yunnan (China) via spatial analysis and modal-adaptive prospectivity modeling. Trans. Nonferrous Metals Soc. China, 30, 1058–1070.

Wang, J.J., Tao, C.H., Wang, H.J., Deng, X.M., Xiong, W., Li, Z., 2018. Study of self-potential observation ways in the seafloor polymetallic sulfide deposits. Haiyang Xuebao.

Wang, X., Xie, H., 2018. A Review on Applications of Remote Sensing and Geographic Information Systems (GIS) in Water Resources and Flood Risk Management. Water, 10, 608.

Wang, C., Yanjiang, Y.U., Kou, B., 2017. Summary of marine in-situ testing methods. Ocean Dev. Manag., 34, 81–86.

Wang, M., Denga, M., Wu, Z., Luo, X., Jing, J., Chen, K., 2017. The deep-tow marine controlled-source electromagnetic transmitter system for gas hydrate exploration. J. Appl. Geophys., 137, 138–144.

Wang, Q., Wang, X.B., Yang, J., Min, G., Guo, J., 2016. Locating of the concealed skarn iron ore deposit based on multivariate information constraints: a case study of the Beiya gold mine in Yunnan province. Chin. J. Geophys., 59, 4771–4781.

Wang, J., Sain, K., Wang, X., Satyavani, N., Wu, S., 2014. Characteristics of bottom simulating reflectors for hydrate filled fractured sediments in Krishna-Godavari basin, eastern Indian margin. Journal of Petroleum Science and Engineering, 122, 515–523.

Wang, P., Li, Q., Li, C., 2014. Hydrocarbon and mineral resources. Geology of the China Seas, Elsevier 571–641.

Wang, Y.J., Fan, W.M., Zhang, G.W., Zhang, Y.H., 2013. Phanerozoic tectonics of the South China block: key observations and controversies. Gondwana Res., 23, 1273–1305.

Wang, X.J., Hutchinson, D.R., Wu, S.G., Yang, S.X., Guo, Y.Q., 2011a. Elevated gas hydrate saturation within silt and silty clay sediments in the Shenhu area, South China Sea. J. Geophys. Res., 116:B05102. doi:10.1029/2010JB007944.

Wang, X.J., Wu, S.G., Lee, M., Guo, Y.Q., Yang, S.X., Liang, J.Q., 2011b. Gas hydrate saturation from acoustic impedance and resistivity logs in the Shenhu area, South China Sea. Mar. Petrol. Geol., 28, 1625–1633. doi:10.1016/j.marpetgeo.2011.07.002.

Wang, X., Zhang, Q., Zhou, G., 2007. National-Scale geochemical mapping projects in China. Geostandards and Geoanalytical Research, 31(4), 311–320.

Wang, X.J., Wu, S.G., Liu, X.W., 2006. Factors affecting the estimation of gas hydrate and free gas saturation. Chin. J. Geophys. 49(2), 505–511. doi:10.1002/cjg2.853.

Wang, H., Liu, G., Xu, W., Wang, G., 2005. GIS-based landslide hazard assessment: an overview. Prog. Phys. Geogr., 29, 548–567.

Wannamaker, P.E., Rose, P.E., Doerner, W.M., Berard, B.C., McCulloch, J., Nurse, K., 2004. Magnetotelluric surveying and monitoring at the Coso geothermal area, California, in support of the enhanced geothermal systems concept: survey parameters initial results. Proceedings of the Twenty-Ninth Workshop on Geothermal Reservoir and Engineering: Stanford University, 287–294.

Wannamaker, P.E., 1997. Tensor CSAMT survey over the Sulphur Springs thermal area, Valles Caldera, New Mexico, United States of America, Part I: implications for structure of the western caldera. Geophysics, 62, 451–465.

Ward, S., 1959. AFMAG – airborne and ground. Geophysics, 24, 761–787.

Ward, S.H., O'Donnell, J., Rivera, R., Ware, G.H., Fraser, D.C., 1966. AFMAG –applications and limitations. Geophysics, 31, 576–605.

Ward, S.H., O'Brien, D.P., Parry, J.R., McKnight, B.K., 1968, AFMAG – interpretation. Geophysics, 33(4), 621–644.

Watzel, R., Ruhlemann, C., Vink, A., 2020. Mining mineral resources from the seabed: opportunities and challenges. Mar. Policy, 114, 103828.

Waxman, M.H., Smits, L.J.M., 1968. Electrical Conductivities in Oil-Bearing Shaly Sands. Soc. Petrol. Eng. J., 8, 107–122.

Webb, S., Edwards, R.N., Yu, L.M., 1993. First measurements from a deep-tow transient electromagnetic sounding system. Mar. Geophys. Res., 15(1), 13–26. 10.1007/BF01204149.

Webb, M., Rowston, P., 1995. The geophysics of the Ernest Henry Cu-Au deposit (N.W.) Qld. Explor. Geophys., 26(2-3), 51–59. 10.1071/EG995051.

Wellmann, J.F., Horowitz, F.G., Schill, E., Regenauer-Lieb, K., 2010. Towards incorporating uncertainty of structural data in 3D geological inversion. Tectonophysics, 490(3-4), 141–151.

Wessel, P., Smith, W.H.F., Scharroo, R., 2018. The integration of geological, geophysical, and geochemical data in resource exploration: case studies from the mining industry. Explor. Geophys., 49(1), 45–60.

Whipple, J., Jeirath, N., Archer, C., Sisk, D.A., Gray, S., Lee, C.J., Gonzalez, J., Wilmes, T., 2019. Aerial drone for well-site and signal survey. U.S. patent application no. 10,192,182. https://patentimages.storage.googleapis.com/2b/3f/ec/0dddba63cb6be5/US10192182.pdf.

White, R.M.S., Collins, S., Denne, R., Hee, R., Brown, P., 2001. A new survey design for 3D IP modelling at Copper Hill. Explor. Geophys., 32, 152–155.

White, D.J., Mwenifumbo, J., Salisbury, M., 2011. Submitted for publication. Seismic properties of rocks from the Flin Flon VMS Camp. Econ. Geol.

White, D.J., Secord, D., Malinowski, M., 2012. 3D seismic imaging of volcanogenic massive sulphide deposits in the Flin Flon mining camp, Canada: part 1 – seismic results. Geophysics, 77, WC47–WC58.

Whiting, T.H., 1986. Aeromagnetics as an aid to geological mapping: a case history from the Arunta inlier, Northern Territory. Aust. J. Earth Sci., 33, 271–286.

Wiese, H., 1962. Geomagnetische tiefensondierung. Teil II: die streichrichtung der Untergrundstrukturen des elektrischen Widerstandes, erschlossen aus geomagnetischen variationen. Geofis Pura Et Appl 52, 83–103.

Wilde, B.J., 2010. Data spacing and uncertainty (Master's thesis), University of Alberta (pp. 20–24).

Williams, R., Chirwa, P., 2021. AI and machine learning in the exploration of rare earth elements in Africa. Miner. Eng., 159, 106670.

Williams, P.F., Goodwin, L.B., Lafrance, B., 1995. Brittle faulting in the Canadian Appalachians and the interpretation of reflection seismic data. J. Struct. Geol., 17, 215–232. doi: 10.1016/0191-8141(94)E0046-2.

Wills, B.A., 2015. Wills Mineral Processing Technology: An Introduction to the Practical Aspects of Ore Treatment and Mineral Recovery (8th ed.). Oxford: Elsevier Science & Technology. ISBN 978-0-08-097054-7. OCLC 920545608.

Wilson, C.R., 2020. Analog signals and digital time series. In: Essentials of Geophysical Data Processing (pp. 11–24). Cambridge University Press.

Wilson, H., 1985. Analysis of the magnetic gradient tensor: defence Research Establishment Pacific. Canada Technical Memorandum, 85-13, 47.

Wolfgram, P.A., Edwards, R.N., Law, L.K., Bone, M.N., 1986. Polymetallic sulphide exploration on the deep seafloor: the feasibility of the MINI-MOSES experiment. Geophysics 51, 1808–1818.

World Bank, 2023. Sustainable mining practices: innovations in geophysical data collection. World Bank. Retrieved from https://www.worldbank.org.

World Bank, 2021. Minerals for the future: demand, supply, and policy considerations for the green transition. World Bank Group.

World Bank, 2020. Mineral resources and global supply chains: policy reforms and challenges. World Bank.

Wright, P.M., 1981. Seismic methods in mineral exploration. Book Chapter, Seventy-fifth Anniversary Volume. In: Skinner, B.J., Society of Economic Geologists, Economic Geology Publishing Company, 10.5382/AV75.

Wright, D.A., Ziolkowski, A., Hobbs, B.A. 2001. Hydrocarbon detection with a multichannel transient electromagnetic survey. In: SEG Technical Program Extented Abstracts (1435–1438). Society of Exploration Geophysicists.

Wright, D., Ziolkowski, A., Hobbs, B., 2002. Hydrocarbon detection and monitoring with a multicomponent transient electromagnetic (MTEM) survey. Lead. Edge, 21, 852–864.

Wu, J., Milkereit, B., Boerner, D.E., 1995. Seismic imaging of the enigmatic Sudbury Structure. J. Geophys. Res., 100, 4117–4130.

Wu, L.P., Shi, K.F., 1996. Exploration of underground thermal water in Songshan area and a discussion on its genetic model. Geophys. Geochem. Explor., 20, 309–315.

Wu, L.P., Shi, K.F., Li, Y.H., Li, S.H., 1996. Application of CSAMT method to the search for ground water. Chinese Journal of Geophysics, 39, 712–717.

Wu, N.Y., Zhang, H.Q., Yang, S.X., Liang, J.Q., Wang, H.B., 2007. Preliminary discussion on natural gas hydrate (NGH) reservoir system of Shenhu area, Northern South China Sea. Nat. Gas. Ind., 27(9), 1–6. (In Chinese with English abstract). doi:10.3321/j.issn:1000-0976.2007.09.001.

Wu, N.Y., Yang, S., Zhang, H., et al., 2008. Preliminary discussion on gas hydrate reservoir system of Shenhu Area, North Slope of South China Sea, Proceedings of the 6th International Conference on Gas Hydrates (ICGH 2008, 5700 Vancouver, British Columbia, Canada.

Wu, S.G., Gong, Y.H., Mi, L.J., Wang, Z.J., Wang, X.J., 2010. Study on hydrocarbon leakage system and associated gas hydrate reservoirs in the deep water basin of Northern South China Sea. Geoscience 24(3), 433–440. (In Chinese with English abstract). doi: 10.3969/j.issn.1000-8527.2010.03.003.

Wu, G., Shen, C., Tan, H., Yang, G., 2016. Gravity anomaly and crustal density structure in Jilantai rift zone and its adjacent region. Earthq. Sci. DOI 10.1007/s11589-016-0158-z.

Wu, M.A., Wang, Q.S., Zheng, G.W., Cai, X.B., Yang, S.X., Di, Q.S., 2011. Discovery of the Nihe iron deposit in Lujiang, Anhui, and its exploration significance. Acta Geol. Sin., 85, 802–809, (In Chinese with English abstract).

Wulder, M.A., 1998. The Prediction of Leaf Area Index from Forest Polygons Decomposed through the Integration of Remote Sensing, GIS, UNIX, and C. Comput. Geosci., 24(2), 151–157.

Wyborn, L.A.I., Heinrich, C.A., Jaques, A.L., 1994. Australian Proterozoic mineral systems: essential ingredients and mappable criteria. Australian Institute of Mining and Metallurgy, Annual conference, Melbourne, 109–115.

Wynn, W.M., Frahm, C.P., Carroll, P.J., Clark, R.H., Wellhoner, J., Wynn, M.J., 1975. Advanced superconducting gradiometer/magnetometer arrays and a novel signal processing technique. IEEE Trans. Magn., MAG-11, 701–707.

Xiang, T.Z., Xia, G.S., Zhang, L., 2018. Mini-UAV-based remote sensing: techniques, applications and prospectives. arXiv, arXiv:1812.07770v1.

Xie, J., Cui, Y.-A., Liu, J.-X., Guo, Y.-J., Zhang, L.-J., Luo, Y.-J., Zhang, P.-F., 2023. A review on theory, modeling, inversion, and application of self-potential in marine mineral exploration. Trans. Nonferrous Met. Soc. China, 33, 1214–1232.

Xiong, Y., Zuo, R., 2017. Effects of misclassification costs on mapping mineral prospectivity. Ore Geol. Rev., 82, 1–9.

Xiong, Y., Zuo, R., Carranza, E.J.M., 2018. Mapping mineral prospectivity through big data analytics and a deep learning algorithm. Ore Geol. Rev., 102, 811–817.

Xiong, Y., Zuo, R., 2021. A positive and unlabeled learning algorithm for mineral prospectivity mapping. Comput. Geosci., 147, 104667.

Xu, Y., Zhang, X., 2024. Geophysical Techniques in Global Mineral Exploration: the Evolution and Future Prospects. Earth Sci. Rev., 150, 98–112.

Xu, S.Z., Liu, B., 1995. A numerical method for calculating MT field on a layered model with continuous change of conductivity in each layer. Chinese Journal of Geophysics, 38, 262–268.

Xue, G.Q., Chen, W.Y., Zhou, N.N., Li, H., 2013. Short-offset TEM technique with a grounded wire source for deep sounding. Chin. J. Geophys., 56(1), 255–261 (in Chinese).

Xue, G.Q., 2018. The development of near-source electromagnetic methods in China. J. Environ. Eng. Geophys., 23, 115–124.

Yadav, A., Nayak, S., Kanta, H., Sangwan, P., Pundir, A., Chatterjee, R., 2019. A case study of azimuthal fracture characterization in Cambay Basin, India. J. Appl. Geophys., 169, 239–248.

Yagi, R., Simili, F.F., Araújo, J.C., Prado, R.M., Sanchez, S.V., Ribeiro, C.E.R., Barreto, V.C.M., 2006. Zinc application in seeds and its effect on germination, nutrition and initial development of sorghum. Pesqui. Agropecu. Bras., 41(4), 655–660.

Yamamoto, M., Nakamura, R., Kasaya, T., Kumagai, H., Suzuki, K., Ken, T.K., 2017. Spontaneous and widespread electricity generation in natural deep sea hydrothermal fields Angew Chem (Int Ed), 56, 5725–5728.

Yamazaki, T., Nakatani, N., Arai, R., Sekimoto, T., Katayama, H., 2021. Combined mining and pulp-lifting of ferromanganese nodules and rare-earth element rich mud around Minamitorishima island in the Western North Pacific: a prefeasibility study. Minerals, 11:310.

Yan, B., Zhu, W.H., Liu, L.S., Fang, G.Y., 2013. An optimization method for induction mag netometer of 0.1 mHz to 1 kHz. IEEE Trans. Magn., 49(10), 5294–5300. 10.1109/TMAG.2013.2264821.

Yang, H., Van der Meer, F., Bakker, W., Tan, Z.J., 1999. A back-propagation neural network for mineralogical mapping from AVIRIS data. Int. J. Remote Sens., 20, 91–110.

Yang, M.J., Song, Y.C., Liu, Y., Chen, Y.J., Li, Q.P., 2010. Influence of pore size, salinity and gas composition upon the hydrate formation conditions. Chin. J. Chem. Eng., 18, 292–296.

Yang, J., Liu, Y., Wu, X.P., 2015. 3D simulation of marine CSEM using vector finite element method on unstructured grids (in Chinese). Chin. J. Geophys., 58, 2827–2838.

Yang, J.M., Liu, L., Lyu, H.N., Lin, Z.Q., 2020. Deep-sea mining equipment in China: current status and prospect. Strategy Study CAE, 22(6):1–9 (in Chinese). 10.15302/J-SSCAE-2020.06.001.

Yasukawa, K., Liu, H., Fujinaga, K., et al., 2014. Geochemistry and mineralogy of REY rich mud in the eastern Indian Ocean. J. Asian Earth Sci., 93:25–36.

Yeats, C.J., Hollis, S.P., Halfpenny, A., Corona, J.C., Laflamme, C., Southam, G., Fiorentini, M., Herrington, R.J., Spratt, J., 2017. Actively forming Kuroko ype volcanic hosted massive sulfide (VHMS) mineralization at Iheya North, Okinawa Trough, Japan, Ore Geol. Rev, 84, 20–41.

Yi, M.J., Kim, J.H., Chung, S.H., 2003. Enhancing the resolving power of least-squares inversion with active constraint balancing. Geophys, 68, 931–941.

Yilmaz, O., 2001. Seismic Data Analysis: Processing, Inversion, and Interpretation of Seismic Data. Society of Exploration Geophysicists.

Yin, C.C., Ben, F., Liu, Y.H., Huang, W., Cai, J., 2014. MCSEM 3D modeling for arbitrarily anisotropic media (in Chinese). Chin. J. Geophys., 57, 4110–4122.

Yincan, Y., et al., 2017. Submarine shallow natural gas. Elsevier, Marine Geohazard in China, 453–521.

Yokoyama, I., Tajima, H., 1957. A gravity survey on volcano Mihara, Ooshima Island by means of a Worden gravimeter. Bulletin of the Earthquake Research Institute, 35, 23–33.

Yousefi, M., Carranza, E.J.M., Kreuzer, O.P., Nykanen, V., Hronsky, J.M.A., Mihalasky, M.J., 2021. Data analysis methods for prospectivity modelling as applied to mineral exploration targeting: state-of-the-art and outlook. Journal of Geochemical Exploration (Science Direct), 229, 106839.

Yousefi, B., Castanedo, C.I., Maldague, X.P., Beaudoin, G., 2020. Assessing the reliability of an automated system for mineral identification using LWIR Hyperspectral Infrared imagery. Miner. Eng., 155, p.106409.

Yousefi, M., Ranjbar, M.S., Gharibi, A.G., 2019. Integration of GIS and exploration information systems for mineral targeting. Journal of Earth Sciences, 30(5), 889–904.

Yousefi, M., Kreuzer, O.P., Nykanen, V., Hronsky, J.M.A., 2019. Exploration information systems – a proposal for the future use of GIS in mineral exploration targeting. Geol. Rev., 111, p.103005.

Yousefi, B., Sojasi, S., Castanedo, C.I., Maldague, X.P., Beaudoin, G., Chamberland, M., 2018. Comparison assessment of low rank sparse-PCA based-clustering/classification for automatic mineral identification in longwave- infrared hyperspectral imagery. Infrared Phys. Technol., 93, 103–111.

Yousefi, M., Carranza, E.J.M., 2017. Union score and fuzzy logic mineral prospectivity mapping using discretized and continuous spatial evidence values. J. Afr. Earth Sci., 128, 47–60.

Yusefi, M.R., Hafizi, M.K., Salari, B., 2017. Gold exploration using induced polarization and resistivity methods in Meyduk-Latala area. In Conference Proceedings, 79th EAGE Conference and Exhibition 2017, Paris, France, 1–3.

Yousefi, M., Carranza, E.J.M., 2016. Data-driven index overlay and Boolean logic mineral prospectivity modeling in greenfields exploration. Nat. Resour. Res., 25, 3–18.

Yousefi, M., Nykanen, V., 2016. Data-driven logistic-based weighting of geochemical and geological evidence layers in mineral prospectivity mapping. J. Geochem. Explor., 164, 94–106.

Yousefi, M., Carranza, E.J.M., 2015a. Fuzzification of continuous-value spatial evidence for mineral prospectivity mapping. Comput. Geosci., 74, 97–109.

Yousefi, M., Carranza, E.J.M., 2015b. Prediction-area (P-A) plot and C-A fractal analysis to classify and evaluate evidential maps for mineral prospectivity modeling. Comput. Geosci. 79, 69–81.

Yousefi, M., Carranza, E.J.M., 2015c. Geometric average of spatial evidence data layers: a GIS-based multi-criteria decision-making approach to mineral prospectivity mapping. Comput. Geosci. 83, 72–79.

Yousefi, M., Carranza, E.J.M., 2015d. Geochemical-geophysical mineral prospectivity mapping using Random Forests. Computers & Geosciences, 79, 69–81.

Yousefi, M., Kamkar-Rouhani, A., Carranza, E.J.M., 2014. Application of staged factor analysis and logistic function to create a fuzzy stream sediment geochemical evidence layer for mineral prospectivity mapping. Geochem. Explor. Environ. Anal., 14, 45–58.

Yousefi, M., Kamkar-Rouhani, A., Carranza, E.J.M., 2012. Geochemical mineralization probability index (GMPI): a new approach to generate enhanced stream sediment geochemical evidential map for increasing probability of success in mineral potential mapping. J. Geochem. Explor., 115, 24–35.

Yu, L.C., King, L.S., Hoon, A.T.C., Yean, P.C.C., 2015. A review study of oil and gas facilities for fixed and floating offshore platforms. Res J Appl Sci Eng Technol., 10(6):672–679.

Yuan, Y., Shen, F., Li, X., GPS multipath and NLOS mitigation for relative positioning in urban environments. Aerospace Science and Technology, 107(2020) 106315.

Yuan, J., Edwards, R.N., 2000. The assessment of marine gas hydrates through electrical remote sounding: hydrate without a BSR? Geophys. Res. Lett., 27(16), 2397–2400. 10.1029/2000GL011585.

Yungul, S., 1950. Interpretation of spontaneous polarization anomalies caused by spheroidal ore bodies. Geophysics, 15, 237–246.

Yuningsih, E.T., Matsueda, H., Setyaraharja, E.P., Rosana, M.F., 2012. The Arinem Te-bearing gold-silver-basemetal deposit, West Java, Indonesia. Resour. Geol., 62, 140–158.

Yusuf, T.U., 2016. Overview of effective geophysical methods used in the study of environmental pollutions by waste dumpsites. African research review. An International Multi-discipline Journal Ethopia. DOI: 10.4314/afrrev.v10i2.8, 10, 134–135.

Zadeh, M.H., Tangestani, M.H., Roldan, F.V., Yusta, I., 2014. Spectral characteristics of minerals in alteration zones associated with porphyry copper deposits in the middle part of Kerman copper belt, SE Iran. Ore Geology Reviews, Elsevier, 62, 191–198.

Zakir, F.A., Qari, M.H.T., Mostafa, M.E., 1999. A new optimizing technique for preparing lineament density maps. Int. J. Remote Sens., 20, 1073–1085.

Zana, M., Okorogbo, E., Ayodele, I., 2019. Geophysical Surveys in Africa: contributions to Sustainable Resource Management. J. Afr. Earth Sci., 156(2), 1–15.

Zawadzki, D., Maciag, L., Blasco, I., et al., 2022. Geochemistry and mineralogy of ferromanganese crusts from the western Cocos-Nazca spreading centre, Pacific. Minerals, 12:538.

Zeviani, W.M., Silva, C.A., Oliveira, W.J., Muniz, J.A., 2012. Non-linear models to potassium release from animals manure in latosols. Ciencia Rural, 42(10), 1789–1796.

Zhang, T., Li, Q., Li, Y., Liu, X., 2023. Underwater optical image restoration method for natural/artificial light. JMSE, 11:470.

Zhang, S.E., Bourdaeu, J.E., Nwaka, G.T., Ghorban, Y., 2022. Advanced geochemical Exploration knowledge using machine learning: prediction of unknown elemental concentrations and operational prioritization of re-analysing campaign. Artificial Intelligence in Geosciences 3, 86–100.

Zhang, H., Liu, X., Chen, A., Li, W., Lu, Y., Guo, X., 2021a. Design and application of an in-situ test device for rheological characteristic measurements of liquefied submarine sediments. J Mar Sci Eng 9(6):639.

Zhang, X., Yang, H., 2020. Machine Learning in Geophysical Data Interpretation: a Review and Prospects. Explor. Geophys., 51(4), 380–392.

Zhang, L., Xu, L., Wang, X., 2020. Artificial intelligence and smart irrigation systems: a review. Agricultural Water Management, 242, 106318.

Zhang, L., Yang, L.Q., Groves, D.I., Sun, S.C., Liu, Y., Wang, J.Y., Li, R.H., Wu, S.G., Gao, L., Guo, J.L., Chen, X.G., Chen, J.H., 2019. An overview of timing and structural geometry of gold, gold-antimony and antimony mineralization in the Jiangnan orogen, southern China. Ore Geol. Rev, 115 103173.

Zhang, H., Jia, Y., Liu, X., Wei, Z., Ji, C., Jiao, X., Zhu, C., 2019. Progress in in-situ measurement of sediment mechanical properties for full ocean depth. Mar. Geol. Front., 35, 1–9.

Zhang, L., Xu, X., Gao, Z., 2018. 4D Seismic Exploration for Dynamic Subsurface Monitoring. Journal of Seismic Exploration, 35(1), 67–84.

Zhang, S., Liu, R., Zhao, T., 2018. Mapping radiation distribution on ground based on the measurement using an unmanned aerial vehicle. J. Environ. Radioact. 193-194, 44–56.

Zhang, Z., Zuo, R., Xiong, Y., 2016. A comparative study of fuzzy weights of evidence and random forests for mapping mineral prospectivity for skarn-type Fe deposits in the southwestern Fujian metallogenic belt, China. Sci. China Earth Sci., 59, 556–572.

Zhang, L., 2015. OBS multi-component seismic data processing technology and its application. The University of Chinese Academy of Sciences, Guangzhou (in Chinese).

Zhang, L.C., Zhai, M.G., Wan, Y.S., Guo, J.H., Dai, Y.P., Wang, C.L., Liu, L., 2012. Study of the Precambrain BIF-iron deposits in the North China Craton: progresses and questions. Acta Petrol. Sin., 28, 3431–3445.

Zhang, H., 2011. Research on magnetic exploration inversion technology and its application. Hebei University of Engineering, Handan (in Chinese).

Zhang, X., Pazner, M., Duke, N., 2007. Lithologic and mineral information extraction for gold exploration using ASTER data in the south Chocolate Mountains (California) ISPRS J. Photogramm. Remote Sens., 62(4), 271–282.

Zhang, J., Zhou, C., Xu, K., Watanabe, M., 2002. Flood disaster monitoring and evaluation in China. Glob. Environ. Chang. Part B Environ. Hazards, 4, 33–43.

Zhang, C., Mushayandevu, M.F., Reid, A.B., Fairhead, J.D., Odegard, M.E., 2000. Euler deconvolution of gravity tensor gradient data. Geophysics, 65, 512–520.

Zhao, D., 1992. Tomographic imaging of P and S wave velocity structure beneath northeastern Japan. J. Geophys. Res 97, 19909–19928.

Zhao, D., 2001. Seismic structure and origin of hotspots and mantle plumes. Earth Planet. Sci. Lett 192, 251–265.

Zhao, D., 2015. Multiscale Tomography, first ed. Springer, Tokyo, pp. 304. doi:10.1007/978-44-431-55360-1.

Zhao, D., Mishra, O.P., Sanda, R., 2002. Influence of fluid and magma on earthquakes: seismological evidence. Phys. Earth Planet. Inter 132, 249–267.

Zhao, W., Li, T., Zhang, H., 2022. Sustainable mineral exploration: the role of geophysical mapping in reducing environmental impact. Environmental Geoscience Reviews, 11(3), 254–271.

Zhao, G., 2021. Digital Transformation of Petroleum Geophysical Exploration: towards Real-Time Data Acquisition, Automatic Processing, and Intelligent Interpretation. Geophysical Prospecting for Petroleum, 60(2), 175–189.

Zhdanov, M.S., 2012. The first practical 3D inversion of towed streamer EM data from the Troll field trial. SEG Las Vegas Annual Meeting.

Zhdanov, M.S., 2018. Marine electromagnetic methods. In: Foundations of Geophysical Electromagnetic Theory and Methods (Second Edition). Elsevier, 978-0-444-63890-8.

Zhong, Y., Chen, Z., Gonzalez, F.J., et al., 2018. Rare earth elements and yttrium in ferromanganese deposits from the South China Sea: distribution, composition and resource considerations. Acta Oceanol. Sin., 37:41–54.

Zhou, B., Yin, C., Ren, Z., Wang, X., 2022. Editorial for the special issue "Electromagnetic exploration: theory, methods and applications." Minerals, 12, 1505.

Zhou, X., Li, Y., 2021. Rare earth element resources: a geographical and geochemical perspective. Journal of Mineral Resources, 29(1), 45–67.

Zhou, J., Li, H., Zhang, W., 2020. New approaches in seismic exploration for deep mineral deposits. Geophys. Prospect., 65(7), 1523–1538.

Zhou, N.N., Xue, G.Q., Chen, W., Cheng, J., 2015. Large-depth hydrogeological detection in the North China-type coalfield through short-offset grounded-wire TEM. Environ. Earth Sci., 74, 2393–2404.

Zhou, J., Revil, A., Karaoulis, M., Hale, D., Doetsch, J., Cuttler, S., 2014. Image-guided inversion of electrical resistivity data. Geophys. J. Int. 197, 292–309. (doi:10.1093/gji/ggu001).

Zhou, W., Chen, G., Li, H., Luo, H., Huang, S.L., 2007. GIS Application in Mineral Resource Analysis – a Case Study of Offshore Marine Placer Gold at Nome, Alaska. Computers & Geosciences, 33(6), 773–788.

Zhu, Z., Tao, C., Shen, J., Revil, A., Deng, X., Liao, S., Zhou, J., Wang, W., Nie, Z., Yu, J., 2020. Self-potential tomography of a deep-sea polymetallic sulfide deposit on Southwest Indian Ridge. Journal of Geophysical Research: Solid Earth, 125, e2020JB019738.

Zhu, Z., Tao, C., Shen, J., Revil, A., Deng, X., Liao, S., Zhou, J., Wang, W., Nie, Z., Yu, J., 2020. Self-potential tomography of a deep sea polymetallic sulfide deposit on Southwest Indian Ridge. Journal of Geophysical Research: Solid Earth, 125, e2020JB019738.

Zhu, Z.M., Shen, J.S., Tao, C.H., Deng, X.M., Wu, T., Nie, Z.F., Wang, W.Y., Su, Z.Y., 2021. Autonomous-underwater-vehicle-based marine multicomponent self-potential method: observation scheme and navigational correction [J]. Geoscientific Instrumentation Methods and Data Systems, 10, 35–43.

Zhu, A.X., Hudson, B., Burt, J., Lubich, K., Simonson, D., Soil mapping using GIS, expert knowledge, and fuzzy logic. Soil Sci. Soc. Am. J. 2001.

Zimmerman, J.E., 1981. Cryocoolers for geophysical SQUID magnetometers. In: (Weinstock, H., Overton, W.C.) SQUID Applications to Geophysics, Proceedings of the Workshop Held 2-4 June 1980, Los Alamos, Soc. Expl. Geophys., Tulsa, Oklahoma, 81–83.

Ziolkowski, A.M., Hobbs, B.A., Wright, D.A., 2007. Multitransient electromagnetic demonstration survey in France. Geophysics, 72(4), F197–F209.

Ziolkowski, A., Parr, R., Wright, D., Nockles, V., Limond, C., Morris, E., Linfoot, J., 2010. Multi-transient electromagnetic repeatability experiment over the North Sea Harding field Geophys. Prospect., 58(6), 1159–1176.

Zlatanova, S., Coltekin, A., Ledoux, H., Stoter, J., Biljecki, F., 2015. Applications of 3D City Models: state of the Art Review. ISPRS Int. J. Geo-Inf., 4, 2842–2889.

Zohdy, A.A.R., 1989. A new method for the automatic interpretation of Schlumberger and Wenner sounding curves. Geophysics, 54, 245–253.

Zoheir, B., Emam, A., 2012. Integrating geologic and satellite imagery data for high resolution mapping and gold exploration targets in the South-Eastern Desert, Egypt. J. Afr. Earth Sci., 66–67,22–34.

Zonge, K.L., Sauck, W.A., Sumner, J.S., 1972. Comparison of time frequency and phase measurements in induced polarisation. Geophys. Prospect., 20, 626–648.

Zonge, K.L., Hughes, L.H., 1991. Controlled-source audio-frequency magnetotellurics, In: Nabighian, M.C., ed., Electromagnetic Methods in Applied Geophysics, 2, Applications, Part B: SEG, 713–809.

Zou, H.F., Liu, S.Y., Cai, G.J., Puppala, A.J., 2020. Probabilistic identification of contaminated soils using resistivity piezocone penetration tests. Acta Geotech. 15(3):761–779.

Zuo, R., Carranza, E.J.M., 2011. Support vector machine: a tool for mapping mineral prospectivity. Comput. Geosci., 37, 1967–1975.

Zuo, R., Xiong, Y., Wang, J., Carranza, E.J.M., 2019. Deep learning and its application in geochemical mapping. Earth Sci. Rev., 192, 1–14.

Zuo, R., 2020. Geodata Science-based mineral prospectivity mapping: a review. Nat. Resour. Res., 29, 3415–3424.

Zuo, R., Wang, Z., 2020. Effects of random negative training samples on mineral prospectivity mapping. Nat. Resour. Res., 29, 3443–3455.

Zurmotai, N.H., 2020. GIS, remote sensing and GPS: their activity, integration and fieldwork. International Journal of Applied Research, 6(9), 328–332.

Index

Note: Page numbers followed by "f" indicate figures, "t" indicate tables, "b" indicate boxes.

B

C

D

N

Printed and bound by CPI Group (UK) Ltd, Croydon, CR0 4YY
14/09/2026
02216593-0002